Environmental Science

FOURTEENTH EDITION

G. TYLER MILLER

SCOTT E. SPOOLMAN

Pecold/Shutterstock.com

BROOKS/COLE
CENGAGE Learning™

Australia • Brazil • Japan • Korea • Mexico • Singapore • Spain • United Kingdom • United States

Environmental Science, 14e
G. Tyler Miller and Scott E. Spoolman

Publisher: Yolanda Cossio

Development Editor: Jake Warde

Assistant Editor: Alexis Glubka

Editorial Assistant: Lauren Crosby

Media Editor: Alexandria Brady

Marketing Manager: Tom Ziolkowski

Marketing Assistant: Jing Hu

Marketing Communications Manager: Linda Yip

Content Project Manager: Hal Humphrey

Design Director: Rob Hugel

Art Director: John Walker

Print Buyer: Karen Hunt

Rights Acquisitions Specialist: Roberta Broyer

Production Service/Compositor: Thompson Steele, Inc.

Photo Researcher: Abigail Reip

Copy Editor: Deborah Thompson

Illustrator: Patrick Lane, ScEYEence Studios; Rachel Ciemma

Cover Designer: John Walker

Cover Image: © Frans Lanting/www.lanting.com. For more information about this photo, see p. iv.

Library of Congress Control Number: 2011934330

ISBN-13: 978-1-111-98893-7

ISBN-10: 1-111-98893-5

Brooks/Cole
20 Davis Drive
Belmont, CA 94002-3098
USA

Cengage Learning is a leading provider of customized learning solutions with office locations around the globe, including Singapore, the United Kingdom, Australia, Mexico, Brazil, and Japan. Locate your local office at **www.cengage.com/global**.

Cengage Learning products are represented in Canada by Nelson Education, Ltd.

To learn more about Brooks/Cole, visit **www.cengage.com/brookscole**.

Purchase any of our products at your local college store or at our preferred online store **www.cengagebrain.com**.

Printed in Canada
2 3 4 5 6 7 17 16 15 14 13

Brief Contents

About the Cover Photo

(a) Tropical rainforest in Peru

(b) Scarlet macaw

This tropical rainforest canopy **(a)** is part of the Tambopata National Reserve in Peru. The tropical forest researcher suspended by a cable in the upper left corner of the photo is installing an artificial nest for use by **(b)** a scarlet macaw or some other type of macaw.

Tropical rainforests are found near the equator and occupy only about 2% of the earth's land surface. However, these storehouses of the earth's biological diversity are home for up to half of the planet's known land-based animal and plant species. These forests and many of their species are under great environmental stress. At least half of the earth's total area of tropical forest has been disturbed or destroyed by human activities, mostly since about 1950. These forests are cleared for building roads, harvesting timber, growing crops, grazing cattle, mining minerals, and building settlements. These activities are likely to increase as the human population grows and uses more of the world's resources. By the end of this century, most mature tropical rainforests and many of their unique species may be gone, unless large areas of these forests are protected in reserves such as the one shown in this photo.

Because these forests hold such an incredible diversity of life forms, scientists study them to learn about how this forest life has sustained itself for many millions of years. To do this work, scientists climb into the top portions of the canopies where most of the forests' species are found. They erect small observation platforms and laboratories and rope walkways between portions of a canopy. In Panama, the Smithsonian Tropical Research Institute has used a construction crane to suspend researchers in a forest canopy.

The clearing of these forests destroys or degrades the habitats for thousands of species such as the scarlet macaw **(b)** and several other macaw species. The large and colorful macaws are found in a vast area covering parts of Mexico, Brazil, Paraguay, Bolivia, and Peru. Because of this large range, most macaw species are not currently endangered. However, certain macaw species have disappeared from many local areas because of destruction and fragmentation of their habitat and because they have been captured for the pet trade. The installation of artificial nests for macaws **(a)** is part of an effort to help build their populations.

Detailed Contents

Mark Edwards/Peter Arnold, Inc.

Photo 1 This child and his family in Katmandu, Nepal, collect beer bottles and sell them for cash to a brewery that will reuse them.

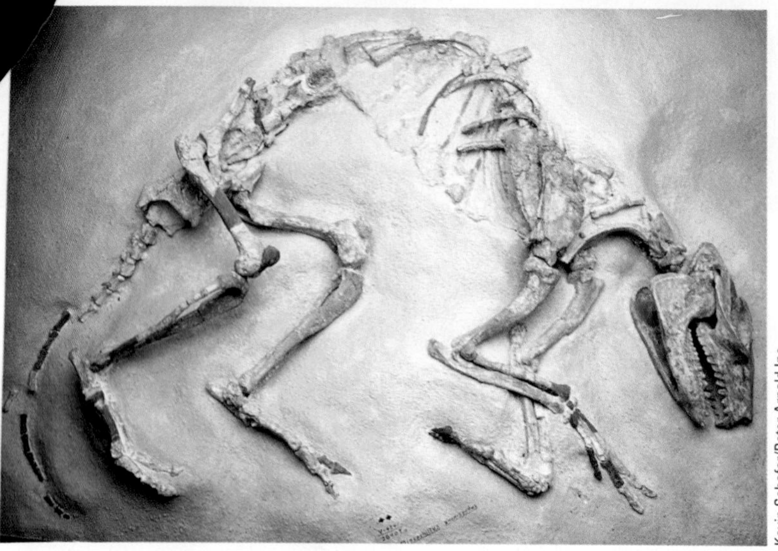

Kevin Schafer/Peter Arnold Inc.

Photo 2 This fossilized skeleton is the mineralized remains of an herbivore that lived 26 to 66 million years ago.

Wolfgang Poelzer/Peter Arnold, Inc.

Photo 3 Population (school) of glassfish in the Red Sea.

Photo 4 This brown bear (the predator) in the U.S. state of Alaska has captured and will feed on this salmon (the prey).

PhotoAlto/SuperStock

Photo 5 Healthy tree (left) and tree infested with parasitic mistletoe (right).

Paul Marcus/Shutterstock.com

Photo 6 Endangered Sumatran corpse flower.

Photo 7 Endangered green sea turtle.

© Reinhard Dirscherl/Alamy

© age fotostock/SuperStock

Photo 8 A coastal marsh in Peru.

Alison Gannett/Peter Arnold, Inc.

Photo 9 Energy efficient *straw bale house* in Crested Butte, Colorado (USA) during construction.

Alison Gannett/Peter Arnold, Inc.

Photo 10 Completed energy efficient *straw bale house* in Crested Butte, Colorado (USA).

Photo 11 Cow dung is collected, dried, and burned as fuel for cooking and heating in some parts of India.

Photo 12 Hubbard glacier in the U.S. state of Alaska stores water for a long time as part of the hydrologic cycle. However, mostly because of recent atmospheric warming, many of the world's glaciers are slowly melting.

SUPPLEMENTS

PREFACE

For Instructors

We wrote this book to help instructors achieve three goals: *first,* to explain to their students the scientific fundamentals of how life on the earth has survived and thrived; *second,* to use this scientific foundation to help students understand the multiple environmental problems that we face and evaluate possible solutions to them; and *third,* to inspire their students to make a difference in how we treat the earth, which supports our lives and economies, and thus in how we treat ourselves and our descendants. To help achieve these goals, we present our view of the earth, the environmental problems we face, and some possible solutions to them through the lens of *sustainability*—the integrating theme of this book.

What's New in This Edition?

Our texts have been widely praised for keeping users up to date in the rapidly changing field of environmental science. We continue to maintain this proven strength, and in this edition, and have added some new features.

- **More than 50 new or expanded topics including our vision of a more sustainable world in 2060 (p. 5).** Sections on energy resources now include coverage of the 2010 BP oil spill in the Gulf of Mexico and more on nuclear reactors. Climate change is emphasized with new coverage on the warming of the world's lakes, climate change tipping points, and innovative efforts to reduce methane and soot emissions. Updated examples of natural capital degradation include the Great Pacific garbage patch and an aluminum plant hazardous waste spill in Hungary. We also stress that environmental science is rife with good news and with promise for a better future. We include new information on how pollution prevention pays, and the greening of college campuses.
- **New *Core Case Studies*. Over 25% of the book's 17 chapters have new *Core Case Studies* that are threaded throughout each chapter.** New Core Case Studies include: A Vision of a More Sustainable World in 2060 (Chapter 1); Why Should We Protect Sharks? (Chapter 4); Slowing Population Growth in China (Chapter 6); A Temperate Deciduous Forest (Chapter 7); The Astounding Potential for Wind Power in the United States (Chapter 13); and Melting Ice in Greenland (Chapter 15). Additional case studies, not published in the book, are available online in ***Explore More*** (see below).
- **Three social science principles of sustainability are introduced.** These three principles based on environmental economics, political science, and ethics complement the three scientific principles of sustainability. (See Figure 1-2, p. 8 and Figure 17-23, p. 461.)
- **Two new features engage students in environmental investigations and research.** We include ***Doing Environmental Science*** exercises at the end of each chapter. The goal is to get students to carry out some environmental investigations in the real world, particularly around their schools or local communities. Also at the end of each chapter is a ***Global Environment Watch*** exercise. These exercises help students learn how to use an extensive database of environmental articles and other materials that are updated several times a day and are available exclusively for users of this book.
- **Enhanced Visual Learning.** As always, we have maintained the highly visual nature of this book by updating our illustration program. We have replaced, upgraded, or updated more than 120 of our photos and diagrams, or about 30% of our figures, carefully keying them to our overarching theme of sustainability and our five subthemes—*natural capital, natural capital degradation, solutions, tradeoffs,* and *individuals matter*. New to this edition is ***Virtual Field Trips in Environmental Issues***. Using a visual case study approach, this new online supplement brings the field to you. Dynamic panoramas, videos, photographs, maps, and quizzes work together to engage students in understanding and interacting with five important environmental issues: keystone species, climate change's role in extinctions, invasive species, the evolution of a species due to its environment, and an ecosystem approach to sustaining biodiversity.
- ***Explore More*: Online Library of Case Studies and Short Articles.** In this edition, we offer a new feature called ***Explore More***, which is a collection of more than 125 brief articles on a variety of environmental topics, written by the authors and available online. These supplementary items are available for both instructors and students on our website at **www.cengagebrain.com**. An abbreviated list is

found on the inside of the back cover of the Instructor version of this book. The articles may be viewed electronically, or used to customize your printed textbook. Each article is also referred to within the Instructor's Manual, along with a website address. This feature adds a new level of flexibility to help instructors customize their courses.

Concept-Centered Approach

To help students focus on the main ideas, we built each major chapter section around one to two *key concepts*, which state the most important take-away messages of each chapter. They are listed at the front of each chapter (see p. 62), and each chapter section begins with a key question and concepts (see pp. 65, 70, and 72), which are highlighted and referenced throughout each chapter. At the end of each chapter, we also summarize and reinforce learning by listing its *three big ideas*.

Sustainability Is the Integrating Theme of This Book

Sustainability, a watchword of the 21st century for those concerned about the environment, is the overarching theme of this textbook. You can see the sustainability emphasis by looking at the Brief Contents (p. iii).

Three scientific **principles of sustainability**—based on our analysis of how life on the earth has been sustained for at least 3.5 billion years—play a major role in carrying out this book's sustainability theme. These principles are introduced in Chapter 1, depicted in Figure 1-2 (p. 8) and on the back cover of the student edition, and used throughout the book, with each reference marked in the margin by [SUSTAINABILITY icon]. (See pp. 8, 69, 79, and 325.)

In this edition we also introduce three *social science principles of sustainability* based on key concepts from environmental economics, political science, and ethics. These principles are depicted in Figure 17-23 (p. 461) and on the back cover of the student edition.

Core Case Studies and the Sustainability Theme

Each chapter opens with a *Core Case Study* (p. 61), which is applied throughout the chapter. These connections to the Core Case Study are indicated in the book's margins by CORE CASE STUDY. (See pp. 64, 73, and 74.) *Thinking About* exercises strategically placed throughout each chapter (see pp. 184, 195, 199, and 212) challenge students to make these and other connections for themselves. Each chapter ends with a *Revisiting* box (p. 234), which connects the Core Case Study and other material in the chapter to the three scientific principles of sustainability.

Five Subthemes Guide the Way toward Sustainability

We use the following five major subthemes to integrate material throughout this book.

- ***Natural capital.*** Sustainability depends on the natural resources and natural services that support all life and economies. Examples of diagrams that illustrate this subtheme are Figures 1-3, p. 9; 3-9, p. 46; and 9-4, p. 177.
- ***Natural capital degradation.*** We describe how human activities can degrade natural capital. Examples of diagrams that illustrate this subtheme are Figures 9-9, p. 180; 10-11, p. 214; and 12-12, p. 285.
- ***Solutions.*** We pay a great deal of attention to the search for *solutions* to natural capital degradation and other environmental problems. We present proposed solutions in a balanced manner and challenge students to use critical thinking to evaluate them. Some figures and many chapter sections and subsections present tested and proposed solutions to various environmental problems. Examples are Figures 9-14, p. 184; 10-25, p. 229; and 13-48, p. 343. We also present a number of technologies and social trends that could soon break out and change the world much more rapidly than most people think. This potentially good news is summarized in Figure 17-24 (p. 462).
- ***Trade-Offs.*** The search for solutions involves *trade-offs*, because any solution requires weighing advantages against disadvantages. We use *Trade-Offs* diagrams to present the advantages and disadvantages of various environmental technologies and the solutions to environmental problems. Examples are Figures 10-17, p. 219; 11-17, p. 244; and 13-37, p. 333.
- ***Individuals Matter.*** Throughout the book *Individuals Matter* boxes and some of the Core Case Studies describe what various scientists and concerned citizens have done to help us work toward living more sustainably. (See pp. 65, 199, and 353). In addition, a number of *What Can You Do?* diagrams describe how readers can deal with the problems we face. Examples are Figures 8-10, p. 161; 11-19, p. 253; and 15-31, p. 403. Eight especially important things individuals can do—the *sustainability eight*—are summarized in Figure 17-21 (p. 459).

Science-Based Global Coverage

Chapters 2–7 discuss how scientists work and introduce the scientific principles (see Brief Contents, p. iii) needed for a basic understanding of how life on earth has survived and thrived for 3.5 billion years. These chapters also describe the variety of environmental problems we face and some methods for evaluating proposed solutions to these problems. Important environmental science topics are explored in depth in 27 *Science Focus* boxes distributed among the chapters (see pp. 194, 232, and 290). Science is also integrated throughout the book in various *Case Studies* (see pp. 73, 162, and 323) and in figures (see Figures 13-2, p. 301; 14-2, p. 349; and 14-11, p. 362). *Green Career* notations in the text point to various green careers on which further information can be found on the website for this book.

This book also provides a *global perspective* on two levels. *First,* ecological principles reveal how all the world's life is connected and sustained within the biosphere (Chapter 3) and these principles are applied throughout the book. *Second,* the book integrates information and images from around the world into its presentation of environmental problems and their possible solutions. This includes more than 40 global maps and U.S. maps in the basic text and in Supplement 6.

Case Studies

In addition to the 17 Core Case Studies, each of which is integrated throughout its chapter, 51 additional *Case Studies* (see pp. 74, 249, and 264) appear throughout the book (see items in **BOLD** type in the Detailed Contents, pp. v–xii). The total of 68 case studies provides an in-depth look at specific environmental problems and their possible solutions.

Critical Thinking

The introduction on *Learning Skills* describes critical thinking skills for students (pp. 2–4). Critical thinking applications appear throughout the book in several forms:

- As more than 60 *Thinking About* exercises. This *interactive approach to learning* reinforces textual and graphic information and concepts by asking students to analyze material immediately after it is presented rather than waiting until the end of chapter (see pp. 184, 195, and 199).
- In all *Science Focus* boxes.
- As 57 *Connections* boxes that stimulate critical thinking by exploring often surprising connections between environmental problems and common human activities (see pp. 75, 181, 218, and 327).
- In the captions of most of the book's figures (see Figures 1-8, p. 14; 3-11, p. 48; and 10-21, p. 224).
- As end-of-chapter exercises (see pp. 92, 202, and 346).

Visual Learning

This is a highly visual book with more than 400 photos and diagrams, 30% of which are new, improved, or updated in this edition. They are designed to present complex ideas in understandable ways relating to the real world (see Figures 3-14, p. 51; 11-30, p. 266; 13-9, p. 307; and 15-15, p. 387). Our illustration program includes more than 170 photos carefully selected over the years and reviewed and updated in each new edition.

To enhance visual learning, more than 200 animations, at least 30 referenced in the text as *Active Figures,* are available online, as well as integrated into an e-book version of this textbook. To find the premium online resources that are right for your students, talk to your Brooks/Cole sales representative.

Three Levels of Flexibility

There are hundreds of ways to organize the content of this course to fit the needs of different instructors with a wide variety of professional backgrounds as well as course lengths and goals. To meet these diverse needs, we have designed a highly flexible book that allows instructors to vary the order of chapters and sections within chapters without exposing students to terms and concepts that could confuse them.

We recommend that instructors start with Chapter 1 because it defines basic terms and gives an overview of sustainability, population issues, pollution, resources, and economic development issues that are treated throughout the book. This provides a springboard for instructors to use other chapters in almost any order.

One often-used strategy is to follow Chapter 1 with Chapters 2–7, which introduce basic science and ecological concepts. Instructors can then use the remaining chapters in any order desired. Some instructors follow Chapter 1 with Chapter 17 on environmental economics, politics, and worldviews before proceeding to the chapters on basic science and ecological concepts.

We provide a *second level of flexibility* in 7 Supplements (see p. xii in the Detailed Contents and p. S1), which instructors can assign as desired to meet the needs of their specific courses. Examples include U.S. environmental history (Supplement 3), basic chemistry (Supplement 4), weather basics (Supplement 5), a collection of 23 maps (Supplement 6), and a set of 16 graphs showing trends in basic environmental data with one or more data analysis exercises accompanying each graph. (Supplement 7).

Our new **Explore More** feature provides a third level of flexibity by providing 125 brief articles on a variety of topics written by the authors.

In-Text Study Aids

Each chapter begins with a list of *Key Questions and Concepts* showing how the chapter is organized and what students will be learning. When a new term is introduced and defined, it is printed in boldface type, and all such terms are summarized in the glossary at the end of the book and highlighted in review questions at the end of each chapter.

Thinking About exercises (65 in all) reinforce learning by asking students to think critically about the implications of various environmental issues and solutions immediately after they are discussed in the text. The captions of many figures contain questions that involve students in thinking about and evaluating their content.

Each chapter ends with a *Review* section containing a detailed set of review questions by chapter section that include all chapter key terms in boldface (p. 91), followed by a set of *Critical Thinking* (p. 92) questions to encourage students to think critically and apply what they have learned to their lives. Each chapter also has one or two *Doing Environmental Science* exercises (new to this edition), an exercise on using the Global Environment Watch database of articles and other information (new to this edition), and a *Data Analysis* or *Ecological Footprint Analysis* problem built around ecological footprint data or some other environmental data set.

Supplements for Students

A multitude of electronic supplements available to students take the learning experience beyond the textbook:

- *CengageNOW* is an online learning tool that helps students assess their unique study needs. Students take a pre-test and a personalized study plan provides them with specific resources for review. A post-test then identifies content that might require further study.
- *WebTutor* on WebCT or Blackboard provides qualified adopters of this textbook with access to a full array of study tools, including flashcards, practice quizzes, animations, exercises, web links, and more.
- *Environmental Science CourseMate for Environmental Science 14e* helps you make the grade.

  Environmental Science CourseMate includes:
  - an interactive eBook, with highlighting, note taking and search capabilities
  - interactive learning tools including:
    - Quizzes
    - Flashcards
    - Videos
    - Animations
    - and more!

  Go to **login.cengagebrain.com** to access these resources.
- *Global Environment Watch.* Updated several times a day, the Global Environment Watch is a focused portal into GREENR—the Global Reference on the Environment, Energy, and Natural Resources—an ideal one-stop site for classroom discussion and research projects. This resource center keeps courses up-to-date with the most current news on the environment. Users get access to information from trusted academic journals, news outlets, and magazines, as well as statistics, an interactive world map, videos, primary sources, case studies, podcasts, and much more. Login or purchase access at **www.cengagebrain.com/shop/isbn/9781423929444** to complete the exercises found at the end of each chapter.
- **New! *Virtual Field Trips in Environmental Issues.*** *Virtual Field Trips in Environmental Issues* brings the field to you, with dynamic panoramas, videos, photographs, maps, and quizzes covering important topics within environmental science. A case study approach covers the issues of *keystone species, climate change's role in extinctions, invasive species, the evolution of a species due to its environment*, and *an ecosystem approach to sustaining biodiversity*. Students are engaged, interacting with real issues to help them think critically about the world around them.
- *Audio Study Tools.* Students can download these useful study aids, which contain valuable information such as reviews of important concepts, key terms, questions, clarifications of common misconceptions, and study tips.
- Access to *InfoTrac® College Edition* for teachers and students using *CengageNOW* and *WebTutor* on WebCT or Blackboard. This fully searchable online library gives users access to complete environmental articles from several hundred current periodicals and to others dating back over 20 years.

The following materials for this textbook are available on the companion website at **www.cengagebrain.com/shop/isbn/9781111988937**:

- *Chapter Summaries* help guide student reading and study of each chapter.
- *Flashcards* and *Glossary* allow students to test their mastery of each chapter's Key Terms.
- *Chapter Tests* provide multiple-choice practice quizzes.
- Information on a variety of *Green Careers* and tips for sustainable living.
- *What Can You Do?* offers students resources for what they can do to effect individual changes relating to key environmental issues.
- *Weblinks* offer an extensive list of websites with news and research related to each chapter.
- *Explore More* library of articles and case studies.

Other student learning tools include:

- *Essential Study Skills for Science Students* by Daniel D. Chiras. This book includes chapters on developing good study habits, sharpening memory, getting the most out of lectures, labs, and reading assignments, improving test-taking abilities, and becoming a critical thinker. Available for students on instructor's request.

- *Lab Manual.* Edited by Edward Wells, this lab manual includes both hands-on and data analysis labs to help your students develop a range of skills. Create a custom version of this Lab Manual by adding labs you have written or ones from our collection with Cengage Custom Publishing. An Instructor's Manual for the labs will be available to adopters.
- *What Can You Do?* This guide presents students with a variety of ways that they can affect the environment, and shows them how to track the effect their actions have on their carbon footprint. Available for students on instructor's request.

Supplements for Instructors

- *PowerLecture.* This DVD, available to adopters, allows you to create custom lectures in Microsoft® PowerPoint using lecture outlines, all of the figures and photos from the text, bonus photos and animations. PowerPoint's editing tools allow use of slides from other lectures, modification or removal of figure labels and leaders, insertion of your own slides, saving slides as JPEG images, and preparation of lectures for use on the web.
- *Instructor's Manual.* Available to adopters. The Instructor's Manual has been thoughtfully revised and updated to make creating your lectures even easier. Some of the features include updated answers to the end-of-chapter questions, including suggested answers for the review questions and the quantitative exercises, and a revised video reference list with web resources. Also available on PowerLecture.
- *Test Bank.* Available to adopters. The Test Bank contains thousands of questions and answers in a variety of formats, including multiple choice, true/false, fill-in-the-blank, critical thinking questions, and essay questions. Also available on PowerLecture.
- *Transparencies.* Featuring most illustrations from the chapters, this set contains 250 printed transparencies of key figures, and 250 electronic masters. These electronic masters will allow you to print, in color, only those additional figures you need.
- *Environmental Science CourseMate.* How do you assess your students' engagement in your course? How do you know if your students have read the material or viewed the resources you've assigned? How can you tell if your students are struggling with a concept? With CourseMate, you can use the included Engagement Tracker to assess student preparation and engagement. Use the tracking tools to see progress for the class as a whole or for individual students. Identify students at risk early in the course. Uncover which concepts are most difficult for your class. Monitor time on task. Keep your students engaged.
- *ABC and BBC Videos for Environmental Science.* The 45 ABC clips and the 27 BBC clips are informative and short video clips that cover current news stories on environmental issues from around the world. These clips are a great way to start a lecture or spark a discussion. Available on DVD with a workbook, on the *PowerLecture* DVD, and within *CourseMate, CengageNOW,* and *WebTutor* with additional Internet activities.
- *ExamView.* This full-featured program helps you create and deliver customized tests (both print and online) in minutes, using its complete word processing capabilities. Available on the PowerLecture DVD.

Other Textbook Options

Instructors wanting a book with a different length and emphasis can use one of the three other books that we have written for various types of environmental science courses: *Living in the Environment,* 17th edition (675 pages, Brooks/Cole 2012), *Sustaining the Earth: An Integrated Approach,* 10th edition (330 pages, Brooks/Cole, 2012) and *Essentials of Ecology,* 6th edition (276 pages, Brooks/Cole, 2012).

Additionally, Cengage Learning offers custom solutions for your course—whether it's making a small modification to *Environmental Science* to match your syllabus or combining multiple sources to create something truly unique. You can pick and choose chapters, and add additional cases, readings, articles, or labs, to create a text that fits the way you teach. Contact your Cengage Learning representative to explore custom solutions for your course.

Help Us Improve This Book or Its Supplements

Let us know how you think this book can be improved. If you find any errors, bias, or confusing explanations, please e-mail us about them at:

mtg89@hotmail.com

spoolman@tds.net

Most errors can be corrected in subsequent printings of this edition, as well as in future editions.

Acknowledgments

We wish to thank the many students and teachers who have responded so favorably to the 13 previous editions of *Environmental Science,* the 17 editions of *Living in the Environment,* the 10 editions of *Sustaining the Earth,* and the 6 editions of *Essentials of Ecology,* and who have corrected errors and offered many helpful suggestions for improvement. We are also deeply indebted to the more than 295 reviewers, who pointed out errors and suggested many important improvements in the various editions of these three books.

It takes a village to produce a textbook, and the members of the talented production team, listed on the copyright page, have made vital contributions. Our special thanks go to development editor Jake Warde, production editors Hal Humphrey and Nicole Barone, copy editor Deborah Thompson, layout expert Judy Maenle, photo researcher Abigail Reip, artist Patrick Lane, media editor Alexandria Brady, assistant editor Alexis Glubka, editorial assistant Lauren Crosby, and Brooks/Cole's hard-working sales staff. Finally, we are very fortunate to have the guidance and inspiration of life sciences publisher Yolanda Cossio and her dedicated team of highly talented people who make this book possible.

G. Tyler Miller

Scott E. Spoolman

Guest Essayists

Guest essays by the following authors are available on *CengageNOW*: **M. Kat Anderson**, ethnoecologist with the National Plant Center of the USDA's Natural Resource Conservation Center; **Lester R. Brown**, president, Earth Policy Institute; **Alberto Ruz Buenfil**, environmental activist, writer, and performer; **Robert D. Bullard**, professor of sociology and director of the Environmental Justice Resource Center at Clark Atlanta University; **Michael Cain**, ecologist and adjunct professor at Bowdoin College; **Herman E. Daly**, senior research scholar at the School of Public Affairs, University of Maryland; **Lois Marie Gibbs**, director, Center for Health, Environment, and Justice; **Garrett Hardin**, professor emeritus (now deceased) of human ecology, University of California, Santa Barbara; **John Harte**, professor of energy and resources, University of California, Berkeley; **Paul G. Hawken**, environmental author and business leader; **Jane Heinze-Fry**, environmental educator; **Paul F. Kamitsuja**, infectious disease expert and physician; **Amory B. Lovins**, energy policy consultant and director of research, Rocky Mountain Institute; **Bobbi S. Low**, professor of resource ecology, University of Michigan; **John J. Magnuson**, Director Emeritus of the Center for Limnology, University of Wisconsin, Madison; **Lester W. Milbrath**, director of the research program in environment and society, State University of New York, Buffalo; **Peter Montague**, director, Environmental Research Foundation; **Norman Myers**, tropical ecologist and consultant in environment and development; **David W. Orr**, professor of environmental studies, Oberlin College; **Noel Perrin**, adjunct professor of environmental studies, Dartmouth College; **David Pimentel**, professor of insect ecology and agricultural sciences, Cornell University; **John Pichtel**, Ball State University; **Andrew C. Revkin**, environmental author and environmental reporter for the *New York Times*; **Vandana Shiva**, physicist, educator, environmental consultant; **Nancy Wicks**, ecopioneer and director of Round Mountain Organics; and **Donald Worster**, environmental historian and professor of American history, University of Kansas.

Pedagogy Contributors

Dr. Dean Goodwin and his colleagues, Berry Cobb, Deborah Stevens, Jeannette Adkins, Jim Lehner, Judy Treharne, Lonnie Miller, and Tom Mowbray provided excellent contributions to the Data Analysis and Ecological Footprint Analysis exercises. Mary Jo Burchart of Oakland Community College wrote the in-text *Global Environment Watch* exercises.

Cumulative List of Reviewers

Barbara J. Abraham, Hampton College; Donald D. Adams, State University of New York at Plattsburgh; Larry G. Allen, California State University, Northridge; Susan Allen-Gil, Ithaca College; James R. Anderson, U.S. Geological Survey; Mark W. Anderson, University of Maine; Kenneth B. Armitage, University of Kansas; Samuel Arthur, Bowling Green State University; Gary J. Atchison, Iowa State University; Thomas W. H. Backman, Lewis Clark State University; Marvin W. Baker, Jr., University of Oklahoma; Virgil R. Baker, Arizona State University; Stephen W. Banks, Louisiana State University in Shreveport; Ian G. Barbour, Carleton College; Albert J. Beck, California State University, Chico; Eugene C. Beckham, Northwood University; Diane B. Beechinor, Northeast Lakeview College; W. Behan, Northern Arizona University; David Belt, Johnson County Community College; Keith L. Bildstein, Winthrop College; Andrea Bixler, Clarke College; Jeff Bland, University of Puget Sound; Roger G. Bland, Central Michigan University; Grady Blount II, Texas A&M University, Corpus Christi; Lisa K. Bonneau, University of Missouri-Kansas City; Georg Borgstrom, Michigan State University; Arthur C. Borror, University of New Hampshire; John H. Bounds, Sam Houston State University; Leon F. Bouvier, Population Reference Bureau; Daniel J. Bovin, Université Laval; Jan Boyle, University of Great Falls; James A. Brenneman, University of Evansville; Michael F. Brewer, Resources for the Future, Inc.; Mark M. Brinson, East Carolina University; Dale Brown, University of Hartford; Patrick E. Brunelle, Contra Costa College; Terrence J. Burgess, Saddleback College North; David Byman, Pennsylvania State University, Worthington–Scranton; Michael L. Cain, Bowdoin College, Lynton K. Caldwell, Indiana University; Faith Thompson Campbell, Natural Resources Defense Council, Inc.; John S. Campbell, Northwest College; Ray Canterbery, Florida State University; Ted J. Case, University of San Diego; Ann Causey, Auburn University; Richard A. Cellarius, Evergreen State University; William U. Chandler, Worldwatch Institute; F. Christman, University of North Carolina, Chapel Hill; Lu Anne Clark, Lansing Community College; Preston Cloud, University of California, Santa Barbara; Bernard C. Cohen, University of Pittsburgh; Richard A. Cooley, University of California, Santa Cruz; Dennis J. Corrigan; George Cox, San Diego State University; John D. Cunningham, Keene State College; Herman E. Daly, University of Maryland; Raymond F. Dasmann, University of California, Santa Cruz; Kingsley Davis, Hoover Institution; Edward E. DeMartini, University of California, Santa Barbara; James Demastes, University of Northern Iowa; Charles E. DePoe, Northeast Louisiana University; Thomas R. Detwyler, University of Wisconsin; Bruce DeVantier, Southern Illinois University Carbondale; Peter H. Diage, University of California, Riverside; Stephanie Dockstader, Monroe Community College; Lon D. Drake, University of Iowa; Michael Draney, University of Wisconsin–Green Bay; David DuBose, Shasta College; Dietrich Earnhart, University of Kansas; Robert East, Washington & Jefferson College; T. Edmonson, University of Washington; Thomas Eisner, Cornell University; Michael Esler, Southern Illinois University; David E. Fairbrothers, Rutgers University; Paul P. Feeny, Cornell University; Richard S. Feldman, Marist College; Vicki Fella-Pleier, La Salle University; Nancy Field, Bellevue Community College; Allan Fitzsimmons, University of Kentucky; Andrew J. Friedland, Dartmouth College; Kenneth O. Fulgham, Humboldt State University; Lowell L. Getz, University of Illinois at Urbana–Champaign; Frederick F. Gilbert, Washington State University; Jay Glassman, Los Angeles Valley College; Harold Goetz, North Dakota State University; Srikanth Gogineni, Axia College of University of Phoenix; Jeffery J. Gordon, Bowling Green State University; Eville Gorham, University of Minnesota; Michael Gough, Resources for the Future; Ernest M. Gould, Jr., Harvard University; Peter Green, Golden West College; Katharine B. Gregg, West Virginia Wesleyan College; Paul K. Grogger, University of Colorado at Colorado Springs; L. Guernsey, Indiana State University; Ralph Guzman, University of California, Santa Cruz; Raymond Hames, University of Nebraska, Lincoln; Robert Hamilton IV, Kent State University, Stark Campus; Raymond E. Hampton, Central Michigan University; Ted L. Hanes, California State University, Fullerton; William S. Hardenbergh, Southern Illinois University at Carbondale; John P. Harley, Eastern Kentucky University; Neil A. Harriman, University of Wisconsin, Oshkosh; Grant A. Harris, Washington State University; Harry S. Hass, San Jose City College; Arthur N. Haupt, Population Reference Bureau; Denis A. Hayes, environmental consultant; Stephen Heard, University of Iowa; Gene Heinze-Fry, Department of Utilities, Commonwealth of Massachusetts; Jane Heinze-Fry, environmental educator; John G. Hewston, Humboldt State University; David L. Hicks, Whitworth College; Kenneth M. Hinkel, University of Cincinnati; Eric Hirst, Oak Ridge National Laboratory; Doug Hix, University of Hartford; S. Holling, University of British Columbia; Sue Holt, Cabrillo College; Donald Holtgrieve, California State University, Hayward; Michelle Homan, Gannon University; Michael H. Horn, California State University, Fullerton; Mark A. Hornberger, Bloomsberg University; Marilyn Houck, Pennsylvania State University; Richard D. Houk, Winthrop College; Robert J. Huggett, College of William and Mary; Donald Huisingh, North Carolina State University; Catherine Hurlbut, Florida Community College at Jacksonville; Marlene K. Hutt, IBM; David R. Inglis, University of Massachusetts; Robert Janiskee, University of South Carolina; Hugo H. John, University of

Connecticut; Brian A. Johnson, University of Pennsylvania, Bloomsburg; David I. Johnson, Michigan State University; Mark Jonasson, Crafton Hills College; Zoghlul Kabir, Rutgers/New Brunswick; Agnes Kadar, Nassau Community College; Thomas L. Keefe, Eastern Kentucky University; David Kelley, University of St. Thomas; William E. Kelso, Louisiana State University; Nathan Keyfitz, Harvard University; David Kidd, University of New Mexico; Pamela S. Kimbrough; Jesse Klingebiel, Kent School; Edward J. Kormondy, University of Hawaii–Hilo/West Oahu College; John V. Krutilla, Resources for the Future, Inc.; Judith Kunofsky, Sierra Club; E. Kurtz; Theodore Kury, State University of New York at Buffalo; Steve Ladochy, University of Winnipeg; Troy A. Ladine, East Texas Baptist University; Anna J. Lang, Weber State University; Mark B. Lapping, Kansas State University; Michael L. Larsen, Campbell University; Linda Lee, University of Connecticut; Tom Leege, Idaho Department of Fish and Game; Maureen Leupold, Genesee Community College; William S. Lindsay, Monterey Peninsula College; E. S. Lindstrom, Pennsylvania State University; M. Lippiman, New York University Medical Center; Valerie A. Liston, University of Minnesota; Dennis Livingston, Rensselaer Polytechnic Institute; James P. Lodge, air pollution consultant; Raymond C. Loehr, University of Texas at Austin; Ruth Logan, Santa Monica City College; Robert D. Loring, DePauw University; Paul F. Love, Angelo State University; Thomas Lovering, University of California, Santa Barbara; Amory B. Lovins, Rocky Mountain Institute; Hunter Lovins, Rocky Mountain Institute; Gene A. Lucas, Drake University; Claudia Luke; David Lynn; Timothy F. Lyon, Ball State University; Stephen Malcolm, Western Michigan University; Melvin G. Marcus, Arizona State University; Gordon E. Matzke, Oregon State University; Parker Mauldin, Rockefeller Foundation; Marie McClune, The Agnes Irwin School (Rosemont, Pennsylvania); Theodore R. McDowell, California State University; Vincent E. McKelvey, U.S. Geological Survey; Robert T. McMaster, Smith College; John G. Merriam, Bowling Green State University; A. Steven Messenger, Northern Illinois University; John Meyers, Middlesex Community College; Raymond W. Miller, Utah State University; Arthur B. Millman, University of Massachusetts, Boston; Sheila Miracle, Southeast Kentucky Community & Technical College; Fred Montague, University of Utah; Rolf Monteen, California Polytechnic State University; Debbie Moore, Troy University Dothan Campus; Michael K. Moore, Mercer University; Ralph Morris, Brock University, St. Catherine's, Ontario, Canada; Angela Morrow, Auburn University; William W. Murdoch, University of California, Santa Barbara; Norman Myers, environmental consultant; Brian C. Myres, Cypress College; A. Neale, Illinois State University; Duane Nellis, Kansas State University; Jan Newhouse, University of Hawaii, Manoa; Jim Norwine, Texas A&M University, Kingsville; John E. Oliver, Indiana State University; Mark Olsen, University of Notre Dame; Carol Page, copy editor; Eric Pallant, Allegheny College; Bill Paletski, Penn State University; Charles F. Park, Stanford University; Richard J. Pedersen, U.S. Department of Agriculture, Forest Service; David Pelliam, Bureau of Land Management, U.S. Department of Interior; Murray Paton Pendarvis, Southeastern Louisiana University; Dave Perault, Lynchburg College; Rodney Peterson, Colorado State University; Julie Phillips, De Anza College; John Pichtel, Ball State University; William S. Pierce, Case Western Reserve University; David Pimentel, Cornell University; Peter Pizor, Northwest Community College; Mark D. Plunkett, Bellevue Community College; Grace L. Powell, University of Akron; James H. Price, Oklahoma College; Marian E. Reeve, Merritt College; Carl H. Reidel, University of Vermont; Charles C. Reith, Tulane University; Roger Revelle, California State University, San Diego; L. Reynolds, University of Central Arkansas; Ronald R. Rhein, Kutztown University of Pennsylvania; Charles Rhyne, Jackson State University; Robert A. Richardson, University of Wisconsin; Benjamin F. Richason III, St. Cloud State University; Jennifer Rivers, Northeastern University; Ronald Robberecht, University of Idaho; William Van B. Robertson, School of Medicine, Stanford University; C. Lee Rockett, Bowling Green State University; Terry D. Roelofs, Humboldt State University; Daniel Ropek, Columbia George Community College; Christopher Rose, California Polytechnic State University; Richard G. Rose, West Valley College; Stephen T. Ross, University of Southern Mississippi; Robert E. Roth, Ohio State University; Dorna Sakurai, Santa Monica College; Arthur N. Samel, Bowling Green State University; Shamili Sandiford, College of DuPage; Floyd Sanford, Coe College; David Satterthwaite, I.E.E.D., London; Stephen W. Sawyer, University of Maryland; Arnold Schecter, State University of New York; Frank Schiavo, San Jose State University; William H. Schlesinger, Ecological Society of America; Stephen H. Schneider, National Center for Atmospheric Research; Clarence A. Schoenfeld, University of Wisconsin, Madison; Madeline Schreiber, Virginia Polytechnic Institute; Henry A. Schroeder, Dartmouth Medical School; Lauren A. Schroeder, Youngstown State University; Norman B. Schwartz, University of Delaware; George Sessions, Sierra College; David J. Severn, Clement Associates; Don Sheets, Gardner-Webb University; Paul Shepard, Pitzer College and Claremont Graduate School; Michael P. Shields, Southern Illinois University at Carbondale; Kenneth Shiovitz; F. Siewert, Ball State University; E. K. Silbergold, Environmental Defense Fund; Joseph L. Simon, University of South Florida; William E. Sloey, University of Wisconsin, Oshkosh; Robert L. Smith, West Virginia University; Val Smith, University of Kansas; Howard M. Smolkin, U.S. Environmental Protection Agency; Patricia M. Sparks, Glassboro State College; John E. Stanley, University of Virginia; Mel Stanley, California State Polytechnic University, Pomona; Richard Stevens, Monroe Community College; Norman R. Stewart, University of Wisconsin, Milwaukee; Frank E. Studnicka, University of Wisconsin, Platteville; Chris Tarp, Contra Costa College; Roger E. Thibault, Bowling Green State University; William L. Thomas, California State University, Hayward; Shari Turney, copy editor; John D. Usis, Youngstown State University; Tinco E. A. van Hylckama, Texas Tech University; Robert R. Van Kirk, Humboldt State University; Donald E. Van Meter, Ball State University; Rick Van Schoik, San Diego State University; Gary Varner, Texas A&M University; John D. Vitek, Oklahoma State University; Harry A. Wagner, Victoria College; Lee B. Waian, Saddleback College; Warren C. Walker, Stephen F. Austin State University; Thomas D. Warner, South Dakota State University; Kenneth E. F. Watt, University of California, Davis; Alvin M. Weinberg, Institute of Energy Analysis, Oak Ridge Associated Universities; Brian Weiss; Margery Weitkamp, James Monroe High School (Granada Hills, California); Anthony Weston, State University of New York at Stony Brook; Raymond White, San Francisco City College; Douglas Wickum, University of Wisconsin, Stout; Charles G. Wilber, Colorado State University; Nancy Lee Wilkinson, San Francisco State University; John C. Williams, College of San Mateo; Ray Williams, Rio Hondo College; Roberta Williams, University of Nevada, Las Vegas; Samuel J. Williamson, New York University; Dwina Willis, Freed-Hardeman University; Ted L. Willrich, Oregon State University; James Winsor, Pennsylvania State University; Fred Witzig, University of Minnesota at Duluth; Martha Wolfe, Elizabethtown Community and Technical College; George M. Woodwell, Woods Hole Research Center; Todd Yetter, University of the Cumberlands; Robert Yoerg, Belmont Hills Hospital; Hideo Yonenaka, San Francisco State University; Brenda Young, Daemen College; Anita Závodská, Barry University; Malcolm J. Zwolinski, University of Arizona.

About the Authors

G. Tyler Miller

G. Tyler Miller has written or co-authored 60 editions of various textbooks for introductory courses in environmental science, basic ecology, energy, and environmental chemistry. Since 1975, Miller's books have been the most widely used textbooks for environmental science in the United States and throughout the world. They have been used by almost 3 million students and have been translated into eight languages.

Miller has a PhD from the University of Virginia and has received two honorary doctorate degrees for his contributions to environmental education. He taught college for 20 years and developed an innovative interdisciplinary undergraduate science program before deciding to write environmental science textbooks full time in 1975.

He describes his hopes for the future as follows:

If I had to pick a time to be alive, it would be the next 75 years. Why? First, there is overwhelming scientific evidence that we are in the process of seriously degrading our own life-support system. In other words, we are living unsustainably. Second, within the next 75 years we have the opportunity to learn how to live more sustainably by working with the rest of nature, as described in this book.

I am fortunate to have three smart, talented, and wonderful sons—Greg, David, and Bill. I am especially privileged to have Kathleen as my wife, best friend, and research associate. It is inspiring to have a brilliant, beautiful (inside and out), and strong woman who cares deeply about nature as a lifemate. She is my hero. I dedicate this book to her and to the earth.

Scott Spoolman

Scott Spoolman is a writer with over 25 years of experience in educational publishing. He has worked with Tyler Miller since 2003, first as a contributing editor and now as coauthor on several editions of *Living in the Environment, Environmental Science,* and *Sustaining the Earth*.

Spoolman holds a master's degree in science journalism from the University of Minnesota. He has authored numerous articles in the fields of science, environmental engineering, politics, and business. He worked as an acquisitions editor on a series of college forestry textbooks. He has also worked as a consulting editor in the development of more than 70 college and high school textbooks in the natural and social sciences fields.

In his free time, he enjoys exploring the forests and waters of his native Wisconsin along with his family—his wife, environmental educator Gail Martinelli, and his children, Will and Katie.

Spoolman has the following to say about his collaboration with Tyler Miller.

I am honored to be working with Tyler Miller as a coauthor to continue the Miller tradition of thorough, clear, and engaging writing about the vast and complex field of environmental science. I share Tyler Miller's passion for ensuring that these textbooks and their multimedia supplements will be valuable tools for students and instructors. To that end, we strive to introduce this interdisciplinary field in ways that will be informative and sobering, but also tantalizing and motivational.

If the flip side of any problem is indeed an opportunity, then this truly is one of the most exciting times in history for students to start an environmental career. Environmental problems are numerous, serious, and daunting, but their possible solutions generate exciting new career opportunities. We place high priorities on inspiring students with these possibilities, challenging them to maintain a scientific focus, pointing them toward rewarding and fulfilling careers, and in doing so, working to help sustain life on the earth.

My Environmental Journey

G. Tyler Miller

My environmental journey began in 1966 when I heard a lecture on population and pollution problems by Dean Cowie, a biophysicist with the U.S. Geological Survey. It changed my life. I told him that if even half of what he said was valid, I would feel ethically obligated to spend the rest of my career teaching and writing to help students learn about the basics of environmental science. After spending six months studying the environmental literature I told him that he had greatly underestimated the seriousness of these problems.

I developed one of the country's first undergraduate environmental studies programs and in 1971 published my first introductory environmental science book, an interdisciplinary study of the connections between energy laws (thermodynamics), chemistry, and ecology. In 1975, I published my first environmental science textbook. And here we are, 60 editions of my various environmental science textbooks later, with the 14th edition of this book.

Beginning in 1985, I spent 10 years living and writing environmental science textbooks in the deep woods in an adapted school bus that I used as an environmental science laboratory. I evaluated the use of passive solar energy design to heat the structure, buried earth tubes to bring in air cooled by the earth (geothermal cooling) at a cost of about $1 per summer, set up active and passive solar systems to provide hot water, supplemented passive solar heating with an energy-efficient instant hot water system powered by LPG, installed energy-efficient windows and appliances, and a composting (waterless) toilet, employed biological pest control, composted food wastes, used natural planting (no grass or lawnmowers), gardened organically, and experimented with a number of other potential solutions to major environmental problems that we face. I learned a lot and had great fun.

I also used this time to learn and think about how nature works by studying the natural plants and animals around me and by thinking about how nature has sustained an incredible variety of life for several billion years on the marvelous planet that is our home. My experience from *living in nature* is reflected in much of the material in this book. It also helped me come up with the three simple principles of sustainability that serve as the integrating theme for this textbook and to apply these principles to living my life more sustainably.

I came out of the woods in 1995 to learn about how to live more sustainably in an urban setting where most people live. Since then I have lived in two urban villages, one in a small town and one within a large urban area where walking was my primary method for getting anywhere.

Since 1970 my goal has been to use a car as little as possible. Since I work at home I have a "low-pollute commute" from my bedroom to a chair and a laptop computer. I usually take one or two airplane trips a year to visit my sister and my publisher.

As you will learn in this book, life involves a series of environmental tradeoffs. I still have a large, harmful environmental impact, but I continue struggling to reduce it. I hope you will join me in striving to live more sustainably and sharing what you learn with others. It is not always easy but it sure is fun.

CENGAGE LEARNING'S COMMITMENT TO SUSTAINABLE PRACTICES

We the authors of this textbook and Cengage Learning, the publisher, are committed to making the publishing process as sustainable as possible. This involves four basic strategies:

- *Using sustainably produced paper.* The book publishing industry is committed to increasing the use of recycled fibers, and Cengage Learning is always looking for ways to increase this content. Cengage Learning works with paper suppliers to maximize the use of paper that contains only wood fibers that are certified as sustainably produced, from the growing and cutting of trees all the way through paper production. (See page 185 in this book for more information on *certification of forest products* such as paper.)
- *Reducing resources used per book.* The publisher has an ongoing program to reduce the amount of wood pulp, virgin fibers, and other materials that go into each sheet of paper used. New, specially designed printing presses also reduce the amount of scrap paper produced per book.
- *Recycling.* Printers recycle the scrap paper that is produced as part of the printing process. Cengage Learning also recycles waste cardboard from shipping cartons, along with other materials used in the publishing process.
- *Process improvements.* In years past, publishing has involved using a great deal of paper and ink for writing and editing of manuscripts, copyediting, reviewing page proofs, and creating illustrations. Almost all of these materials are now saved through use of electronic files. Except for our review of page proofs, very little paper and ink were used in the preparation of this textbook.

Learning Skills

Students who can begin early in their lives to think of things as connected, even if they revise their views every year, have begun the life of learning.

MARK VAN DOREN

Why Is It Important to Study Environmental Science?

Welcome to **environmental science**—an *interdisciplinary* study of how the earth works, how we interact with the earth, and how we can deal with the environmental problems we face. Because environmental issues affect every part of your life, the concepts, information, and issues discussed in this book and in the course you are taking will be useful to you now and throughout your life.

Understandably, we are biased, but *we strongly believe that environmental science will be the single most important course in your education*. What could be more important than learning how the earth works, how we affect its life-support system, and how we can reduce our environmental impact?

We live in an incredibly challenging era. We are becoming increasingly aware that during this century we need to make a new cultural transition in which we learn how to live more sustainably by sharply reducing our degradation of our life-support system. We hope this book will inspire you to become involved in this change in the way we view and treat the earth, which sustains us, our economies, and all other living things.

You Can Improve Your Study and Learning Skills

Maximizing your ability to learn should be one of your most important lifetime educational goals. It involves continually trying to *improve your study and learning skills*. Here are some suggestions for doing so:

Develop a passion for learning. As the famous physicist and philosopher Albert Einstein put it, "I have no special talent. I am only passionately curious."

Get organized. Becoming more efficient at studying gives you more time for other interests.

Make daily to-do lists in writing. Put items in order of importance, focus on the most important tasks, and assign a time to work on these items. Because life is full of uncertainties, you might be lucky to accomplish half of the items on your daily list. Shift your schedule as needed to accomplish the most important items.

Set up a study routine in a distraction-free environment. Develop a written daily study schedule and stick to it. Study in a quiet, well-lighted space. Work while sitting at a desk or table—not lying down on a couch or bed. Take breaks every hour or so to help you stay focused. During each break, take several deep breaths and move around.

Avoid procrastination. Avoid putting work off until another time. Do not fall behind on your reading and other assignments. Set aside a particular time for studying each day.

Do not eat dessert first. Otherwise, you may never get to the main meal (studying). When you have accomplished your study goals, reward yourself with dessert (play or leisure).

Make hills out of mountains. It is psychologically difficult to climb a mountain, which is what reading an entire book, reading a chapter in a book, writing a paper, or cramming to study for a test can feel like. Instead, break these large tasks (mountains) down into a series of small tasks (hills). Each day, read a few pages of a book or chapter, write a few paragraphs of a paper, and review what you have studied and learned. As American automobile designer and builder Henry Ford put it, "Nothing is particularly hard if you divide it into small jobs."

Look at the big picture first. Get an overview of an assigned reading in this book by looking at the *Key Questions and Concepts* box at the beginning of each chapter. It lists both the key questions explored in the chapter sections and their corresponding key concepts. Use this list as a chapter roadmap. When you finish a chapter you can also use the list to review.

Ask and answer questions as you read. For example, "What is the main point of a particular subsection or paragraph?" Relate your own questions to the key questions and key concepts addressed in each major chapter section. In this way, you can write out a chapter outline to help you understand the chapter material.

Focus on key terms. Use the glossary in your textbook to look up the meanings of terms or words you do not understand. This book shows all key terms in **bold** type and lesser, but still important, terms in *italicized* type. The review questions at the end of each chapter also include the chapter's key terms in bold. Flash cards for testing your mastery of key terms for each chapter are available on the website for this book, or you can make your own by putting a term on one side of an index card or piece of paper and its meaning on the other side.

Interact with what you read. We suggest that you mark key sentences and paragraphs with a highlighter or pen. You might put an asterisk in the margin next to material you think is important and double asterisks next to material you think is especially important. Write comments in the margins such as *beautiful, confusing, misleading,* or *wrong.* You could fold down the top corners of pages on which you highlighted passages. Then you can flip through a chapter or book and quickly review the key ideas.

Review to reinforce learning. Before each class session, review the material you learned in the previous session.

Become a good note taker. Do not try to take down everything your instructor says. Instead, write down main points and key facts using your own shorthand system. Review, fill in, and organize your notes as soon as possible after each class.

Check what you have learned. At the end of each chapter you will find review questions that cover all of the key material in each chapter section. We suggest that you try to answer each of these questions after studying each chapter section.

Write out answers to questions to focus and reinforce learning. Write down your answers to the critical thinking questions found in the *Thinking About* boxes throughout the chapters, in many figure captions, and at the end of each chapter. These questions are designed to inspire you to think critically about key ideas and connect them to other ideas and your own life. Also, write down your answers to the end-of-chapter review questions. The website for each chapter has an additional detailed list of review questions for that chapter. Save your answers for review and test preparation.

Use the buddy system. Study with a friend or become a member of a study group to compare notes, review material, and prepare for tests. Explaining something to someone else is a great way to focus your thoughts and reinforce your learning. Attend any review sessions offered by instructors or teaching assistants.

Learn your instructor's testing style. Does your instructor emphasize multiple-choice, fill-in-the-blank, true-or-false, factual, or essay questions? How much of the test will come from the textbook and how much from lecture material? You may not like your instructor's testing style or feel that it works well, but the reality is that your instructor is in charge.

Become a good test taker. Avoid cramming. Eat well and get plenty of sleep before a test. Arrive on time or early. Calm yourself and increase your oxygen intake by taking several deep breaths. (Do this also about every 10–15 minutes while taking the test.) Look over the test and answer the questions you know well first. Then work on the harder ones. Use the process of elimination to narrow down the choices for multiple-choice questions. Paring them down to two choices gives you a 50% chance of guessing the right answer. For essay questions, organize your thoughts before you start writing. If you have no idea what a question means, make an educated guess. You might earn some partial credit by writing something like this: "If this question means so and so, then my answer is ________."

Develop an optimistic but realistic outlook. Try to be a "glass is half full" rather than a "glass is half empty" person. Pessimism, fear, anxiety, and excessive worrying (especially over things you cannot control) are destructive and can lead to inaction. Try to nurture energizing feelings of realistic optimism instead of immobilizing feelings of pessimism.

Take time to enjoy life. Every day, take time to laugh, to enjoy nature, beauty, and friendship, and to pursue the things you enjoy.

You Can Improve Your Critical Thinking Skills

Critical thinking involves developing skills to analyze information and ideas, judge their validity, and make decisions. Critical thinking helps you to distinguish between facts and opinions, evaluate evidence and arguments, take and defend informed positions on issues, integrate information and see relationships, and apply your knowledge to dealing with new and different problems, and to your own lifestyle choices. Here are some basic skills for learning how to think more critically.

Question everything and everybody. Be skeptical, as any good scientist is. Do not believe everything you hear and read, including the content of this textbook, without evaluating the information you receive. Seek other sources and opinions.

Identify and evaluate your personal biases and beliefs. Each of us has biases and beliefs taught to us by our parents, teachers, friends, and role models. What are your basic beliefs, values, and biases? Where did they come from? What assumptions are they based on? How sure are you that they are right and why? According to the American psychologist and philosopher William James, "A great many people think they are thinking when they are merely rearranging their prejudices."

Be open-minded and flexible. Be open to considering different points of view. Suspend judgment until you gather more evidence, and be willing to change your mind. Recognize that there may be a number of useful and acceptable solutions to a problem. There are trade-offs involved in dealing with any environmental issue, as you will learn in this book. One way to evaluate divergent views is to try to take the viewpoints of other people. How do they see the world? What are their basic assumptions and beliefs? Are their positions logically consistent with their assumptions and beliefs?

Be humble about what you know. Some people are so confident in what they know that they stop thinking and questioning. Always be willing to question your own thinking.

Evaluate how the information related to an issue was obtained. Are the statements you heard or read based on firsthand knowledge and research or on hearsay? Are unnamed sources used? Is the information based on reproducible and widely accepted scientific studies or on preliminary scientific results that may be valid but need further testing? Is the information based on a few isolated stories or experiences or on carefully controlled studies with the results reviewed by experts in the field involved?

Question the evidence and conclusions presented. What are the conclusions or claims for the information? What evidence is presented to support them? Does the evidence support them? Is there a need to gather more evidence to test the conclusions? Are there other, more reasonable conclusions?

Try to uncover differences in basic beliefs and assumptions. On the surface, most arguments or disagreements involve differences of opinion about the validity or meaning of certain facts or conclusions. Scratch a little deeper and you will find that most disagreements are based on different (and often hidden) basic assumptions concerning how we look at and interpret the world. Try to identify such differences.

Try to identify and assess any motives on the part of those presenting evidence and drawing conclusions. What is their expertise in this area? Do they have any unstated assumptions, beliefs, biases, or values? Do they have a personal agenda? Can they benefit financially or politically from acceptance of their evidence and conclusions? Would investigators with different basic assumptions or beliefs take the same data and come to different conclusions?

Expect and tolerate uncertainty. Recognize that scientists cannot establish absolute proof or certainty about anything. However, the reliable results of science can have a high degree of certainty.

Check the arguments you hear and read for logical fallacies and debating tricks. Here are six examples of such debating tricks. *First,* attack the presenter of an argument rather than the argument itself. *Second,* appeal to emotion rather than facts and logic. *Third,* claim that if one piece of evidence or one conclusion is false, then all other related pieces of evidence and conclusions are false. *Fourth,* say that a conclusion is false because it has not been scientifically proven. (Scientists never prove anything absolutely, but they can often establish high degrees of certainty.) *Fifth,* use irrelevant or misleading information to divert attention from important points. *Sixth,* present only either/or alternatives when there may be a number of options.

Do not believe everything you read on the Internet. The Internet is a wonderful and easily accessible source of information that provides explanations and opinions on almost any subject or issue. Web logs, or blogs, have become a major source of information. However, because the Internet is so open, anyone can post anything they want to some blogs and other websites with no editorial control or review by experts. As a result, evaluating information on the Internet is one of the best ways to put into practice your critical thinking skills. Use the Internet, but think critically and proceed with caution.

Develop principles or rules for evaluating evidence. Develop a written list of principles to serve as guidelines for evaluating evidence and claims. Continually evaluate and modify this list on the basis of your experience.

Become a seeker of wisdom, not a vessel of information. Many people believe that the main goal of education is to learn as much as you can by gathering more and more information. We believe that the primary goal is to learn how to sift through mountains of facts and ideas to find the few *nuggets of wisdom* that are the most useful for understanding the world and for making decisions. This book is full of facts and numbers, but they are useful only to the extent that they lead to an understanding of key ideas, scientific laws, theories, concepts, and connections. The major goals of the study of environmental science are to find out how nature works and sustains itself (*environmental wisdom*) and to use *principles of environmental wisdom* to help make human societies and economies more sustainable, more just, and more beneficial and enjoyable for all. As writer Sandra Carey observed, "Never mistake knowledge for wisdom. One helps you make a living; the other helps you make a life." Or as American writer Walker Percy suggested, "Some individuals with a high intelligence but lacking wisdom can get all A's and flunk life."

To help you practice critical thinking, we have supplied questions throughout this book, found within each chapter in brief boxes labeled *Thinking About,* in the captions of many figures, and at the end of each chapter. There are no right or wrong answers to many of these questions.

Use the Learning Tools We Offer in this Book

We have included a number of tools throughout this textbook that can help you improve your learning skills. First, use the *Key Questions and Concepts* list at the beginning of each chapter to preview and review a chapter.

Next, note that we use three different logos throughout the text. The *Core Case Study* logos show you how material throughout the chapter connects to the chapter-opening core case study. When you see the *sustainability* logo, you will know that you have just read something that relates directly to the overarching theme of this text, summarized by the three *principles of sustainability,* which are introduced in Figure 1-3 and which appear on the back

CORE CASE STUDY

SUSTAINABILITY

cover of the student edition. The *Good News* logos GOOD NEWS are intended to show you examples of successes that people have had in dealing with the environmental challenges we face.

We also include *Connections* boxes to show you some of the often surprising connections between environmental problems or processes and some of the products and services we use every day. These, along with the *Thinking About* boxes scattered throughout the text, are intended to get you to think about your environmental impact.

At the end of each chapter, we list what we consider to be the *three big ideas* that you should take away from the chapter. And following that list in each chapter, we have a *Revisiting* box that reviews the Core Case Study and explains how the principles of sustainability can be applied to the Core Case Study and other key material.

Finally, we have a *Review* section at the end of each chapter with questions listed for each chapter section. These questions cover all of the key material and key terms in each chapter. More exercises and projects follow this review section at the end of each chapter.

Know Your Own Learning Style

People have different ways of learning and it can be helpful to know your own learning style. *Visual learners* learn best from reading and viewing illustrations and diagrams. They can benefit from using flash cards (available on the website for this book) to memorize key terms and ideas. This is a highly visual book with many carefully selected photographs and diagrams.

Auditory learners learn best by listening and discussing. They might benefit from reading aloud while studying and taping lectures for study and review. *Logical learners* learn best by using concepts and logic to understand a subject.

Part of what determines your learning style is how your brain works. According to the *split-brain hypothesis*, the left hemisphere of your brain is good at logic, analysis, and evaluation, and the right half of your brain is good at visualizing, synthesizing, and creating. We provide material that stimulates both sides of your brain.

The study and critical thinking skills encouraged in this book and in most courses largely involve the left brain. However, you can also learn by letting your creative right brain loose. You can do this by brainstorming ideas with classmates with the rule that no left-brain criticism is allowed until the session is over.

When you are trying to solve a problem, try to rest, meditate, take a walk, exercise, or do something to shut down your controlling left-brain activity and allow the right side of your brain to work on the problem.

This Book Presents a Positive and Realistic Environmental Vision of the Future

Making and implementing environmental decisions always involves *trade-offs*. Our goal is to give balanced presentations of different viewpoints, and the advantages and disadvantages of various technologies and proposed solutions to environmental problems, as well as the good and bad news about environmental problems, without injecting personal bias.

Studying a subject as important as environmental science and ending up with no conclusions, opinions, and beliefs means that both teacher and student have failed. However, any conclusions one does reach must be based on using critical thinking to evaluate different ideas and to understand the trade-offs involved. Our goal is to present a positive vision of our environmental future based on realistic optimism.

Help Us Improve This Book

Researching and writing a book that covers and connects ideas in such a wide variety of disciplines is a challenging and exciting task. Almost every day, we learn about some new connection in nature.

In a book this complex, there are bound to be some errors—some typographical mistakes that slip through and some statements that you might question, based on your knowledge and research. We invite you to contact us and point out any bias, correct any errors you find, and suggest ways to improve this book. Please e-mail your suggestions to Tyler Miller at **mtg89@hotmail.com** or Scott Spoolman at **spoolman@tds.net**.

Now start your journey into this fascinating and important study of how the earth works and how we can leave the planet in a condition at least as good as what we found. Have fun.

Study nature, love nature, stay close to nature. It will never fail you.

FRANK LLOYD WRIGHT

Environmental Problems, Their Causes, and Sustainability

1

A Vision of a More Sustainable World in 2060

CORE CASE STUDY

Emily Briggs and Michael Rodriguez graduated from college in 2014. Michael earned a masters degree in environmental education, became a middle-school teacher, and loved teaching environmental science. Emily, meanwhile, went to law school and later established a thriving practice as an environmental lawyer.

In 2022, Michael and Emily met when they were doing volunteer work for an environmental organization. They later got married, had a child, and taught her about some of the world's environmental problems (Figure 1-1, left) and about the joys of nature that they had experienced as children (Figure 1-1, right). As a result, their daughter also became deeply involved in working to promote a more sustainable world and eventually passed this goal on to her own child.

When Michael and Emily ware growing up, there had been increasing signs of stress on the earth's life support system—its land, air, water, and wildlife—due to the harmful environmental impacts of more people consuming more resources. But a major transition in environmental awareness began around 2010 when a growing number of people began to transform their lifestyles and economies to be more in tune with the ways in which nature had sustained itself for billions of years before humans walked the earth. Over several decades, this combination of environmental awareness and action paid off.

In January of 2060, Emily and Michael celebrated the birth of their grandchild. He was born into a world that was still rich with a great diversity of plants, animals, and ecosystems. The loss of this biological diversity, which had been a looming threat when Michael and Emily were young adults, had slowed to a trickle. And the atmosphere, oceans, lakes, and rivers were gradually cleansing themselves.

Energy waste had been cut in half. Renewable energy from the sun, wind, flowing water, underground heat, and fuels produced from prairie grasses and algae had largely replaced nonrenewable energy from highly polluting oil and coal and from nuclear power with its dangerous, long-lived radioactive wastes. By 2050, significant atmospheric warming and the resulting climate change had occurred as many climate scientists had projected in the 1990s. But the threat of further climate change had begun to decrease, as the use of cleaner energy resources and efforts to reduce energy waste became the norm.

By 2060, farmers producing most of the world's food had shifted to farming practices that helped to conserve water and renew depleted soils. In addition, the human population had peaked at about 8 billion in 2040, instead of at the projected 9.6 billion, and then had begun a slow decline.

In 2060, Emily and Michael felt a great sense of pride, knowing that they and their child and countless others had helped to bring about these improvements so that future generations could live more sustainably on this marvelous planet that is our only home.

Sustainability is the capacity of the earth's natural systems and human cultural systems to survive, flourish, and adapt to changing environmental conditions into the very long-term future. It is about people caring enough to pass on a better world to all the generations to come. And it is the overarching theme of this textbook. Here, we describe the environmental problems we face, and we explore possible solutions. Our goal is to present to you a realistic and hopeful vision of what could be.

Mostovyi Sergii Igorevich/Shutterstock.com

Colin Hawkins/Getty Images

Figure 1-1 These parents—like Emily and Michael in our fictional vision of a possible world in 2060—are teaching their children about some of the world's environmental problems (left) and helping them to enjoy the wonders of nature (right). Their goal is to teach their children to care for the earth in hopes of passing on a better world to future generations.

Key Questions and Concepts*

1-1 What are three principles of sustainability?

CONCEPT 1-1A Nature has sustained itself for billions of years by relying on solar energy, biodiversity, and nutrient cycling.

CONCEPT 1-1B Our lives and economies depend on energy from the sun and on natural resources and natural services (*natural capital*) provided by the earth.

1-2 How are our ecological footprints affecting the earth?

CONCEPT 1-2 As our ecological footprints grow, we are depleting and degrading more of the earth's natural capital.

1-3 Why do we have environmental problems?

CONCEPT 1-3A Major causes of environmental problems are population growth, wasteful and unsustainable resource use, poverty, and not including the harmful environmental costs of resource use in the market prices of goods and services.

CONCEPT 1-3B Our environmental worldview plays a key role in determining whether we live unsustainably or more sustainably.

1-4 What is an environmentally sustainable society?

CONCEPT 1-4 Living sustainably means living off the earth's natural income without depleting or degrading the natural capital that supplies it.

Note: Supplements 2 (p. S3), 4 (p. S10), and 6 (p. S22) can be used with this chapter.

*This is a *concept-centered* book, with each major section of each chapter built around one or two key concepts derived from the natural or social sciences. Key questions and concepts are summarized at the beginning of each chapter. You can use this summary as a preview and as a review of the key ideas in each chapter.

Alone in space, alone in its life-supporting systems, powered by inconceivable energies, mediating them to us through the most delicate adjustments, wayward, unlikely, unpredictable, but nourishing, enlivening, and enriching in the largest degree—is this not a precious home for all of us? Is it not worth our love?

BARBARA WARD AND RENÉ DUBOS

1-1 What Are Three Principles of Sustainability?

▶ **CONCEPT 1-1A Nature has sustained itself for billions of years by relying on solar energy, biodiversity, and nutrient cycling.**

▶ **CONCEPT 1-1B Our lives and economies depend on energy from the sun and on natural resources and natural services (*natural capital*) provided by the earth.**

Environmental Science Is a Study of Connections in Nature

The **environment** is everything around us, or as the famous physicist Albert Einstein put it, "The environment is everything that isn't me." It includes the living and the nonliving things (air, water, and energy) with which we interact in a complex web of relationships that connect us to one another and to the world we live in.

Despite our many scientific and technological advances, we are utterly dependent on the environment for clean air and water, food, shelter, energy, and everything else we need to stay alive and healthy. As a result, we are part of, and not apart from, the rest of nature.

This textbook is an introduction to **environmental science**, an *interdisciplinary* study of how humans interact with the living and nonliving parts of their environment. It integrates information and ideas from the *natural sciences* such as biology, chemistry, and geology; the *social sciences* such as geography, economics, and political science; and the *humanities* such as philosophy and ethics. The three goals of environmental science are **(1)** to learn how life on the earth has survived and thrived, **(2)** to understand how we interact with the environment, and **(3)** to find ways to deal with environmental problems and live more sustainably.

A key component of environmental science is **ecology**, the biological science that studies how **organisms**, or living things, interact with one another and with their environment. Every organism is a member of a certain **species**, a group of organisms that have a unique set of characteristics that distinguish them from all other organisms and, for organisms that reproduce sexually, can mate and produce fertile offspring.

A major focus of ecology is the study of ecosystems. An **ecosystem** is a set of organisms within a defined area or volume that interact with one another and with their environment of nonliving matter and energy. For example, a forest ecosystem consists of plants (especially trees), animals, and mostly tiny micro-organisms that decompose organic materials and recycle their chemicals, all interacting with one another and with solar energy and the chemicals in the forest's air, water, and soil.

We should not confuse environmental science and ecology with **environmentalism**, a social movement dedicated to protecting the earth's life-support systems for all forms of life. Environmentalism is practiced more in the political and ethical arenas than in the realm of science.

Nature's Survival Strategies Follow Three Principles of Sustainability

Nature has been dealing with significant changes in environmental conditions that have affected life on the earth since it first appeared about 3.5 billion years ago. This is why many environmental experts say that when we face an environmental change that becomes a problem for us or other species, we should begin by learning how nature has dealt with such changes and then mimic nature's solutions.

In our study of environmental science, the most important question we can ask is how did the incredible variety of life on the earth sustain itself for at least 3.5 billion years in the face of catastrophic changes in environmental conditions? Such changes had various causes, including gigantic meteorites impacting the earth, ice ages lasting for hundreds of millions of years, and long warming periods during which melting ice raised sea levels by hundreds of feet.

Our species has been around for only about 200,000 years—less than the blink of an eye relative to the billions of years that life has existed on the earth. We named ourselves *Homo sapiens sapiens* (Latin for "wise man"). With our large and complex brains and language ability, we are a very *smart* species. Within only a few hundred years, we have learned how to take over most of the earth to support our basic needs and rapidly growing wants. But it remains to be seen whether we are the *wise* species that we claim to be. Many argue that a species in the process of degrading its own life-support system could not be considered wise.

Our research leads us to believe that in the face of drastic environmental changes, there are three overarching science-based themes to the long-term sustainability of life on this planet: *solar energy, biodiversity*, and *chemical cycling*, as summarized below and in Figure 1-2 (p. 8) (**Concept 1-1A**). In other words, we must rely on the sun (see photo on title page), promote multiple options for life, and minimize waste. These powerful and simple ideas make up three **principles of sustainability** or *lessons from nature* that we use throughout the book to guide us in living more sustainably and moving toward a more sustainable future such as the one we outlined in the **Core Case Study*** that opened this chapter.

SUSTAINABILITY

CORE CASE STUDY

- **Reliance on solar energy:** The sun warms the planet and provides energy that plants use to produce **nutrients**, or the chemicals necessary for life, for themselves and for us and most other animals. The energy contained in the sun's radiation is called **solar energy**. Without it, life as we know it would not exist. The sun also powers indirect forms of solar energy such as wind and flowing water, which would not exist without the sun's energy, and which we can use to produce electricity.
- **Biodiversity** (short for *biological diversity*): **Biodiversity** is the astounding variety of different organisms, the natural systems in which they exist and interact (such as deserts, grasslands, forests [see front cover photo], and oceans), and the natural services that these organisms and living systems provide free of charge (such as renewal of the *topsoil* that makes up the top layer of the earth's crust, pest control, and air and water purification). The feeding relationships and other interactions among species also provide population control that limits the ultimate population size of any species. Biodiversity also provides countless ways for life to adapt to changing environmental conditions. Without it, most life forms would have been wiped out long ago.
- **Chemical cycling:** Also referred to as **nutrient cycling**, **chemical cycling** is the circulation of chemicals from the environment (mostly from soil and water) through organisms and back to the environment is necessary for life. Natural processes keep this cycle going, and the earth receives no new supplies of these chemicals. Thus, for life to sustain itself, these chemicals must cycle in this way indefinitely. Without chemical cycling, there would be no air, no water, no soil, no food, and no life. This also means that there is little waste in nature, other than in the human world, because the wastes of organisms become nutrient raw materials for other organisms.

*We use the opening Core Case Study as a theme to connect and integrate much of the material in each chapter. The logo indicates these connections.

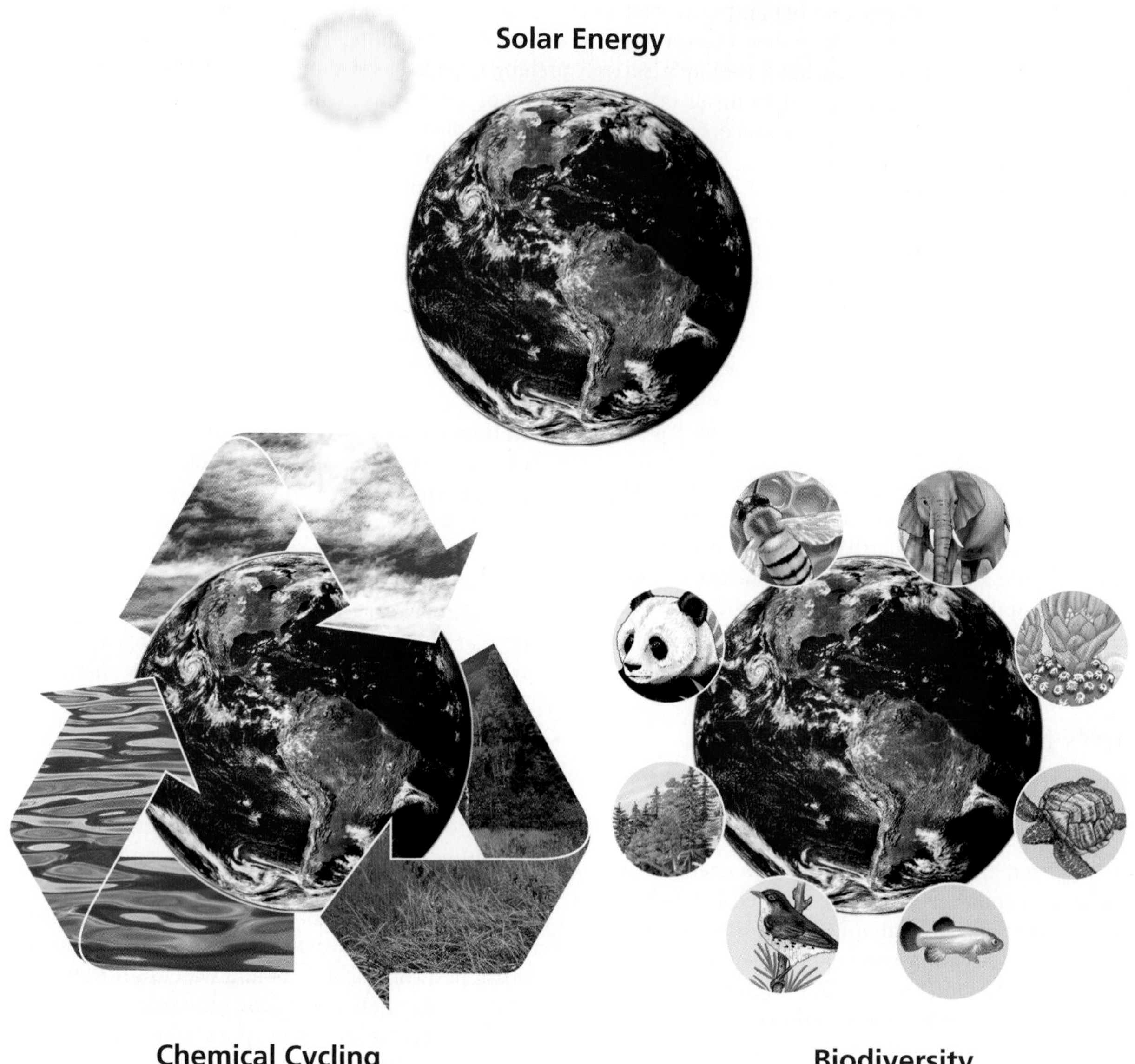

Figure 1-2 Three principles of sustainability: We derive these three interconnected principles of sustainability from learning how nature has sustained a huge variety of life on the earth for at least 3.5 billion years, despite drastic changes in environmental conditions (**Concept 1-1A**).

Sustainability Has Certain Key Components

Sustainability, the central integrating theme of this book, has several critical components that we use as subthemes. One such component is **natural capital**—the natural resources and natural services that keep us and other forms of life alive and support human economies (Figure 1-3).

Natural resources are materials and energy in nature that are essential or useful to humans. They are often classified as *renewable resources* (such as air, water, soil, plants, and wind) or *nonrenewable resources* (such as copper, oil, and coal). **Natural services** are processes in nature such as purification of air and water and renewal of topsoil, which support life and human economies.

One vital natural service is nutrient cycling (Figure 1-4, p. 10). An important component of nutrient cycling is topsoil—a vital natural resource that provides us and the world's other land-dwelling species with food. Without nutrient cycling in topsoil, life as we know it could not exist on the earth's land. Hence, we consider nutrient cycling to be the basis for one of the three **principles of sustainability**.

Natural capital is supported by energy from the sun—another of the **principles of sustainability** (Figure 1-3). Thus, our lives and economies depend on energy from the sun, and on natural resources and natural services (*natural capital*) provided by the earth (**Concept 1-1B**).

A second component of sustainability—and another subtheme of this text—is to recognize that many human

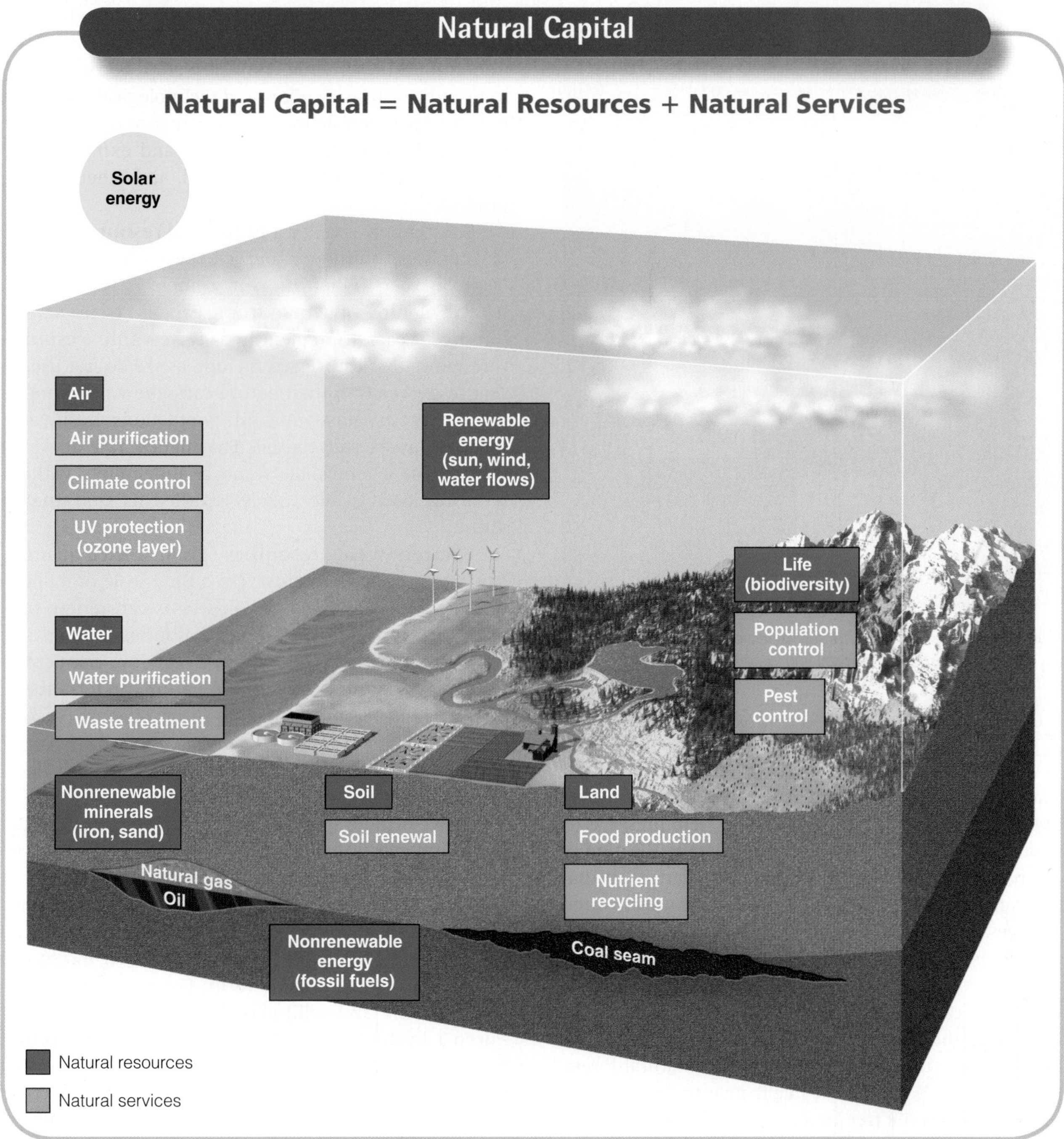

Figure 1-3 These key *natural resources* (blue) and *natural services* (orange) support and sustain the earth's life and human economies (**Concept 1-1A**).

activities can *degrade natural capital* by using normally renewable resources faster than nature can restore them, and by overloading natural systems with pollution and wastes. For example, in some parts of the world, we are clearing mature forests much faster than they can grow back and eroding topsoil faster than nature can renew it. We are also loading some rivers, lakes, and oceans with chemical and animal wastes faster than these bodies of water can cleanse themselves.

This leads us to a third component of sustainability: *solutions*. While environmental scientists search for solutions to problems such as the unsustainable degradation of forests and other forms of natural capital, their work is limited to finding the *scientific* solutions; the political solutions are left to political processes. For example, a scientific solution to the problems of depletion of forests might be to stop burning or cutting down biologically diverse, mature forests and to allow nature to replenish them. A scientific solution to the problem of pollution of rivers might be to prevent the excessive dumping of chemicals and wastes into streams and to allow them to recover naturally. However, to implement such solutions, governments would probably have to enact and enforce laws and regulations.

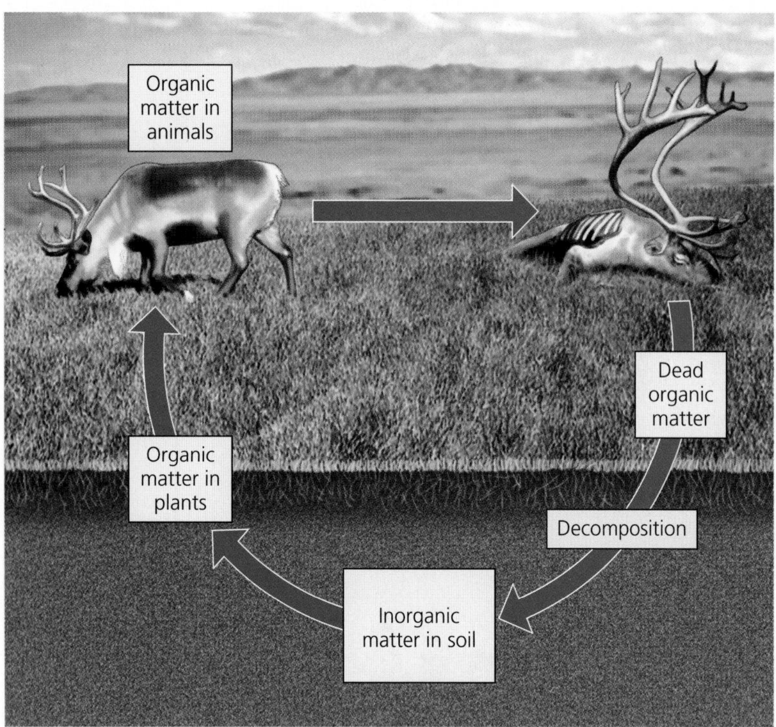

Figure 1-4 *Nutrient cycling:* This important natural service recycles chemicals needed by organisms from the environment (mostly from soil and water) through those organisms and back to the environment.

The search for solutions often involves conflicts. For example, when a scientist argues for protecting a natural forest on publicly owned land to help preserve its important diversity of plants and animals, the timber company that had planned to harvest the trees in that forest might protest. Dealing with such conflicts often involves making *trade-offs*, or compromises—another component of sustainability. For example, the timber company might be persuaded to plant a tree farm—consisting of neat rows of a rapidly growing tree species—in an area that had already been cleared or degraded, instead of clearing the trees in a diverse natural forest. In return, the government might give the company a subsidy to plant the tree farm.

A shift toward environmental sustainability should be based on scientific concepts and results that are widely accepted by experts in a particular field, as discussed in more detail in Chapter 2. In making such a shift, *individuals matter*—another subtheme of this book. Society's shift toward sustainability ultimately depends on the actions of individuals (**Core Case Study**), CORE CASE STUDY beginning with the daily choices we all make. Thus, *sustainability begins at personal and local levels.*

Some Resources Are Renewable and Some Are Not

From a human standpoint, a **resource** is anything obtained from the environment to meet our needs and wants. Some resources such as solar energy, fresh air, fertile topsoil, and edible wild plants are directly available for use. Other resources such as petroleum, iron, underground water, and cultivated crops become useful to us only with some effort and technological ingenuity. For example, petroleum was merely a mysterious, oily fluid until we learned how to find and extract it, and convert it into gasoline, heating oil, and other saleable products.

Solar energy is called a **perpetual resource** because its supply is continuous and is expected to last at least 6 billion years, when the sun completes its life cycle. It takes nature anywhere from several days to several hundred years to replenish a **renewable resource** through natural processes, as long as we do not use up that resource faster than nature can renew it. Examples include forests, grasslands, fish populations, freshwater, fresh air, and fertile topsoil. The highest rate at which we can use a renewable resource indefinitely without reducing its available supply is called its **sustainable yield**.

Nonrenewable resources exist in a fixed quantity, or *stock*, in the earth's crust. On a time scale of millions to billions of years, geologic processes can renew such resources. However, on the much shorter human time scale of hundreds to thousands of years, we can deplete these resources much faster than nature can form them. Such exhaustible stocks include *energy resources* (such as coal and oil), *metallic mineral resources* (such as copper and aluminum), and *nonmetallic mineral resources* (such as salt and sand).

As we deplete such resources, human ingenuity can often find substitutes. However, sometimes there is no acceptable or affordable substitute for a resource.

We can recycle or reuse some nonrenewable resources, such as copper and aluminum, to extend their supplies. **Reuse** is the practice of using a resource over and over in the same form. For example, we can collect, wash, and refill glass bottles many times (see Photo 1 in the Detailed Table of Contents). **Recycling** involves collecting waste materials and processing them into new materials. For example, we can crush and melt discarded aluminum to make new aluminum cans or other aluminum products. Reuse and recycling are two ways to live more sustainably by following one of nature's three **principles of sustainability** (Figure 1-2). However, we cannot recycle or reuse SUSTAINABILITY energy resources such as oil and coal. Once burned, their concentrated energy is no longer available to us.

Recycling nonrenewable metallic resources uses much less energy, water, and other resources and produces much less pollution and environmental degradation than exploiting virgin metallic resources. And reusing such resources has a lower environmental impact than recycling. From an environmental and sustainability viewpoint, the priorities for more sustainable use of nonrenewable resources such as metals and plastics should be: **R**educe (use less), **R**euse, and **R**ecycle. According to a number of environmental scientists, we already know how to reuse or recycle 80–90% of the nonrenewable metal and plastic resources that we use. GOOD NEWS

Countries Differ in their Resource Use and Environmental Impact

The United Nations (UN) classifies the world's countries as economically more developed or less developed, based primarily on their average income per person. High-income, **more-developed countries** include the United States, Canada, Japan, Australia, New Zealand, and most European countries. The more-developed countries have 18% of the world's population, use about 88% of the world's resources, and produce about 75% of the world's pollution and waste, according to UN and World Bank data.

All other nations, in which 82% of the world's people live, are classified as **less-developed countries**, most of them in Africa, Asia, and Latin America. Some are *middle-income, moderately-developed countries* such as China, India, Brazil, Thailand, and Mexico. Others are *low-income, least-developed countries* such as Congo, Haiti, Nigeria, and Nicaragua. Figure 6, p. S27, in Supplement 6 is a map of high-, upper-middle-, lower-middle-, and low-income countries.

1-2 How Are Our Ecological Footprints Affecting the Earth?

▶ **CONCEPT 1-2** **As our ecological footprints grow, we are depleting and degrading more of the earth's natural capital.**

We Are Living Unsustainably

The bad news is that according to a massive and growing body of scientific evidence, we are living unsustainably by wasting, depleting, and degrading the earth's natural capital at an accelerating rate. The entire process is known as **environmental degradation**, summarized in Figure 1-5. We also refer to this as **natural capital degradation**.

In many parts of the world, renewable forests are shrinking, deserts are expanding, topsoil is eroding, and suburbs are replacing croplands. In addition, the lower

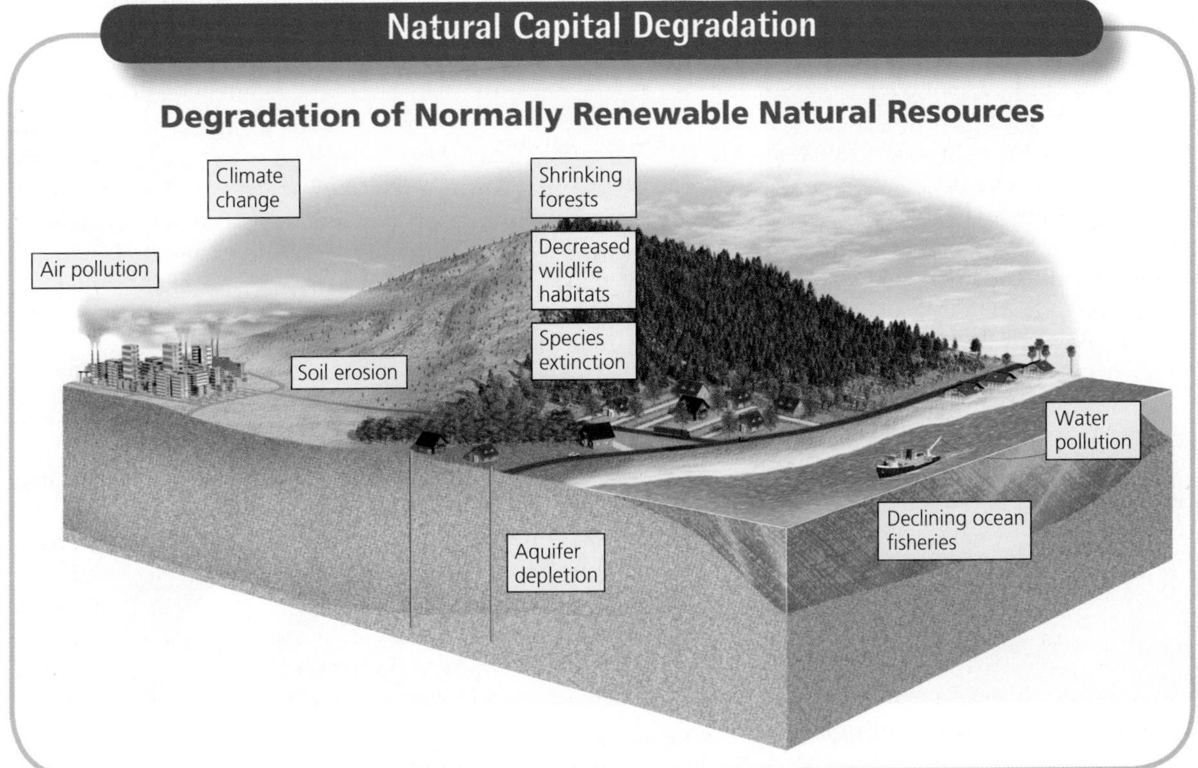

Figure 1-5 These are examples of the degradation of normally renewable natural resources and services in parts of the world, mostly as a result of growing populations and rising rates of resource use per person.

atmosphere is warming, glaciers are melting, sea levels are rising, and floods, droughts, severe weather, and forest fires are increasing in some areas. In a number of regions, rivers are running dry, harvests of many species of edible fish are dropping sharply, and coral reefs are disappearing. Species are becoming extinct at least 100 times faster than in pre-human times, and extinction rates are expected to increase to at least 1,000 times faster during this century.

In 2005, the UN released its *Millennium Ecosystem Assessment*, a 4-year study by 1,360 experts from 95 countries. According to this study, human activities have degraded about 60% of the earth's natural or ecosystem services (Figure 1-3, orange boxes), mostly since 1950. In its summary statement, the report warned that "human activity is putting such a strain on the natural functions of Earth that the ability of the planet's ecosystems to sustain future generations can no longer be taken for granted."

GOOD NEWS

The good news, also included in the UN report, is that we have solutions to these problems that we could implement within a few decades in order to make the transition to a more sustainable future within your lifetime (**Core Case Study**), as you will learn in reading this book.

CORE CASE STUDY

Pollution Comes from a Number of Sources

A major environmental problem is **pollution**, which is contamination of the environment by a chemical or other agent such as noise or heat to a level that is harmful to the health, survival, or activities of humans or other organisms. Polluting substances, or *pollutants*, can enter the environment naturally, such as from volcanic eruptions, or through human activities, such as the burning of coal and gasoline, and the dumping of chemicals into rivers and oceans. At a high enough concentration in the air, in water, or in our bodies, almost any chemical can cause harm and be classified as a pollutant.

The pollutants we produce come from two types of sources. **Point sources** are single, identifiable sources. Examples are the smokestack of a coal-burning power or industrial plant (Figure 1-6), the drainpipe of a factory, and the exhaust pipe of an automobile. **Nonpoint sources** are dispersed and often difficult to identify. Examples are pesticides blown from the land into the air and the runoff of fertilizers, pesticides, and trash from the land into streams and lakes (Figure 1-7). It is much easier and cheaper to identify and control or prevent pollution from point sources than from widely dispersed nonpoint sources.

We have tried to deal with pollution in two very different ways. One method is **pollution cleanup**, which involves cleaning up or diluting pollutants after we have produced them. The other method is **pollution prevention**, which reduces or eliminates the production of pollutants.

So far, we have relied mostly on pollution cleanup. Environmental scientists have identified three problems with this approach. *First,* cleanup is only a temporary fix as long as population and resource consumption levels grow without compensating improvements in pollution-control technology.

Second, cleanup often removes a pollutant from one part of the environment only to cause pollution in another. For example, we can collect garbage, but the garbage is then *burned* (possibly causing air pollution and leaving toxic ash that must be put somewhere),

Ray Pfortner/Peter Arnold, Inc.

Figure 1-6 This *point-source air pollution* rises from a pulp mill in New York State (USA).

Igor Jandric/Shutterstock.com

Figure 1-7 The trash in this river came from a large area of land and is an example of *nonpoint-source water pollution*.

dumped on the land (possibly causing water pollution through runoff or seepage into groundwater), or *buried* (possibly causing soil and groundwater pollution).

Third, once pollutants become dispersed into the environment at harmful levels, it usually costs too much to reduce them to acceptable levels.

We need both pollution prevention (front-of-the-pipe) and pollution cleanup (end-of-the-pipe) solutions. But a number of environmental scientists and economists urge us to put more emphasis on prevention because it works better and in the long run is cheaper than cleanup. Pollution prevention is another key to a more sustainable future (**Core Case Study**). CORE CASE STUDY

The Tragedy of the Commons: Overexploiting Commonly Shared Renewable Resources

Some renewable resources can be used by almost anyone. Examples are the atmosphere and the open ocean and its fishes.

Many open-access renewable resources have been environmentally degraded. In 1968, biologist Garrett Hardin (1915–2003) called such degradation the *tragedy of the commons.* It occurs because each user of a shared common resource or open-access resource reasons, "If I do not use this resource, someone else will. The little bit that I use or pollute is not enough to matter, and anyway, it's a renewable resource."

When the number of users is small, this logic works. Eventually, however, the cumulative effect of many people trying to exploit a shared resource can degrade it and eventually exhaust or ruin it. Then no one can benefit from it. That is the tragedy.

There are two major ways to deal with this difficult problem. One is to use a shared renewable resource at a rate well below its estimated sustainable yield by using less of the resource, regulating access to the resource, or doing both. For example, governments can establish laws and regulations limiting the annual harvests of various types of ocean fishes that we are harvesting at unsustainable levels, and regulating the amount of pollutants we add to the atmosphere or the oceans.

The other way is to convert open-access renewable resources to private ownership. The reasoning is that if you own something, you are more likely to protect your investment. That sounds good, but this approach is not practical for global open-access resources such as the atmosphere and the oceans, which cannot be divided up and sold as private property.

Ecological Footprints: Our Environmental Impacts

Supplying people with renewable resources results in wastes and pollution. We can think of it as an **ecological footprint**—the amount of biologically productive land and water needed to supply a person or a country with the renewable resources that they need and to absorb and recycle the wastes and pollution produced by such resource use. (The developers of this tool for measuring environmental impacts chose to focus on renewable resources, although the use of nonrenewable resources also contributes to environmental impacts.) The **per capita ecological footprint** is the average ecological footprint of an individual in a given country or area.

If a country's (or the world's) total ecological footprint is larger than its current *biological capacity* to replenish its renewable resources and absorb the resulting wastes and pollution, it is said to have an *ecological deficit.* In other words, its people are living unsustainably by depleting their natural capital instead of living off the renewable supply or income provided by such capital. In 2008, the World Wildlife Fund (WWF) and the Global Footprint Network estimated that humanity's global ecological footprint exceeded the earth's current ecological capacity to support humans and other forms of life indefinitely by at least 30% (Figure 1-8, p. 14, bottom) and by 88% in the United States.

In other words, humanity is living unsustainably. According to the ecological footprint model, to sustain indefinitely the world's current population and average renewable resource use per person, and to dispose of the resulting wastes and pollution, we would need the equivalent of 1.3 planet earths. And if we continue on our current path of renewable resource use and population growth using existing technology, by around 2035, we will need 2 planet earths.

According to this model, we would need about five planet earths for everyone in the world to reach the current U.S. level of per-person use of renewable resources using existing technology. Put another way, if everyone consumed as much in the way of renewable resources as the average American does today, the earth could indefinitely support only about 1.3 billion people—not today's 7 billion. (In Supplement 6, see Figure 2, p. S24, for a map of the human ecological footprints for the world, and Figure 5, p. S27, for a map of countries that are either ecological debtors or ecological creditors. For more on ecological footprints, see the Guest Essay by Michael Cain at **www.cengagebrain.com**.)

Ecological footprint data are imperfect estimates. But even if such estimates are too high by a factor of two, it appears that we are in deep trouble if the world's population growth and consumption of renewable resources continue at their current rates. We can reduce the size of our ecological footprints if we use existing and emerging technologies and economic tools to make a shift to more sustainable societies (**Core Case Study**) over the next few decades (Figure 1-8, p. 14, bottom curve). CORE CASE STUDY

Proposed ways to do this include slowing population growth, reducing resource waste, sharply reducing poverty, and shifting from fossil fuels to renewable energy sources, as discussed throughout this book.

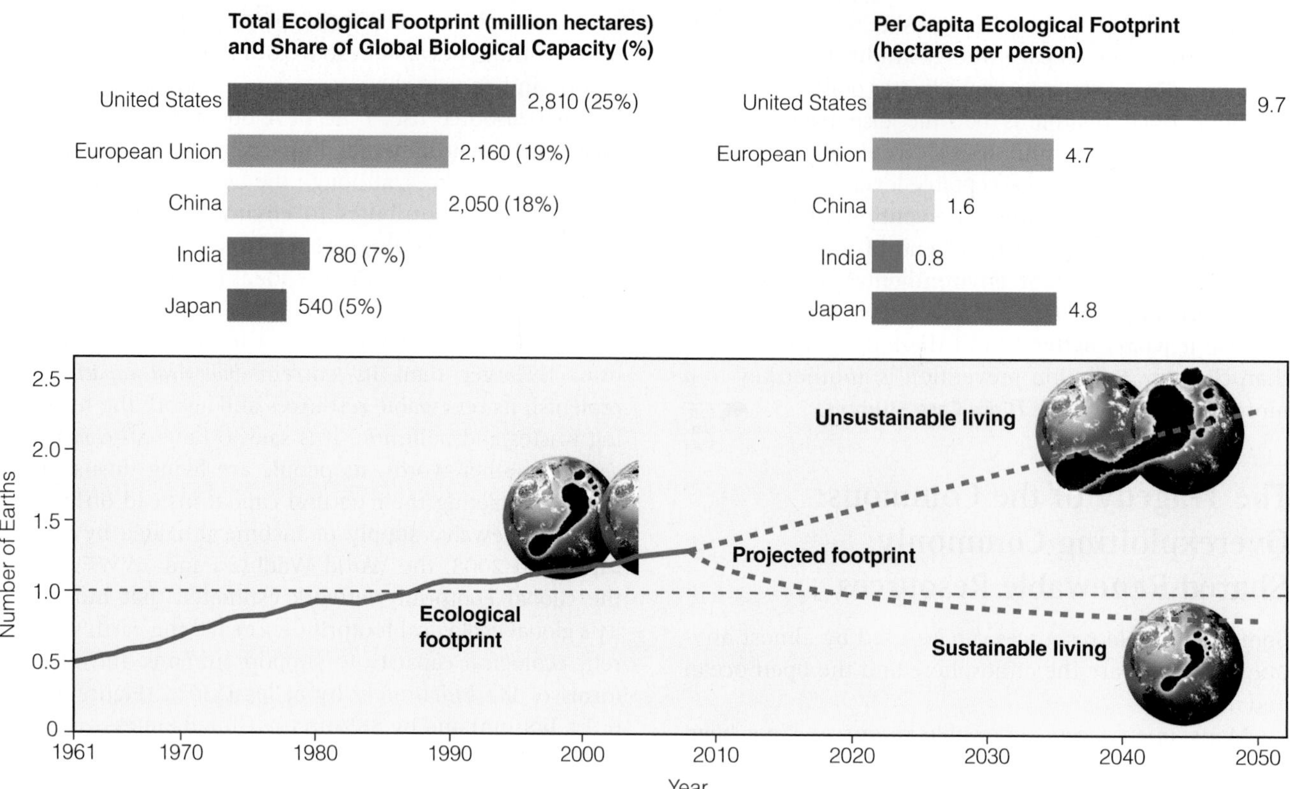

Figure 1-8 Natural capital use and degradation: These graphs show the total and per capita ecological footprints of selected countries (top). In 2008, humanity's total, or global, estimated ecological footprint was at least 30% higher than the earth's ecological capacity (bottom) and is projected to be twice the planet's ecological capacity by around 2035. **Question:** If we are living beyond the earth's renewable biological capacity, why do you think the human population and per capita resource consumption are still growing rapidly? (Data from Worldwide Fund for Nature, Global Footprint Network, *Living Planet Report* 2008.

IPAT Is Another Environmental Impact Model

In the early 1970s, scientists Paul Ehrlich and John Holdren developed a simple model showing how population size (P), affluence, or resource consumption per person (A), and the beneficial and harmful environmental effects of technologies (T) help to determine the environmental impact (I) of human activities—a rough estimate of how much humanity is degrading the natural capital it depends upon. We can summarized this model by the simple equation $I = P \times A \times T$.

Impact (I) = Population (P) × Affluence (A) × Technology (T)

Figure 1-9 shows the relative importance of these three factors in less-developed and more-developed countries. While the ecological footprint model emphasizes the use of renewable resources, this model includes the per capita use of both renewable and nonrenewable resources.

Some forms of technology such as polluting factories, coal-burning power plants, and gas-guzzling motor vehicles increase environmental impact by raising the T factor in the equation. But other technologies reduce environmental impact by decreasing the T factor. Examples are pollution control and prevention technologies, wind turbines and solar cells that generate electricity without polluting, and fuel-efficient cars. In other words, some forms of technology are *environmentally harmful* and some are *environmentally beneficial.*

In most less-developed countries, the key factors in total environmental impact (Figure 1-9, top) are population size and the degradation of renewable resources as a growing number of poor people struggle to stay alive. In more-developed countries, high rates of per capita resource use and the resulting high per capita levels of pollution and resource depletion and degradation usually are the key factors determining overall environmental impact (Figure 1-9, bottom).

■ CASE STUDY
China's New Affluent Consumers

More than a billion super-affluent consumers, mostly in more-developed countries are putting immense pressure on the earth's potentially renewable natural capital and its nonrenewable resources. And well over half a billion new consumers are attaining middle-class, affluent lifestyles with a disposable income of at least $3,000 annually in 20 rapidly developing middle-income countries such as China, India, Indonesia, Brazil, South Korea, and Mexico. In China, India, and Indonesia, the

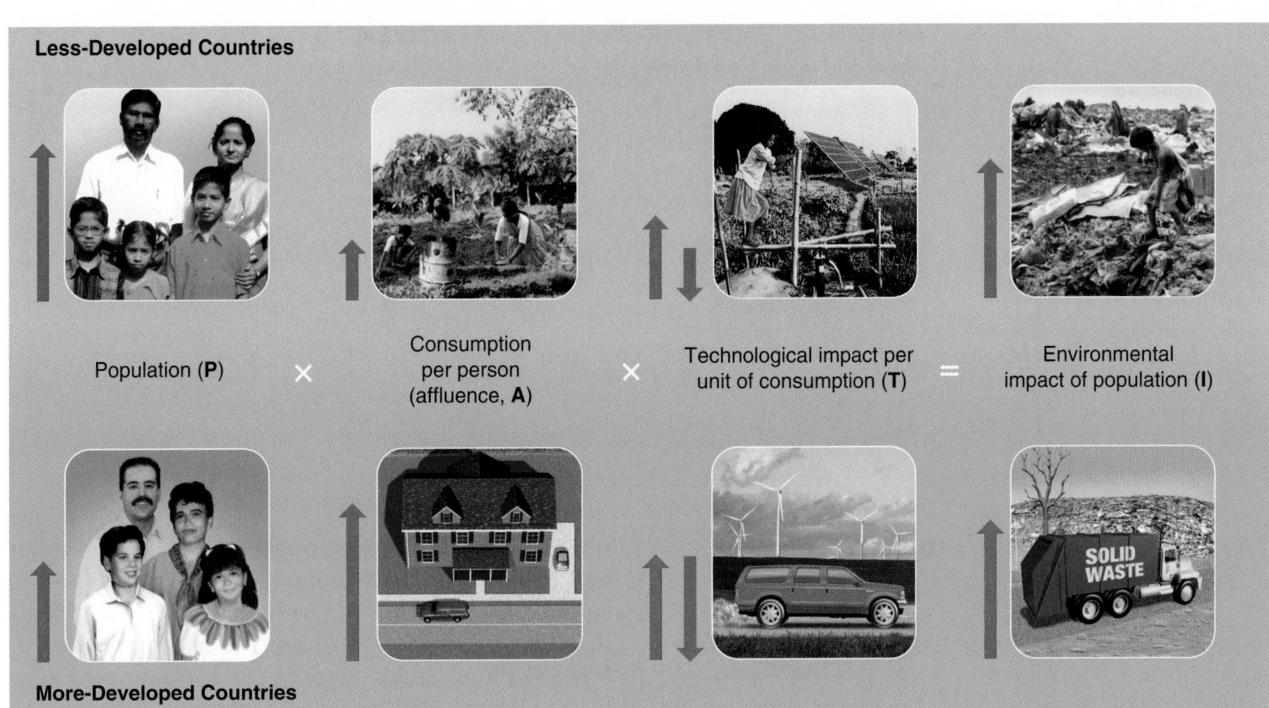

Figure 1-9 *Connections:* This simple model demonstrates how three factors—population size, affluence (resource use per person), and technology—help to determine the environmental impacts of populations in less-developed countries (top) and more-developed countries (bottom).

number of middle-class consumers is approaching 600 million—almost twice the current size of the U.S. population. This number is projected to rise to 945 million by 2015, with China having about two-thirds of these new middle-class consumers.

China has the world's largest population and second-largest economy. It is the world's leading consumer of wheat, rice, meat, coal, fertilizer, steel, cement, and oil. China also leads the world in consumption of goods such as televisions, cell phones, and refrigerators. It has built the world's largest building, the fastest train, and the biggest dam. It has produced more wind turbines than any other country and will soon become the world's largest producer of solar cells. By 2015, China is projected to be the world's largest producer and consumer of cars, most of them more fuel-efficient than those produced in the United States and Europe.

On the other hand, after 20 years of industrialization, China now contains two-thirds of the world's most polluted cities. Some of its major rivers are choked with waste and pollution, and some areas of its coastline are basically devoid of fishes and other ocean life. A massive cloud of air pollution, largely generated in China, affects other Asian countries, the Pacific Ocean, and the West Coast of North America. For more details on China's growing ecological footprint, see the Guest Essay by Norman Myers for this chapter at **www.cengagebrain.com**.

1-3 Why Do We Have Environmental Problems?

▶ **CONCEPT 1-3A Major causes of environmental problems are population growth, wasteful and unsustainable resource use, poverty, and not including the harmful environmental costs of resource use in the market prices of goods and services.**

▶ **CONCEPT 1-3B Our environmental worldview plays a key role in determining whether we live unsustainably or more sustainably.**

Experts Have Identified Four Basic Causes of Environmental Problems

According to a number of environmental and social scientists, the major causes of the environmental problems we face are **(1)** population growth, **(2)** wasteful and unsustainable resource use, **(3)** poverty, and **(4)** failure to include in their market prices the harmful environmental costs of goods and services (Figure 1-10, p. 16).

We discuss all of these causes in detail in later chapters of this book. Let us begin with a brief overview of them.

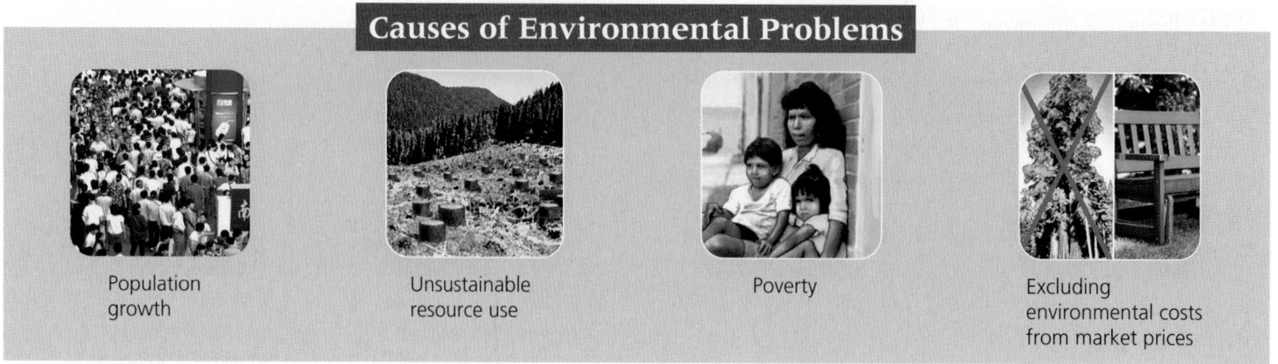

Figure 1-10 Environmental and social scientists have identified four basic causes of the environmental problems we face (**Concept 1-3**). **Question:** For each of these causes, what are two environmental problems that result?

The Human Population Is Growing at a Rapid Rate

Exponential growth occurs when a quantity such as the human population increases at a fixed percentage per unit of time, such as 2% per year. Exponential growth starts off slowly. But after only a few doublings, it grows to enormous numbers because each doubling is twice the total of all earlier growth.

Here is an example of the immense power of exponential growth. Fold a piece of paper in half to double its thickness. If you could continue doubling the thickness of the paper 50 times, it would be thick enough to almost reach the sun—149 million kilometers (93 million miles) away! Hard to believe, isn't it?

The human population has been growing exponentially (Figure 1-11). Collectively, the world's people consume vast amounts of food, water, raw materials, and energy, and they produce huge amounts of pollution and wastes in the process. There are about 7 billion people on the earth with about 83 million more people added each year. There may be 9.6 billion of us by 2050.

No one knows how many people the earth can support indefinitely, and at what level of average resource consumption per person, without seriously degrading the ability of the planet to support us, our economies, and other forms of life. However, the world's expanding total and per capita ecological footprints (Figure 1-9) are disturbing warning signs.

We can slow population growth with the goal of having it level off at around 8 billion by 2040, as suggested in the **Core Case Study**. Some ways to do this include reducing poverty through economic development, promoting family planning, and elevating the status of women, as discussed in Chapter 6.

CORE CASE STUDY

Affluence Has Harmful and Beneficial Environmental Effects

The lifestyles of many consumers in more-developed countries and in less-developed countries such as India and China (see Case Study, p. 14) are built upon grow-

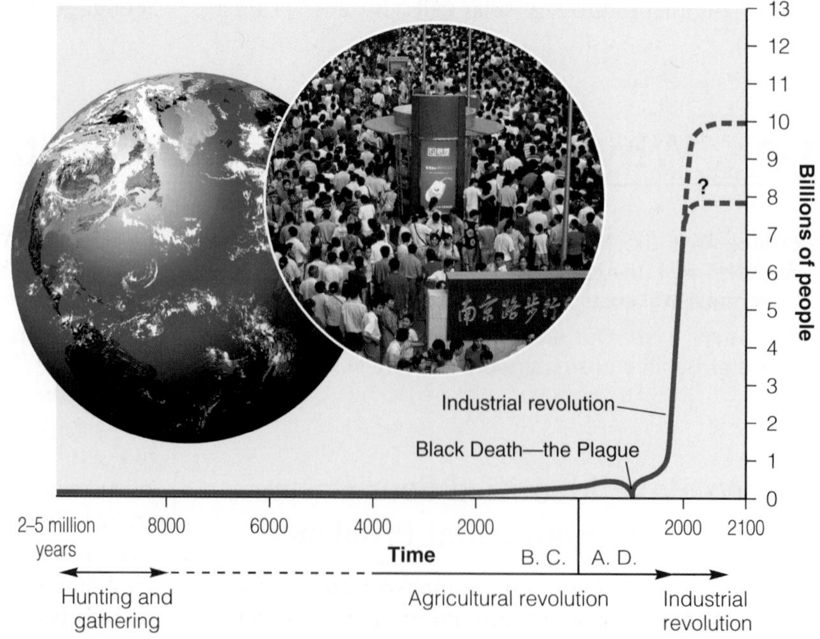

Figure 1-11 *Exponential growth:* The J-shaped curve represents past exponential world population growth, with projections to 2100 showing possible population stabilization as the J-shaped curve of growth changes to an S-shaped curve. (This figure is not to scale.) (Data from the World Bank and United Nations, 2008; photo L. Young/UNEP/Peter Arnold, Inc.)

ing **affluence**, or wealth, which results in high levels of consumption and unnecessary waste of resources. Such affluence is based mostly on the assumption—fueled by mass advertising—that buying more and more material goods will bring fulfillment and happiness.

The harmful environmental effects of affluence are dramatic. Although the U.S. population is only about one-fourth that of India, the average American consumes about 30 times as much as the average Indian and 100 times as much as the average person in the world's poorest countries. According to some ecological footprint estimates, it takes about 27 large tractor-trailer loads of resources per year to support one typical American or 8.4 billion truckloads per year to support the entire U.S. population. Lined up end-to-end, this convoy of trucks would be longer than the distance from the earth to the sun.

Some analysts say that many affluent consumers in the United States and other more-developed countries are afflicted with *affluenza*, an eventually unsustainable addiction to buying more and more stuff, even though numerous studies show that, beyond a certain level, more wealth and more consumption do not increase happiness. Another downside to wealth is that it allows the affluent to obtain the resources they need from almost anywhere in the world without seeing the harmful environmental impacts of their high-consumption, high-waste lifestyles.

On the other hand, affluence can allow for better education, which can lead people to become more concerned about environmental quality. It also provides money for developing technologies to reduce pollution, environmental degradation, and resource waste. As a result, in the United States and most other affluent countries, the air is clearer, drinking water is purer, and most rivers and lakes are cleaner than they were in the 1970s. In addition, the food supply is more abundant and safer, the incidence of life-threatening infectious diseases has been greatly reduced, life spans are longer, and some endangered species are being rescued from extinction hastened by human activities.

GOOD NEWS

These improvements in environmental quality were achieved because of greatly increased scientific research and technological advances financed by affluence. Education also spurred many citizens to insist that businesses and governments work toward improving environmental quality.

Poverty Has Harmful Environmental and Health Effects

Poverty is a condition in which people are unable to fulfill their basic needs for adequate food, water, shelter, healthcare, and education. According to a 2008 study by the World Bank, 1.4 billion people—almost five times the number of people in the United States—live in *extreme poverty* (Figure 1-12) and struggle to live on the equivalent of less than $1.25 a day. And at least 50% of the world's people struggle to live on less than $2.25 a day. Could you do this?

Sean Sprague/Peter Arnold, Inc.

Figure 1-12 *Extreme poverty:* This boy is searching through an open dump in Rio de Janeiro, Brazil, for items to sell. Many children of poor families who live in makeshift shantytowns in or near such dumps often scavenge most of the day for food and other items to help their families survive.

Poverty causes a number of harmful environmental and health effects (Figure 1-13, p. 18). The daily lives of the world's poorest people are focused on getting enough food, water, and cooking and heating fuel to survive. Desperate for short-term survival, some of these individuals unintentionally degrade forests, soils, grasslands, fisheries, and wildlife at an ever-increasing rate. They do not have the luxury of worrying about long-term environmental quality or sustainability. Even though the poor in less-developed countries have no

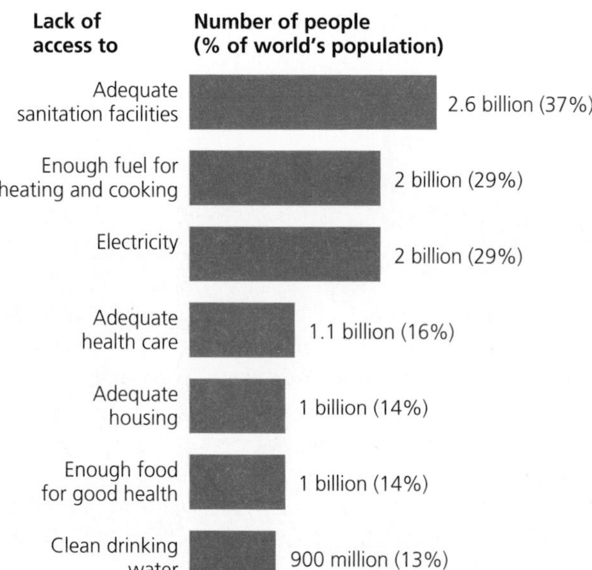

Figure 1-13 These are some of the harmful effects of poverty. **Questions:** Which two of these effects do you think are the most harmful? Why? (Data from United Nations, World Bank, and World Health Organization)

choice but to use very few resources per person, their large population size leads to a high overall environmental impact (Figure 1-9, top).

CONNECTIONS

Poverty and Population Growth

To many poor people, having more children is a matter of survival. Their children help them gather fuel (mostly wood and animal dung), haul drinking water, and tend crops and livestock. The children also help to care for their parents in their old age (their 40s or 50s in the poorest countries) because they do not have social security, health care, and retirement funds. This is largely why populations in some less-developed countries continue to grow at high rates.

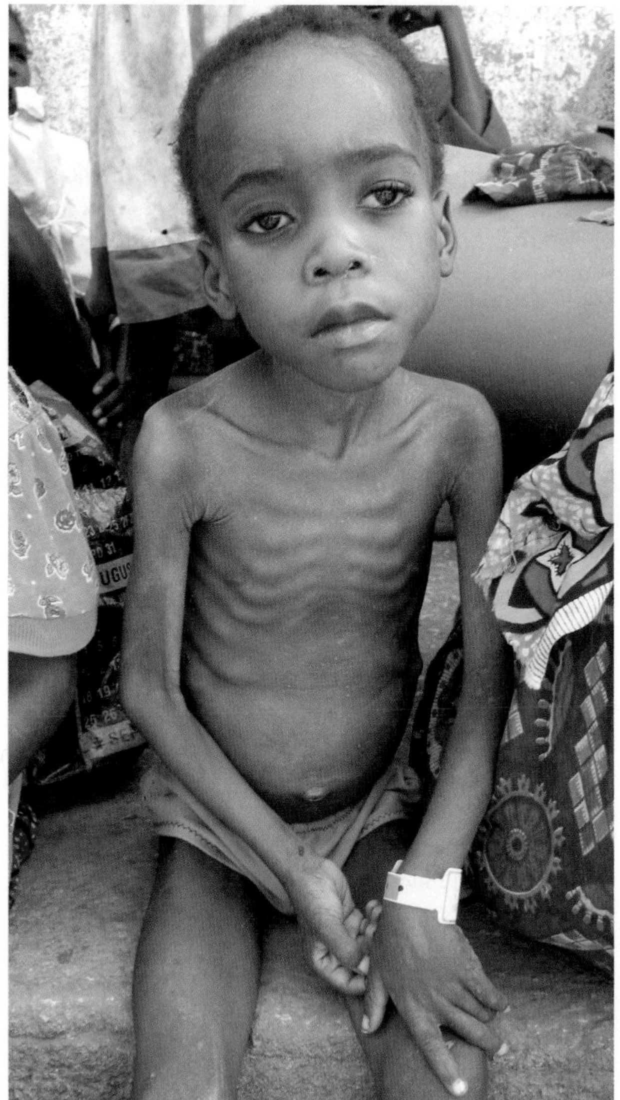

Figure 1-14 *Global outlook:* One of every three children younger than age 5 in less-developed countries, such as this child in Lunda, Angola, suffers from severe malnutrition caused by a lack of calories and protein.

While poverty can increase some types of environmental degradation, the reverse is also true. Pollution and environmental degradation have a severe impact on the poor and can increase their poverty. Consequently, many of the world's poor people die prematurely from several preventable health problems. One such problem is *malnutrition* caused by a lack of protein and other nutrients needed for good health (Figure 1-14). The resulting weakened condition can increase an individual's chances of death from normally nonfatal ailments such as diarrhea and measles.

A second health problem is limited access to adequate sanitation facilities and clean drinking water. About 37% of the world's people have no decent bathroom facilities. They are forced to use backyards, alleys, ditches, and streams. As a result, about one of every seven of the world's people get water for drinking, washing, and cooking from sources polluted by human and animal feces. A third health problem is severe respiratory disease that people get from breathing the smoke of open fires or poorly vented stoves used for heating and cooking inside their homes. This indoor air pollution kills about 2.4 million people a year.

In 2010, the World Health Organization estimated that one or more of these factors, mostly related to poverty, cause premature death for about 7.5 million children under the age of 5 each year. Some hopeful news is that this number of annual deaths is down from about 12 million in 1990. Even so, every day an average of at least 20,500 young children die prematurely from these causes. This is equivalent to *102 fully loaded 200-passenger airliners crashing every day with no survivors!* The daily news rarely covers this ongoing human tragedy.

THINKING ABOUT
The Poor, the Affluent, and Rapidly Increasing Population Growth

Some see the rapid population growth of the poor in less-developed countries as the primary cause of our environmental problems. Others say that the much higher resource use per person in more-developed countries is a more important factor. Which factor do you think is more important? Why?

Prices of Goods and Services Do Not Include Harmful Environmental Costs

Another basic cause of environmental problems has to do with how goods and services are priced in the marketplace.

Companies using resources to provide goods for consumers generally are not required to pay for the harmful environmental costs of supplying such goods. For example, fishing companies pay the costs of catching fish but do not pay for the depletion of fish stocks. Timber companies pay the cost of clear-cutting forests but do not pay for the resulting environmental degradation and loss of wildlife habitat. The primary goal of a company is to maximize profits for its owners or stockholders, which is how capitalism works. Indeed, it would be economic suicide for a company to add these costs to their prices unless government regulations created a level economic playing field by using taxes or regulations to require all businesses to pay for the environmental costs of producing their products.

As a result, the prices of goods and services do not include their harmful environmental and human health costs. Thus, consumers have no effective way to evaluate the harmful effects, on their own health and on the earth's life-support systems, of producing and using these goods and services. For example, scientists and economists have estimated that the real cost of gasoline to consumers in the United States is about \$3.7 per liter (\$14 per gallon) when its estimated short and long-term harmful environmental and health costs are included (as discussed in more detail in Chapter 13).

Another problem arises when governments (taxpayers) give companies *subsidies* such as tax breaks and payments to assist them with using resources to run their businesses. This helps to create jobs and stimulate economies. But environmentally harmful subsidies encourage the depletion and degradation of natural capital. (See the Guest Essay for this chapter about perverse subsidies by Norman Myers at **www.cengagebrain.com**.)

GOOD NEWS

We can live more sustainably by finding ways to include in market prices the harmful environmental and health costs of the goods and services that we use. Two ways to do this over the next two decades are to shift from environmentally harmful government subsidies to environmentally beneficial subsidies, and to tax pollution and waste heavily while reducing taxes on income and wealth. We discuss such *subsidy shifts* and *tax shifts* in Chapter 17.

People Have Different Views about Environmental Problems and Their Solutions

Another challenge we face is that people differ over the seriousness of the world's environmental problems and what we should do to help solve them. Differing opinions about environmental problems arise mostly out of differing environmental worldviews. Your **environmental worldview** is your set of assumptions and values reflecting how you think the world works and what you think your role in the world should be. **Environmental ethics**, which are beliefs about what is right and wrong with how we treat the environment, are an important element in our worldviews. Here are some important *ethical questions* relating to the environment:

- Why should we care about the environment?
- Are we the most important beings on the planet or are we just one of the earth's millions of different forms of life?
- Do we have an obligation to see that our activities do not cause the extinction of other species? Should we try to protect all species or only some? How do we decide which to protect?
- Do we have an ethical obligation to pass on to future generations the extraordinary natural world in a condition that is at least as good as what we inherited?
- Should every person be entitled to equal protection from environmental hazards regardless of race, gender, age, national origin, income, social class, or any other factor? This is the central ethical and political issue for what is known as the *environmental justice* movement. (See the Guest Essay by Robert D. Bullard at **www.cengagebrain.com**.)
- How do we promote sustainability?

THINKING ABOUT
Our Responsibilities

How would you answer each of the questions above? Compare your answers with those of your classmates. Record your answers and, at the end of this course, return to these questions to see if your answers have changed.

People with widely differing environmental worldviews can take the same data, be logically consistent with it, and arrive at quite different answers to such questions because they start with different assumptions and moral, ethical, or religious beliefs. Environmental worldviews are discussed in detail in Chapter 17, but here is a brief introduction.

The **planetary management worldview** holds that we are separate from and in charge of nature, that nature exists mainly to meet our needs and increasing wants, and that we can use our ingenuity and technology to manage the earth's life-support systems, mostly for our benefit, into the distant future.

The **stewardship worldview** holds that we can and should manage the earth for our benefit, but that we have an ethical responsibility to be caring and responsible managers, or *stewards,* of the earth. It says we should encourage environmentally beneficial forms of economic growth and development and discourage environmentally harmful forms.

The **environmental wisdom worldview** holds that we are part of, and dependent on, nature and that nature exists for all species, not just for us. According to this view, our success depends on learning how the earth sustains itself (Figure 1-2 and back cover of this book) and integrating such *environmental wisdom* into the ways we think and act.

1-4 What Is an Environmentally Sustainable Society?

▶ **CONCEPT 1-4 Living sustainably means living off the earth's natural income without depleting or degrading the natural capital that supplies it.**

Environmentally Sustainable Societies Protect Natural Capital and Live Off Its Income

According to most environmental scientists, our ultimate goal should be to achieve an **environmentally sustainable society**—one that meets the current and future basic resource needs of its people in a just and equitable manner without compromising the ability of future generations to meet their basic needs (**Core Case Study**). CORE CASE STUDY

Imagine that you win $1 million in a lottery. Suppose you invest this money (your capital) and earn 10% interest per year. If you live on just the interest, or the income made by your capital, you will have a sustainable annual income of $100,000 that you can spend each year indefinitely without depleting your capital. However, if you spend $200,000 per year, while still allowing interest to accumulate, all of your money will be gone early in the seventh year. Even if you spend only $110,000 per year and allow the interest to accumulate, you will be bankrupt early in the eighteenth year.

The lesson here is an old one: *Protect your capital and live on the income it provides.* Deplete or waste your capital and you will move from a sustainable to an unsustainable lifestyle.

The same lesson applies to our use of the earth's natural capital—the global trust fund that nature has provided for us, for future generations (Figure 1-1), and for the earth's other species. *Living sustainably* means living on **natural income**, the renewable resources such as plants, animals, and soil provided by the earth's natural capital. It also means not depleting or degrading the earth's natural capital, which supplies this income, and providing the human population with adequate and equitable access to this natural income for the foreseeable future (**Concept 1-4**).

There is considerable and growing evidence that we are living unsustainably. A glaring example of this is our growing total and per capita ecological footprints (Figure 1-8).

We Can Work Together to Solve Environmental Problems

Solutions to environmental problems are not black and white, but rather are all shades of gray, because proponents of all sides of these issues have some legitimate and useful insights. In addition, any proposed solution has short- and long-term advantages and disadvantages that we must evaluate. This means that citizens need to work together to find *trade-off solutions* to environmental problems—another important theme of this book. In other words, *individuals matter*—one more important theme of this book.

GOOD NEWS

Here are two pieces of good news: First, research by social scientists suggests that it takes only 5–10% of the population of a community, a country, or the world to bring about major social change. Second, such research also shows that significant social change can occur in a much shorter time than most people think. Anthropologist Margaret Mead summarized our potential for social change: "Never doubt that a small group of thoughtful, committed citizens can change the world. Indeed, it is the only thing that ever has."

Scientific evidence indicates that we have perhaps 50 years and no more than 100 years to make a new cultural shift from unsustainable living to more sustainable living, if we start now. One of the goals of this book is to provide a realistic vision of a more environmentally sustainable future (**Core Case Study**) based on energizing, realistic hope rather than on immobilizing fear, gloom, and doom. CORE CASE STUDY

Figure 1-15 Capturing wind power both on- and offshore is one of the world's most rapidly growing and least environmentally harmful ways to produce electricity.

Varina and Jay Patel/Shutterstock.com

Based on the three **principles of sustainability**, we can derive three strategies for reducing our ecological footprints, helping to sustain the earth's natural capital, and making a transition to more sustainable lifestyles and economies. Those strategies are summarized in the *three big ideas* of this chapter:

- Rely more on renewable energy from the sun (see photo on title page), including indirect forms of solar energy such as wind (Figure 1-15), to meet most of our heating and electricity needs.
- Protect biodiversity by preventing the degradation of the earth's species, ecosystems, and natural processes, and by restoring areas we have degraded.
- Help to sustain the earth's natural chemical cycles by reducing the production of wastes and pollution, not overloading natural systems with harmful chemicals, and not removing natural chemicals faster than the earth's chemical cycles can replace them.

REVISITING A Vision of a More Sustainable Earth

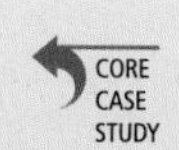

We face an array of serious environmental problems. This book is about *solutions* to these problems. A key to most solutions is to apply the three **principles of sustainability** (Figure 1-2 and the *three big ideas* listed above) to the design of our economic and social systems, and to our individual lifestyles. We can use such strategies to try to slow the rapidly increasing losses of biodiversity, to switch to more sustainable sources of energy, and to promote more sustainable forms of agriculture and other uses of land and water. We can also use them to sharply reduce poverty and slow human population growth.

Suppose that we make good environmental choices during this century, as the fictional characters Emily and Michael and people like them did in the **Core Case Study** that opens this chapter. Then, chances are that we will help to create an extraordinary and more sustainable future for ourselves, for future generations, and for most other forms of life on our planetary home. If we get it wrong, we face irreversible ecological disruption that could set humanity back for centuries and wipe out as many as half of the world's species as well as much of the human population.

You have the good fortune to be a member of the 21st century's *transition generation* that will decide which path humanity takes. *This means confronting the urgent challenges presented by the major environmental problems discussed in this book.* GOOD NEWS

What's the use of a house if you don't have a decent planet to put it on?

HENRY DAVID THOREAU

REVIEW

CORE CASE STUDY

1. Summarize the authors' vision of a more sustainable world, attainable by 2060 (**Core Case Study**).

SECTION 1-1

2. What are the two key concepts for this section? Define **sustainability**. Define **environment**. Distinguish among **environmental science**, **ecology**, and **environmentalism**. Distinguish between an **organism** and a **species**. What is an **ecosystem**? What are **three principles of sustainability** derived from natural processes? What is **solar energy** and why is it important to life on the earth? What is **biodiversity** and why is it important to life on the earth? Define **nutrients**. Define **chemical** or **nutrient cycling** and explain why it is important to life on the earth.

3. Define **natural capital**. Define **natural resources** and **natural services** and give two examples of each. Describe how we can degrade natural capital and how finding solutions to environmental problems involves making trade-offs. Explain why individuals matter in dealing with the environmental problems we face.

4. What is a **resource**? Distinguish between a **perpetual resource** and a **renewable resource** and give an example of each. What is the **sustainable yield** of a renewable resource? Define and give two examples of a **nonrenewable resource**. Distinguish between **recycling** and **reuse** and give an example of each. What percentage of the nonrenewable metals and plastics that we use could be recycled or reused? Distinguish between **more-developed countries** and **less-developed countries** and give an example of a high-income, middle-income, and low-income country.

SECTION 1-2

5. What is the key concept for this section? Define and give three examples of **environmental degradation (natural capital degradation)**. About what percentage of the earth's natural or ecosystem services have been degraded by human activities? Define **pollution**. Distinguish between **point sources** and **nonpoint sources** of pollution and give an example of each. Distinguish between **pollution cleanup** and **pollution prevention**. Describe three drawbacks to solutions that rely mostly on pollution cleanup. What is the *tragedy of the commons* and what are two ways to deal with it?

6. What is an **ecological footprint**? What is a **per capita ecological footprint**? Compare the total and per capita ecological footprints of the United States and China. Use the ecological footprint concept to explain how we are living unsustainably in terms of the number of planet earths that we need to sustain ourselves now and in the future.

7. What is the IPAT model for estimating our environmental impact? Explain how we can use this model to estimate the impacts of the human populations in less-developed and more-developed countries. Describe the environmental impacts of China's new affluent consumers.

SECTION 1-3

8. What are the two key concepts for this section? Identify four basic causes of the environmental problems that we face. What is **exponential growth**? What is the current size of the human population? How many people are added each year? How many people may be here by 2050? What is **affluence**? How do Americans, Indians, and the average people in the poorest countries compare in terms of average consumption per person? What are two types of environmental damage resulting from growing affluence? How can affluence help us to solve environmental problems? What is **poverty** and what are three of its harmful environmental and health effects? Describe the connection between poverty and population growth. Describe three major health problems suffered by many of the world's poor.

9. Explain how excluding the harmful environmental costs of production from the prices of goods and services affects the environmental problems we face. What is the connection between government subsidies, resource use, and environmental degradation? What are two ways to include the harmful environmental and health costs of the goods and services that we use in their market prices? What is an **environmental worldview**? What are **environmental ethics**? Distinguish among the **planetary management**, **stewardship**, and **environmental wisdom worldviews**.

SECTION 1-4

10. What is the key concept for this section? What is an **environmentally sustainable society**? What is **natural income** and what does it mean to live off of natural income? What are two pieces of good news about making the transition to a more sustainable society? Based on the three **principles of sustainability**, what are the three best ways to make a transition sustainability as summarized in this chapter's *three big ideas*? Explain how we can use these three principles to get us closer to the vision of a more sustainable world described in the **Core Case Study** that opens this chapter.

Note: Key terms are in **bold** type. Knowing the meanings of these terms will help you in the course you are taking.

CRITICAL THINKING

1. Do you think you are living unsustainably? Explain. If so, what are the three most environmentally unsustainable components of your lifestyle? List two ways in which you could apply each of the three **principles of sustainability** (Figure 1-2) to making your lifestyle more environmentally sustainable.
2. Do you believe that a vision such as the one described in the **Core Case Study** that opens this chapter is possible? Why or why not? What, if anything, do you believe will be different from that vision of the future? Explain. If your vision of what it will be like in 2060 is sharply different from that in the Core Case Study, write a description of your vision. Compare your answers to this question with those of your classmates.
3. For each of the following actions, state one or more of the three **principles of sustainability** (Figure 1-2) that are involved: **(a)** recycling aluminum cans; **(b)** using a rake instead of leaf blower; **(c)** walking or bicycling to class instead of driving; **(d)** taking your own reusable bags to the grocery store to carry your purchases home; **(e)** volunteering to help restore a prairie; and **(f)** lobbying elected officials to require that at least 20% of your country's electricity be produced with renewable wind power by 2020.
4. Explain why you agree or disagree with the following propositions:
   a. Stabilizing population is not desirable because, without more consumers, economic growth would stop.
   b. The world will never run out of resources because we can use technology to find substitutes and to help us reduce resource waste.
5. What do you think when you read that the average American consumes 30 times more resources than the average citizen of India? Are you happy, skeptical, indifferent, sad, helpless, guilty, concerned, or outraged by this fact? Do you think that these differences in consumption have led to problems? If so, describe them and propose some possible solutions.
6. When you read that at least 20,500 children age 5 and younger die each day (14 per minute) from preventable malnutrition and infectious disease, how does it make you feel? Can you think of something that you and others could do to address this problem? What might that be?
7. Explain why you agree or disagree with each of the following statements: **(a)** humans are superior to other forms of life; **(b)** humans are in charge of the earth; **(c)** the value of other forms of life depends only on whether they are useful to humans; **(d)** based on past extinctions and the history of life on the earth over the last 3.5 billion years, biologists hypothesize that all forms of life eventually become extinct, and we should not worry about whether our activities hasten their extinction; **(e)** all forms of life have an inherent right to exist; **(f)** all economic growth is good; **(g)** nature has an almost unlimited storehouse of resources for human use; **(h)** technology can solve our environmental problems; **(i)** I do not believe I have any obligation to future generations; and **(j)** I do not believe I have any obligation to other forms of life.
8. What are the basic beliefs within your environmental worldview (pp. 19–20)? Record your answer. Then, at the end of this course, return to your answer to see if your environmental worldview has changed. Are the beliefs included in your environmental worldview consistent with the answers you gave above? Are your actions that affect the environment consistent with your environmental worldview? Explain.

DOING ENVIRONMENTAL SCIENCE

Estimate your own ecological footprint by visiting the website **www.myfootprint.org/**. Is it larger or smaller than you thought it would be, according to this estimate? Why do you think this is so? List three ways in which you could reduce your ecological footprint. Try one of them for a week, and write a report on this change.

GLOBAL ENVIRONMENT WATCH EXERCISE

Using the *World Map*, choose one more-developed country and one less-developed country to compare their ecological footprints (found under Quick Facts on the country portal). Click on the ecological footprint number to view a graph of both the ecological footprint and biocapacity of each country. Using those graphs, determine whether these countries are living sustainably or not. What would be some reasons for these trends?

ECOLOGICAL FOOTPRINT ANALYSIS

If the *ecological footprint per person* of a country or the world (Figure 1-8) is larger than its *biological capacity per person* to replenish its renewable resources and absorb the resulting waste products and pollution, the country or the world is said to have an *ecological deficit*. If the reverse is true, the country or the world has an *ecological credit* or *reserve*. Use the data below to calculate the ecological deficit or credit for the countries listed and for the world. (For a map of ecological creditors and debtors see Figure 5, p. S27, in Supplement 6.)

Place	Per Capita Ecological Footprint (hectares per person)	Per Capita Biological Capacity (hectares per person)	Ecological Credit (+) or Debit (–) (hectares per person)
World	2.2	1.8	– 0.4
United States	9.8	4.7	
China	1.6	0.8	
India	0.8	0.4	
Russia	4.4	0.9	
Japan	4.4	0.7	
Brazil	2.1	9.9	
Germany	4.5	1.7	
United Kingdom	5.6	1.6	
Mexico	2.6	1.7	
Canada	7.6	14.5	

Source: Data from WWF *Living Planet Report 2006.*

1. Which two countries have the largest ecological deficits? Why do you think they have such large deficits?
2. Which two countries have an ecological credit? Why do you think each of these countries has an ecological credit?
3. Rank the countries in order from the largest to the smallest per capita ecological footprint.

LEARNING ONLINE

The website for *Environmental Science 14e* contains helpful study tools and other resources to take the learning experience beyond the textbook. Examples include flashcards, practice quizzing, information on green careers, tips for more sustainable living, activities, and a wide range of articles and further readings. To access these materials and other companion resources for this text, please visit **www.cengagebrain.com**. For further details, see the preface, p. xvi.

Science, Matter, and Energy

2

How Do Scientists Learn about Nature? A Story about a Forest

CORE CASE STUDY

Imagine that you learn of a logging company's plans to cut down all of the trees on a hillside in back of your house. You are very concerned and want to know the possible harmful environmental effects of this action on the hillside, the stream at the bottom of the hillside, and your backyard.

One way to learn about such effects is to conduct a *controlled experiment*, just as environmental scientists do. They begin by identifying key *variables* such as water loss and soil nutrient content that might change after the trees are cut down. Then, they set up two groups. One is the *experimental group*, in which a chosen variable is changed in a known way. The other is the *control group*, in which the chosen variable is not changed. Their goal is to compare the two groups after the variable has been changed and to look for differences resulting from the change.

In 1963, botanist F. Herbert Bormann, forest ecologist Gene Likens, and their colleagues began carrying out such a controlled experiment. The goal was to compare the loss of water and soil nutrients from an area of uncut forest (the *control site*) with one that had been stripped of its trees (the *experimental site*).

They built V-shaped concrete dams across the creeks at the bottoms of several forested valleys in the Hubbard Brook Experimental Forest in New Hampshire (Figure 2-1). The dams were designed so that all surface water leaving each forested valley had to flow across a dam, where scientists could measure its volume and dissolved nutrient content.

First, the researchers measured the amounts of water and dissolved soil nutrients flowing from an undisturbed forested area in one of the valleys (the control site) (Figure 2-1, left). These measurements showed that an undisturbed mature forest is very efficient at storing water and retaining chemical nutrients in its soils.

Next, they set up an experimental forest area in another of the valleys. One winter, they cut down all the trees and shrubs in that valley, left them where they fell, and sprayed the area with herbicides to prevent the regrowth of vegetation. Then, they compared outflow of water and nutrients in this experimental site with those in the control site for 3 years.

With no plants to help absorb and retain water, the amount of water flowing out of the deforested valley increased by 30–40%. As this excess water ran rapidly over the ground, it eroded soil and carried dissolved nutrients out of the deforested site. Overall, the loss of key nutrients from the experimental forest was six to eight times that in the nearby uncut control forest.

This controlled experiment revealed one of the ways in which scientists can learn about the effects of our actions on natural systems such as forests. In this chapter, you will learn more about how scientists study nature and about the matter and energy that make up the world within and around us. You will also learn about three *scientific laws*, or rules of nature, that govern the changes that matter and energy undergo.

Figure 2-1 This controlled field experiment measured the effects of deforestation on the loss of water and soil nutrients from a forest. The forested valley (left) was the control site; the cutover valley (right) was the experimental site.

Key Questions and Concepts

2-1 What do scientists do?

CONCEPT 2-1 Scientists collect data and develop theories, models, and laws about how nature works.

2-2 What is matter and what happens when it undergoes change?

CONCEPT 2-2A Matter consists of elements and compounds, which in turn are made up of atoms, ions, or molecules.

CONCEPT 2-2B Whenever matter undergoes a physical or chemical change, no atoms are created or destroyed (the law of conservation of matter).

2-3 What is energy and what happens when it undergoes change?

CONCEPT 2-3A Whenever energy is converted from one form to another in a physical or chemical change, no energy is created or destroyed (first law of thermodynamics).

CONCEPT 2-3B Whenever energy is converted from one form to another in a physical or chemical change, we end up with lower quality or less usable energy than we started with (second law of thermodynamics).

Note: Supplements 1 (p. S2), 2 (p. S3), 4 (p. S10), and 7 (p. S38) can be used with this chapter.

Science is built up of facts, as a house is built of stones; but an accumulation of facts is no more a science than a heap of stones is a house.

HENRI POINCARÉ

2-1 What Do Scientists Do?

▶ **CONCEPT 2-1 Scientists collect data and develop theories, models, and laws about how nature works.**

Science Is a Search for Order in Nature

Science is an attempt to discover how nature works and to use that knowledge to make predictions about what is likely to happen in nature. It is based on the assumption that events in the physical world follow orderly cause-and-effect patterns that can be understood through careful observation, measurements, experimentation, and modeling. Figure 2-2 summarizes the scientific process.

There is no prescribed scientific method. Scientists use a variety of methods to study nature, although these methods tend to fall within the general process described in Figure 2-2. There is nothing mysterious about this process. You use it all the time in making decisions. As the famous physicist Albert Einstein put it, "The whole of science is nothing more than a refinement of everyday thinking."

Scientists Use Observations, Experiments, and Models to Answer Questions about How Nature Works

Here is a more formal outline of the steps scientists often take in trying to understand the natural world, although they do not always follow the steps in the order listed. The outline is based on the scientific experiment carried out by Bormann and Likens (**Core Case Study**) that illustrates the nature of the scientific process shown in Figure 2-2. CORE CASE STUDY

- *Identify a problem.* Bormann and Likens identified the loss of water and soil nutrients from cutover forests as a problem worth studying.
- *Find out what is known about the problem.* Bormann and Likens searched the scientific literature to find out what scientists knew about both the retention and the loss of water and soil nutrients in forests.

Links: CORE CASE STUDY refers to the Core Case Study. SUSTAINABILITY refers to the book's sustainability theme. GOOD NEWS refers to good news about the environmental challenges we face.

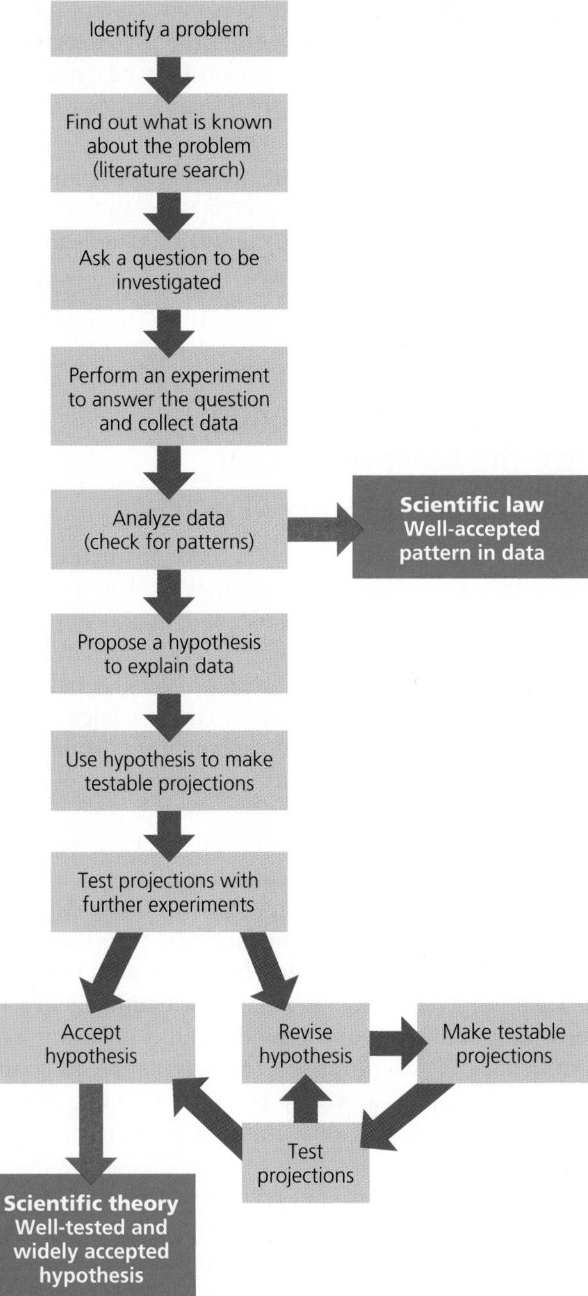

Figure 2-2 This diagram illustrates what scientists do. Scientists use this overall process for testing ideas about how the natural world works. However, they do not necessarily follow the order of steps shown here. For example, a scientist might start by coming up with a hypothesis to answer the initial question and then running experiments to test the hypothesis.

- *Ask a question to investigate*. The scientists asked, "How does clearing forested land affect its ability to store water and retain soil nutrients?"
- *Perform an experiment and collect and analyze data to answer the question*. To collect **data**—information needed to answer their questions—scientists often perform experiments and make observations and measurements. (Sometimes, they simply observe and measure natural phenomena to collect data without doing an experiment.) Bormann and Likens collected and analyzed data on the water and soil nutrients flowing from a valley with an undisturbed forest (Figure 2-1, left) and from a nearby valley where they had cleared the forest for their experiment (Figure 2-1, right).
- *Propose a hypothesis to explain the data*. Scientists suggest a **scientific hypothesis**—a possible and testable explanation of what scientists observe in nature or in the results of their experiments. Bormann and Likens came up with the following hypothesis to explain their data: When a forest is cleared of its vegetation and exposed to rain and melting snow, it retains less water than it did before it was cleared and loses large quantities of its soil nutrients.
- *Use the hypothesis to make testable projections*. Scientists make projections about what should happen if their hypothesis is correct and then run experiments to test the projections. Bormann and Likens projected that if their hypothesis was valid for nitrogen, then a cleared forest should also lose other soil nutrients such as phosphorus over a similar time period and under similar weather conditions.
- *Test the projections with further experiments, models, or observations*. To test their projection, Bormann and Likens repeated their controlled experiment and measured the phosphorus content of the soil. Another way to test projections is to develop a **model**, an approximate representation or simulation of a system, such as a forested or deforested valley. Data from the research carried out by Bormann and Likens and from other scientists' research can be fed into such models and used to project the loss of phosphorus and other types of soil nutrients. Scientists can then compare these projections with the actual measured losses to test the validity of the models.
- *Accept or revise the hypothesis*. If their new data do not support their hypothesis, scientists come up with other testable explanations. This process of proposing and testing various hypotheses goes on until there is general agreement among the scientists in this field of study that a particular hypothesis is the best explanation of the data. After Bormann and Likens confirmed that the soil in a cleared forest also loses phosphorus, they measured losses of other soil nutrients, which further supported their hypothesis. Research and models by other scientists also supported the hypothesis. A well-tested and widely accepted scientific hypothesis or a group of related hypotheses is called a **scientific theory**. Thus, Bormann and Likens and other scientists developed a theory that trees and other plants hold soil in place and help it to retain water and nutrients needed by the plants for their growth.

Scientists Are Curious and Skeptical, and Demand Lots of Evidence

Four important features of the scientific process are *curiosity, skepticism, reproducibility,* and *peer review.* Good scientists are extremely curious about how nature works. But they tend to be highly skeptical of new data, hypotheses, and models until they can test and verify them with lots of evidence. Scientists say, "Show me your evidence and explain the reasoning behind the scientific ideas or hypotheses that you propose to explain your data." They also require that any evidence gathered must be reproducible. In other words, other scientists should be able to get the same results when they run the same experiments.

Science is a community effort, and an important part of the scientific process is **peer review**. It involves scientists openly publishing details of the methods and models they used, the results of their experiments, and the reasoning behind their hypotheses for other scientists working in the same field (their peers) to evaluate.

For example, Bormann and Likens (**Core Case Study**) CORE CASE STUDY submitted the results of their forest experiments to a respected scientific journal. Before publishing this report, the journal's editors asked other soil and forest experts to review it. Other scientists have repeated the measurements of soil content in undisturbed and cleared forests of the same type and also for different types of forests. Their results have been subjected to peer review as well. In addition, computer models of forest systems have been used to evaluate this problem, with the results also subjected to peer review.

Scientific knowledge advances in this self-correcting way, with scientists continually questioning the measurements and data produced by their peers. They also collect new data and sometimes come up with new and better hypotheses. Skepticism and debate among peers in the scientific community is essential to the scientific process.

Critical Thinking and Creativity Are Important in Science

Scientists use logical reasoning and critical thinking skills (pp. 2–3) to learn about the natural world. Thinking critically involves four important steps:

1. Be skeptical about everything you read or hear.
2. Look at the evidence and evaluate it and any related information, along with opinions from a variety of sources. Validating information and identifying opinions are especially important steps to take in this Internet age when we are often exposed to unreliable data and to opinions from uninformed amateurs posing as experts.
3. Be open to many viewpoints and evaluate each one before coming to a conclusion.
4. Identify and evaluate your personal assumptions, biases, and beliefs. As the American psychologist and philosopher William James observed, "A great many people think they are thinking when they are merely rearranging their prejudices."

Logic and critical thinking are very important tools in science, but imagination, creativity, and intuition are just as vital. According to physicist Albert Einstein, "There is no completely logical way to a new scientific idea."

Scientific Theories and Laws Are the Most Important and Certain Results of Science

The real goal of scientists is to develop theories and laws, based on facts and data that explain how the physical world works, as illustrated in the quotation that opens this chapter. *Any scientific conclusion that is deemed a theory should not be taken lightly.* It has been tested widely, is supported by extensive evidence, and is accepted as being a useful explanation of some phenomenon by most scientists in a particular field or related fields of study.

Because of this rigorous testing process, scientific theories are rarely overturned unless new evidence discredits them or scientists come up with better explanations. So when you hear someone say, "Oh, that's just a theory," you will know that he or she does not have a clear understanding of what a scientific theory is, how important it is, and how difficult it is for any scientific idea to reach this level of acceptance. In sports terms, developing a widely accepted scientific theory is roughly equivalent to winning a gold medal in the Olympics.

Another important and reliable outcome of science is a **scientific law**—a well-tested and accepted description of what we find happening repeatedly in nature in the same way. An example is the *law of gravity.* After making many thousands of observations and measurements of objects falling from different heights, scientists developed the following scientific law: all objects fall to the earth's surface at predictable speeds.

We can break a society's law, for example, by driving faster than the speed limit. But *we cannot break a scientific law,* unless we discover new data that lead to changes in the law.

For a superb look at how the scientific process is applied to expanding our understanding of the natural world, see the Annenberg Video series, *The Habitable Planet: A Systems Approach to Environmental Science* (see the website at **www.learner.org/resources/series209.html**). Each of the 13 videos describes how scientists working on two different problems related to each subject are learning about how nature works. We regularly cross-reference material in this book to these videos.

The Results of Science Can Be Tentative, Reliable, or Unreliable

Sometimes, preliminary scientific results that capture news headlines have not been widely tested and accepted by peer review. They are not yet considered reliable, and can be thought of as **tentative science** or **frontier science**. Some of these results will be validated and classified as reliable and some will be discredited and classified as unreliable. At the frontier stage, it is normal for scientists to disagree about the meaning and accuracy of data and the validity of hypotheses and results. This is how scientific knowledge advances.

By contrast, **reliable science** consists of data, hypotheses, models, theories, and laws that are widely accepted by all or most of the scientists who are considered experts in the field under study, in what is sometimes referred to as a *scientific consensus*. The results of reliable science are based on the self-correcting process of testing, open peer review, reproducibility, and debate. New evidence and better hypotheses may discredit or alter accepted views. But until that happens, those views are considered to be the results of reliable science.

Scientific hypotheses and results that are presented as reliable without having undergone the rigors of widespread peer review, or that have been discarded as a result of peer review, are considered to be **unreliable science**. Here are some critical thinking questions you can use to uncover unreliable science:

- Was the experiment well designed? Did it involve a control group? (**Core Case Study**) CORE CASE STUDY
- Have other scientists reproduced the results?
- Does the proposed hypothesis explain the data? Have scientists made and verified projections based on the hypothesis?
- Are there no other, more reasonable explanations of the data?
- Are the investigators unbiased in their interpretations of the results? Were all the investigators' funding sources unbiased?
- Have the data and conclusions been subjected to peer review?
- Are the conclusions of the research widely accepted by other experts in this field?

If "yes" is the answer to each of these questions, then you can classify the results as reliable science. Otherwise, the results may represent tentative science that needs further testing and evaluation, or you can classify them as unreliable science.

Science Has Some Limitations

Environmental science and science in general have three important limitations. *First,* scientists cannot prove or disprove anything absolutely, because there is always some degree of uncertainty in scientific measurements, observations, and models. Instead, scientists try to establish that a particular scientific theory or law has a very high *probability* or *certainty* (at least 90%) of being useful for understanding some aspect of nature.

Many scientists don't use the word *proof* because this implies "absolute proof" to people who don't understand how science works. For example, most scientists will rarely say something like, "Cigarettes cause lung cancer." Rather, they might say, "Overwhelming evidence from thousands of studies indicates that people who smoke regularly for many years have a greatly increased chance of developing lung cancer."

Suppose someone tells you that some statement is not true because it has not been scientifically proven. When this happens, you can conclude either that he or she does not understand how science works, or that the person is trying to trick you by telling you something that is true but irrelevant and misleading.

THINKING ABOUT
Scientific Proof

Does the fact that science can never prove anything absolutely mean that its results are not valid or useful? Explain.

A *second* limitation of science is that scientists are human and thus are not totally free of bias about their own results and hypotheses. However, the high standards of evidence required through peer review can usually uncover or greatly reduce personal bias and expose occasional cheating by scientists who falsify their results.

A *third* limitation—especially important to environmental science—is that many systems in the natural world involve a huge number of variables with complex interactions. This makes it difficult, too costly, and too time consuming to test one variable at a time in controlled experiments such as the one described in the **Core Case Study** that opens this chapter. CORE CASE STUDY To try to deal with this problem, scientists develop *mathematical models* that can take into account the interactions of many variables. Running such models on high-speed computers can sometimes overcome the limitations of testing each variable individually, saving both time and money. In addition, scientists can use computer models to simulate global experiments on phenomena like climate change that cannot be done in a controlled physical experiment.

Despite these limitations, science is the most useful way that we have of learning about how nature works and of projecting how it might behave in the future. But we still know too little about how the earth works, about its present state of environmental health, and about the current and future environmental impacts of our activities. These knowledge gaps point to important *research frontiers,* listed for each chapter on this book's website at **www.cengagebrain.com**.

2-2 What Is Matter and What Happens When It Undergoes Change?

▶ **CONCEPT 2-2A Matter consists of elements and compounds, which in turn are made up of atoms, ions, or molecules.**

▶ **CONCEPT 2-2B Whenever matter undergoes a physical or chemical change, no atoms are created or destroyed (the law of conservation of matter).**

Matter Consists of Elements and Compounds

To begin our study of environmental science, we look at matter—the stuff that makes up life and its environment. **Matter** is anything that has mass and takes up space. It can exist in three *physical states*—solid, liquid, and gas—and two *chemical forms*—elements and compounds.

An **element** is a fundamental type of matter that has a unique set of properties and cannot be broken down into simpler substances by chemical means. For example, the elements gold (Figure 2-3, right) and mercury (Figure 2-3, left) cannot be broken down chemically into any other substance.

Some matter is composed of one element, such as gold or mercury. However, most matter consists of **compounds**, combinations of two or more different elements held together in fixed proportions. For example, water is a compound made of the elements hydrogen and oxygen that have chemically combined with one another. (For more details, see Supplement 4, p. S10, on the basics of chemistry.)

To simplify things, chemists represent each element by a one- or two-letter symbol. Table 2-1 lists the elements and their symbols that you need to know to understand the material in this book.

Morgan Lane Photography/Shutterstock.com

Andraž Cerar/Shutterstock.com

Figure 2-3 Gold (right) and mercury (left) are chemical elements; each has a unique set of properties and cannot be broken down into simpler substances.

Table 2-1 Chemical Elements Used in This Book

Element	Symbol
arsenic	As
bromine	Br
calcium	Ca
carbon	C
chlorine	Cl
fluorine	F
gold	Au
lead	Pb
lithium	Li
mercury	Hg
nitrogen	N
phosphorus	P
sodium	Na
sulfur	S
uranium	U

Atoms, Molecules, and Ions Are the Building Blocks of Matter

The most basic building block of matter is an **atom**, the smallest unit of matter into which an element can be divided and still have its characteristic chemical properties. The idea that all elements are made up of atoms is called the **atomic theory** and it is the most widely accepted scientific theory in chemistry.

Atoms are incredibly small. For example, more than 3 million hydrogen atoms could sit side by side on the period at the end of this sentence. If you could view atoms with a supermicroscope, you would find that each different type of atom contains a certain number of three types of *subatomic particles*: **neutrons (n)** with no electrical charge, **protons (p)** with a positive electrical charge (+), and electrons (e) with a negative electrical charge (−).

Each atom has an extremely small but dense center called the **nucleus**, which contains one or more protons (p) and, in most cases, one or more neutrons (n). Outside of the nucleus we find one or more electrons (e) in rapid motion (see Figure 1, p. S10, in Supplement 4). Each atom has equal numbers of positively charged protons and negatively charged electrons. Because these electrical charges cancel one another, *atoms as a whole have no net electrical charge*.

Each element has a unique **atomic number** equal to the number of protons in the nucleus of its atom.

Carbon (C), with 6 protons in its nucleus, has an atomic number of 6, whereas uranium (U), a much larger atom, has 92 protons in its nucleus and thus an atomic number of 92.

Because electrons have so little mass compared to protons and neutrons, *most of an atom's mass is concentrated in its nucleus.* The mass of an atom is described by its **mass number**, the total number of neutrons and protons in its nucleus. For example, a carbon atom with 6 protons and 6 neutrons in its nucleus has a mass number of 12, and a uranium atom with 92 protons and 143 neutrons in its nucleus has a mass number of 235 (92 + 143 = 235).

Each atom of a particular element has the same number of protons in its nucleus. But the nuclei of atoms of a particular element can vary in the number of neutrons they contain, and therefore, in their mass numbers. The forms of an element having the same atomic number but different mass numbers are called **isotopes** of that element. Scientists identify isotopes by attaching their mass numbers to the name or symbol of the element. For example, the three most common isotopes of carbon are carbon-12 (with six protons and six neutrons), carbon-13 (with six protons and seven neutrons), and carbon-14 (with six protons and eight neutrons). Carbon-12 makes up about 98.9% of all naturally occurring carbon.

A second building block of matter is a **molecule**, a combination of two or more atoms of the same or different elements held together by forces called *chemical bonds*. Molecules are the basic building blocks of many compounds (see Figure 5, p. S12, in Supplement 4 for examples). An example of a molecule is that of water, or H_2O, which consists of two atoms of hydrogen and one atom of oxygen held together by chemical bonds. Another example is methane, or CH_4 (the major component of natural gas), which consists of four atoms of hydrogen and one atom of carbon.

A third building block of some types of matter is an **ion**—an atom or a group of atoms with one or more net positive or negative electrical charges. Like atoms, ions are made up of protons, neutrons, and electrons. (See p. S10 in Supplement 4 for details on how ions form.) Chemists use a superscript after the symbol of an ion to indicate how many positive or negative electrical charges it has, as shown in Table 2-2.

The nitrate ion (NO_3^-) is a nutrient essential for plant growth. Figure 2-4 shows measurements of the loss of nitrate ions from the deforested area (Figure 2-1, right) in the controlled experiment run by Bormann and Likens (**Core Case Study**). Numerous chemical analyses of the water flowing through the dam at the cleared forest site showed an average 60-fold rise in the concentration of NO_3^- compared to water running off the forested site. After a few years, however, vegetation began growing back on the cleared valley and nitrate levels in its runoff returned to normal levels.

CORE CASE STUDY

Table 2-2 Chemical Ions Used in This Book

Positive Ion	Symbol	Components
hydrogen ion	H^+	One H atom, one positive charge
sodium ion	Na^+	One Na atom, one positive charge
calcium ion	Ca^{2+}	One Ca atom, two positive charges
aluminum ion	Al^{3+}	One Al atom, three positive charges
ammonium ion	NH_4^+	One N atom, four H atoms, one positive charge
Negative Ion	**Symbol**	**Components**
chloride ion	Cl^-	One Cl atom, one negative charge
hydroxide ion	OH^-	One O atom, one H atom, one negative charge
nitrate ion	NO_3^-	One N atom, three O atoms, one negative charge
carbonate ion	CO_3^{2-}	One C atom, three O atoms, two negative charges
sulfate ion	SO_4^{2-}	One S atom, four O atoms, two negative charges
phosphate ion	PO_4^{3-}	One P atom, four O atoms, three negative charges

Ions are also important for measuring a substance's **acidity** in a water solution, a chemical characteristic that helps determine how a substance dissolved in water will interact with and affect its environment. The acidity of a water solution is based on the comparative amounts of hydrogen ions (H^+) and hydroxide ions (OH^-) contained in a particular volume of the solution. Scientists use **pH** as a measure of acidity. Pure water (not tap water or rainwater) has an equal number of H^+ and OH^- ions. It is called a neutral solution and has a pH of 7. An acidic solution has more hydrogen ions than hydroxide ions and has a pH less than 7. A basic solution has more hydroxide ions than hydrogen ions and has a pH greater than 7. (See Figure 6, p. S13, in Supplement 4 for more details.)

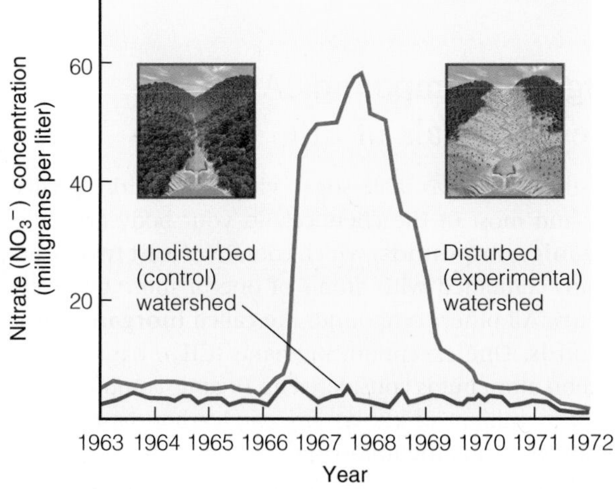

Figure 2-4 This graph shows the loss of nitrate ions (NO_3^-) from a deforested watershed in the Hubbard Brook Experimental Forest in New Hampshire (Figure 2-1, right). (Data from F. H. Bormann and Gene Likens)

Table 2-3 Compounds Used in This Book

Compound	Formula
sodium chloride	NaCl
sodium hydroxide	NaOH
carbon monoxide	CO
oxygen	O_2
nitrogen	N_2
chlorine	Cl_2
carbon dioxide	CO_2
nitric oxide	NO
nitrogen dioxide	NO_2
nitrous oxide	N_2O
nitric acid	HNO_3
methane	CH_4
glucose	$C_6H_{12}O_6$
water	H_2O
hydrogen sulfide	H_2S
sulfur dioxide	SO_2
sulfuric acid	H_2SO_4
ammonia	NH_3
calcium carbonate	$CaCO_3$

Chemists use a **chemical formula** to show the number of each type of atom or ion in a compound. This shorthand contains the symbol for each element present (Table 2-1) and uses subscripts to show the number of atoms or ions of each element in the compound's basic structural unit. Examples of compounds and their formulas encountered in this book are sodium chloride (NaCl) and water (H_2O, read as "H-two-O"). These and other compounds important to our study of environmental science are listed in Table 2-3.

You might want to mark these pages containing Tables 2-1, 2-2, and 2-3, because they show the key elements, ions, and compounds used in this book. Think of them as lists of the main chemical characters in the story of matter that makes up the natural world.

Organic Compounds Are the Chemicals of Life

Plastics, as well as table sugar, vitamins, aspirin, penicillin, and most of the chemicals in your body are called **organic compounds**, which contain at least two carbon atoms combined with atoms of one or more other elements. All other compounds are called **inorganic compounds**. One exception, methane (CH_4), has only one carbon atom but is considered an organic compound.

The millions of known organic (carbon-based) compounds include the following:

- *Hydrocarbons:* compounds of carbon and hydrogen atoms. One example is methane (CH_4), the main component of natural gas and the simplest organic compound. Another is octane (C_8H_{18}), a major component of gasoline.
- *Chlorinated hydrocarbons:* compounds of carbon, hydrogen, and chlorine atoms. An example is the insecticide DDT ($C_{14}H_9Cl_5$).
- *Simple carbohydrates (simple sugars):* certain types of compounds of carbon, hydrogen, and oxygen atoms. An example is glucose ($C_6H_{12}O_6$), which most plants and animals break down in their cells to obtain energy. (For more details, see Figure 8, p. S14, in Supplement 4.)

Larger and more complex organic compounds, essential to life, are composed of *macromolecules*. Some of these molecules are called *polymers*, formed when a number of simple organic molecules (*monomers*) are linked together by chemical bonds—somewhat like rail cars linked in a freight train. The three major types of organic polymers are:

- *complex carbohydrates* such as cellulose and starch, which consist of two or more monomers of simple sugars such as glucose (see Figure 8, p. S14, in Supplement 4),
- *proteins* formed by monomers called *amino acids* (see Figure 9, p. S14, in Supplement 4), and
- *nucleic acids* (DNA and RNA) formed by monomers called *nucleotides* (see Figures 10 and 11, pp. S14 and S15, in Supplement 4).

Lipids, which include fats and waxes, are not made of monomers but are a fourth type of macromolecule essential for life (see Figure 12, p. S15, in Supplement 4).

Matter Comes to Life through Cells, Genes, and Chromosomes

All organisms are composed of one or more **cells**—the fundamental structural and functional units of life. They are minute compartments covered with a thin membrane, and within them, the processes of life occur. The idea that all living things are composed of cells is called the *cell theory* and it is the most widely accepted scientific theory in biology.

Earlier, we mentioned nucleotides in DNA (see Figures 10 and 11, pp. S14 and S15, in Supplement 4). Within some DNA molecules are certain sequences of nucleotides called **genes**. Each of these distinct pieces of DNA contains instructions, or codes, called *genetic information*, for making specific proteins. Each of these coded units of genetic information leads to a specific **trait**, or characteristic, passed on from parents to offspring during reproduction in an animal or plant.

In turn, thousands of genes make up a single **chromosome**, a double helix DNA molecule (see Figure 11, p. S15, in Supplement 4) wrapped around some proteins. Genetic information coded in your chromosomal DNA is what makes you different from an oak leaf, an

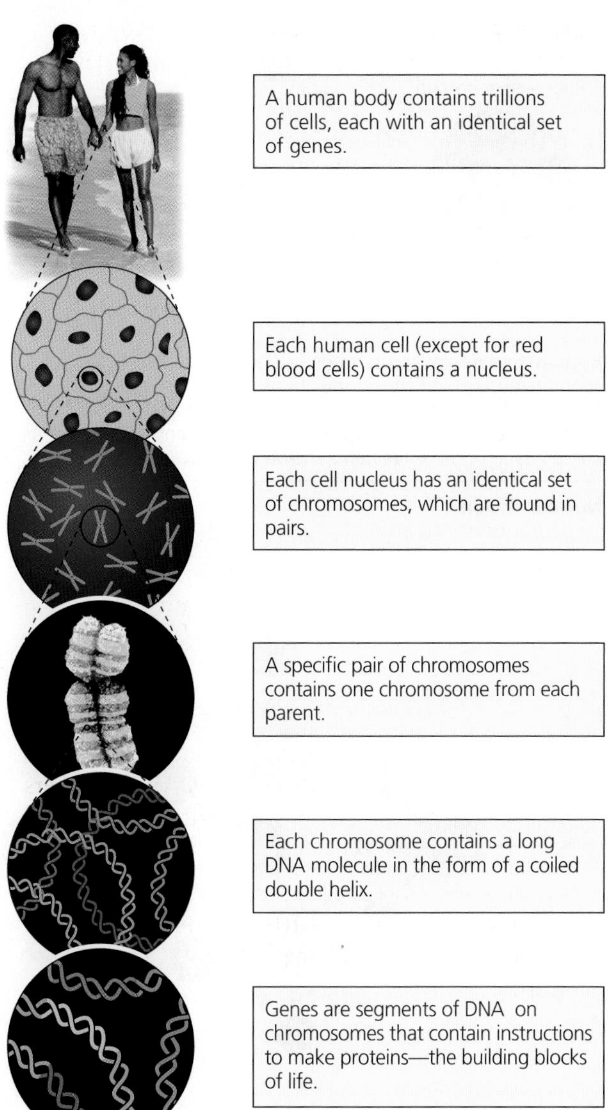

Figure 2-5 This diagram shows the relationships among cells, nuclei, chromosomes, DNA, and genes.

alligator, or a mosquito, and from your parents. The relationships of genetic material to cells are depicted in Figure 2-5.

Some Forms of Matter Are More Useful than Others

Matter quality is a measure of how useful a form of matter is to humans as a resource, based on its availability and *concentration*—the amount of it that is contained in a given area or volume. **High-quality matter** is highly concentrated, is typically found near the earth's surface, and has great potential for use as a resource. **Low-quality matter** is not highly concentrated, is often located deep underground or dispersed in the ocean or atmosphere, and usually has little potential for use as a resource. Figure 2-6 illustrates examples of differences in matter quality.

Figure 2-6 These examples illustrate the differences in matter quality. *High-quality matter* (left column) is fairly easy to extract and is highly concentrated; *low-quality matter* (right column) is not highly concentrated and is more difficult to extract than high-quality matter.

Matter Undergoes Physical, Chemical, and Nuclear Changes

When a sample of matter undergoes a **physical change**, there is no change in its *chemical composition*. A piece of aluminum foil cut into small pieces is still aluminum foil. When solid water (ice) melts and when liquid water boils, the resulting liquid water and water vapor are still made up of H_2O molecules.

When a **chemical change**, or **chemical reaction**, takes place, there is a change in the chemical composition of the substances involved. Chemists use a *chemical equation* to show how chemicals are rearranged in a chemical reaction. For example, coal is made up almost entirely of the element carbon (C). When coal is burned completely in a power plant, the solid carbon (C) in the coal combines with oxygen gas (O_2) from the atmosphere to form the gaseous compound carbon dioxide (CO_2). Chemists use the following shorthand chemical equation to represent this chemical reaction:

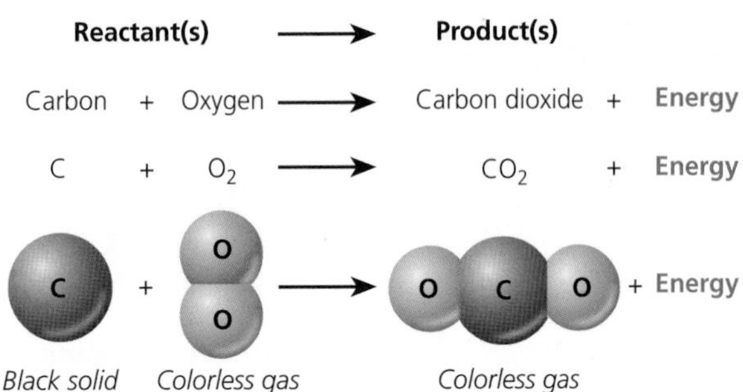

In addition to physical and chemical changes, matter can undergo three types of **nuclear change**, or change in the nuclei of its atoms: radioactive decay, nuclear fission, and nuclear fusion, which are described and defined in Figure 2-7.

We Cannot Create or Destroy Atoms: The Law of Conservation of Matter

We can change elements and compounds from one physical or chemical form to another, but we can never create or destroy any of the atoms involved in any physical or chemical change. All we can do is rearrange the atoms, ions, or molecules into different spatial patterns (physical changes) or chemical combinations (chemical changes). These facts, based on many thousands of measurements, describe a scientific law known as the **law of conservation of matter**: Whenever matter undergoes a physical or chemical change, no atoms are created or destroyed (**Concept 2-2B**).

CONNECTIONS

Waste and the Law of Conservation of Matter

The law of conservation of matter means we can never really throw anything away because the atoms in any form of matter cannot be destroyed as it undergoes physical or chemical changes. Stuff that we put out in the trash may be buried in a sanitary landfill, but we have not really thrown it away because the atoms in this waste material will always be around in one form or another. We can burn trash, but we then end up with ash that must be put somewhere, and with gases emitted by the burning that can pollute the air. We can reuse or recycle some materials and chemicals, but the law of conservation of matter means we will always face the problem of what to do with some quantity of the wastes and pollutants we produce because their atoms cannot be destroyed.

Radioactive decay

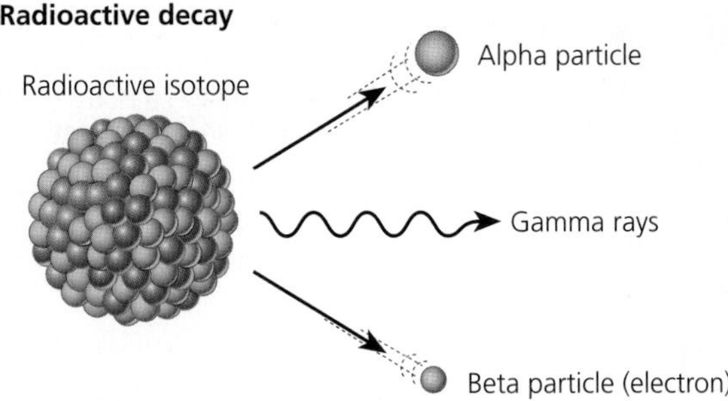

Radioactive decay occurs when nuclei of unstable isotopes spontaneously emit fast-moving chunks of matter (alpha particles or beta particles), high-energy radiation (gamma rays), or both at a fixed rate. A particular radioactive isotope may emit any one or a combination of the three items shown in the diagram.

Nuclear fission

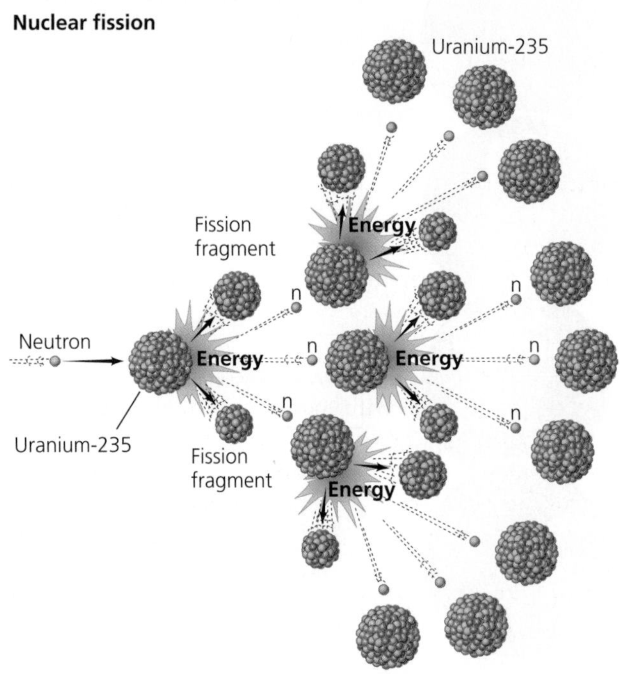

Nuclear fission occurs when the nuclei of certain isotopes with large mass numbers (such as uranium-235) are split apart into lighter nuclei when struck by a neutron and release energy plus two or three more neutrons. Each neutron can trigger an additional fission reaction and lead to a *chain reaction*, which releases an enormous amount of energy very quickly.

Nuclear fusion

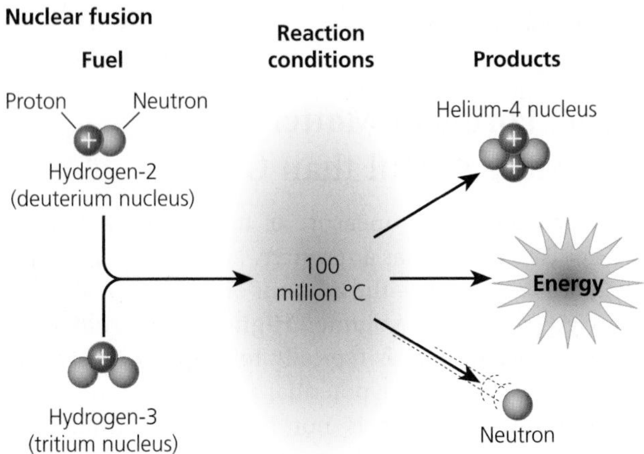

Nuclear fusion occurs when two isotopes of light elements, such as hydrogen, are forced together at extremely high temperatures until they fuse to form a heavier nucleus and release a tremendous amount of energy.

Figure 2-7 There are three types of nuclear changes: natural radioactive decay (top), nuclear fission (middle), and nuclear fusion (bottom).

2-3 What Is Energy and What Happens When It Undergoes Change?

▶ **CONCEPT 2-3A Whenever energy is converted from one form to another in a physical or chemical change, no energy is created or destroyed (first law of thermodynamics).**

▶ **CONCEPT 2-3B Whenever energy is converted from one form to another in a physical or chemical change, we end up with lower-quality or less-usable energy than we started with (second law of thermodynamics).**

Energy Comes in Many Forms

Suppose you find this book on the floor and you pick it up and put it on your desktop. In doing this you have to use a certain amount of muscular force to move the book from one place to another. In scientific terms, work is done when any object is moved a certain distance (work = force × distance). And whenever you touch a hot object such as a stove, heat flows from the stove to your finger. Both of these examples involve **energy**: the capacity to do work or to transfer heat.

There are two major types of energy: *moving energy* (called kinetic energy) and *stored energy* (called potential energy). Matter in motion has **kinetic energy**, which is energy associated with motion. Examples are flowing water, electricity (electrons flowing through a wire or other conducting material), and wind (a mass of moving air that we can use to produce electricity).

Another form of kinetic energy is **heat**, the total kinetic energy of all moving atoms, ions, or molecules within a given substance. When two objects at different temperatures come in contact with one another, heat flows from the warmer object to the cooler object. You learned this the first time you touched a hot stove.

In another form of kinetic energy, called **electromagnetic radiation**, energy travels in the form of a *wave* as a result of changes in electrical and magnetic fields. There are many different forms of electromagnetic radiation (see Figure 2-8), each having a different *wavelength* (the distance between successive peaks or troughs in the wave), and *energy content*. Forms of electromagnetic radiation with short wavelengths, such as gamma rays, X rays, and ultraviolet (UV) radiation, have more energy than do forms with longer wavelengths, such as visible light and infrared (IR) radiation. Visible light makes up most of the spectrum of electromagnetic radiation emitted by the sun.

The other major type of energy is **potential energy**, which is stored and potentially available for use. Examples of this type of energy include a rock held in your hand, the water in a reservoir behind a dam, and the chemical energy stored in the carbon atoms of coal.

We can change potential energy to kinetic energy. If you hold this book in your hand, it has potential energy. However, if you drop it on your foot, the book's potential energy changes to kinetic energy. When a car engine burns gasoline, the potential energy stored in the chemical bonds of the gasoline molecules changes into kinetic energy that propels the car, and into heat that flows into the environment. When water in a reservoir flows through channels in a dam, its potential energy becomes kinetic energy that we can use to spin turbines in the dam to produce electricity—another form of kinetic energy.

About 99% of the energy that heats the earth and our buildings and supports plants (through a process called *photosynthesis*) that form the basis of our food supply comes from the sun at no cost to us. This is in keeping with the solar energy **principle of sustainability** (see back cover). Without this essentially inexhaustible solar energy, the earth's average temperature would be −240°C (−400°F), and life as we know it would not exist.

SUSTAINABILITY

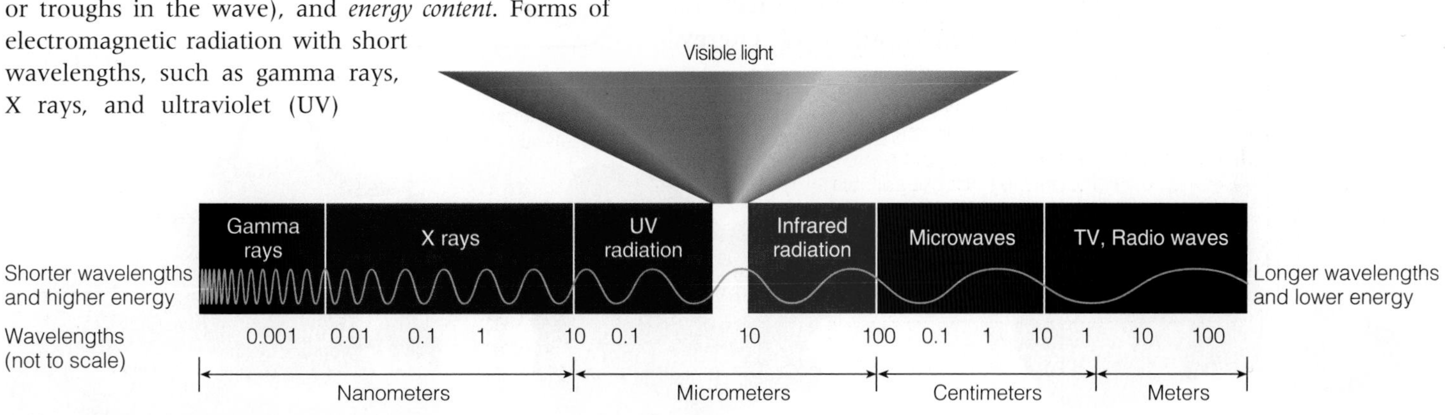

Active Figure 2-8 The *electromagnetic spectrum* consists of a range of electromagnetic waves, which differ in wavelength (the distance between successive peaks or troughs) and energy content. *See an animation based on this figure at* www.cengagebrain.com.

This direct input of solar energy produces several indirect forms of renewable solar energy. Examples are *wind, hydropower* (falling and flowing water), and *biomass* (solar energy converted to chemical energy and stored in the chemical bonds of organic compounds in trees and other plants) that can be burned to provide heat.

Commercial energy sold in the marketplace makes up the remaining 1% of the energy we use to supplement the earth's direct input of solar energy. About 81% of the commercial energy used in the world and 83% of the commercial energy that is used in the United States comes from burning **fossil fuels** (oil, coal, and natural gas, Figure 2-9). They are called fossil fuels because they were formed over millions of years as layers of the decaying remains of ancient plants and animals were exposed to intense heat and pressure within the earth's crust.

Some Types of Energy Are More Useful Than Others

Energy quality is a measure of the capacity of a type of energy to do useful work. **High-quality energy** has a great capacity to do useful work because it is concentrated. Examples are very high-temperature heat, concentrated sunlight, high-speed wind, and the energy released when we burn gasoline or coal.

By contrast, **low-quality energy** is so dispersed that it has little capacity to do useful work. For example, the enormous number of moving molecules in the atmosphere or in an ocean together have such low-quality energy and such a low temperature, that we cannot use them to move things or to heat things to high temperatures.

Energy Changes Are Governed by Two Scientific Laws

After observing and measuring energy being changed from one form to another in millions of physical and chemical changes, scientists have summarized their results in the **first law of thermodynamics**, also known as the **law of conservation of energy**. According to this scientific law, whenever energy is converted from one form to another in a physical or chemical change, no energy is created or destroyed (**Concept 2-3A**).

This scientific law tells us that no matter how hard we try or how clever we are, we cannot get more energy out of a physical or chemical change than we put in. This is one of nature's basic rules that we cannot violate.

Because the first law of thermodynamics states that energy cannot be created or destroyed, but only converted from one form to another, you may be tempted to think we will never have to worry about running out of energy. Yet if you fill a car's tank with gasoline and drive around or use a flashlight battery until it is dead, something has been lost. What is it? The answer is *energy quality,* the amount of energy available for performing useful work.

Thousands of experiments have shown that whenever energy is converted from one form to another in a physical or chemical change, we end up with lower-quality or less useable energy than we started with (**Concept 2-3B**). This is a statement of the **second law of thermodynamics**. The resulting low-quality energy usually takes the form of heat that flows into the environment. In the environment, the random motion of air or water molecules further disperses this heat, decreasing its temperature to the point where its energy quality is too low to do much useful work.

In other words, *when energy is changed from one form to another, it always goes from a more useful to a less useful form*. No one has ever found a violation of this fundamental scientific law. See examples of how the first and second laws of thermodynamics apply in our world at **www.cengagebrain.com**.

CONNECTIONS

Can We Recycle or Reuse Energy?

We can recycle and reuse various forms of matter such as paper and aluminum. However, because of the second law of thermodynamics, we can never recycle or reuse high-quality energy to perform useful work. Once the concentrated or high-quality energy in a serving of food, a full tank of gasoline, or a chunk of uranium is released, it is degraded to low-quality heat that is dispersed into the environment at a low temperature.

Andrea Danti/Shutterstock.com

Tom Mc Nemar/Shutterstock.com

Olga Utlyakova/Shutterstock.com

Figure 2-9 *Fossil fuels:* Oil, coal, and natural gas (left, center, and right, respectively) supply most of the commercial energy that we use to supplement energy from the sun.

Scientists estimate that about 84% of the energy used in the United States is either unavoidably wasted because of the second law of thermodynamics (41%) or unnecessarily wasted (43%). For example, two widely used technologies—the incandescent light bulb and the internal combustion engine found in most motor vehicles—waste enormous amounts of energy. Thus, thermodynamics teaches us an important lesson: the cheapest and quickest way to get more energy is to stop wasting almost half the energy we use.

Three Scientific Laws Govern What We Can and Cannot Do With Matter and Energy

The three scientific laws of matter and energy are the *three big ideas* of this chapter:

- **There is no away.** According to the *law of conservation of matter*, no atoms are created or destroyed whenever matter undergoes a physical or chemical change. Thus, we cannot do away with matter; we can only change it from one physical state or chemical form to another.
- **You cannot get something for nothing.** According to the *first law of thermodynamics*, or the *law of conservation of energy*, whenever energy is converted from one form to another in a physical or chemical change, no energy is created or destroyed. This means that in such changes, we cannot get more energy out than we put in.
- **You cannot break even.** According to the *second law of thermodynamics*, whenever energy is converted from one form to another in a physical or chemical change, we always end up with lower-quality or less usable energy than we started with.

No matter how smart we are or how hard we try, we cannot violate these three basic scientific laws, or rules of nature, that place limits on what we can do with matter and energy resources.

REVISITING The Hubbard Brook Experimental Forest and Sustainability

The controlled experiment discussed in the **Core Case Study** that opened this chapter revealed that clearing a mature forest degrades some of its natural capital (see Figure 1-3, p. 9). Specifically, the loss of trees and vegetation altered the ability of the forest to retain and recycle water and other critical plant nutrients—a crucial ecological function based on one of the three **principles of sustainability** (see Figure 1-2, p. 8, or the back cover). In other words, the uncleared forest was a more sustainable system than a similar area of cleared forest (Figure 2-1).

This clearing of vegetation also violated the other two principles of sustainability. For example, the cleared forest lost most of its plants that had used solar energy to produce food for animals. And the loss of plants and the resulting loss of animals reduced the life-sustaining biodiversity of the cleared forest.

Humans clear forests to harvest timber, grow crops, build settlements, and expand cities. The key question is how far can we go in expanding our ecological footprints (see Figure 1-8, p. 14) without threatening the quality of life for our own species and for the other species that keep us alive and support our economies? To live more sustainably, we need to find and maintain a balance between preserving undisturbed natural systems and modifying other natural systems for our use.

Logic will get you from A to B. Imagination will take you everywhere.

ALBERT EINSTEIN

REVIEW

CORE CASE STUDY

1. Describe the *controlled scientific experiment* carried out in the Hubbard Brook Experimental Forest (**Core Case Study**).

SECTION 2-1

2. What is the key concept for this section? What is **science**? Describe the steps involved in a scientific process. What is **data**? What is a **model**? Distinguish among a **scientific hypothesis**, a **scientific theory**, and a **scientific law**. What is **peer review** and why is it important? Explain why scientific theories are not to be taken lightly and why people often use the term *theory* incorrectly.
3. Explain why scientific theories and laws are the most important and most certain results of science.

4. Distinguish among **tentative science (frontier science)**, **reliable science**, and **unreliable science**. What are three limitations of science in general and environmental science in particular?

SECTION 2-2

5. What are the two key concepts for this section? What is **matter**? Distinguish between an **element** and a **compound** and give an example of each. Distinguish among **atoms**, **molecules**, and **ions** and give an example of each. What is the **atomic theory**? Distinguish among **protons**, **neutrons**, and **electrons**. What is the **nucleus** of an atom? Distinguish between the **atomic number** and the **mass number** of an element. What is an **isotope**? What is **acidity**? What is **pH**?

6. What is a **chemical formula**? Distinguish between **organic compounds** and **inorganic compounds** and give an example of each. Distinguish among complex carbohydrates, proteins, nucleic acids, and lipids. What is a **cell**? Distinguish among a **gene**, a **trait**, and a **chromosome**. What is **matter quality**? Distinguish between **high-quality matter** and **low-quality matter** and give an example of each.

7. Distinguish between a **physical change** and a **chemical change (chemical reaction)** and give an example of each. What is a **nuclear change**? Explain the differences among **radioactive decay**, **nuclear fission**, and **nuclear fusion**. What is the **law of conservation of matter** and why is it important?

SECTION 2-3

8. What are the two key concepts for this section? What is **energy**? Distinguish between **kinetic energy** and **potential energy** and give an example of each. What is **heat**? Define and give two examples of **electromagnetic radiation**. What are **fossil fuels** and what three fossil fuels do we use to provide most of the energy that we use to supplement energy from the sun? What is **energy quality**? Distinguish between **high-quality energy** and **low-quality energy** and give an example of each.

9. What is the **first law of thermodynamics (law of conservation of energy)** and why is it important? What is the **second law of thermodynamics** and why is it important? Explain why the second law means that we can never recycle or reuse high-quality energy.

10. What are this chapter's *three big ideas*? Relate the three **principles of sustainability** to the Hubbard Brook Experimental Forest controlled experiment described in the **Core Case Study** that opens this chapter. Which two of these principles were violated by the clearing of the experimental forest, and what were the results?

SUSTAINABILITY

CORE CASE STUDY

Note: Key terms are in **bold** type.

CRITICAL THINKING

1. What ecological lesson can we learn from the controlled experiment on the clearing of forests described in the **Core Case Study** that opened this chapter?

CORE CASE STUDY

2. You observe that all of the fish in a pond have died. Describe how you might use the scientific process described in the **Core Case Study** and summarized in Figure 2-2 to determine the cause of this fish kill.

CORE CASE STUDY

3. Respond to the following statements:
   a. Scientists have not absolutely proven that anyone has ever died from smoking cigarettes.
   b. The natural greenhouse theory—that certain gases such as water vapor and carbon dioxide warm the lower atmosphere—is not a reliable idea because it is just a scientific theory.

4. A tree grows and increases its mass. Explain why this is not a violation of the law of conservation of matter.

5. If there is no "away" where organisms can get rid of their wastes because of the law of conservation of matter, why is the world not filled with waste matter?

6. Someone wants you to invest money in an automobile engine, claiming that it will produce more energy than the energy in the fuel used to run it. What is your response? Explain.

7. Use the second law of thermodynamics to explain why we can use oil only once as a fuel, or in other words, why we cannot recycle its high-quality energy.

8. a. Imagine you have the power to revoke the law of conservation of matter for one day. What are three things you would do with this power? Explain your choices.
   b. Imagine you have the power to violate the first law of thermodynamics for one day. What are three things you would do with this power? Explain your choices.

DOING ENVIRONMENTAL SCIENCE

Find a newspaper or magazine article or a report on the Web that attempts to discredit a scientific hypothesis because it has not been proven. Analyze the piece by doing the following: **(a)** determine its source (author or organization); **(b)** detect an alternative hypothesis, if any, that is offered by the author; **(c)** determine the primary objective of the author (e.g., to debunk the original hypothesis, to state an alternative hypothesis, or to raise new questions; **(d)** summarize the evidence given by the author(s) for his or her position; and **(e)** compare the authors' evidence with the evidence for the original hypothesis. Write a report summarizing your analysis and compare it with those of other members of your class.

GLOBAL ENVIRONMENT WATCH EXERCISE

Search for *Climate Change* and use this portal to research this controversy. Evaluate an argument on each side of the issue and the credentials of the individuals presenting the information. Then decide whether you think climate change research is an example of reliable or unreliable science and write a short report. Revisit your conclusion after you have studied more on this issue later in your course.

DATA ANALYSIS

Consider the graph below that shows loss of calcium from the experimental cutover site of the Hubbard Brook Experimental Forest compared with that of the control site. Note that this figure is very similar to Figure 2-4, which compares loss of nitrates from the two sites. After studying this graph, answer the questions below.

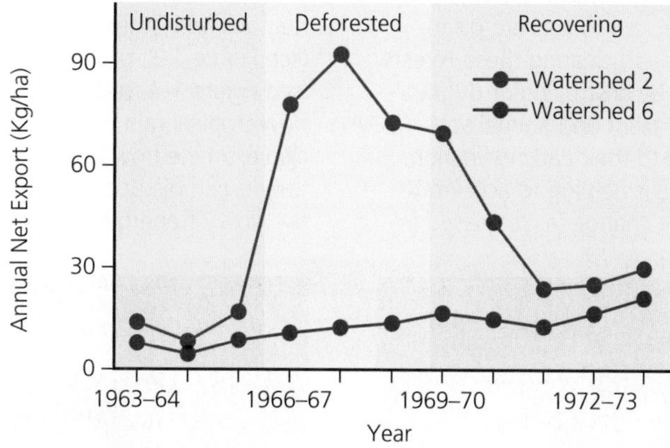

1. In what year did the calcium loss from the experimental site begin a sharp increase? In what year did it peak? In what year did it again level off?
2. In what year were the calcium losses from the two sites closest together? In the span of time between 1963 and 1972, did they ever get that close again?
3. Does this graph support the hypothesis that cutting the trees from a forested area causes the area to lose nutrients more quickly than leaving the trees in place? Explain.

LEARNING ONLINE

The website for *Environmental Science 14e* contains helpful study tools and other resources to take the learning experience beyond the textbook. Examples include flashcards, practice quizzing, information on green careers, tips for more sustainable living, activities, and a wide range of articles and further readings. To access these materials and other companion resources for this text, please visit **www.cengagebrain.com**. For further details, see the preface, p. xvi.

3 Ecosystems: What Are They and How Do They Work?

CORE CASE STUDY Tropical Rain Forests Are Disappearing

Tropical rain forests (see front cover photo) are found near the earth's equator and contain an incredible variety of life. These lush forests are warm year round and have high humidity and heavy rainfall almost daily. Although they cover only about 2% of the earth's land surface, studies indicate that they contain up to half of the world's known terrestrial plant and animal species. For these reasons, they make an excellent natural laboratory for the study of *ecosystems*—communities of organisms interacting with one another and with the physical environment of matter and energy in which they live.

So far, at least half of these forests have been destroyed or disturbed by humans cutting down trees, growing crops, grazing cattle, and building settlements (Figure 3-1), and the degradation of these centers of biodiversity is increasing. Ecologists warn that without strong protective measures, most of these forests will probably be gone or severely degraded within your lifetime.

So why should we care that tropical rain forests are disappearing? Scientists give three reasons. *First*, clearing these forests will reduce the earth's vital biodiversity by destroying or degrading the habitats of many of the unique plant and animal species that live in them, which could lead to their early extinction. *Second*, the destruction of these forests is helping to accelerate atmospheric warming, and thus climate change (as discussed in more detail in Chapter 15). The reason is that eliminating large areas of trees faster than they can grow back decreases the forests' ability to remove human-generated emissions of the greenhouse gas carbon dioxide (CO_2)—most of them from the burning of carbon-containing fossil fuels—from the atmosphere.

Third, rain forest loss will change regional weather patterns in ways that could prevent the return of diverse tropical rain forests in cleared or severely degraded areas. If this irreversible *ecological tipping point* is reached, tropical rain forests in such areas will become much less diverse tropical grasslands.

Ecologists study an ecosystem to learn how its variety of organisms interact with their living (*biotic*) environment of other organisms and with their nonliving (*abiotic*) environment of soil, water, other forms of matter, and energy, mostly from the sun. In effect, ecologists study *connections in nature*. Tropical rain forests and other ecosystems recycle nutrients and provide humans and other organisms with essential natural services (see Figure 1-3, p. 9) and natural resources such as nutrients (see Figure 1-4, p. 10). In this chapter, we look more closely at how tropical rain forests, and ecosystems in general, work. We also examine how human activities such as the clear-cutting of forests can disrupt the cycling of nutrients within ecosystems and the flow of energy through them.

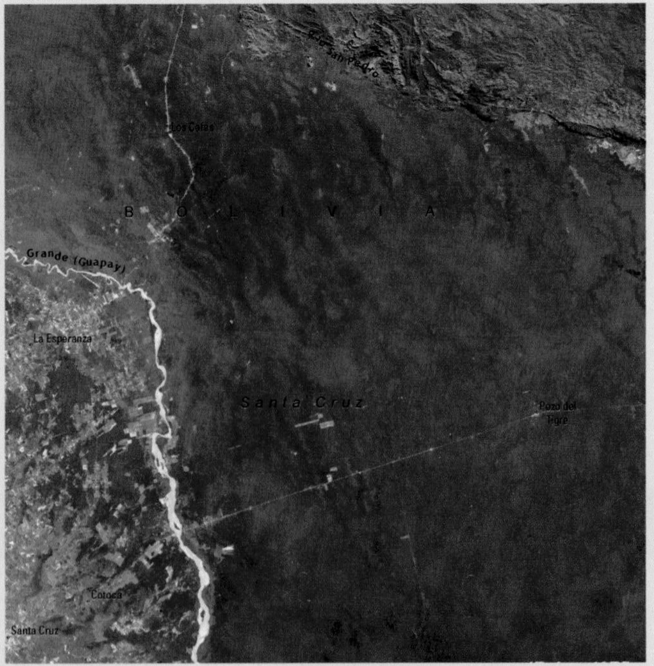

Figure 3-1 Natural capital degradation and depletion: These satellite images show the loss of topical rain forest, cleared for farming, cattle grazing, and settlements, near the Bolivian city of Santa Cruz between June 1975 (left) and May 2003 (right).

Key Questions and Concepts

3-1 What keeps us and other organisms alive?

CONCEPT 3-1A The four major components of the earth's life-support system are the atmosphere (air), the hydrosphere (water), the geosphere (rock, soil, and sediment), and the biosphere (living things).

CONCEPT 3-1B Life is sustained by the flow of energy from the sun through the biosphere, the cycling of nutrients within the biosphere, and gravity.

3-2 What are the major components of an ecosystem?

CONCEPT 3-2 Some organisms produce the nutrients they need, others get the nutrients they need by consuming other organisms, and some recycle nutrients back to producers by decomposing the wastes and remains of other organisms.

3-3 What happens to energy in an ecosystem?

CONCEPT 3-3 As energy flows through ecosystems in food chains and webs, the amount of chemical energy available to organisms at each successive feeding level decreases.

3-4 What happens to matter in an ecosystem?

CONCEPT 3-4 Matter, in the form of nutrients, cycles within and among ecosystems and the biosphere, and human activities are altering these chemical cycles.

3-5 How do scientists study ecosystems?

CONCEPT 3-5 Scientists use both field research and laboratory research, as well as mathematical and other models to learn about ecosystems.

Note: Supplements 1 (p. S2), 2 (p. S3), 4 (p. S10), and 5 (p. S18) can be used with this chapter.

To halt the decline of an ecosystem, it is necessary to think like an ecosystem.

DOUGLAS WHEELER

3-1 What Keeps Us and Other Organisms Alive?

▶ **CONCEPT 3-1A The four major components of the earth's life-support system are the atmosphere (air), the hydrosphere (water), the geosphere (rock, soil, and sediment), and the biosphere (living things).**

▶ **CONCEPT 3-1B Life is sustained by the flow of energy from the sun through the biosphere, the cycling of nutrients within the biosphere, and gravity.**

Earth's Life-Support System Has Four Major Components

The earth's life-support system consists of four main spherical systems (Figure 3-2, p. 42) that interact with one another—the atmosphere (air), the hydrosphere (water), the geosphere (rock, soil, and sediment), and the biosphere (living things) (**Concept 3-1A**).

The **atmosphere** is a thin spherical envelope of gases surrounding the earth's surface. Its inner layer, the **troposphere**, extends only about 17 kilometers (11 miles) above sea level at the tropics and about 7 kilometers (4 miles) above the earth's north and south poles. It contains air that we breathe, consisting mostly of nitrogen (78% of the total volume) and oxygen (21%). Most of the remaining 1% of the air consists of water vapor, carbon dioxide, and methane, all of which are called **greenhouse gases**, which absorb and release energy that warms the inner layer of the atmosphere. Without these gases, the earth would be too cold for the existence of life as we know it.

The next layer, reaching from 17 to 50 kilometers (11–31 miles) above the earth's surface, is called the **stratosphere**. Its lower portion holds enough ozone (O_3) gas to filter out about 95% of the sun's harmful *ultraviolet (UV) radiation*. This global sunscreen allows life to exist on the surface of the planet.

The **hydrosphere** consists of all of the water on or near the earth's surface. It is found as *water vapor* in the atmosphere, as *liquid water* on the surface and underground, and as *ice*—polar ice, icebergs, glaciers (See Photo 12 in the Detailed Contents), and ice in frozen soil layers called *permafrost*. The oceans, which cover about 71% of the globe, contain about 97% of the earth's water.

Figure 3-2 Natural capital: This diagram illustrates the general structure of the earth, showing that it consists of a land sphere (*geosphere*), an air sphere (*atmosphere*), a water sphere (*hydrosphere*), and a life sphere (*biosphere*) (**Concept 3-1A**).

The **geosphere** consists of the earth's intensely hot *core*, a thick *mantle* composed mostly of rock, and a thin outer *crust*. Most of the geosphere is located in the earth's interior. Its upper portion contains nonrenewable *fossil fuels*—coal, oil, and natural gas—and minerals that we use, as well as renewable soil chemicals (nutrients) that organisms need in order to live, grow, and reproduce.

The **biosphere** consists of the parts of the atmosphere, hydrosphere, and geosphere where life is found. If the earth were an apple, the biosphere would be no thicker than the apple's skin. *One important goal of environmental science is to understand the interactions that occur within this thin layer of air, water, soil, and organisms.*

Three Factors Sustain the Earth's Life

Life on the earth depends on three interconnected factors (**Concept 3-1B**):

1. The *one-way flow of high-quality energy* from the sun, through living things in their feeding interactions, into the environment as low-quality energy (mostly heat dispersed into air or water at a low temperature), and eventually to outer space as heat (Figure 3-3). No round trips are allowed because high-quality energy cannot be recycled, according to the two laws of thermodynamics (see Chapter 2, p. 36). As this solar energy interacts with carbon dioxide (CO_2) and other gases in the troposphere, it warms the troposphere—a process known as the **greenhouse effect**. Without this process, the earth would be too cold to support the forms of life we find here today (see *The Habitable Planet*, Video 2, at **www.learner.org/resources/series209.html**).
2. The *cycling of nutrients* (the atoms, ions, and molecules needed for survival by living organisms) through parts of the biosphere. Because the earth is closed to significant inputs of matter from space, its essentially fixed supply of nutrients must be continually recycled to support life (see Figure 1-4, p. 10).

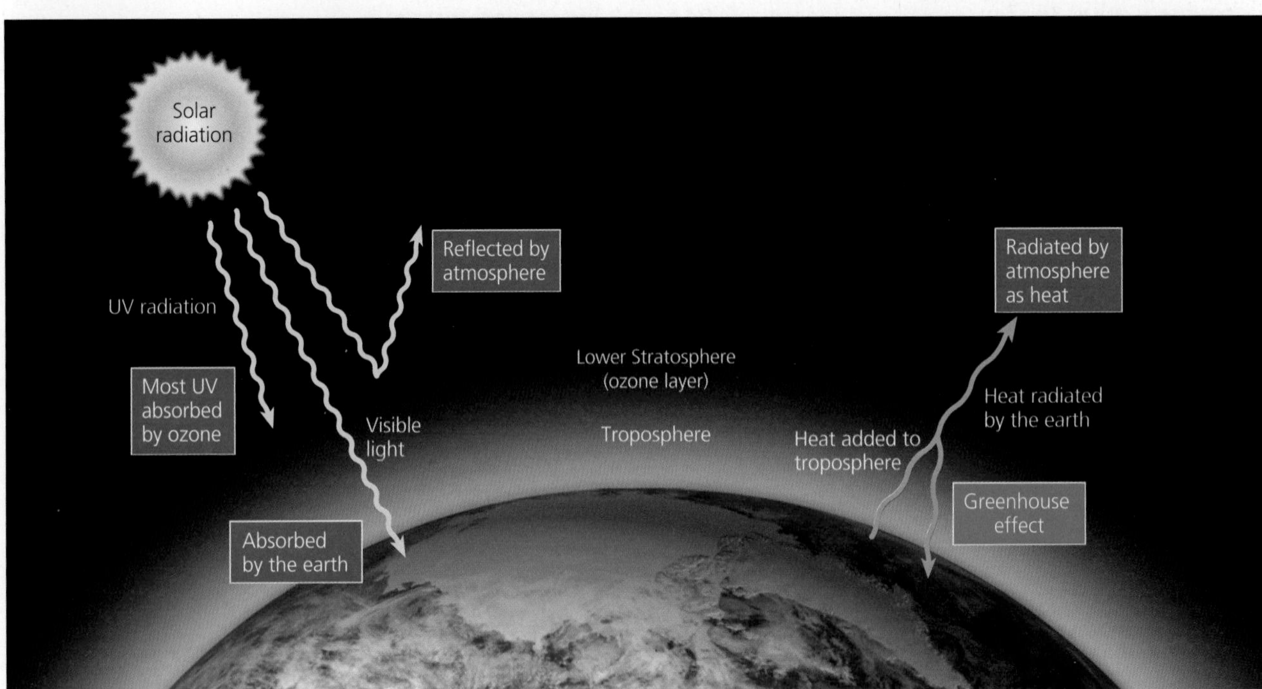

Active Figure 3-3 High-quality solar energy flows from the sun to the earth, is degraded to lower quality energy (mostly heat) as it interacts with the earth's air, water, soil, and life forms, and eventually flows into space. *See an animation based on this figure at* www.cengagebrain.com.

Nutrient cycles in ecosystems and in the biosphere are round trips, which can take from seconds to centuries to complete. The law of conservation of matter (see Chapter 2, p. 34) governs this nutrient cycling process.

3. *Gravity*, which allows the planet to hold onto its atmosphere and helps to enable the movement and cycling of chemicals through air, water, soil, and organisms.

3-2 What Are the Major Components of an Ecosystem?

▶ **CONCEPT 3-2 Some organisms produce the nutrients they need, others get the nutrients they need by consuming other organisms, and some recycle nutrients back to producers by decomposing the wastes and remains of other organisms.**

Ecologists Study Interactions in Nature

Ecology is the science that focuses on how organisms interact with one another and with their nonliving environment of matter and energy. Scientists classify matter into levels of organization ranging from atoms to galaxies. Ecologists study interactions within and among five of these levels—**organisms**, **populations**, **communities**, **ecosystems**, and the **biosphere**, which are illustrated and defined in Figure 3-4.

The biosphere and its ecosystems are made up of living (*biotic*) and nonliving (*abiotic*) components. Examples of nonliving components are water, air, nutrients, rocks, heat, and solar energy. Living components include plants, animals, microbes, and all other organisms. Figure 3-5 (p. 44) is a greatly simplified diagram of some of the living and nonliving components of a terrestrial ecosystem.

Ecosystems Have Several Important Components

Ecologists assign every type of organism in an ecosystem to a *feeding level*, or **trophic level**, depending on its source of food or nutrients. We can broadly classify the living organisms that transfer energy and nutrients from one trophic level to another within an ecosystem as producers and consumers.

Producers, sometimes called **autotrophs** (self-feeders), make the nutrients they need from compounds and energy obtained from their environment (**Concept 3-2**). In a process called **photosynthesis**,

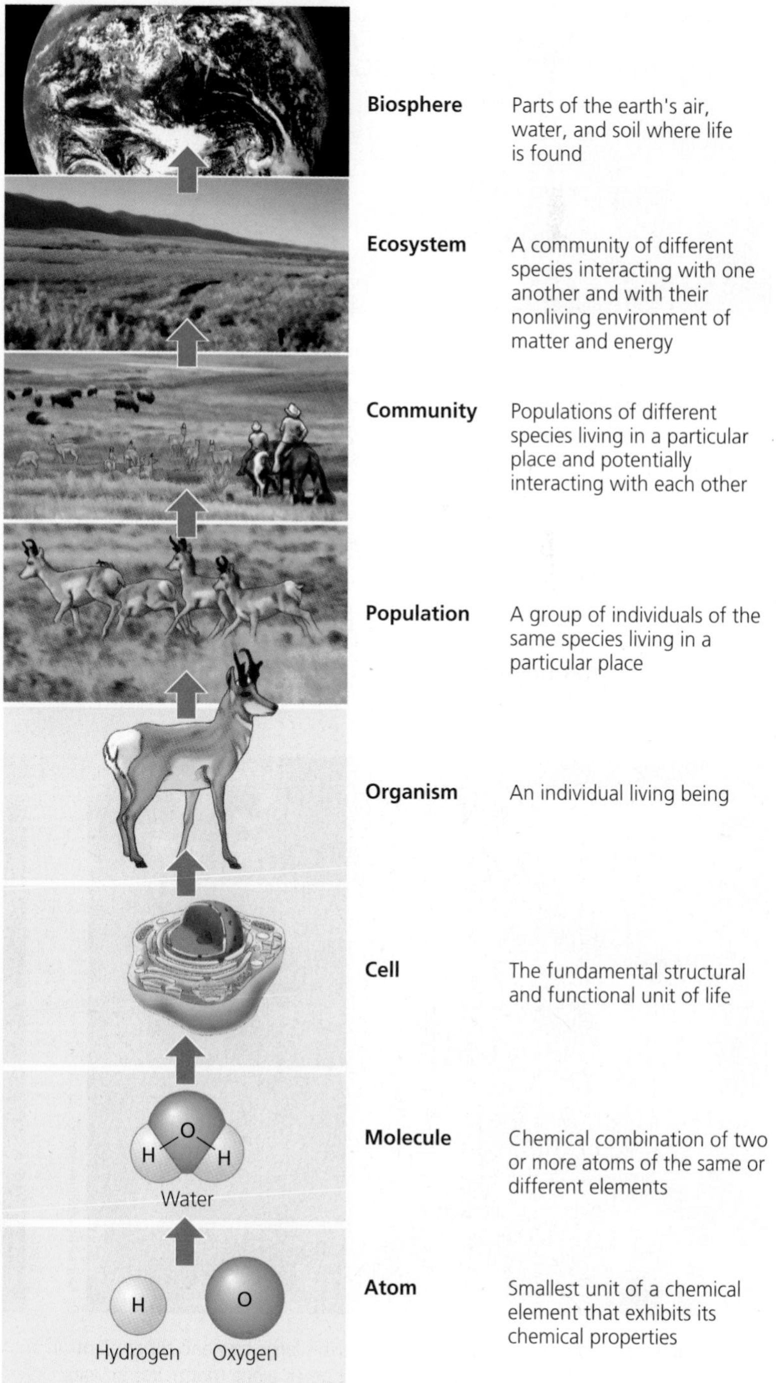

Active Figure 3-4 This diagram illustrates some levels of the organization of matter in nature. Ecology focuses on the top five of these levels. *See an animation based on this figure at* www.cengagebrain.com.

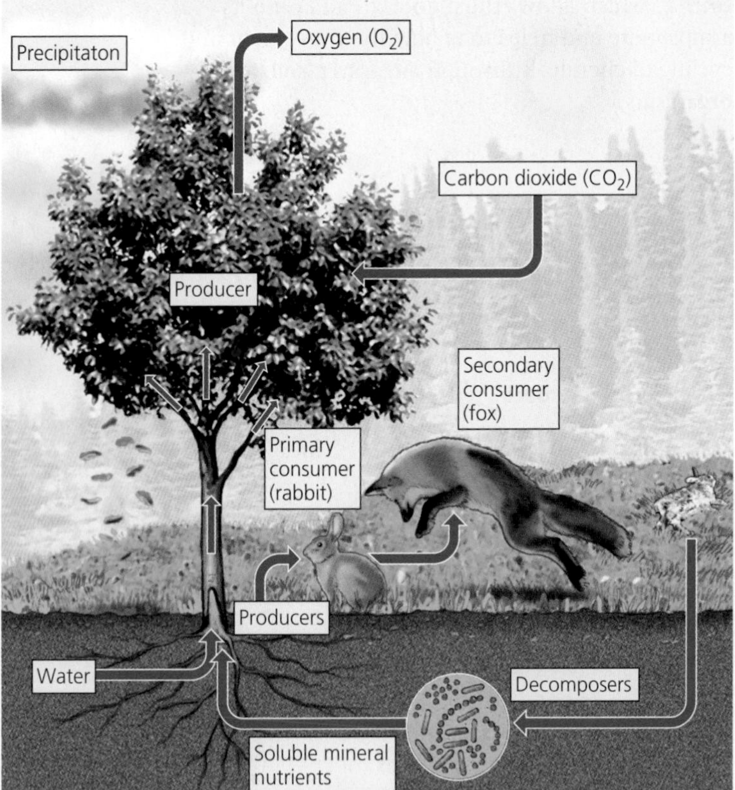

Active Figure 3-5 Key living and nonliving components of an ecosystem in a field are shown in this diagram. *See an animation based on this figure at* **www.cengagebrain.com**.

plants typically capture about 1% of the solar energy that falls on their leaves and use it in combination with carbon dioxide and water to form organic molecules, including energy-rich carbohydrates (such as glucose, $C_6H_{12}O_6$), which store the chemical energy plants need. Although hundreds of chemical changes take place during photosynthesis, we can summarize the overall reaction as follows:

$$\text{carbon dioxide} + \text{water} + \textbf{solar energy} \longrightarrow \text{glucose} + \text{oxygen}$$

$$6\,CO_2 + 6\,H_2O + \textbf{solar energy} \longrightarrow C_6H_{12}O_6 + 6\,O_2$$

(In Supplement 4, see p. S16 for information on how to balance chemical equations such as this one.)

On land, most producers are green plants (Figure 3-6, left). In freshwater and ocean ecosystems, algae and aquatic plants growing near shorelines are the major producers (Figure 3-6, right). In open water, the dominant producers are *phytoplankton*—mostly microscopic organisms that float or drift in the water (see *The Habitable Planet*, Video 3, at **http://www.learner.org/resources/series209.html**).

All other organisms in an ecosystem are **consumers**, or **heterotrophs** ("other-feeders"), that cannot produce the nutrients they need through photosynthesis or other processes (**Concept 3-2**). They get their nutrients by feeding on other organisms (producers or other consumers) or their remains. In other words, all consumers (including humans) depend on producers for their nutrients.

There are several types of consumers. **Primary consumers**, or **herbivores** (plant eaters), are animals that eat mostly green plants. Examples are caterpillars, giraffes (Figure 3-7, left), and *zooplankton*, or tiny sea animals that feed on phytoplankton. **Carnivores** (meat eaters) are animals that feed on the flesh of other animals. Some carnivores such as spiders, lions (Figure 3-7, right), and most small fishes are **secondary consumers** that feed on the flesh of herbivores. Other carnivores such as tigers, hawks, and killer whales (orcas) are **tertiary** (or higher-level) **consumers** that feed on the flesh of other carnivores. Some of these relationships are shown in Figure 3-5.

Dr. Morley Read/Shutterstock.com

Lance Rider/Shutterstock.com

Figure 3-6 Some producers live on land, such as this large tree and other plants in an Amazon rain forest (**Core Case Study**) in Brazil (left). Others, such as green algae (right), live in water.

Figure 3-7 The giraffe (left) feeding on the leaves of a tree is an herbivore. The lions (right) are carnivores feeding on the dead body of a giraffe that they have killed.

Omnivores such as pigs, rats, and humans eat plants and other animals.

THINKING ABOUT
What You Eat

When you ate your most recent meal, were you an herbivore, a carnivore, or an omnivore?

Decomposers are consumers that, in the process of obtaining their own nutrients, release nutrients from the wastes or remains of plants and animals and then return those nutrients to the soil, water, and air for reuse by producers (**Concept 3-2**). Most decomposers are bacteria and fungi. Other consumers, called **detritus feeders**, or **detritivores**, feed on the wastes or dead bodies of other organisms; these wastes are called *detritus* (dee-TRI-tus), which means debris. Examples are earthworms, some insects, and vultures.

Hordes of detritus feeders and decomposers can transform a fallen tree trunk into wood particles and, finally, into simple inorganic molecules that plants can absorb as nutrients (Figure 3-8). Thus, in natural ecosystems the wastes and dead bodies of organisms serve

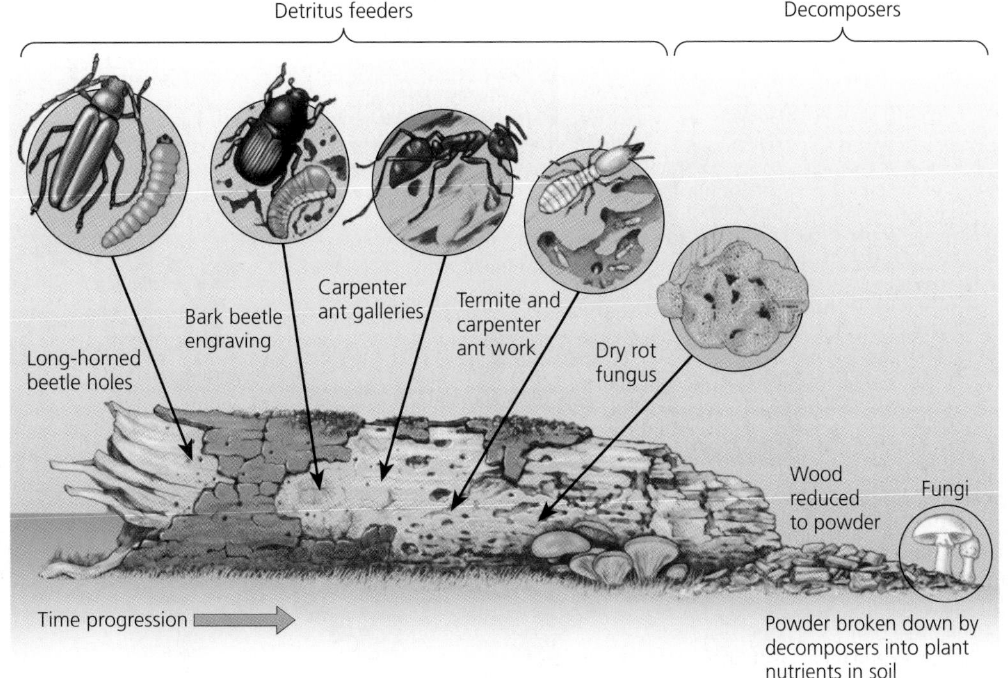

Figure 3-8 Various detritivores and decomposers (mostly fungi and bacteria) can "feed on" or digest parts of a log and eventually convert its complex organic chemicals into simpler inorganic nutrients that can be taken up by producers.

SCIENCE FOCUS

Many of the World's Most Important Organisms Are Invisible to Us

They are everywhere. Billions can be found inside your body, on your body, in a handful of soil, and in a cup of ocean water.

These mostly invisible rulers of the earth are *microbes*, or *microorganisms*, catchall terms for many thousands of species of bacteria, protozoa, fungi, and floating phytoplankton—most too small to be seen with the naked eye.

Microbes do not get the respect they deserve. Most of us view them primarily as threats to our health in the form of infectious bacteria or fungi that cause athlete's foot and other skin diseases, and protozoa that cause diseases such as malaria. But these harmful microbes are in the minority.

We are alive largely because of multitudes of microbes toiling away completely out of sight. Bacteria in our intestinal tracts help break down the food we eat, and microbes in our noses help prevent harmful bacteria from reaching our lungs.

Bacteria and other microbes help to purify the water we drink by breaking down plant and animal wastes that may be in the water. Bacteria and fungi (such as yeast) also help to produce foods such as bread, cheese, yogurt, soy sauce, beer, and wine. Bacteria and fungi in the soil decompose organic wastes into nutrients that can be taken up by plants that are then eaten by humans and many other plant eaters. Without these tiny creatures, we would go hungry and be up to our necks in waste matter.

Microbes, particularly phytoplankton in the ocean, provide much of the planet's oxygen, and help to regulate the earth's temperature by removing some of the carbon dioxide produced when we burn coal, natural gas, and gasoline (see The Habitable Planet, Video 3, at **www.learner.org/resources/series209.html**). Scientists are working on using microbes to develop new medicines and fuels. Genetic engineers are inserting genetic material into existing microorganisms to convert them to microbes that can be used to clean up polluted water and soils.

Some microorganisms assist us in controlling diseases that affect plants and in controlling populations of insect species that attack our food crops. Relying more on these microbes for pest control could reduce the use of potentially harmful chemical pesticides. In other words, microbes are a vital part of the earth's natural capital.

Critical Thinking

What are two advantages that microbes have over humans for thriving in the world?

as resources for other organisms, as the nutrients that make life possible are continually recycled, in keeping with one of the three **principles of sustainability** (see back cover). As a result, *there is very little waste of nutrients in nature*.

Decomposers and detritus feeders, many of which are microscopic organisms (Science Focus, above), are the key to nutrient cycling. Without them, the planet would be overwhelmed with plant litter, animal wastes, dead animal bodies, and garbage.

Producers, consumers, and decomposers use the chemical energy stored in glucose and other organic compounds to fuel their life processes. In most cells, this energy is released by **aerobic respiration**, which uses oxygen to convert glucose (or other organic nutrient molecules) back into carbon dioxide and water. The net

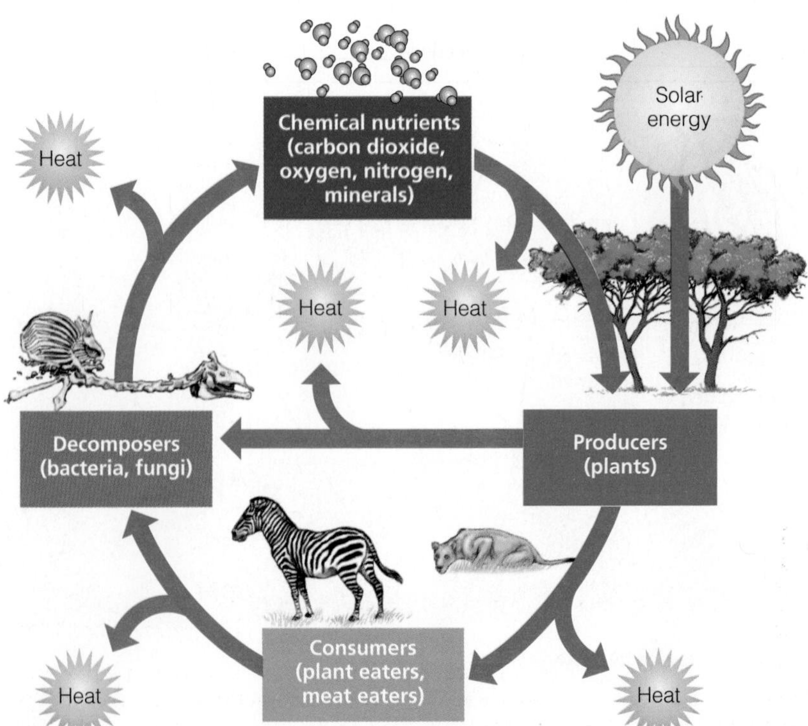

Active Figure 3-9 Natural capital: This diagram shows the main structural components of an ecosystem (energy, chemicals, and organisms). Nutrient cycling and the flow of energy—first from the sun, then through organisms, and finally into the environment as low-quality heat—link these components. *See an animation based on this figure at* **www.cengagebrain.com**.

effect of the hundreds of steps in this complex process is represented by the following chemical reaction:

glucose + oxygen ⟶ carbon dioxide + water + energy

$$C_6H_{12}O_6 + 6\,O_2 \longrightarrow 6\,CO_2 + 6\,H_2O + \text{energy}$$

Notice that, although the detailed steps differ, the net chemical change for aerobic respiration is the opposite of that for photosynthesis (p. 44).

To summarize, ecosystems and the biosphere are sustained through a combination of *one-way energy flow* from the sun through these systems and the *nutrient cycling* of key materials within them (**Concept 3-1B**)—in keeping with two of the **principles of sustainability** (Figure 3-9). Let us look more closely at the flow of energy and the cycling of chemicals in ecosystems.

3-3 What Happens to Energy in an Ecosystem?

▶ **CONCEPT 3-3 As energy flows through ecosystems in food chains and webs, the amount of chemical energy available to organisms at each successive feeding level decreases.**

Energy Flows through Ecosystems in Food Chains and Food Webs

The chemical energy stored as nutrients in the bodies and wastes of organisms flows through ecosystems from one trophic (feeding) level to another. For example, a plant uses solar energy to store chemical energy in a leaf. A caterpillar eats the leaf, a robin eats the caterpillar, and a hawk eats the robin. Decomposers and detritus feeders consume the wastes and remains of all members of this and other food chains and return their nutrients to the soil for reuse by producers.

A sequence of organisms, each of which serves as a source of nutrients or energy for the next, is called a **food chain** (Figure 3-10). Every use and transfer of energy by organisms involves a loss of some degraded high-quality energy to the environment as heat, in accordance with the second law of thermodynamics.

In natural ecosystems, most consumers feed on more than one type of organism, and most organisms are eaten or decomposed by more than one type of consumer. Because of this, organisms in most ecosystems form a complex network of interconnected food chains called a **food web** (Figure 3-11, p. 48). We can assign trophic levels in food webs just as we can in food chains. Food chains and webs show how producers, consumers, and decomposers are connected to one another as energy flows through trophic levels in an ecosystem.

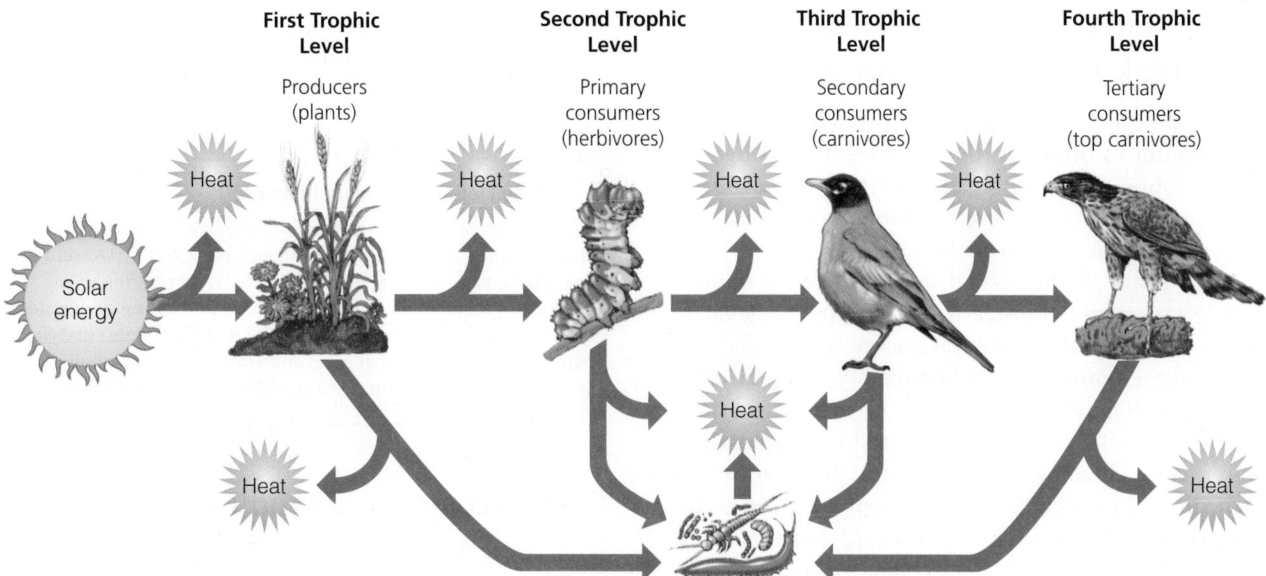

Active Figure 3-10 This diagram illustrates a *food chain*, which shows how the chemical energy in nutrients flows through various *trophic levels*. *See an animation based on this figure at* www.cengagebrain.com. **Question:** Think about what you ate for breakfast. At what level or levels on a food chain were you eating?

Active Figure 3-11 This diagram illustrates a greatly simplified *food web* in the southern hemisphere. The shaded middle area shows a simple food chain that is part of these complex interacting feeding relationships. Many more participants in the web, including an array of decomposer and detritus feeder organisms, are not shown here. *See an animation based on this figure at* www.cengagebrain.com. **Question:** Can you imagine a food web of which you are a part? Try drawing a simple diagram of it.

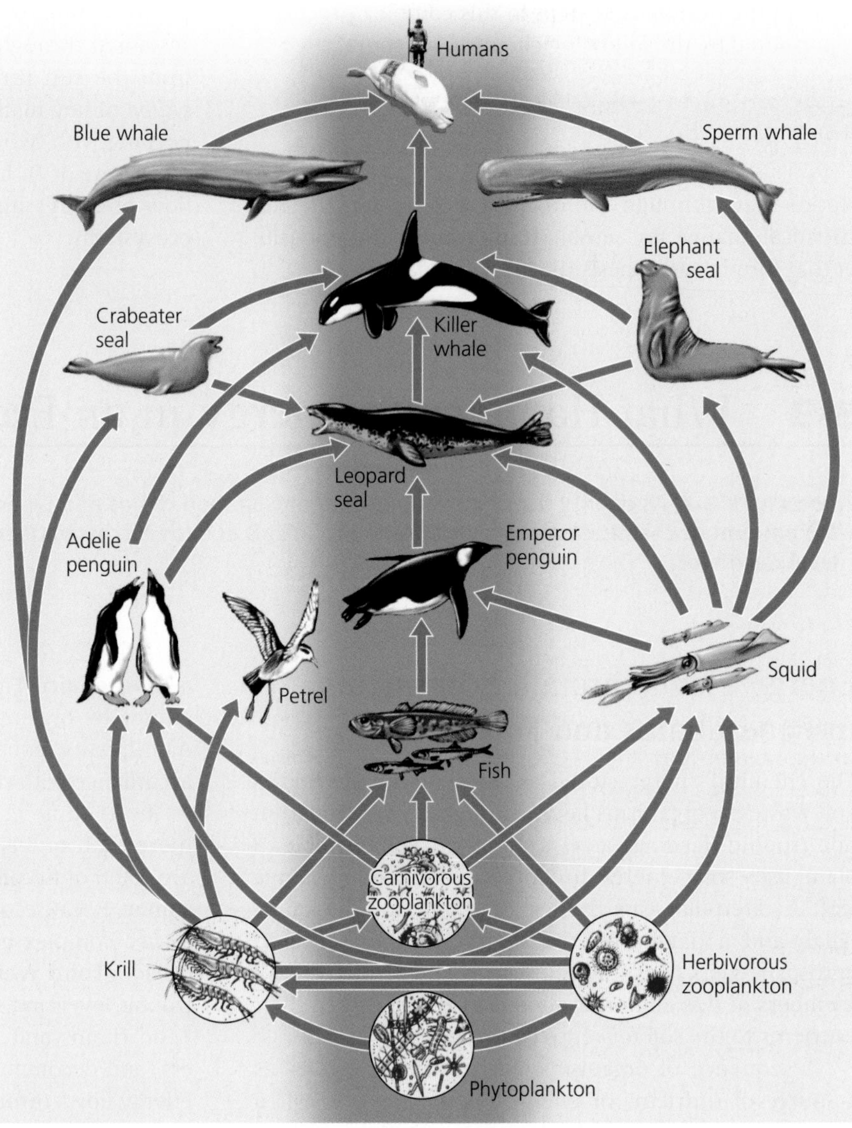

Usable Energy Decreases with Each Link in a Food Chain or Web

Each trophic level in a food chain or web contains a certain amount of **biomass**, the dry weight of all organic matter contained in its organisms. In a food chain or web, chemical energy stored in biomass is transferred from one trophic level to another.

As a result of the second law of thermodynamics, with each energy transfer from one trophic level to another, some usable chemical energy is degraded and lost to the environment as low-quality heat. In other words, as energy flows through ecosystems in food chains and webs, there is a decrease in the amount of high-quality chemical energy available to organisms at each succeeding feeding level (**Concept 3-3**). The **pyramid of energy flow** in Figure 3-12 illustrates this energy loss for a simple food chain, assuming a 90% energy loss with each transfer.

The large loss in chemical energy between successive trophic levels explains why food chains and webs rarely have more than four or five trophic levels. Thus, there are far fewer tigers in tropical rain forests (**Core Case Study**) and other areas than there are insects. CORE CASE STUDY

CONNECTIONS

Energy Flow and Feeding People

Energy flow pyramids explain why the earth could support more people if they all ate at lower trophic levels by consuming grains, vegetables, and fruits directly rather than passing such crops through another trophic level and eating herbivores such as cattle. About two-thirds of the world's people survive primarily by eating wheat, rice, and corn at the first trophic level because most of them cannot afford to eat much meat.

Some Ecosystems Produce Plant Matter Faster Than Others Do

The amount, or mass, of living organic material (biomass) that a particular ecosystem can support is determined by how much solar energy its producers can capture and store as chemical energy and by how rapidly

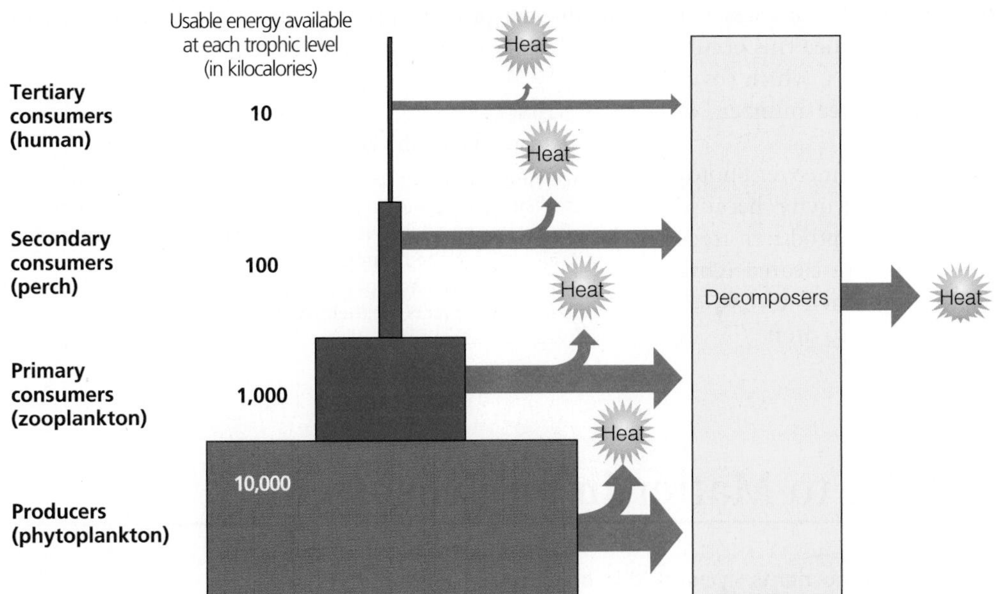

Active Figure 3-12 This model is a generalized *pyramid of energy flow* that shows the decrease in usable chemical energy available at each succeeding trophic level in a food chain or web. *See an animation based on this figure at* www.cengagebrain.com. **Question:** Why is a vegetarian diet more energy efficient than a meat-based diet?

they can do so. **Gross primary productivity (GPP)** is the *rate* at which an ecosystem's producers (usually plants) convert solar energy into chemical energy in the form of biomass found in their tissues. It is usually measured in terms of energy production per unit area over a given time span, such as kilocalories per square meter per year (kcal/m^2/yr).

To stay alive, grow, and reproduce, producers must use some of the chemical energy stored in the biomass they make for their own respiration. **Net primary productivity (NPP)** is the *rate* at which producers use photosynthesis to produce and store chemical energy *minus* the *rate* at which they use some of this stored chemical energy through aerobic respiration. NPP measures how fast producers can make the chemical energy that is stored in their tissues and that is potentially available to other organisms (consumers) in an ecosystem. Thus, *the planet's NPP ultimately limits the number of consumers (including humans) that can survive on the earth.* This is an important lesson from nature.

Ecosystems and aquatic life zones differ in their NPP as illustrated in Figure 3-13. Despite its low NPP, the

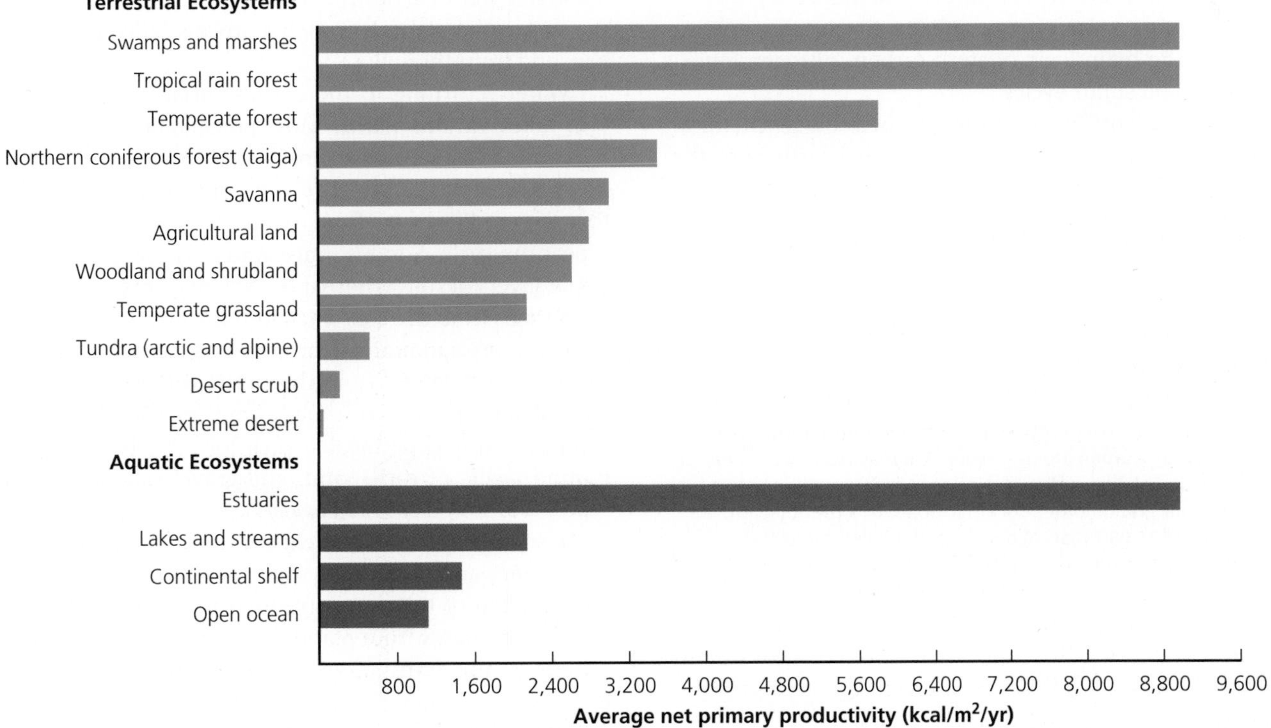

Figure 3-13 The estimated annual average *net primary productivity* in major life zones and ecosystems is expressed in this graph as kilocalories of energy produced per square meter per year (kcal/m^2/yr). **Question:** What are the three most productive and three least productive systems? (Data from R. H. Whittaker, *Communities and Ecosystems*, 2nd ed., New York: Macmillan, 1975)

open ocean produces more of the earth's biomass per year than any other ecosystem or life zone. This occurs because the volume of the world's oceans, which cover 71% of the earth's surface, contain huge numbers of producers such as tiny phytoplankton.

On land, tropical rain forests (see front cover photo) have a very high net primary productivity because of their large number and variety of producer trees and other plants. When such forests are cleared (**Core Case Study**, Figure 3-1, right) or burned to plant crops or graze cattle, there is a sharp drop in the net primary productivity and a loss of many of the diverse array of plant and animal species.

CORE CASE STUDY

CONNECTIONS
Humans and Earth's Net Primary Productivity

Peter Vitousek, Stuart Rojstaczer, and other ecologists estimate that humans now use, waste, or destroy 10–55% of the earth's total potential NPP. This is a remarkably high value, considering that the human population makes up less than 1% of the total biomass of all of the earth's consumers that depend on producers for their nutrients.

3-4 What Happens to Matter in an Ecosystem?

▶ **CONCEPT 3-4 Matter, in the form of nutrients, cycles within and among ecosystems and the biosphere, and human activities are altering these chemical cycles.**

Nutrients Cycle within and among Ecosystems

The elements and compounds that make up nutrients move continually through air, water, soil, rock, and living organisms within ecosystems, as well as in the biosphere in cycles called **biogeochemical cycles** (literally, life-earth-chemical cycles), or **nutrient cycles**. This is in keeping with one of the three **principles of sustainability** (see back cover). These cycles, which are driven directly or indirectly by incoming solar energy and by the earth's gravity, include the hydrologic (water), carbon, nitrogen, phosphorus, and sulfur cycles.

SUSTAINABILITY

As nutrients move through their biogeochemical cycles, they may accumulate in certain portions of the cycles and remain there for different periods of time. These temporary storage sites such as the atmosphere, the oceans and other bodies of water, and underground deposits are called *reservoirs*.

CONNECTIONS
Nutrient Cycles and Life

Nutrient cycles connect past, present, and future forms of life. Some of the carbon atoms in your skin may once have been part of an oak leaf, a dinosaur's skin, or a layer of limestone rock. Your grandmother, rock star Bono, or a hunter–gatherer who lived 25,000 years ago may have inhaled some of the nitrogen molecules you just inhaled.

The Water Cycle

Water is an amazing substance (Science Focus, p. 52) that is necessary for life on the earth. The **hydrologic cycle**, or **water cycle**, collects, purifies, and distributes the earth's fixed supply of water, as shown in Figure 3-14.

The water cycle is powered by energy from the sun and involves three major processes—evaporation, precipitation, and transpiration. Incoming solar energy causes *evaporation* of water from the earth's oceans, lakes, rivers, and soil. Evaporation changes liquid water into water vapor in the atmosphere, and gravity draws the water back to the earth's surface as *precipitation* (rain, snow, sleet, and dew). Over land, about 90% of the water that reaches the atmosphere evaporates from the surfaces of plants, through a process called *transpiration*, and from the soil.

Water returning to the earth's surface as precipitation takes various paths. Most precipitation falling on terrestrial ecosystems becomes *surface runoff*. This water flows into streams, which eventually carry water back to lakes and oceans, from which it can evaporate to repeat the cycle. Some surface water also seeps into the upper layers of soils where it is used by plants, and some evaporates from the soils back into the atmosphere.

Some precipitation is converted to ice that is stored in *glaciers* (see Photo 12 in the Detailed Contents), usually for long periods of time. Some precipitation sinks through soil and permeable rock formations to underground layers of rock, sand, and gravel called *aquifers*, where it is stored as *groundwater*.

A small amount of the earth's water ends up in the living components of ecosystems. As producers, plants absorb some of this water through their roots, most of which evaporates from plant leaves back into the atmosphere during transpiration; some of the water combines with carbon dioxide during photosynthesis to produce high-energy organic compounds such as carbohydrates. Eventually these compounds are broken down in plant cells, which release the water back into

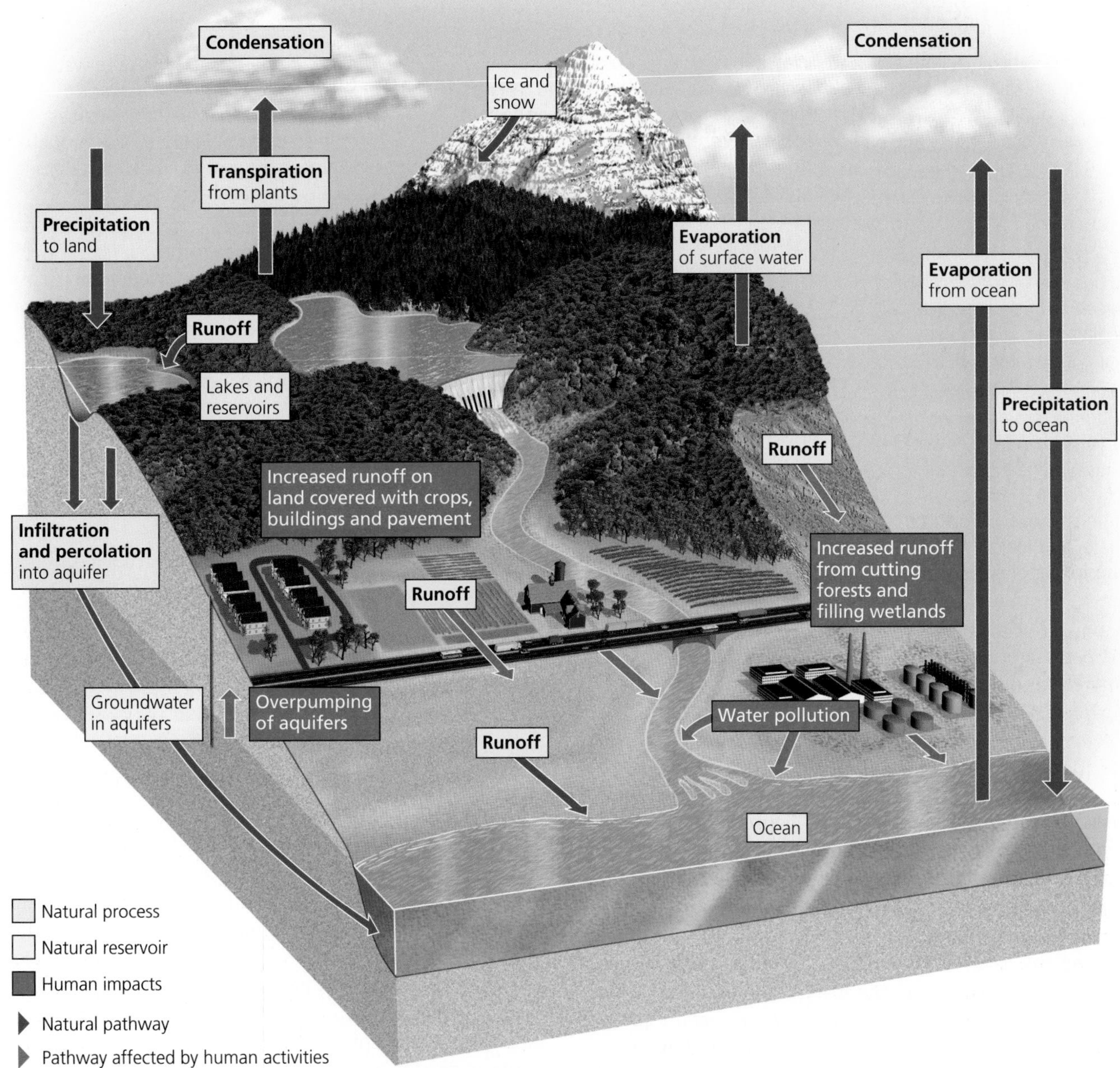

Active Figure 3-14 Natural capital: This diagram is a simplified model of the *water cycle*, or *hydrologic cycle*, in which water circulates in various physical forms within the biosphere. Major harmful impacts of human activities are shown by the red arrows and boxes. *See an animation based on this figure at* www.cengagebrain.com. **Question:** What are three ways in which your lifestyle directly or indirectly affects the hydrologic cycle?

the environment. Consumers get their water from their food and by drinking it.

Because water dissolves many nutrient compounds, it is a major medium for transporting nutrients within and between ecosystems. Water is also the primary sculptor of the earth's landscape as it flows over and wears down rock over millions of years.

Throughout the hydrologic cycle, many natural processes purify water. Evaporation and subsequent precipitation act as a natural distillation process that removes impurities dissolved in water. Water flowing aboveground through streams and lakes, and underground in aquifers is naturally filtered and partially purified by chemical and biological processes—mostly by the actions of decomposer bacteria—as long as these natural processes are not overloaded. Thus, *the hydrologic cycle can be viewed as a cycle of natural renewal of water quality.*

Only about 0.024% of the earth's vast water supply is available to humans and other species as liquid freshwater in accessible groundwater deposits and in lakes, rivers, and streams. The rest is too salty for us to use, is stored as ice, or is too deep underground to extract at affordable prices using current technology.

SCIENCE FOCUS

Water's Unique Properties

Water is a remarkable substance with a unique combination of properties:

- *Forces of attraction*, called *hydrogen bonds* (see Figure 7, p. S13, in Supplement 4), *hold water molecules together*—the major factor determining water's distinctive properties.
- *Water exists as a liquid over a wide temperature range because of the forces of attraction between its molecules.* Without water's high boiling point, the oceans would have evaporated long ago.
- *Liquid water changes temperature slowly because it can store a large amount of heat without a large change in its own temperature.* This high heat storage capacity helps protect living organisms from temperature changes, moderates the earth's climate, and makes water an excellent coolant for car engines and power plants.
- *It takes a large amount of energy to evaporate water because of the forces of attraction between its molecules.* Water absorbs large amounts of heat as it changes into water vapor and releases this heat as the vapor condenses back to liquid water. This helps to distribute heat throughout the world and to determine regional and local climates. It also makes evaporation a cooling process—explaining why you feel cooler when perspiration evaporates from your skin.
- *Liquid water can dissolve a variety of compounds* (see Figure 4, p. S12, in Supplement 4). It carries dissolved nutrients into the tissues of living organisms, flushes waste products out of those tissues, serves as an all-purpose cleanser, and helps to remove and dilute the water-soluble wastes of civilization. This property also means that water-soluble wastes can easily pollute water.
- *Water filters out wavelengths of the sun's ultraviolet radiation that would harm some aquatic organisms.* However, down to a certain depth, it is transparent to sunlight that is necessary for photosynthesis.
- *The forces of attraction between water molecules also allow liquid water to adhere to a solid surface.* This enables narrow columns of water to rise through a plant from its roots to its leaves (a process called *capillary action*).
- *Unlike most liquids, water expands when it freezes.* This means that ice floats on water because it has a lower density (mass per unit of volume) than liquid water has. Otherwise, lakes and streams in cold climates would freeze solid, losing most of their aquatic life. Because water expands upon freezing, it can break pipes, crack a car's engine block (if it doesn't contain antifreeze), break up pavement, and fracture rocks.

Critical Thinking

What are three ways in which your life would be different if there were no special forces of attraction (hydrogen bonds) between water molecules?

Humans alter the water cycle in three major ways (see the red arrows and boxes in Figure 3-14). *First,* we withdraw large quantities of freshwater from streams, lakes, and aquifers sometimes at rates faster than nature can replace it. As a result, some aquifers are being depleted and some rivers no longer flow to the ocean. *Second,* we clear vegetation from land for agriculture, mining, road building, and other activities, and cover much of the land with buildings, concrete, and asphalt. This increases runoff and reduces infiltration that would normally recharge groundwater supplies. And *third,* we drain and fill wetlands for farming and urban development, which interferes with the flood control provided by wetlands that act like sponges to absorb and hold overflows of water from heavy rains and melting snow.

CONNECTIONS

Clearing a Rain Forest Can Affect Local Weather

Clearing vegetation can alter weather patterns by reducing transpiration, especially in dense tropical rain forests (**Core Case Study** and Figure 3-1). Because so many plants in such a forest transpire water into the atmosphere, vegetation is the primary source of local rainfall. Cutting down the forest raises ground temperatures (because it reduces shade) and can reduce local rainfall so much that the forest cannot grow back. When such an *ecological tipping point* is reached, these biologically diverse forests are converted into much less diverse tropical grasslands.

The Carbon Cycle

Carbon is the basic building block of the carbohydrates, fats, proteins, DNA, and other organic compounds necessary for life. Various compounds of carbon circulate through the biosphere, the atmosphere, and parts of the hydrosphere, in the carbon cycle shown in Figure 3-15.

The carbon cycle is based on carbon dioxide (CO_2) gas, which makes up about 0.039% of the volume of the earth's atmosphere and is also dissolved in water. Carbon dioxide (along with water vapor in the water cycle) is a key component of the atmosphere's thermostat. If the carbon cycle removes too much CO_2 from the atmosphere, the atmosphere will cool, and if it generates too much CO_2, the atmosphere will get warmer. Thus, even slight changes in this cycle caused by natural or human factors can affect the earth's climate and ultimately help to determine the types of life that can exist in various places.

Terrestrial producers remove CO_2 from the atmosphere and aquatic producers remove it from the water. (See *The Habitable Planet,* Video 3, at **www.learner.org/resources/series209.html** for information on the effects of phytoplankton on the carbon cycle and on the earth's climate.) These producers then use photosynthesis to convert CO_2 into complex carbohydrates such as glucose ($C_6H_{12}O_6$).

The cells in oxygen-consuming producers, consumers, and decomposers then carry out aerobic res-

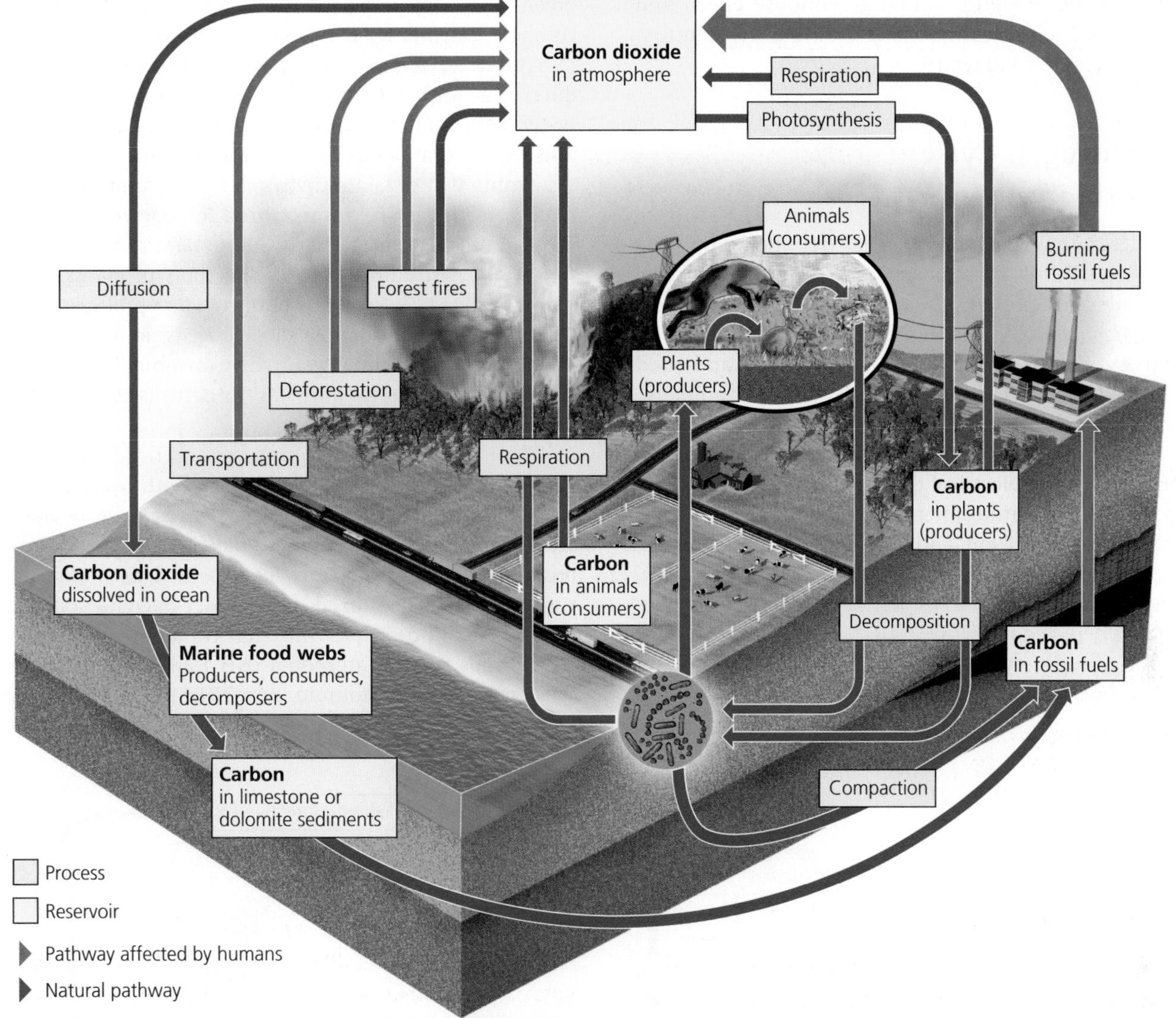

Active Figure 3-15 **Natural capital:** This simplified model illustrates the circulation of various chemical forms of carbon in the global *carbon cycle*, with major harmful impacts of human activities shown by the red arrows. *See an animation based on this figure at* **www.cengagebrain.com**. **Question:** What are three ways in which you directly or indirectly affect the carbon cycle?

piration. This process breaks down glucose and other complex organic compounds to produce CO_2 in the atmosphere and water for reuse by producers. This linkage between *photosynthesis* in producers and *aerobic respiration* in producers, consumers, and decomposers circulates carbon in the biosphere. Oxygen and hydrogen—the other elements in carbohydrates—cycle almost in step with carbon.

Some carbon atoms take a long time to recycle. Decomposers release the carbon stored in the bodies of dead organisms on land back into the air as CO_2. However, in water, decomposers release carbon that can be stored as insoluble carbonates in bottom sediment. Indeed, marine sediments are the earth's largest store of carbon. Over millions of years, buried deposits of dead plant matter and bacteria are compressed between layers of sediment, where high pressure and heat convert them to carbon-containing *fossil fuels* such as coal, oil, and natural gas (see Figure 2-9, p. 36, and Figure 3-15). This carbon is not released to the atmosphere as CO_2 for recycling until these fuels are extracted and burned, or until long-term geological processes expose these deposits to air. In only a few hundred years, we have extracted and burned huge quantities of fossil fuels that took millions of years to form. This is why, on a human time scale, fossil fuels are nonrenewable resources.

We are altering the carbon cycle (see the red arrows in Figure 3-15), mostly by adding large amounts of carbon dioxide to the atmosphere (see Figure 14, p. S44, Supplement 7) when we burn carbon-containing fossil fuels and clear carbon-absorbing vegetation from forests, especially tropical forests, faster than it can grow back (**Core Case Study**).

CORE CASE STUDY

Computer models of the earth's climate systems indicate that increased concentrations of atmospheric CO_2 and other greenhouse gases such as methane (CH_4) are very likely to warm the atmosphere by enhancing the planet's natural greenhouse effect, and thus to change the earth's climate during this century, as we discuss in Chapter 15.

The Nitrogen Cycle: Bacteria in Action

Chemically unreactive nitrogen gas (N_2) makes up 78% of the volume of the atmosphere. Nitrogen is a crucial component of proteins, many vitamins, and nucleic acids such as DNA (see Figure 11, p. S15, in Supplement 4). However, N_2 cannot be absorbed and used directly as a nutrient by multicellular plants or animals.

Fortunately, two natural processes convert, or *fix*, N_2 into compounds that plants and animals can use as nutrients. One is electrical discharges, or lightning, taking place in the atmosphere. The other takes place in aquatic systems, in soil, and in the roots of some plants, where specialized bacteria, called *nitrogen-fixing bacteria*, complete this conversion as part of the **nitrogen cycle**, which is depicted in Figure 3-16.

The nitrogen cycle consists of several major steps. In *nitrogen fixation*, specialized bacteria in soil as well as blue-green algae (cyanobacteria) in aquatic environments combine gaseous N_2 with hydrogen to make ammonia (NH_3). The bacteria use some of the ammonia they produce as a nutrient and excrete the rest into the soil or water. Some of the ammonia is converted to ammonium ions (NH_4^+) that plants can use as a nutrient.

Ammonia not taken up by plants may undergo *nitrification*. In this process, specialized soil bacteria convert most of the NH_3 and NH_4^+ in soil to *nitrate ions* (NO_3^-), which are easily taken up by the roots of plants. The plants then use these forms of nitrogen to produce various amino acids, proteins, nucleic acids, and vitamins. Animals that eat plants eventually consume these nitrogen-containing compounds, as do detritus feeders and decomposers.

Plants and animals return nitrogen-rich organic compounds to the environment as both wastes and cast-off particles of tissues such as leaves, skin, or hair, and through their bodies when they die and are decomposed or eaten by detritus feeders. In *ammonification*, vast armies of specialized decomposer bacteria convert this detritus into simpler nitrogen-containing inorganic compounds such as ammonia (NH_3) and water-soluble salts containing ammonium ions (NH_4^+).

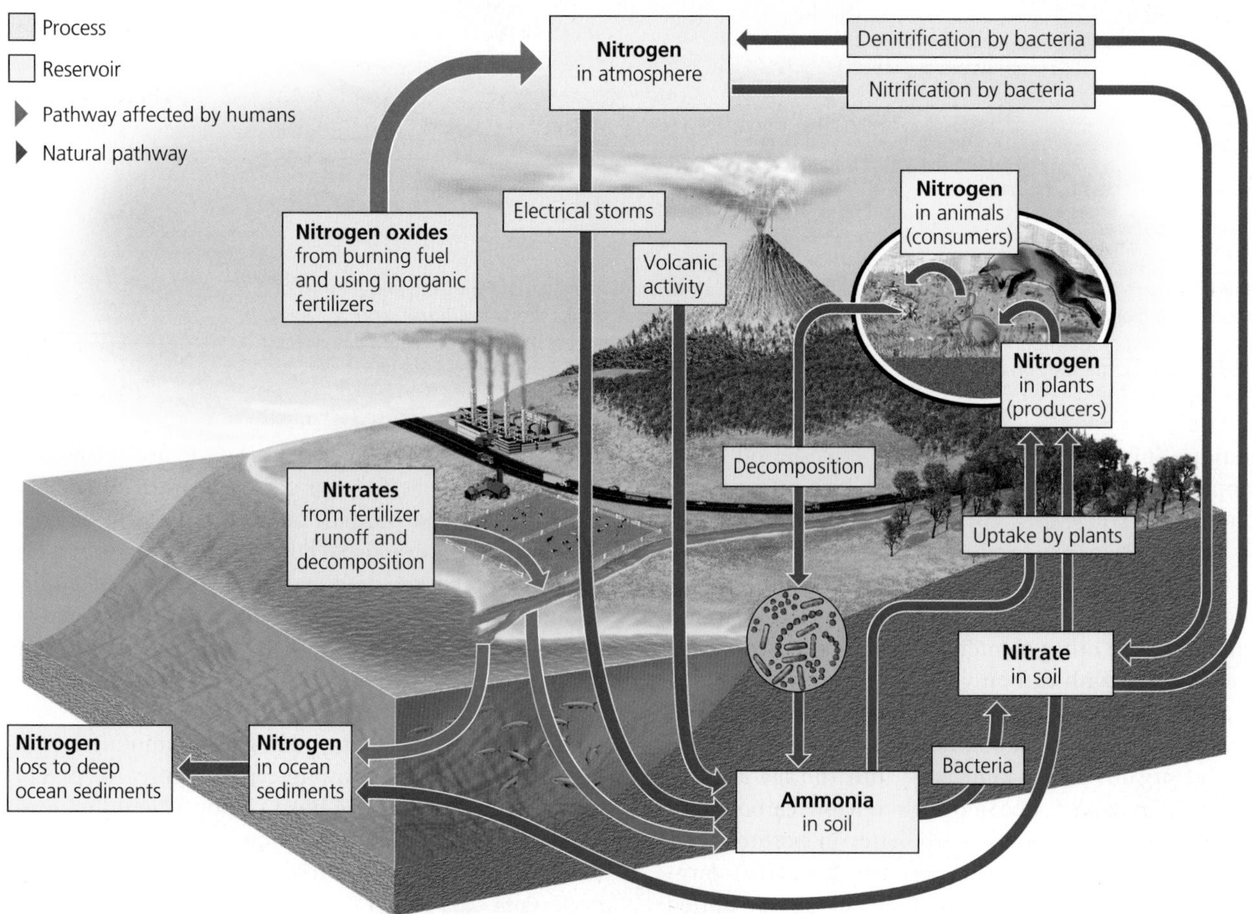

Active Figure 3-16 Natural capital: This diagram is a simplified model of the circulation of various chemical forms of nitrogen in the *nitrogen cycle* in a terrestrial ecosystem, with major harmful human impacts shown by the red arrows. *See an animation based on this figure at* **www.cengagebrain.com**. **Question:** What are three ways in which you directly or indirectly affect the nitrogen cycle?

In *denitrification*, specialized bacteria in waterlogged soil and in the bottom sediments of lakes, oceans, swamps, and bogs convert NH_3 and NH_4^+ back into nitrate ions, and then into nitrogen gas (N_2) and nitrous oxide gas (N_2O)—a greenhouse gas with 300 times more warming potential per molecule than carbon dioxide (CO_2). These gases are released to the atmosphere to begin the nitrogen cycle again.

We intervene in the nitrogen cycle in several ways (see the red arrows in Figure 3-16). According to the 2005 Millennium Ecosystem Assessment, since 1950, human activities have more than doubled the annual release of nitrogen from the land into the rest of the environment—most of this from the greatly increased use of inorganic fertilizers to grow crops—and the amount released is projected to double again by 2050 (see Figure 16, p. S44, in Supplement 7).

This excessive input of nitrogen into the air and water contributes to pollution and other problems to be discussed in later chapters. Nitrogen overload is a serious and growing local, regional, and global environmental problem that has attracted little attention.

THINKING ABOUT
The Nitrogen Cycle and Tropical Deforestation

CORE CASE STUDY

What effects might the clearing and degrading of tropical rain forests (**Core Case Study**) have on the nitrogen cycle in these forest ecosystems and on any nearby aquatic systems? (See Figure 2-1, p. 25, and Figure 2-4, p. 31.)

The Phosphorus Cycle

Compounds of phosphorous (P) circulate through water, the earth's crust, and living organisms in the **phosphorus cycle**, depicted in Figure 3-17. Most of these compounds contain *phosphate* ions (PO_4^{3-}), which serve as an important nutrient. In contrast to the cycles of water, carbon, and nitrogen, the phosphorus cycle does not include the atmosphere. The major reservoir for phosphorous is phosphate salts containing PO_4^{3-} in terrestrial rock formations and ocean bottom sediments. The phosphorus cycle is slow compared to the water, carbon, and nitrogen cycles.

As water runs over exposed rocks, it slowly erodes away inorganic compounds that contain phosphate ions. The running water carries these phosphate ions into the soil where they can be absorbed by the roots of plants and by other producers. Phosphate compounds are also transferred by food webs from producers to consumers, eventually including detritus feeders and decomposers. In both producers and consumers, phosphates are a component of biologically important molecules such as nucleic acids (see Figure 10, p. S14, in Supplement 4) and energy transfer molecules such as ADP and ATP (see Figure 14, p. S16, in Supplement 4). Phosphate is also a major component of vertebrate bones and teeth.

Phosphate can be lost from the cycle for long periods of time when it is washed from the land into streams and rivers and is carried to the ocean. There it can be

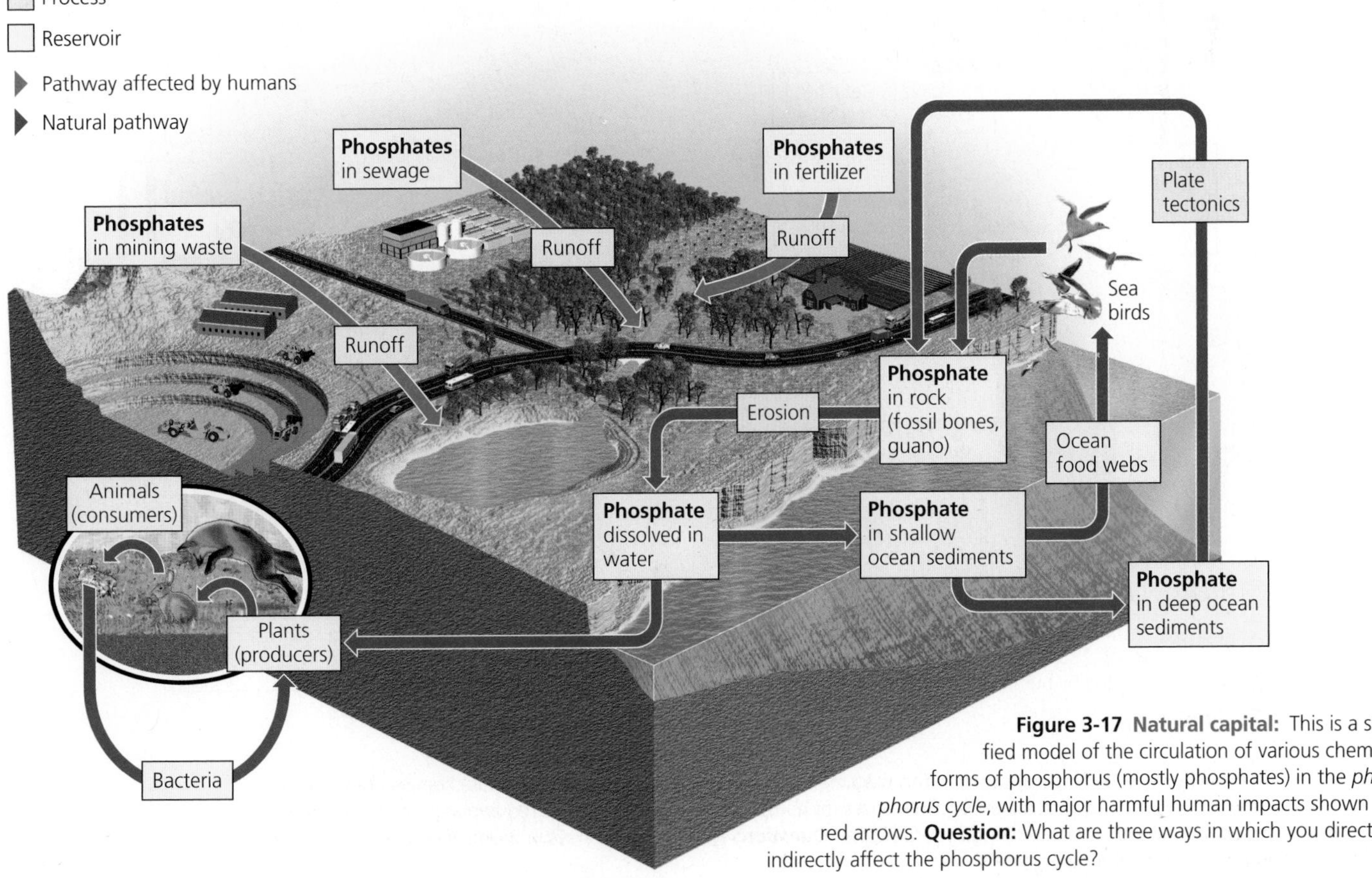

Figure 3-17 Natural capital: This is a simplified model of the circulation of various chemical forms of phosphorus (mostly phosphates) in the *phosphorus cycle*, with major harmful human impacts shown by the red arrows. **Question:** What are three ways in which you directly or indirectly affect the phosphorus cycle?

deposited as marine sediment and remain trapped for millions of years. Someday, geological processes may uplift and expose these seafloor deposits, from which phosphate can be eroded to start the cycle again.

Because most soils contain little phosphate, the lack of it often limits plant growth on land unless phosphorus (as phosphate salts mined from the earth) is applied to the soil as a fertilizer. Lack of phosphorus also limits the growth of producer populations in many freshwater streams and lakes because phosphate salts are only slightly soluble in water and thus do not release many phosphate ions that producers need as nutrients.

Human activities are affecting the phosphorous cycle (as shown by the red arrows in Figure 3-17). This includes removing large amounts of phosphate from the earth to make fertilizer and reducing phosphate levels in tropical soils by clearing forests (**Core Case Study**). CORE CASE STUDY Topsoil that is eroded from fertilized crop fields, lawns, and golf courses carries large quantities of phosphate ions into streams, lakes, and oceans. There they stimulate the growth of producers such as algae and various aquatic plants. Phosphate-rich runoff from the land can produce huge populations of algae (Figure 3-6, right), which can upset chemical cycling and other processes in lakes.

The Sulfur Cycle

Sulfur circulates through the biosphere in the **sulfur cycle**, shown in Figure 3-18. Much of the earth's sulfur is stored underground in rocks and minerals and in the form of sulfate (SO_4^{2-}) salts buried deep under ocean sediments.

Sulfur also enters the atmosphere from several natural sources. Hydrogen sulfide (H_2S)—a colorless, highly poisonous gas with a rotten-egg smell—is released from active volcanoes and from organic matter broken down by anaerobic decomposers in flooded swamps, bogs, and tidal flats. Sulfur dioxide (SO_2), a colorless and suffocating gas, also comes from volcanoes.

Particles of sulfate (SO_4^{2-}) salts, such as ammonium sulfate, enter the atmosphere from sea spray, dust storms, and forest fires. Plant roots absorb sulfate ions and incorporate the sulfur as an essential component of many proteins.

In the oxygen-deficient environments of flooded soils, freshwater wetlands, and tidal flats, specialized bacteria convert sulfate ions to sulfide ions (S^{2-}). The sulfide ions can then react with metal ions to form insoluble metallic sulfides, which are deposited as rock or metal ores (often extracted by mining and converted to various metals), and the cycle continues.

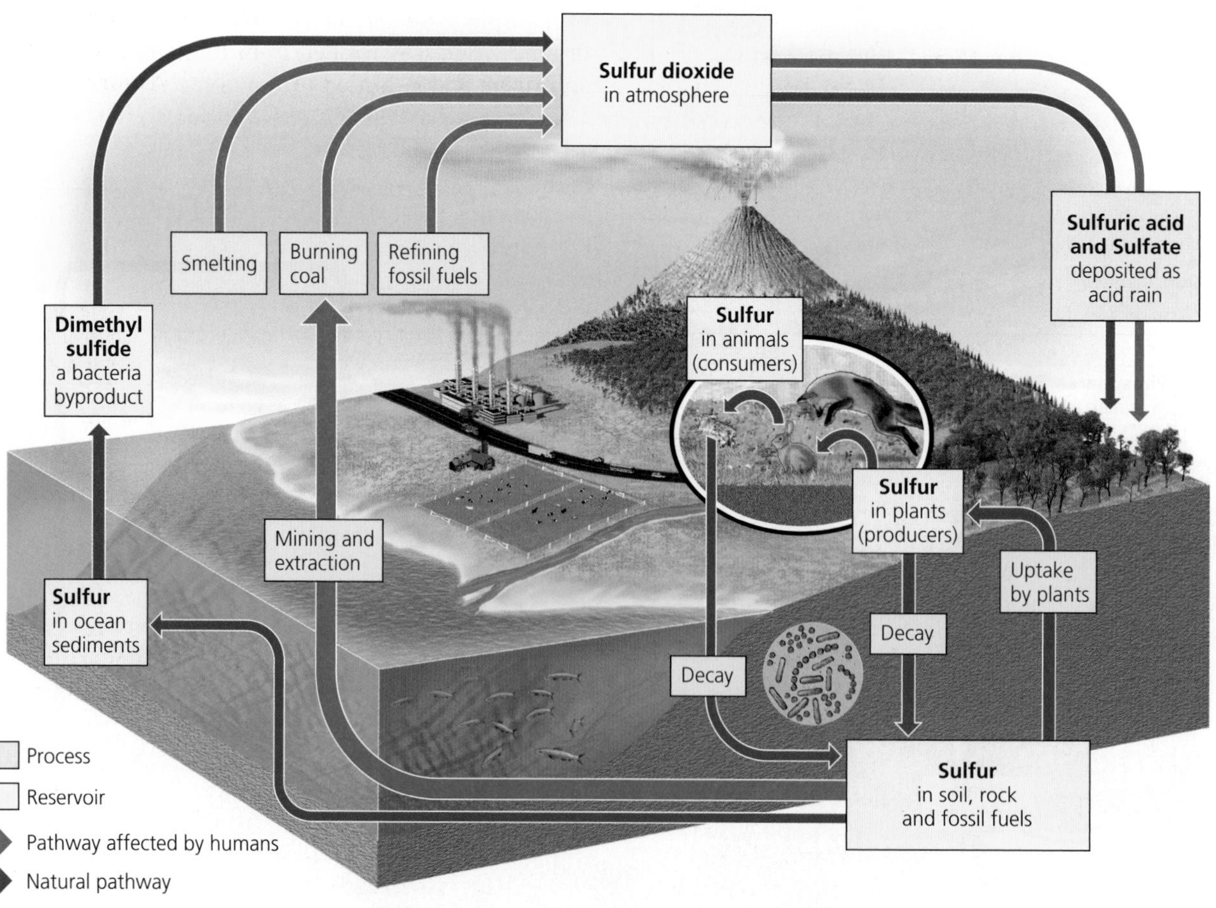

Active Figure 3-18 Natural capital: This is a simplified model of the circulation of various chemical forms of sulfur in the *sulfur cycle*, with major harmful impacts of human activities shown by the red arrows. *See an animation based on this figure at* www.cengagebrain.com. **Question:** What are three ways in which your lifestyle directly or indirectly affects the sulfur cycle?

Human activities have affected the sulfur cycle primarily by releasing large amounts of sulfur dioxide (SO_2) into the atmosphere (as shown by the red arrows in Figure 3-18). We release sulfur to the atmosphere in three ways. *First,* we burn sulfur-containing coal and oil to produce electric power. *Second,* we refine sulfur-containing oil (petroleum) to make gasoline, heating oil, and other useful products. *Third,* we extract metals such as copper, lead, and zinc from sulfur-containing compounds in rocks that are mined for these metals. In the atmosphere, SO_2 is converted to droplets of sulfuric acid (H_2SO_4) and particles of sulfate (SO_4^{2-}) salts, which return to the earth as acid deposition, which in turn can damage ecosystems.

3-5 How Do Scientists Study Ecosystems?

▶ **CONCEPT 3-5 Scientists use both field research and laboratory research, as well as mathematical and other models to learn about ecosystems.**

Some Scientists Study Nature Directly

Scientists use field and laboratory research and mathematical and other models to learn about ecosystems (**Concept 3-5**). *Field research,* sometimes called "muddy-boots biology," involves going into forests (see front cover photo) and other natural settings to observe and measure the structure of ecosystems and what happens in them. Most of what we know about ecosystems has come from such research. **GREEN CAREER:** ecologist

Sometimes ecologists carry out controlled experiments by isolating and changing a variable in part of an area and comparing the results with nearby unchanged areas. A good example of this is reported in the Chapter 2 Core Case Study, p. 25. (See also *The Habitable Planet,* Videos 4 and 9, at **www.learner.org/resources/series209.html**).

Scientists also use aircraft and satellites equipped with sophisticated cameras and other *remote sensing* devices to scan and collect data on the earth's surface. Then they use *geographic information system (GIS)* software to capture, store, analyze, and display such information. Such software can electronically store geographic and ecological spatial data as numbers or as images. For example, a GIS can convert digital satellite images into global, regional, and local maps showing variations in vegetation, gross primary productivity, air pollution emissions, and many other variables. **GREEN CAREERS:** GIS analyst; remote sensing analyst

Some Scientists Study Ecosystems in the Laboratory

Since the 1960s, ecologists have increasingly supplemented field research by using *laboratory research,* setting up, observing, and making measurements of model ecosystems and populations under laboratory conditions. They have created such simplified systems in containers such as culture tubes, bottles, aquariums, and greenhouses, and in indoor and outdoor chambers where they can control temperature, light, CO_2, humidity, and other variables.

Such systems make it easier for scientists to carry out controlled experiments. In addition, laboratory experiments often are quicker and less costly than similar experiments in the field. But there is a catch: scientists must consider how well their scientific observations and measurements in a simplified, controlled system under laboratory conditions reflect what actually takes place under more complex and dynamic conditions found in nature. Thus, the results of laboratory research must be coupled with and supported by field research.

THINKING ABOUT
Greenhouse Experiments and Tropical Rain Forests

How would you design an experiment that includes an experimental group and a control group and uses a greenhouse to determine how clearing a patch of tropical rain forest vegetation (**Core Case Study**) affects the temperature above the cleared patch?

We Need to Learn More about the Health of the World's Ecosystems

Since the late 1960s, ecologists have developed mathematical and other models that simulate ecosystems, and that can be run on high-speed computers. Such computer simulations help scientists understand large and very complex systems, such as lakes, oceans, forests, and the earth's climate system, that cannot be adequately studied and modeled in field and laboratory research (see *The Habitable Planet,* Videos 2, 3, and 12, at **www.learner.org/resources/series209.html**). **GREEN CAREER:** ecosystem modeler

Of course, simulations and projections made with ecosystem models are no better than the data and assumptions used to develop the models. Ecologists must do careful field and laboratory research to get *baseline data,* or beginning measurements, of the variables they are studying. They also must determine the relationships among key variables that they will use to develop and test ecosystem models.

According to the 2005 Millennium Ecosystem Assessment, scientists have less than half of the basic ecological data they need in order to evaluate the status of ecosystems in the United States, to see how they are changing, and to develop effective strategies for preventing or slowing their degradation. Even fewer data are available for most other parts of the world. Ecologists have called for a massive program to develop baseline data for the world's ecosystems.

Here are this chapter's *three big ideas*:

- Life is sustained by the flow of energy from the sun through the biosphere, the cycling of nutrients within the biosphere, and gravity.
- Some organisms produce the nutrients they need, others survive by consuming other organisms, and still others recycle nutrients back to producer organisms.
- Human activities are altering the flow of energy through food chains and webs, and the cycling of nutrients within ecosystems and the biosphere.

REVISTING Tropical Rain Forests and Sustainability

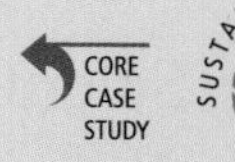

This chapter applied two of the **principles of sustainability** (see back cover) by which the biosphere and the ecosystems it contains have been sustained over the long term. *First*, the source of energy for the biosphere and almost all of its ecosystems is *solar energy*, which flows through these systems. *Second*, the ecosystems continually *recycle the chemical nutrients* that their organisms need for survival, growth, and reproduction. These two major processes support and are enhanced by *biodiversity*, in keeping with the third sustainability principle.

This chapter started with a discussion of the importance of incredibly diverse tropical rain forests (**Core Case Study**), which showcase the functioning of the three **principles of sustainability**, which also apply to the world's other ecosystems. Producers within rain forests rely on solar energy to produce a vast amount of biomass through photosynthesis. Species living in the forests take part in, and depend on, the cycling of nutrients in the biosphere and the flow of energy through the biosphere. Tropical rain forests contain a huge and vital part of the earth's biodiversity, and interactions among species living in these forests help to sustain these complex ecosystems. We explore biodiversity and these important species interactions more deeply in the next two chapters.

All things come from earth, and to earth they all return.

MENANDER (342–290 BC)

REVIEW

CORE CASE STUDY

1. What are three harmful effects of the clearing and degradation of tropical rain forests (**Core Case Study**)?

SECTION 3-1

2. What are the two key concepts for this section? Define and distinguish among the **atmosphere**, **troposphere**, **stratosphere**, **hydrosphere**, **geosphere**, and **biosphere**. What are **greenhouse gases** and why are they important? What three interconnected factors sustain life on earth? Describe the flow of energy to and from the earth. What is the **greenhouse effect** and why is it important?

SECTION 3-2

3. What is the key concept for this section? Define **ecology**. Define **organism**, **population**, and **community** and give an example of each. Define and distinguish between an **ecosystem** and the **biosphere**.
4. Distinguish between the living and nonliving components in ecosystems and give two examples of each. What are two other terms used to describe these ecosystem components?
5. What is a **trophic level**? Distinguish among **producers (autotrophs)**, **consumers (heterotrophs)**, **decomposers** and **detritus feeders (detritivores)**, and give

an example of each in an ecosystem. Distinguish among **primary consumers (herbivores)**, **carnivores**, **secondary consumers**, **tertiary consumers**, and **omnivores**, and give an example of each.

6. Describe the processes of **photosynthesis** and **aerobic respiration**. How are these two processes related? What two processes sustain ecosystems and the biosphere and how are they linked? Explain the importance of microbes.

SECTION 3-3

7. What is the key concept for this section? Define and distinguish between a **food chain** and a **food web**. Explain what happens to energy as it flows through food chains and webs. What is **biomass**? What is the **pyramid of energy flow**? Why are there more insects than tigers in the world?

8. Distinguish between **gross primary productivity (GPP)** and **net primary productivity (NPP)**, and explain their importance.

SECTION 3-4

9. What is the key concept for this section? What happens to matter in an ecosystem? What is a **biogeochemical cycle (nutrient cycle)**? Describe the **hydrologic**, or **water cycle**. What are its three major processes? Summarize the unique properties of water. Explain how clearing a rain forest can affect local weather (**Core Case Study**). CORE CASE STUDY Explain how human activities are affecting the water cycle. Describe the **carbon**, **nitrogen**, **phosphorus**, and **sulfur cycles**, and explain how human activities are affecting each cycle. Explain how nutrient cycles connect past, present, and future life.

SECTION 3-5

10. What is the key concept for this section? Describe three ways in which scientists study ecosystems. Explain why we need much more basic data about the structure and condition of the world's ecosystems. What are this chapter's *three big ideas*? How are the three **principles of sustainability** showcased in tropical rain forests? SUSTAINABILITY

Note: Key terms are in **bold** type.

CRITICAL THINKING

1. How would you explain the importance of tropical rain forests (**Core Case Study**) to people who think that such forests have no connection to their lives? CORE CASE STUDY

2. Explain **(a)** why the flow of energy through the biosphere depends on the cycling of nutrients, and **(b)** why the cycling of nutrients depends on gravity (**Concept 3-1B**).

3. Explain why microbes are so important. List two beneficial effects and two harmful effects of microbes on your health and lifestyle. Write a brief description of what you think would happen to you if microbes were eliminated from the earth.

4. Make a list of the foods you ate for lunch or dinner today. Trace each type of food back to a particular producer species. Describe the sequence of feeding levels that led to your feeding.

5. Use the second law of thermodynamics (see Chapter 2, p. 36) to explain why many poor people in less-developed countries live on a mostly vegetarian diet.

6. Why do farmers not need to apply carbon to grow their crops but often need to add fertilizer containing nitrogen and phosphorus?

7. What changes might take place in the hydrologic cycle if the earth's climate becomes **(a)** hotter or **(b)** cooler? In each case, what are two ways in which these changes might affect your lifestyle?

8. What would happen to an ecosystem if **(a)** all of its decomposers and detritus feeders were eliminated, **(b)** all of its producers were eliminated, and **(c)** all of its insects were eliminated? Could a balanced ecosystem exist with only producers and decomposers and no consumers such as humans and other animals? Explain.

DOING ENVIRONMENTAL SCIENCE

Visit a nearby terrestrial ecosystem or aquatic life zone and try to identify major producers, primary and secondary consumers, detritus feeders, and decomposers. Take notes and describe at least one example of each of these types of organisms. Make a simple sketch showing how these organisms might be related to each other or to other organisms in a food chain.

GLOBAL ENVIRONMENT WATCH EXERCISE

Search for *Nitrogen Cycle* and look for information on how humans are affecting the nitrogen cycle. Specifically look for impacts on the atmosphere and on human health from emissions of nitrogen oxides, and look for the harmful ecological effects of the runoff of nitrate fertilizers into rivers and lakes. Make a list of these impacts and use this information to review your daily activities. Find three things that you do regularly that contribute to these impacts.

ECOLOGICAL FOOTPRINT ANALYSIS

Based on the following carbon dioxide emissions data and 2010 population data, answer the questions below.

Country	Total Carbon Footprint—Carbon Dioxide Emissions in Metric Gigatons per Year*	Population in Billions (2010)	Per Capita Carbon Footprint—Per Capita Carbon Dioxide Emissions per Year
China	5.0	1.3	
India	1.3	1.2	
Japan	1.3	0.13	
Russia	1.5	0.14	
United States	6.0	0.31	
WORLD	29	6.9	

(Data from the World Resources Institute and the International Energy Agency)

*The prefix *giga-* stands for 1 billion.

1. Calculate the per capita carbon footprint for each country and for the world, and complete the table.
2. It has been suggested that a sustainable average worldwide carbon footprint per person should be no more than 2.0 metric tons per person per year. How many times larger is the U.S. carbon footprint per person than **(a)** the sustainable level, and **(b)** the world average?
3. By what percentage will China, Japan, Russia, the United States, and the world each have to reduce their carbon footprints per person to achieve the estimated maximum sustainable carbon footprint per person of 2.0 metric tons per person per year?

LEARNING ONLINE

The website for *Environmental Science 14e* contains helpful study tools and other resources to take the learning experience beyond the textbook. Examples include flashcards, practice quizzing, information on green careers, tips for more sustainable living, activities, and a wide range of articles and further readings. To access these materials and other companion resources for this text, please visit **www.cengagebrain.com**. For further details, see the preface, p. xvi.

Biodiversity and Evolution

4

Why Should We Protect Sharks?

CORE CASE STUDY

More than 400 known species of sharks (Figure 4-1) inhabit the world's oceans. They vary widely in size and behavior, from the goldfish-sized dwarf dog shark to the whale shark (Figure 4-1, left), which can grow to a length roughly equal to that of a typical city bus and weigh as much as two full-grown African elephants.

Many people, influenced by movies, popular novels, and widespread media coverage of shark attacks, think of sharks as people-eating monsters. In reality, the three largest species—the whale shark (Figure 4-1, left), basking shark, and megamouth shark—are gentle giants. These plant-eating sharks swim through the water with their mouths open, filtering out and swallowing huge quantities of phytoplankton.

Media coverage of shark attacks greatly exaggerates the danger from sharks. Every year, members of a few species such as the great white, bull, tiger, oceanic whitetip, and hammerhead sharks (Figure 4-1, right) injure 60–75 people worldwide, with the majority of the attacks occurring in U.S. waters. Between 1998 and 2009, there were of six deaths per year, on average, from such attacks. Some of these sharks feed on sea lions and other marine mammals and sometimes mistake swimmers, surfers, and scuba divers for their usual prey. Despite the publicity and fear, sharks caused only 49 known deaths in U.S. waters during the 339 years between 1670 and 2009, according to the International Shark Attack File.

However, *for every shark that injures or kills a person every year, people kill at least 1 million sharks. This amounts to 73–97 million shark deaths each year*, according to Australia's Shark Research Institute. Many sharks are caught for their fins and then thrown back alive into the water, fins removed, to bleed to death or drown because they can no longer swim. This practice is called *finning*. The fins are widely used in Asia as an ingredient in shark fin soup and as a pharmaceutical cure-all. According to the World Wildlife Fund, shark fin soup can cost as much as $100 a bowl in Hong Kong, which imports about half of the world's shark fins. We also kill sharks for their livers, hides, and jaws, and because we fear them. And the long fishing lines and massive nets used by fishing fleets trap and kill some sharks.

According to a 2009 study by the International Union for Conservation of Nature (IUCN), about 32% of the world's open-ocean shark species are threatened or nearly threatened with extinction. One of the most endangered species is the scalloped hammerhead shark (Figure 4-1, right). Sharks are especially vulnerable to population declines because they grow slowly, take a long time to reach sexual maturity, and have only a few offspring per generation. Today, they are among the earth's most vulnerable and least protected animals.

Sharks have been around for more than 400 million years. As *keystone species*, some shark species play crucial roles in helping to keep their ecosystems functioning. Feeding at or near the tops of food webs, they remove injured and sick animals from the ocean. Without this ecosystem service provided by sharks, the oceans would be teeming with dead and dying fish and marine mammals.

In addition to playing their important ecological roles, sharks could help to save human lives. If we can learn why they almost never get cancer, we could possibly use this information to fight cancer in our own species. Scientists are also studying sharks' highly effective immune systems, which allow their wounds to heal without becoming infected.

Some argue that we should protect sharks simply because they, like any other species, have a right to exist. Another reason is that some sharks are keystone species, which means that we and other species need the free ecosystem services they provide. In this chapter we explore the keystone species' role and other special roles played by species in the story of the earth's vital biodiversity.

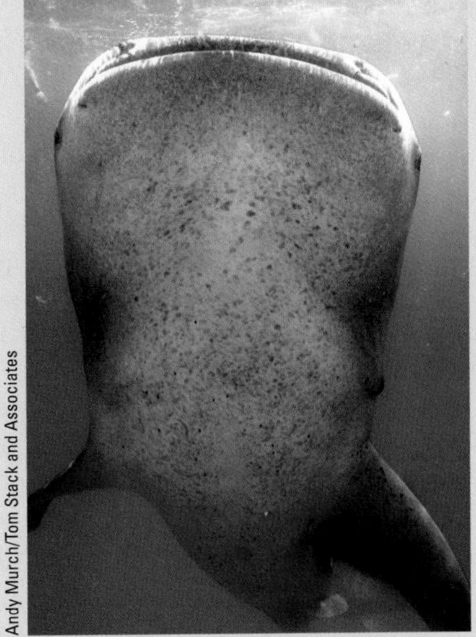

Andy Murch/Tom Stack and Associates

Westend61/SuperStock

Figure 4-1 The threatened whale shark (left), which feeds on plankton, is the largest fish in the ocean. The endangered scalloped hammerhead shark (right) is swimming in waters near the Galapagos Islands, Ecuador. They are known for eating stingrays, which are related to sharks.

Key Questions and Concepts

4-1 What is biodiversity and why is it important?

CONCEPT 4-1 The biodiversity found in genes, species, ecosystems, and ecosystem processes is vital to sustaining life on earth.

4-2 How does the earth's life change over time?

CONCEPT 4-2A The scientific theory of evolution explains how life on earth changes over time due to changes in the genes of populations.

CONCEPT 4-2B Populations evolve when genes mutate and give some individuals genetic traits that enhance their abilities to survive and to produce offspring with these traits (natural selection).

4-3 How do geological processes and climate change affect evolution?

CONCEPT 4-3 Tectonic plate movements, volcanic eruptions, earthquakes, and climate change have shifted wildlife habitats, wiped out large numbers of species, and created opportunities for the evolution of new species.

4-4 How do speciation, extinction, and human activities affect biodiversity?

CONCEPT 4-4A As environmental conditions change, the balance between the formation of new species and the extinction of existing species determines the earth's biodiversity.

CONCEPT 4-4B Human activities are decreasing biodiversity by causing the extinction of many species and by destroying or degrading habitats needed for the development of new species.

4-5 What roles do species play in ecosystems?

CONCEPT 4-5A Each species plays a specific ecological role called its niche.

CONCEPT 4-5B Any given species may play one or more of four important roles—native, nonnative, indicator, or keystone—in a particular ecosystem.

Note: Supplements 2 (p. S3), 3 (p. S6), 4 (p. S10), and 6 (p. S22) can be used with this chapter.

There is grandeur to this view of life . . . that, whilst this planet has gone cycling on . . . endless forms most beautiful and most wonderful have been, and are being, evolved.

CHARLES DARWIN

4-1 What Is Biodiversity and Why Is It Important?

▶ **CONCEPT 4-1 The biodiversity found in genes, species, ecosystems, and ecosystem processes is vital to sustaining life on earth.**

Biodiversity Is a Crucial Part of the Earth's Natural Capital

Biological diversity, or **biodiversity**, is the variety of the earth's *species*, the genes they contain, the ecosystems in which they live, and the ecosystem processes such as energy flow and nutrient cycling that sustain all life (Figure 4-2).

For a group of sexually reproducing organisms, a **species** is a set of individuals that can mate and produce fertile offspring. Every organism is a member of a certain species with certain distinctive *traits*, or characteristics.

We do not know how many species there are on the earth. Estimates range from 8 million to 100 million. The best guess is that there are 10–14 million species. So far, biologists have identified almost 2 million species. Up to half of the world's plant and animal species live in tropical rain forests that humans are rapidly clearing (see Figure 3-1, p. 40). Insects make up most of the world's known species (see Science Focus, p. 64). Scientists believe that most of the unidentified species live in the planet's rain forests and the largely unexplored oceans.

Species diversity, the number and variety of the species present in any biological community (Figure 4-3), is the most obvious component of biodiversity. Another important component is **genetic diversity**, the variety of genes found in a population or in a species (Figure 4-3). The earth's variety of species contains an even greater variety of genes, which enable life on the earth to survive and adapt to dramatic environmental changes.

Links: CORE CASE STUDY refers to the Core Case Study. SUSTAINABILITY refers to the book's sustainability theme. GOOD NEWS refers to good news about the environmental challenges we face.

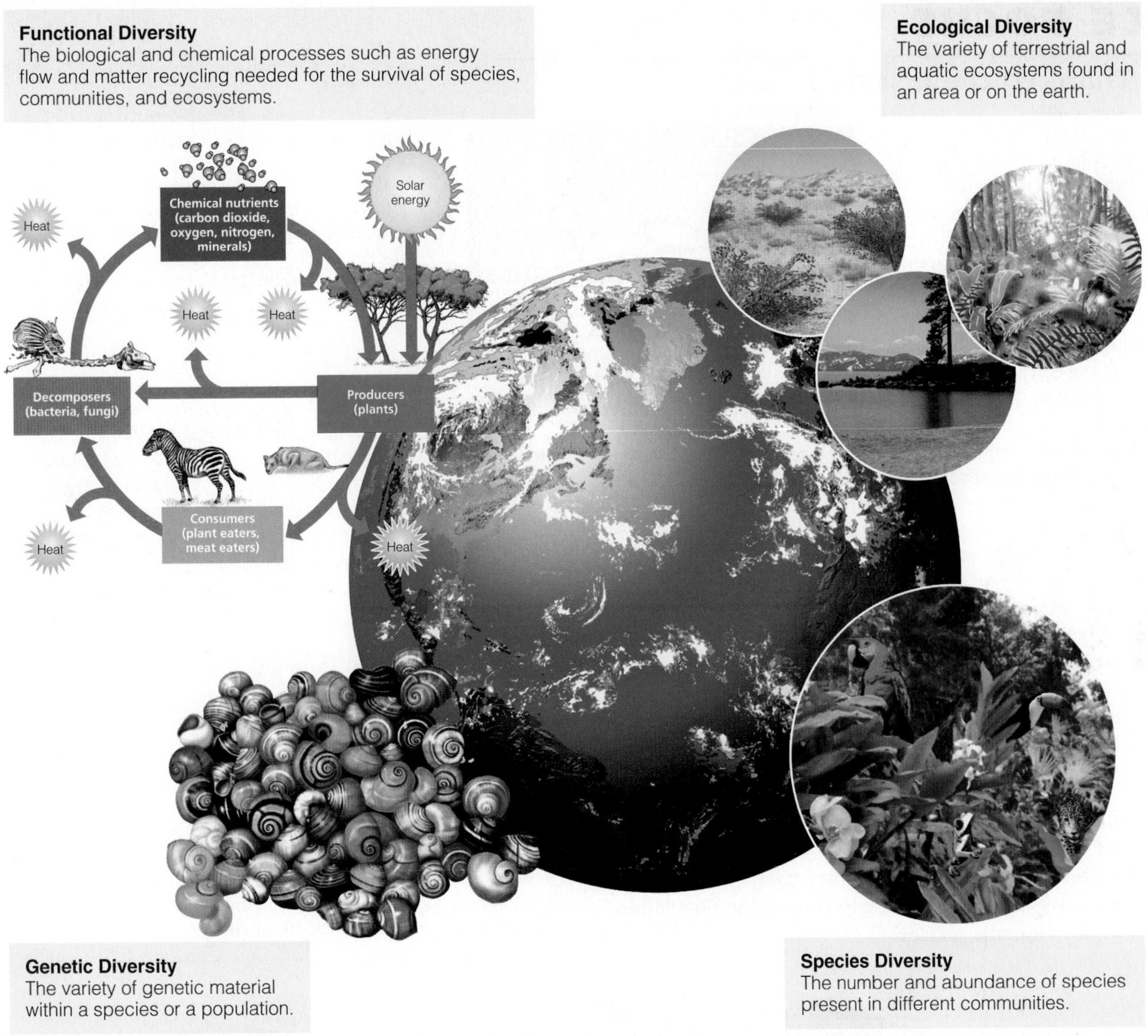

Functional Diversity
The biological and chemical processes such as energy flow and matter recycling needed for the survival of species, communities, and ecosystems.

Ecological Diversity
The variety of terrestrial and aquatic ecosystems found in an area or on the earth.

Heat
Chemical nutrients (carbon dioxide, oxygen, nitrogen, minerals)
Solar energy
Heat
Heat
Decomposers (bacteria, fungi)
Producers (plants)
Consumers (plant eaters, meat eaters)
Heat
Heat

Genetic Diversity
The variety of genetic material within a species or a population.

Species Diversity
The number and abundance of species present in different communities.

Active Figure 4-2 Natural capital: This diagram illustrates the major components of the earth's biodiversity—one of the earth's most important renewable resources and a key component of the planet's natural capital (see Figure 1-3, p. 9). *See an animation based on this figure at* www.cengagebrain.com. **Question:** Why do you think we should protect the earth's biodiversity from our actions?

Ecosystem diversity—the earth's variety of deserts, grasslands, forests, mountains, oceans, lakes, rivers, and wetlands is another major component of biodiversity. Each of these ecosystems is a storehouse of genetic and species diversity.

Biologists have classified the terrestrial (land) portion of the biosphere into **biomes**—large regions such as forests, deserts, and grasslands with distinct climates and certain species (especially vegetation) adapted to them. Figure 4-4 (p. 64) shows different major biomes along the 39th parallel spanning the United States. We discuss biomes in more detail in Chapter 7.

Yet another important component of biodiversity is **functional diversity**—the variety of processes such as energy flow and matter cycling that occur within ecosystems (see Figure 3-9, p. 46) as species interact with

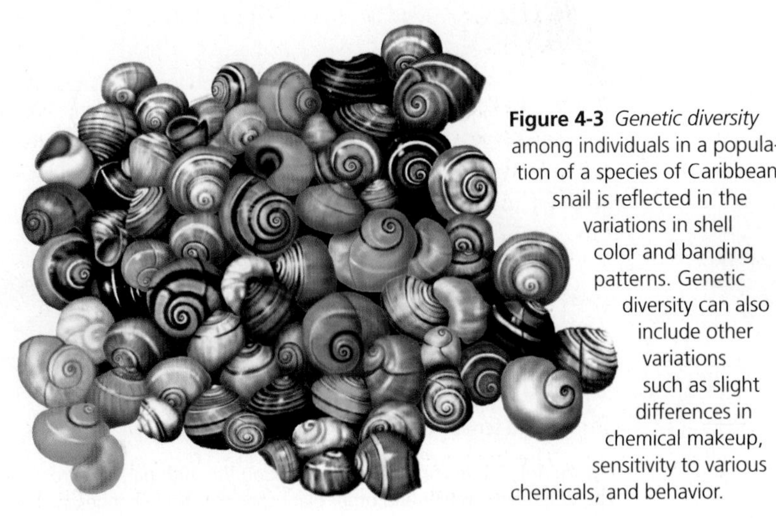

Figure 4-3 *Genetic diversity* among individuals in a population of a species of Caribbean snail is reflected in the variations in shell color and banding patterns. Genetic diversity can also include other variations such as slight differences in chemical makeup, sensitivity to various chemicals, and behavior.

SCIENCE FOCUS

Have You Thanked the Insects Today?

Insects are an important part of the earth's natural capital, although they generally have a bad reputation. We classify many insect species as *pests* because they compete with us for food, spread human diseases such as malaria, bite or sting us, and invade our lawns, gardens, and houses. Some people fear insects and fail to recognize the vital roles insects play in helping to sustain life on earth. For example, many of the earth's plant species depend on insects to pollinate their flowers (Figure 4-A, top).

Insects, which have been around for at least 2,000 times longer than the current version of the human species (*Homo sapiens sapiens*), are phenomenally successful forms of life. Some reproduce at an astounding rate and can rapidly develop new genetic traits such as resistance to pesticides. They also have an exceptional ability to evolve into new species when faced with changing environmental conditions, and they are very resistant to extinction. This is fortunate because, according to ant specialist and biodiversity expert E. O. Wilson, if all insects disappeared, the life-support systems for us and other species would be seriously disrupted.

The environmental lesson: although insects do not need newcomer species such as humans, we and most other land organisms need them.

John Henry Williams/Bruce Coleman USA

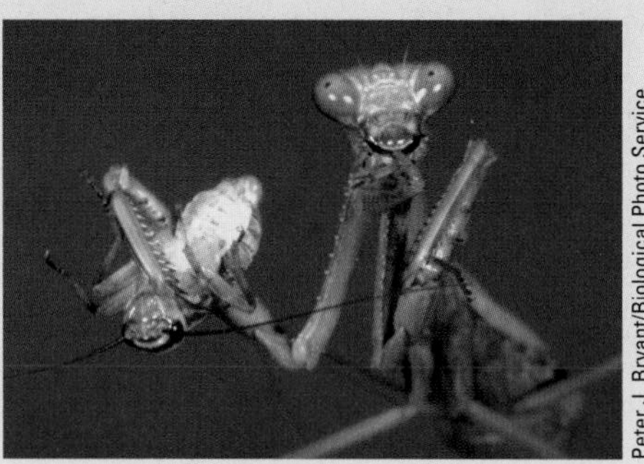
Peter J. Bryant/Biological Photo Service

Figure 4-A *Importance of insects:* The monarch butterfly, like bees and numerous other insects, feeds on pollen in a flower (top), and pollinates flowering plants that serve as food for many plant eaters, including humans. The praying mantis, which is eating a house cricket (bottom), and many other insect species help to control the populations of most of the insect species we classify as pests.

Critical Thinking

Identify three insect species not discussed above that benefit your life.

one another in food chains and webs. Part of the importance of some shark species (**Core Case Study**) is their role in supporting these processes within their ecosystems and helping to control the populations of various aquatic species.

> **THINKING ABOUT**
> **Sharks and Biodiversity**
> What are three ways in which sharks (**Core Case Study**) support one or more of the four components of biodiversity within their environment?

Figure 4-4 The major biomes found along the 39th parallel across the United States are shown in this diagram. The differences in tree and other plant species reflect changes in climate, mainly differences in average annual precipitation and temperature.

INDIVIDUALS MATTER

Edward O. Wilson: A Champion of Biodiversity

As a boy growing up in the southeastern United States, Edward O. Wilson (Figure 4-B) became interested in insects at the age of nine. He has said, "Every kid has a bug period. I never grew out of mine."

Before entering college, Wilson had decided he would specialize in the study of ants and is now recognized as one of the world's experts on ants. Starting with these tiny organisms, and throughout his long career, he has steadily widened his focus to include the entire biosphere.

He has applied the results of his ant research to the study and understanding of other social organisms, including humans. He has also proposed the hypothesis that humans have a natural affinity for wildlife and wild places—a concept he calls *biophilia* (or love of life).

In the 1960s, Wilson and other scientists developed the *theory of island biogeography*, which examines how species diversity on islands is affected by the sizes and locations of the islands. It has been applied to areas that resemble islands, such as mountain forests surrounded by developed land. It has also been important in the creation of wildlife preserves. One of Wilson's landmark works is *The Diversity of Life*, published in 1992, in which he put together the principles and practical issues of biodiversity more completely than anyone had to that point. Wilson is now deeply involved in writing and lecturing about the need for global conservation efforts and is promoting the goal of completing a global survey of biodiversity. He has won more than 100 national and international awards and has written 25 books, two of which won the Pulitzer Prize for General Nonfiction. About the importance of biodiversity, he writes:

> "Until we get serious about exploring biological diversity . . . science and humanity at large will be flying blind inside the biosphere. How can we fully understand the ecology of a pond or forest patch without knowledge of the thousands of species . . . the principal channels of materials and energy flow? . . . How can we save Earth's life forms from extinction if we don't even know what most of them are?"

Jim Harrison

Figure 4-B Edward O. Wilson

The earth's biodiversity is a vital part of the natural capital that helps to keep us alive and supports our economies. With the help of technology, we use biodiversity to provide us with food, wood, fibers, energy from wood and biofuels, and medicines. Biodiversity also plays critical roles in preserving the quality of the air and water, maintaining the fertility of topsoil, decomposing and recycling waste, and controlling populations of species that we call pests. We owe much of what we know about biodiversity to a fairly small number of researchers, such as Edward O. Wilson (Individuals Matter, above).

4-2 How Does the Earth's Life Change over Time?

▶ **CONCEPT 4-2A The scientific theory of evolution explains how life on earth changes over time due to changes in the genes of populations.**

▶ **CONCEPT 4-2B Populations evolve when genes mutate and give some individuals genetic traits that enhance their abilities to survive and to produce offspring with these traits (natural selection).**

Biological Evolution by Natural Selection Explains How Life Changes over Time

Most of what we know about the long history of life on the earth comes from **fossils**: mineralized or petrified replicas of skeletons, bones, teeth, shells, leaves, and seeds, or impressions of such items found in rocks (see Photo 2 in the Detailed Contents). Scientists also drill core samples from glacial ice at the earth's poles and on mountaintops, and examine the signs of ancient life found at different layers in these cores.

The entire body of evidence gathered using these methods, which is called the *fossil record*, is uneven and incomplete. Some forms of life left no fossils, and some fossils have decomposed. The fossils found so far represent probably only 1% of all species that have ever lived. Trying to reconstruct the development of life with so little evidence is the work of *paleontology*—a challenging scientific detective game. **GREEN CAREER:** paleontologist

How did we end up with such an amazing array of species? The scientific answer involves **biological evolution** (or simply **evolution**): the process whereby earth's life changes over time through changes in the genes of populations of organisms in succeeding generations (**Concept 4-2A**). One of the most important scientific theories is the **theory of evolution by natural selection**, which is a scientific explanation of how the process of evolution takes place. According to this theory, because of differences in their genes, some individuals in a population can have one or more traits that give them a better chance of surviving and reproducing under existing environmental conditions, and this eventually leads to a population with a higher proportion of individuals possessing those traits (**Concept 4-2B**).

In other words, over time, natural selection leads to changes in the genetic makeup of populations of organisms. Also, at some point, it can lead to the development of new species (a process we discuss further in Section 4-4). This important scientific theory provides an explanation of how life has changed over the past 3.5 billion years and why life is so diverse today. However, there are still many unanswered questions that generate scientific debate about the details of evolution by natural selection.

Mutations and Changes in the Genetic Makeup of Populations Lead to Biological Evolution by Natural Selection

The process of biological evolution by natural selection involves changes in a population's genetic makeup through successive generations. Note that *populations—not individuals—evolve by becoming genetically different.*

The first step in this process is the development of *genetic variability*, or variety in the genetic makeup of individuals in a population. This occurs through **mutations**: random changes in the DNA molecules (see Figure 11, p. S15, in Supplement 4) of a gene in any cell that can be inherited by offspring. Some mutations also occur from exposure to external agents such as radioactivity and natural and human-made chemicals (called *mutagens*).

Mutations can occur in any cell, but only those taking place in genes of reproductive cells are passed on to offspring. Sometimes, such a mutation can result in a new genetic trait, called a *heritable trait*, which can be passed from one generation to the next. In this way, populations develop differences among individuals, including genetic variability.

The next step in biological evolution is *natural selection*, in which environmental conditions favor some individuals over others. The favored individuals possess heritable traits that give them some advantage over other individuals in a given population. Such a trait is called an **adaptation**, or **adaptive trait**—any heritable trait that improves the ability of an individual organism to survive and to reproduce at a higher rate than other individuals in a population are able to do under prevailing environmental conditions.

For example, in the face of snow and cold, a few gray wolves in a population that have thicker fur might live longer and thus produce more offspring than do those without thicker fur. As those longer-lived wolves mate, genes for thicker fur spread throughout the population and individuals with those genes increase in number and pass this helpful trait on to more offspring. Thus, the scientific concept of natural selection explains how populations adapt to changes in environmental conditions.

Another important example of natural selection at work is the evolution of *genetic resistance* in certain bacteria to widely used antibacterial drugs, or antibiotics. Genetic resistance is the ability of one or more organisms in a population to tolerate a chemical designed to kill the population. For example, because of genetic diversity, one or more organisms in a population of bacteria might have a gene that allows the organisms to break the chemical down into other harmless chemicals. They can then pass these genes on to members of succeeding generations.

Antibiotics have become a force of natural selection. When such drugs are used, the few bacteria that are genetically resistant to them (because of some trait the bacteria possess) survive and rapidly produce more offspring than the bacteria that were killed by the drug could have produced. Thus, the antibiotic eventually loses its effectiveness as genetically resistant bacteria rapidly reproduce and those that are susceptible to the drug die off (Figure 4-5). Such evolution by natural selection also allows rapidly producing insect pests to become genetically resistant to the pesticides we spray on crop fields.

One way to summarize the process of biological evolution by natural selection is: *Genes mutate, individuals are selected, and populations evolve such that they are better adapted to survive and reproduce under existing environmental conditions* (**Concept 4-2B**).

A remarkable example of species evolution by natural selection is *Homo sapiens sapiens*. We have evolved certain traits that have allowed us to take over much of the world (see Science Focus, at right).

Adaptation through Natural Selection Has Limits

In the not-too-distant future, will adaptations to new environmental conditions through natural selection allow our skin to become more resistant to the harmful effects of UV radiation, our lungs to cope with air pol-

(a)
A group of bacteria, including genetically resistant ones, are exposed to an antibiotic.

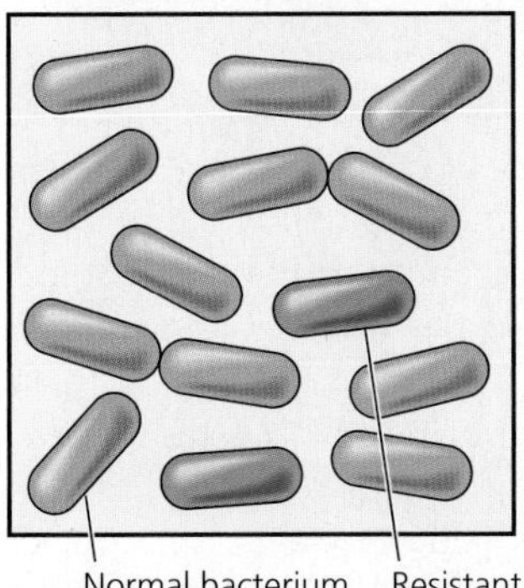

(b)
Most of the normal bacteria die.

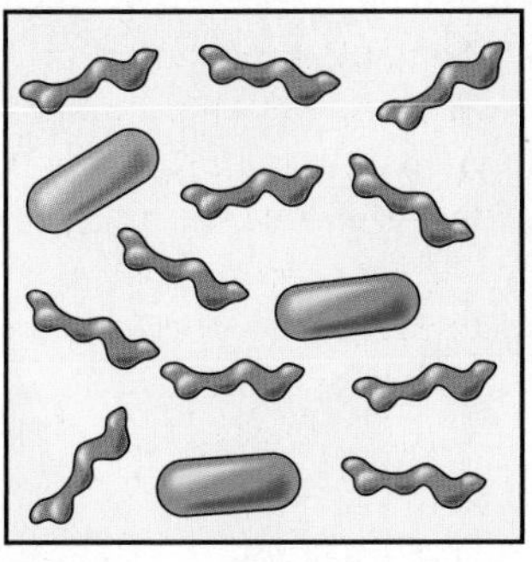

(c)
The genetically resistant bacteria start multiplying.

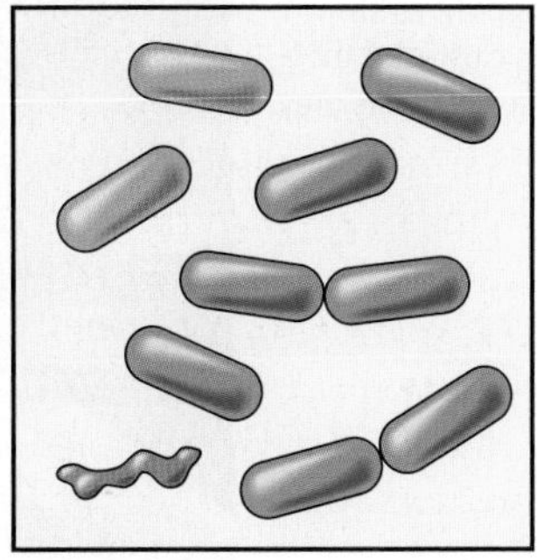

(d)
Eventually the resistant strain replaces all or most of the strain affected by the antibiotic.

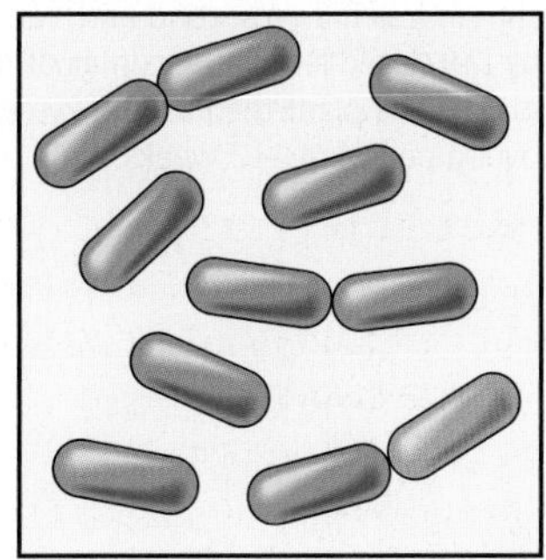

Figure 4-5 *Evolution by natural selection:* **(a)** A population of bacteria is exposed to an antibiotic, which **(b)** kills all individuals except those possessing a trait that makes them resistant to the drug. **(c)** The resistant bacteria multiply and eventually **(d)** replace all or most of the nonresistant bacteria.

lutants, and our livers to better detoxify pollutants in our bodies?

According to scientists in this field, the answer is *no* because of two limitations on adaptation through natural selection. *First,* a change in environmental conditions can lead to such an adaptation only for genetic traits already present in a population's gene pool or for traits resulting from mutations, which occur randomly.

Second, even if a beneficial heritable trait is present in a population, the population's ability to adapt may be limited by its reproductive capacity. Populations of genetically diverse species that reproduce quickly—such as weeds, mosquitoes, rats, bacteria, and cockroaches—often adapt to a change in environmental conditions in a short time (days to years). By contrast, species that cannot produce large numbers of offspring rapidly—such as elephants, tigers, sharks, and humans—take a much longer time (typically thousands or even millions of years) to adapt through natural selection.

There Are Three Incorrect Ideas about Evolution through Natural Selection

According to evolution experts, there are three common misconceptions about biological evolution through natural selection. One is that "survival of the fittest" means "survival of the strongest." To biologists, *fitness* is a measure of reproductive success, not strength. Thus, the fittest individuals are those that leave the most descendants.

SCIENCE FOCUS

How Did Humans Become Such a Powerful Species?

Like many other species, humans have survived and thrived because we have certain traits that allow us to adapt to and modify parts of the environment to increase our survival chances.

Evolutionary biologists attribute our success to three adaptations: *strong opposable thumbs* that allowed us to grip and use tools better than the few other animals that have thumbs could do; an *ability to walk upright,* which gave us agility and freed up our hands for many uses; and a *complex brain,* which allowed us to develop many skills, including the ability to use speech and to read and write to transmit complex ideas.

These adaptations have helped us to develop tools, weapons, protective devices, and technologies that extend our limited senses of sight, hearing, and smell, and make up for some of our physical deficiencies. Thus, in an eye-blink of the 3.5-billion-year history of life on earth, we have developed powerful technologies and taken over much of the earth's net primary productivity for our own use. At the same time, we have degraded much of the planet's life-support system as our ecological footprints have grown (see Figure 1-8, p. 14).

However, adaptations that make a species successful during one period of time may not be enough to ensure the species' survival when environmental conditions change. This is no less true for humans, and some environmental conditions are now changing rapidly, largely due to our own actions. (We focus on several such changes in later chapters.)

One of our adaptations—our powerful brain—may enable us to live more sustainably by understanding and copying the ways in which nature has sustained itself for billions of years, despite major changes in environmental conditions. GOOD NEWS

Critical Thinking

An important adaptation of humans is strong opposable thumbs, which allow us to grip and manipulate things with our hands. Make a list of the things you could not do if you did not have opposable thumbs.

Figure 4-6 One type of carnivorous plant is the Venus flytrap. A fly or other small insect, entering a hinged opening on any one of the plant's leaves, touches trigger hairs that cause the opening to snap shut in less than a second and trap its prey, as shown by the closed leaf in the center of this cluster of the plants. The plant then uses enzymes to digest its prey over a period of 1–2 weeks.

Mary Lane/Shutterstock.com

Another misconception is that organisms develop certain traits because they need them. Certain plants, called carnivorous plants (Figure 4-6), feed on insects not because they once needed to in order to survive. Rather, some ancestors of these plants had characteristics that enabled them to trap insects and to draw nutrients from them. This gave them an advantage over other plants that could not trap insects for food. These plants then produced more offspring that had such characteristics, and their populations grew and continued to evolve in this way. (Go on a virtual field trip to learn more about carnivorous plants at **www.cengagebrain.com**.)

A third misconception is that evolution by natural selection involves some grand plan of nature in which species become more perfectly adapted. From a scientific standpoint, no plan or goal for genetic perfection has been identified in the evolutionary process.

4-3 How Do Geological Processes and Climate Change Affect Evolution?

CONCEPT 4-3 Tectonic plate movements, volcanic eruptions, earthquakes, and climate change have shifted wildlife habitats, wiped out large numbers of species, and created opportunities for the evolution of new species.

Geological Processes Affect Natural Selection

The earth's surface has changed dramatically over its long history. Scientists have discovered that huge flows of molten rock within the earth's interior have broken its surface into a series of gigantic solid plates, called *tectonic plates*. For hundreds of millions of years, these plates have drifted slowly on the planet's mantle (Figure 4-7).

This process has played a role in the extinction of species, as continental areas split apart, and also in the rise of new species when isolated island areas such as the Hawaiian Islands and the Galapagos Islands were created. Rock and fossil evidence indicates that 200–250 million years ago, all of the earth's present-day continents were connected in a supercontinent called Pangaea (Figure 4-7, left). About 135 million years ago, Pangaea began splitting apart as the earth's tectonic plates moved, eventually resulting in the present-day locations of the continents (Figure 4-7, right).

This fact, that tectonic plates drift, has had two important effects on the evolution and distribution of life on the earth. *First*, the locations (latitudes) of continents and oceanic basins have greatly influenced the earth's climate and thus have helped to determine where plants and animals can live. *Second*, the move-

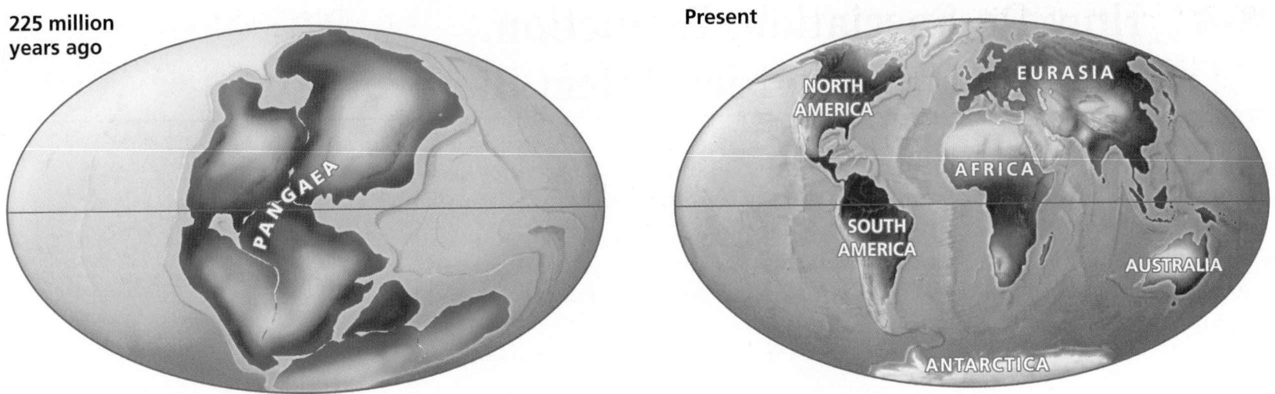

Figure 4-7 Over millions of years, the earth's continents have moved very slowly on several gigantic tectonic plates. **Question:** How might an area of land splitting apart cause the extinction of a species?

ment of continents has allowed species to move, adapt to new environments, and form new species through natural selection.

The shifting of tectonic plates can cause *earthquakes,* which can also affect biological evolution by causing fissures in the earth's crust that can separate and isolate populations of species. Over long periods of time, this can lead to the formation of new species as each isolated population changes genetically in response to new environmental conditions. *Volcanic eruptions* that occur along the boundaries of tectonic plates can also affect biological evolution by destroying habitats and reducing, isolating, or wiping out populations of species (**Concept 4-3**).

Climate Change and Catastrophes Affect Natural Selection

Throughout its long history, the earth's climate has changed drastically. Sometimes it has cooled and covered much of the earth with glacial ice. At other times it has warmed, melted that ice, and drastically raised sea levels, which in turn increased the total area covered by the oceans and decreased the earth's total land area. Such alternating periods of cooling and heating have led to the advance and retreat of ice sheets at high latitudes over much of the northern hemisphere, most recently about 18,000 years ago (Figure 4-8).

These long-term climate changes have a major effect on biological evolution by determining where different types of plants and animals can survive and thrive, and by changing the locations of different types of ecosystems such as deserts, grasslands, and forests (**Concept 4-3**). Some species became extinct because the climate changed too rapidly for them to adapt and survive, and new species evolved to take over their ecological roles.

Another force affecting natural selection has been catastrophic events such as collisions between the earth and large asteroids. There have probably been many of these collisions during the 3.5 billion years of life on earth. Such impacts have caused widespread destruction of ecosystems and wiped out large numbers of species. On the other hand, they have also caused shifts in the locations of ecosystems and created opportunities for the evolution of new species.

On a long-term basis, the three **principles of sustainability** (see back cover), especially the biodiversity principle (Figure 4-2), have enabled life on earth to adapt to drastic changes in environmental conditions (see *The Habitable Planet,* Video 1, at **www.learner.org/resources/series209.html**).

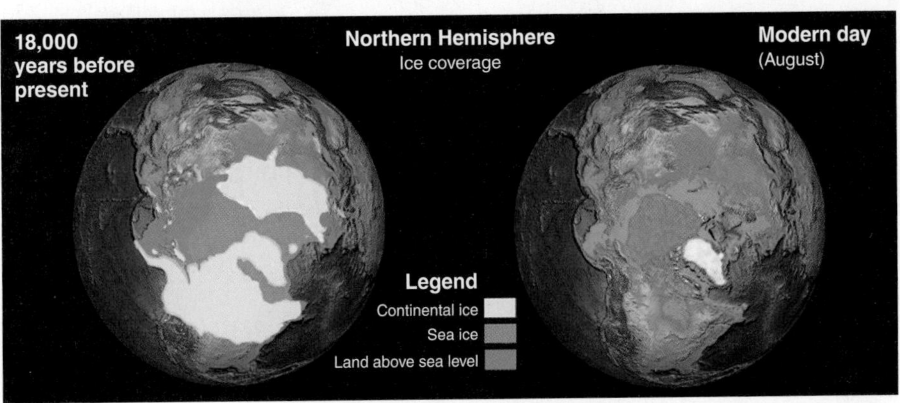

Figure 4-8 These maps of the northern hemisphere show the large-scale changes in glacial ice coverage during the past 18,000 years. Other smaller changes in glacial ice on mountain ranges such as the European Alps are not shown. **Question:** What are two characteristics of an animal and two characteristics of a plant that natural selection would have favored as these ice sheets (left) advanced? (Data from the National Oceanic and Atmospheric Administration)

4-4 How Do Speciation, Extinction, and Human Activities Affect Biodiversity?

▶ **CONCEPT 4-4A As environmental conditions change, the balance between the formation of new species and the extinction of existing species determines the earth's biodiversity.**

▶ **CONCEPT 4-4B Human activities are decreasing biodiversity by causing the extinction of many species and by destroying or degrading habitats needed for the development of new species.**

How Do New Species Evolve?

Under certain circumstances, natural selection can lead to an entirely new species. In this process, called **speciation**, one species splits into two or more different species. For sexually reproducing organisms, a new species forms when one population of a species has evolved to the point where its members can no longer breed and produce fertile offspring with members of another population that did not change or that evolved differently.

The most common way in which speciation occurs, especially among sexually reproducing species, is when a barrier or distant migration separates two or more populations of a species and prevents the flow of genes between them. This happens in two phases: first, geographic isolation and then reproductive isolation.

Geographic isolation occurs when different groups of the same population of a species become physically isolated from one another for a long period of time. For example, part of a population may migrate in search of food and then begin living as a separate population in another area with different environmental conditions. Populations can also be separated by a physical barrier (such as a mountain range, stream, or road), a volcanic eruption, tectonic plate movements, or winds or flowing water that carry a few individuals to a distant area.

In **reproductive isolation**, mutation and change by natural selection operate independently in the gene pools of geographically isolated populations. If this process continues long enough, members of the geographically and reproductively isolated populations of sexually reproducing species may become so different in genetic makeup that they cannot produce live, fertile offspring if they are rejoined and attempt to interbreed. Then, one species has become two, and speciation has occurred (Figure 4-9).

Humans are playing an increasing role in the process of speciation. We have learned to shuffle genes from one species to another through artificial selection and, more recently, through genetic engineering (see Science Focus, at right).

Sooner or Later All Species Become Extinct

Another process affecting the number and types of species on the earth is **extinction**, a process in which an entire species ceases to exist (*biological extinction*) or

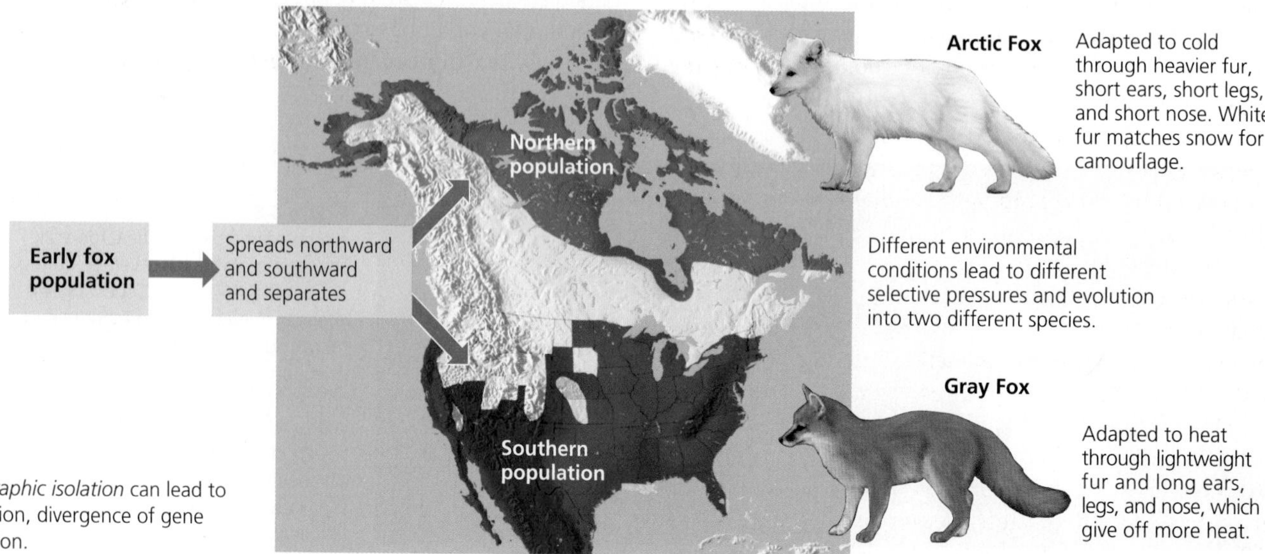

Figure 4-9 *Geographic isolation* can lead to reproductive isolation, divergence of gene pools, and speciation.

SCIENCE FOCUS

Changing the Genetic Traits of Populations

We have used **artificial selection** to change the genetic characteristics of populations with similar genes. In this process, we select one or more desirable genetic traits in the population of a plant or animal such as a type of wheat, fruit, or dog. Then we use *selective breeding*, or *crossbreeding*, to generate populations of the species containing large numbers of individuals with the desired traits.

Note that artificial selection involves crossbreeding between genetic varieties of the same species or between species that are genetically close to one another, and thus it is not a form of speciation. Most of the grains, fruits, and vegetables we eat are produced by artificial selection. Artificial selection has given us food crops with higher yields, cows that give more milk, trees that grow faster, and many different types of dogs and cats. But traditional crossbreeding is a slow process and it can be used only on species that are close to one another genetically.

Now scientists are using **genetic engineering** to speed up our ability to manipulate genes. In this process, scientists alter an organism's genetic material by adding, deleting, or changing segments of its DNA to produce desirable traits or to eliminate undesirable ones. It enables scientists to transfer genes between different species that would not interbreed in nature. For example, we can put genes from a cold-water fish species into a tomato plant to give it properties such as resistance to cold.

Scientists have used genetic engineering to develop modified crop plants, new drugs, pest-resistant plants, and animals that grow rapidly (Figure 4-C). They have also created genetically engineered bacteria to extract minerals such as copper from their underground ores and to clean up spills of oil and other toxic pollutants.

Critical Thinking

What might be some beneficial and harmful effects on the evolutionary process if genetic engineering is widely applied to plants and animals?

R. L. Brinster and R. E. Hammer/School of Veterinary Medicine, University of Pennsylvania

Figure 4-C These mice are an example of genetic engineering. The 6-month-old mouse on the left is normal; the same-age mouse on the right had a human growth hormone gene inserted into its cells. Mice with this gene grow two to three times faster than, and twice as large as, mice without it. **Question:** How do you think the creation of such species might change the process of evolution by natural selection?

a population of a species becomes extinct over a large region, but not globally (*local extinction*). When environmental conditions change, a population of a species faces three possible futures: *adapt* to the new conditions through natural selection, *move* (if possible) to an area with more favorable conditions, or *become extinct*.

Species that are found in only one area are called **endemic species** and are especially vulnerable to extinction. They exist on islands and in other unique areas, especially in tropical rain forests where most species have highly specialized roles. For these reasons, they are unlikely to be able to migrate or adapt in the face of rapidly changing environmental conditions.

Fossils and other scientific evidence indicate that all species eventually become extinct, but drastic changes in environmental conditions can eliminate large groups of species. Throughout most of the earth's long history, species have disappeared at a low rate, called **background extinction**. Based on the fossil record and analysis of ice cores, biologists estimate that the average annual background extinction rate has been about 0.0001% of all species per year, which amounts to 1 species lost for every million species on the earth. At this rate, if there were 10 million species on the earth, about 10 them, on average, would go extinct every year. And if there were 100 million species, then an average of 100 of them would disappear naturally every year.

There Have Been Several Mass Extinctions of Life on the Earth

In contrast to ongoing background extinctions, a **mass extinction** is a significant rise in extinction rates above the background level. In such a catastrophic, widespread, and often global event, large groups of species (25–95% of all species) are wiped out. Fossil and geological evidence indicate that there have probably been five mass extinctions (20–60 million years apart) during the past 500 million years.

A mass extinction provides an opportunity for the evolution of new species that can fill unoccupied ecological roles or newly created ones. As a result, evidence indicates that each occurrence of mass extinction has been followed by an increase in species diversity over several million years as new species have arisen to occupy new habitats or to exploit newly available resources. As environmental conditions change, the balance between formation of new species (speciation) and extinction of existing species determines the earth's biodiversity (**Concept 4-4A**). The existence of millions of species today means that speciation, on average, has kept ahead of extinction.

There is growing evidence that we are experiencing the beginning of a new mass extinction. There is also considerable evidence that much of the current increase in extinctions and the resulting loss of biodiversity are primarily due to human activities (**Concept 4-4B**), as our ecological footprints spread over the planet (see Figure 1-8 and Figure 2, p. S24, in Supplement 6).

Research indicates that the largest cause of the growing rate of species extinctions and ecosystem disruptions is the loss, fragmentation, and degradation of terrestrial and aquatic habitats. These losses occur as we cultivate more land to grow crops, clear more forest land for farming, ranching, and settlement (Figure 3-1, p. 40), and harvest many ocean and freshwater species of fish faster than they can replenish their populations. We further examine these and other causes of the growing rate of extinction in Chapters 8 and 9.

4-5 What Roles Do Species Play in Ecosystems?

▶ **CONCEPT 4-5A Each species plays a specific ecological role called its niche.**

▶ **CONCEPT 4-5B Any given species may play one or more of four important roles—native, nonnative, indicator, or keystone—in a particular ecosystem.**

Each Species Plays a Role in Its Ecosystem

An important principle of ecology is that *each species has a specific role to play in the ecosystems where it is found* (**Concept 4-5A**). Scientists describe the role that a species plays in its ecosystem as its **ecological niche**, or simply **niche** (often pronounced "nitch"). It is a species' way of life in a community and includes everything that affects its survival and reproduction, such as how much water and sunlight it needs, how much space it requires, what it feeds on, what feeds on it, and the temperatures and other conditions it can tolerate. A species' niche should not be confused with its **habitat**, which is the place where it lives. Its niche is its pattern of living.

Scientists use the niches of species to classify them mostly as *generalists* or *specialists*. **Generalist species** have broad niches. They can live in many different places, eat a variety of foods, and often tolerate a wide range of environmental conditions. Flies, cockroaches, mice, rats, white-tailed deer, and humans are generalist species.

In contrast, **specialist species** occupy narrow niches. They may be able to live in only one type of habitat, use just one or only a few types of food, or tolerate a narrow range of climatic and other environmental conditions. For example, some shorebirds occupy specialized niches, feeding on crustaceans, insects, and other organisms found on sandy beaches and their adjoining coastal wetlands (Figure 4-10).

Because of their narrow niches, specialists are more prone to extinction when environmental conditions change. For example, China's *giant panda* is highly endangered because of a combination of habitat loss, low birth rate, and its specialized diet consisting mostly of bamboo.

Is it better to be a generalist or a specialist? It depends. When environmental conditions are fairly constant, as in a tropical rain forest, specialists have an advantage because they have fewer competitors. But

Figure 4-10 This diagram illustrates the specialized feeding niches of various bird species in a coastal wetland. This specialization reduces competition and allows sharing of limited resources.

under rapidly changing environmental conditions, the more adaptable generalist usually is better off than the specialist.

> **THINKING ABOUT**
> **A Shark's Niche**
> CORE CASE STUDY
> Do you think that most shark species (**Core Case Study**) occupy specialist or generalist niches? Explain.

Species Can Play Four Major Roles within Ecosystems

Niches can be classified further in terms of specific roles that certain species play within ecosystems. Ecologists describe *native, nonnative, indicator,* and *keystone* roles. Any given species may play one or more of these four roles in a particular ecosystem (**Concept 4-5B**).

Native species are those species that normally live and thrive in a particular ecosystem. Other species that migrate into, or are deliberately or accidentally introduced into, an ecosystem are called **nonnative species**, also referred to as *invasive, alien,* and *exotic species*.

Some people tend to think of nonnative species as threatening. In fact, most introduced and domesticated plant species such as food crops and flowers and animals such as chickens, cattle, and fish from around the world are beneficial to us. However, some nonnative species can compete with and reduce a community's native species, causing unintended and unexpected consequences. In 1957, for example, Brazil imported wild African honeybees to help increase honey production. Instead, the bees displaced domestic honeybees and reduced the honey supply.

Since then, these nonnative honeybee species—popularly known as "killer bees"—have moved northward into Central America and parts of the southwestern and southeastern United States. These wild African bees are are aggressive and unpredictable and have killed thousands of domesticated animals and an estimated 1,000 people in the western hemisphere, many of whom were allergic to bee stings.

Nonnative species can spread rapidly if they find a new location with favorable conditions. In their new niches, these species often do not face the predators and diseases they face in their native niches, or they may be able to out-compete some native species in their new locations. We will examine this environmental threat in greater detail in Chapter 8.

Indicator Species Serve as Biological Smoke Alarms

Species that provide early warnings of damage to a community or an ecosystem are called **indicator species**. Birds are excellent biological indicators because they are found almost everywhere and are affected quickly by environmental changes such as the loss or fragmentation of their habitats and the introduction of chemical pesticides. The populations of many bird species are declining. Some amphibians are also classified as indicator species (see the following Case Study).

■ CASE STUDY
Why Are Amphibians Vanishing?

Amphibians (frogs, toads, and salamanders) live part of their lives in water and part on land. Populations of some amphibians, also believed to be indicator species, are declining throughout the world.

Amphibians were the first vertebrates (animals with backbones) to set foot on the earth. Historically, they have also been better at adapting to environmental changes through evolution than many other species have been. Many amphibian species (Figure 4-11) are having difficulty adapting to some of the rapid environmental changes that have taken place in the air and water and on the land during the past few decades—changes resulting mostly from human activities.

Since 1980, populations of hundreds of the world's almost 6,000 amphibian species have been vanishing or declining in almost every part of the world, even in protected wildlife reserves and parks. According to a 2008 assessment by the International Union for the Conservation of Nature and Natural Resources (IUCN), about 32% of all known amphibian species (and more than 80% of those in the Caribbean) are threatened with extinction, and populations of another 43% of the species are declining.

Frogs are especially sensitive and vulnerable to environmental disruption at various points in their life

Abhindia/Shutterstock.com

Figure 4-11 Some species of the poison dart frog, found in the tropical forests of Brazil and southern Suriname, are threatened with extinction, mostly because of an infectious fungus and habitat loss due to logging and farming. This frog's bright blue color warns predators that it is poisonous to eat. The toxic secretions of at least three of these species are used to poison the tips of blow darts that native peoples in the frog's tropical habitat use for hunting.

cycle. As juveniles (tadpoles), frogs live in water and eat plants; as adults, they live mostly on land and eat insects that can expose them to pesticides. The eggs of frogs have no protective shells to block harmful UV radiation or pollution. As adults, they take in water and air through their thin, permeable skins, which can readily absorb pollutants from water, air, or soil. And they have no hair, feathers, or scales to protect them.

No single cause has been identified to explain these amphibian declines. However, scientists have identified a number of factors that can affect frogs and other amphibians at various points in their life cycles:

- *Habitat loss and fragmentation*, especially from the draining and filling of freshwater wetlands, deforestation, farming, and urban development.
- *Prolonged drought*, which can dry up breeding pools so that few tadpoles survive.
- *Increases in UV radiation* resulting from reductions in stratospheric ozone during the past few decades, which were caused by chemicals we put into the lower atmosphere that ended up in the stratosphere. Higher doses of UV radiation can harm embryos of amphibians in shallow ponds and adults basking in the sun for warmth.
- *Parasites* such as flatworms, which feed on the amphibian eggs laid in water, apparently have caused an increase in the number of births of amphibians with missing or extra limbs.
- *Pollution*, especially exposure to pesticides in ponds and in the bodies of insects consumed by frogs, which can make them more vulnerable to bacterial, viral, and fungal diseases and to some parasites.
- *Viral and fungal diseases*, especially the chytrid fungus that attacks the skin of frogs, apparently reducing their ability to take in water, thus leading to death from dehydration. Such diseases can spread because adults of many amphibian species congregate in large numbers to breed.
- *Climate change.* A 2005 study found an apparent correlation between climate change caused by atmospheric warming and the extinction of about two-thirds of the 110 known species of harlequin frog in tropical forests in Central and South America.
- *Overhunting*, especially in Asia and France, where frog legs are a delicacy.
- *Natural immigration of, or deliberate introduction of, nonnative predators and competitors* (such as certain fish species).

A combination of such factors, which vary from place to place, probably is responsible for most of the decline and disappearances among amphibian species.

Why should we care if some amphibian species become extinct? Scientists give three reasons. *First,* amphibians are sensitive biological indicators of changes in environmental conditions such as habitat loss and degradation, air and water pollution, UV radiation, and climate change. The growing number threats to their survival indicate that environmental health is deteriorating in many parts of the world.

Second, adult amphibians play important ecological roles in biological communities. For example, amphibians eat more insects (including mosquitoes) than do birds. In some habitats, the extinction of certain amphibian species could lead to extinction of other species, such as reptiles, birds, aquatic insects, fish, mammals, and other amphibians that feed on them or their larvae.

Third, amphibians are a genetic storehouse of pharmaceutical products waiting to be discovered. For example, compounds in secretions from amphibian skin have been isolated and used as painkillers and antibiotics, and in treatments for burns and heart disease.

Many scientists believe that the rapidly increasing chances for global extinction of a variety of amphibian species is a warning about the harmful effects of an array of environmental threats to biodiversity, mostly resulting from human activities.

Keystone Species Play Critical Roles in Their Ecosystems

Keystone species are species whose roles have a large effect on the types and abundance of other species in an ecosystem. Such species often exist in relatively limited numbers in their ecosystems, but the effects that they have there are often much larger than their numbers would suggest. And because of their often-smaller numbers, some keystone species are more vulnerable to extinction than other species are. Examples are the wolf, leopard, lion, some shark species (**Core Case Study**), and the American alligator (see the following Case Study). CORE CASE STUDY

The loss of a keystone species can lead to population crashes and extinctions of other species in a community that depends on them for certain ecological services. This is why it so important for scientists to identify and protect keystone species.

THINKING ABOUT
Sharks and Biodiversity

Why are some shark species (**Core Case Study**) considered to be keystone species?

■ CASE STUDY
The American Alligator—A Keystone Species That Almost Went Extinct

The American alligator (Figure 4-12) has no natural predators except for humans. It is a keystone species because it plays a number of important roles in

Figure 4-12 *Keystone species:* The American alligator plays an important ecological role in its marsh and swamp habitats in the southeastern United States.

the ecosystems where it is found. This species has outlived the dinosaurs and survived many challenges to its existence.

However, starting in the 1930s, hunters in the United States began killing large numbers of these animals for their exotic meat and their soft belly skin, used to make expensive shoes, belts, and pocketbooks. Other people hunted alligators for sport or out of dislike for the large reptile. By the 1960s, hunters and poachers had wiped out 90% of the alligators in the U.S. state of Louisiana, and the alligator population in the Florida Everglades was also near extinction.

Those who did not care much for the alligator were probably not aware of its important ecological role—its keystone niche—in subtropical wetland ecosystems. Alligators dig deep depressions, or gator holes, using their jaws and claws. These depressions hold freshwater during dry spells, serve as refuges for aquatic life, and supply freshwater and food for fishes, insects, snakes, turtles, birds, and other animals.

Large alligator nesting mounds also provide nesting and feeding sites for some herons and egrets, and red-bellied turtles use old gator nests for incubating their eggs. In addition, alligators eat large numbers of gar, a predatory fish, which helps to maintain populations of game fish such as bass and bream that the gar eat.

As alligators create gator holes and nesting mounds, they help to keep shore and open water areas free of invading vegetation. Without this free ecosystem service, freshwater ponds and coastal wetlands where alligators live would be filled in with shrubs and trees, and dozens of species would disappear from these ecosystems.

For these reasons, most ecologists classify the American alligator as a keystone species because of its important ecological role in helping to maintain the sustainability of the ecosystems in which it is found.

In 1967, the U.S. government placed the American alligator on the endangered species list. By 1977, because it was protected, its populations had made a strong enough comeback in its key habitats to be removed from the endangered species list.

Today, there are well over a million alligators in Florida, and the state now allows property owners to kill alligators that stray onto their land. To conservation biologists, the comeback of the American alligator is an important success story in wildlife conservation. Go on a virtual video field trip to learn more about the role of the American alligator as a keystone species in its habitat at **www.cengagebrain.com**.

THINKING ABOUT
The American Alligator and Biodiversity

What are two ways in which the American alligator supports one or more of the four components of biodiversity (Figure 4-2) within its environment?

CONNECTIONS
Sharks and the Fishing Business

CORE CASE STUDY

Various shark species (**Core Case Study**) act as keystone species by helping to keep the populations of some species in check, while supporting the populations of other species. In 2007, scientists reported that the decline in certain shark populations along the U.S. Atlantic coast led to an explosion in populations of rays and skates, which sharks normally feed on. Now the rays and skates are feasting on bay scallops, resulting in a sharp decline in their population and in the area's bay scallop fishing business.

Here are this chapter's *three big ideas*.

- Populations evolve when genes mutate and give some individuals genetic traits that enhance their abilities to survive and to produce offspring with these traits (natural selection).
- Human activities are degrading the earth's vital biodiversity by causing the extinction of species and by disrupting habitats needed for the development of new species.
- Each species plays a specific ecological role (ecological niche) in the ecosystem where it is found.

REVISITING Sharks and Sustainability

The **Core Case Study** on sharks at the beginning of this chapter shows that the importance of a species does not always match the public's perception of it. While many people worry about the fates of panda bears and polar bears (both facing threats of extinction), most people fear sharks and do not view them as cute and cuddly or worthy of protection. Yet they are no less important to their ecosystems. If certain species of sharks disappeared, ocean ecosystems would be severely damaged and some might even collapse. This could greatly disrupt the earth's climates and food supplies for billions of people. And like the polar bears, almost a third of the world's sharks face serious threats of extinction due to human activities.

In this chapter, we studied the importance of biodiversity, especially the numbers and varieties of species found in different parts of the world, along with the other forms of biodiversity—genetic, ecosystem, and functional diversity. We also studied the process whereby all species came to be, according to the scientific theory of biological evolution through natural selection.

Taken together, biodiversity and evolution represent two vital and irreplaceable forms of natural capital. Each depends upon the other and upon whether humans can respect and help to preserve this natural capital.

Finally, we examined the variety of roles played by species in ecosystems. For example, we saw that some shark species and other species are keystone species on which their ecosystems depend for survival and health. Human activities that threaten the survival of keystone species can threaten the survival of other species that depend on them and can threaten entire ecosystems.

Ecosystems and the variety of species they contain are functioning examples of the three **principles of sustainability** (see back cover) in action. They depend on solar energy and on the chemical cycling of nutrients. Ecosystems also help to sustain biodiversity in all its forms. In the next chapter, we delve further into how species interact and how their interactions result in the natural regulation of populations and thus help to maintain biodiversity.

All we have yet discovered is but a trifle in comparison with what lies hid in the great treasury of nature.

ANTOINE VAN LEEUWENHOCK

REVIEW

CORE CASE STUDY

1. Describe the threats to many of the world's shark species (**Core Case Study**) and explain why we should avoid hastening the extinction of any shark species through our activities. CORE CASE STUDY

SECTION 4-1

2. What is the key concept for Section 4-1? List and define the four major components of **biodiversity (biological diversity)**. What is the importance of biodiversity? Define **species**. Summarize the importance of insects. Define and give three examples of **biomes**. Summarize the scientific contributions of Edward O. Wilson.

SECTION 4-2

3. What are the two key concepts for Section 4-2? What is a **fossil** and why are fossils important for understanding the history of life? What is **biological evolution**? State the **theory of evolution by natural selection**. What is a **mutation** and what role do mutations play in evolution by natural selection? What is an **adaptation (adaptive trait)**? How did humans become such a powerful species? What are two limits to evolution by natural selection? What are three myths about evolution through natural selection?

SECTION 4-3

4. What is the key concept for Section 4-3? Describe how geologic processes can affect natural selection. How can climate change and catastrophes such as asteroid impacts affect natural selection?

SECTION 4-4

5. What are the two key concepts for Section 4-4? What is **speciation**? Distinguish between **geographic isolation** and **reproductive isolation**, and explain how they can lead to the formation of a new species. Distinguish between **artificial selection** and **genetic engineering** and give an example of each.

6. What is **extinction**? What is an **endemic species** and why can such a species be vulnerable to extinction? Distinguish between **background extinction** and **mass extinction**.

SECTION 4-5

7. What are the two key concepts for Section 4-5? What is an **ecological niche**? Distinguish between **specialist species** and **generalist species** and give an example of each.

8. Define and distinguish among **native**, **nonnative**, and **indicator species** and give an example of each type. What major ecological roles do amphibian species play? List nine factors that contribute to the threat of extinction for frogs and other amphibians. What are three reasons for protecting amphibians?

9. Define and give an example of a **keystone species**. Describe the role of some sharks as a keystone species. Describe the role of the American alligator as a keystone species and how it was brought back from near extinction.

10. What are this chapter's *three big ideas*? How are ecosystems and the variety of species they contain related to the three principles of sustainability?

Note: Key terms are in **bold** type.

CRITICAL THINKING

1. How might we and other species be affected if all of the world's sharks (Core Case Study) were to go extinct?

2. What role does each of the following processes play in helping to implement the three principles of sustainability: **(a)** natural selection, **(b)** speciation, and **(c)** extinction?

3. How would you respond to someone who tells you that:
   a. he or she does not believe in biological evolution because it is "just a theory"?
   b. we should not worry about air pollution because natural selection will enable humans to develop lungs that can detoxify pollutants?

4. **(a)** Explain how you would experimentally determine whether or not an organism is an indicator species. **(b)** Do the same exercise for a keystone species.

5. Is the human species a keystone species? Explain. If humans were to become extinct, what are three species that might also become extinct and three species whose populations would probably grow?

6. How would you respond to someone who says that because extinction is a natural process, we should not worry about the loss of biodiversity when species become extinct largely as a result of our activities?

7. List three ways in which you could apply **Concept 4-4B** to making your lifestyle more environmentally sustainable.

8. Congratulations! You are in charge of the future evolution of life on the earth. What are the three things that you would consider to be the most important to do?

DOING ENVIRONMENTAL SCIENCE

Design and carry out an experiment, as discussed in Question 4 above, in which you would try to determine whether an organism is **(a)** a keystone species and **(b)** an indicator species. Take careful notes on what you observe. State your hypothesis and explain why your observations do or do not support it.

GLOBAL ENVIRONMENT WATCH EXERCISE

Search for *Shark Finning* to find out what actions are being taken by various nations to reduce the killing of sharks, particularly the practice of finning (Core Case Study). Write a short summary report on your research.

DATA ANALYSIS

This graph shows data collected by scientists who were investigating biodiversity on islands as it relates to island size and distance from the nearest mainland. One of the hypotheses developed by these scientists was that larger islands tend to have higher species densities.

The graph shows measurements taken for two variables: the area on an island within which measurements were taken, and the population density of a certain species of lizard found in each area. Note that the areas studied varied from 100 square meters (m^2) (1,076 square feet) to 300 m^2 (3,228 square feet), as shown on the x-axis. Population densities measured by the scientists varied from 0.1 to 0.4 individuals per m^2 (y-axis). Study the data and answer the following questions.

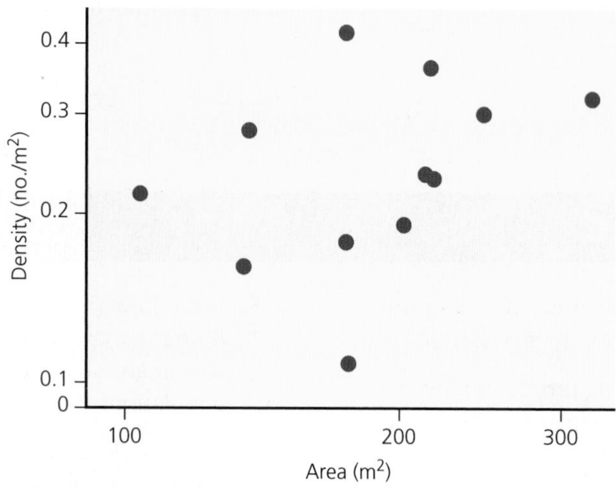

1. How many measurements below 0.2 individuals per m^2 were made? How many of these measurements were made in areas smaller than 200 m^2? How many were made in areas larger than 200 m^2?
2. How many measurements above 0.2 individuals per m^2 were made? How many of these measurements were made in areas smaller than 200 m^2? How many were made in areas larger than 200 m^2?
3. Do your answers support the hypothesis that larger areas tend to have higher species densities? Explain.

LEARNING ONLINE

The website for *Environmental Science 14e* contains helpful study tools and other resources to take the learning experience beyond the textbook. Examples include flashcards, practice quizzing, information on green careers, tips for more sustainable living, activities, and a wide range of articles and further readings. To access these materials and other companion resources for this text, please visit **www.cengagebrain.com**. For further details, see the preface, p. xvi.

Biodiversity, Species Interactions, and Population Control

5

The Southern Sea Otter: A Species in Recovery

CORE CASE STUDY

Southern sea otters (Figure 5-1, left) live in giant kelp forests (Figure 5-1, right) in shallow waters along part of the Pacific coast of North America. Most remaining members of this endangered species are found off the western coast of the United States between the cities of Santa Cruz and Santa Barbara, California. Scientific studies indicate that the otters act as a keystone species in a kelp forest system by helping to control the populations of sea urchins and other kelp-eating species.

Southern sea otters are fast and agile swimmers that dive to the ocean bottom looking for shellfish and other prey. These marine mammals use stones as tools to pry shellfish off rocks under water. When they return to the surface they swim on their backs, and using their bellies as a table, break open the shells by cracking them against a stone (Figure 5-1, left). Each day, a sea otter consumes about a fourth of its weight in clams, mussels, crabs, sea urchins, abalone, and about 40 other species of bottom-dwelling organisms.

It is estimated that between 16,000 and 17,000 southern sea otters once lived in the waters of their habitat areas along the Pacific coast of North America. By the early 1900s, however, the otters were hunted almost to extinction in this region by fur traders who killed them for their thick, luxurious fur. Commercial fisherman also killed otters because they were competing with them for valuable abalone and other shellfish.

The population of southern sea otters off the California segment of the coast has generally increased over time from a low of about 50 in 1938. In 2010, it was estimated to be 2,711. This otter population's partial recovery had gotten a boost in 1977 when the U.S. Fish and Wildlife Service declared the species endangered in most of its range, with a population of only 1,850 individuals. Despite some recovery, this species has a long way to go before its population will be large enough to justify removing it from the endangered species list.

Why should we care about the southern sea otters of California? One reason is *ethical*: many people believe it is wrong to allow human activities to cause the extinction of a species. Another reason is that people love to look at these appealing and highly intelligent animals as they play in the water. As a result, they help to generate millions of dollars a year in tourism revenues in coastal areas where they are found.

A third reason to care about otters—and a key reason in our study of environmental science—is that biologists classify them as a *keystone species* (p. 74). Without southern sea otters, scientists hypothesize that sea urchins and other kelp-eating species would probably destroy the kelp forests and much of the rich biodiversity of species they support.

Biodiversity is an important part of the earth's natural capital and is the focus of one of the three **principles of sustainability** (see back cover). In this chapter, we will look at two factors that affect biodiversity: how species interact and help control one another's population sizes and how biological communities and populations respond to changes in environmental conditions.

SUSTAINABILITY

Tom and Pat Leeson, Ardea London Ltd

Paul Whitted/Shutterstock.com

Figure 5-1 An endangered southern sea otter in Monterey Bay, California (USA) uses a stone to crack the shell of a clam (left). It lives in a giant kelp bed (right).

Key Questions and Concepts

5-1 How do species interact?

CONCEPT 5-1 Five types of interactions among species—interspecific competition, predation, parasitism, mutualism, and commensalism—affect the resource use and population sizes of the species in an ecosystem.

5-2 What limits the growth of populations?

CONCEPT 5-2 No population can continue to grow indefinitely because of limitations on resources and because of competition among species for those resources.

5-3 How do communities and ecosystems respond to changing environmental conditions?

CONCEPT 5-3 The structure and species composition of communities and ecosystems change in response to changing environmental conditions through a process called ecological succession.

Note: Supplements 2 (p. S3), 4 (p. S10), and 6 (p. S22) can be used with this chapter.

In looking at nature, never forget that every single organic being around us may be said to be striving to increase its numbers.

CHARLES DARWIN

5-1 How Do Species Interact?

▶ **CONCEPT 5-1 Five types of interactions among species—interspecific competition, predation, parasitism, mutualism, and commensalism—affect the resource use and population sizes of the species in an ecosystem.**

Most Species Compete with One Another for Certain Resources

Ecologists have identified five basic types of interactions between species as they share limited resources such as food, shelter, and space. These types of interactions are called *interspecific competition, predation, parasitism, mutualism,* and *commensalism,* and they all have significant effects on the resource use and population sizes of the species in an ecosystem (**Concept 5-1**).

The most common interaction between species is **interspecific competition**, which occurs when members of two or more species interact to use the same limited resources such as food, water, light, and space. While fighting for resources does occur, most interspecific competition involves the ability of one species to become more efficient than another species in getting food, space, light, or other resources.

Recall that each species plays a role in its ecosystem called its *ecological niche* (see Chapter 4, p. 72). When two species compete with one another for the same resources, their niches *overlap*. The greater this overlap, the more intense their competition for key resources. If one species can take over the largest share of one or more key resources, each of the other competing species must move to another area (if possible), adapt by shifting its feeding habits or behavior through natural selection to reduce or alter its niche, suffer a sharp population decline, or become extinct in that area.

Humans compete with many other species for space, food, and other resources. As our ecological footprints grow and spread (see Figure 1-8, p. 14), and we convert more of the earth's land and aquatic resources to our uses, we are taking over the habitats of many other species and depriving them of resources that they need in order to survive.

Over a time scale long enough for natural selection to occur, populations of some species develop adaptations that allow them to reduce or avoid competition with other species for resources. One way this happens is through **resource partitioning**. It occurs when species competing for similar scarce resources evolve specialized traits that allow them to share resources by using parts of them, using them at different times, or using them in different ways.

Figure 5-2 shows resource partitioning by some insect-eating bird species. In this case, their adaptations allow them to reduce competition by feeding in different portions of certain spruce trees and by feeding on different insect species. Figure 4-10 (p. 72) shows how

Links: CORE CASE STUDY refers to the Core Case Study. SUSTAINABILITY refers to the book's sustainability theme. GOOD NEWS refers to good news about the environmental challenges we face.

Figure 5-2 *Sharing the wealth:* This diagram illustrates resource partitioning among five species of insect-eating warblers in the spruce forests of the U.S. state of Maine. Each species spends at least half its feeding time in its associated yellow-highlighted areas of these spruce trees. (After R. H. MacArthur, "Population Ecology of Some Warblers in Northeastern Coniferous Forests," *Ecology* 36 (1958): 533–536.)

the evolution of specialized feeding niches has reduced competition for resources among bird species in a coastal wetland.

Most Consumer Species Feed on Live Organisms of Other Species

In **predation**, a member of one species (the **predator**) feeds directly on all or part of a living organism (the **prey**) as part of a food web. Together, the two different species, such as a brown bear (the predator, or hunter) and a salmon (the prey, or hunted; see Photo 4 in the Detailed Contents), form a **predator–prey relationship**. Such relationships are also shown in Figures 3-10 (p. 47) and 3-11 (p. 48).

In giant kelp forest ecosystems, sea urchins prey on kelp, a form of seaweed (Science Focus, p. 82). However, as a keystone species, southern sea otters (**Core Case Study**, Figure 5-1, left) prey on the sea urchins and thus keep them from destroying the kelp forests. CORE CASE STUDY

Predators have a variety of methods that help them to capture prey. *Herbivores* can simply walk, swim, or fly up to the plants they feed on. *Carnivores* feeding on mobile prey have two main options: *pursuit* and *ambush*. Some, such as the cheetah, catch prey by running fast; others, such as the American bald eagle, can fly and have keen eyesight; still others cooperate in capturing their prey. For example, African lions capture giraffes by hunting in packs.

Other predators use *camouflage* to hide in plain sight and ambush their prey. For example, praying mantises (see Figure 4-A, right, p. 64) sit on flowers or plants of a color similar to their own and ambush visiting insects. White ermines (a type of weasel) and snowy owls hunt their prey in snow-covered areas. People camouflage themselves to hunt wild game and use camouflaged traps to capture wild animals.

Some predators use *chemical warfare* to attack their prey. For example, some spiders and poisonous snakes use venom to paralyze their prey and to deter their predators.

Prey species have evolved many ways to avoid predators, including abilities to run, swim, or fly fast, and some have highly developed senses of sight or smell that alert them to the presence of predators. Other avoidance adaptations include protective shells (as on armadillos and turtles), thick bark (on giant sequoias), spines (on porcupines), and thorns (on cacti and rose bushes). Many lizards have brightly colored tails that break off when they are attacked, often giving them enough time to escape.

Other prey species use the camouflage of certain shapes or colors. Chameleons and cuttlefish have the ability to change color. Some insect species have shapes that look like twigs (Figure 5-3a, p. 82), bark, thorns, or even bird droppings on leaves. A leaf insect can be almost invisible against its background (Figure 5-3b), as can an arctic hare in its white winter fur.

Chemical warfare is another common strategy. Some prey species discourage predators with chemicals that are *poisonous* (oleander plants), *irritating* (stinging nettles and bombardier beetles, Figure 5-3c, p. 82), *foul smelling* (skunks, skunk cabbages, and stinkbugs), or *bad tasting* (buttercups and monarch butterflies,

SCIENCE FOCUS

Threats to Kelp Forests

A kelp forest is composed of large concentrations of a seaweed called *giant kelp* whose long blades grow straight to the surface (Figure 5-1, right). Under good conditions, the blades can grow 0.6 meter (2 feet) a day. A gas-filled bladder at its base holds up each blade. The blades are very flexible and can survive all but the most violent storms and waves.

Kelp forests are one of the most biologically diverse ecosystems found in marine waters, supporting large numbers of marine plants and animals. These forests help reduce shore erosion by blunting the force of incoming waves and trapping some of the outgoing sand. People harvest kelp as a renewable resource, extracting a substance called algin from its blades. This substance is used as a thickening or stabilizing agent in toothpaste, cosmetics, ice cream, and hundreds of other products.

Sea urchins and pollution are major threats to kelp forests. Acting as predators, large populations of sea urchins (Figure 5-A) can rapidly devastate a kelp forest because they eat the bases of young kelp plants. Southern sea otters, a keystone species, help to control populations of sea urchins. An adult southern sea otter (Figure 5-1, left) can eat up to 50 sea urchins a day—equivalent to a 68-kilogram (150-pound) person eating 160 quarter-pound hamburgers a day. Scientific studies indicate that without southern sea otters, giant kelp-forest ecosystems off the coast of California would collapse, thereby reducing aquatic biodiversity.

CORE CASE STUDY

Figure 5-A This purple sea urchin inhabits the coastal waters of the U.S. state of California.

A second threat to kelp forests is polluted water running off of the land and into the coastal waters where kelp forests grow. The pollutants in this runoff include pesticides and herbicides, which can kill kelp plants and other kelp forest species and upset the food webs in these forests. Another runoff pollutant is fertilizer. Its plant nutrients (mostly nitrates) can cause excessive growth of algae and other plants, which block some of the sunlight needed to support the growth of giant kelp.

A looming threat to kelp forests is the warming of the world's oceans. Giant kelp forests require fairly cool water. If coastal waters get warmer during this century, as projected by climate models, many—perhaps most—of the kelp forests off the coast of California will disappear. If this happens, the southern sea otter and many other species will also disappear, unless they can migrate to other suitable locations, which are few and far between.

Critical Thinking

List three ways to protect giant kelp forests and southern sea otters.

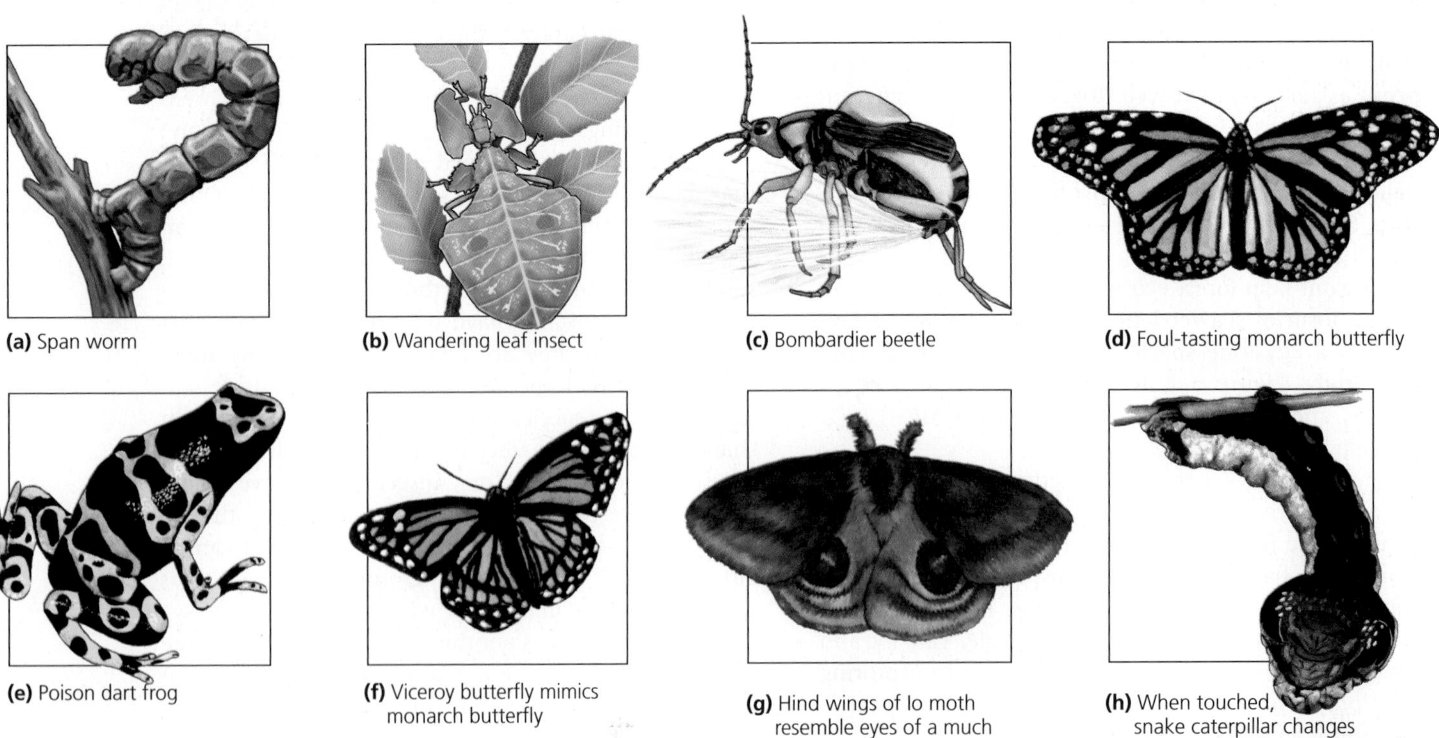

Figure 5-3 These prey species have developed specialized ways to avoid their predators: **(a, b)** *camouflage*, **(c, d, e)** *chemical warfare*, **(d, e)** *warning coloration*, **(f)** *mimicry*, **(g)** *deceptive looks*, and **(h)** *deceptive behavior*.

Figure 5-3d). When attacked, some species of squid and octopus emit clouds of black ink, allowing them to escape by confusing their predators.

Many bad-tasting, bad-smelling, toxic, or stinging prey species have evolved *warning coloration*, brightly colored advertising that helps experienced predators to recognize and avoid them. They flash a warning: "Eating me is risky." Examples are brilliantly colored poisonous frogs (Figure 5-3e and Figure 4-11, p. 73) and foul-tasting monarch butterflies (Figure 5-3d). When a bird such as a blue jay eats a monarch butterfly, it usually vomits and learns to avoid them.

CONNECTIONS

Coloration and Dangerous Species

Biologist Edward O. Wilson gives us two rules, based on coloration, for evaluating possible danger from any unknown animal species we encounter in nature. *First*, if it is small and strikingly beautiful, it is probably poisonous. *Second*, if it is strikingly beautiful and easy to catch, it is probably deadly.

Some butterfly species such as the nonpoisonous viceroy (Figure 5-3f) gain protection by looking and acting like the monarch, a protective device known as *mimicry*. Other prey species use *behavioral strategies* to avoid predation. Some attempt to scare off predators by puffing up (blowfish), spreading their wings (peacocks), or mimicking a predator (Figure 5-3h). Some moths have wings that look like the eyes of much larger animals (Figure 5-3g). Other prey species gain some protection by living in large groups such as schools of fish and herds of antelope.

THINKING ABOUT

Predation and the Southern Sea Otter

Describe **(a)** a trait possessed by the southern sea otter (**Core Case Study**) that helps it to catch prey and **(b)** a trait that helps it to avoid being preyed upon.

At the individual level, members of the predator species benefit and members of the prey species are harmed. At the population level, predation plays a role in evolution by natural selection (see Chapter 4, p. 66). Animal predators, for example, tend to kill the sick, weak, aged, and least fit members of a prey population because they are the easiest to catch. This leaves behind individuals in that population with better defenses against predation. Such individuals tend to survive longer and leave more offspring with adaptations that can help them avoid predation.

Some people tend to view certain animal predators with contempt. When a hawk tries to capture and feed on a rabbit, some root for the rabbit. Yet the hawk, like all predators, is merely trying to get enough food for itself and its young. In doing so, it plays an important ecological role in controlling rabbit populations.

Interactions between Predator and Prey Species Can Drive Each Other's Evolution

Predator and prey populations can exert intense natural selection pressures on one another. Over time, as a prey species develops traits that make it more difficult to catch, its predators face selection pressures that favor traits increasing their ability to catch their prey. Then the prey species must get better at eluding the more effective predators.

When populations of two different species interact in such a way over a long period of time, changes in the gene pool of one species can lead to changes in the gene pool of the other. Such changes can help both sides to become more competitive or to avoid or reduce competition. Biologists call this natural selection process **coevolution**.

Consider the species interaction between bats (the predator) and certain species of moths (the prey). Bats like to eat moths, and they hunt at night using *echolocation* to navigate and to locate their prey. This means that they emit pulses of extremely high frequency and high-intensity sound that bounce off objects, and then they capture the returning echoes that tell them where their prey is located.

As a countermeasure to this effective prey-detection system, certain moth species have evolved ears that are especially sensitive to the sound frequencies that bats use to find them. When the moths hear the bat frequencies, they try to escape by dropping to the ground or flying evasively. Some bat species have evolved ways to counter this defense by switching the frequency of their sound pulses. In turn, some moths have evolved their own high-frequency clicks to jam the bats' echolocation systems. Some bat species have then adapted by turning off their echolocation systems and using the moths' clicks to locate their prey.

Some Species Feed off Other Species by Living On or Inside Them

Parasitism occurs when one species (the *parasite*) feeds on another organism (the *host*), usually by living on or inside the host. In this relationship, the parasite benefits and the host is often harmed but not immediately killed.

A parasite usually is much smaller than its host and rarely kills it. However, most parasites remain closely associated with their hosts, draw nourishment from them, and may gradually weaken them over time.

Some parasites such as tapeworms live inside their hosts. Other parasites such as mistletoe plants (see Photo 5 in the Detailed Contents) and sea lampreys (Figure 5-4, p. 84) attach themselves to the outsides of

Figure 5-4 *Parasitism:* This blood-sucking, parasitic sea lamprey has attached itself to an adult lake trout from the Great Lakes (USA, Canada).

their hosts. Some parasites such as fleas and ticks move from one host to another while others, such as tapeworms, spend their adult lives within a single host.

From the host's point of view, parasites are harmful. But from the population perspective, parasites can promote biodiversity by contributing to species richness, and in some cases, they help to keep the populations of their hosts in check.

In Some Interactions, Both Species Benefit

In **mutualism**, two species behave in ways that benefit both by providing each with food, shelter, or some other resource. One example is pollination of flowering plants by species such as honeybees and butterflies that feed on the nectar of flowers (see Figure 4-A, left, p. 64).

Figure 5-5 shows two examples of mutualistic relationships that combine *nutrition* and *protection*. One involves birds that ride on the backs of large animals like African buffalo, elephants, and rhinoceroses (Figure 5-5a). The birds remove and eat parasites and pests (such as ticks and flies) from the animals' bodies and often make noises warning the larger animals when predators are approaching.

A second example involves the clownfish species (Figure 5-5b). Clownfish usually live in a group within sea anemones, whose tentacles sting and paralyze most fish that touch them. The clownfish, which are not harmed by the tentacles, gain protection from predators and feed on the detritus left from the anemones' meals. The sea anemones benefit because the clownfish protect them from some of their predators and parasites.

In *gut inhabitant mutualism*, vast armies of bacteria in the digestive systems of animals help to break down (digest) the animals' food. In turn, the bacteria receive a sheltered habitat and food from their hosts. Hundreds of millions of bacteria in your gut secrete enzymes that help you digest the food you eat. Cows and termites are able to digest cellulose in the plant tissues they eat because of the large number of microorganisms, mostly certain types of bacteria, that live in their guts.

It is tempting to think of mutualism as an example of cooperation between species. In reality, each species benefits by unintentionally exploiting the other as a result of traits they obtained through natural selection. Both species in a mutualistic pair are in it for themselves.

In Some Interactions, One Species Benefits and the Other Is Not Harmed

Commensalism is an interaction that benefits one species but has little, if any, beneficial or harmful effect on the other. One example involves plants called

(a) Oxpeckers and black rhinoceros

(b) Clownfish and sea anemone

Figure 5-5 Examples of *mutualism:* **(a)** Oxpeckers (or tickbirds) feed on parasitic ticks that infest large, thick-skinned animals such as the endangered black rhinoceros. **(b)** A clownfish gains protection and food by living among deadly, stinging sea anemones and helps to protect the anemones from some of their predators.

Luiz C. Marigo/Peter Arnold, Inc.

Figure 5-6 In an example of *commensalism*, this bromeliad—an epiphyte, or air plant—in Brazil's Atlantic tropical rain forest roots on the trunk of a tree, rather than in soil, without penetrating or harming the tree.

epiphytes (air plants) such as certain types of orchids and bromeliads that attach themselves to the trunks or branches of large trees in tropical and subtropical forests (Figure 5-6). Epiphytes benefit by having a solid base on which to grow. They also live in an elevated spot that gives them better access to sunlight, water from the humid air and rain, and nutrients falling from the tree's upper leaves and limbs. Their presence apparently does not harm the tree. Similarly, birds benefit by nesting in trees, generally without harming them.

5-2 What Limits the Growth of Populations?

▶ **CONCEPT 5-2 No population can continue to grow indefinitely because of limitations on resources and because of competition among species for those resources.**

Populations Can Grow, Shrink, or Remain Stable

A **population** is a group of individuals of the same species living in a particular place (see Photo 3 in Detailed Contents). Over time, the number of individuals in a population may increase, decrease, remain about the same, or go up and down in cycles that are a response to changes in environmental conditions.

Four variables—*births, deaths, immigration,* and *emigration*—govern changes in population size. A population increases by birth and immigration (arrival of individuals from outside the population) and decreases by death and emigration (departure of individuals from the population):

$$\text{Population change} = (\text{Births} + \text{Immigration}) - (\text{Deaths} + \text{Emigration})$$

The presence or absence of certain physical and chemical factors, known as **limiting factors**, can help to determine the number of organisms in a population. For example, the amount of water available to organisms in a forest limits the number of plants and animals that can live there. If the system dries up and gets little or no moisture, then most of the organisms living there either die or must move away (if they can) in order to survive. Similarly, if flooding occurs, many of the organisms must move or die. Such factors are more important than other factors in regulating population growth.

Another factor that can affect population size is **population density**, the number of individuals in a population found within a defined area or volume. Some limiting factors become more important as a population's density increases. For example, in a dense population, parasites and diseases can spread more easily and have the effect of controlling population growth. On the other hand, a higher population density can help sexually reproducing individuals to find mates more easily in order to produce offspring and increase the size of a population. Such density-dependent factors tend to regulate a population by keeping it at a fairly constant size.

Species Have Different Reproductive Patterns

Species use different reproductive patterns to help ensure their long-term survival. Some species have many, usually small, offspring and give them little or no parental care or protection. Examples include algae, bacteria, and most insects. These species reproduce at an early age and overcome typically massive losses of offspring by producing so many offspring that a few will likely survive to reproduce many more offspring and keep this reproductive pattern going.

At the other extreme are species that tend to reproduce later in life and have a small number of offspring with longer life spans. Typically, the offspring of mammals with this reproductive strategy develop inside their mothers (where they are safe), and are born fairly large. After birth, they mature slowly and are cared for and protected by one or both parents, and in some cases by living in herds or groups, until they reach reproductive age and begin the cycle again.

Most large mammals (such as elephants, whales, and humans) follow this reproductive pattern. Many of these species—especially those with long times between generations and with low reproductive rates like elephants, rhinoceroses, and sharks—are vulnerable to extinction. Most organisms have reproductive patterns between these two extremes.

No Population Can Grow Indefinitely: J-Curves and S-Curves

Some species have an incredible ability to increase their numbers. Members of such populations typically reproduce at an early age, have many offspring each time they reproduce, and reproduce many times, with short intervals between successive generations. For example, with no controls on its population growth, a species of bacteria that can reproduce every 20 minutes would generate enough offspring to form a layer 0.3 meter (1 foot) deep over the entire earth's surface in only 36 hours!

Fortunately, this will not happen. Research reveals that regardless of their reproductive strategy, no population of a species can grow indefinitely because of limitations on resources and competition with populations of other species for those resources (**Concept 5-2**). In the real world, a rapidly growing population of any species eventually reaches some size limit imposed by the availability of one or more limiting factors such as light, water, temperature, space, or nutrients, or by exposure to predators or infectious diseases.

There are always limits to population growth in nature. For example, one reason California's southern sea otters (**Core Case Study**) face extinction is that they cannot reproduce rapidly (Science Focus, at right).

CORE CASE STUDY

Environmental resistance is the combination of all factors that act to limit the growth of a population. It largely determines an area's **carrying capacity**: the maximum population of a given species that a particular habitat can sustain indefinitely. The growth rate of a population decreases as its size nears the carrying capacity of its environment because resources such as food, water, and space begin to dwindle.

A population with few, if any, limitations on its resource supplies can grow exponentially at a fixed rate such as 1% or 2% per year. *Exponential growth* (see Chapter 1, p. 16) starts slowly but then accelerates as the population increases, because the base size of the population is increasing. Plotting the number of individuals against time yields a J-shaped growth curve (Figure 5-7, left portion of curve).

Exponential growth (left third of the curve in Figure 5-7) occurs when a population has essentially unlimited resources to support its growth. Such exponential growth is eventually converted to *logistic growth* (center of the curve in Figure 5-7), in which the growth rate decreases as the population becomes larger and faces environmental resistance. Over time, the population size stabilizes at or near the *carrying capacity* of its environment, which results in a sigmoid (S-shaped) population growth curve. Depending on resource availability, the size of a population often fluctuates around its carrying capacity (right third of the curve in Figure 5-7). However, a population may temporarily exceed its carrying capacity and then suffer a sharp decline or crash in its numbers (see next subsection).

Changes in the population sizes of keystone species such as the southern sea otter (**Core Case Study**) can alter the species composition and biodiversity of an ecosystem. For example, a decline in a population of California southern sea otters causes a decline in the populations of species dependent on them, including the giant kelp (Science Focus, at right). This

CORE CASE STUDY

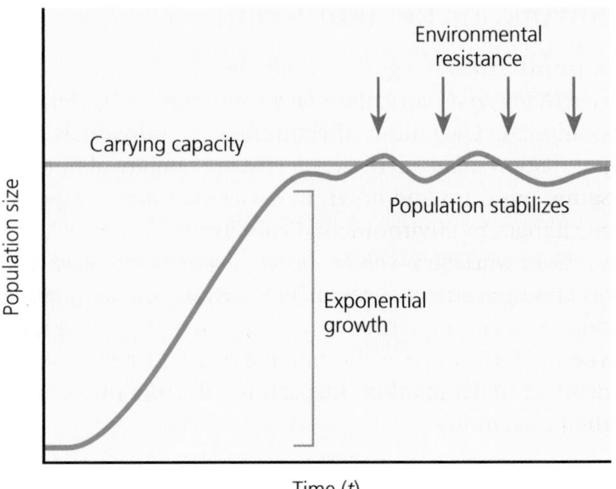

Active Figure 5-7 No population can continue to grow indefinitely (**Concept 5-2**). *See an animation based on this figure at* **www.cengagebrain.com**. **Question:** What is an example of environmental resistance that humans have not been able to overcome?

SCIENCE FOCUS

Why Do California's Southern Sea Otters Face an Uncertain Future?

The southern sea otter (**Core Case Study**) cannot rapidly increase its numbers for several reasons. Female southern sea otters reach sexual maturity between 2 and 5 years of age, can reproduce until age 15, and typically each produce only one pup a year.

The population size of southern sea otters has fluctuated in response to changes in environmental conditions. One such change has been a rise in populations of the orcas (killer whales) that feed on them. Scientists hypothesize that orcas began feeding more on southern sea otters when populations of their normal prey, sea lions and seals, began declining.

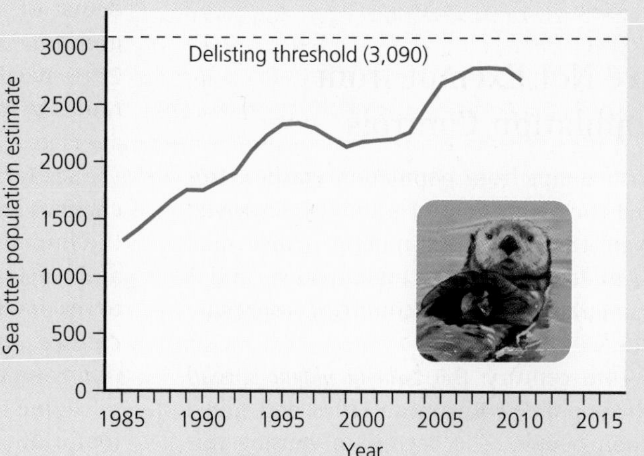

Figure 5-B This graph tracks the average annual population size of southern sea otters off the coast of the U.S. state of California, 1985–2010. (Data from U.S. Geological Survey)

Another factor may be parasites known to breed in cats. Scientists hypothesize that some southern sea otters may be dying because coastal area cat owners flush feces-laden cat litter down their toilets or dump it in storm drains that empty into coastal waters. The feces contain the parasites that then infect the otters.

Thorny-headed worms from seabirds also are known to be killing sea otters, as are toxic algae blooms triggered by urea, a key ingredient in fertilizer that washes into coastal waters. PCBs and other fat-soluble toxic chemicals released by human activities can accumulate in the tissues of the shellfish on which otters feed, and this proves fatal to otters. The facts that southern sea otters feed at high trophic levels and live close to the shore makes them vulnerable to these and other pollutants in coastal waters. In other words, as an *indicator* species, southern sea otters reveal the condition of coastal waters in their habitat.

Some southern sea otters die when they encounter oil spilled from ships. The entire California southern sea otter population could be wiped out by a large oil spill from a single tanker off the state's central coast or from an offshore oil well that ruptures, should drilling for oil be allowed off this coast. These factors, mostly resulting from human activities, plus a fairly low reproductive rate, have hindered the ability of the endangered southern sea otter to rebuild its population (Figure 5-B). According to the U.S. Geological Survey, the California southern sea otter population would have to reach at least 3,090 animals for 3 years in a row before it could be considered for removal from the endangered species list.

Critical Thinking

How would you design a controlled experiment to test the hypothesis that cat litter flushed down toilets may be killing southern sea otters?

reduces the species diversity of the kelp forest and alters its functional biodiversity by upsetting its food web and reducing energy flows and nutrient cycling within the forest.

When a Population Exceeds Its Carrying Capacity It Can Crash

Some species do not make a smooth transition from exponential growth to logistic growth (Figure 5-8). Such populations use up their resource supplies and temporarily *overshoot,* or exceed, the carrying capacity of their environment. This occurs because of a *reproductive time lag*: the period of time needed for the birth rate to fall and for the death rate to rise in response to resource overconsumption.

In such cases, the population suffers a sharp decline, called *dieback,* or **population crash**, unless the excess individuals can switch to new resources or move to an area that has more resources. Such a crash occurred when reindeer were introduced onto a small island in the Bering Sea (Figure 5-8).

However, the carrying capacity of any given area is not fixed. In some areas, it can increase or decrease seasonally and from year to year because of variations

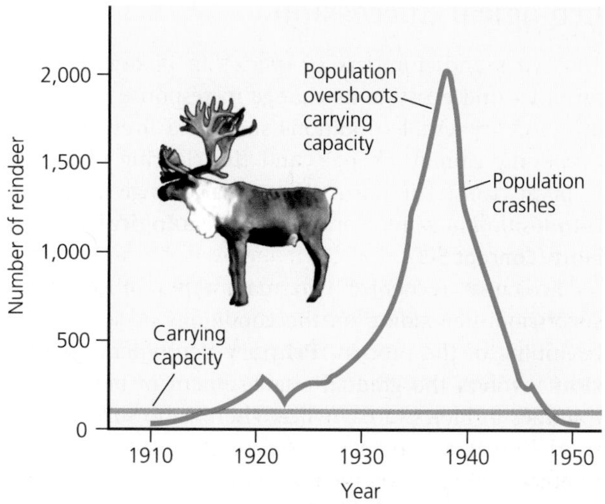

Figure 5-8 This graph tracks the exponential growth, overshoot, and population crash of reindeer introduced onto the small Bering Sea island of St. Paul. **Question:** Why do you think the sizes of some populations level off while others such as the reindeer in this example exceed their carrying capacities and crash?

in weather, such as a drought that causes decreases in available vegetation. Other factors include the presence or absence of predators and an abundance or scarcity of competitors.

Humans Are Not Exempt from Nature's Population Controls

Humans are not exempt from population crashes. Ireland experienced such a crash after a fungus destroyed its potato crop in 1845. About 1 million people died from hunger or diseases related to malnutrition, and 3 million people migrated to other countries, especially the United States.

During the 14th century, the *bubonic plague* spread through densely populated European cities and killed at least 25 million people. The bacterium causing this disease normally lives in rodents. It was transferred to humans by fleas that fed on infected rodents and then bit humans. The disease spread like wildfire through crowded cities, where sanitary conditions were poor and rats were abundant. Today, several antibiotics, not available until recently, can be used to treat bubonic plague.

Currently, the world is experiencing a global epidemic of AIDS, caused by infection with the human immunodeficiency virus (HIV). Between 1981 and 2010, AIDS killed more than 27 million people and continues to claim another 1.8 million lives each year—an average of 3 deaths per minute.

So far, technological, social, and other cultural changes have expanded the earth's carrying capacity for the human species. We have increased food production and used large amounts of energy and matter resources to occupy formerly uninhabitable areas, to expand agriculture, and to control the populations of other species that compete with us for resources.

Some say we can keep expanding our ecological footprint (see Figure 2, p. S24, in Supplement 6) indefinitely, mostly because of our technological ingenuity. Others say that sooner or later, we will reach the limits that nature always imposes on all populations.

5-3 How Do Communities and Ecosystems Respond to Changing Environmental Conditions?

▶ **CONCEPT 5-3 The structure and species composition of communities and ecosystems change in response to changing environmental conditions through a process called ecological succession.**

Communities and Ecosystems Change over Time: Ecological Succession

The types and numbers of species in biological communities and ecosystems change in response to changing environmental conditions such as a fires, volcanic eruptions, climate change, and the clearing of forests to plant crops. The normally gradual change in species composition in a given area is called **ecological succession** (Concept 5-3).

Ecologists recognize two main types of ecological succession, depending on the conditions present at the beginning of the process. **Primary ecological succession** involves the gradual establishment of biotic communities in lifeless areas where there is no soil in a terrestrial ecosystem or no bottom sediment in an aquatic ecosystem. Examples include bare rock exposed by a retreating glacier (Figure 5-9), newly cooled lava, an abandoned highway or parking lot, and a newly created shallow pond or reservoir. Primary succession usually takes hundreds to thousands of years because of the need to build up fertile soil or aquatic sediments to provide the nutrients needed to establish a plant community.

The other, more common type of ecological succession is called **secondary ecological succession**, in which a series of communities or ecosystems with different species develop in places containing soil or bottom sediment. This type of succession begins in an area where an ecosystem has been disturbed, removed, or destroyed, but some soil or bottom sediment remains. Candidates for secondary succession include abandoned farmland (Figure 5-10, p. 90), burned or cut forests, heavily polluted streams, and land that has been flooded. Because some soil or sediment is present, new vegetation can begin to grow, usually within a few weeks. It begins with the germination of seeds already in the soil and seeds imported by wind or in the droppings of birds and other animals.

Primary and secondary ecological succession are important natural services that tend to increase biodiversity, and thus the sustainability of communities and ecosystems, by increasing species richness and interactions among species. Such interactions in turn

Figure 5-9 *Primary ecological succession:* Over almost a thousand years, these plant communities developed, starting on bare rock exposed by a retreating glacier on Isle Royal, Michigan (USA) in western Lake Superior. The details of this process vary from one site to another. **Question:** What are two ways in which lichens, mosses, and plants might get started growing on bare rock?

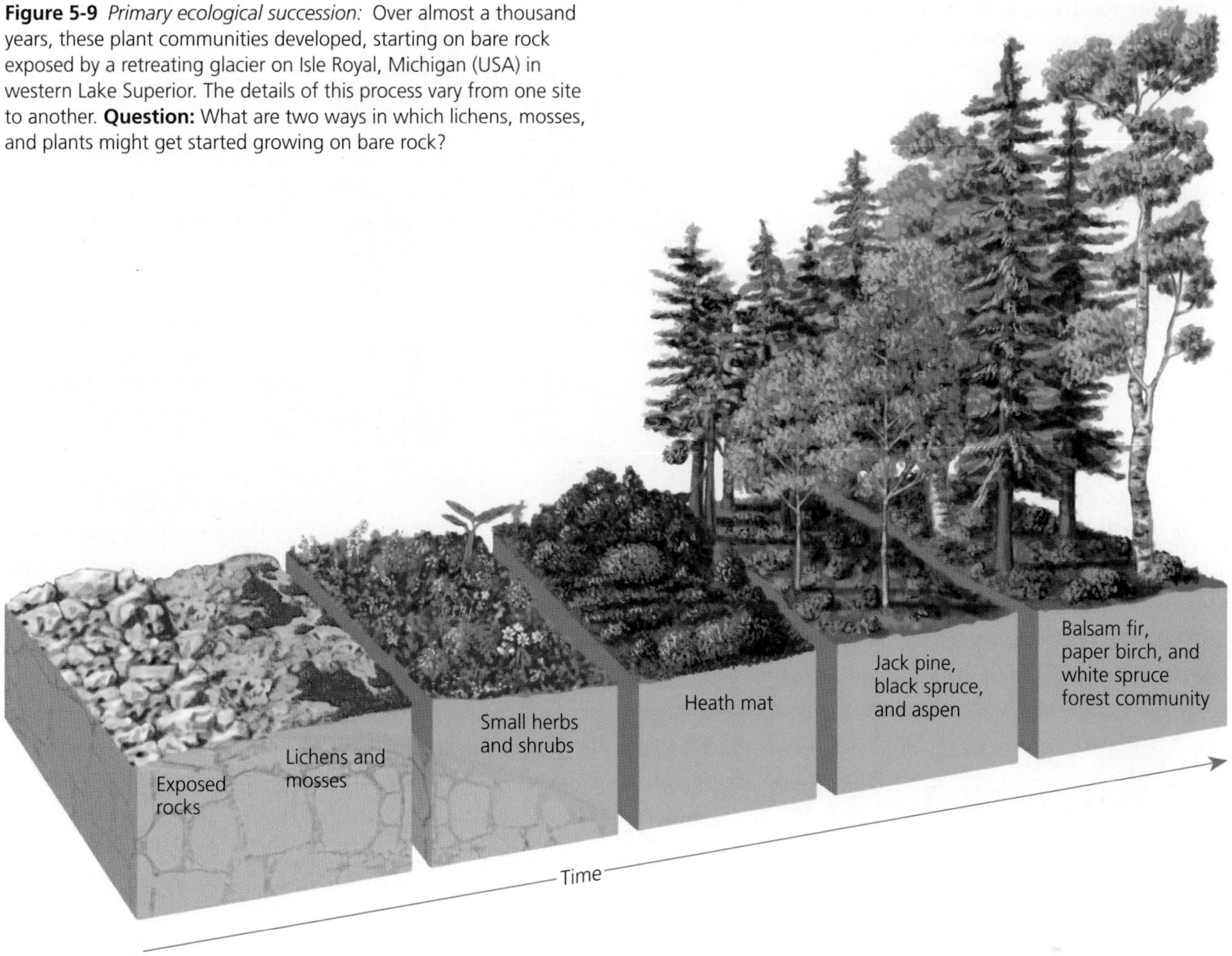

enhance sustainability by promoting population control and by increasing the complexity of food webs. This then enhances the energy flow and nutrient cycling, which are functional components of biodiversity (see Figure 4-2, p. 63). As part of the earth's natural capital, both types of succession are examples of *natural ecological restoration*.

Succession Does Not Follow a Predictable Path

According to the traditional view, succession proceeds in an orderly sequence along an expected path until a certain stable type of *climax community* occupies an area. Such a community is dominated by a few long-lived plant species and is in balance with its environment. This equilibrium model of succession is what ecologists once meant when they talked about the *balance of nature*.

Over the last several decades, many ecologists have changed their views about balance and equilibrium in nature. Under the balance-of-nature view, a large terrestrial community or ecosystem undergoing succession eventually became covered with an expected type of climax vegetation such as a mature forest (Figures 5-9 and 5-10). There is a general tendency for succession to lead to more complex, diverse, and presumably stable ecosystems. But a close look at almost any terrestrial community or ecosystem reveals that it consists of an ever-changing mosaic of patches of vegetation in different stages of succession.

The current view is that we cannot predict a given course of succession or view it as inevitable progress toward an ideally adapted climax plant community or ecosystem. Rather, succession reflects the ongoing struggle by different species for enough light, water, nutrients, food, and space. Most ecologists now recognize that mature, late-successional ecosystems are not in a state of permanent equilibrium. Rather, they are in a state of continual disturbance and change.

Living Systems Are Sustained through Constant Change

All living systems, from a cell to the biosphere, are constantly changing in response to changing environmental conditions. Continents move, climates change, and disturbances and succession change the composition of communities and ecosystems.

Active Figure 5-10 *Natural ecological restoration of disturbed land:* This diagram shows the undisturbed secondary ecological succession of plant communities on an abandoned farm field in the U.S. state of North Carolina. It took 150–200 years after the farmland was abandoned for the area to become covered with a mature oak and hickory forest. A new disturbance such as deforestation or fire would create conditions favoring pioneer species such as annual weeds. In the absence of new disturbances, secondary succession would recur over time, but not necessarily in the same sequence shown here. *See an animation based on this figure at* www.cengagebrain.com. **Questions:** Do you think the annual weeds (left) would continue to thrive in the mature forest (right)? Why or why not?

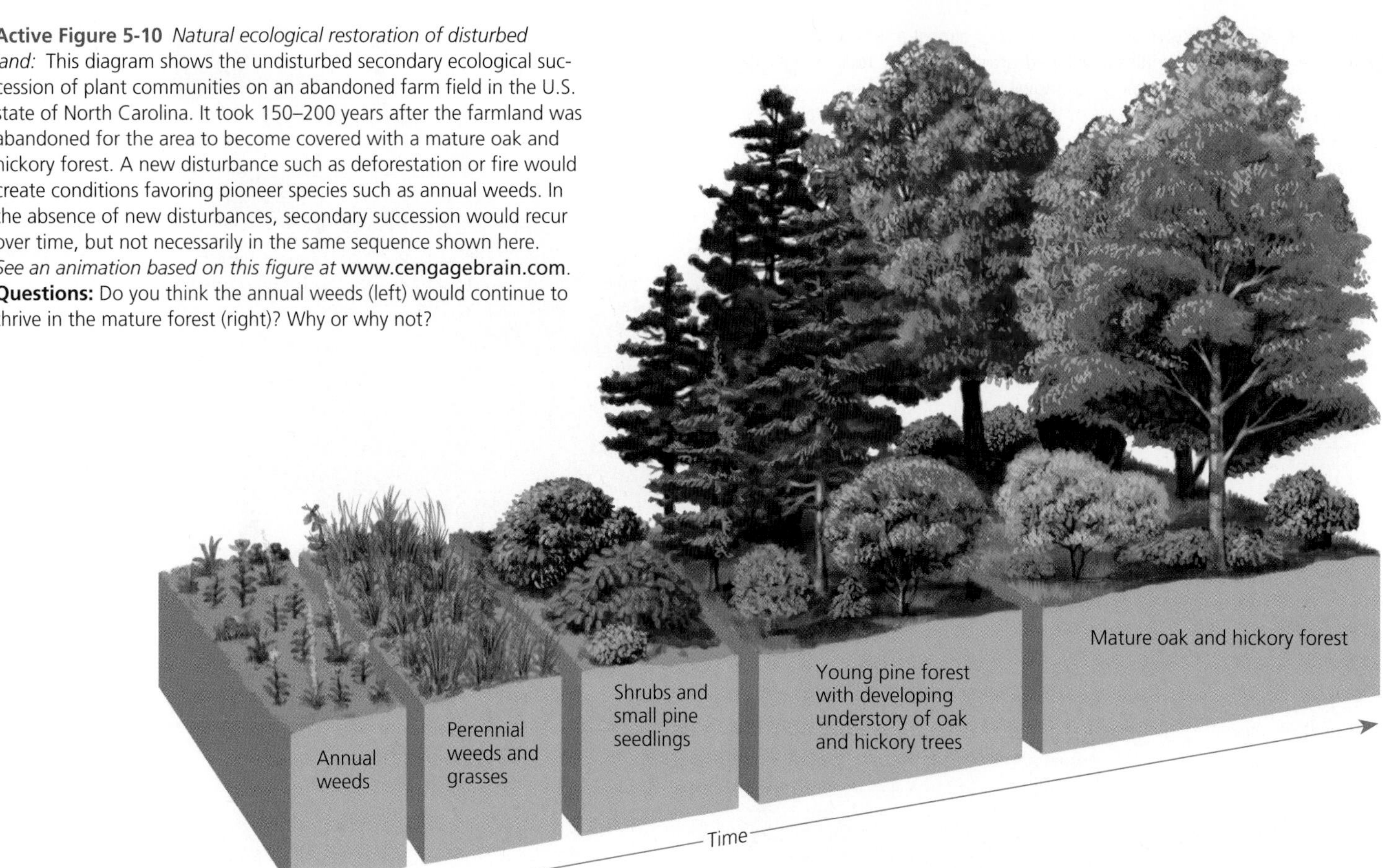

Living systems contain complex processes that interact to provide some degree of stability, or sustainability, over each system's expected life span. This *stability*, or capacity to withstand external stress and disturbance, is maintained only by constant change in response to changing environmental conditions. For example, in a mature tropical rain forest (see front cover photo), some trees die and others take their places. However, unless the forest is cut, burned, or otherwise destroyed, you would still recognize it as a tropical rain forest 50 or 100 years from now.

It is useful to distinguish between two aspects of stability in living systems. One is **inertia**, or **persistence**: the ability of a living system such as a grassland or a forest to survive moderate disturbances. A second factor is **resilience**: the ability of a living system to be restored through secondary succession after a more severe disturbance.

Evidence suggests that some ecosystems have one of these properties but not the other. For example, tropical rain forests have high species richness and high inertia and thus are resistant to significant change or damage. But once a large tract of tropical rain forest is cleared or severely damaged, the resilience of the resulting degraded forest ecosystem may be so low that it reaches an ecological tipping point after which it may not be restored by secondary ecological succession. One reason for this is that most of the nutrients in a typical rain forest are stored in its vegetation, not in the topsoil, as in most other terrestrial ecosystems. Once the nutrient-rich vegetation is gone, daily rains can remove most of the remaining soil nutrients and thus prevent a tropical rain forest from regrowing on a large cleared area.

By contrast, grasslands are much less diverse than most forests, and consequently they have low inertia and can burn easily. However, because most of their plant matter is stored in underground roots, these ecosystems have high resilience and can recover quickly after a fire, as their root systems produce new grasses. Grassland can be destroyed only if its roots are plowed up and something else is planted in its place, or if it is severely overgrazed by livestock or other herbivores.

This variation among species in inertia and resilience is yet another example of how biodiversity has helped life on earth to sustain itself for billions of years. It illustrates one aspect of the biodiversity **principle of sustainability** (see back cover).

Here are this chapter's *three big ideas*:

- Certain interactions among species affect their use of resources and their population sizes.
- There are always limits to population growth in nature.
- Changes in environmental conditions cause communities and ecosystems to gradually alter their species composition and population sizes (ecological succession).

REVISITING Southern Sea Otters and Sustainability

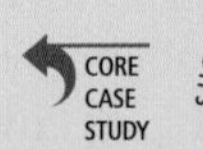

Before the arrival of European settlers on the western coast of North America, the southern sea otter population was part of a complex ecosystem made up of kelp, bottom-dwelling creatures, otters, whales, and other species depending on one another for survival. Giant kelp forests served as food and shelter for sea urchins. Sea otters ate the sea urchins and other kelp eaters. Some species of whales and sharks ate the otters. And detritus from all these species helped to maintain the giant kelp forests. Each of these interacting populations was kept in check by—and helped to sustain—all the others.

When European settlers arrived and began hunting the otters for their pelts, they probably didn't know much about the intricate web of life beneath the ocean surface. But with the effects of overhunting, people realized they had done more than simply take sea otters. They had torn the web, disrupted an entire ecosystem, and triggered a loss of valuable natural resources and services, including biodiversity.

Populations of most plants and animals depend, directly or indirectly, on solar energy and all populations play roles in the cycling of nutrients in the ecosystems where they live. In addition, the biodiversity found in the variety of species in different terrestrial and aquatic ecosystems provides alternative paths for energy flow and nutrient cycling, and better opportunities for natural selection as environmental conditions change. When we disrupt these paths, we violate all three **principles of sustainability**. In this chapter, we looked more closely at two effects of one of those principles: first, *biodiversity promotes sustainability*, and second, *there are always limits to population growth in nature*, mostly because of biodiversity and diverse species interactions.

We cannot command nature except by obeying her.

SIR FRANCIS BACON

REVIEW

CORE CASE STUDY

1. Explain how southern sea otters act as a keystone species in their environment (**Core Case Study**). Explain why we should care about protecting this species from extinction that could result primarily from human activities.

SECTION 5-1

2. What is the key concept for Section 5-1? Define **interspecific competition** and give an example of it. Explain how this type of species interaction can affect the population sizes of species in ecosystems. Describe and give an example of **resource partitioning** and explain how it can increase species diversity.

3. Define **predation** and distinguish between a **predator** species and a **prey** species and give an example of each. What is a **predator–prey relationship**? Explain why we should help to preserve kelp forests. Describe three ways in which prey species can avoid their predators and three ways in which predators can increase their chances of feeding on their prey. Define and give an example of **coevolution**.

4. Define **parasitism** and give two examples of it. Explain how parasites can be helpful from a population point of view. Define **mutualism** and **commensalism** and give an example of each. For each of these interactions, note which species are harmed and which are benefited or not affected by the interaction.

SECTION 5-2

5. What is the key concept for Section 5-2? Define **population** and describe four variables that govern changes in population size. Write an equation showing how these variables interact. Define **limiting factor** and give an example. Define **population density** and explain how some limiting factors can become more important as a population's density increases. Describe two different reproductive strategies that can enhance the long-term survival of a species.

6. Explain why no population can grow indefinitely. Distinguish between the **environmental resistance** and the **carrying capacity** of an environment, and use these concepts to explain why there are always limits to population growth in nature. Why are southern sea otters making a slow comeback and what factors are threatening this recovery?

7. Define and give an example of a **population crash**. Explain why humans are not exempt from nature's population controls.

SECTION 5-3

8. What is the key concept for Section 5-3? What is **ecological succession**? Distinguish between **primary ecological succession** and **secondary ecological succession** and give an example of each. Explain why succession does not follow a predictable path.

9. Explain how living systems achieve some degree of sustainability by undergoing constant change in response to changing environmental conditions. In terms of the stability of ecosystems, distinguish between **inertia (persistence)** and **resilience** and give an example of each.

10. What are this chapter's *three big ideas*? Explain how changes in the nature and size of populations are related to the three **principles of sustainability**.

Note: Key terms are in **bold** type.

CRITICAL THINKING

1. What difference would it make if the southern sea otter (**Core Case Study**) became extinct primarily because of human activities? What are three things we could do to help prevent the extinction of this species?

CORE CASE STUDY

2. Use the second law of thermodynamics (see Chapter 2, p. 36) to help explain why predators are generally less abundant than their prey. In your explanation, make use of the pyramid of energy flow (see Figure 3-12, p. 49).

3. Explain why most species with a high capacity for population growth (such as bacteria, flies, and cockroaches) tend to have small individuals, while those with a low capacity for population growth (such as humans, elephants, and whales) tend to have large individuals.

4. Which reproductive strategy do most species of insect pests and harmful bacteria use? Why does this make it difficult for us to control their populations?

5. List three factors limiting human population growth in the past that we have overcome. Describe how we overcame each of these factors. List two factors that may limit human population growth in the future. Do you think that we are close to reaching those limits? Explain.

6. If the human species were to suffer a population crash, what are three species that might move in to occupy part of our ecological niche? Explain why this might happen.

7. How would you reply to someone who argues that we should not worry about the effects that human activities have on natural systems because natural succession will heal the wounds of such activities and restore the balance of nature?

8. How would you reply to someone who contends that efforts to preserve natural systems are not worthwhile because nature is largely unpredictable?

DOING ENVIRONMENTAL SCIENCE

Visit a nearby land area, such as a partially cleared or burned forest, a grassland, or an abandoned crop field, and record signs of secondary ecological succession. Take notes on your observations and formulate a hypothesis about what sort of disturbance led to this succession. Include your thoughts about whether this disturbance was natural or caused by humans. Study the area carefully to see whether you can find patches that are at different stages of succession and record your thoughts about what sorts of disturbances have caused these differences. You might want to research the topic of ecological succession in such an area.

GLOBAL ENVIRONMENT WATCH EXERCISE

Search for *Climate Change* and use this portal to find sources of information about how a warmer ocean might affect California's coastal kelp beds on which the southern sea otters depend (**Core Case Study**). Write a report on what you found. Try to include information on current effects of warmer water on the kelp beds as well as projections about future effects. Also, summarize any information you might find on possible ways to prevent harm to these kelp beds.

CORE CASE STUDY

DATA ANALYSIS

The graph below shows changes in the size of an emperor penguin population in terms of breeding pairs on the island of Terre Adelie in the Antarctic. Use the graph to answer the questions below.

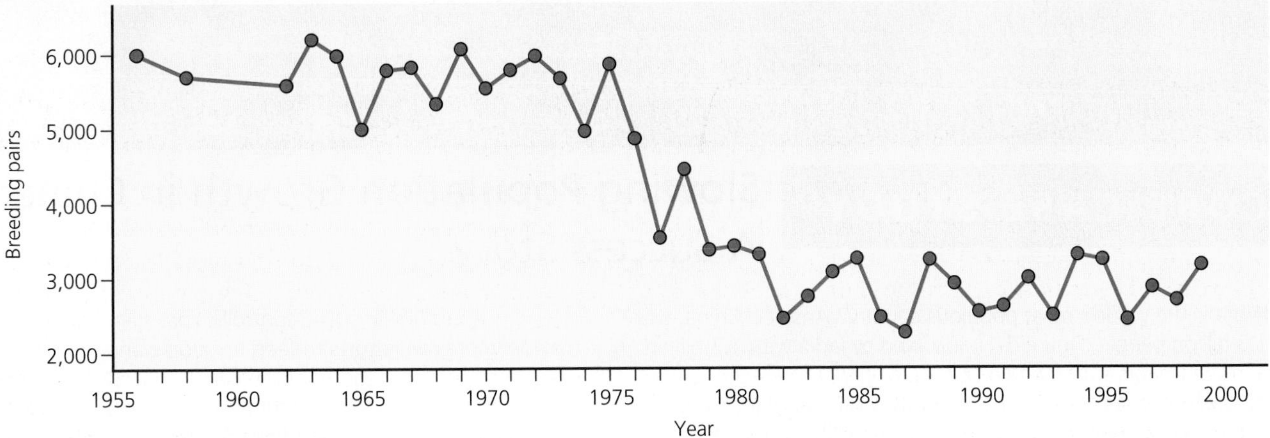

Source: Data from *Nature*, May 10, 2001.

1. What was the approximate carrying capacity of the island for the penguin population from 1960 to 1975? What was the approximate carrying capacity of the island for the penguin population from 1980 to 2000? (Hint: see Figure 5-7.)
2. What was the percentage decline in the penguin population from 1975 to 2000?

LEARNING ONLINE

The website for *Environmental Science 14e* contains helpful study tools and other resources to take the learning experience beyond the textbook. Examples include flashcards, practice quizzing, information on green careers, tips for more sustainable living, activities, and a wide range of articles and further readings. To access these materials and other companion resources for this text, please visit **www.cengagebrain.com**. For further details, see the preface, p. xvi.

6 The Human Population and Urbanization

CORE CASE STUDY Slowing Population Growth in China: A Success Story

What is the world's most populous country? Answer: China, with 1.3 billion people (Figure 6-1), followed by India with 1.2 billion and the United States with 312 million. In 2011, the U.S. Census Bureau projected that if current trends continue, China's population will increase to about 1.4 billion by 2026 and then will begin a slow decline to as low as 750 million by the end of this century.

Since the 1960s, which country has done the most to reduce its population growth: China or the United States? Answer: China. In the 1960s, China's large population was growing so rapidly that there was a serious threat of mass starvation. To avoid this, government officials decided to take measures that eventually led to the establishment of the world's most extensive, intrusive, and strict family planning and birth control program.

China's goal has been to sharply reduce population growth by promoting one-child families. The government provides contraceptives, sterilizations, and abortions for married couples. In addition, married couples pledging to have no more than one child receive a number of benefits including better housing, more food, free health care, salary bonuses, and preferential job opportunities for their child. Couples who break their pledge lose such benefits.

L. Young/UNEP/Peter Arnold, Inc.

Figure 6-1 Thousands of people crowd a street in China, where almost one of every five persons on the planet lives.

Since this government-controlled program began, China has made impressive efforts to feed its people and bring its population growth under control. Between 1972 and 2011, the country cut its birth rate in half and trimmed the average number of children born to its women from 5.7 to 1.5, compared to 2.0 children per woman in the United States. As a result, the U.S. population is growing faster than China's population.

Since 1980, China has undergone rapid industrialization and economic growth. According to the Earth Policy Institute, between 1990 and 2010, this reduced the number of people living in extreme poverty by almost 500 million. It also helped at least 300 million Chinese—a number roughly equal to the entire U.S. population—to become middle-class consumers. Over time, China's rapidly growing middle class will consume more resources per person, increasing China's ecological footprint (see Figure 1-8, p. 14) within its own borders and in other parts of the world that provide it with resources. This will put a strain on the earth's natural capital unless China steers a course toward more sustainable economic development.

So why should we care about population growth in the world, in China, in the United States, or anywhere else? The primary reason is that each of us depends on the earth's life-support systems to meet our basic needs for air, water, food, land, shelter, and energy as well as our needs for a number of other natural resources that we use to produce an incredible variety of manufactured goods.

As our population grows and incomes rise, we use more of the earth's natural resources to satisfy our growing wants and this increases our ecological footprints. The result can be degradation and depletion of the natural capital (see Figure 1-3, p. 9 and Figure 1-5, p. 11) that keeps us alive and supports our lifestyles and economies. No one knows how long we can continue degrading and depleting the earth's natural capital by living off of nature's credit card as our population and resource use continue to grow. But there are warning signs that nature may begin calling in some of our ecological debts.

Key Questions and Concepts

6-1 How many people can the earth support?

CONCEPT 6-1 We do not know how long we can continue increasing the earth's carrying capacity for humans without seriously degrading the life-support system that keeps us and many other species alive.

6-2 What factors influence the size of the human population?

CONCEPT 6-2A Population size increases through births and immigration, and decreases through deaths and emigration.

CONCEPT 6-2B The average number of children born to the women in a population (*total fertility rate*) is the key factor that determines population size.

6-3 How does a population's age structure affect its growth or decline?

CONCEPT 6-3 The numbers of males and females in young, middle, and older age groups determine how fast a population grows or declines.

6-4 How can we slow human population growth?

CONCEPT 6-4 We can slow human population growth by reducing poverty, elevating the status of women, and encouraging family planning.

6-5 What are the major urban resource and environmental problems?

CONCEPT 6-5 Most cities are unsustainable because of high levels of resource use, waste, pollution, and poverty.

6-6 How does transportation affect urban environmental impacts?

CONCEPT 6-6 In some countries, many people live in widely dispersed urban areas and depend mostly on motor vehicles for their transportation, which greatly expands their ecological footprints.

6-7 How can cities become more sustainable and livable?

CONCEPT 6-7 An *ecocity* allows people to choose walking, biking, or mass transit for most transportation needs; to recycle or reuse most of their wastes; to grow much of their food; and to protect biodiversity by preserving surrounding land.

Note: Supplements 2 (p. S3), 6 (p. S22), and 7 (p. S38) can be used with this chapter.

The problems to be faced are vast and complex, but come down to this: 7 billion people are breeding exponentially. The process of fulfilling their wants and needs is stripping earth of its biotic capacity to support life; a climactic burst of consumption by a single species is overwhelming the skies, earth, waters, and fauna.

PAUL HAWKEN

6-1 How Many People Can the Earth Support?

▶ **CONCEPT 6-1 We do not know how long we can continue increasing the earth's carrying capacity for humans without seriously degrading the life-support system that keeps us and many other species alive.**

Human Population Growth Continues but It Is Unevenly Distributed

For most of history, the human population grew slowly (see Figure 1-11, p. 16, left part of curve). But for the past 200 years, the human population has grown rapidly, resulting in the characteristic J-curve of exponential growth (see Figure 1-11, p. 16, right part of curve).

Three major factors account for this population increase. *First,* humans developed the ability to expand into almost all of the planet's climate zones and habitats. *Second,* the emergence of early and modern agriculture allowed us to grow more food for each unit of land area farmed. *Third,* death rates dropped sharply because of improved sanitation and health care, and development of antibiotics and vaccines to help control infectious diseases. Thus, *most of the increase in the world's population during the last 100 years took place because of a sharp drop in death rates—not a sharp rise in birth rates*.

About 10,000 years ago, when agriculture began, there were roughly 5 million humans on the planet; now there are about 7 billion and by 2050 we may be trying to support 9.6 billion people. (See Figure 3, p. S39, in Supplement 7 for a timeline of key events related to human population growth.)

The rate of population growth has slowed (Figure 6-2), but the world's population is still growing at a rate that added about 83 million people during 2011—an average of 227,000 more people each day, or 2 more people every time your heart beats. (See *The Habitable Planet,* Video 5, at **www.learner.org/resources/series209.html** for a discussion of how demographers measure population size and growth.)

Geographically, this growth is unevenly distributed and this pattern is expected to continue. About 2% of the 83 million new arrivals on the planet in 2011 were added to the world's more-developed countries; the other 98% were added to the world's middle- and low-income, less-developed countries. At least 95% of the 2.6 billion people likely to be added to the world's population between 2011 and 2050 will be born into the least-developed countries, which are not equipped to deal with the pressures of such rapid growth.

SCIENCE FOCUS

How Long Can the Human Population Keep Growing?

To survive and provide resources for growing numbers of people, humans have modified, cultivated, built on, and degraded a large and increasing portion of the earth's natural systems. Our activities have directly affected, to some degree, about 83% of the earth's land surface, excluding Antarctica and Greenland (see Figure 2, p. S24, in Supplement 6), as our ecological footprints have spread across the globe (see Figure 1-8, p. 14).

Scientific studies of the populations of other species tell us that *no population can continue growing indefinitely*. How long can we continue to avoid the reality of the earth's carrying capacity for our species by sidestepping many of the factors that sooner or later limit the growth of any population?

According to one view, the planet has too many people collectively degrading the earth's life-support system. To some scientists and other analysts, the key problem is *overpopulation* because of the sheer number of people in less-developed countries (see Figure 1-9, top, p. 15), which have 82% of the world's population. To others, the key factor is *overconsumption* in affluent, more-developed countries because of their high rates of resource use per person (see Figure 1-9, bottom, p. 15). Such resource consumption increases the environmental impact, or ecological footprint, of each person (see Figure 1-8, bottom, p. 14). People who hold the general view that overpopulation is contributing to our major environmental problems argue that slowing human population growth should be an important priority.

Another view of population growth is that, so far, technological advances have allowed us to overcome the environmental limits that all populations of other species face. This has the effect of increasing the earth's carrying capacity for our species.

Some analysts believe that because of our technological ingenuity, there are few, if any, limits to human population growth and resource use per person. They point out that average life expectancy in most of the world has been steadily increasing despite dire warnings from some environmental scientists that we are seriously degrading our life support system.

To these analysts, adding more people means having more workers, consumers, and creative people to support ever-increasing economic growth. They believe that we can avoid serious damage to our life-support systems through technological advances, especially by increasing food production per unit of cropland and by continuing to make major advances in health care. As a result, they see no need to slow the world's population growth.

Proponents of slowing and eventually stopping population growth point out that we are not providing the basic necessities for about one of every five people—who struggle to survive on the equivalent of about $1.25 per day. The number of people living in such extreme poverty today—1.4 billion—is larger than China's entire population and almost 5 times the population of the United States. This raises a serious question: If we fail to meet the basic needs for 1.4 billion people today, what will happen in 2050 when there may be 2.6 billion more of us?

They also warn of two potentially serious consequences if we do not sharply lower birth rates. First, death rates may increase because of declining health and environmental conditions as well as increasing social disruption in some areas, as is already happening in parts of Africa. Second, resource use and degradation of normally renewable resources may intensify as more consumers increase their already large ecological footprints in more-developed countries and in rapidly developing countries such as China and India.

So far, advances in food production and health care have staved off widespread environmental collapse. But there is extensive and growing evidence that we are steadily depleting and degrading much of the natural capital that supports us. We can get away with this for awhile, because the earth's life support system is very complex and resilient, and because of built-in time delays between disturbances to the system and responses to them from within the system. But like unseen termites eating away the foundation of a house, at some point, such disturbances can reach a *tipping* or *breaking point* beyond which there could be damaging and irreversible change.

No one knows how close we are to environmental limits that many scientists say sooner or later will control the size of the human population. However, these scientists argue that this is a vital scientific, political, economic, and ethical issue that we must confront.

Critical Thinking

How close do you think we are to the environmental limits of human population growth—very close, moderately close, or far away? Explain.

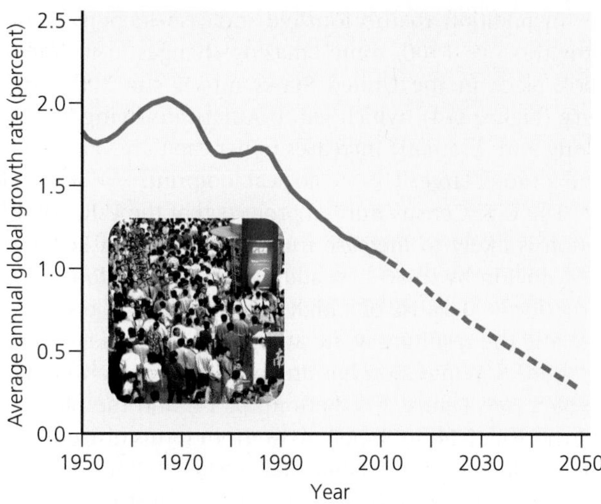

Figure 6-2 This graph tracks the annual growth rate of world population, 1950–2011, with projections to 2050. (Data from United Nations Population Division and U.S. Census Bureau)

There is little, if any, chance for stabilizing the size of the human population in the near future, barring global nuclear war or some other major catastrophe. However, during this century, the human population may level off as it moves from a J-shaped curve of exponential growth to an S-shaped curve of logistic growth (see Figure 5-7, p. 86).

This raises a question: How many people can the earth support indefinitely? Some analysts believe this is the wrong question. Instead, they argue that we should ask, what is the planet's **cultural carrying capacity**? That would be the maximum number of people who could live in reasonable freedom and comfort indefinitely, without decreasing the ability of the earth to sustain future generations. (See the Guest Essay by Garrett Hardin on this topic at **www.cengagebrain.com**.) This issue has long been a topic of scientific and political debate (Science Focus, at left).

6-2 What Factors Influence the Size of the Human Population?

▶ **CONCEPT 6-2A Population size increases through births and immigration, and decreases through deaths and emigration.**

▶ **CONCEPT 6-2B The average number of children born to the women in a population (*total fertility rate*) is the key factor that determines population size.**

The Human Population Can Grow, Decline, or Remain Fairly Stable

The basics of global population change are quite simple. If there are more births than deaths during a given period of time, the earth's population increases, and when the opposite is true, it decreases. When the number of births equals the number of deaths during a particular time period, population size does not change.

Instead of using the total numbers of births and deaths per year, demographers use the **birth rate**, or **crude birth rate** (the number of live births per 1,000 people in a population in a given year), and the **death rate**, or **crude death rate** (the number of deaths per 1,000 people in a population in a given year).

Human populations grow or decline in particular countries, cities, or other areas through the interplay of three factors: *births (fertility)*, *deaths (mortality)*, and *migration*. We can calculate the **population change** of an area by subtracting the number of people leaving a population (through death and emigration) from the number entering it (through birth and immigration) during a specified period of time (usually one year) (**Concept 6-2A**).

$$\text{Population change} = (\text{Births} + \text{Immigration}) - (\text{Deaths} + \text{Emigration})$$

When births plus immigration exceed deaths plus emigration, a population increases; when the reverse is true, a population declines. (Figure 7, p. S28, in Supplement 6 is a map showing the percentage rates of population increase in the world's countries in 2011.)

Women Are Having Fewer Babies but the World's Population Is Still Growing

A key factor affecting human population growth and size is the **total fertility rate (TFR)**: the average number of children born to the women in a population during their reproductive years (**Concept 6-2B**). (See Figure 8, p. S28, in Supplement 6 for a map showing how TFRs vary throughout the world.)

Between 1955 and 2011, the average global lifetime number of births of live babies per woman dropped from 5 to 2.5. Those who support slowing the world's population growth view this as good news. However, to eventually halt population growth, the global average lifetime number of births per woman will have to drop to 2.1.

CONNECTIONS

Population Growth and Fertility Rates in China CORE CASE STUDY

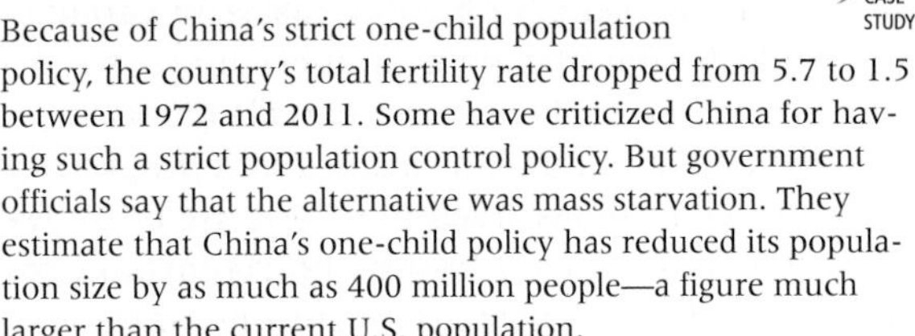

Because of China's strict one-child population policy, the country's total fertility rate dropped from 5.7 to 1.5 between 1972 and 2011. Some have criticized China for having such a strict population control policy. But government officials say that the alternative was mass starvation. They estimate that China's one-child policy has reduced its population size by as much as 400 million people—a figure much larger than the current U.S. population.

■ **CASE STUDY**

The U.S. Population Is Growing

The population of the United States grew from 76 million in 1900 to 312 million in 2011, despite oscillations in the country's TFR and birth rates (Figure 6-3). During the period of high birth rates between 1946 and 1964, known as the *baby boom,* 79 million people were added to the U.S. population. At the peak of the baby boom in 1957, the average TFR was 3.7 children per woman. In 2011, as in most years since 1972, it has been at or below 2.1 children per woman, compared to 1.5 in China in 2011.

The drop in the TFR has slowed the rate of population growth in the United States. But the country's population is still growing faster than those of all other more-developed countries as well as that of China, and it is not close to leveling off. According to the U.S. Census Bureau, about 2.5 million people were added to the U.S. population in 2011. About 1.6 million (almost two-thirds of the total) were added because there were that many more births than deaths, and about 900,000 (a third of the total) were immigrants.

In addition to the fourfold increase in population growth since 1900, some amazing changes in lifestyles took place in the United States during the 20th century (Figure 6-4), which led to Americans living longer along with dramatic increases in per capita resource use and a much larger U.S. ecological footprint.

The U.S. Census Bureau projects that the U.S. population is likely to increase from 312 million in 2011 to 423 million by 2050—an addition of 111 million more Americans. Because of a high per capita rate of resource use and the resulting waste and pollution, each addition to the U.S. population has an enormous environmental impact (see Figure 1-8, bottom, p. 14, and the map in Figure 2, p. S24, in Supplement 6). In terms of environmental impact per person, many analysts consider the United States to be by far the world's most overpopulated country, mostly because of its high rate of resource use per person and its fairly high population growth rate compared to most other more-developed countries and to China. However, in the not-too-distant future China may become the world's most overpopulated country because of its large population and its greatly increased resource use per person (Figure 1-9, p. 15).

THINKING ABOUT

U.S. Population

Do you believe that the United States is the world's most overpopulated country, given its very high rate of resource consumption per person? Explain.

Several Factors Affect Birth Rates and Fertility Rate

Many factors affect a country's average birth rate and TFR. One is the *importance of children as a part of the labor force,* especially in less-developed countries. This is why it makes sense for many poor couples in those countries to have a large number of children. They need help with hauling daily drinking water, gathering wood for heating and cooking, and tending crops and livestock.

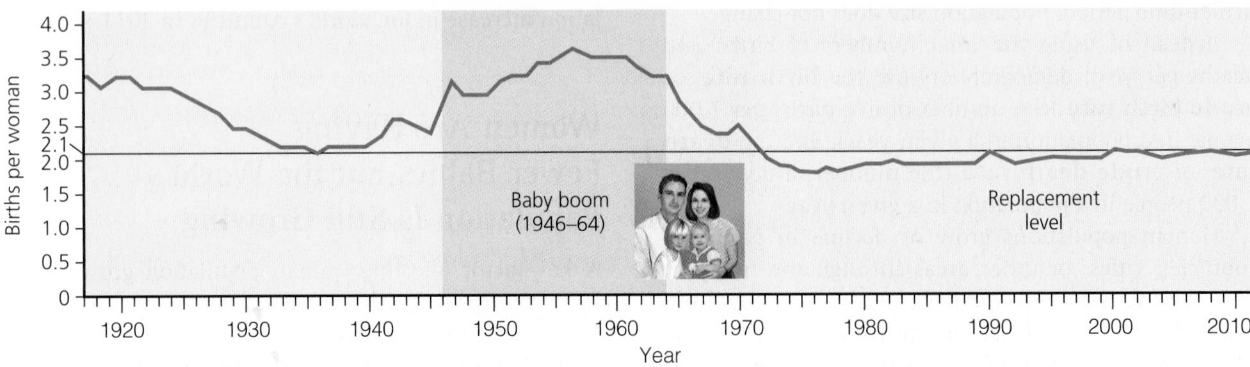

Figure 6-3 This graph shows the total fertility rates for the United States between 1917 and 2011. **Question:** The U.S. fertility rate has declined and remained at or below replacement levels since 1972. So why is the population of the United States still increasing? (Data from Population Reference Bureau and U.S. Census Bureau)

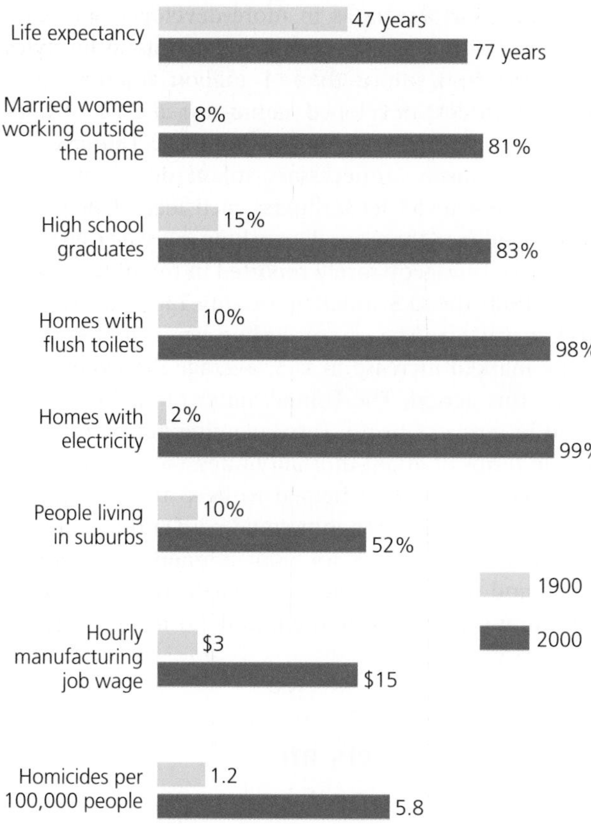

Figure 6-4 Some major changes took place in the United States between 1900 and 2000. **Question:** Which two of these changes do you think had the biggest impacts on the U.S. ecological footprint? (Data from U.S. Census Bureau and Department of Commerce)

Another economic factor is the *cost of raising and educating children*. Birth and fertility rates tend to be lower in more-developed countries, where raising children is much more costly because they do not enter the labor force until they are in their late teens or twenties. In the United States, it costs more than $286,000 to raise a middle-class child from birth to age 18. By contrast, many children in poor countries receive little education and instead have to work to help their families survive (Figure 6-5).

The *availability of, or lack of, private and public pension systems* can influence the decisions of some couples on how many children to have, especially the poor in less-developed countries. Pensions reduce a couple's need to have several children to help support them in old age.

Urbanization plays a role. People living in urban areas usually have better access to family planning services and tend to have fewer children than do those living in rural areas (especially in less-developed countries) where children are often needed to help raise crops and carry daily water and fuelwood supplies.

Another important factor is the *educational and employment opportunities available for women*. Total fertility rates tend to be low when women have access to education and paid employment outside the home. In less-developed countries, a woman with no education typically has two more children than does a woman with a high school education.

Average age at marriage (or, more precisely, the average age at which a woman has her first child) also plays a role. Women normally have fewer children when their average age at marriage is 25 or older. In 2011, the average age at marriage for U.S. women was 26, up from 25 in 2000.

Birth rates and TFRs are also affected by the *availability of legal abortions*. Each year, about 190 million women become pregnant. The United Nations and the Alan Guttmacher Institute estimate that at least 40 million of these women get abortions—about 20 million of them legal and the other 20 million illegal (and often

Mark Edwards/Peter Arnold, Inc.

Figure 6-5 These young girls are child laborers in the state of Rajasthan in India. They are weaving wool on looms to make carpets for export, and they receive very little money for their work.

unsafe). The *availability of reliable birth control methods* also allows women to control the number and spacing of the children they have.

Religious beliefs, traditions, and cultural norms also play a role. In some countries, these factors favor large families as many people there strongly oppose abortion and some forms of birth control.

CONNECTIONS

China, Male Children, and the Bride Shortage

CORE CASE STUDY

In China, there is a strong preference for male children, because unlike sons, daughters are likely to marry and leave their parents, and tradition dictates that the families of brides are obligated to provide expensive dowries. Some pregnant Chinese women use ultrasound to determine the gender of their fetuses and get an abortion if the child is female. This has led to problems. Some thieves steal baby boys and sell them to families that want a boy. Also, the government estimates that by 2030, approximately 30 million Chinese men will not be able to find wives. Because of this rapidly growing "bride shortage," young girls in some parts of rural China are being kidnapped and sold as brides for single men in other parts of the country.

Several Factors Affect Death Rates

The rapid growth of the world's population over the past 100 years has been caused largely by a decline in death rates, especially in less-developed countries. More people in these countries started living longer and fewer infants died because of increased food supplies and distribution, better nutrition, medical advances such as immunizations and antibiotics, improved sanitation, and safer water supplies.

Two useful indicators of the overall health of people in a country or region are **life expectancy** (the average number of years a newborn infant can be expected to live) and the **infant mortality rate** (the number of babies out of every 1,000 born who die before their first birthday). (See a map of infant mortality by country in Figure 9, p. S29, of Supplement 6.) Between 1955 and 2011, the average global life expectancy increased from 48 years to 70 years. Between 1900 and 2011, the average U.S. life expectancy increased from 47 years to 78 years. However, the United States ranks 49th among nations in life expectancy, down from 5th in 1950. Research studies indicate that a key factor in this ranking is poor health, even though the United States leads the world in health care costs per person.

GOOD NEWS

Infant mortality is viewed as one of the best measures of a society's quality of life because it reflects a country's general level of nutrition and health care. A high infant mortality rate usually indicates insufficient food (*undernutrition*), poor nutrition (*malnutrition*), and a high incidence of infectious disease (usually from drinking contaminated water and having weakened disease resistance due to undernutrition and malnutrition). Infant mortality also affects the TFR. In areas with low infant mortality rates, women tend to have fewer children because fewer children die at an early age.

Infant mortality rates in more-developed and less-developed countries have declined dramatically since 1965. Even so, more than 4 million infants (most of them in less-developed countries) die of *preventable* causes during their first year of life. This average of 11,000 mostly unnecessary infant deaths per day is equivalent to 55 jet airliners, each loaded with 200 infants younger than age 1, crashing *every day* with no survivors—a tragedy rarely reported in the news.

In 1900 the U.S. infant mortality rate was 165. In 2011 it was 6.1. This sharp decline was a major factor in the marked increase in U.S. average life expectancy during this period. The United States ranks first in the world in terms of health care spending per person, but 54th in terms of infant mortality rates.

Three factors have helped to keep the U.S. infant mortality rate higher than it could be: **(1)** the generally inadequate health care for poor women during pregnancy and for their babies after birth, **(2)** drug addiction among pregnant women, and **(3)** a high teenage pregnancy rate.

Migration Affects an Area's Population Size

The third factor in population change is **migration**: the movement of people into (*immigration*) and out of (*emigration*) specific geographic areas. Most people migrating from one area or country to another seek jobs and economic improvement. Some also migrate because of religious persecution, ethnic conflicts, political oppression, wars, and certain types of environmental degradation and depletion, such as soil erosion and water and food shortages. Those in the last category can be considered *environmental refugees*. One UN study estimated that a million people are added to this category every year.

CONNECTIONS

Climate Disruption and Environmental Refugees

Environmental scientist Norman Myers warns that, as the world experiences the climate disruption in this century as projected by key climate scientists, conditions that create environmental crises such as drought and flooding will worsen. With more such crises, the number of environmental refugees could soar to as many as 250 million before the end of this century. (See more on this in the Guest Essay by Norman Myers at **www.cengagebrain.com**.)

■ **CASE STUDY**

The United States: A Nation of Immigrants

Since 1820, the United States has admitted almost twice as many immigrants and refugees as all other countries combined. The number of legal immigrants (including refugees) has varied during different periods because of changes in immigration laws and rates of economic

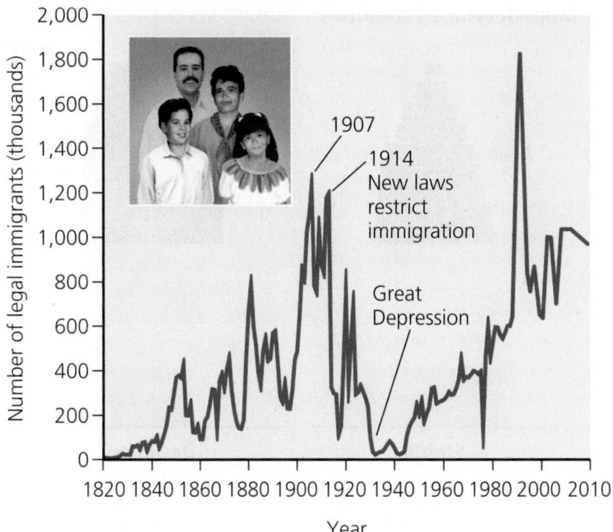

Figure 6-6 This graph shows legal immigration to the United States, 1820–2006 (the last year for which data are available). The large increase in immigration since 1989 resulted mostly from the Immigration Reform and Control Act of 1986, which granted legal status to certain illegal immigrants who could show they had been living in the country prior to January 1, 1982. (Data from U.S. Immigration and Naturalization Service and the Pew Hispanic Center)

growth (Figure 6-6). Currently, legal and illegal immigration account for about 36% of the country's annual population growth.

Between 1820 and 1960, most legal immigrants to the United States came from Europe. Since 1960, most have come from Latin America (53%) and Asia (25%), followed by Europe (14%). In 2010, Hispanics (2 out of 3 from nearby Mexico) made up 15% of the U.S. population, and by 2050, are projected to make up 30% of the population.

There is controversy over whether to reduce legal immigration to the United States. Proponents of reducing legal immigration argue that it would allow the United States to stabilize its population sooner and help to reduce the country's enormous environmental impact. Those opposed to reducing current levels of legal immigration argue that it would diminish the historical role of the United States as a place of opportunity for the world's poor and oppressed. They also argue that it would take away from the cultural diversity and innovative spirit that have been hallmarks of American culture since the country's beginnings.

In addition, according to several studies that include a 2006 study by the Pew Hispanic Center, most immigrants and their descendants start new businesses, create jobs, add cultural vitality, and help the United States to succeed in the global economy. Many immigrants also take menial and low-paying jobs that most other Americans shun.

In 2011, there were an estimated 11 million illegal immigrants in the United States, about 62% of them from nearby Mexico. There is intense political controversy over what to do about illegal immigration. Some want to deport all illegal immigrants, even though law enforcement officials warn that this is essentially impossible. Others want to set up programs that allow illegal immigrants (except for those with criminal records) to remain in the country as long as they are working their way to full citizenship.

6-3 How Does a Population's Age Structure Affect Its Growth or Decline?

▶ **CONCEPT 6-3 The numbers of males and females in young, middle, and older age groups determine how fast a population grows or declines.**

A Population's Age Structure Helps Us to Make Projections

An important factor determining whether the population of a country increases or decreases is its **age structure**: the numbers or percentages of males and females in young, middle, and older age groups in that population (**Concept 6-3**).

Population experts construct a population age-structure diagram by plotting the percentages or numbers of males and females in the total population in each of three age categories: *prereproductive* (ages 0–14), consisting of individuals normally too young to have children; *reproductive* (ages 15–44), consisting of those normally able to have children; and *postreproductive* (ages 45 and older), with individuals normally too old to have children. Figure 6-7 (p. 102) presents generalized age-structure diagrams for countries with rapid, slow, zero, and negative population growth rates. A population with a large proportion of its people in the prereproductive age group (far left) has a significant potential for rapid population growth.

A country with a large percentage of its people younger than age 15 (represented by a wide base in Figure 6-7, far left) will experience rapid population growth unless death rates rise sharply. Because of this

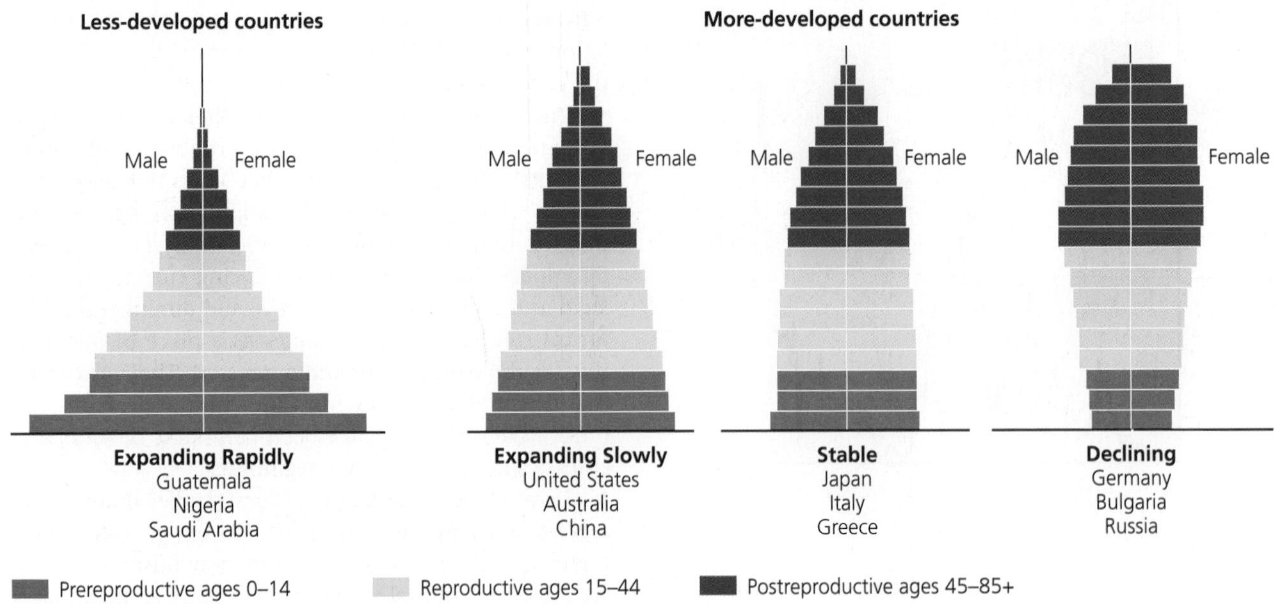

Active Figure 6-7 This chart represents the generalized population age-structure diagrams for countries with rapid (1.5–3%), slow (0.3–1.4%), zero (0–0.2%), and negative (declining) population growth rates. *See an animation based on this figure at* www.cengagebrain.com. **Question:** Which of these diagrams best represents the country where you live? (Data from Population Reference Bureau)

demographic momentum, the number of births in such a country will rise for several decades even if women have an average of only one or two children, due to the large number of girls entering their prime reproductive years.

In 2011, about 27% of the world's population—29% in the less-developed countries and 16% in more-developed countries—was under age 15. In 2011, the five countries with the largest percentage of their population under age 15 were, in order: Niger, Uganda, Burkina Faso, Democratic Republic of the Congo, and Zambia—all less-developed African countries. By 2025, the world's current 1.9 billion people under age 15—more than 1 of every 4 persons on the planet—will move into their prime reproductive years. The dramatic differences in population age structure between less-developed countries (Figure 6-7, far left) and more-developed countries (Figure 6-7, far right) show why most future human population growth will take place in less-developed countries.

The global population of seniors—people who are 65 and older—is projected to triple by 2050, when one of every six people will be a senior. (See the Case Study that follows.) This graying of the world's population is due largely to declining birth rates and medical advances that have extended life spans. In 2011, the five nations with the largest percentage of their population age 65 or older were, in order: Japan, Germany, Italy, Sweden, and Greece. In such countries, the number of working adults is shrinking in proportion to the number of seniors, which in turn is slowing the growth of these countries' tax revenues. Some analysts worry about how such societies will support their growing populations of seniors.

■ CASE STUDY

The American Baby Boom

Changes in the distribution of a country's age groups have long-lasting economic and social impacts. For example, consider the American baby boom, which added 79 million people to the U.S. population between 1946 and 1964 (Figure 6-3). Over time, this group looks like a bulge moving up through the country's age structure, as shown in Figure 6-8.

For decades, members of the baby-boom generation have strongly influenced the U.S. economy because they make up about 36% of all adult Americans. Baby boomers created the youth market in their teens and twenties and are now creating the late middle age and senior markets. In addition to having this economic impact, the baby-boom generation plays an increasingly important role in deciding who gets elected to public office and what laws are passed.

After 2011, when the first baby boomers began turning 65, the number of Americans older than age 65 will grow at the rate of about 10,000 a day through 2030, when their numbers will amount to one of every five people in the country. This process has been called the *graying of America*. As the number of working adults declines in proportion to the number of seniors, so will the tax revenues necessary for supporting the growing senior population. Some analysts say that this will prompt calls for admitting more immigrants into the country in order to swell the workforce and the income tax revenues. A recent study by the U.N. Population Division estimated that the U.S. will need to add an average of 10.8 million immigrants a year to maintain its current ratio of tax-paying workers to retirees.

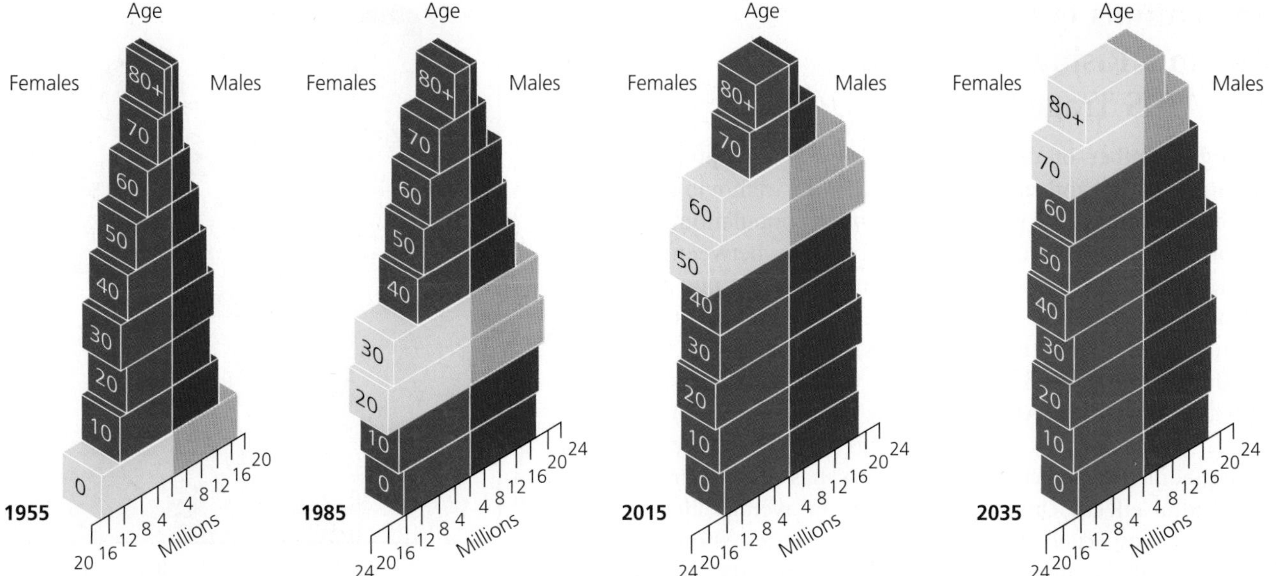

Active Figure 6-8 These charts track the baby-boom generation in the United States, showing the U.S. population by age and sex for 1955, 1985, 2015 (projected), and 2035 (projected). *See an animation based on this figure at* **www.cengagebrain.com**. (Data from U.S. Census Bureau)

Populations Made Up Mostly of Older People Can Decline Rapidly

As the age structure of the world's population changes and the percentage of people age 65 or older increases, more countries will begin experiencing population declines. If population decline is gradual, its harmful effects usually can be managed. However, some countries are experiencing fairly rapid declines and feeling such effects more severely.

Japan has the world's highest percentage of elderly people and the world's lowest percentage of young people. In 2011, Japan's population was 128 million. By 2050, its population is projected to be 95 million, a 26% drop. As its population declines, there will be fewer adults working and paying taxes to support an increasing elderly population. Because Japan discourages immigration, it may face a bleak economic future. As a result, some have called for the country to rely more on robots to do its manufacturing jobs and on selling robots in the global economy to help support its aging population.

In China, the growth in numbers of children has slowed because of the one-child policy (**Core Case Study**). As a result, the average age of China's population has been increasing over the past two decades at one of the fastest rates ever recorded. While China's population is not yet declining, the UN estimates that by 2025, China is likely to have too few young workers to support its rapidly aging population. This graying of the Chinese population could lead to a declining work force, higher wages for workers, limited funds for supporting continued economic development, and fewer children and grandchildren to care for the growing number of elderly people. These concerns and other factors may slow economic growth and have led to some relaxation of China's one-child population control policy.

Figure 6-9 lists some of the problems associated with rapid population decline. Countries currently faced with rapidly declining populations include Japan, Russia, Germany, Bulgaria, Hungary, Ukraine, Serbia, Greece, Portugal, and Italy.

Some Problems with Rapid Population Decline

- Can threaten economic growth
- Labor shortages
- Less government revenues with fewer workers
- Less entrepreneurship and new business formation
- Less likelihood for new technology development
- Increasing public deficits to fund higher pension and health-care costs
- Pensions may be cut and retirement age increased

Figure 6-9 Rapid population decline can cause several problems. **Question:** Which two of these problems do you think are the most important?

Populations Can Decline Due to a Rising Death Rate: The AIDS Tragedy

A large number of deaths from AIDS can disrupt a country's social and economic structure by removing significant numbers of young adults from its population. According to the World Health Organization, between 1981 and 2010, AIDS killed more than 27 million people and it takes about 1.8 million more lives each year (including 18,000 in the United States).

Unlike hunger and malnutrition, which kill mostly infants and children, AIDS kills many young adults and leaves many children orphaned. This change in the young-adult age structure of a country has a number of harmful effects. One is a sharp drop in average life expectancy, especially in several African countries where 15–26% of the adult population is infected with HIV.

Another effect of the AIDS pandemic is the loss of productive young-adult workers and trained personnel such as scientists, farmers, engineers, and teachers, as well as government, business, and health-care workers. This means there are fewer taxpayers and fewer workers available to support the very young and the elderly. Worldwide, AIDS is the leading cause of death for people ages 15–49. This loss of productive working adults can affect the age structure of a population (Figure 6-10) and result in large numbers of orphans, many of them also infected with HIV.

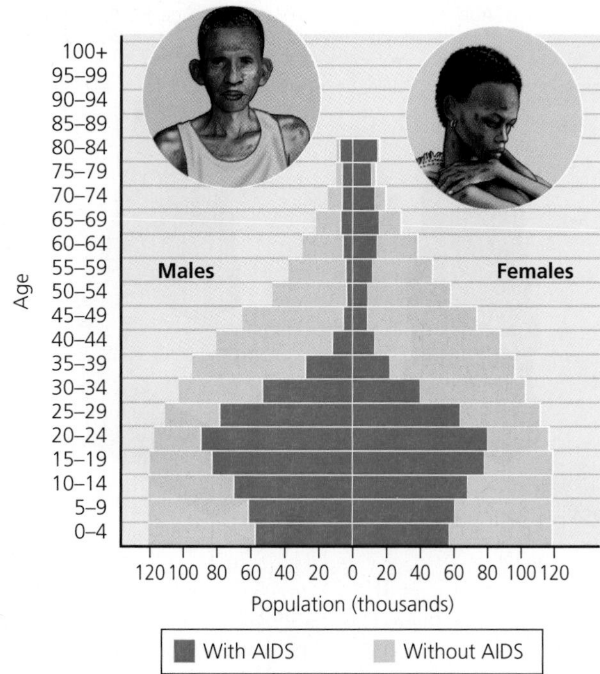

Figure 6-10 *Global outlook:* In Botswana, more than 25% of this age group was infected with HIV in 2009. This figure shows two projected age structures for Botswana's population in 2020—one including the possible effects of the AIDS epidemic (red bars), and the other not including those effects (yellow bars). See the Data Analysis Exercise at the end of this chapter for further analysis of this problem. (Data from the U.S. Census Bureau) **Question:** How might this affect Botswana's economic development?

6-4 How Can We Slow Human Population Growth?

▶ **CONCEPT 6-4 We can slow human population growth by reducing poverty, elevating the status of women, and encouraging family planning.**

There Are Three Effective Ways to Slow Population Growth

Scientific studies and experience have shown that the three most effective ways to slow or stop population growth are **(1)** to reduce poverty, primarily through economic development and universal primary education, **(2)** to elevate the status of women, and **(3)** to encourage family planning and reproductive health care (**Concept 6-4**). Let's look more closely at each of these strategies.

Promote Economic Development

Demographers, examining the birth and death rates of western European countries that became industrialized during the 19th century, developed a hypothesis of population change known as the **demographic transition**: As countries become industrialized and economically developed, their populations tend to grow more slowly. According to the hypothesis based on such data, this transition takes place in four stages (Figure 6-11).

Some analysts believe that most of the world's less-developed countries will make a demographic transition over the next few decades, mostly because modern technology can raise per capita incomes by bringing economic development and family planning to such countries. Other analysts fear that rapid population growth, extreme poverty, and increasing environmental degradation and resource depletion in some low-income, less-developed countries could leave these countries stuck in stage 2 of the demographic transition.

Other factors that could hinder the demographic transition in some less-developed countries are shortages of scientists, engineers, and skilled workers; insuf-

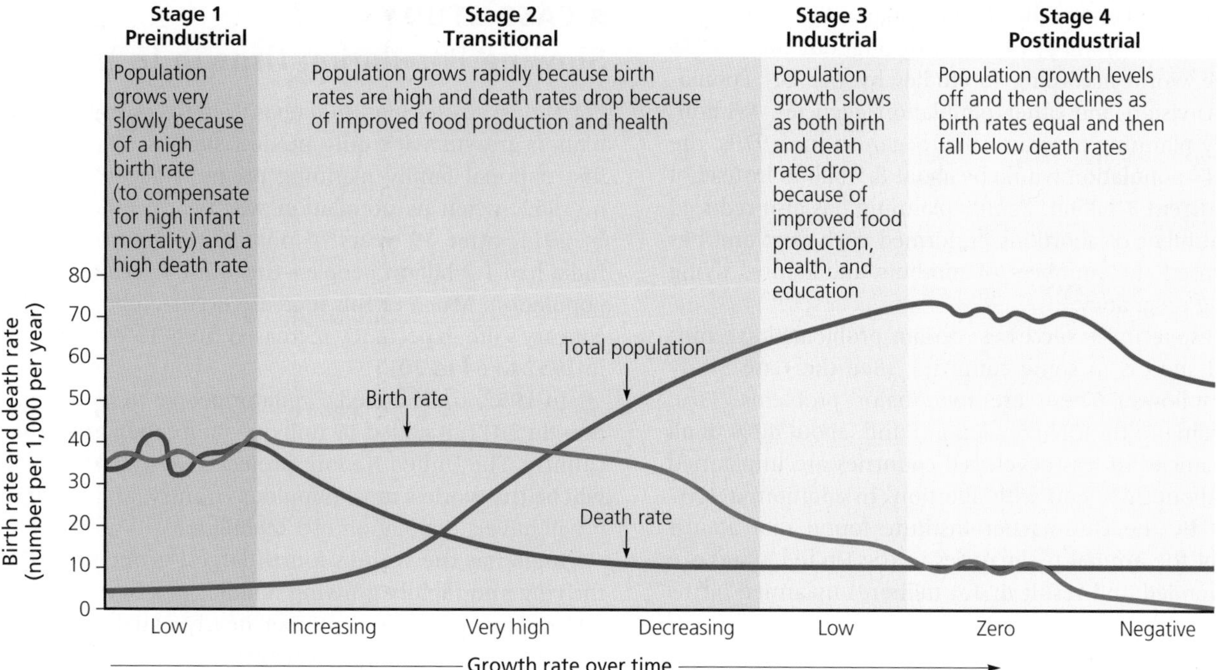

Active Figure 6-11 The *demographic transition*, which a country can experience as it becomes industrialized and more economically developed, can take place in four stages. *See an animation based on this figure at* **www.cengagebrain.com. Question:** At what stage is the country where you live?

ficient financial capital; large foreign debt to more-developed countries; and a drop in economic assistance from more-developed countries since 1985. This could leave large numbers of people trapped in poverty.

Empower Women

A number of studies show that women tend to have fewer children if they are educated, have the ability to control their own fertility, earn an income, and live in societies that do not suppress their rights. Although women make up roughly half of the world's population, in most societies they have fewer rights and educational and economic opportunities than men have.

Women do almost all of the world's domestic work and child care for little or no pay and provide more unpaid health care (within their families) than do all of the world's organized health-care services combined. They also do 60–80% of the work associated with growing food, gathering, and hauling wood (Figure 6-12) and animal dung for use as fuel, and hauling water in rural areas of Africa, Latin America, and Asia. As one Brazilian woman observed, "For poor women, the only holiday is when you are asleep."

While women account for 66% of all hours worked, they receive only 10% of the world's income and own just 2% of the world's land. They also make up 70% of the world's poor and 64% of its 800 million illiterate adults. Poor women who cannot read often have an average of 5–7 children, compared to 2 or fewer children in societies where almost all women can read.

Promote Family Planning

Family planning provides educational and clinical services that help couples choose how many children to have and when to have them. Such programs vary from culture to culture, but most of them provide information on birth spacing, birth control, and health care for pregnant women and infants.

Mark Edwards/Peter Arnold, Inc.

Figure 6-12 These women from a village in the West African country of Burkina Faso are bringing home fuelwood. Typically, they spend two hours a day, two or three times a week, searching for and hauling fuelwood.

Family planning has been a major factor in reducing the number of births throughout most of the world according to studies by the UN Population Division and other population agencies. Without family planning programs that began in the 1970s, the world's population would be about 8.5 billion instead of the current 7 billion. Family planning has also reduced the number of abortions performed each year and has decreased the numbers of mothers and fetuses dying during pregnancy. GOOD NEWS

Despite these successes, certain problems have hindered success in some countries. (See the Case Study that follows.) There are two major problems. *First*, according to the UN Population Fund, about 42% of all pregnancies in less-developed countries are unplanned and about 26% end with abortion. In addition, a 2007 study by the Guttmacher Institute found that almost half of the annual pregnancies in the United States are unintended and result in 1.4 million unplanned births and 1.3 million abortions.

Second, an estimated 201 million couples in less-developed countries want to limit their number of children and determine their spacing, but they lack access to family planning services. According to the UN Population Fund and the Alan Guttmacher Institute, meeting women's current unmet needs for family planning and contraception could prevent about 52 million unwanted pregnancies, 22 million induced abortions, 1.4 million infant deaths, and 142,000 pregnancy-related deaths of women *each year*. This could reduce the projected global population size by more than 1 billion people, at an average cost of $20 per couple per year. The experiences of countries such as Japan, Thailand, Bangladesh, South Korea, Taiwan, Iran, and China (**Core Case Study**) show that a country can achieve or come close to replacement-level fertility within a decade or two. CORE CASE STUDY

■ CASE STUDY

Slowing Population Growth in India

For almost six decades, India has tried to control its population growth with only modest success. The world's first national family planning program began in India in 1952, when its population was nearly 400 million. In 2011, after 59 years of population control efforts, India had 1.2 billion people—the world's second largest population. Much of this increase occurred because the country's life expectancy increased from 38 years of age in 1952 to 64 in 2011.

In 1952, India added 5 million people to its population. In 2011, it added 19 million—more than any other country. The United Nations projects that by 2030, India will be the world's most populous country, and by 2050 it will have a population of 1.69 billion.

India has the world's fourth largest economy and a thriving and rapidly growing middle class of more than 100 million people—a number nearly equal to a third of the U.S. population. However, the country faces a number of serious poverty, malnutrition, and environmental problems that could worsen as its population continues to grow rapidly. About one of every four of India's urban population lives in slums, and prosperity and progress have not touched many of the nearly 650,000 rural villages where more than two-thirds of India's population lives. Nearly half of the country's labor force is unemployed or underemployed, and 42% of its population lives in extreme poverty (Figure 6-13). By comparison, in China, the percentage of the population living in extreme poverty is about 16%, down from 60% in 1990.

For decades, the Indian government has provided family planning services throughout the country and has strongly promoted a smaller average family size. Even so, Indian women have an average of 2.6 children.

ullstein-Leber/Peter Arnold, Inc.

Figure 6-13 Four of every ten people in India, like this homeless mother and her child, struggle to live on the equivalent of less than $1.25 per day.

Two factors help account for larger families in India. First, most poor couples believe they need several children to work and care for them in old age. Second, as in China (**Core Case Study**), the strong cultural preference in India for male children means that some couples keep having children until they produce one or more boys. The result: even though 9 of every 10 Indian couples have access to at least one modern birth control method, only 47% actually use one (compared to 84% in China). India's efforts to slow its population growth have been less successful than those in China because India is the world's largest democracy and, unlike China, cannot impose such a program on its people.

Like China, India also faces critical resource and environmental problems. With 17% of the world's people, India has just 2.3% of the world's land resources and 2% of its forests. About half the country's cropland is degraded as a result of soil erosion and overgrazing. In addition, more than two-thirds of its water is seriously polluted, sanitation services often are inadequate, and many of its major cities suffer from serious air pollution.

India is undergoing rapid economic growth, which is expected to accelerate. As members of its growing middle class increase their use of resources, raising their rate of resource use per person, India's ecological footprint will expand and increase the pressure on the country's and the earth's natural capital. On the other hand, economic growth may help India to slow its population growth by accelerating its demographic transition.

6-5 What Are the Major Urban Resource and Environmental Problems?

CONCEPT 6-5 Most cities are unsustainable because of high levels of resource use, waste, pollution, and poverty.

Scientists See Three Important Urban Trends

The world's first cities emerged about 6,000 years ago. Today, 51% of the world's people, and 79% of Americans, live in urban areas. (See the Case Study that follows.)

Urban areas grow in two ways—by *natural increase* (more births than deaths) and by *immigration*, mostly from rural areas. Rural people are *pulled* to urban areas in search of jobs, food, housing, educational opportunities, and better health care. Some are also *pushed* from rural to urban areas by factors such as poverty, lack of land for growing food, declining agricultural jobs, famine, war, and religious, and political conflicts.

Here are three major trends in urban population dynamics:

1. *The proportion of the global population living in urban areas has increased dramatically.* While 2% of the world's people lived in urban areas in 1850, the figure is now 55% (75% in more-developed countries) and is projected to be 70% by 2050. Urban growth in some countries is very rapid, with the percent of people in China living in urban areas growing from 28% to 50% between 1990 and 2011.
2. *The numbers and sizes of urban areas are mushrooming.* There are more than 400 urban areas with 1 million or more people. In addition, in 2011 there were 22 *megacities* or *megalopolises*—cities with 10 million or more people—19 of which are in less-developed countries (Figure 6-14, p. 108). Such megacities will soon be eclipsed by *hypercities* with more than 20 million people. For example, greater Tokyo, Japan, contains 36 million people—more than the entire population of Canada. In 2010, the United Nations reported that some of the world's megacities and hypercities are merging into urban *megaregions* that spread across huge areas and across entire countries.
3. *Poverty is becoming increasingly urbanized, mostly in less-developed countries.* The United Nations estimates that at least 1 billion people in less-developed countries (more than three times the current U.S. population) live in crowded and unsanitary slums and shantytowns in cities or on their outskirts.

THINKING ABOUT
Urban Trends

If you could reverse one of the three urban trends discussed here, which one would it be? Explain.

CASE STUDY

Urbanization in the United States

Between 1800 and 2011, the percentage of the U.S. population living in urban areas increased from 5% to 79%. This population shift has occurred in four phases.

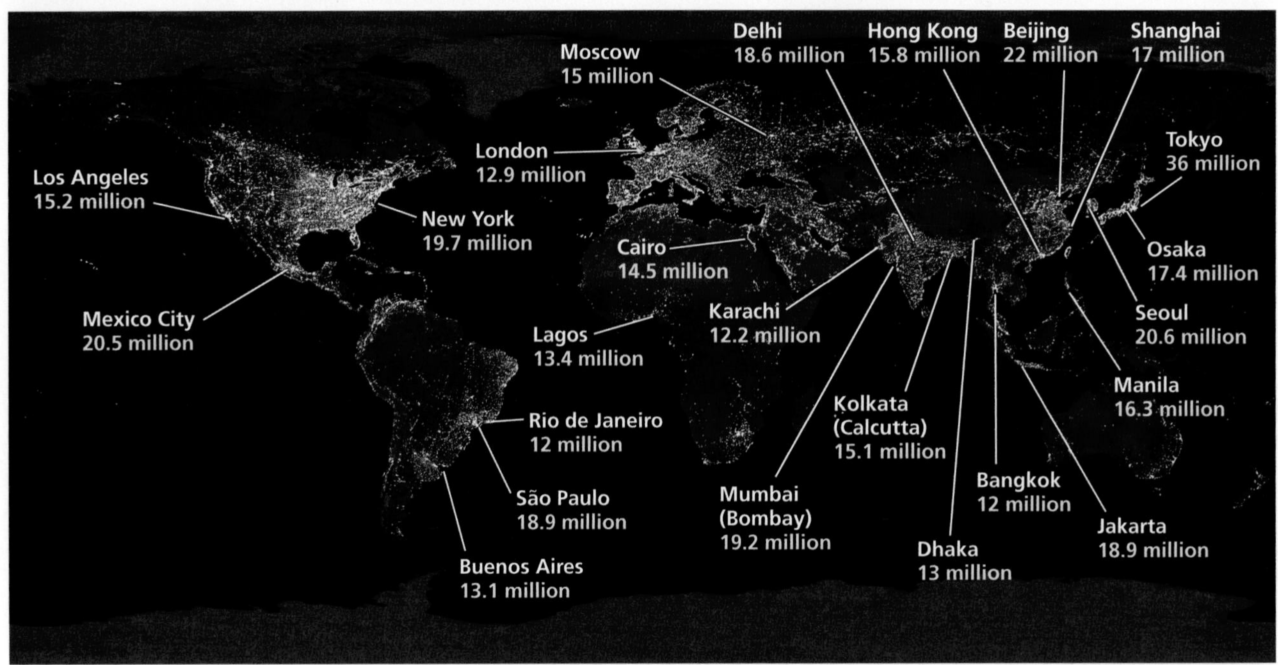

Figure 6-14 *Global outlook:* Major urban areas throughout the world are revealed in these satellite images of the earth at night, showing city lights. **Question:** In order, what were the world's five most populous cities in 2011? (Data from National Geophysics Data Center, National Oceanic and Atmospheric Administration, and United Nations)

First, *people migrated from rural areas to large central cities*. Currently, about three-fourths of Americans live in cities with at least 50,000 people, and nearly half live in urban areas with 1 million or more residents (Figure 6-15, shaded areas).

Second, *many people migrated from large central cities to smaller cities and suburbs*. Currently, about half of urban Americans live in the suburbs, nearly a third in central cities, and the rest in rural housing developments beyond suburbs.

Third, *many people migrated from the North and East to the South and West*. Since 1980, about 80% of the U.S. population increase has occurred in the South and West. This trend is likely to continue.

Fourth, since the 1970s, and especially since 1990, *some people have fled both cities and suburbs, and migrated to developed areas outside of suburbs*. The result is rapid growth of *exurbs*—housing developments scattered over vast areas that lie beyond suburbs and have no socioeconomic centers.

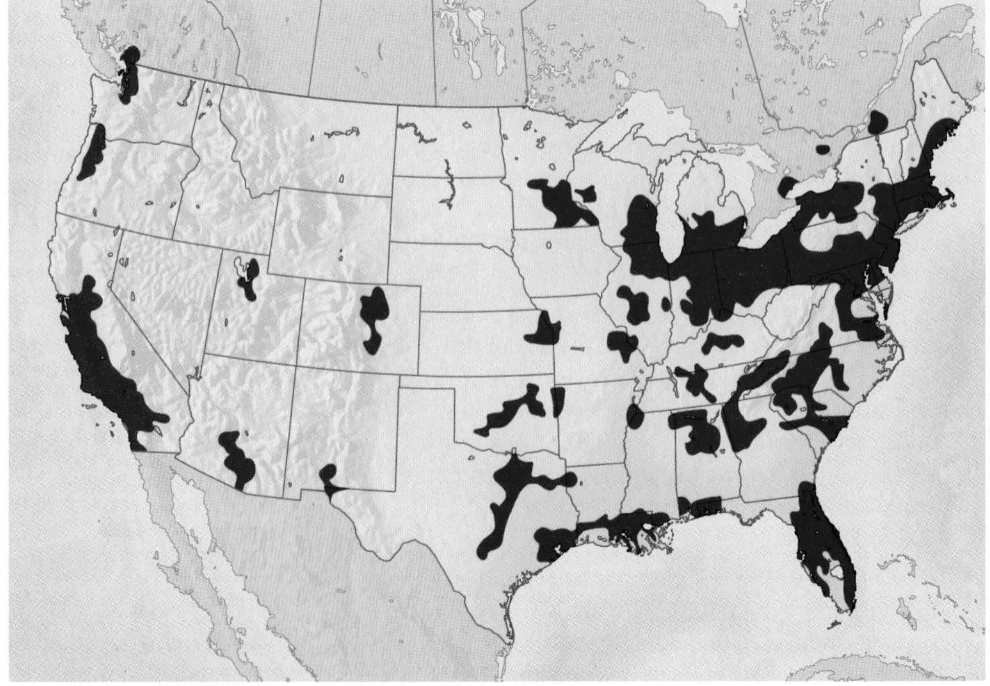

Figure 6-15 The shaded areas are urban areas in the United States with more than 1 million people, where nearly half (48%) of Americans live. **Question:** Why are many of the largest urban areas located near water? (Data from National Geophysical Data Center/National Oceanic and Atmospheric Administration, U.S. Census Bureau)

Since 1920, many of the worst urban environmental problems in the United States have GOOD NEWS been reduced significantly. Most people have better working and housing conditions, while air and water quality have improved. Better sanitation, clean public water supplies, and medical care have slashed death rates and incidences of sickness from infectious diseases. Concentrating most of the population in urban areas also has helped protect the country's biodiversity by reducing the destruction and degradation of wildlife habitat.

However, a number of U.S. cities—especially older ones—have deteriorating services and aging *infrastructures* (streets, bridges, dams, power lines, schools, waste management, water supply pipes, and sewers). Funds for repairing and upgrading urban infrastructure have declined as the flight of people to the suburbs and exurbs has decreased city revenues from property taxes.

Urban Sprawl Gobbles Up the Countryside

In the United States and some other countries, **urban sprawl**—the growth of low-density development on the edges of cities and towns—is eliminating surrounding agricultural and wild lands (Figure 6-16). It results in a dispersed jumble of housing developments, shopping malls, parking lots, and office complexes that are loosely connected by multilane highways and freeways.

Urban sprawl is the product of ample affordable land, automobiles, and gasoline, and little or no urban planning. Many people prefer living in unplanned suburbs and exurbs. Compared to central cities, these areas provide lower-density living and access to larger lot sizes and single-family homes. Often these areas also have newer public schools and lower crime rates.

On the other hand, urban sprawl has caused or contributed to a number of environmental problems. Because of nonexistent or inadequate mass transportation in most such areas, sprawl forces people to drive everywhere, emitting climate-changing greenhouse gases and air pollutants in the process. Sprawl has decreased energy efficiency, increased traffic congestion, and destroyed prime cropland, forests, and wetlands. It has also led to the economic deaths of many central cities as people and businesses move out of these areas. Figure 6-17 (p. 110) summarizes some of the undesirable consequences of urban sprawl.

Urbanization Has Advantages

Urbanization has many benefits. From an *economic standpoint,* cities are centers of industry, GOOD NEWS commerce, transportation, innovation, education, technological advances, and jobs. Urban residents in many

Image courtesy of the U.S. Geological Survey

1973

Image courtesy of the U.S. Geological Survey

2009

Active Figure 6-16 These satellite images show *urban sprawl* in and around the U.S. city of Las Vegas, Nevada, between 1973 and 2009. Between 2000 and 2010, Las Vegas was the fastest-growing urban area in the United States with a 42% increase in its population. *See an animation at* www.cengagebrain.com *to examine how the San Francisco Bay area in the U.S. state of California grew in population between 1900 and 1990.* **Question:** What might be a limiting factor on population growth in Las Vegas?

Figure 6-17 These are some of the undesirable impacts of urban sprawl, or car-dependent development. **Question:** Which five of these effects do you think are the most harmful?

Natural Capital Degradation

Urban Sprawl

Land and Biodiversity	**Water**	**Energy, Air, and Climate**	**Economic Effects**
Loss of cropland	Increased use and pollution of surface water and groundwater	Increased energy use and waste	Decline of downtown business districts
Loss and fragmentation of forests, grasslands, wetlands, and wildlife habitat	Increased runoff and flooding	Increased emissions of carbon dioxide and other air pollutants	More unemployment in central cities

parts of the world also tend to live longer than do rural residents and to have lower infant mortality and fertility rates. They also have better access to medical care, family planning, education, and social services than do their rural counterparts.

Urban areas also have some environmental advantages. Recycling is more economically feasible because the high concentrations of recyclable materials in urban areas. Concentrating people in cities helps to preserve biodiversity by reducing the stress on wildlife habitats. Central cities also can save energy if residents rely more on energy-efficient mass transportation, walking, and bicycling.

Urbanization Has Disadvantages

Most Urban Areas Are Unsustainable Systems. Urban populations occupy only about 2% of the earth's land area but consume about 75% of its resources and produce about 75% of the world's pollution and wastes. Because of this high resource input of food, water, and materials, and the resulting high waste output (Figure 6-18), *most of the world's urban areas have huge ecological footprints that extend far beyond their boundaries, and such cities are not self-sustaining systems* (**Concept 6-5**). For example, according to an analysis by Mathis Wackernagel and William Rees, developers

Figure 6-18 Natural capital degradation: Urban areas are rarely sustainable systems. The typical city depends on large nonurban areas for huge inputs of matter and energy resources, while it generates large outputs of waste matter and heat. **Question:** How would you apply the three **principles of sustainability** (see back cover) to lessen some of these impacts?

of the ecological footprint concept, London, England, requires an area 58 times as large as the city to supply its residents with resources. They estimate that if all of the world's people used resources at the same rate as Londoners do, it would take at least three more planet Earths to meet their needs.

CONNECTIONS
Urban Living and Biodiversity Awareness

Recent studies reveal that most urban dwellers live most or all of their lives in an artificial environment that isolates them from forests, grasslands, streams, and other natural areas that make up the world's biodiversity. As a result, many urban residents tend to be unaware of the importance of protecting not only the earth's increasingly threatened biodiversity but also its other forms of natural capital that support their lives and the cities in which they live.

Most Cities Lack Vegetation. In urban areas, most trees, shrubs, grasses, and other plants are destroyed to make way for buildings, roads, parking lots, and housing developments. Thus, most cities do not benefit from vegetation that would absorb air pollutants, give off oxygen, provide shade, reduce soil erosion, provide wildlife habitats, and offer aesthetic pleasure. As one observer remarked, "Cities are places where they cut down most of the trees and then name the streets after them."

Many Cities Have Water Problems. As cities grow and their water demands increase, expensive reservoirs and canals must be built and deeper wells must be drilled. This can deprive rural and wild areas of surface water and can deplete underground water supplies.

Flooding also tends to be greater in some cities that are built on floodplains near rivers or along low-lying coastlines subject to natural flooding. In addition, covering land with buildings, asphalt, and concrete causes precipitation to run off quickly and overload storm drains. Further, urban development has often destroyed or degraded large areas of wetlands that have served as natural sponges to help absorb excess storm water. Many of the world's largest coastal cities (Figure 6-14) face a new threat of flooding some time in this century as sea levels rise because of projected climate change due to a warmer atmosphere.

Some cities in arid areas depend on water withdrawn from rivers and reservoirs that are fed by mountaintop glaciers. Such cities will face severe water shortages if projected atmospheric warming melts these mountaintop glaciers.

Cities Tend to Concentrate Pollution and Health Problems. Because of their high population densities and high resource consumption, cities produce most of the world's air pollution, water pollution, and solid and hazardous wastes. Pollutant levels are generally higher because pollution is produced in a smaller area and cannot be dispersed and diluted as readily as pollution produced in rural areas can. In addition, high population densities in urban areas can increase the spread of infectious diseases, especially if adequate drinking water and sewage systems are not available. Excessive exposure to noise is another problem.

Cities Affect Local Climates. On average, cities tend to be warmer, rainier, foggier, and cloudier than suburbs and nearby rural areas. In cities, the enormous amount of heat generated by cars, factories, furnaces, lights, air conditioners, and heat-absorbing dark roofs and streets creates an *urban heat island* that is surrounded by cooler suburban and rural areas. As cities grow and merge, their heat islands merge, which can reduce the natural dilution and cleansing of polluted air.

The artificial light created by cities also affects some plant and animal species. For example, some endangered sea turtles lay their eggs on beaches at night and require darkness to do so. In addition, each year large numbers of migrating birds, lured off course by the lights of high-rise buildings, fatally collide with these structures.

THINKING ABOUT
Disadvantages of Urbanization

Which two of the disadvantages discussed here for living in urban areas do you think are the most serious? Explain.

Life Is a Desperate Struggle for the Urban Poor in Less-Developed Countries

Poverty is a way of life for many urban dwellers in less-developed countries. At least 1 billion people live under crowded and unsanitary conditions in cities in these countries, and according to a 2006 UN study, that number could reach 1.4 billion by 2020. (See the Case Study that follows.)

Some of these people live in *slums*—areas dominated by tenements and rooming houses where several people might live in a single room. Others live in *squatter settlements* and *shantytowns* on the outskirts of these cities. They build shacks from corrugated metal, plastic sheets, scrap wood, cardboard, and other scavenged building materials, or they live in rusted shipping containers and junked cars. Still others live or sleep on the streets (Figure 6-13).

Poor people living in shantytowns and squatter settlements usually lack clean water supplies, sewers, electricity, and roads, and are subject to severe air and water pollution and hazardous wastes from nearby factories. Many of these settlements are in locations especially prone to landslides, flooding, or earthquakes. Some city governments regularly bulldoze squatter shacks and send police to drive illegal settlers out. The people usually move back in or develop another shantytown elsewhere.

■ CASE STUDY
Mexico City

Mexico City—the world's second most populous city—is an urban area in crisis. It has 20 million residents, and each year, at least 400,000 new residents arrive.

Mexico City suffers from severe air pollution, close to 50% unemployment, deafening noise, overcrowding, traffic congestion, inadequate public transportation, and a soaring crime rate. More than one-third of its residents live in slums called *barrios* or in squatter settlements that lack running water and electricity.

At least 3 million people in the barrios have no sewage facilities. As a result, large amounts of human waste are deposited in gutters, vacant lots, and open ditches every day, attracting armies of rats and swarms of flies. When the winds pick up dried excrement, a *fecal snow* blankets parts of the city. This bacteria-laden fallout leads to widespread salmonella and hepatitis infections, especially among children.

Mexico City has serious air pollution problems because of a combination of factors: too many cars and polluting factories, more smog because of its warm and sunny climate, and topographical bad luck. The city sits in a high-elevation, bowl-shaped valley surrounded on three sides by mountains—conditions that can trap air pollutants at ground level. Breathing the city's air is said to be roughly equivalent to smoking three packs of cigarettes per day, and respiratory diseases are rampant. (See *The Habitable Planet,* Video 11 at **www.learner.org/resources/series209.html** to learn how scientists are measuring air pollution levels in Mexico City.)

Over the past two decades, Mexico City has made progress in reducing the severity of some of its air pollution problems. GOOD NEWS The percentage of days each year in which air pollution standards are violated has fallen from 50% to 20%. Private cars must pass emissions tests every six months and are banned from the city one day a week. Its most polluting factories have been relocated, its subway system has been greatly expanded, and it has a large fleet of low-emission buses. The city also has rent-a-bike stations in the central city and has planted more than 25 million trees, which absorb carbon dioxide and some air pollutants.

While writer Carlos Fuentes once dubbed this city "Makesicko City," many of its leaders and citizens are working hard to make it a more healthful, enjoyable, and sustainable place to live.

6-6 How Does Transportation Affect Urban Environmental Impacts?

▶ **CONCEPT 6-6 In some countries, many people live in widely dispersed urban areas and depend mostly on motor vehicles for their transportation, which greatly expands their ecological footprints.**

Cities Can Grow Outward or Upward

If a city cannot spread outward, it must grow vertically—upward and downward (below ground)—so that it occupies a small land area with a high population density. Most people living in *compact cities* such as Hong Kong, China, and Tokyo, Japan, get around by walking, biking, or using mass transit such as rail or buses.

In other parts of the world, a combination of plentiful land and networks of highways have produced *dispersed cities* whose residents depend on motor vehicles for most travel (**Concept 6-6**). Such car-centered cities are found in the United States, Canada, Australia, and other countries where ample land often is available for these cities to expand outward. The resulting urban sprawl (Figure 6-16) can have a number of undesirable effects (Figure 6-17).

The United States is a prime example of a car-centered nation. With 4.5% of the world's people, the United States has almost 31% of all motor vehicles and uses about 43% of the world's gasoline, according to the U.S. Department of Transportation. In its dispersed urban areas, passenger vehicles are used for 98% of all transportation, and 75% of residents drive alone to work every day. Largely because of urban sprawl, the total annual distance driven by all Americans combined is about the same as the total distance driven each year by all other drivers in the world.

Motor Vehicles Have Advantages and Disadvantages

Motor vehicles provide mobility and offer a convenient and comfortable way to get from one place to another. For many people, they are also symbols of power, sex appeal, social status, and success. Much of the world's economy is also built on producing motor vehicles and supplying fuel, roads, services, and repairs for them.

Despite their important benefits, motor vehicles have many harmful effects on people and on the environment. Globally, each year automobile accidents kill approximately 1.2 million people—an average of 3,300 deaths per day—and injure another 15 million people. They also kill about 50 million wild animals and family pets every year.

In the United States, motor vehicle accidents kill about 33,000 people per year and injure another 5 million, at least 300,000 of them severely. *Car accidents have killed more Americans than have all the wars in the country's history.*

Motor vehicles are the world's largest source of outdoor air pollution, which causes 30,000–60,000 premature deaths per year in the United States, according to the Environmental Protection Agency. They are also the fastest-growing source of climate-changing CO_2 emissions. At least a third of the world's urban land and half of that in the United States is devoted to roads, parking lots, gasoline stations, and other automobile-related uses.

Another problem is congestion. U.S. motorists spend an increasing amount of their time—totaling about a week a year per driver—in traffic jams, as streets and freeways often resemble parking lots. Traffic congestion in some cities in less-developed countries is much worse. Building more roads may not be the answer. Many analysts agree with economist Robert Samuelson that "the number of cars expands to fill available concrete."

Reducing Automobile Use Is Not Easy, but It Can Be Done

Some environmental scientists and economists suggest that we can reduce the harmful effects of automobile use by making drivers pay directly for most of the environmental and health costs caused by their automobile use—a *user-pays* approach.

One way to phase in such *full-cost pricing* would be to charge a tax or fee on gasoline to cover the estimated harmful environmental and health costs of driving. According to a study by the International Center for Technology Assessment, such a tax would amount to about $3.18 per liter ($12 per gallon) of gasoline in the United States. Gradually phasing in such a tax, as has been done in many European nations, would spur the use of more energy-efficient motor vehicles and mass transit, decrease dependence on imported oil, and thus increase economic and national security. It would also reduce pollution and environmental degradation and help to slow projected climate disruption.

Proponents of this approach urge governments to do two major things. *First,* fund programs to educate people about the hidden environmental and health costs they are paying for gasoline. *Second,* use gasoline tax revenues to help finance mass-transit systems, bike lanes, and sidewalks as alternatives to cars, and to reduce taxes on income, wages, and wealth to offset the increased taxes on gasoline. Such a *tax shift* would help to make higher gasoline taxes more politically and economically acceptable.

However, taxing gasoline heavily would be difficult in the United States, for three reasons. *First,* it faces strong opposition from two groups. One group is made up of those people who feel they are already overtaxed, many of whom are largely unaware of the huge hidden costs they are paying for gasoline. The other group is made up of the economically and politically powerful transportation-related industries such as carmakers, oil and tire companies, road builders, and many real estate developers. *Second,* the dispersed nature of most U.S. urban areas makes people dependent on cars. *Third,* fast, efficient, reliable, and affordable mass-transit options, bike lanes, and sidewalks are not widely available in the United States.

U.S. taxpayers might accept sharp increases in gasoline taxes if a tax shift were employed, as mentioned above. Another way to reduce automobile use and urban congestion is to raise parking fees and to charge tolls on roads, tunnels, and bridges leading into cities—especially during peak traffic times. Densely populated Singapore is rarely congested because it auctions the rights to buy a car, and its cars carry electronic sensors that automatically charge the drivers a large fee every time they enter the city. Several European cities have also imposed stiff fees for motor vehicles entering their central cities.

In Germany, Austria, Italy, Switzerland, and the Netherlands, more than 300 cities have *car-sharing* networks. Members reserve a car in advance or contact the network and are directed to the closest car. They are billed every month for the time they use a car and the distance they travel. In Berlin, Germany, car sharing has cut car ownership by 75%. According to the Worldwatch Institute, car sharing in Europe has reduced the average driver's CO_2 emissions by 40–50%. Car-sharing companies have sprouted up in a number of U.S. cities.

Some Cities Promote Alternatives to Cars

There are several alternatives to motor vehicles, each with its own advantages and disadvantages. They include *bicycles, mass-transit rail systems in urban areas, bus systems in urban areas,* and *high-speed rail systems between urban areas* (Figures 6-19 through 6-22, p. 114).

THINKING ABOUT
Bicycles

Do you, or would you, use a conventional or an electric bicycle to go to and from work or school? Explain.

Bicycling and walking account for about a third of all urban trips in the Netherlands and in Copenhagen, Denmark, compared to only 1% in the United States. Some cities actually provide bicycles that people can rent at a very low cost. In addition, Japan and the Netherlands strive to integrate the use of bicycles and commuter rail by providing secure bicycle parking at rail stations.

Western Europe and Japan have high-speed *bullet trains* that travel between cities at up to 306 kilometers (190 miles) per hour. China has the world's fastest bullet train, and by 2010, will have completed a vast network of 42 high-speed train routes throughout the country, which is roughly the same size as the United States.

Many analysts in the United States have talked about building high-speed bullet-train routes between cities for decades, but so far, none have been built. Such projects involve controversy over routes, extremely high land acquisition costs, and very high construction and operating costs. Unless they are built along popular routes, they do not attract enough riders. Also, building such systems usually requires large federal government subsidies, and operating them usually requires local government subsidies.

Trade-Offs

Bicycles

Advantages	Disadvantages
Are quiet and non-polluting	Provide little protection in an accident
Take few resources to make	Provide no protection from bad weather
Burn no fossil fuels	Are impractical for long trips
Require little parking space	Secure bike parking not yet widespread

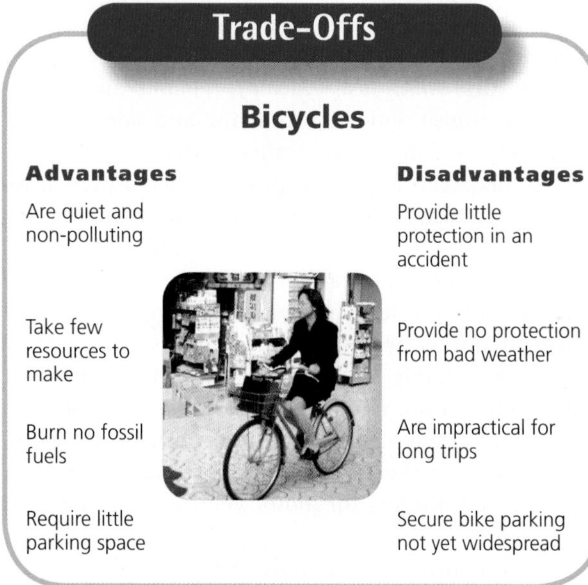

Figure 6-19 Bicycle use has advantages and disadvantages. The key to increasing bicycle use is the creation of bicycle-friendly systems, including bike lanes. **Question:** Which single advantage and which single disadvantage do you think are the most important?

Trade-Offs

Buses

Advantages	Disadvantages
Reduce car use and air pollution	Can lose money because they require affordable fares
Can be rerouted as needed	Can get caught in traffic and add to noise and pollution
Cheaper than heavy-rail system	Commit riders to transportation schedules

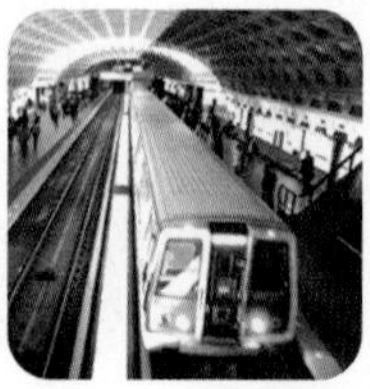

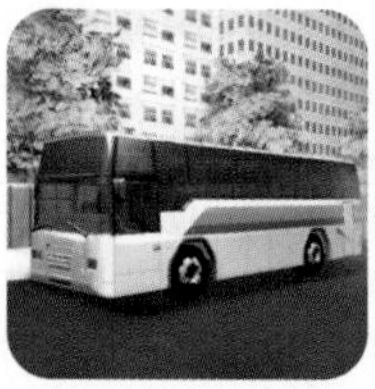

Figure 6-21 Bus rapid-transit systems (where several buses running in express lanes can be coupled together) and conventional bus systems in urban areas have advantages and disadvantages. **Question:** Which single advantage and which single disadvantage do you think are the most important?

Trade-Offs

Mass Transit Rail

Advantages	Disadvantages
Uses less energy and produces less air pollution than cars do	Expensive to build and maintain
Use less land than roads and parking lots use	Cost-effective only along a densely populated corridor
Causes fewer injuries and deaths than cars	Commits riders to transportation schedules

Figure 6-20 Mass-transit rail systems in urban areas have advantages and disadvantages. **Question:** Which single advantage and which single disadvantage do you think are the most important?

Trade-Offs

High-Speed Rail

Advantages	Disadvantages
Much more energy efficient per rider than cars and planes are	Cost effective only along heavily travelled routes
Less air pollution than cars and planes	Causes noise and vibration for nearby residents
Can reduce need for air travel, cars, roads, and parking areas	Usually requires large government subsidies to build and run

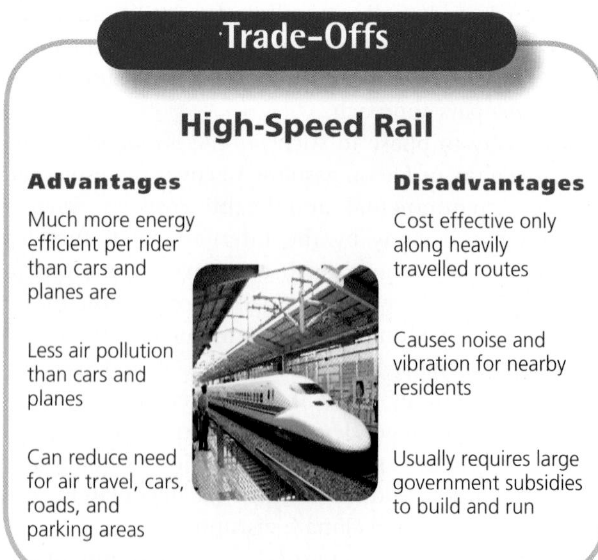

Figure 6-22 High-speed rail systems between urban areas have advantages and disadvantages. **Question:** Which single advantage and which single disadvantage do you think are the most important?

6-7 How Can Cities Become More Sustainable and Livable?

▶ **CONCEPT 6-7 An *ecocity* allows people to choose walking, biking, or mass transit for most transportation needs; to recycle or reuse most of their wastes; to grow much of their food; and to protect biodiversity by preserving surrounding land.**

We Can Make Urban Areas More Environmentally Sustainable and Enjoyable Places to Live

Many environmental scientists and urban planners call for us to make new and existing urban areas more sustainable and enjoyable places to live through good ecological design. (See the Guest Essay on this topic by David Orr at **www.cengagebrain.com**.)

Smart growth is one way to encourage more environmentally sustainable development that allows for less dependence on cars, controls and directs sprawl, and reduces wasteful resource use (see the following Case Study). It recognizes that urban growth will occur, but uses zoning laws and other tools to channel that growth into areas where it might cause less harm.

Smart growth can discourage sprawl, reduce traffic, protect ecologically sensitive and important lands and waterways, and develop neighborhoods that are more enjoyable places in which to live. Figure 6-23 lists popular smart growth tools that we can use to prevent and control urban growth and sprawl.

Another approach to making cities more sustainable, called *new urbanism,* involves developing villages within cities. The idea is to give people the chance to live within walking distance of where they work, shop, and go for entertainment. (See the following Case Study.)

■ CASE STUDY
The New Urban Village of Vauban

Since the early 1990s, the city of Freiberg, Germany, with a population of about 216,000, has given high priority to controlling urban sprawl. As part of that effort, the city planned a new suburb called Vauban, completed in 2006 and designed to be virtually free of cars.

A small minority of homeowners in Vauban own cars. Street parking, driveways, and garages are generally forbidden in the village. Car owners have two places to park—two large garages on the edge of town—and a parking space in either of them costs $40,000.

Vauban is designed so that each of its 5,000 homes is located within easy walking distance of trains going to and from surrounding communities. Stores, banks, restaurants, and schools are also within walking distance of all homes. The town has numerous bike paths and a car-sharing club. Mass transit allows its 15,000 residents to work or shop in the city of Freiburg.

The community has no single-family homes. People live in stylish row houses within four- and five-story buildings that were designed with an emphasis on use of passive solar energy. The shared walls between these

Solutions

Smart Growth Tools

Limits and Regulations

Limit building permits

Draw urban growth boundaries

Create greenbelts around cities

Zoning

Promote mixed use of housing and small businesses

Concentrate development along mass transportation routes

Planning

Ecological land-use planning

Environmental impact analysis

Integrated regional planning

Protection

Preserve open space

Buy new open space

Prohibit certain types of development

Taxes

Tax land, not buildings

Tax land on value of actual use instead of on highest value as developed land

Tax Breaks

For owners agreeing not to allow certain types of development

For cleaning up and developing abandoned urban sites

Revitalization and New Growth

Revitalize existing towns and cities

Build well-planned new towns and villages within cities

Figure 6-23 We can use these *smart growth* or *new urbanism tools* to prevent and control urban growth and sprawl. **Questions:** Which five of these tools do you think would be best for preventing or controlling urban sprawl? Which, if any, of these tools are used in your community?

homes help them to be highly energy-efficient. Vauban's planners and designers hope that some of the best features of this urban ecovillage will be widely used in other planned communities around the world.

THINKING ABOUT
New Urbanism

If you had the opportunity, would you live in Vauban? Why or why not? If you could change something to make the village more attractive to you, what would that be?

A more environmentally sustainable city, called an *ecocity* or *green city*, emphasizes the following goals, most of which apply one or more of the three **principles of sustainability** (See back cover):

SUSTAINABILITY

- Use solar and other locally available renewable energy resources, and design buildings to be heated and cooled as much as possible by natural processes.
- Build and redesign cities for people, not for cars.
- Greatly reduce the waste of matter and energy.
- Prevent pollution.
- Reuse, recycle, and compost 60–85% of all municipal solid waste.
- Protect and encourage biodiversity by preserving undeveloped land and protecting and restoring natural systems and wetlands in and around cities.
- Promote urban gardens and farmers markets.
- Use zoning and other tools to keep urban sprawl at environmentally sustainable levels.

An ecocity is a people-oriented city, not a car-oriented city. Its residents are able to walk, bike, or use low-polluting mass transit for most of their travel. Its buildings, vehicles, and appliances meet high energy-efficiency standards. Trees and plants adapted to the local climate and soils are planted throughout the city to provide shade, beauty, and wildlife habitats, and to reduce air pollution, noise, and soil erosion. Small organic gardens and a variety of plants adapted to local climate conditions often replace conventional grass lawns.

In an ecocity, abandoned lots and industrial sites are cleaned up and restored. Nearby forests, grasslands, wetlands, and farms are preserved. Much of the food that people eat comes from nearby organic farms, solar greenhouses, community gardens, and small gardens on rooftops, in yards, and in windows boxes. Parks are easily available to everyone. People who design or live in ecocities take seriously the advice that U.S. urban planner Lewis Mumford gave more than 3 decades ago: "Forget the damned motor car and build cities for lovers and friends."

The ecocity is not a futuristic dream, but a reality in many cities that are striving to become more environmentally sustainable and livable. Examples include Curitiba, Brazil (see the following Case Study); Bogotá, Colombia; Waitakere City, New Zealand; Stockholm, Sweden; Helsinki, Finland; Leicester, England; Neerlands, the Netherlands; and in the United States, Portland, Oregon; Davis, California; Olympia, Washington; and Chattanooga, Tennessee.

GOOD NEWS

■ CASE STUDY
The Ecocity Concept in Curitiba, Brazil

An example of an ecocity is Curitiba ("koor-i-TEE-ba"), a city of 3.2 million people, known as the "ecological capital" of Brazil. In 1969, planners in this city decided to focus on an inexpensive and efficient mass-transit system rather than on the car. Curitiba now has what some experts consider to be the world's best bus system, in which clean and modern buses transport about 72% of all commuters every day throughout the city using express lanes dedicated to buses (Figure 6-24). Only high-rise apartment buildings are allowed near major bus routes, and each building must devote its bottom two floors to stores—a practice that reduces the need for residents to travel.

GOOD NEWS

Cars are banned from 49 blocks in the center of the downtown area, which has a network of pedestrian walkways connected to bus stations, parks, and bicycle paths running throughout most of the city. Consequently, Curitiba uses less energy per person and has lower emissions of greenhouse gases and air pollutants and less traffic congestion than do most comparable cities.

The city transformed flood-prone areas along its six rivers into a series of interconnected parks. Volunteers have planted more than 1.5 million trees throughout the city, none of which can be cut down without a permit, which also requires that two trees must be planted for each one that is cut down.

Curitiba recycles roughly 70% of its paper and 60% of its metal, glass, and plastic. Recovered materials are sold mostly to the city's more than 500 major industries, which must meet strict pollution standards.

The city's poor people receive free medical and dental care, child care, and job training, and 40 feeding centers are available for street children. Poor people who live in areas not served by garbage trucks, can collect garbage and exchange filled garbage bags for surplus food, bus tokens, and school supplies. The city uses old buses as roving classrooms to train its poor in basic job hunting skills and job training. Other retired buses have become health clinics, soup kitchens, and day care centers that are free to low-income parents.

About 95% of Curitiba's citizens can read and write, and 83% of its adults have at least a high school education. All school children study ecology. Polls show that 99% of the city's inhabitants would not want to live anywhere else.

John Maier, Jr./Peter Arnold, Inc.

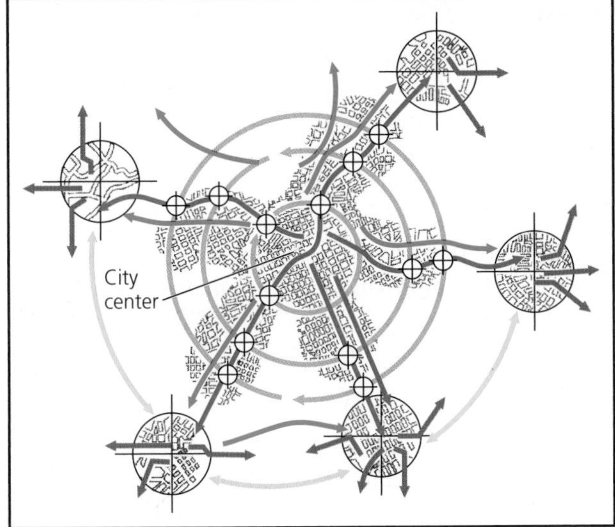

Route
Express Interdistrict Direct Feeder Workers

Figure 6-24 **Solutions:** This bus rapid-transit (BRT) system in Curitiba, Brazil, moves large numbers of passengers around rapidly because each of the system's five major "spokes," connecting the city center with outlying districts, has two express lanes used only by buses. Double- and triple-length bus sections are coupled together as needed to carry up to 300 passengers. Boarding is speeded up by the use of extra-wide bus doors and boarding platforms under glass tubes where passengers can pay before getting on the bus (top left).

Curitiba now faces new challenges, as do all cities, mostly due to a fivefold increase in its population since 1965. Curitiba's once-clear streams are often overloaded with pollutants. The bus system is nearing capacity, and car ownership is on the rise. The city is considering building a light-rail system to relieve some of the pressure on the bus system.

This internationally acclaimed model of urban planning and sustainability is the brainchild of architect and former college professor Jaime Lerner, who has served as the city's mayor three times since 1969. In the face of new challenges, Lerner and other leaders in Curitiba argue that education is still a key to making cities more sustainable, and they want Curitiba to continue serving as an educational example for that great purpose.

Here are this chapter's *three big ideas*:

- The human population is increasing rapidly and may soon bump up against environmental limits.
- We can slow human population growth by reducing poverty, encouraging family planning, and elevating the status of women.
- Most urban areas, home to half of the world's people, are unsustainable but they can be made more sustainable and livable within your lifetime.

REVISITING Population Growth in China, Urbanization, and Sustainability

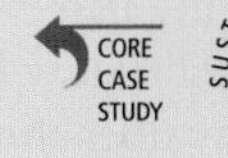

This chapter began by considering the case of China's growing population (**Core Case Study**), which serves as a good application for the concepts that organize this chapter. China's story shows how population growth can be affected by a key factor such as China's strict national policy to promote one-child families. China's age structure tells us about challenges the country will soon face, as well as those it is currently facing, and it explains why the government is beginning to ease its one-child policy. China is also facing the problems of rapid urbanization. It provides a unique example of how a society can attempt to slow its population growth while experiencing rapid urbanization, and how unexpected consequences might occur as a result.

In the first six chapters of this book, you have learned how ecosystems and species have been sustained throughout the earth's history, in keeping with three **principles of sustainability**, by relying on solar energy, nutrient recycling, and biodiversity (see back cover). These three principles can guide us in dealing with the problems of population growth and decline, the rapid shift of the world's population to urban areas, and the challenges of making cities more sustainable and livable.

For example, by employing solar energy technologies more widely, we can cut pollution and greenhouse gas emissions that are worsening as the population grows. By reusing and recycling more materials, we could cut waste and reduce the ecological footprints of urban areas. And in focusing on preserving biodiversity, we could preserve more undeveloped areas around cities, making them more sustainable places to live.

Our numbers expand but Earth's natural systems do not.

LESTER R. BROWN

REVIEW

CORE CASE STUDY

1. Summarize the story of population growth in China and the Chinese government's efforts to regulate it (**Core Case Study**).

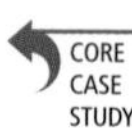

SECTION 6-1

2. What is the key concept for this section? List three factors that account for the rapid increase in the world's human population over the past 200 years. What are the world's three most populous countries? About how many people are added to the world's population each year? How many of us are likely to be on the planet in 2050? Define **cultural carrying capacity**. Summarize the debate over whether and how long the human population can keep growing.

SECTION 6-2

3. What are the two key concepts for this section? Distinguish between **crude birth rate** and **crude death rate**. List four variables that affect the **population change** of an area and write an equation showing how they are related. What is **total fertility rate (TFR)**? How has the world's TFR changed since 1955? Discuss population growth and fertility rates in China.

4. Describe population growth in the United States and explain why it is high compared to population growth in most other more-developed countries and in China (**Core Case Study**). CORE CASE STUDY

5. List nine factors that affect birth rates and fertility rates. Explain why there is a bride shortage in China. Define **life expectancy** and **infant mortality rate** and explain how they affect the population size of a country. Why does the United States have a lower life expectancy and higher infant mortality rate than a number of other more-developed countries? What is **migration**? Describe immigration into the United States and the benefits and issues it raises.

SECTION 6-3

6. What is the key concept for this section? What is the **age structure** of a population? Explain how age structure affects population growth and economic growth. Describe the American baby boom and some of its economic and social effects. Describe and compare the aging of the U. S. and Chinese populations. What are some problems related to rapid population decline due to an aging population? How has the AIDS epidemic affected the age structure of some countries in Africa?

SECTION 6-4

7. What is the key concept for this section? What is the **demographic transition** and what are its four stages? What factors could hinder some less-developed countries from making this transition? What is **family planning**? Describe the roles of promoting economic development, elevating the status of women, and family planning in slowing population growth. Describe India's efforts to control its population growth.

SECTION 6-5

8. What is the key concept for this section? List three trends in global urban growth. Describe four phases of urban growth in the United States. What is **urban sprawl**? List five undesirable effects of urban sprawl. What are four advantages of urbanization? List six disadvantages of urbanization. Explain why most cities and urban areas are not sustainable. Describe some of the problems faced by the poor who live in urban areas. Summarize the urban problems of Mexico City, Mexico along with the city's efforts to deal with these problems.

SECTION 6-6

9. What is the key concept for this section? What are the major advantages and disadvantages of motor vehicles? List three ways to reduce dependence on motor vehicles. Describe the major advantages and disadvantages of relying more on **(a)** bicycles, **(b)** mass-transit light-rail systems, **(c)** bus rapid-transit systems within urban areas, and **(d)** rapid-rail systems between urban areas.

SECTION 6-7

10. What is the key concept for this section? What is **smart growth**? List five tools used to promote smart growth. Describe the new urban village of Vauban, Germany. List eight goals of ecocity design. Summarize the efforts to make Curitiba, Brazil, a sustainable ecocity. What are this chapter's *three big ideas*? Describe how the story of human population growth in China serves as a good application of the key concepts of this chapter. Give an example of how the three **principles of sustainability** can guide us in dealing with the problems that stem from population growth and urbanization. SUSTAINABILITY

Note: Key terms are in **bold** type.

CRITICAL THINKING

1. Do you think that the problems resulting from China's one-child policy (**Core Case Study**) outweigh the problems of overpopulation that likely would have resulted without some sort of regulation of population growth in China? Can you think of other ways in which China could try to regulate its population growth? Explain. Should China do away with its one-child policy? Explain.

2. If you could greet a new person every second without taking a break, how many people could you greet in one day? How many in a year? How many in a lifetime of 80 years? How long would you have to live to greet all 7 billion people on earth at a rate of one every second working around the clock?

3. Identify a major local, national, or global environmental problem, and describe the role of population growth in this problem.

4. Do you believe that the population is too high in **(a)** China (**Core Case Study**), **(b)** the world, **(c)** your own country, and **(d)** the area where you live? Explain.

5. Some people believe the most important environmental goal is to sharply reduce the rate of population growth in less-developed countries, where at least 92% of the world's population growth is expected to take place between 2011 and 2050. Others argue that the most serious environmental problems stem from high levels of resource consumption per person in more-developed countries, which use 88% of the world's resources and have much larger ecological footprints per person (see Figure 1-8, p. 14) than do people in less-developed countries. What is your view on this issue? Explain.

6. Experts have identified population growth as one of the major causes of the environmental problems we face (Figure 1-10, p. 16). The population of United States is growing faster than that of any of the world's other more-developed countries as well as that of China. However, this problem is rarely mentioned, and the U.S. government has no official policy to slow U.S. population growth. Why do think this is so? Do you agree with this approach? If not, list three things you would do to slow U.S. population growth.

7. List three reasons why you **(a)** enjoy living in a large city, **(b)** would like to live in a large city, or **(c)** do not wish to live in a large city.

8. If you own a car or hope to own one, what conditions, if any, would encourage you to rely less on the automobile and to travel to school or work by bicycle, on foot, by mass transit, or by carpool?

9. Congratulations! You are in charge of the world. List the three most important features of your policy for dealing with **(a)** global population growth and **(b)** urban growth and development.

DOING ENVIRONMENTAL SCIENCE

Prepare an age structure diagram for your community. You will need to estimate how many people belong in each age category (see p. 101). To do this, interview a randomly drawn sample of the population to find out their ages and then divide your sample into age groups. (Sample equal numbers of males and females.) Then find out the total population of your community in order to make your estimates. Create your diagram and then use it to project future population trends. Write a report in which you discuss some economic and social problems that might arise as a result of these trends.

GLOBAL ENVIRONMENT WATCH EXERCISE

Research the status of China's one-child policy and answer the following questions: How is China changing its policy, if at all (**Core Case Study**)? How are people responding to these changes? Have there been any changes in the population trends in China in recent years? How has the Chinese government's policy affected such changes, if at all?

ECOLOGICAL FOOTPRINT ANALYSIS

The chart below shows selected population data for two different countries, A and B. Study the chart and answer the questions that follow.

	Country A	Country B
Population (millions)	144	82
Crude birth rate	43	8
Crude death rate	18	10
Infant mortality rate	100	3.8
Total fertility rate	5.9	1.3
% of population under 15 years old	45	14
% of population older than 65 years	3.0	19
Average life expectancy at birth	47	79
% urban	44	75

(Data from Population Reference Bureau 2010. *World Population Data Sheet*)

1. Calculate the rates of natural increase (due to births and deaths, not counting immigration) for the populations of country A and country B. Based on these calculations and the data in the table, for each of the countries, suggest whether it is a more-developed country or a less-developed country and explain the reasons for your answers.
2. Describe where each of the two countries may be in terms of its stage in the demographic transition (Figure 6-11). Discuss factors that could hinder either country from progressing to later stages in the demographic transition.
3. Explain how the percentages of people under 15 years of age in each country could affect its per capita and total ecological footprints.

LEARNING ONLINE

The website for *Environmental Science 14e* contains helpful study tools and other resources to take the learning experience beyond the textbook. Examples include flashcards, practice quizzing, information on green careers, tips for more sustainable living, activities, and a wide range of articles and further readings. To access these materials and other companion resources for this text, please visit **www.cengagebrain.com**. For further details, see the preface, p. xvi.

Climate and Biodiversity

7

A Temperate Deciduous Forest

CORE CASE STUDY

The earth hosts a great diversity of species and *habitats*, or places where these species can live. Some species live in land, or *terrestrial*, habitats such as deserts, grasslands, and forests—the three major types of terrestrial ecosystems, also called *biomes*. Other species live in water-covered habitats in *aquatic life zones*, such as rivers, lakes, and oceans.

Why do forests grow on some areas of the earth's land while deserts form in other areas and still other areas are covered with water? The answers lie largely in differences in *climate*, the average atmospheric conditions, primarily average annual precipitation and temperature, in a given region over a period of time ranging from at least three decades to thousands of years. These differences lead to three major types of climate—*tropical* (areas near the equator, receiving the most intense sunlight), *polar* (areas near the earth's poles, receiving the least intense sunlight), and *temperate* (areas between the tropic and polar regions).

Throughout these regions, we find different types of ecosystems, vegetation, and animals adapted to the various climate conditions. For example, scattered throughout the temperate areas of the globe, we find *temperate deciduous forests* (Figure 7-1). Such forests typically see warm summers, cold winters, and abundant precipitation—rain in summer and snow in winter months. They are dominated by a few species of *broadleaf deciduous trees* such as oak, hickory, maple, aspen, and birch. Animal species living in these forests include predators such as bears, wolves, foxes, and wildcats. They feed on herbivores such as white-tailed deer, squirrels, rabbits, opossums, and mice. Warblers, robins, and other bird species live in these forests during the spring and summer, mating and raising their young.

These species are adapted to the conditions of temperate forests. For example, most of the trees' leaves, after developing their vibrant colors in the fall (Figure 7-1, left), drop off the trees. This allows the trees to survive the cold winters by becoming dormant (Figure 7-1, right). Each spring, they sprout new leaves and spend their summers growing and producing until the cold weather returns. Some forest mammals spend their summers storing fat and then they hibernate during winter, sleeping the coldest months away. Many bird species migrate to warmer climates in the late fall to avoid the cold and snow.

Temperate deciduous forests once covered the eastern half of the United States and western Europe. But as these areas became urbanized, most of the original forests were cleared. Today, on a worldwide basis, this terrestrial system has been disturbed by human activity more than any other.

In this chapter, we examine the key role that climate plays in the formation and location of forests and all other major terrestrial ecosystems. We also look at the nature of the earth's major aquatic ecosystems.

Paul W. Johnson/Biological Photo Service

Paul W. Johnson/Biological Photo Service

Figure 7-1 The earth has three major climate zones: *tropical*, where the climate is generally warm throughout the year; *temperate*, where the climate is not extreme and typically changes through four different annual seasons; and *polar*, where it is fiercely cold during winter months and cool or cold during summer months. The photos show a temperate deciduous forest (left) in the U.S. state of Rhode Island in the fall and the same forest in winter (right).

Key Questions and Concepts

7-1 What factors influence climate?

CONCEPT 7-1 Key factors that influence an area's climate are incoming solar energy, the earth's rotation, global patterns of air and water movement, gases in the atmosphere, and the earth's surface features.

7-2 How does climate affect the nature and location of biomes?

CONCEPT 7-2 Differences in long-term average annual precipitation and temperature lead to the formation of tropical, temperate, and cold deserts, grasslands, and forests, and largely determine their locations.

7-3 How have human activities affected the world's terrestrial ecosystems?

CONCEPT 7-3 In many areas, human activities are disrupting ecological and economic services provided by the earth's deserts, grasslands, forests, and mountains.

7-4 What are the major types of aquatic systems?

CONCEPT 7-4 Saltwater and freshwater aquatic life zones cover almost three-fourths of the earth's surface, and oceans dominate the planet.

7-5 Why are marine aquatic systems important and how have human activities affected them?

CONCEPT 7-5 Saltwater ecosystems are irreplaceable reservoirs of biodiversity, providing major ecological and economic services that are being threatened by human activities.

7-6 What are the major types of freshwater systems and how have human activities affected them?

CONCEPT 7-6 Freshwater lakes, rivers, and wetlands provide important ecological and economic services that are being disrupted by human activities.

Note: Supplements 2 (p. S2), 5 (p. S18), and 6 (p. S22) can be used with this chapter.

To do science is to search for repeated patterns, not simply to accumulate facts, and to do the science of geographical ecology is to search for patterns of plant and animal life that can be put on a map.

ROBERT H. MACARTHUR

7-1 What Factors Influence Climate?

▶ **CONCEPT 7-1 Key factors that influence an area's climate are incoming solar energy, the earth's rotation, global patterns of air and water movement, gases in the atmosphere, and the earth's surface features.**

The Earth Has Many Different Climates

The first step in understanding issues related to climate, such as how climate affects terrestrial biodiversity (**Core Case Study**), is to be sure that we understand the difference between weather and climate. **Weather** is a set of physical conditions of the lower atmosphere such as temperature, precipitation, humidity, wind speed, cloud cover, and other factors in a given area over a period of hours or days. (Supplement 5, p. S18, introduces you to the basics of weather.)

CORE CASE STUDY

Weather differs from **climate**, which is an area's general pattern of atmospheric conditions over periods ranging from at least three decades to thousands of years. In other words, climate is weather, averaged over a long time. Figure 7-2 depicts the earth's current major climate zones and ocean currents, which are key components of the earth's natural capital (see Figure 1-3, p. 9).

Climate varies in different parts of the earth primarily because, over long periods of time, patterns of global air circulation and ocean currents distribute heat and precipitation unevenly between the tropics and other parts of the world (Figure 7-3). Three major factors that affect the circulation of air in the lower atmosphere are

1. *Uneven heating of the earth's surface by the sun.* Air is heated much more at the equator, where the sun's rays strike directly, than at the poles, where sunlight strikes at an angle and spreads out over a much greater area (Figure 7-3, right). These differences in the input of solar energy to the atmosphere help

Links: CORE CASE STUDY refers to the Core Case Study.  refers to the book's sustainability theme. GOOD NEWS refers to good news about the environmental challenges we face.

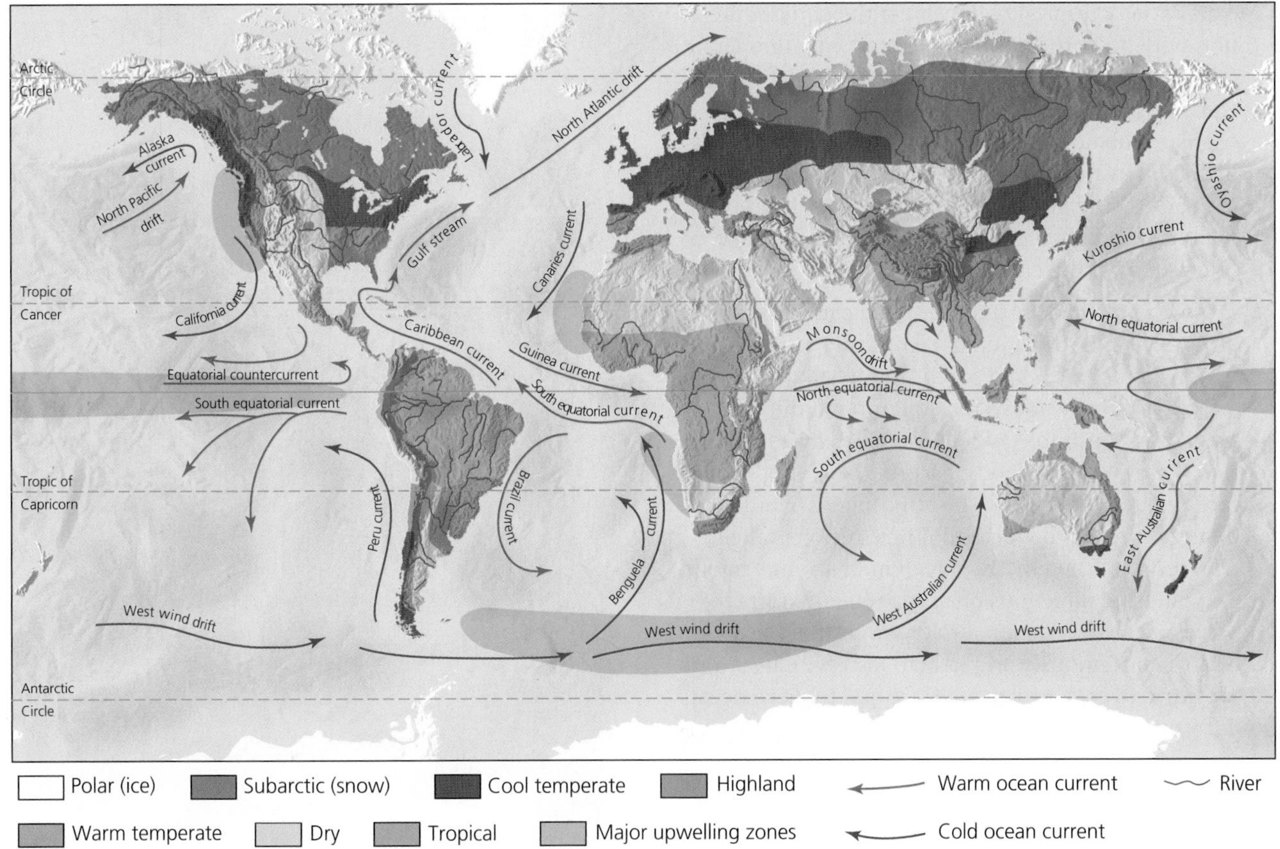

Active Figure 7-2 Natural capital: This generalized map of the earth's current climate zones also shows the major ocean currents and upwelling areas (where currents bring nutrients from the ocean bottom to the surface). *See an animation based on this figure at* www.cengagebrain.com. **Question:** Based on this map, what is the general type of climate where you live?

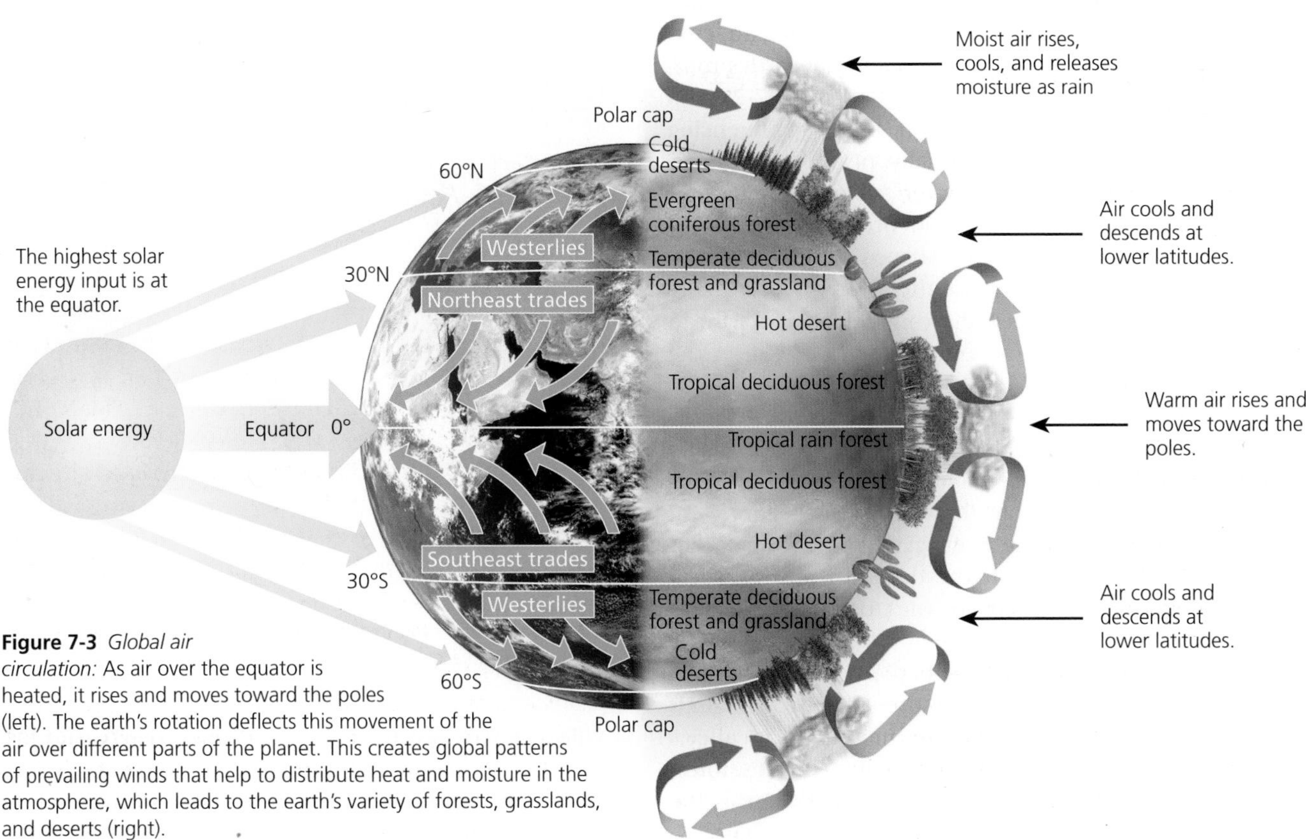

Figure 7-3 *Global air circulation:* As air over the equator is heated, it rises and moves toward the poles (left). The earth's rotation deflects this movement of the air over different parts of the planet. This creates global patterns of prevailing winds that help to distribute heat and moisture in the atmosphere, which leads to the earth's variety of forests, grasslands, and deserts (right).

explain why tropical regions near the equator are hot, why polar regions are cold, and why temperate regions in between generally have both warm and cool temperatures (Figure 7-2). The intense input of solar radiation in tropical regions leads to greatly increased evaporation of moisture from forests, grasslands, and bodies of water. As a result, tropical regions normally receive more precipitation than do other areas of the earth.

2. *Rotation of the earth on its axis.* As the earth rotates around its axis, the equator spins faster than the regions to its north and south. As a result, heated air masses, rising above the equator and moving north and south to cooler areas, are deflected in different ways over different parts of the planet's surface (Figure 7-3, right). The atmosphere over these different areas is divided into huge regions called *cells,* distinguished by the direction of air movement. The differing directions of air movement are called *prevailing winds*—major surface winds that blow almost continuously and help to distribute heat and moisture over the earth's surface and to drive ocean currents (Figure 7-3, left).

3. *Properties of air, water, and land.* Heat from the sun evaporates ocean water and transfers heat from the oceans to the atmosphere, especially near the hot equator. This evaporation of water creates giant cyclical convection cells that circulate air, heat, and moisture both vertically and from place to place in the atmosphere, as shown in Figure 7-4.

Prevailing winds blowing over the oceans produce mass movements of surface water called **ocean currents**. Driven by prevailing winds and the earth's rotation, the earth's major ocean currents (Figure 7-2) help to redistribute heat from the sun, thereby influencing climate and vegetation, especially near coastal areas. This heat and differences in water *density* (mass per unit volume) create warm and cold ocean currents. Prevailing winds and irregularly shaped continents interrupt these currents and cause them to flow in roughly circular patterns between the continents, clockwise in the northern hemisphere and counterclockwise in the southern hemisphere.

Water also moves vertically in the oceans as denser water sinks while less dense water rises. This creates a connected loop of deep and shallow ocean currents (which are separate from those shown in Figure 7-2). This loop acts somewhat like a giant conveyer belt that moves heat to and from the deep sea and transfers warm and cold water between the tropics and the poles (Figure 7-5).

The ocean and the atmosphere are strongly linked in two ways: ocean currents are affected by winds in the atmosphere, and heat from the ocean affects atmospheric circulation. One example of the interactions between the ocean and the atmosphere is the *El Niño–Southern Oscillation,* or *ENSO.* (See Figure 4, p. S19, in Supplement 5 and *The Habitable Planet,* Video 3, at **www.learner.org/resources/series209.html**.) This large-scale weather phenomenon occurs every few years when prevailing winds in the tropical Pacific Ocean weaken and change direction. The resulting above-average warming of Pacific waters alters the

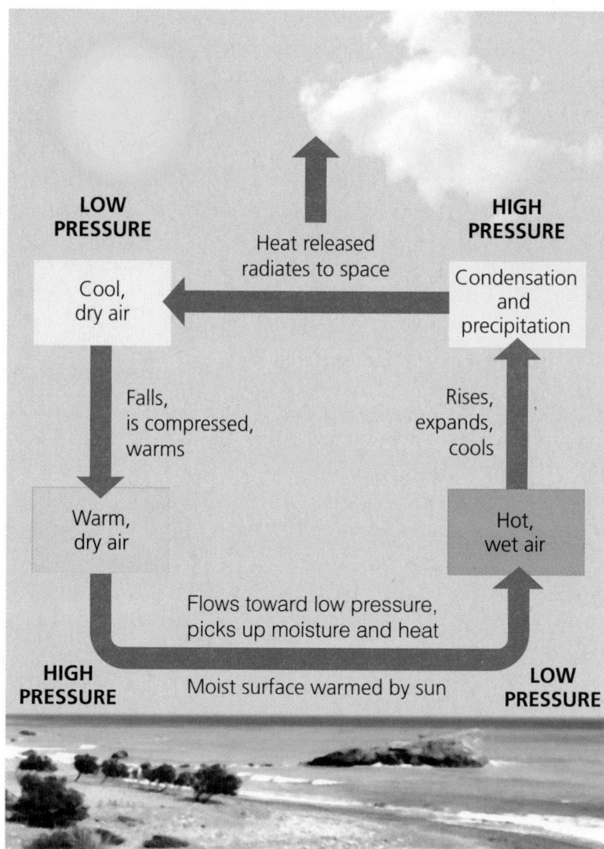

Figure 7-4 Energy is transferred by *convection* in the atmosphere—the process by which warm, wet air rises, then cools and releases heat and moisture as precipitation (right side and top, center). Then the cooler, denser, and drier air sinks, warms up, and absorbs moisture as it flows across the earth's surface (left side and bottom) to begin the cycle again.

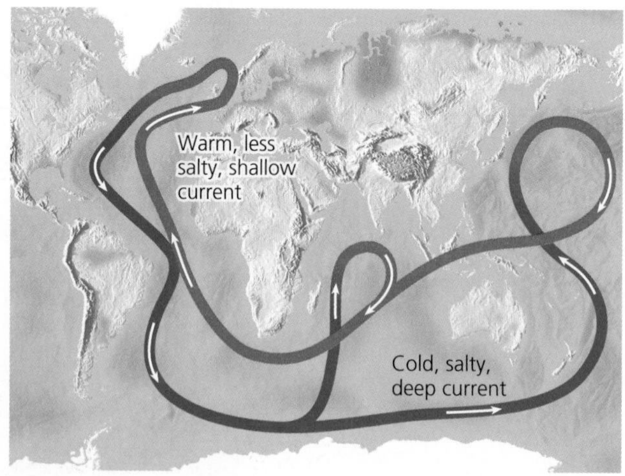

Figure 7-5 A connected loop of deep and shallow ocean currents transports warm and cool water to various parts of the earth. **Question:** How do you think this loop affects the climates of the coastal areas adjacent to it?

weather over at least two-thirds of the earth for 1 or 2 years (see Figure 5, p. S20, in Supplement 5).

The earth's air circulation patterns, prevailing winds, and configuration of continents and oceans are all factors in the formation of six giant convection cells, three of them south of the equator and three north of the equator. These cells lead to an irregular distribution of climates and of the resulting deserts, grasslands, and forests, as shown in Figure 7-3, right (**Concept 7-1**).

Greenhouse Gases Warm the Lower Atmosphere

As energy flows from the sun to the earth, some of it is reflected by the earth's surface back into the atmosphere (see Figure 3-3, p. 42). Molecules of certain gases in the atmosphere, including water vapor (H_2O), carbon dioxide (CO_2), methane (CH_4), and nitrous oxide (N_2O), absorb some of this solar energy and release a portion of it as infrared radiation (heat) that warms the atmosphere. Thus, these gases, called **greenhouse gases**, play a role in determining the earth's average temperatures and its climates.

The earth's surface also absorbs much of the solar energy that strikes it and transforms it into longer-wavelength infrared radiation, which then rises into the lower atmosphere. Some of this heat escapes into space, but some is absorbed by molecules of greenhouse gases and emitted into the lower atmosphere as even longer-wavelength infrared radiation (see Figure 2-8, p. 35). Some of this released energy radiates into space, and some adds to the warming of the lower atmosphere and the earth's surface. Together, these processes result in a natural warming of the troposphere, called the **natural greenhouse effect** (see Figure 3-3, p. 42, and *The Habitable Planet*, Video 2, at **www.learner.org/resources/series209.html**). Without this natural heating effect, the earth would be a very cold and mostly lifeless planet.

Human activities such as the burning of fossil fuels, clearing of forests, and growing of crops release carbon dioxide, methane, and nitrous oxide into the atmosphere. A large and growing body of scientific evidence, combined with climate model projections, indicate that the large inputs of these greenhouse gases into the atmosphere from human activities very likely will significantly enhance the earth's natural greenhouse effect and change the earth's climate during this century. If this occurs, it will likely alter precipitation patterns, raise average sea levels, and shift areas where we can grow crops and where some types of plants and animals (including humans) can live, as discussed more fully in Chapter 15. Some of these changes could last for centuries to thousands of years.

The Earth's Surface Features Affect Local Climates

Heat is absorbed and released more slowly by water than by land. This difference creates land and sea breezes. As a result, the world's oceans and large lakes help to moderate the weather and climates of nearby lands.

Various other topographic features of the earth's surface can create local and regional climatic conditions that differ from the general climate of some regions. For example, mountains interrupt the flow of prevailing surface winds and the movement of storms. When moist air blowing inland from an ocean reaches a mountain range, it is forced upward. As it rises, it cools and expands, and then loses most of its moisture as rain and snow that fall on the windward slope of the mountain (Figure 7-6).

As the drier air mass passes over the mountaintops, it flows down the leeward slopes (facing away from the wind), and warms up. This increases its ability to hold moisture, but the air releases little moisture and instead tends to dry out plants and soil below. This process is called the **rain shadow effect** (Figure 7-6), and

Figure 7-6 The *rain shadow effect* is a reduction of rainfall and loss of moisture from the landscape on the leeward side of a mountain. Warm, moist air in onshore winds loses most of its moisture as rain and snow that fall on the windward slopes of a mountain range. This leads to semiarid and arid conditions on the leeward side of the mountain range and on the land beyond.

over many decades, it results in *semiarid* or *arid* conditions on the leeward side of a high mountain range. Sometimes this effect leads to the formation of deserts such as Death Valley, a part of the Mojave Desert found in parts of the U.S. states of California, Nevada, Utah, and Arizona. Death Valley lies in the rain shadow of Mount Whitney, the highest mountain in the Sierra Nevada range.

Cities also create distinct microclimates. Bricks, concrete, asphalt, and other building materials absorb and hold heat, and buildings block wind flow. Motor vehicles and the heating and cooling systems of buildings release large quantities of heat and pollutants. As a result, cities on average tend to have more haze and smog, higher temperatures that make them *heat islands,* and lower wind speeds than the surrounding countryside.

7-2 How Does Climate Affect the Nature and Location of Biomes?

CONCEPT 7-2 Differences in long-term average annual precipitation and temperature lead to the formation of tropical, temperate, and cold deserts, grasslands, and forests, and largely determine their locations.

Climate Helps to Determine Where Organisms Can Live

Differences in climate (Figure 7-2) help to explain why one area of the earth's land surface is a desert, another a grassland, and another a forest (**Core Case Study**). CORE CASE STUDY They also help to explain why global air circulation patterns (Figure 7-3) account for different types of deserts, grasslands, and forests. Furthermore, different combinations of varying average annual precipitation and temperatures lead to the formation of tropical (hot), temperate (moderate), and polar (cold) deserts, grasslands, and forests, as summarized in Figure 7-7 (**Concept 7-2**).

Another way in which climate and vegetation vary, in addition to *latitude,* is with changing *elevation,* or height above sea level. If you climb a tall mountain, from its base to its summit, you can observe changes in plant life similar to those you would encounter in traveling from the equator to the earth's northern polar region (Figure 7-8). For example, if you hike up a tall mountain in Ecuador's Andes range, your trek will begin in a tropical rain forest and end up on a glacier at the summit.

Figure 7-9 (p. 128) shows how scientists have divided the world into several major **biomes**—large terrestrial regions, each characterized by certain types of climate and dominant plant life. The variety of terrestrial biomes and aquatic systems is one of the four components of the earth's biodiversity (see Figure 4-2, p. 63)—a vital part of the earth's natural capital.

Each biome contains many ecosystems whose communities have adapted to differences in climate, soil, and other environmental factors (**Core Case Study**). CORE CASE STUDY People have removed or altered much of the natural vegetation in some areas for farming, livestock grazing, obtaining timber and fuelwood, mining, and construction of towns and cities. (Figure 3, p. S26, in Supplement 6 shows the major biomes of North America.)

By comparing Figure 7-9 with Figure 7-2, you can see how the world's major biomes vary with climate. Figure 4-4 (p. 64) shows how major biomes along the 39th parallel in the United States are related to its different climates.

On maps such as the one in Figure 7-9, biomes are shown with sharp boundaries, and each biome is covered with one general type of vegetation. In reality, *biomes are not uniform*. They consist of a *mosaic of patches,* each with somewhat different biological communities but with similarities typical of the biome. These patches occur primarily because of the irregular distribution of the resources needed by plants and animals and because human activities have removed or altered the natural vegetation in many areas.

THINKING ABOUT
Biomes, Climate, and Human Activities

Use Figure 7-2 to determine the general type of climate where you live and Figure 7-9 to determine the general type of biome that should exist where you live. Then use Figure 2, pp. S24–25, in Supplement 6 to determine how human ecological footprints have affected the biome where you live.

There Are Three Major Types of Deserts

In a *desert,* annual precipitation is low and often scattered unevenly throughout the year. During the day, the baking sun warms the ground and evaporates water

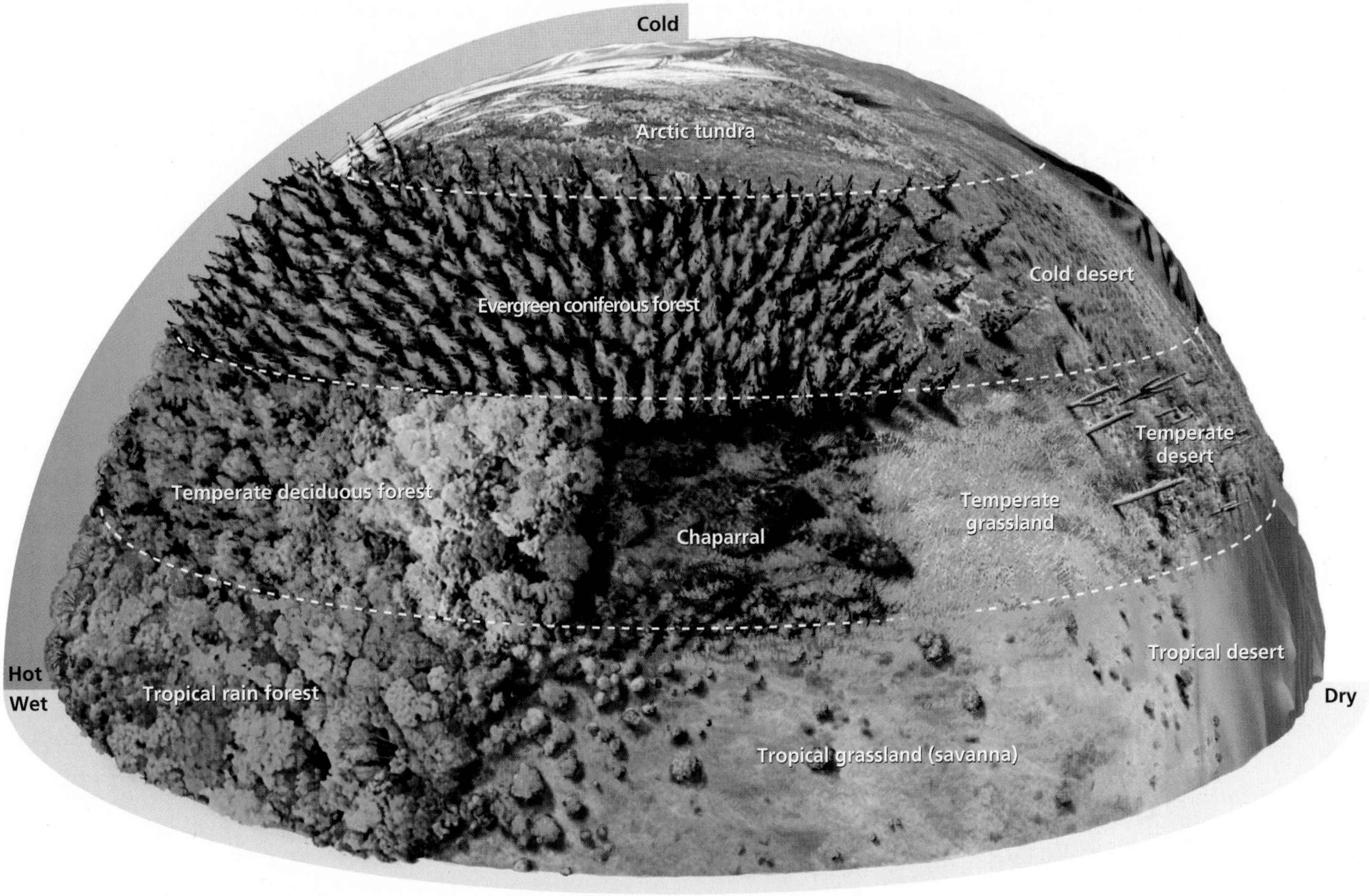
Cold
Arctic tundra
Evergreen coniferous forest
Cold desert
Temperate desert
Temperate deciduous forest
Temperate grassland
Chaparral
Tropical desert
Hot
Wet
Tropical rain forest
Dry
Tropical grassland (savanna)

Figure 7-7 **Natural capital:** Average precipitation and average temperature, acting together as limiting factors over a long time, help to determine the type of desert, grassland, or forest in any particular area, and thus the types of plants, animals, and decomposers found in that area (assuming it has not been disturbed by human activities).

Elevation
Mountain ice and snow
Tundra (herbs, lichens, mosses)
Coniferous Forest
Deciduous Forest
Tropical Forest
Latitude (south to north)
Tropical Forest
Deciduous Forest
Coniferous Forest
Tundra (herbs, lichens, mosses)
Polar ice and snow

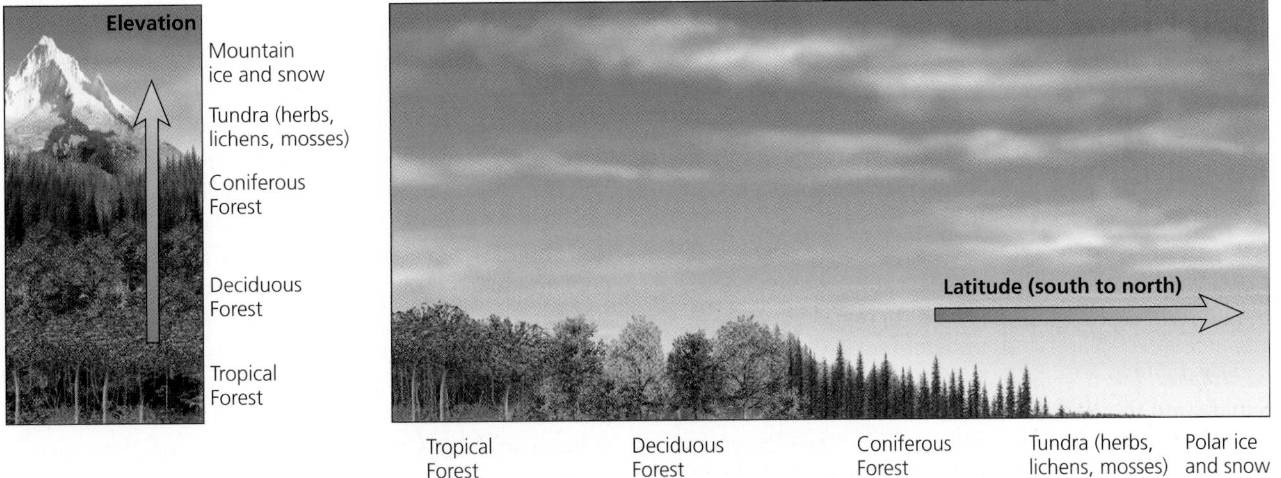

Figure 7-8 Biomes and climate both change with elevation (left), as well as with latitude (right). Parallel changes in vegetation type occur when we travel from the equator toward the north pole and from lowlands to mountaintops. **Question:** How might the components of the left diagram change as the earth warms during this century? Explain.

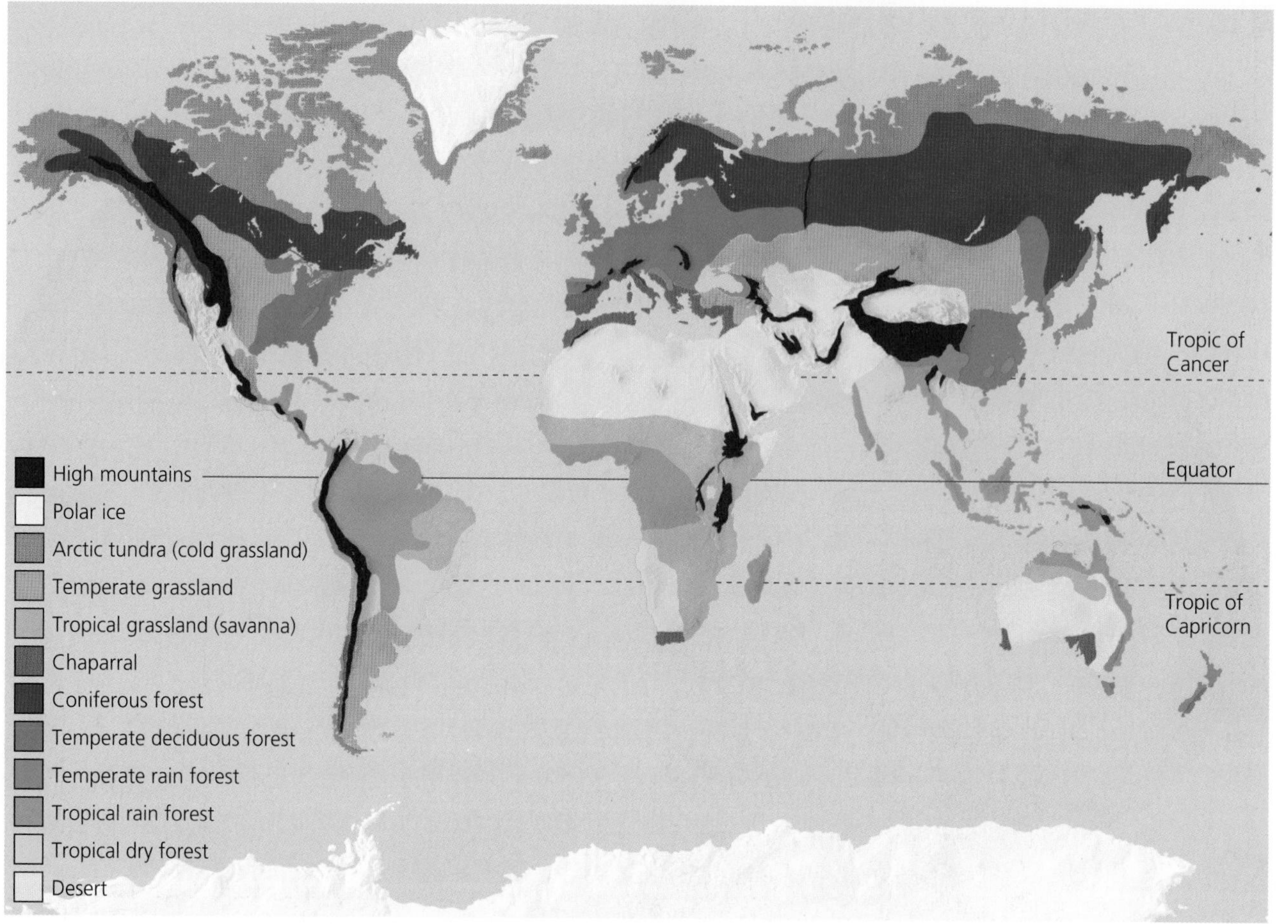

Active Figure 7-9 Natural capital: The earth's major *biomes*—each characterized by a certain combination of climate and dominant vegetation—result primarily from differences in climate. *See an animation based on this figure at* **www.cengagebrain.com**. **Question:** If you take away human influences such as farming and urban development, what kind of biome do you live in?

from plant leaves and from the soil. But at night, most of the heat stored in the ground radiates quickly into the atmosphere. Desert soils have little vegetation and moisture to help store the heat and the skies above deserts are usually clear. This explains why in a desert you may roast during the day but shiver at night.

The lack of vegetation, especially in tropical and polar deserts, makes them vulnerable to sandstorms driven by winds that can spread sand from one area to another. Desert surfaces are also vulnerable to disruption from vehicles such as SUVs (Figure 7-10, top photo).

A combination of low rainfall and varying average temperatures creates tropical, temperate, and cold deserts (Figures 7-7 and 7-10 and **Concept 7-2**).

Tropical deserts (Figure 7-10, top photo) such as the Sahara and the Namib of Africa are hot and dry most of the year (Figure 7-10, top graph). They have few plants and a hard, windblown surface strewn with rocks and some sand. They are the deserts we often see in the movies.

In *temperate deserts* (Figure 7-10, center photo) such as the Sonoran Desert in southeastern California, southwestern Arizona, and northwestern Mexico, daytime temperatures are high in summer and low in winter and there is more precipitation than in tropical deserts (Figure 7-10, center graph). The sparse vegetation primarily consists of widely dispersed, drought-resistant shrubs and cacti or other succulents adapted to the lack of water and temperature variations.

In *cold deserts* such as the Gobi Desert in Mongolia, vegetation is sparse (Figure 7-10, bottom photo). Winters are cold, summers are warm or hot, and precipitation is low (Figure 7-10, bottom graph). Desert plants and animals have adaptations that help them to stay cool and to get enough water to survive (Science Focus, p. 130).

Desert ecosystems are fragile. Their soils take from decades to centuries to recover from disturbances such as off-road vehicle traffic (Figure 7-10, top photo). This is because deserts have slow plant growth, low species diversity, slow nutrient cycling (due to low bacterial activity in the soils), and very little water. Also, off-road vehicle traffic in deserts can destroy the habitats for a variety of animals that live underground in this biome.

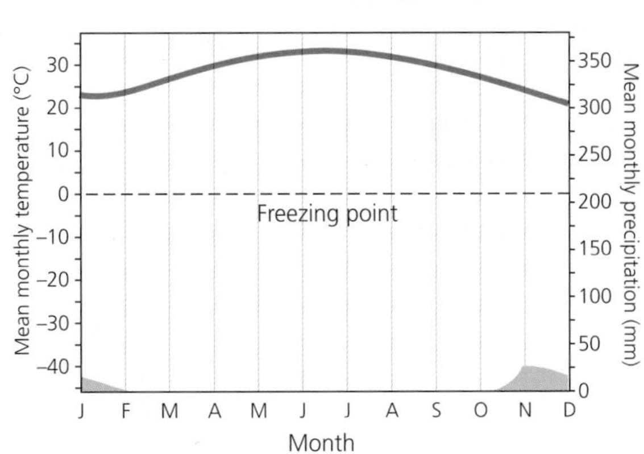

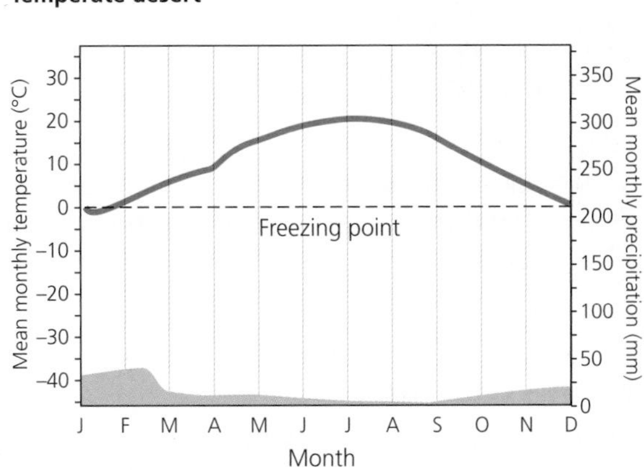

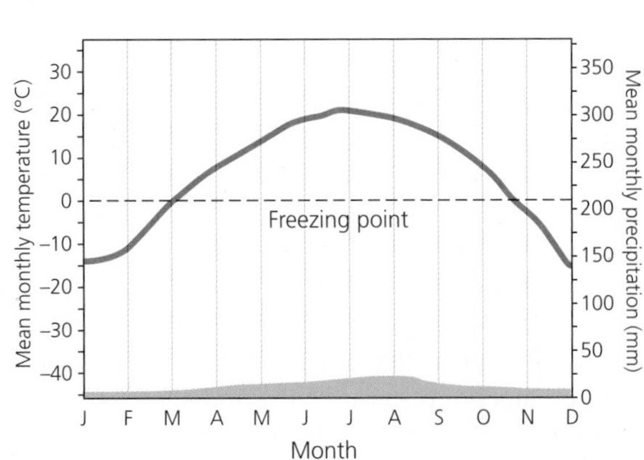

Figure 7-10 These climate graphs track the typical variations in annual temperature (red) and precipitation (blue) in tropical, temperate, and cold deserts. Top photo: a *tropical desert* in the United Arab Emirates, in which a sport utility vehicle (SUV) participates in a popular but environmentally destructive SUV rodeo. Center photo: a *temperate desert* in southeastern California, with saguaro cactus, a prominent species in this ecosystem. Bottom photo: a *cold desert*, Mongolia's Gobi Desert, where Bactrian camels live. **Question:** Which month of the year has the highest temperature and which month has the lowest rainfall for each of the three types of deserts?

SCIENCE FOCUS

Staying Alive in the Desert

Adaptations for survival in the desert have two themes: *beat the heat* and *every drop of water counts*.

Desert plants have evolved a number of strategies based on such adaptations. During long hot and dry spells, plants such as mesquite and creosote drop their leaves to survive in a dormant state. *Succulent* (fleshy) *plants* such as the saguaro ("sah-WAH-ro") cactus (Figure 7-10, middle photo) have three adaptations: they have no leaves, which can lose water to the atmosphere through *transpiration*; they store water and synthesize food in their expandable, fleshy tissue; and they reduce water loss by opening their pores only at night to take up carbon dioxide (CO_2). The spines of these and many other desert plants guard them from being eaten by herbivores seeking the precious water they hold.

Some desert plants use deep roots to tap into groundwater. Others such as prickly pear and saguaro cacti use widely spread shallow roots to collect water after brief showers and store it in their spongy tissues.

Some plants found in deserts conserve water by having wax-coated leaves that reduce water loss. Others such as annual wildflowers and grasses store much of their biomass in seeds that remain inactive, sometimes for years, until they receive enough water to germinate. Shortly after a rain, these seeds germinate, grow, and carpet some deserts with dazzling arrays of colorful flowers that last for up to a few weeks.

Most desert animals are small. Some beat the heat by hiding in cool burrows or rocky crevices by day and coming out at night or in the early morning. Others become dormant during periods of extreme heat or drought. Some larger animals such as camels can drink massive quantities of water when it is available and store it in their fat for use as needed. Also, the camel's thick fur actually helps it to keep cool because the air spaces in the fur insulate the camel's skin against the outside heat. And camels do not sweat, which reduces their water loss through evaporation. Kangaroo rats never drink water. They get the water they need by breaking down fats in seeds that they consume.

Insects and reptiles such as rattlesnakes have thick outer coverings to minimize water loss through evaporation, and their wastes are dry feces and a dried concentrate of urine. Many spiders and insects get their water from dew or from the food they eat.

Critical Thinking

What are three steps you would take to survive in the open desert if you had to?

There Are Three Major Types of Grasslands

Grasslands occur primarily in the interiors of continents in areas that are too moist for deserts to form and too dry for forests to grow (Figure 7-9). Grasslands persist because of a combination of seasonal drought, grazing by large herbivores, and occasional fires—all of which keep shrubs and trees from growing in large numbers.

The three main types of grassland—tropical, temperate, and cold (arctic tundra)—result from combinations of low average precipitation and varying average temperatures (Figures 7-7 and 7-11 and **Concept 7-2**).

One type of tropical grassland, called a *savanna*, contains widely scattered clumps of trees such as acacia (Figure 7-11, top photo), which are covered with thorns that keep some herbivores away. This biome usually has warm temperatures year-round and alternating dry and wet seasons (Figure 7-11, top graph).

Tropical savannas in East Africa are home to *grazing* (primarily grass-eating) and *browsing* (twig- and leaf-nibbling) hoofed animals, including wildebeests (Figure 7-11, top photo), gazelles, zebras, giraffes, and antelopes, as well as their predators such as lions, hyenas, and humans. Herds of these grazing and browsing animals migrate to find water and food in response to seasonal and year-to-year variations in rainfall (Figure 7-11, blue region in top graph) and food availability. Savanna plants, like those in deserts, are adapted to survive drought and extreme heat; many have deep roots that can tap into groundwater.

> CONNECTIONS
>
> **Grassland Niches and Feeding Habits**
>
> As an example of differing niches, some large herbivores have evolved specialized eating habits that minimize competition among species for the vegetation found on the savanna. For example, giraffes eat leaves and shoots from the tops of trees, elephants eat leaves and branches farther down, wildebeests prefer short grasses, and zebras graze on longer grasses and stems.

In a *temperate grassland*, winters can be bitterly cold, summers are hot and dry, and annual precipitation is fairly sparse and falls unevenly throughout the year (Figure 7-11, center graph). Because the aboveground parts of most of the grasses die and decompose each year, organic matter accumulates to produce a deep, fertile topsoil. This topsoil is held in place by a thick network of the drought-tolerant grasses' intertwined roots (unless the topsoil is plowed up, which exposes it to being blown away by high winds found in these biomes). The natural grasses are also adapted to fires that burn the plant parts above the ground but do not harm the roots, from which new grass can grow.

In the mid-western and western areas of the United States, we find two types of temperate grasslands depending primarily on average rainfall: *short-grass prairies* (Figure 7-11, center photo) and the *tall-grass prairies* (which get more rain). Many of the world's natural temperate grasslands have been converted to farmland, because their fertile soils are useful for growing crops (Figure 7-12, p. 132) and grazing cattle.

Martin Harvey/Peter Arnold, Inc.

Tropical grassland (savanna)

Mean monthly temperature (°C): 30, 20, 10, 0, −10, −20, −30, −40
Mean monthly precipitation (mm): 350, 300, 250, 200, 150, 100, 50, 0
Freezing point
J F M A M J J A S O N D
Month

Kent & Charlene Krone/SuperStock

Temperate grassland (prairie)

Mean monthly temperature (°C): 30, 20, 10, 0, −10, −20, −30, −40
Mean monthly precipitation (mm): 350, 300, 250, 200, 150, 100, 50, 0
Freezing point
J F M A M J J A S O N D
Month

M. Leder/Peter Arnold, Inc.

Cold grassland (arctic tundra)

Mean monthly temperature (°C): 30, 20, 10, 0, −10, −20, −30, −40
Mean monthly precipitation (mm): 350, 300, 250, 200, 150, 100, 50, 0
Freezing point
J F M A M J J A S O N D
Month

Figure 7-11 These climate graphs track the typical variations in annual temperature (red) and precipitation (blue) in tropical, temperate, and cold (arctic tundra) grasslands. Top photo: *savanna* (*tropical grassland*) in Maasai Mara National Park in Kenya, Africa, with wildebeests grazing. Center photo: *prairie* (*temperate grassland*) near East Glacier Park in the U.S. state of Montana, with wildflowers in bloom. Bottom photo: *arctic tundra* (*cold grassland*) in autumn in the U.S. state of Alaska. **Question:** Which month of the year has the highest temperature and which month has the lowest rainfall for each of the three types of grassland?

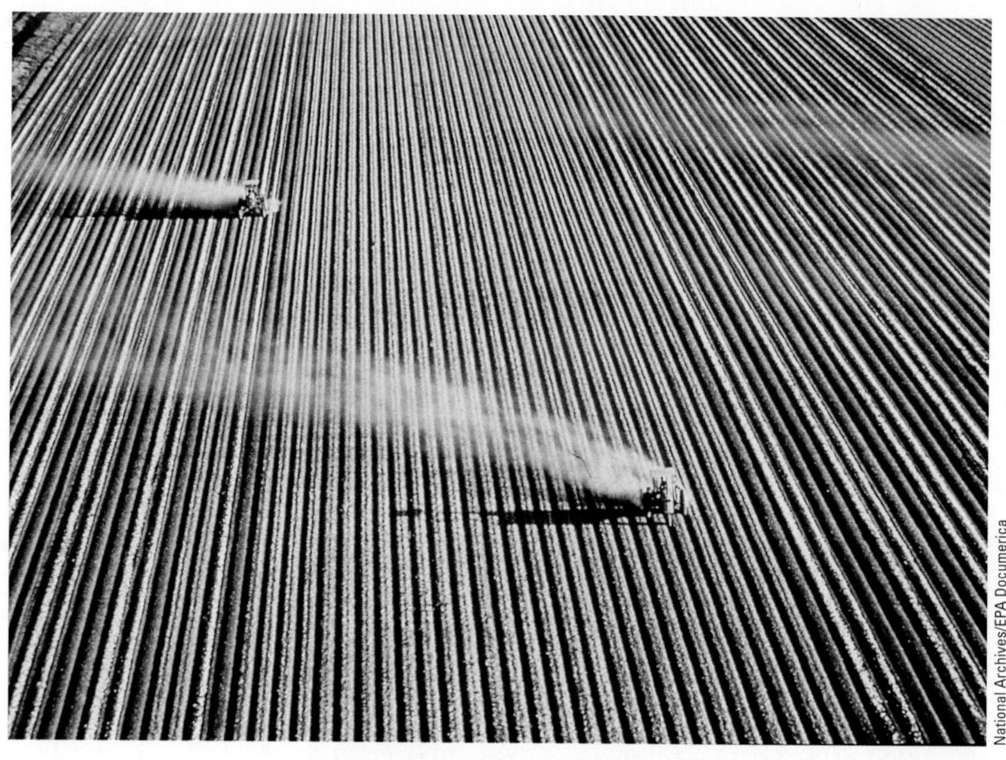

Figure 7-12 Natural capital degradation: This intensively cultivated cropland is an example of the replacement of biologically diverse temperate grasslands with a monoculture crop in the U.S. state of California. When humans remove the tangled root network of natural grasses, the fertile topsoil becomes subject to severe wind erosion unless it is covered with some type of vegetation.

Cold grasslands, or *arctic tundra* (Russian for "marshy plain"), lie south of the arctic polar ice cap (Figure 7-9). During most of the year, these treeless plains are bitterly cold (Figure 7-11, bottom graph), swept by frigid winds, and covered with ice and snow. Winters are long with few hours of daylight, and scant precipitation falls primarily as snow.

Under the snow, this biome is carpeted with a thick, spongy mat of low-growing plants, primarily grasses, mosses, lichens, and dwarf shrubs (Figure 7-11, bottom photo). Trees or tall plants cannot survive in the cold and windy tundra because they would lose too much of their heat. Most of the annual growth of the tundra's plants occurs during the 7- to 8-week summer, when there is daylight almost around the clock. Figure 7-13 describes some components and food-web interactions in an arctic tundra ecosystem.

One outcome of the extreme cold is the formation of **permafrost**, underground soil in which captured water stays frozen for more than two consecutive years. During the brief summer, the permafrost layer keeps melted snow and ice from draining into the ground. As a consequence, many shallow lakes, marshes, bogs, ponds, and other seasonal wetlands form when snow and frozen surface soil melt on the waterlogged tundra (Figure 7-11, bottom photo). Hordes of mosquitoes, black flies, and other insects thrive in these shallow surface pools. They serve as food for large colonies of migratory birds (especially waterfowl) that return from the south to nest and breed in the bogs and ponds.

Animals in this biome survive the intense winter cold through adaptations such as thick coats of fur (arctic wolf, arctic fox, and musk oxen) and feathers (snowy owl) and living underground (arctic lemming). In the summer, caribou migrate to the tundra to graze on its vegetation.

Tundra is a fragile biome. Most tundra soils formed about 17,000 years ago when glaciers began retreating after the last Ice Age (see Figure 4-8, p. 69). These soils usually are nutrient poor and have little organic detritus. Because of the short growing season, tundra soil and vegetation recover very slowly from damage or disturbance. Human activities in the arctic tundra—primarily on and around oil drilling sites, pipelines, mines, and military bases—leave scars that persist for centuries.

Another type of tundra, called *alpine tundra,* occurs above the limit of tree growth but below the permanent snow line on high mountains (Figure 7-8, left). The vegetation is similar to that found in arctic tundra, but it receives more sunlight than arctic vegetation gets. During the brief summer, alpine tundra can be covered with an array of beautiful wildflowers.

There Are Three Major Types of Forests

Forests are lands that are dominated by trees. The three main types of forest—*tropical, temperate* (**Core Case Study**), and *cold* (northern coniferous, or boreal)—result from combinations of varying precipitation levels and varying average temperatures (Figures 7-7 and 7-14, p. 134, and **Concept 7-2**). CORE CASE STUDY

Tropical rain forests (Figure 7-14, top photo, and front cover photo) are found near the equator (Figure 7-9), where hot, moisture-laden air rises and dumps its mois-

Figure 7-13 Some components and interactions in an *arctic tundra (cold grassland) ecosystem*. When these organisms die, decomposers break down their organic matter into minerals that plants use. Organisms are not drawn to scale. **Question:** If the arctic fox were eliminated from this ecosystem, what species might increase and what species might decrease in population size?

ture (Figure 7-3). These lush forests have year-round, uniformly warm temperatures, high humidity, and almost daily heavy rainfall (Figure 7-14, p. 134, top graph). This fairly constant warm and wet climate is ideal for a wide variety of plants and animals.

Tropical rain forests are dominated by *broadleaf evergreen plants*, which keep most of their leaves year-round. The tops of the trees form a dense canopy (Figure 7-14, top photo) that blocks most light from reaching the forest floor. For this reason, there is little vegetation on the forest floor. Many of the plants that do live at the ground level have enormous leaves to capture what little sunlight filters through to the dimly lit forest floor.

Some trees are draped with vines (called lianas) that reach for the treetops to gain access to sunlight. Once up into the *canopy*, the vines grow from one tree to another, providing walkways for many species living there. When a large tree is cut down, its network of lianas can pull down other trees.

Tropical rain forests have a very high net primary productivity (see Figure 3-13, p. 49). They are teeming with life and possess incredible biological diversity. Although tropical rain forests cover only about 2% of the earth's land surface, ecologists estimate that they contain at least 50% of the earth's known terrestrial plant and animal species. For example, a single tree in these forests may support several thousand different insect species. Plants from tropical rain forests are a source of chemicals used as blueprints for making most of the world's prescription drugs.

Rain forest species occupy a variety of specialized niches in distinct layers, which help to enable these forests' great biodiversity (high species richness). For example, vegetation layers are structured, for the most

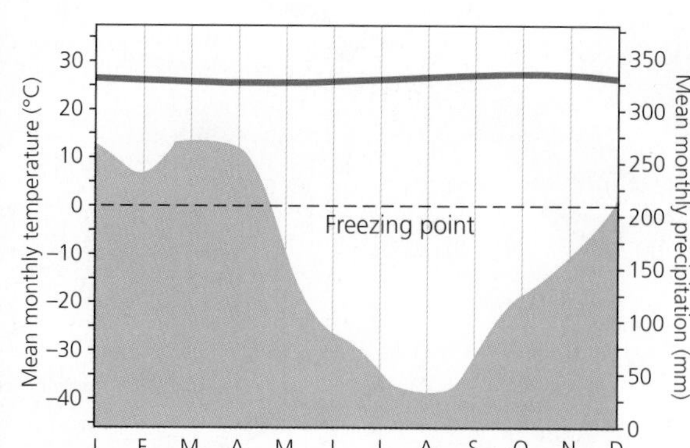

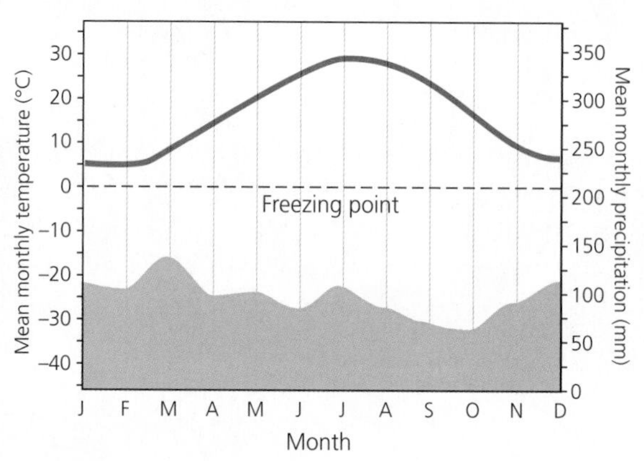

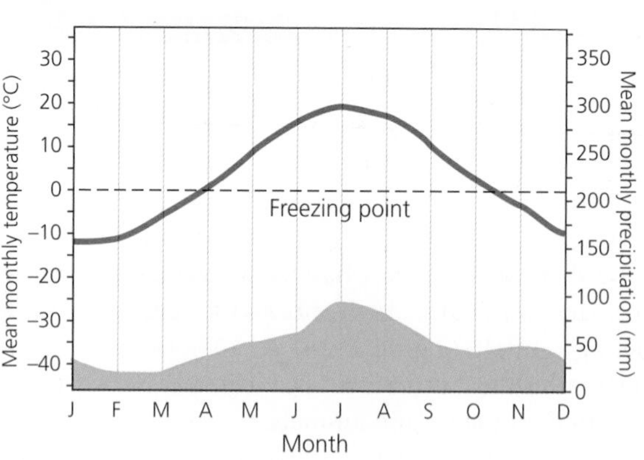

Figure 7-14 These climate graphs track the typical variations in annual temperature (red) and precipitation (blue) in tropical, temperate, and cold (northern coniferous, or boreal) forests. Top photo: the closed canopy of a *tropical rain forest* in the western Congo Basin of Gabon, Africa. Middle photo: a *temperate deciduous forest* in the autumn near Hamburg, Germany. Bottom photo: a *northern coniferous forest* in Canada's Jasper National Park. **Question:** Which month of the year has the highest temperature and which month has the lowest rainfall for each of the three types of forest?

Figure 7-15 Specialized plant and animal niches are *stratified*, or arranged roughly in layers, in a tropical rain forest. Filling such specialized niches enables species to avoid or minimize competition for resources and results in the coexistence of a great variety of species.

part, according to the plants' needs for sunlight, as shown in Figure 7-15. Much of the animal life, particularly insects, bats, and birds, lives in the sunny canopy layer, with its abundant shelter and supplies of leaves, flowers, and fruits. To study life in the canopy, ecologists climb trees and build platforms and boardwalks in the upper canopy. (See *The Habitable Planet*, Videos 4 and 9, at **www.learner.org/resources/series209.html** for information on how scientists gather information about tropical rain forests and the effects of human activities on such forests.)

Dropped leaves, fallen trees, and dead animals decompose quickly in tropical rain forests because of the warm, moist conditions and the hordes of decomposers. About 90% of the nutrients released by this rapid decomposition are quickly taken up and stored by trees, vines, and other plants. Nutrients that are not taken up are soon leached from the thin topsoil by the almost daily rainfall. As a result, very little plant litter builds up on the ground. The resulting lack of fertile soil helps to explain why rain forests are not good places to clear and grow crops or graze cattle on a sustainable basis.

So far, at least half of these forests have been destroyed or disturbed by human activities such as growing crops and raising cattle, and the pace of this destruction and degradation is increasing (see Chapter 3 Core Case Study, p. 40). Ecologists warn that without strong protective measures, most of these forests, along with their rich biodiversity and other very valuable ecological services, could be gone within your lifetime. (See *The Habitable Planet*, Video 9, at **www.learner.org/resources/series209.html**.)

THINKING ABOUT
Tropical Rain Forest Destruction

What harmful effects might the loss of most of the world's remaining tropical rain forests have on your lifestyle and that of any child or grandchild that you might have? What are two things you could do to help reduce this loss?

The second major type of forest, the *temperate deciduous forest*, is the subject of this chapter's **Core Case Study** (see middle photo of Figure 7-14). CORE CASE STUDY

Figure 7-16 Some of the components and species interactions in a *temperate deciduous forest ecosystem* in North America are illustrated here. When these organisms die, decomposers break down their organic matter into minerals that plants can use. Organisms are not drawn to scale. **Question:** If the broad-winged hawk were eliminated from this ecosystem, which species might increase and which species might decrease in population size?

Because they have cooler temperatures and fewer decomposers than tropical forests have, these forests also have a slower rate of decomposition. As a result, they accumulate a thick layer of slowly decaying leaf litter, which becomes a storehouse of nutrients. Figure 7-16 shows some of the major components and species interactions of this type of forest.

On a global basis, temperate forests have been degraded by various human activities, especially logging and urban expansion, more than any other terrestrial biome. However, within 100–200 years, forests of this type that have been cleared can return through secondary ecological succession (see Figure 5-10, p. 90).

Evergreen coniferous forests (Figure 7-14, bottom photo) are also called *boreal forests* or *taigas* ("TIE-guhs"). These cold forests are found just south of the arctic tundra in northern regions across North America, Asia, and Europe (Figure 7-9) and above certain altitudes in the Sierra Nevada and Rocky Mountain ranges of the United States. In this subarctic climate, winters are long and extremely cold; in the northernmost taigas, winter sunlight is available only 6–8 hours per day. Summers are short, with cool to warm temperatures (Figure 7-14, bottom graph), and the sun shines as long as 19 hours a day during mid-summer.

Most boreal forests are dominated by a few species of *coniferous* (cone-bearing) *evergreen trees* such as spruce, fir, cedar, hemlock, and pine that keep most of their leaves (or needles) year-round. Most of these species have small, needle-shaped, wax-coated leaves that can withstand the intense cold and drought of winter, when snow blankets the ground. Plant diversity is low because few species can survive the winters when soil moisture is frozen.

Beneath the stands of trees in these forests is a deep layer of partially decomposed conifer needles. Decomposition is slow because of low temperatures, the waxy coating on the needles, and high soil acidity. The decomposing conifer needles make the thin, nutrient-poor topsoil acidic, which prevents most other plants (except certain shrubs) from growing on the forest floor.

This biome contains a variety of wildlife. Year-round residents include bears, wolves, moose, lynx, and many burrowing rodent species. Caribou spend the winter in taiga and the summer in arctic tundra (Figure 7-11, bottom). During the brief summer, warblers and other insect-eating birds feed on hordes of flies, mosquitoes, and caterpillars.

Coastal coniferous forests or *temperate rain forests* are found in scattered coastal temperate areas with ample rainfall or moisture from dense ocean fogs. Thick stands of these forests with large conifers such as Sitka spruce, Douglas fir, and redwoods once dominated undisturbed areas of these biomes along the coast of North America, from Canada to northern California in the United States.

Mountains Play Important Ecological Roles

Some of the world's most spectacular environments are high on *mountains* (Figure 7-17), steep or high-elevation lands which cover about one-fourth of the earth's land surface (Figure 7-9). Mountains are places where dramatic changes in altitude, slope, climate, soil, and vegetation take place over a very short distance (Figure 7-8, left).

Mark Hamblin/WWI/Peter Arnold, Inc.

Figure 7-17 Mountains such as this one in the U.S. state of Washington play important ecological roles.

About 1.2 billion people (17% of the world's population) live in mountain ranges or in their foothills, and 4 billion people (57% of the world's population) depend on mountain systems for all or some of their water. Because of the steep slopes, mountain soils are easily eroded when the vegetation holding them in place is removed by natural disturbances such as landslides and avalanches, or by human activities such as timber cutting and agriculture. Many mountains are *islands of biodiversity* surrounded by a sea of lower-elevation landscapes transformed by human activities.

Mountains play important ecological roles. They contain the majority of the world's forests, which are habitats for much of the planet's terrestrial biodiversity. They often are habitats for *endemic species,* those that are found nowhere else on earth. They also serve as sanctuaries for animal species that are capable of migrating to higher altitudes and surviving in such environments if they are driven from lowland areas by human activities or by a warming climate.

CONNECTIONS

Mountains and Climate

Mountains help to regulate the earth's climate. Many mountaintops are covered with glacial ice and snow that reflect some solar radiation back into space, which helps to cool the earth and offset atmospheric warming. However, many of the world's mountain glaciers are melting, primarily because of atmospheric warming. While glaciers reflect solar energy, the darker rocks exposed by melting glaciers absorb that energy. This helps to warm the atmosphere above them, which melts more ice and warms the atmosphere more—in an escalating cycle of change.

Finally, mountains play a critical role in the hydrologic cycle (see Figure 3-14, p. 51) by serving as major storehouses of water. During winter, precipitation is stored as ice and snow. In the warmer weather of spring and summer, much of this snow and ice melts and is released to streams for use by wildlife and by humans for drinking and for irrigating crops. With atmospheric warming, mountaintop snow packs and glaciers are melting earlier in the spring each year. This is lowering food production in certain areas, because much of the water needed throughout the summer to irrigate crops gets released too quickly.

Scientific measurements and climate models indicate that a large number of the world's mountaintop glaciers may disappear during this century if the atmosphere keeps getting warmer as projected. This could force many people to move from their homelands in search of new water supplies and places to grow their crops. Despite the ecological, economic, and cultural importance of mountain ecosystems, protecting them has not been a high priority for governments or for many environmental organizations.

7-3 How Have Human Activities Affected the World's Terrestrial Ecosystems?

CONCEPT 7-3 In many areas, human activities are disrupting ecological and economic services provided by the earth's deserts, grasslands, forests, and mountains.

Humans Have Disturbed Most of the Earth's Land

According to the 2005 Millennium Ecosystem Assessment, about 62% of the world's major terrestrial ecosystems are being degraded or used unsustainably, as the human ecological footprint gets bigger and spreads across the globe (see Figure 2, pp. S24–25, in Supplement 6). Figure 7-18 summarizes some of the human impacts on the world's deserts, grasslands, forests, and mountains (**Concept 7-3**).

How long can we keep eating away at these terrestrial forms of natural capital without threatening our economies and the long-term survival of our own and many other species? No one knows. But there are increasing signs that we need to come to grips with this vital issue.

Many environmental scientists now call for a global effort to protect the world's remaining wild areas from development. In addition, they call for us to restore many of the land areas that have been degraded, especially in areas that are rich in biodiversity. However, such efforts are highly controversial because of timber, mineral, fossil fuel, and other resources found on or under many of the earth's remaining wild land areas. These issues are discussed in Chapter 9.

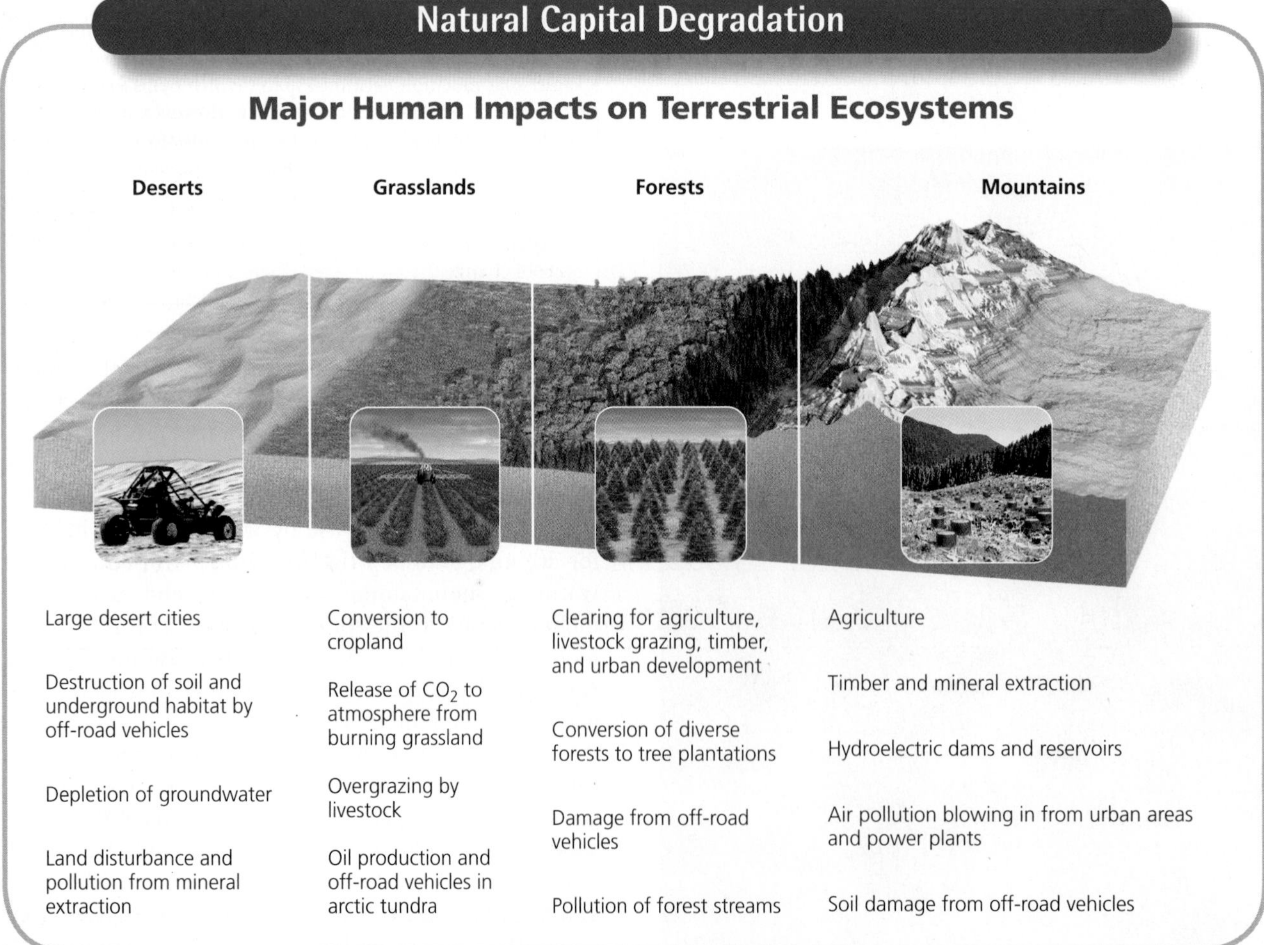

Figure 7-18 The major human impacts on the world's deserts, grasslands, forests, and mountains (**Concept 7-3**) are summarized here. **Question:** For each of these biomes, which two of the impacts listed do you think are the most harmful?

7-4 What Are the Major Types of Aquatic Systems?

▶ **CONCEPT 7-4 Saltwater and freshwater aquatic life zones cover almost three-fourths of the earth's surface, and oceans dominate the planet.**

Most of the Earth Is Covered with Water

When viewed from a certain point in outer space, the earth appears to be almost completely covered with water (Figure 7-19). Without water—which makes the earth a blue planet—life as we know it would not exist.

Saltwater in the oceans covers about 71% of the earth's surface, holds almost 98% of its water, and makes up 97% of the biosphere where life is found. Each of us is connected to, and utterly dependent on, the earth's global ocean through the water cycle (Figure 3-14, p. 51) with every breath, every swallow of water, and every bite of food that we take.

Although the *global ocean* is a single and continuous body of water, geographers divide it into four large areas—the Atlantic, Pacific, Arctic, and Indian Oceans—separated by the continents. The largest ocean is the Pacific, which contains more than half of the earth's water and covers one-third of the earth's surface.

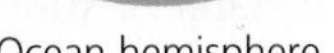

Figure 7-19 *The ocean planet:* The salty oceans cover 90% of the planet's ocean hemisphere (left) and nearly half of its land-ocean hemisphere (right). (**Concept 7-4**).

The aquatic counterparts of biomes are called **aquatic life zones**—saltwater and freshwater portions of the biosphere that can support life. The distribution of many aquatic organisms is determined largely by the water's *salinity*—the amounts of various salts such as sodium chloride (NaCl) dissolved in a given volume of water. As a result, aquatic life zones are classified into two major types: *saltwater* or *marine life zones* (oceans and their bays, estuaries, and other coastal systems) and *freshwater life zones* (lakes, rivers, streams, and inland wetlands). Although some systems such as estuaries are a mix of saltwater and freshwater, we classify them as marine systems for purposes of discussion.

Saltwater and freshwater life zones contain four major types of organisms. The first is weakly swimming, free-floating *plankton,* which include *phytoplankton* (tiny plants including many types of algae); *zooplankton* (plankton that feed on other plankton); and *ultraplankton* (huge populations of photosynthetic bacteria). Second is *nekton,* strongly swimming consumers such as fish, turtles, and whales. Third is *benthos,* or bottom-dwellers such as oysters, clams, and lobsters. Fourth is *decomposers* (mostly bacteria), which break down organic compounds in the dead bodies and wastes of aquatic organisms. In most aquatic systems, the key factors that determine the types and numbers of organisms found at various depths are *water temperature, dissolved oxygen content, availability of food,* and *availability of light and nutrients required for photosynthesis.*

7-5 Why Are Marine Aquatic Systems Important and How Have Human Activities Affected Them?

▶ **CONCEPT 7-5 Saltwater ecosystems are irreplaceable reservoirs of biodiversity, providing major ecological and economic services that are being threatened by human activities.**

Oceans Provide Vital Ecological and Economic Services

Why should we care about the oceans? The answer is that they are an irreplaceable part of the earth's life-support system. They provide enormously valuable ecological and economic services (Figure 7-20, p. 140) that keep us and other species alive and that support our economies.

Marine aquatic systems are enormous reservoirs of biodiversity. Marine life is found in three major *life*

Figure 7-20 Marine systems provide a number of important ecological and economic services (**Concept 7-5**). **Questions:** Which two ecological services and which two economic services do you think are the most important? Why?

Natural Capital

Marine Ecosystems

Ecological Services	Economic Services
Climate moderation	Food
CO_2 absorption	Animal and pet feed
Nutrient cycling	Pharmaceuticals
Waste treatment	Harbors and transportation routes
Reduced storm impact (mangroves, barrier islands, coastal wetlands)	Coastal habitats for humans
Habitats and nursery areas	Recreation
Genetic resources and biodiversity	Employment
Scientific information	Oil and natural gas
	Minerals

zones: the coastal zone, the open sea, and the ocean bottom (Figure 7-21).

The **coastal zone** is the warm, nutrient-rich, shallow water that extends from the high-tide mark on land to the gently sloping, shallow edge of the *continental shelf* (the submerged part of the continents). The coastal zone makes up less than 10% of the world's ocean area, but it contains 90% of all marine species and is the site of most large commercial marine fisheries.

Most coastal zone aquatic systems such as estuaries, coastal marshes, mangrove forests, and coral reefs have a high net primary productivity (see Figure 3-13, p. 49). This is the result of the zone's ample supplies of sunlight and plant nutrients that flow from land and are distributed by wind and ocean currents. Here, we look at some of these systems in more detail.

Estuaries and Coastal Wetlands Are Highly Productive

An **estuary** is where a river meets the sea (Figure 7-22). It is a partially enclosed body of water where seawater mixes with freshwater as well as nutrients and pollutants from streams, rivers, and runoff from the land.

Estuaries are associated with **coastal wetlands**—coastal land areas covered with water all or part of the year. These wetlands include coastal marshes (called *salt marshes* in temperate zones; see Photo 8 in Detailed Contents), and **mangrove forests**, which can contain any of 69 different tree species that can grow in salt water (Figure 7-23, p. 142). Coastal wetlands contain some of the earth's most productive ecosystems because of high nutrient inputs from rivers and nearby land, rapid circulation of nutrients by tidal flows, and ample sunlight penetrating their shallow waters.

Sea grass beds are another component of coastal marine biodiversity (Figure 7-24, p. 142). They consist of combinations of as many as 60 species of plants that grow underwater in shallow marine and estuarine areas along most continental coastlines. These highly productive and physically complex systems support a variety of marine species. They also help to stabilize shorelines and reduce wave impacts.

Life in these coastal ecosystems is harsh. It must adapt to significant daily and seasonal changes in tidal and river flows, salinity, and water temperature, and is subject to runoff of eroded soil sediment and other pollutants from the land.

These coastal aquatic systems provide vital ecological and economic services. They help to maintain water quality in tropical coastal zones by filtering toxic pollutants, excess plant nutrients, and sediments, and by absorbing other pollutants. They provide food, habitats, and nursery sites for a variety of aquatic and terrestrial species. They also reduce storm damage and coastal erosion by absorbing waves and storing excess water produced by storms and tsunamis such as those that devastated many coastal areas of Indonesia and several other countries in 2004, as well as the coastal area of northeast Japan in March of 2011.

According to a 2008 UN Food and Agriculture Organization report, despite their ecological and economic importance, at least one-fifth of the world's mangrove forests were eliminated between 1980 and 2005, mostly to make way for human coastal developments.

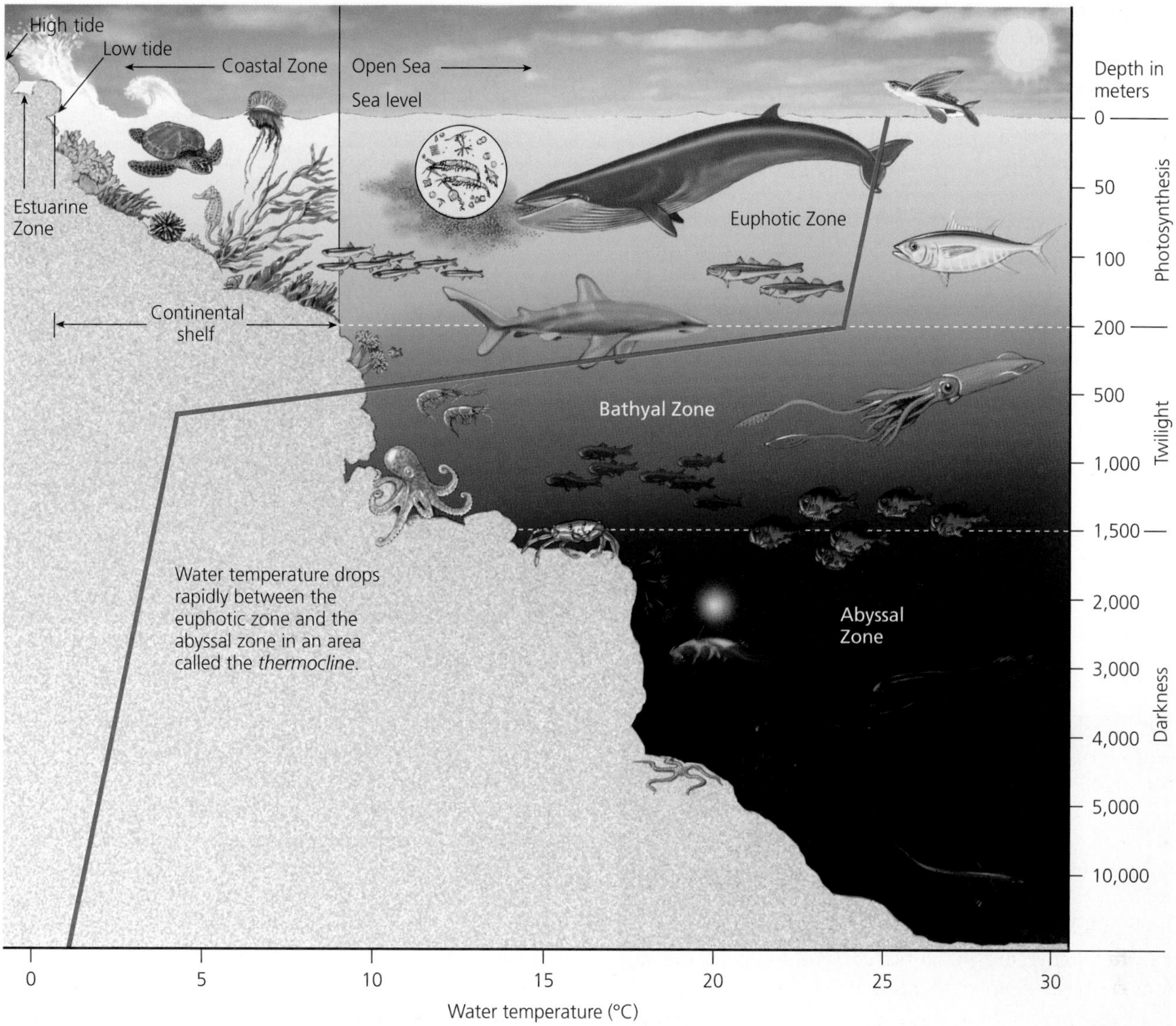

Figure 7-21 This diagram illustrates the major life zones and vertical zones (not drawn to scale) in an ocean. The actual depths of these zones vary by location. Available light determines the euphotic, bathyal, and abyssal zones. Temperature zones also vary with depth, shown here by the red line. **Question:** How is an ocean like a rain forest? (Hint: see Figure 7-15.)

NASA

Figure 7-22 This satellite photo shows an *estuary* in the African nation of Madagascar. The plume of cloudy water, formed by eroded sediments, is located at the mouth of the Betsiboka River, which flows through the estuary and into the Mozambique Channel. Because of its topography, heavy rainfall, and the clearing of its forests for agriculture, Madagascar is the world's most eroded country.

Figure 7-23 This mangrove forest is in Daintree National Park in Queensland, Australia. The tangled roots and dense vegetation in these coastal forests act like shock absorbers to reduce damage from storms and tsunamis. They also provide a highly complex habitat for a diversity of invertebrates and fishes.

Theo Allofs/Visuals Unlimited

Rich Carey/Shutterstock.com

Figure 7-24 Sea-grass beds support a variety of marine species. Since 1980, about 29% of the world's sea-grass beds have been lost to pollution and other disturbances.

Coral Reefs Are Storehouses of Biodiversity

Another important coastal ecosystem is a **coral reef** (Figure 7-25), an underwater structure, built primarily of limestone, that hosts a large variety of marine species. It forms in the clear, warm coastal waters of the tropics and subtropics. Coral reefs are dazzling centers of biodiversity—among the world's oldest, most diverse, and most productive ecosystems. They are being damaged and destroyed at an alarming rate by a variety of human activities.

These structures are formed by massive colonies of tiny animals called *polyps* (close relatives of jellyfish). They slowly build reefs by secreting a protective crust of limestone (calcium carbonate) around their soft bodies. When the polyps die, their empty crusts remain behind as a platform for more reef growth. The resulting elaborate network of crevices, ledges, and holes serve as natural "condominiums" for a variety of marine animals.

Coral reefs are the result of a mutually beneficial relationship between the polyps and tiny single-celled algae called *zooxanthellae* ("zoh-ZAN-thel-ee") that live in the tissues of the polyps. The algae provide the polyps with food and oxygen through photosynthesis, and they also give the reefs their stunning coloration. The polyps, in turn, provide the algae with a well-protected home and some of their nutrients.

Although coral reefs occupy only about 0.2% of the ocean floor, they provide important ecological and economic services. For example, they act as natural barriers that help to protect about 15% of the world's coastlines from erosion caused by battering waves and storms. And they provide habitats for about 25% of all marine organisms.

Economically, coral reefs produce about 10% of the global fish catch—25% of the catch in less-developed countries—and they provide fishing and ecotourism jobs for some of the world's poorest countries. These biological treasures give us an underwater world to study and enjoy. Each year, more than 1 million scuba divers and snorkelers visit coral reefs to experience these wonders of aquatic biodiversity.

Coral reefs are vulnerable to damage because they grow slowly and are disrupted easily. They thrive only in clear and fairly shallow water of constant high salin-

Figure 7-25 A healthy coral reef in the Red Sea (left) is covered by colorful algae (left), while a bleached coral reef (right) has lost most of its algae because of changes in the environment (such as cloudy water, high water temperatures, and increased acidity). With the algae gone, the white limestone of the corals' skeletons becomes visible. If the environmental stress is not removed and no other algae fill the abandoned niche, the corals die.

ity. Runoff of soil and other materials from the land can cloud the water and block the sunlight needed by the reefs' producer organisms. The water in which they live must have a temperature of 18–30°C (64–86°F), and if the water is too acidic, it can dissolve their calcium carbonate structures and shells. This explains why the biggest long-term threat to coral reefs may be projected climate change, which could raise the water temperature above this limit in most reef areas and increase the acidity of ocean water to harmful levels.

The increasing acidity of ocean water is a problem that many scientists say is not getting enough attention. It happens when ocean water absorbs some of the excess carbon dioxide (CO_2) that we are adding to the atmosphere, mostly by burning carbon-containing fossils fuels. The CO_2 reacts with ocean water to form a weak acid, which can slowly dissolve the calcium carbonate that makes up the corals.

Another problem is *coral bleaching* (Figure 7-25, upper right). It occurs when stresses such as increased temperature cause the algae, upon which corals depend for food, to die off. Without food, the coral polyps die, leaving behind a white skeleton of calcium carbonate.

Studies by the Global Coral Reef Monitoring Network and other scientist groups estimate that since the 1950s, some 45–53% of the world's shallow coral reefs have been destroyed or degraded and another 25–33% could be lost within 20–40 years. In the Caribbean Sea, 80% of the shallow reefs are dead. And deep coral reefs that are thousands of years old are being destroyed by large numbers of trawler fishing boats that drag huge weighted nets across the ocean bottom.

The Open Sea and the Ocean Floor Host a Variety of Species

The sharp increase in water depth at the edge of the continental shelf separates the coastal zone from the vast volume of the ocean called the **open sea**. Primarily on the basis of the penetration of sunlight, this deep blue sea is divided into three *vertical zones* (Figure 7-21), or layers of water. Temperatures also change with depth and such changes help to define zones of varying species diversity in these layers.

The *euphotic zone* is the brightly lit upper zone, where drifting phytoplankton carry out about 40% of the world's photosynthetic activity (see *The Habitable Planet*, Video 3, at **www.learner.org/resources/series209.html**). Large, fast-swimming predatory fishes such as swordfish, sharks, and bluefin tuna populate this zone.

The *bathyal zone* is the dimly lit middle zone, which receives little sunlight and therefore does not contain photosynthesizing producers. Zooplankton and smaller fishes, many of which migrate to feed on the surface at night, populate this zone.

The deepest zone, called the *abyssal zone*, is dark and very cold. There is no sunlight to support photosynthesis, and this zone has little dissolved oxygen. Nevertheless, the deep ocean floor is teeming with life—so much so, that it is considered a major life zone—because it contains enough nutrients to support a large number of species. Most organisms of the deep waters and ocean floor get their food from showers of dead and decaying organisms, called *marine snow*, drifting down from the upper, lighted levels of the ocean.

Net primary productivity (NPP) is quite low in the open sea, except in upwelling areas, where currents bring up nutrients from the ocean bottom. However, because the open sea covers so much of the earth's surface, it makes the largest contribution to the earth's overall NPP.

Human Activities Are Disrupting and Degrading Marine Ecosystems

In their desire to live near a coast, some people are unwittingly helping to destroy or degrade the aquatic biodiversity and the ecological and economic services (Figure 7-20) that make coastal areas so valuable. In 2010, about 45% of the world's population and more than 50% of the U.S. population lived along or near coasts, and these percentages are increasing rapidly.

Major threats to marine systems from human activities include:

- Coastal development, which can destroy and pollute coastal habitats (see *The Habitable Planet*, Video 5, at **www.learner.org/resources/series209.html**)
- Runoff from nonpoint sources of pollutants such as silt, fertilizers, pesticides, and livestock wastes (see *The Habitable Planet*, Videos 7 and 8)
- Point-source pollution such as sewage from cruise ships and spills from oil tankers
- Overfishing, which depletes populations of commercial fish species
- Use of fishing trawlers, which drag weighted nets across the ocean bottom, degrading and destroying its habitats
- Invasive species (often introduced by humans) that can deplete populations of native aquatic species and cause economic damage
- Climate change, enhanced by human activities, which is warming the oceans and making them more acidic; this could cause a rise in sea levels during this century that would destroy coral reefs and flood coastal marshes and coastal cities (see *The Habitable Planet*, Videos 7 and 8)

Human activities are disrupting parts of the ocean's vital fabric of life at breathtaking speed. Figure 7-26 shows some of the effects of these human impacts on marine systems, which we examine more closely in Chapter 9. The world's oceans are in trouble and thus so are we. One reason for our overuse and abuse of the oceans' resources and ecological services is that we know little about what life they contain and how ocean systems work. Oceans are the least explored regions of our planet. We know much more about the surface of the moon than we do about the depths of the world's oceans.

THINKING ABOUT
Coral Reef Destruction

How might the loss of most of the world's remaining tropical coral reefs affect your life and the lives of any children or grandchildren you might have? What are two things you could do to help reduce this loss?

Natural Capital Degradation

Major Human Impacts on Marine Ecosystems and Coral Reefs

Marine Ecosystems	Coral Reefs
Half of coastal wetlands lost to agriculture and urban development	Ocean warming
Over one-fifth of mangrove forests lost to agriculture, development, and shrimp farms since 1980	Rising ocean acidity
Beaches eroding because of coastal development and rising sea levels	Soil erosion
Ocean bottom habitats degraded by dredging and trawler fishing	Algae growth from fertilizer runoff
At least 20% of coral reefs severely damaged and 25–33% more threatened	Bleaching
	Rising sea levels
	Increased UV exposure
	Damage from anchors
	Damage from fishing and diving

Figure 7-26 Major threats to marine ecosystems resulting from human activities are summarized here. **Questions:** Which three of these threats do you think are the most serious? Why?

7-6 What Are the Major Types of Freshwater Systems and How Have Human Activities Affected Them?

> **CONCEPT 7-6 Freshwater lakes, rivers, and wetlands provide important ecological and economic services that are being disrupted by human activities.**

Water Stands in Some Freshwater Systems and Flows in Others

Freshwater life zones include *standing* bodies of freshwater such as lakes, ponds, and inland wetlands, and *flowing* systems such as streams and rivers. Although these freshwater systems cover less than 2.2% of the earth's surface, they provide a number of important ecological and economic services (Figure 7-27).

Lakes are large natural bodies of standing freshwater formed when precipitation, runoff, streams, rivers, and groundwater seepage fill depressions in the earth's surface. Causes of such depressions include glaciation (Lake Louise in Alberta, Canada), displacement of the earth's crust (Lake Nyasa in East Africa), and volcanic activity (Crater Lake in the U.S. state of Oregon).

Freshwater lakes vary tremendously in size, depth, and nutrient content. Deep lakes normally consist of four distinct zones that are defined by their depth and distance from shore (Figure 7-28, p. 146).

Ecologists classify lakes according to their nutrient content and primary productivity. Lakes that have a small supply of plant nutrients are called **oligotrophic** (poorly nourished) **lakes** (Figure 7-29, left, p. 146). This type of lake is often deep and has steep banks.

Glaciers and mountain streams supply water to many such lakes, bringing little in the way of sediment or microscopic life to cloud the water. These lakes usually have crystal-clear water and small populations of phytoplankton and fish species (such as smallmouth bass and trout). Because of their low levels of nutrients, these lakes have a low net primary productivity. Over time, sediments, organic material, and inorganic nutrients wash into most oligotrophic lakes, and plants grow and decompose to form bottom sediments.

A lake with a large supply of the nutrients needed by producers is called a **eutrophic** (well-nourished) **lake** (Figure 7-29, right). Such lakes typically are shallow and have murky brown or green water. Because of their high levels of nutrients, these lakes have a high net primary productivity. Excessive inputs of nutrients from the atmosphere and from human sources in nearby urban and agricultural areas can accelerate the eutrophication of lakes by putting excessive amounts of nutrients into them—a process called **cultural eutrophication**.

Freshwater Streams and Rivers Carry Water from the Mountains to the Oceans

Precipitation that does not sink into the ground or evaporate is called **surface water**. It becomes **runoff** when it flows into streams or lakes. A **watershed**, or **drainage basin**, is the land area that delivers runoff, sediments, and dissolved substances to a stream or lake. Small streams join to form rivers, and rivers flow down

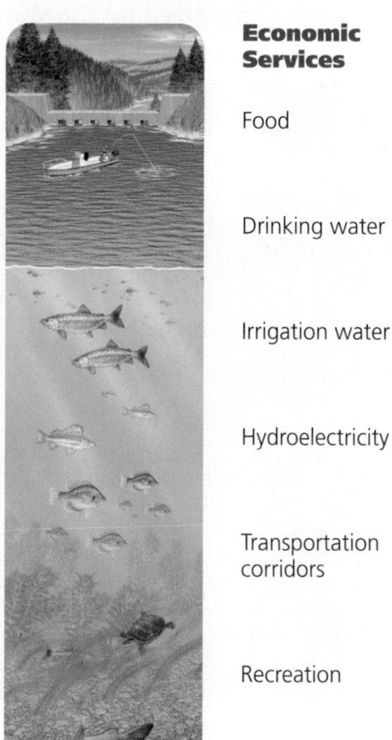

Figure 7-27 Freshwater systems provide many important ecological and economic services (**Concept 7-6**). **Questions:** Which two ecological services and which two economic services do you think are the most important? Why?

Active Figure 7-28 There are typically distinct zones of life in any fairly deep temperate-zone lake. *See an animation based on this figure at* www.cengagebrain.com. **Question:** How are deep lakes like tropical rain forests? (Hint: See Figure 7-15)

across sloping land, eventually flowing into a lake or an ocean (Figure 7-30).

In many areas, streams begin in mountainous or hilly areas, which collect and release water falling to the earth's surface as rain or as snow that melts during warm seasons. The downward flow of surface water and groundwater from mountain highlands to the sea typically takes place in three aquatic life zones characterized by different environmental conditions: the *source zone*, the *transition zone*, and the *floodplain zone* (Figure 7-30). Rivers and streams can differ somewhat from this generalized model.

As streams flow, they shape the land through which they pass. Over millions of years, the friction of moving water has leveled mountains and cut deep canyons, and rocks and soil removed by the water have been depos-

Figure 7-29 These photos show the effect of nutrient enrichment on a lake. Crater Lake in the U.S. state of Oregon (left) is an example of an *oligotrophic lake*, which is low in nutrients. Because of the low density of plankton, its water is quite clear. The lake on the right, found in western New York State, is a *eutrophic lake*. Because of an excess of plant nutrients, its surface is covered with mats of algae.

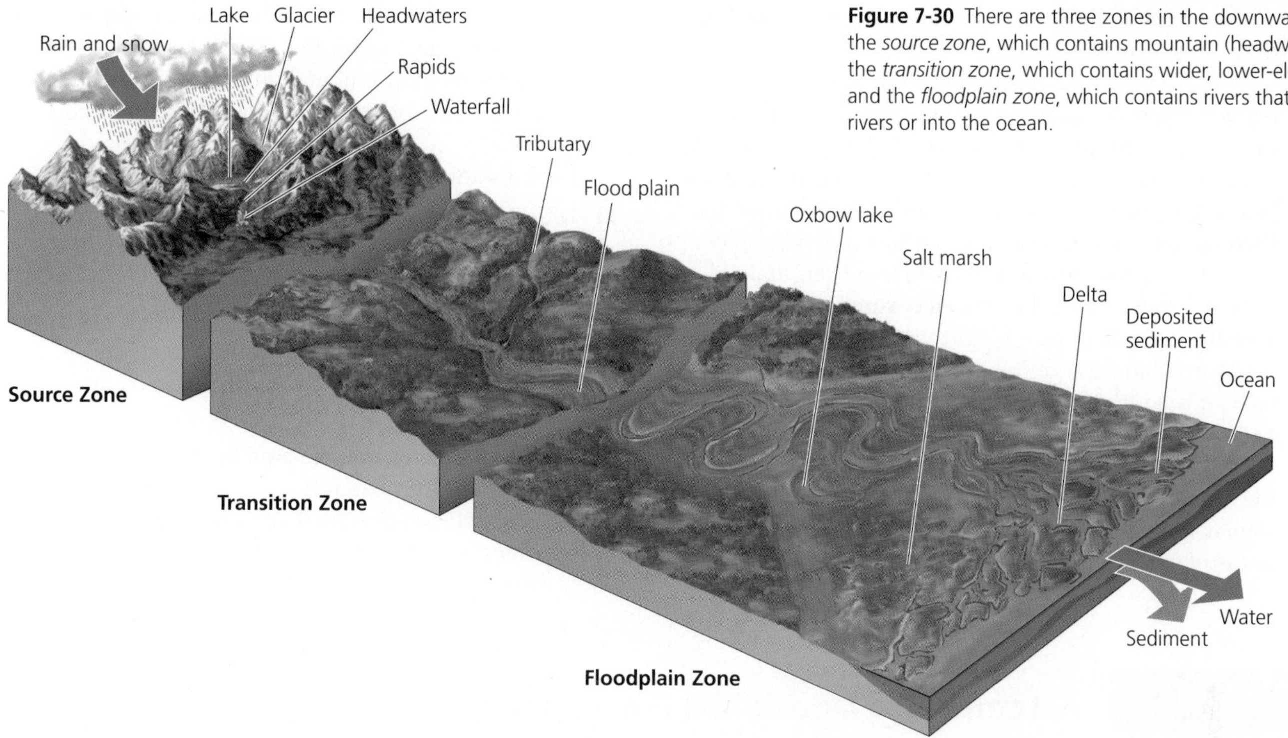

Figure 7-30 There are three zones in the downward flow of water: the *source zone*, which contains mountain (headwater) streams; the *transition zone*, which contains wider, lower-elevation streams; and the *floodplain zone*, which contains rivers that empty into larger rivers or into the ocean.

ited as sediment in low-lying areas. Where rivers meet the sea, such deposits can form coastal deltas and wetlands, which are important examples of natural capital. They absorb and slow the velocity of floodwaters from coastal storms, hurricanes, and tsunamis.

CONNECTIONS
Streams and Bordering Land

Streams receive many of their nutrients from bordering land ecosystems. Such nutrients come from falling leaves, animal feces, insects, and other forms of biomass washed into streams during heavy rainstorms or by melting snow. Thus, the levels and types of nutrients in a stream depend on what is happening in the stream's watershed.

Freshwater Inland Wetlands Are Vital Sponges

Inland wetlands are lands covered with freshwater all or part of the time (excluding lakes, reservoirs, and streams) and located away from coastal areas. They include *marshes* (dominated by grasses and reeds with few trees), *swamps* (dominated by trees and shrubs), and *prairie potholes* (depressions carved out by ancient glaciers). Other examples are *floodplains*, which receive excess water during heavy rains and floods, and the wet *arctic tundra* in summer. Some wetlands are covered with water year-round and others are under water for only a short time each year.

Inland wetlands provide a number of free ecological and economic services, which include:

- filtering and degrading toxic wastes and pollutants,
- reducing flooding and erosion by absorbing storm water and releasing it slowly, and by absorbing overflows from streams and lakes,
- maintaining stream flows during dry periods,
- helping to recharge groundwater aquifers,
- helping to maintain biodiversity by providing habitats for a variety of species,
- supplying valuable products such as fishes and shellfish, blueberries, cranberries, wild rice, and timber, and
- providing recreation for birdwatchers, nature photographers, boaters, anglers, and waterfowl hunters.

THINKING ABOUT
Inland Wetlands

Which two of the ecological and economic services provided by inland wetlands do you believe are the most important? Why?

Human Activities Are Disrupting and Degrading Freshwater Systems

Human activities are disrupting and degrading many of the ecological and economic services provided by freshwater rivers, lakes, and wetlands (**Concept 7-6**) in four major ways. *First*, dams and canals alter and destroy terrestrial and aquatic wildlife habitats along rivers and in their coastal deltas and estuaries by reducing water flow and increasing damage from coastal storms.

Second, flood control levees and dikes built along rivers disconnect the rivers from their floodplains, destroy aquatic habitats, and alter or reduce the functions of nearby wetlands.

Third, cities and farms add pollutants and excess plant nutrients to nearby streams, rivers, and lakes. This can cause explosions in the populations of algae and cyanobacteria (Figure 7-29, right), which deplete the lake's dissolved oxygen. Fishes and other species may then die off, which causes a major loss in biodiversity.

Fourth, many inland wetlands have been drained or filled to grow crops or have been covered with concrete, asphalt, and buildings.

More than 50% of the inland wetlands estimated to have existed in the continental United States during the 1600s no longer exist. This loss of natural capital has been an important factor in increasing flood damage in the United States. Many other countries have suffered similar losses. For example, 80% of all inland wetlands in Germany and France have been destroyed.

When we look further into human impacts on aquatic systems in Chapter 9, we will also explore possible solutions to environmental problems that result from these impacts, as well as ways to help sustain aquatic biodiversity.

Here are this chapter's *three big ideas*:

- Differences in climate, based mostly on long-term differences in average temperature and precipitation, largely determine the types and locations of the earth's deserts, grasslands, and forests.
- Saltwater and freshwater aquatic systems cover almost three-fourths of the earth's surface, and oceans dominate the planet.
- The earth's terrestrial and aquatic systems provide important ecological and economic services, which are being degraded and disrupted by human activities.

REVISITING A Temperate Deciduous Forest and Sustainability

In this chapter's **Core Case Study**, we considered the influence of climate on terrestrial biodiversity in the formation of biomes—deserts, grasslands, and forests—and aquatic ecosystems. We focused on the *temperate deciduous forest* biome, noting how the plants and animals that have evolved within such forests are adapted to climatic conditions there. Thus we saw that *climate*—the weather conditions in a given area averaged over periods of time ranging from three decades to thousands of years—plays a key role in determining the nature of terrestrial and aquatic systems as well as the life forms that live in those systems.

These relationships are in keeping with the three **principles of sustainability** (see back cover). The earth's dynamic climate system helps to distribute heat from solar energy and to recycle the earth's nutrients. In turn, this helps to generate and support the terrestrial and aquatic biodiversity found in the earth's biomes, oceans, lakes, rivers, and wetlands. Through these global processes, life has sustained itself for at least 3.5 billion years.

Scientists have made some progress in understanding the ecology of the world's terrestrial and aquatic systems and how the vital ecological and economic services they provide are being degraded and disrupted. One of the major lessons from their research is: *in nature, everything is connected*. According to these scientists, we urgently need more research on the components and workings of the world's biomes and aquatic life zones, on how they are interconnected, and on which of them are in the greatest danger of being disrupted by human activities. With such vital information, we will have a clearer picture of how our activities affect the natural capital that supports the earth's life and what we can do to help sustain it.

When we try to pick out anything by itself, we find it hitched to everything else in the universe.

JOHN MUIR

REVIEW

CORE CASE STUDY

1. Describe a temperate deciduous forest (**Core Case Study**) and explain why it serves as an example of how differences in climate lead to the formation of different types of ecosystems.

SECTION 7-1

2. What is the key concept for this section? Distinguish between **weather** and **climate**. Describe three major factors that determine how air circulates in the lower atmosphere. Describe how the properties of air, water, and land affect global air circulation. Define **ocean currents** and explain how they, along with global air circulation, support the formation of forests, grasslands, and deserts. Define **greenhouse gas** and the **natural greenhouse effect**. Why are they important to the earth's life and climate? What is the **rain shadow effect** and how can it lead to the formation of deserts? Why do cities tend to have more haze and smog, higher temperatures, and lower wind speeds than the surrounding countryside?

SECTION 7-2

3. What is the key concept for this section? What is a **biome**? Explain why there are three major types of each of the major biomes (deserts, grasslands, and forests). Explain why biomes are not uniform. Describe how climate and vegetation vary with latitude and elevation.

4. Describe how the three major types of deserts differ in their climate and vegetation. How do desert plants and animals survive? Describe how the three major types of grasslands differ in their climate and vegetation. What is **permafrost**? Describe how the three major types of forests differ in their climate and vegetation. What important ecological roles do mountains play?

SECTION 7-3

5. What is the key concept for this section? About what percentage of the world's major terrestrial ecosystems are being degraded or used unsustainably? Summarize the ways in which human activities have affected the world's deserts, grasslands, forests, and mountains.

SECTION 7-4

6. What is the key concept for this section? What percentage of the earth's surface is covered with water? What is an **aquatic life zone**? Distinguish between a saltwater (marine) life zone and a fresh-water life zone, and give two examples of each.

SECTION 7-5

7. What is the key concept for this section? What major ecological and economic services are provided by marine systems? Define and distinguish between the **coastal zone** and the **open sea**. Distinguish between an **estuary** and a **coastal wetland** and explain why they each have high net primary productivity. Define and explain the importance of **sea grass beds**. What is a **mangrove forest** and what is its ecological and economic importance? What is a **coral reef**? How do they form and what major ecological and economic services do they provide? Why does the open sea have a low net primary productivity?

8. Describe seven major threats to marine systems resulting from human activities. List the ways in which each of the major marine systems is threatened.

SECTION 7-6

9. What is the key concept for this section? What major ecological and economic services do freshwater systems provide? What is a **lake**? What four zones are found in most deep lakes? Distinguish among **oligotrophic** and **eutrophic lakes**. What is **cultural eutrophication**? Define **surface water**, **runoff**, and **watershed** (**drainage basin**). Describe the three zones that streams and rivers pass through as they flow from highlands to the sea. Give three examples of **inland wetlands** and explain their ecological importance. What are four ways in which human activities are disrupting and degrading freshwater systems?

10. What are this chapter's *three big ideas*? Describe the connections between climate and differences among the terrestrial and aquatic systems. How are these relationships in keeping with the three **principles of sustainability** (see back cover)?

Note: Key terms are in **bold** type.

CRITICAL THINKING

1. Why do you think temperate deciduous forests (**Core Case Study**) are among the biomes most extensively disturbed by human activities?

2. What would be likely to happen to the earth's climate **(a)** if most of the world's oceans disappeared and **(b)** if most of the world's land disappeared?

3. Describe the roles of temperature and precipitation in determining what parts of the earth's land are covered with: **(a)** desert, **(b)** arctic tundra, **(c)** temperate grassland, **(d)** tropical rain forest, and **(e)** temperate deciduous forest (**Core Case Study**).

4. How might the distribution of the world's forests, grasslands, and deserts shown in Figure 7-9 differ if the prevailing winds shown in Figure 7-3 did not exist?

5. Which biomes are best suited for **(a)** raising crops and **(b)** grazing livestock? Use the three **principles of sustainability** to come up with three guidelines for growing crops and grazing livestock more sustainably in these biomes.

6. What type of biome do you live in? (If you live in a developed area, what type of biome was the area before it was developed?) List three ways in which your lifestyle could be contributing to the degradation of this biome. What changes could you make in order to reduce your contribution, if any?

7. You are a defense attorney arguing in court for sparing a tropical rain forest from being cut down. Give your three best arguments for the defense of this ecosystem. Do the same for the case of temperate decidious forest (**Core Case Study**).

8. Congratulations! You are in charge of the world. What are the three most important features of your plan for helping to sustain the earth's **(a)** terrestrial biodiversity and ecosystems services and **(b)** aquatic biodiversity and ecosystem services?

DOING ENVIRONMENTAL SCIENCE

Using Google Earth, find an undeveloped ecosystem somewhere on the planet that you can zoom in on to observe some features. For example, find a mountain system or a river system that is undeveloped. Write a description of what you see. Then find such an example of a biome that has been developed and write a detailed comparison of your undeveloped and developed biome samples. If possible, check in on these systems once during your course term and again at the end of the term and describe any changes to the systems that you can observe. Suggest a hypothesis to explain major changes.

GLOBAL ENVIRONMENT WATCH EXERCISE

Search for *Coral reefs* and use the topic portal to find information on **(a)** trends in the global rate of coral destruction; **(b)** what areas of the world are seeing rising rates and what areas are seeing falling rates; and **(c)** what is being done to protect them in various areas. Write a report on your findings.

DATA ANALYSIS

In this chapter, you learned how long-term variations in average temperatures and average precipitation play a major role in determining the types of deserts, forests, and grasslands found in different parts of the world. Below are typical annual climate graphs for a tropical grassland (savanna) in Africa and a temperate grassland in the mid-western United States.

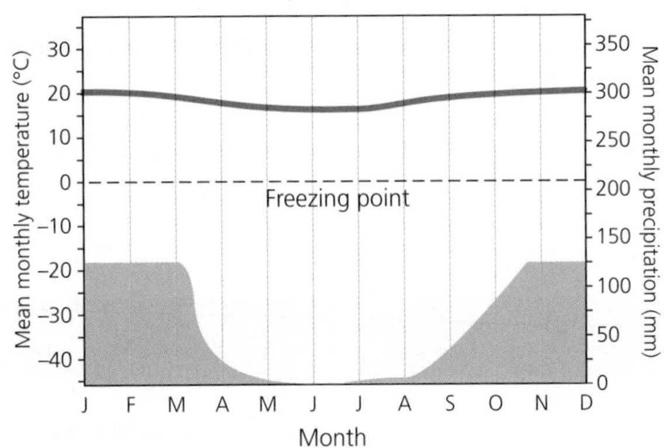

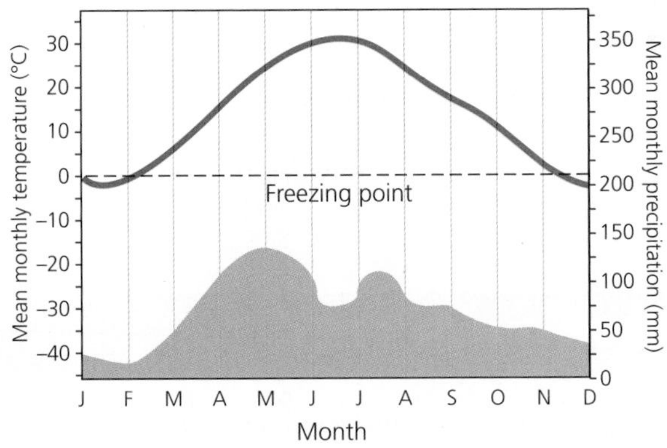

1. In what month (or months) does the most precipitation fall in each of these areas?
2. What are the driest months in each of these areas?
3. What is the coldest month in the tropical grassland?
4. What is the warmest month in the temperate grassland?

LEARNING ONLINE

The website for *Environmental Science 14e* contains helpful study tools and other resources to take the learning experience beyond the textbook. Examples include flashcards, practice quizzing, information on green careers, tips for more sustainable living, activities, and a wide range of articles and further readings. To access these materials and other companion resources for this text, please visit **www.cengagebrain.com**. For further details, see the preface, p. xvi.

Sustaining Biodiversity: The Species Approach

8

Polar Bears and Climate Change

CORE CASE STUDY

The world's 20,000–25,000 polar bears are found in 19 populations scattered across the frozen Arctic. About 60% of them are in Canada, and the rest live in arctic areas of Denmark, Norway, Russia, and the U.S. state of Alaska.

Throughout the winter, polar bears hunt for seals on floating sea ice (Figure 8-1) that expands southward each winter and contracts as the temperature rises during summer. By eating the seals, the bears build up their body fat. In the summer and fall, they live off this fat until hunting resumes when the ice expands again during winter.

Scientific measurements reveal that the earth's atmosphere is getting warmer and that this warming is occurring twice as fast in the Arctic as in the rest of the world. Thus, arctic ice is melting faster and the average annual area of floating sea ice in the Arctic during the summer is decreasing. The floating winter ice is also breaking up earlier each year, shortening the polar bears' hunting season. And much of the remaining ice is getting too thin to support the weight of a polar bear.

These changes mean that polar bears have less time to feed and store fat. As a result, they must fast longer, which weakens them. As females become weaker, their ability to reproduce and keep their young cubs alive declines. Polar bears are strong swimmers, but ice shrinkage has forced them to swim longer distances to find enough food and to spend more time during winter hunting on land, where it is nearly impossible for them to find enough prey. As the bears grow hungrier, they are more likely to go to human settlements looking for food, which gives some people the mistaken impression that their populations are growing. There are also reports that some hungry adult polar bears have been eating polar bear cubs.

In 2008, the U.S. Fish and Wildlife Service placed the Alaskan polar bear on its list of threatened species. Alaska state government officials are trying to have this decision repealed because they say it will hurt economic growth in their state. Oil and coal industry leaders also want the listing decision overturned because they fear it might be used as a way to regulate carbon dioxide, the climate-changing gas that is released into the atmosphere when we burn oil, coal, and natural gas.

According to a 2006 study by the International Union for the Conservation of Nature, the world's total polar bear population is likely to decline by 30–35% by 2050. By the end of this century, polar bears might be found only in zoos.

Scientists project that during this century, human activities, especially those that cause habitat loss and contribute to projected climate change, are likely to lead to the extinction of one-fourth to one-half of the world's known plant and animal species. Many biologists consider the rapid loss of the earth's vital biodiversity—largely resulting from human activities—to be one of the most serious and long-lasting environmental problem that the world faces. In this chapter, we discuss the causes of the rising rate of species extinction and possible ways to slow it down.

Figure 8-1 On floating ice in Svalbard, Norway, a polar bear feeds on its seal prey. Polar bears in the Arctic could become extinct sometime during this century if projected atmospheric warming melts much of the floating sea ice on which they hunt seals. **Question:** Do you think it matters that the polar bear may become extinct primarily because of human activities? Explain.

Key Questions and Concepts

8-1 What role do humans play in the extinction of species?

CONCEPT 8-1 Species are becoming extinct 100 to 1,000 times faster than they were before modern humans evolved, and by the end of this century, the extinction rate is expected to be 10,000 times higher than it was before humans arrived.

8-2 Why should we care about the rising rate of species extinction?

CONCEPT 8-2 We should avoid speeding up the extinction of wild species because of the economic and ecological services they provide, and because wild species have a right to exist regardless of their usefulness to us.

8-3 How do humans accelerate species extinction?

CONCEPT 8-3 The greatest threats to any species are (in order) loss or degradation of habitat, harmful invasive species, human population growth, pollution, climate change, and overexploitation.

8-4 How can we protect wild species from extinction?

CONCEPT 8-4 We can reduce the rising rate of species extinction and help to protect overall biodiversity by establishing and enforcing national environmental laws and international treaties, creating a variety of protected wildlife sanctuaries, and taking precautionary measures to prevent such harm.

Note: Supplements 2 (p. S3), 6 (p. S22), and 7 (p. S38) can be used with this chapter.

The last word in ignorance is the person who says of an animal or plant: "What good is it?" . . . If the land mechanism as a whole is good, then every part of it is good, whether we understand it or not. Harmony with land is like harmony with a friend; you cannot cherish his right hand and chop off his left.

ALDO LEOPOLD

8-1 What Role Do Humans Play in the Extinction of Species?

▶ **CONCEPT 8-1 Species are becoming extinct 100 to 1,000 times faster than they were before modern humans evolved, and by the end of this century, the extinction rate is expected to be 10,000 times higher than it was before humans arrived.**

Extinctions Are Natural but Sometimes They Increase Sharply

When a species can no longer be found anywhere on the earth, it has suffered **biological extinction**. The disappearance of any species, and especially those that play keystone roles (see Chapter 4, p. 74), can weaken or break some of the connections in the ecosystems where they had existed and thus can threaten various ecosystem services. This can lead to additional extinctions of species with strong connections to species that have already gone extinct.

The balance between formation of new species and extinction of existing species determines the earth's biodiversity. The extinction of many species in a relatively short period of geologic time is called a **mass extinction**. Geological and other records indicate that the earth has experienced perhaps five mass extinctions, when 50–95% of the world's species appear to have become extinct. After each mass extinction, the earth's overall biodiversity eventually returned to equal or higher levels, but each recovery required several million years.

Some Human Activities Are Causing Extinction Rates to Rise

Extinction is a natural process. But a growing body of scientific evidence indicates that it has accelerated as human populations have spread over most of the earth's land, consuming huge quantities of resources and creating large and growing ecological footprints (see Figure 1-8, p. 14). According to biodiversity expert Edward O. Wilson (Individuals Matter, p. 65), "The natural world is everywhere disappearing before our eyes—cut to pieces, mowed down, plowed under, gobbled up, replaced by human artifacts."

According to the 2005 Millennium Ecosystem Assessment and other studies, humans have taken over, disturbed, or polluted about 60%–80% of the earth's land surface (see Figure 2, pp. S24–25, in Supplement 6). Most of these disturbances involved filling in wetlands or converting grasslands and forests to crop fields and urban areas. Human activities have also polluted and disturbed almost 50% of the oceans and other surface waters that cover about 71% of the earth's surface.

Using the methods described in the Science Focus box below, scientists from around the world who conducted the 2005 Millennium Ecosystem Assessment estimated that the current annual rate of species extinction is at least 100 to 1,000 times the estimated *background extinction rate* (see p. 71) of about 0.0001%, which existed before modern humans evolved some 200,000 years ago.

Biodiversity researchers project that during this century, the extinction rate is likely to rise to at least 10,000 times the background rate (**Concept 8-1**)—mostly because of habitat loss and degradation, climate change due primarily to projected atmospheric warming, and other environmentally harmful effects of human activities. If this estimate is correct, the annual extinction rate would rise to about 1% per year.

So why is this a big deal? A 1% per year species extinction rate might not seem like much. However, according to biodiversity researchers Edward O. Wilson and Stuart Pimm, with a 1% extinction rate, at least 25% and as many as 50% of the world's current animal and plant species could vanish by the end of this century. In the chilling words of biodiversity expert Norman Myers, "Within just a few human generations, we shall—in the absence of greatly expanded conservation efforts—impoverish the biosphere to an extent that will persist for at least 200,000 human generations."

If such estimates are only half correct, we can see why many biologists consider such a massive loss of the biodiversity within the span of a single human lifetime to be the most important and long-lasting environmental problem we face. This is even more worrisome when we consider that this biodiversity supports all life on earth and all human economies.

THINKING ABOUT
Extinction

How might your lifestyle change if human activities contribute to the extinction of up to half of the world's species during this century? How might this affect the lives of any children or grandchildren you might have? List two aspects of your lifestyle that contribute to this threat to the earth's natural capital.

In fact, most extinction experts consider a projected extinction rate of 1% a year to be on the low side, for several reasons. *First,* both the rate of species loss and the extent of biodiversity losses are likely to increase sharply during the next 50–100 years because of the projected growth of the human population and the growing rate of resource use per person, and because of the projected human influence on climate change.

Second, current and projected extinction rates are much higher than the global average in parts of the world that are already highly endangered centers of biodiversity. Norman Myers and other biodiversity researchers urge us to focus our efforts on slowing the much higher rates of extinction in such *biodiversity hotspots.* They see this emergency action as the best and quickest way to protect much of the earth's biodiversity

SCIENCE FOCUS

Estimating Extinction Rates

Scientists who try to catalog extinctions, estimate past extinction rates, and project future rates face three problems. *First*, because the natural extinction of a species typically takes a very long time, it is not easy to document. *Second*, we have identified only about 2 million of the world's estimated 8 million to 100 million species. *Third*, scientists know little about the nature and ecological roles of most of the species that have been identified.

One approach to estimating future extinction rates is to study records documenting the rates at which mammals and birds (the easiest to observe) have become extinct since humans began their rapidly increasing domination of the planet about 10,000 years ago; around that time, we began the shift from getting our food by hunting and gathering in the wild to growing our own food. This information can be compared with fossil records of extinctions that occurred before the development of agriculture.

Another approach is to observe how decreases in habitat size affect extinction rates. The *species–area relationship*, studied by Edward O. Wilson (Individuals Matter, p. 65) and Robert MacArthur, suggests that, on average, a 90% loss of land habitat in a given area can cause the extinction of about 50% of the species living in that area.

Scientists also use mathematical models to estimate the risk of a particular species becoming endangered or extinct within a certain period of time. These *population viability analysis* (PVA) models include factors such as trends in population size, changes in habitat availability, interactions with other species, and genetic factors.

Researchers know that their estimates of extinction rates are based on inadequate data and sampling, and on incomplete models. Thus, they are continually striving to get better data and to improve the models they use to estimate extinction rates.

At the same time, they point to clear evidence that human activities have accelerated the rate of species extinction and that this rate is increasing. According to these biologists, arguing over the numbers and waiting to get better data and models should not delay our acting now to help prevent the extinction of species resulting mostly from human activities.

Critical Thinking

Does the fact that extinction rates can only be estimated (not proven absolutely) make them unreliable? Why or why not?

(a) Mexican gray wolf: about 42 in the forests of Arizona and New Mexico

(b) Sumatran tiger: less than 60 in Sumatra, Indonesia

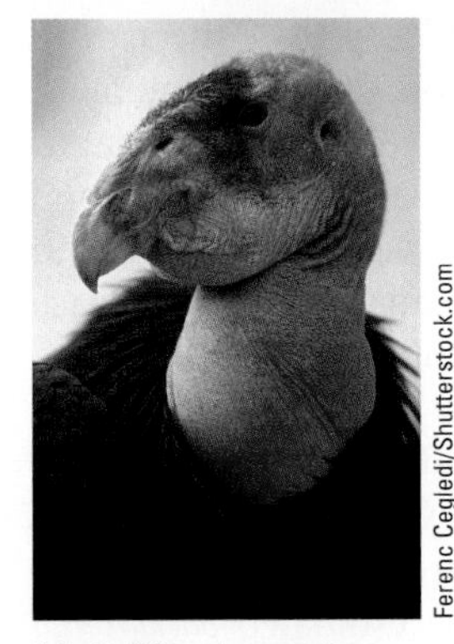

(c) California condor: 187 in the southwestern United States

(d) Whooping crane: 407 in North America

Figure 8-2 *Endangered natural capital:* These four critically endangered species are threatened with extinction, largely because of human activities. The number below each photo indicates the estimated total number of individuals of that species remaining in the wild. **Question:** What kinds of human activities do you think are putting these four species in danger?

from being lost during this century. (We discuss this further in Chapter 9.)

Third, we are eliminating, degrading, fragmenting, and simplifying many biologically diverse environments—such as tropical forests, tropical coral reefs, wetlands, and estuaries—that serve as potential sites for the emergence of new species. Thus, in addition to increasing the rate of extinction, we may be limiting the long-term recovery of biodiversity by reducing the rate of speciation for some species. In other words, we are also creating a *speciation crisis.* (See the Guest Essay by Normal Myers on this topic at **www.cengagebrain.com**.) Based on what we know about the recovery of biodiversity after past mass extinctions, it will take 5–10 million years for the earth's processes to replace the projected number of species that are likely to go extinct during this century.

In addition, Philip Levin, Donald Levin, and other biologists argue that, while our activities are likely to reduce the speciation rates for some species, they might help to increase the speciation rates for other rapidly reproducing opportunist species such as weeds and rodents, as well as cockroaches and other species of insects. Such rapidly expanding populations of generalist species could crowd and compete with certain other species, further accelerating their extinction.

Endangered and Threatened Species Are Ecological Smoke Alarms

Biologists classify species that are heading toward biological extinction as either *endangered* or *threatened.* An **endangered species** has so few individual survivors that the species could soon become extinct. A **threatened species** (also known as a *vulnerable species*) still has enough remaining individuals to survive in the short term, but because of declining numbers, it is likely to become endangered in the near future.

An example of a threatened species is the polar bear (**Core Case Study**). Figure 8-2 shows four highly endangered species. CORE CASE STUDY

Some species have characteristics that increase their chances of becoming extinct (Figure 8-3). As Edward O.

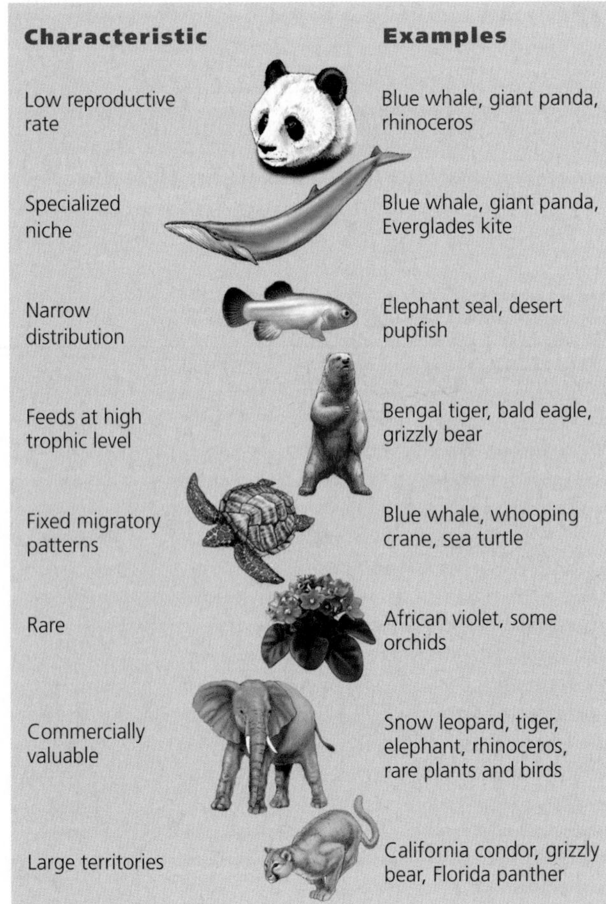

Figure 8-3 This diagram shows some characteristics that can put certain species in greater danger of becoming extinct. **Question:** Which of these characteristics might possibly contribute to the extinction of the polar bear (**Core Case Study**) during this century?

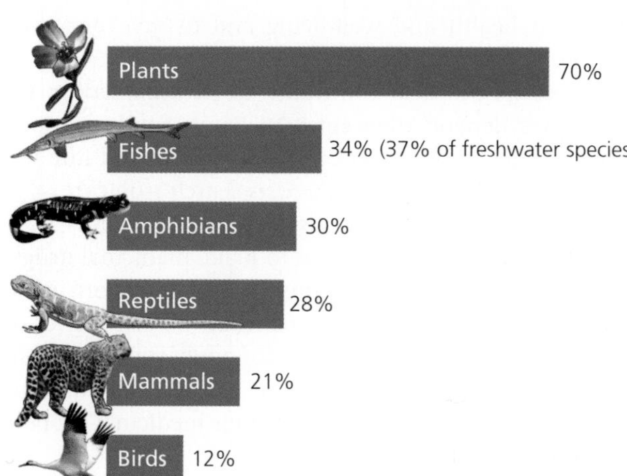

Figure 8-4 Endangered natural capital: This diagram shows the percentages of various types of known species that are threatened with extinction because of human activities. **Question:** Why do you think plants and fish species top this list? (Data from International Union for Conservation of Nature, Conservation 2009)

Wilson, biodiversity expert, puts it, "The first animal species to go are the big, the slow, the tasty, and those with valuable parts such as tusks and skins."

Some species have *behavioral characteristics* that make them prone to extinction. The passenger pigeon and the Carolina parakeet, both extinct, nested in large flocks that made them easy to kill. Some types of species are more threatened with extinction resulting from human activities than others are (Figure 8-4).

8-2 Why Should We Care about the Rising Rate of Species Extinction?

▶ **CONCEPT 8-2 We should avoid speeding up the extinction of wild species because of the economic and ecological services they provide, and because wild species have a right to exist regardless of their usefulness to us.**

Species Are a Vital Part of the Earth's Natural Capital

If all species eventually become extinct, why should we worry about the rate of extinction? Does it matter that the passenger pigeon became extinct because of human activities, or that the remaining polar bears (**Core Case Study**), orangutans, and some unknown plant or insect in a tropical forest might suffer the same fate? In 1900, there were more than 315,000 wild orangutans (Figure 8-5), which are found only in the tropical forests of Indonesia and Malaysia. According to

CORE CASE STUDY

age fotostock/SuperStock

Figure 8-5 Natural capital degradation: These endangered orangutans are shown in their rapidly disappearing tropical forest habitat. **Question:** What difference will it make if human activities cause the extinction of the orangutan?

the WWF, today there are fewer than 30,000 left in the wild (90% of them in Indonesia). These highly intelligent animals are disappearing at a rate of more than 1,000–2,000 per year because of illegal smuggling and the clearing of their tropical forest habitat to make way for plantations that supply palm oil used in cosmetics, cooking, and the production of biodiesel fuel. An illegally smuggled, live orangutan sells for a street price of up to $10,000. Without urgent protective action, the endangered orangutan may be the first great ape species to become extinct primarily because of human activities.

New species eventually evolve to take the places of those species lost through mass extinctions. So why should we care if we speed up the extinction rate over the next 50–100 years? According to biologists, there are three major reasons why we should work to prevent our activities from causing or speeding the extinction of other species.

First, the world's species provide natural resources and natural services (see Figure 1-3, p. 9) that help to keep us alive and support human economies. Each species also has ecological value because it plays a role in the key ecosystem functions of energy flow and chemical cycling (see Figure 3-9, p. 46), in keeping with one of the three **principles of sustainability**.

SUSTAINABILITY

Thus, by eliminating species, especially those that play keystone roles (see Chapter 4, p. 74), we can upset ecosystems and speed up the extinction of other species that depend on those systems. Eventually, such degradation of the earth's natural capital can threaten human health and lifestyles. It follows that, by protecting species from extinction caused by human activities, and by protecting the vital habitats of those species from environmental degradation (as we discuss in the next chapter), we are helping to sustain our own health and well-being and to save our own species, cultures, and economies.

Most species also contribute to *economic services* on which we depend (**Concept 8-2**). For example, various plant species provide food crops, fuelwood and lumber, paper, and medicine. *Bioprospectors* search tropical forests and other ecosystems to find plants and animals that scientists can use to make medicinal drugs. According to a United Nations University report, 62% of all cancer drugs were derived from the discoveries of bioprospectors. Despite their economic and medicinal potential, less than 0.5 % of the world's known plant species have been examined for their medicinal properties. **GREEN CAREER:** bioprospecting

GOOD NEWS

Preserving species also provides economic benefits through wildlife tourism, or *ecotourism*, which generates more than $1 million per minute in tourist revenues, worldwide. Conservation biologist Michael Soulé estimates that a male lion living to age 7 generates about $515,000 in tourist dollars in Kenya, but only about $1,000 if killed for its skin. To biologist Edward O. Wilson, carelessly and rapidly eliminating species that make up an essential part of the world's biodiversity is like burning millions of books that we have never read.

A *second* reason for preventing extinctions caused by human activities is that analysis of past mass extinctions indicates that it is likely to take 5–10 million years for natural speciation to rebuild the biodiversity that is likely to be lost during this century.

Third, many people believe that each wild species has a right to exist, regardless its usefulness to us (**Concept 8-2**). According to this stewardship view, we have an ethical responsibility to prevent species from becoming extinct as a result of human activities and to prevent the degradation of the world's ecosystems and their overall biodiversity.

8-3 How Do Humans Accelerate Species Extinction?

▶ **CONCEPT 8-3 The greatest threats to any species are (in order) loss or degradation of habitat, harmful invasive species, human population growth, pollution, climate change, and overexploitation.**

Loss of Habitat Is the Single Greatest Threat to Species: Remember HIPPCO

Biodiversity researchers summarize the most important direct causes of extinction resulting from human activities using the acronym **HIPPCO: H**abitat destruction, degradation, and fragmentation; **I**nvasive (nonnative) species; **P**opulation growth and increasing use of resources; **P**ollution; **C**limate change; and **O**verexploitation (**Concept 8-3**).

According to biodiversity researchers, the greatest threat to wild species is habitat loss (Figure 8-6), degradation, and fragmentation. A recent example of this is the loss of habitat for polar bears (**Core Case Study**). Because the atmosphere above the Arctic has been getting warmer during the past several decades, the floating sea ice that is a vital part of the bears' habitat is melting away beneath their feet, which is causing a decline in their numbers.

CORE CASE STUDY

Deforestation in tropical areas (see Figure 3-1, p. 40) is the greatest eliminator of species, followed

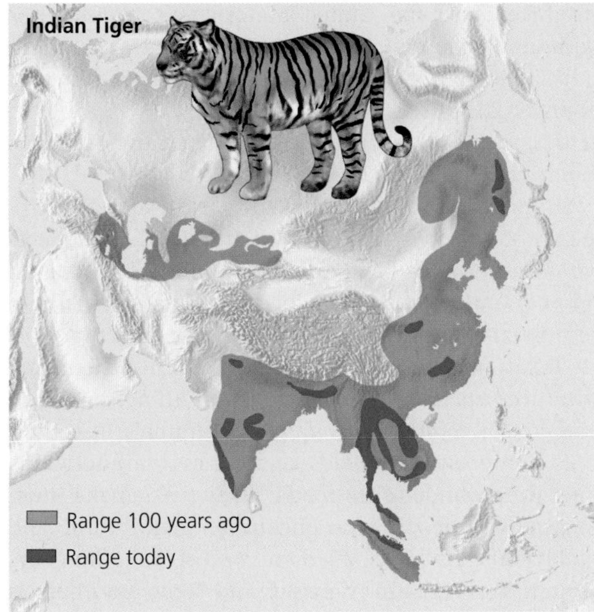

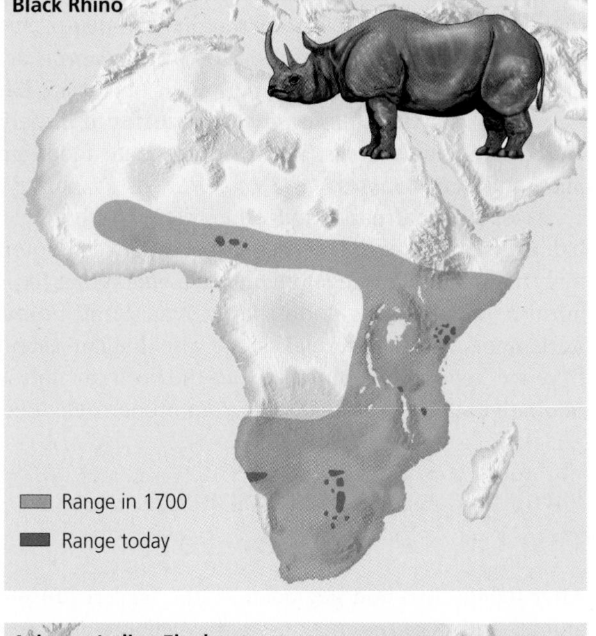

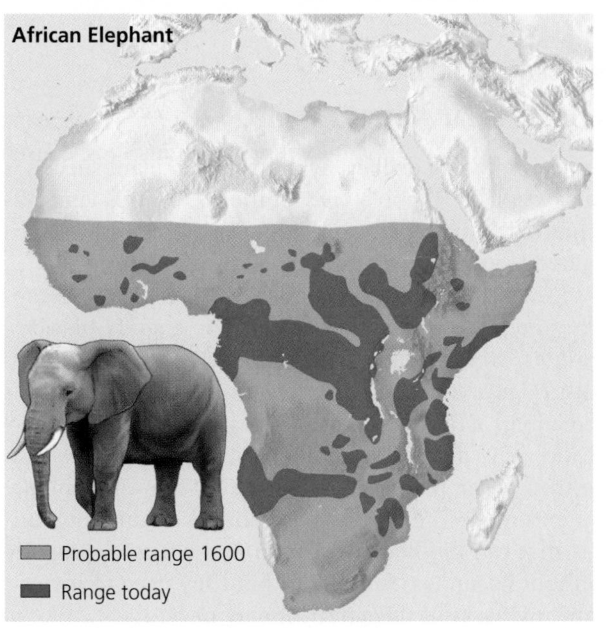

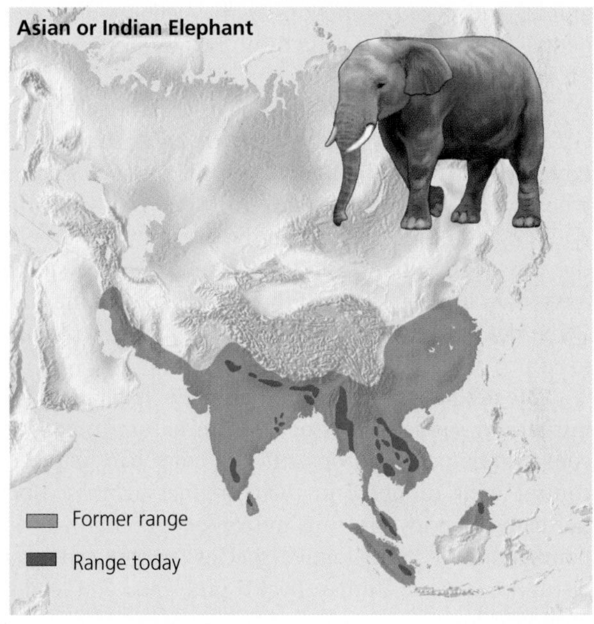

Active Figure 8-6 Natural capital degradation: These maps reveal the reductions in the ranges of four wildlife species, primarily as the result of severe habitat loss and fragmentation, and illegal hunting for some of their valuable body parts. What will happen to these and millions of other species during the next few decades when the human population grows by at least 2 billion—the equivalent of 6 times the current U.S. population and almost twice the current population of China—as is projected by scientists? *See an animation based on this figure at* **www.cengagebrain.com**. **Question:** Would you support expanding these ranges even though this would reduce the land available for human habitation and farming? Explain. (Data from International Union for the Conservation of Nature and World Wildlife Fund)

by the destruction and degradation of coral reefs (see Figure 7-25, right, p. 143) and wetlands, the replacement of biologically diverse grasslands with monoculture crops (see Figure 7-12, p. 132), and the pollution of streams, lakes, and oceans.

Island species—many of them found nowhere else on earth—are especially vulnerable to extinction when their habitats are destroyed, degraded, or fragmented and they have nowhere else to go. This is why the collection of islands that make up the U.S. state of Hawaii is America's "extinction capital"—with 63% of its species at risk.

Habitat fragmentation occurs when a large, intact area of habitat such as a forest or natural grassland is divided, typically by roads, logging, crop fields, and urban development, into smaller, isolated patches or "habitat islands." This process can decrease tree cover in forests (see The Habitable Planet, Video 9, at **www.learner.org/resources/series209.html**), and block animal migration routes. It can also divide populations of a species into smaller, increasingly isolated groups that are more vulnerable to predators, competitor species, disease, and catastrophic events such as storms and fires. In addition, habitat fragmentation can create barriers that limit the abilities of some species to disperse and colonize new areas, to locate adequate food supplies, and to find mates.

Most national parks and other nature reserves are habitat islands, many of them surrounded by potentially damaging logging and mining operations, coal-burning power plants, industrial activities, and human settlements. Freshwater lakes are also habitat islands that are especially vulnerable to the introduction of nonnative species and pollution from human activities.

We Have Introduced Species That Can Disrupt Ecosystems

After habitat loss and degradation, the biggest cause of animal and plant extinctions is the deliberate or accidental introduction of harmful invasive species into ecosystems (**Concept 8-3**).

Many species introductions have been beneficial to us. According to a study by ecologist David Pimentel, introduced domesticated species such as corn, wheat, rice, and other food crops, as well as cattle, poultry, and other livestock, provide more than 98% of the U.S. food supply. Similarly, nonnative tree species are grown in about 85% of the world's tree plantations. Some deliberately introduced species have also helped to control pests.

GOOD NEWS

The problem is that, in their new habitats, some introduced species face none of the natural predators, competitors, parasites, or pathogens that had helped to control their numbers in their original habitats. Such nonnative species can thus outcompete and crowd out populations of many native species, trigger ecological disruptions, cause human health problems, and lead to economic losses.

> CONNECTIONS
>
> **Giant Snails and Meningitis**
>
> In 1988, a giant African land snail was imported into Brazil as a cheap substitute for conventional escargot (snails), used as a source of food. It grows to the size of a human fist and weighs 1 kilogram (2.2 pounds) or more. When export prices for escargot fell, breeders dumped the imported snails into forests and other natural systems. Since then, it has spread to 23 of Brazil's states and devoured many native plants and food crops such as lettuce. It also can carry rat lungworm, a parasite that burrows into the human brain and causes meningitis (a potentially lethal swelling of the membranes that cover the brain and spinal cord), and it carries another parasite that can rupture human intestines. Authorities eventually banned the snail, but it was too late. So far, the snail has been unstoppable.

Figure 8-7 shows a few of the estimated 7,100 invasive species that, after being deliberately or accidentally introduced into the United States, have caused ecological and economic harm. According to the U.S. Fish and Wildlife Service, about 40% of the species listed as endangered in the United States and 95% of those in the U.S. state of Hawaii are on the list because of threats from invasive species.

Some deliberately introduced species, such as *kudzu* (see the Case Study that follows) and the *European wild boar* (Figure 8-7) have caused significant ecological and economic damage. Biologists estimate that there are now about 4 million European wild boars in Florida, Texas, and 22 other U.S. states. These deliberately introduced species will eat almost anything, and they compete for food with endangered animals, use their noses to root up farm fields, and cause traffic accidents when they wander onto roads. Their tusks make them dangerous to people who encounter them. Game and wildlife officials have failed to control their numbers through hunting and trapping, and some say there is no way to stop them.

■ CASE STUDY
The Kudzu Vine

An example of a deliberately introduced plant species is the *kudzu* ("CUD-zoo") *vine*. In the 1930s, this vine was imported from Japan and planted in the southeastern United States in an attempt to control soil erosion.

Kudzu does control erosion. But it is grows so rapidly and is so difficult to kill that it engulfs hillsides, gardens, trees, stream banks, and anything else in its path (Figure 8-8). Dig it up or burn it and it still keeps spreading. Grazing goats and repeated doses of herbicides can destroy it, but goats and herbicides also destroy other plants, and herbicides can contaminate water supplies. Scientists have found a common fungus that can kill kudzu within a few hours, apparently without harming other plants, but they need to investigate any harmful side effects it may have.

This plant—sometimes called "the vine that ate the South"—has spread throughout much of the southeastern United States. It could spread to the north if the climate gets warmer as scientists project. Visit **www.cengagebrain.com** and go on a virtual field trip to learn more about the spread of the kudzu vine in the United States.

In 2009, scientists reported that kudzu plants contribute to photochemical smog in the eastern and southern United States by emitting nitric oxide (NO) into the atmosphere, especially during the summer months. This increases concentrations of harmful ground-level ozone (O_3) gas.

Kudzu is considered a menace in the United States, but Asians use a powdered kudzu starch in beverages, gourmet confections, and herbal remedies for a range of diseases. A Japanese firm has built a large kudzu farm and processing plant in the U.S. state of Alabama and ships the extracted starch to Japan. Almost every part

Deliberately Introduced Species

Accidentally Introduced Species

Figure 8-7 These are some of the more than 7,100 harmful invasive (nonnative) species that have been deliberately or accidentally introduced into the United States.

of the kudzu plant is edible. Its deep-fried leaves are delicious and contain high levels of vitamins A and C. Stuffed kudzu leaves, anyone?

GOOD NEWS

Although kudzu can engulf and kill trees, it might eventually help to save some of them. Researchers at the Georgia Institute of Technology have found that kudzu could be used in place of trees as a source of fiber for making paper. Also, ingesting small amounts of kudzu powder can lessen one's desire for alcohol, and thus it could be used to reduce alcoholism and binge drinking.

Some Accidentally Introduced Species Can Disrupt Ecosystems

Many unwanted nonnative invaders arrive from other continents as stowaways on aircraft, in the ballast water of tankers and cargo ships, and as hitchhikers on imported products such as wooden packing crates. Cars and trucks can also spread the seeds of nonnative plant species embedded in their tire treads. Many tourists return home with living plants that can multiply

Chuck Pratt/Bruce Coleman, Inc.

Figure 8-8 Kudzu has taken over this abandoned house in the U.S. state of Mississippi.

and become invasive. Some of these plants might also contain insects that can escape, multiply rapidly, and threaten crops.

In the 1930s, the extremely aggressive Argentina fire ant (Figure 8-7) was accidentally introduced into the United States in Mobile, Alabama. The ants may have arrived on shiploads of lumber or coffee imported from South America. They can float on water and have no natural predators in the southern United States where they have spread rapidly. They are also found in Puerto Rico, New Mexico, and California. Now the insect has stowed away on imported goods and shipping containers and has invaded other countries, including China, Taiwan, Malaysia, and Australia.

When these ants invade an area, they can wipe out as much as 90% of native ant populations. Mounds containing fire ant colonies cover many fields and invade yards in the southeastern United States. Walk on one of these mounds, and as many as 100,000 ants may swarm out of their nest to attack you with painful, burning stings. They have killed deer fawns, birds, livestock, pets, and at least 80 people who were allergic to their venom.

Widespread pesticide spraying in the 1950s and 1960s temporarily reduced fire ant populations. But this chemical warfare actually hastened the advance of the rapidly multiplying fire ants by reducing populations of many native ant species. Even worse, it promoted development of genetic resistance to pesticides in the fire ants through natural selection (see Figure 4-5, p. 67). In other words, we helped wipe out the fire ants' competitors and made them more genetically resistant to pesticides by relying on a chemical rather than an ecological strategy for reducing the spread of this species.

GOOD NEWS

In 2009, pest management scientist Scott Ludwig reported some success in using tiny parasitic flies to reduce fire ant populations. The flies dive-bomb the fire ants and lay eggs inside them. Maggots then hatch and eat away the brains of the ants. After about two weeks, the ants become staggering zombies, and after about a month, their heads fall off. Then the parasitic fly emerges looking for more fire ants to attack and kill. The researchers say that the flies do not attack native ant species. More research is needed to see how well this biological approach works.

Burmese and African pythons and several species of boa constrictors have accidentally ended up in Everglades in the U.S. state of Florida. About a million of these snakes, imported from Africa and Asia, have been sold as pets. After learning that these reptiles do not make good pets, some owners have dumped them into the wetlands in the Everglades.

Some of these snakes can live 25–30 years, reach 6 meters (20 feet) in length, weigh more than 90 kilograms (200 pounds), and be as big around as a telephone pole. They are hard to find and kill, and they reproduce rapidly. They seize their prey with their sharp teeth, wrap themselves around the prey, and squeeze it to death before feeding on it. They have huge appetites and have devoured a variety of birds, raccoons, pet cats and dogs, and full-grown deer. Pythons have been known to eat American alligators—a keystone species in the Everglades ecosystem (see Chapter 4, Case Study, p. 74).

According to wildlife officials, tens of thousands of these snakes now live in the Everglades and their numbers are increasing rapidly. It is feared that they will spread to other swampy wetlands in the southern half of the United States by the end of this century.

Bioinvaders also affect aquatic systems and have been blamed for about two-thirds of all known fish extinctions in the United States between 1900 and 2009. Many of these invaders arrive in the ballast water stored in tanks in large cargo ships to keep them stable. These ships take in ballast water—along with whatever microorganisms and tiny aquatic organisms it contains—from one harbor and dump it into another. This is an environmentally harmful effect of globalized trade.

The Great Lakes of North America have been invaded by more than 185 alien species. At least 13 of these invading species threaten some native species and each year cause billions of dollars in damages. One such invader is the fish-killing sea lamprey (see Figure 5-4, p. 84). Another is a thumbnail-sized mollusk called the *zebra mussel* (Figure 8-7), which reproduces rapidly and has no known natural enemies in the Great Lakes. It has displaced other mussel species and depleted the food supply for some native species. The mussels have also clogged irrigation pipes, shut down water intake pipes for power plants and city water supplies, jammed ship rudders, and grown in huge masses on boat hulls, piers, rocks, and almost any exposed aquatic surface (Figure 8-9).

Prevention Is the Best Way to Reduce Threats from Invasive Species

Once a harmful nonnative species becomes established in an ecosystem, its removal is almost impossible—somewhat like trying to collect smoke after it has come out of a chimney. Clearly, the best way to limit the harmful impacts of nonnative species is to prevent them from being introduced and becoming established.

Scientists suggest several ways to do this:

- Fund a massive research program to identify the key characteristics that enable some species to become successful invaders; the types of ecosystems that are vulnerable to invaders; and the natural predators, parasites, bacteria, and viruses that might be used to control populations of established invaders.
- Greatly increase ground surveys and satellite observations to track invasive plant and animal species, and develop better models for predicting how they will spread and what harmful effects they might have.

NOAA Great Lakes Environmental Research Laboratory

Figure 8-9 These *Zebra mussels* are attached to a water current meter in Lake Michigan. The invaders entered the Great Lakes through ballast water dumped from a European ship. They have become a major nuisance and a threat to commerce as well as to biodiversity in the Great Lakes.

- Identify major harmful invader species and establish international treaties banning their transfer from one country to another, as is now done for endangered species, while stepping up inspection of imported goods to enforce such bans.
- Require cargo ships to discharge their ballast water and replace it with saltwater at sea before entering ports, or require them to sterilize such water or to pump nitrogen into the water to displace dissolved oxygen and kill most invader organisms.
- Educate the public about the environmentally harmful effects of releasing exotic plants and pets into the environment.

Figure 8-10 shows some of the things you can do to help prevent or slow the spread of harmful invasive species. GOOD NEWS

What Can You Do?

Controlling Invasive Species

- Do not capture or buy wild plants and animals.
- Do not remove wild plants from their natural areas.
- Do not release wild pets back into nature.
- Do not dump the contents of an aquarium into waterways, wetlands, or storm drains.
- When camping, use wood found near your campsite instead of bringing firewood from somewhere else.
- Do not dump unused bait into waterways.
- After dogs visit woods or the water, brush them before taking them home.
- After each use, clean your mountain bike, canoe, boat, motor, and trailer, all fishing tackle, hiking boots, and other gear before heading for home.

Figure 8-10 Individuals matter: Here is a list of some ways to prevent or slow the spread of harmful invasive species. **Questions:** Which two of these actions do you think are the most important? Why? Which of these actions do you plan to take?

Population Growth, Overconsumption, Pollution, and Climate Change Can Cause Species Extinctions

Past and projected *human population growth* and excessive and wasteful consumption of resources have greatly expanded the human ecological footprint (see Figure 1-8, p. 14, and Figure 2, pp. S24–25, in Supplement 6). This has eliminated, degraded, and fragmented vast areas of wildlife habitat (Figure 8-6). Acting together, these two factors have caused the extinction of many species (**Concept 8-3**). (See *The Habitable Planet*, Video 13, at **www.learner.org/resources/series209.html**.)

Pollution also threatens some species with extinction (**Concept 8-3**), as has been shown by the unintended effects of certain pesticides. According to the U.S. Fish and Wildlife Service, each year pesticides kill a large number of the honeybee colonies that pollinate almost 33% of U.S. food crops (see the Case Study that follows). They also kill more than 67 million birds and 6–14 million fish each year, and they threaten about 20% of the country's endangered and threatened species.

During the 1950s and 1960s, populations of fish-eating birds such as ospreys, brown pelicans, and bald eagles plummeted. A chemical derived from the pesticide DDT, when biologically magnified in food webs (Figure 8-11, p. 162), made the birds' eggshells so fragile they could not reproduce successfully. Also hard hit were such predatory birds as the prairie falcon, sparrow hawk, and peregrine falcon, which help to control populations of rabbits, ground squirrels, and other crop eaters. Since the U.S. ban on DDT in 1972, most of these bird species have made a comeback. GOOD NEWS

Finally, according to a 2004 study by Conservation International, projected *climate change* could help to drive a quarter to half of all land animals and plants to extinction by the end of this century. Scientific studies indicate that polar bears (**Core Case Study**) and 10 of the world's 17 penguin species are already threatened because of higher temperatures and melting sea ice in their polar habitats. CORE CASE STUDY

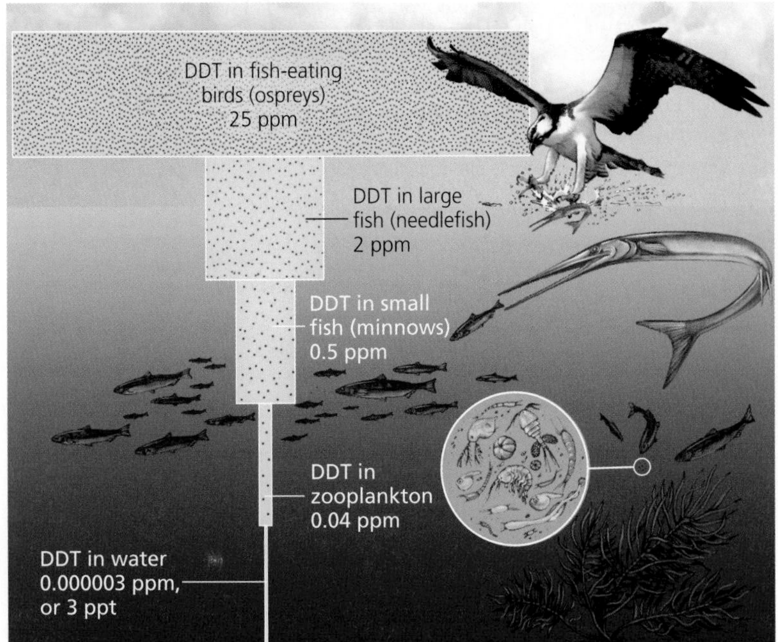

Figure 8-11 *Bioaccumulation* and *biomagnification:* DDT is a fat-soluble chemical that can accumulate in the fatty tissues of animals. In a food chain or web, the accumulated DDT is biologically magnified in the bodies of animals at each higher trophic level. (Dots in this figure represent DDT.) The concentration of DDT in the fatty tissues of organisms was biomagnified about 10 million times in this food chain in an estuary on Long Island Sound in the U.S. state of New York. If each phytoplankton organism takes up and retains one unit of DDT, a small fish eating thousands of zooplankton (which feed on the phytoplankton) will store thousands of units of DDT in its fatty tissue. Each large fish that eats ten of the smaller fish will ingest and store tens of thousands of units, and each bird (or human) that eats several large fish will ingest hundreds of thousands of units. **Question:** How does this story demonstrate the value of pollution prevention?

THINKING ABOUT
Polar Bears

CORE CASE STUDY

What difference would it make if most or all of the world's polar bears (**Core Case Study**) disappeared? List two things you could do to help protect the world's remaining polar bears from extinction, largely due to climate change.

■ CASE STUDY
Where Have All the Honeybees Gone?

About 33% of the U.S. food supply comes from insect-pollinated plants, and honeybees are responsible for 80% of that pollination, according to the U.S. Department of Agriculture (USDA). Because of their importance in sustaining global food security, as well as their key role in a number of food webs, some scientists classify honeybees as a keystone species (see Chapter 4, p. 74). Since the 1980s, 30%–40% of all honeybee populations in the United States have declined in numbers, with all of the bees in some colonies mysteriously disappearing.

So far, no single culprit has been found. One cause of this decline could be pesticide exposure. Bees pick up traces of more than 120 different pesticides during pollination, and the wax in beehives absorbs these and other toxins. Other possible causes include attacks by certain parasitic mites that can wipe out a colony in hours, invasion by Africanized honeybees (killer bees; Figure 8-7), a virus traced to Israel, and a certain fungus. Another factor contributing to their decline is poor nutrition caused by a decrease in the natural diversity of flowers and other plants on which bees feed.

In 2010, about 34% of all commercial honeybee colonies in the United States (each with 30,000 to 100,000 individual bees) were lost, according to the USDA. Almost one-third of the deaths were due to *colony collapse disorder* (CCD) in which most or all of the adult worker bees vanished from their hives. Suspected causes of CCD include parasites, a fungus, viruses, bacteria, and pesticides. For commercial bee colonies that are trucked around the country and rented out for pollination, another possible cause of CCD is stress and poor nutrition from the artificial diet they are fed. Research in 2010 indicated that a virus and a fungus might be interacting to cause some colonies to collapse.

Another possible cause might be microwave radiation from cell phones and cell towers. The microwaves could be disrupting the navigation systems of some worker bees, disorienting them and preventing them from returning to their hives with the nectar and pollen they have collected. Alternatively, it may be decreasing their ability to fly by damaging their nervous systems. In an experiment performed to test these hypotheses, cell phones were placed near beehives, and the hives collapsed within 5 to 10 days after the worker bees failed to return from their foraging.

CONNECTIONS
Pesticides, Honeybees, and Food Prices

China, where some scientists warn that pesticides are overused, gives us a glimpse of a future without enough honeybees. Individual trees in pear orchards in China's Sichuan province are now largely pollinated by hand at great cost. Without honeybees, many fruits, vegetables, and nuts that depend on bees for pollination could become too expensive for most people worldwide. Prices of meat and dairy products would also rise, because honeybees pollinate forage crops such as alfalfa and clover hay that are fed to farm animals. This shows what can happen when human activities impair one of nature's free ecological services such as pollination.

So what can we do to reduce threats to honeybees? Many beekeepers are having some success in reducing CCD by practicing stringent hygiene, improving the diets of the bees, and trying to reduce viral infections. The USDA suggests that we can help by cutting back on our use of pesticides, especially at midday when honeybees are most likely to be searching for nectar. We can also make our yards and gardens into buffets for honey bees by planting native plants that they like, such as bee balm, foxglove, red clover, and joe-pye weed. Bees also need places to live, so some homeowners are purchasing bee houses from their local garden centers.

> **THINKING ABOUT**
> **Honeybees**
> What difference would it make to you if most of the honeybees disappeared? What are two things you could do to reduce the loss of honeybees?

The Illegal Killing, Capturing, and Selling of Wild Species Threatens Biodiversity

Some protected species are illegally killed (poached) for their valuable parts or are sold live to collectors. The global illegal trade in wildlife brings in an average of at least $600,000 an hour and at least two of every three live animals smuggled around the world die in transit. Organized crime has moved into illegal wildlife smuggling because of the huge profits involved. Few of the smugglers are caught or punished.

To poachers, a highly endangered, live mountain gorilla (Figure 8-12) is worth $150,000 and a pelt of a critically endangered giant panda (less than 1,600 are left in the wild in China) can bring $100,000. A poached rhinoceros horn (Figure 8-13) can be worth as much as $55,500 per kilogram ($25,000 per pound). Rhinoceros are killed for their horns, which are used to make dagger handles in the Middle East. Powdered rhino horn also serves as a fever reducer and an alleged aphrodisiac in China and other parts of Asia.

Elephants continue to lose habitat in Africa and Asia (Figure 8-6). In addition, each year about 25,000 African elephants are killed illegally for their valuable ivory tusks, despite an international ban on the sale of poached ivory since 1989.

In 1900, there were an estimated 100,000 wild adult tigers. Today only about 3,200 tigers are left, in a rapidly shrinking range (Figure 8-6, top left), according to a 2010 study by the World Wildlife Fund (WWF). The Indian, or Bengal, tiger is at risk because a coat made from its fur can sell for as much as $100,000 in Tokyo. Because the body parts of a single tiger are worth as much as $70,000—and because few of its poachers are caught or punished—it is not surprising that the illegal hunting of tigers has skyrocketed and that all tiger species have become highly endangered. Without emergency action to curtail poaching and preserve their habitat, few if any tigers may be left in the wild by 2022 according to the WWF.

> **THINKING ABOUT**
> **Tigers**
> What do you think would happen to their ecosystems if all the world's tigers disappeared? What are two things you would do to help protect the world's remaining tigers from extinction?

Jiri Haureljuk/ Shutterstock.com

Figure 8-12 This male mountain gorilla inhabits the bush of the African country of Rwanda. Mountain gorillas feed primarily on the leaves, shoots, and stems of a variety of plants. Only about 700 individuals of this critically endangered species remain in the wild.

Martin Harvey/Peter Arnold, Inc

Figure 8-13 A poacher in South Africa killed this white rhinoceros for its horns. With only about 170 individuals left in the wild, this species is close to extinction. **Question:** What would you say if you could talk to the poacher who killed this animal?

Across the globe, the legal and illegal trade in wild species for use as pets is also a huge and very profitable business. Many owners of wild pets do not know that, for every live animal captured and sold in the pet market, many others are killed or die in transit. More than 60 bird species, mostly parrots, are endangered or threatened because of this wild-bird trade, according to the IUCN.

> **CONNECTIONS**
>
> **The Pet Trade and Infectious Diseases**
>
> Most people are unaware that some imported exotic animals carry diseases such as hantavirus, Ebola virus, Asian bird flu, herpes B virus (carried by most adult macaques), and salmonella (from pets such as hamsters, turtles, and iguanas). These are diseases that can spread quite easily from pets to their owners and then to other people.

Other wild species whose populations are depleted because of the pet trade include many amphibians, various reptiles, some mammals, and many tropical fishes (taken mostly from the coral reefs of Indonesia and the Philippines). Divers catch tropical fish by using plastic squeeze bottles of poisonous cyanide to stun them. For each fish caught alive, many more die. In addition, the cyanide solution kills the coral polyps that build reefs.

Some exotic plants, especially orchids and cacti (see Figure 7-10, center, p. 129), are endangered when they are gathered (often illegally) and sold to collectors to decorate houses, offices, and landscapes. A collector might pay $5,000 for a single rare orchid. A mature crested saguaro cactus can earn cactus rustlers as much as $15,000.

Wild species in their natural habitats have ecological value because of the roles they play in their ecosystems. They can also have great economic value if left in the wild. According to the U.S. Fish and Wildlife Service, collectors of exotic birds may pay $10,000 for an endangered hyacinth macaw parrot smuggled out of Brazil; however, during its lifetime, a single hyacinth macaw left in the wild might account for as much as $165,000 in tourist revenues. Some scientists are using this fact to help in their efforts to preserve biodiversity (Individuals Matter, below).

> **THINKING ABOUT**
>
> **Collecting Wild Species**
>
> Some people believe it is wrong to collect wild animals and plants for display and personal pleasure. They believe we should leave most exotic wild species in the wild. Explain why you agree or disagree with this view.

INDIVIDUALS MATTER

A Scientist Who Confronted Poachers

In Thailand, biologist Pilai Poonswad (Figure 8-A) decided to do something about poachers taking rhinoceros hornbills (Figure 8-B) from a rain forest. This bird species is one of the world's largest hornbills and its large beak amplifies its peculiar, loud squawking.

Dr. Poonswad visited the poachers in their villages and showed them why the birds are worth more alive than dead. Today, some former poachers earn money by taking ecotourists into the forest to see these magnificent birds. Because of their vested financial interest in preserving the hornbills, these former poachers now help to protect the birds from further poaching.

The rhinoceros hornbill's population in this area of Thailand was in steady decline, but partly because of Dr. Poonswad's work, it is now gradually recovering. It is considered to be an indicator species in some of its tropical rain forest habitats.

©Rolex Awards/Tomas Bertelsen

Figure 8-A Professor Pilai Poonswad, a biologist at Mahidol University in Thailand, decided to confront poachers who were a threat to the rare rhinoceros hornbill.

Karen McGougan/Bruce Coleman USA

Figure 8-B The rare rhinoceros hornbill is found in tropical and subtropical forest habitats in parts of Asia. It emits a loud honking squawk and uses its long bill to defend itself against predators such as snakes and monkeys. Agricultural development and logging threaten its habitat. Some local tribesmen kill hornbills for food and for their feathers. These birds are also captured and sold live as part of the illegal wildlife trade.

Rising Demand for Bush Meat Threatens Some African Species

For centuries, indigenous people in much of West and Central Africa have sustainably hunted wildlife for *bush meat* as a source of food. But in the last two decades, bush-meat hunting in some areas has skyrocketed as hunters try to provide food for rapidly growing populations or make a living by supplying restaurants with exotic meats from gorillas and other species.

Throughout most of our history, humans have survived by hunting and gathering wild species. The problem is that today, because of the growing human population, bush-meat hunting has led to the local extinction of many wild animals in parts of West and Central Africa. It has driven at least one species—Miss Waldron's red colobus monkey—to complete extinction. It is also a factor in reducing some populations of orangutans (Figure 8-5), gorillas (Figure 8-12), chimpanzees, elephants, and hippopotamuses. Another problem is that butchering and eating some forms of bush meat has helped to spread fatal diseases such as HIV/AIDS and the Ebola virus from animals to humans.

The U.S. Agency for International Development (USAID) is trying to reduce unsustainable hunting for bush meat in some areas of Africa by introducing alternative sources of food such as farmed fish. They are also showing villagers how to breed large rodents such as cane rats as a source of food.

■ CASE STUDY
A Disturbing Message from the Birds

Approximately 70% of the world's nearly 10,000 known bird species are declining in numbers, and much of this decline is clearly related to human activities, summarized by HIPPCO.

First, roughly one of every eight (12%) of all bird species is threatened with extinction mostly by habitat loss, degradation, and fragmentation (the H in HIPPCO), according to the 2010 *Red List of Endangered Species* published by the IUCN. About three-fourths of the threatened bird species live in forests, many of which are being cleared at a rapid rate, especially in the tropical areas of Asia and Latin America.

According to a 2009 study, the U.S. Fish and Wildlife Service and the U.S Geological Survey, nearly 33% of the more than 800 bird species in the United States are endangered or threatened, mostly because of habitat loss and degradation, invasive species, and climate change.

The greatest declines have occurred among long-distance migrant songbird species such as tanagers, orioles, thrushes, vireos, and warblers. These birds nest deep in North American woodlands in the summer and spend their winters in Central or South America, or on the Caribbean Islands.

The primary causes of these population declines appear to be habitat loss and fragmentation of the birds' breeding habitats. In North America, woodlands are fragmented or cleared for road construction and housing developments. In Central and South America, tropical forest habitats, mangroves, and wetland forests are suffering the same fate.

After habitat loss, the intentional or accidental introduction of nonnative species such as bird-eating rats is the second greatest danger, affecting about 28% of the world's threatened birds. Other such invasive species (the I in HIPPCO) are snakes, mongooses, and both domestic and feral cats, which kill hundreds of millions of birds each year.

Millions of migrating birds are killed every year when they collide with power lines, communications towers, and skyscrapers that have been erected within their migration routes. As many as 1 billion birds in the United States die each year when they fly into glass windows, especially those in tall city buildings that are lit up at night. This is the number-one cause of U.S. bird mortality.

Population growth, the first P in HIPPCO, also threatens some bird species, as more and more people spread out over the landscape and increase their use of timber, food, and other resources, destroying or disturbing some bird habitats. The second P in HIPPCO is for pollution, another major threat to birds. Countless birds are exposed to oil spills, pesticides, and herbicides that destroy their habitats. And birds sometimes eat lead shotgun pellets that fall into wetlands and lead sinkers left by anglers. Lead poisoning is a severe threat to many birds, especially waterfowl.

The greatest new threat to birds is climate change. A study done for the WWF, found that climate change is causing declines of some bird populations in every part of the globe. This is expected to increase sharply during this century.

Finally, overexploitation of many bird species is a major threat to bird populations. Fifty-two of the world's 388 parrot species are threatened, partly because so many parrots are captured (often illegally) for the pet trade. They are taken from tropical areas and sold, usually to buyers in Europe and the United States.

Industrialized fishing fleets also pose a threat. At least 23 species of seabirds face extinction. Many of these are diving birds that drown after becoming hooked on baited lines or trapped in huge nets that are set out by fishing boats.

Biodiversity scientists view this decline of bird species with alarm for two reasons. One reason is that birds are excellent *environmental indicators* because they live in every climate and biome, respond quickly to environmental changes in their habitats, and are relatively easy to track and count.

A second reason is that birds perform critically important economic and ecological services in ecosystems throughout the world. For example, many birds

SCIENCE FOCUS

Vultures, Wild Dogs, and Rabies: Some Unexpected Scientific Connections

In 2004, the World Conservation Union placed three species of vultures found in India and South Asia on the critically endangered list. During the early 1990s, there were more than 40 million of these carcass-eating vultures. But within a few years, their populations had fallen by more than 97%.

This is an interesting scientific mystery, but should anyone care if various vulture species disappear? The answer is yes.

Scientists were puzzled, but they eventually discovered that the vultures were being poisoned by *diclofenac*. This anti-inflammatory drug was given to cows to help increase their milk production by reducing inflammation in their bodies. But it caused kidney failure in vultures that fed on the carcasses of such cows.

As the vultures died off, huge numbers of cow carcasses, normally a source of food for the vultures, were now consumed by wild dogs and rats whose populations the vultures had helped to control by reducing their food supply. As wild dog populations exploded due to a greatly increased food supply, the number of dogs with rabies also increased. This increased the risks to people bitten by rabid dogs. In 1997 alone, more than 30,000 people in India died of rabies—more than half the world's total number of rabies deaths that year.

Thus, protecting vulture species from extinction can end up protecting millions of people from a life-threatening disease. Unraveling often-unexpected ecological connections in nature is not only fascinating but also vital to our own lives and health.

Some critics of efforts to protect species and ecosystems from harmful human activities frame the issue as one of choosing between protecting people and protecting wildlife. Many conservation biologists reject this as a misleading conclusion, arguing that it is important to protect both wildlife and people because their fates and well-being are interconnected.

Critical Thinking

What would happen to your life and lifestyle if most of the world's vultures disappeared?

play specialized roles in pollination and seed dispersal, especially in tropical areas. Extinctions of these bird species might lead to extinctions of plants that depend on the birds for pollination. Then, some specialized animals that feed on these plants might also become extinct. Such a cascade of extinctions, in turn, could affect our own food supplies and well-being. Thus, in addition to being a wildlife conservation issue, protecting birds and their habitats is an important issue for human health as well (Science Focus, above).

Biodiversity scientists urge us to listen more carefully to what the birds are telling us about the state of the environment, for their sake, as well as for ours.

THINKING ABOUT
Bird Extinctions

How does your lifestyle directly or indirectly contribute to the extinction of some bird species? What are two things that you think should be done to reduce the extinction of birds?

8-4 How Can We Protect Wild Species from Extinction?

▶ **CONCEPT 8-4 We can reduce the rising rate of species extinction and help to protect overall biodiversity by establishing and enforcing national environmental laws and international treaties, creating a variety of protected wildlife sanctuaries, and taking precautionary measures to prevent such harm.**

International Treaties and National Laws Can Help to Protect Species

Several international treaties and conventions help to protect endangered or threatened wild species (**Concept 8-4**). One of the most far reaching is the 1975 *Convention on International Trade in Endangered Species (CITES)*. This treaty, signed by 175 countries, bans the hunting, capturing, and selling of threatened or endangered species. It lists some 900 species that cannot be commercially traded as live specimens or for their parts or products because they are in danger of extinction. It also restricts international trade in roughly 5,000 species of animals and 28,000 species of plants that are at risk of becoming threatened.

CITES has helped to reduce the international trade in many threatened animals, including elephants, crocodiles, cheetahs, and chimpanzees. But the effects of this treaty are limited because enforcement varies from country to country, and convicted violators often pay only small fines. Also, member countries can exempt themselves from protecting any listed species, and much of the highly profitable illegal trade in wildlife and wildlife products goes on in countries that have not signed the treaty.

The *Convention on Biological Diversity (CBD)*, ratified by 191 countries (but as of 2011, not by the United States), legally commits participating governments to reducing the global rate of biodiversity loss and to sharing equitably in the benefits from use of the world's genetic resources. This includes efforts to prevent or control the spread of ecologically harmful invasive species.

This convention is a landmark in international law because it focuses on ecosystems rather than on individual species, and it links biodiversity protection to issues such as the traditional rights of indigenous (native) peoples. However, because some key countries, including the United States, have not ratified the CBD, implementation has been slow.

The U.S. Endangered Species Act

The *Endangered Species Act of 1973* (ESA; amended in 1982, 1985, and 1988) was designed to identify and protect endangered species in the United States and abroad (**Concept 8-4**). This act is probably the most far-reaching environmental law ever adopted by any nation, which has made it controversial.

Under the ESA, the National Marine Fisheries Service (NMFS) is responsible for identifying and listing endangered and threatened ocean species, while the U.S. Fish and Wildlife Service (USFWS) is to identify and list all other endangered and threatened species. Any decision by either agency to list or delist a species must be based on biological factors alone, without consideration of economic or political factors. However, the two agencies can use economic factors in deciding whether and how to protect endangered habitat and in developing recovery plans for listed species. The ESA also forbids federal agencies (except the Defense Department) to carry out, fund, or authorize projects that would jeopardize an endangered or threatened species, or destroy or modify its critical habitat.

For offenses committed on private lands, fines as high as $100,000 and 1 year in prison can be imposed to ensure protection of the habitats of endangered species, although this has rarely been done. This part of the act has been controversial because at least 90% of the listed species live on private land. Since 1982, however, the ESA has been amended to give private landowners economic incentives to help save endangered species living on their lands. The ESA also makes it illegal for Americans to sell or buy any product made from an endangered or threatened species, or to hunt, kill, collect, or injure such species.

Between 1973 and 2011, the number of U.S. species on the official endangered and threatened species lists increased from 92 to 1,371. According to a study by the Nature Conservancy, about 33% of the country's species are at risk of extinction, and 15% are at high risk—far more than the current number listed.

In 2010, 83% of the protected species were covered by active recovery plans. Successful recoveries include those for the American alligator (see Chapter 4, p. 74), the gray wolf, the peregrine falcon, the bald eagle, and the brown pelican. GOOD NEWS

The ESA also governs commercial shipments of wildlife and wildlife products to and from the United States. The 120 full-time USFWS inspectors can inspect only a small fraction of the more than 200 million wild animals brought legally into the United States annually. Each year, tens of millions of wild animals are also brought in illegally, but few of such shipments are confiscated (Figure 8-14). Even when they are caught, many violators are not prosecuted, and convicted violators often pay only a small fine.

Since 1995, there have been numerous efforts to weaken the ESA and to reduce its already meager annual budget, which is less than what a beer company typically spends on two 30-second TV commercials during the Super Bowl. Other critics would go further and do away with this act.

Most conservation biologists and wildlife scientists agree that the ESA needs to be simplified and streamlined. But they contend that it has not been a failure. To its supporters, it is amazing that the ESA, on a very small budget, has managed to stabilize or improve the conditions of more than half of its listed critically endangered species.

The ESA, along with certain international agreements, has also been used to protect endangered and threatened marine species such as seals, sea lions, sea turtles, and whales (see the Case Studies that follow).

Steve Hillebrand/U.S. Fish and Wildlife Service

Figure 8-14 These confiscated products were made from endangered species. Because of a scarcity of funds and inspectors, probably no more than one-tenth of the illegal wildlife trade in the United States is discovered. The situation is even worse in most other countries.

■ CASE STUDY

Protecting Endangered Sea Turtles

Six of the world's seven species of sea turtle are critically endangered or threatened. One is the endangered leatherback sea turtle, named for its leathery shell.

Two major threats to sea turtles are loss or degradation of beach habitat (where the leatherback and other species come ashore to lay their eggs) and the legal and illegal taking of their eggs. Another threat is the increased use of the turtles as sources of food, medicinal ingredients, tortoiseshell (for jewelry), and leather from their flippers. Pollution also threatens sea turtles, which can mistake discarded plastic bags for jellyfish and choke to death on them. Some turtles drown or starve after becoming entangled in fishing lines and nets (Figure 8-15), as well as in lobster and crab traps.

Many people are working to protect the leatherbacks. On some beaches, nesting areas are roped off. Since 1991, the U.S. government has required offshore shrimp trawlers to use turtle excluder devices (TEDs) that either prevent sea turtles from being caught in their nets or allow netted turtles to escape. TEDs have been adopted in 15 countries that export shrimp to the United States. This has led to a significant decline in the number of sea turtles killed in fishing nets. In 2004, the United States banned long-lining as a method for catching swordfish off the Pacific coast to help save dwindling sea turtle populations there.

GOOD NEWS

Renatura Congo. www.renatura.asso.eu.org

Figure 8-15 This critically endangered leatherback sea turtle was entangled in a fishing net and could have starved to death had it not been rescued.

■ CASE STUDY

Protecting Whales: A Success Story . . . So Far

Whales are fairly easy to kill because of their large size, in some cases, and their need to come to the surface to breathe. Modern hunters have become efficient at hunting and killing whales using radar, spotters in airplanes, fast ships, and harpoon guns. Whale harvesting has followed the classic pattern of a tragedy of the commons (see Chapter 1, p. 13), with whalers killing an estimated 1.5 million whales between 1925 and 1975.

This overharvesting drove 8 of the 11 major species to commercial extinction, and it drove the blue whale, the world's largest animal, to the brink of biological extinction. Fully grown, the blue whale is longer than two city buses and weighs more than 25 adult elephants. An adult has a heart the size of a compact car, and some of its arteries are big enough for a child to swim through.

Before commercial whaling began, an estimated 350,000 blue whales roamed the oceans. In 2010, according to the American Cetacean Society, there were 8,000 to 14,000 blue whales. They take 25 years to mature sexually and have only one offspring every 2–5 years. This low reproductive rate is making it difficult for the species to recover.

In 1946, the International Whaling Commission (IWC) was established to regulate the whaling industry by setting annual quotas for various whale species to prevent overharvesting. But IWC quotas often were based on insufficient data or were ignored by whaling countries. Without enforcement powers, the IWC was not able to stop the decline of most commercially hunted whale species.

In 1970, the United States stopped all commercial whaling and banned all imports of whale products. Under pressure from conservationists and the governments of many nonwhaling nations, the IWC began imposing a moratorium on commercial whaling in 1986. It worked. The estimated number of whales killed commercially, worldwide, dropped from 42,480 in 1970 to about 1,400 in 2010.

GOOD NEWS

Despite the ban on whaling, more than 28,000 whales were hunted and killed between 1986 and 2010, mostly by the nations of Japan, Norway, and Iceland, which have openly defied the ban. Japan hunts and kills at least 1,000 whales a year, for what it claims are scientific purposes. Critics see this as poorly disguised commercial whaling, because most of the whale meat ends up in restaurants and grocery stores.

Japan, Norway, Iceland, Russia, and a growing number of small tropical island countries—which Japan brought into the IWC to support its position—hope to overthrow the IWC ban on commercial whaling and to reverse the international ban on the buying and selling of whale products. They contend that the ban is emotionally motivated and not supported by current scientific estimates of populations of sperm, pilot, and minke whales, which have grown since the moratorium.

Most conservationists disagree. Some argue that whales are intelligent and highly social mammals that should be protected for ethical reasons. Others question IWC estimates of the allegedly recovered whale species, and the inability of the IWC to enforce quotas.

We Can Establish Wildlife Refuges and Other Protected Areas

In 1903, President Theodore Roosevelt established the first U.S. federal wildlife refuge at Pelican Island, Florida (Figure 8-16), to help protect birds such as the brown pelican (Figure 8-16, inset photo) from extinction. It took more than a century, but this protection worked. In 2009, the brown pelican was removed from the U.S. Endangered Species list. By 2011 there were 553 refuges in the National Wildlife Refuge System. Each year, more than 40 million Americans visit these refuges to hunt, fish, hike, and watch birds and other wildlife. (Some hunting is allowed in many refuges.)

GOOD NEWS

More than three-fourths of the refuges serve as wetland sanctuaries that are vital for protecting migratory waterfowl. One-fifth of U.S. endangered and threatened species have habitats in the refuge system, and some refuges have been set aside for specific endangered species (**Concept 8-4**). Such areas have helped Florida's Key deer, the brown pelican, and the trumpeter swan to recover. National parks and other government sanctuaries have also been used to help protect species such as the American bison.

There is also bad news about refuges. According to a General Accounting Office study, activities considered harmful to wildlife such as mining, oil drilling, and use of off-road vehicles occur in nearly 60% of the nation's wildlife refuges. Also, a 2008 study prepared for Congress found that, for years, the wildlife refuges have received so little funding that a third of them have no staff, and boardwalks and public buildings in some refuges are in disrepair.

Biodiversity scientists are urging the U.S. government to set aside more refuges for endangered plants and to significantly increase the long-underfunded budget for the refuge system. They are also calling on the Congress and state legislatures to allow abandoned military lands that contain significant wildlife habitat to become wildlife refuges.

Gene Banks, Botanical Gardens, and Wildlife Farms Can Help to Protect Species

Gene or *seed banks* preserve genetic information and endangered plant species by storing their seeds in refrigerated, low-humidity environments. More than 100 seed banks around the world collectively hold about 3 million samples.

George Gentry/U.S. Fish and Wildlife Service
Jeremy Woodhouse/Peter Arnold, Inc.

Figure 8-16 The Pelican Island National Wildlife Refuge in Florida was America's first National Wildlife Refuge. It was established in 1903 to help protect the brown pelican (see inset photo) and other birds from extinction. The white spots on and around the island are groups of birds using the sanctuary. In 2009, the brown pelican was removed from the U.S. endangered species list.

Some species cannot be preserved in gene banks. The banks are of varying quality, expensive to operate, and can be destroyed by fires and other mishaps. However, a new underground vault on a remote Norwegian island in the Arctic will eventually contain 100 million of the world's seeds. It is not vulnerable to power losses, fires, storms, or war.

The world's 1,600 *botanical gardens* and *arboreta* contain living plants that represent almost one-third of the world's known plant species. But they contain only about 3% of the world's rare and threatened plant species and have too little space and funding to preserve most of those species.

We can take pressure off some endangered or threatened species by raising individuals of these species on *farms* for commercial sale. In Florida, for example, alligators are raised on farms for their meat and hides. Butterfly farms established to raise and protect endangered species flourish in Papua New Guinea, where many butterfly species are threatened by development activities. These farms are also used to educate visitors about the need to protect butterfly species.

Zoos and Aquariums Can Protect Some Species

Zoos, aquariums, game parks, and animal research centers are being used to preserve some individuals of critically endangered animal species, with the long-term goal of reintroducing the species into protected wild habitats.

Two techniques for preserving endangered terrestrial species are egg pulling and captive breeding. *Egg pulling* involves collecting wild eggs laid by critically endangered bird species and then hatching them in zoos or research centers. In *captive breeding*, some or all of the wild individuals of a critically endangered species are collected for breeding in captivity, with the aim of reintroducing the offspring into the wild. Captive breeding has been used to save the peregrine falcon and the California condor (Figure 8-2c).

The ultimate goal of captive breeding programs is to build up populations to a level where they can be reintroduced into the wild. However, most reintroductions fail because of lack of suitable habitat, inability of individuals bred in captivity to survive in the wild, and renewed overhunting or poaching.

Limited space and budgets restrict efforts to maintain breeding populations of endangered animal species in zoos and research centers. The captive population of an endangered species must number 100–500 individuals in order for it to avoid extinction through accident, disease, or loss of genetic diversity due to inbreeding. Recent genetic research indicates that 10,000 or more individuals are needed for an endangered species to maintain its capacity for biological evolution. Zoos and research centers do not have the funding or space to house such large populations.

Public aquariums that exhibit unusual and attractive species of fish and some marine animals such as seals and dolphins help to educate the public about the need to protect such species. But mostly because of limited funds, public aquariums have not served as effective gene banks for endangered marine species, especially marine mammals such as the endangered southern sea otter that need large volumes of water (see Chapter 5, Core Case Study, p. 79).

Thus, zoos, aquariums, and botanical gardens are not feasible solutions for the growing problem of species extinction. Individuals play key roles in helping to save various species from extinction (see Individuals Matter, p. 164). Figure 8-17 lists some things you can do to help deal with this problem. GOOD NEWS

What Can You Do?

Protecting Species

- Do not buy furs, ivory products, or other items made from endangered or threatened animal species.
- Do not buy wood or paper products produced by cutting old-growth forests in the tropics.
- Do not buy birds, snakes, turtles, tropical fish, and other animals that are taken from the wild.
- Do not buy orchids, cacti, or other plants that are taken from the wild.
- Spread the word. Talk to your friends and relatives about this problem and what they can do about it.

Figure 8-17 **Individuals matter:** You can help to prevent the extinction of species. **Questions:** Which two of these actions do you believe are the most important? Why?

The Precautionary Principle

Biodiversity scientists call for us to take precautionary action to avoid hastening species extinctions. This approach is based on the **precautionary principle**: When significant preliminary evidence indicates that an activity can harm human health or the environment, we should take precautionary measures to prevent or reduce such harm even if some of the cause-and-effect relationships have not been fully established scientifically. It is based on the commonsense idea behind many adages such as "Look before you leap."

Scientists use the precautionary principle to argue for both the preservation of species and the protection of entire ecosystems, which is the focus of the next chapter. The precautionary principle is also used as a strategy for dealing with other challenges such as preventing exposure to harmful chemicals in the air we breathe, the water we drink, and the food we eat.

Using limited financial and human resources to protect biodiversity based on the precautionary principle involves dealing with three important questions:

1. How do we allocate limited resources between protecting species and protecting their habitats?
2. How do we decide which species should get the most attention in our efforts to protect as many species as possible? For example, should we focus on protecting the most threatened species or on protecting keystone species? Protecting species that are appealing to humans, such as tigers (Figure 8-2b) and orangutans (Figure 8-5), can increase awareness of the need for wildlife conservation. But some argue that we should instead protect more ecologically important endangered species.
3. How do we determine which habitat areas are the most critical to protect?

Here are this chapter's *three big ideas*:

- We are greatly increasing the extinction of wild species by destroying and degrading their habitats, introducing harmful invasive species, and increasing human population growth, pollution, climate change, and overexploitation.
- We should avoid causing the extinction of wild species because of the economic and ecological services they provide, and because their existence should not depend primarily on their usefulness to us.
- We can work to prevent the extinction of species and to protect overall biodiversity by using laws and treaties, protecting wildlife sanctuaries, and making greater use of the precautionary principle.

REVISITING Polar Bears and Sustainability

We have learned a lot about how to protect species from extinction that results from our activities. We also know about the importance of wild species as key components of the earth's vital biodiversity.

Yet, despite these efforts, there is overwhelming evidence that as many as half of the world's land-based wild species could go extinct during this century, largely as a result of human activities. Inadequate ecological knowledge is part of the problem. But many argue that the real cause of this failure is that we lack the political and ethical will to act on what we know.

In keeping with the three **principles of sustainability** (see back cover), acting to prevent the extinction of species as a result of human activities helps to preserve the earth's biodiversity, energy flow, and matter cycling in ecosystems. Thus, it is not only for these species that we ought to act, but also for the overall long-term health of the biosphere, on which we all depend, as well as for the health and well-being of our own species.

Protecting biodiversity is no longer simply a matter of passing and enforcing endangered species laws and setting aside parks and preserves. It will also require slowing projected climate change, which severely threatens the polar bear and millions of other species and their habitats. In addition, it will require reducing the size and impact of our ecological footprints (see Figure 1-8, p. 14), for individuals as well as for communities and nations.

The great challenge of the twenty-first century is to raise people everywhere to a decent standard of living while preserving as much of the rest of life as possible.

EDWARD O. WILSON

REVIEW

CORE CASE STUDY

1. Describe how human activities threaten polar bears in the Arctic (**Core Case Study**). CORE CASE STUDY

SECTION 8-1

2. What is the key concept for this section? What is **biological extinction**? What is a **mass extinction**? How can the extinction of a species affect other species and ecosystem services? Describe how scientists estimate extinction rates. Give three reasons why many extinction experts believe that projected extinction rates are probably on the low side. What percentage of the world's species are likely to go extinct, largely as a result of human activities, during this century? Distinguish between **endangered species** and **threatened species** and give an example of each. List four characteristics that make some species especially vulnerable to extinction.

SECTION 8-2

3. What is the key concept for this section? What are three reasons for trying to avoid causing the extinction of wild species? Describe two economic and two ecological benefits of species diversity. Explain how saving other species helps save our own species and our cultures and economies.

SECTION 8-3

4. What is the key concept for this section? What is **HIPPCO**? What are the six largest causes of extinction of species resulting from human activities? What is the greatest threat to wild species? What is **habitat fragmentation**? Describe the major effects of habitat loss and fragmentation.

5. Give two examples of the benefits of introducing some nonnative species. Give two examples of the harmful effects of nonnative species that have been introduced **(a)** deliberately and **(b)** accidentally. Describe the harmful and beneficial effects of introducing the kudzu vine. List four ways to limit the harmful impacts of nonnative species. Explain why prevention is the best way to reduce threats from invasive species and list five ways to implement this strategy. Summarize the roles of population growth, overconsumption, pollution, and climate change in the extinction of wild species. Explain how pesticides such as DDT can be biomagnified in food chains and webs. Describe what is happening to some honeybee populations in the United States.

6. Describe the poaching of wild species and give three examples of species that are threatened by this illegal activity. Why are wild tigers likely to disappear within a few decades? What is the connection between infectious diseases in humans and the pet trade? Explain how Pilai Poonswad helped to protect the rare rhinoceros hornbills in Thailand. Describe the threat to some forms of wildlife from increased hunting for bush meat.

7. Summarize the major threats to bird species. List two reasons why we should be alarmed by the decline of many bird species. Describe the relationships between vultures, wild dogs, and rabies in parts of India.

SECTION 8-4

8. What is the key concept for this section? Name two international treaties that are used to help protect species. What is the U.S. Endangered Species Act? How successful has it been, and why is it controversial? Describe efforts to protect sea turtles and whales from extinction hastened by human activities.

9. Summarize the roles and limitations of wildlife refuges, gene banks, botanical gardens, wildlife farms, zoos, and aquariums in protecting some species. What is the **precautionary principle** and how can we use it to help protect wild species and overall biodiversity?

10. What are this chapter's *three big ideas*? Describe how the three **principles of sustainability** are related to protecting the polar bear and other wild species from extinction, and how they are related to protecting the earth's overall biodiversity.

Note: Key terms are in **bold** type.

CRITICAL THINKING

1. What are three aspects of your lifestyle that might be directly or indirectly contributing to the extinction of the polar bear (**Core Case Study**)?

2. Describe your gut-level reaction to the following statement: "Eventually, all species become extinct. So it does not really matter that the passenger pigeon is extinct, or that the polar bear (**Core Case Study**) or the world's remaining tiger species are endangered primarily because of human activities." Be honest about your reaction, and give arguments to support your position.

3. Do you accept the position that each species has the inherent right to survive without human interference, regardless of whether it serves any useful purpose for humans? Explain. Would you extend this right to the *Anopheles* mosquito, which transmits malaria, and to infectious bacteria?

4. Wildlife ecologist and environmental philosopher Aldo Leopold wrote this with respect to preventing the extinction of wild species: "To keep every cog and wheel is the first precaution of intelligent tinkering." Explain how this statement relates to the material in this chapter.

5. What would you do if fire ants invaded your yard and house? Explain your reasoning behind your course of action. How might your actions affect other species or the ecosystem in which you live?

6. Which of the following statements best describes your feelings toward wildlife?
   a. As long as it stays in its space, wildlife is okay.
   b. As long as I do not need its space, wildlife is okay.
   c. I have the right to use wildlife habitat to meet my own needs.
   d. When you have seen one redwood tree, elephant, or orangutan, you have seen them all, so lock up a few of each species in a zoo or wildlife park and do not worry about protecting the rest.
   e. Wildlife should be protected in its current ranges.

7. Write an argument for **(a)** preserving a weed species in your yard, and for **(b)** not exterminating a colony of wood-damaging carpenter ants in your home.

8. Congratulations! You are in charge of preventing the extinction, caused by human activities, of a large share of the world's existing species. List the three most important steps you would take to accomplish this goal.

DOING ENVIRONMENTAL SCIENCE

Identify examples of habitat destruction or degradation in the area in which you live or go to school. Try to determine and record any harmful effects that these activities have had on the populations of one wild plant and one animal species. (Name each of these species and describe how they have been affected.) Do some research on the Internet and/or in a school library on management plans, and then develop a management plan for rehabilitating the habitats and species you have studied. Try to determine whether trade-offs are necessary with regard to the human activities you have observed, and account for these trade-offs in your management plan. Compare your plan with those of your classmates.

GLOBAL ENVIRONMENT WATCH EXERCISE

Search for *Extinction*, and scroll to statistics on the portal's page. Click on "Known Causes of Animal Extinction since 1600." You will find four general categories of causes. Thinking about history from 1600 through today, how do you think humans have changed their impact on species in each of these categories? Has the impact increased or decreased over this time period? Give specific examples of changes in this timeframe to support your answers.

DATA ANALYSIS

Examine these data released by the World Resources Institute and answer the following questions.

Country	Total Land Area in Square Kilometers (Square Miles)	Protected Area as Percent of Total Land Area (2003)	Total Number of Known Breeding Bird Species (1992–2002)	Number of Threatened Breeding Bird Species (2002)	Threatened Breeding Bird Species as Percent of Total Number of Known Breeding Bird Species
Afghanistan	647,668 (250,000)	0.3	181	11	
Cambodia	181,088 (69,900)	23.7	183	19	
China	9,599,445 (3,705,386)	7.8	218	74	
Costa Rica	51,114 (19,730)	23.4	279	13	
Haiti	27,756 (10,714)	0.3	62	14	
India	3,288,570 (1,269,388)	5.2	458	72	
Rwanda	26,344 (10,169)	7.7	200	9	
United States	9,633,915 (3,718,691)	15.8	508	55	

Source of data: World Resources Institute, *Earth Trends, Biodiversity and Protected Areas, Country Profiles*; **http://earthtrends.wri.org/country_profiles/index.php?theme=7**

1. Complete the table by filling in the last column. For example, to calculate this value for Costa Rica, divide the number of threatened breeding bird species by the total number of known breeding bird species and multiply the answer by 100 to get the percentage.
2. Arrange the countries from largest to smallest according to total land area. Does there appear to be any correlation between the size of country and the percentage of threatened breeding bird species? Explain your reasoning.

LEARNING ONLINE

The website for *Environmental Science 14e* contains helpful study tools and other resources to take the learning experience beyond the textbook. Examples include flashcards, practice quizzing, information on green careers, tips for more sustainable living, activities, and a wide range of articles and further readings. To access these materials and other companion resources for this text, please visit **www.cengagebrain.com**. For further details, see the preface, p. xvi.

9 Sustaining Biodiversity: The Ecosystem Approach

CORE CASE STUDY Wangari Maathai and the Green Belt Movement

In the mid-1970s, Wangari Maathai (Figure 9-1) took a hard look at environmental conditions in her native African country of Kenya. Tree-lined streams she had known as a child had dried up. Farms and plantations had displaced vast areas of forest and were draining the watersheds and degrading the soil. Clean drinking water, firewood, and nutritious foods were all in short supply.

Something inside her told Maathai that she had to do something about this environmental degradation. Starting with a small tree nursery in her backyard in 1977, she founded the Green Belt Movement, which continues today. The main goal of this highly regarded women's self-help group is to organize poor women in rural Kenya to plant and protect millions of trees in order to combat deforestation and provide fuelwood. Since 1977, the 50,000 members of this grassroots group have planted and protected more than 45 million trees.

United Nations Environment Programme

Figure 9-1 Wangari Maathai was the first Kenyan woman to earn a PhD and to head an academic department at the University of Nairobi. In 1977, she organized the internationally acclaimed Green Belt Movement to plant millions of trees throughout Kenya. For her work in protecting the environment and in promoting democracy, human rights, and women's rights, she has received many honors, including the 2004 Nobel Peace Prize. In 2005, *Time* magazine (USA) named her as one of the 100 Most Influential People in the World. This photo shows her receiving news of her Nobel Peace Prize award in 2004.

The women are paid a small amount for each seedling they plant that survives. This gives them an income to help them break out of the cycle of poverty. The trees provide fruits, building materials, and fodder for livestock. They also provide more fuelwood, so that women and children do not have to walk so far to find fuel for cooking and heating. The trees also improve the environment by reducing soil erosion and providing shade and beauty. In addition, they help to slow projected climate change by removing CO_2 from the atmosphere.

GOOD NEWS

The success of the Green Belt Movement has sparked the creation of similar programs in more than 30 other African countries. Her efforts inspired the United Nations Environment Programme (UNEP) to implement a global effort to plant at least 1 billion trees a year beginning in 2006. By 2010, about 10.6 billion trees had been planted by participants in 178 countries.

In 2004, Maathai became the first African woman and the first environmentalist to be awarded the Nobel Peace Prize for her lifelong efforts. Within an hour of learning that she had won the prize (Figure 9-1), Maathai planted a tree, telling onlookers it was "the best way to celebrate." She urged everyone in the world to plant a tree as a symbol of commitment and hope. Maathai tells her story in her book, *The Green Belt Movement: Sharing the Approach and the Experience* (Lantern Books, 2003) and in *Unbowed: A Memoir* (Alfred A. Knopf, 2006).

Since 1980, *biodiversity* (see Figure 4-2, p. 63) has emerged as a key concept of biology, and it is the focus of one of the three **principles of sustainability** (see back cover). Biologists warn that human population growth, economic development, and poverty are exerting increasing pressure on terrestrial and aquatic ecosystems and on the services they provide that help to sustain biodiversity. In 2010, a report by two United Nations environmental bodies warned that unless radical and creative action is taken now to conserve the earth's biodiversity, many natural systems that support lives and livelihoods are at risk of collapsing.

SUSTAINABILITY

This chapter is devoted to helping us understand the threats to the earth's forests, grasslands, and other storehouses of terrestrial biodiversity, so that we can seek ways to help sustain these vital ecosystems. To many scientists, sustaining the world's vital biodiversity is one of our most important challenges.

Key Questions and Concepts

9-1 What are the major threats to forest ecosystems?

CONCEPT 9-1 Ecologically valuable forest ecosystems are being cut and burned at unsustainable rates in many parts of the world.

9-2 How should we manage and sustain forests?

CONCEPT 9-2 We can sustain forests by emphasizing the economic value of their ecological services, removing government subsidies that hasten their destruction, protecting old-growth forests, harvesting trees no faster than they are replenished, and planting trees.

9-3 How should we manage and sustain grasslands?

CONCEPT 9-3 We can sustain the productivity of grasslands by controlling the numbers and distribution of grazing livestock, and by restoring degraded grasslands.

9-4 How should we manage and sustain parks and nature reserves?

CONCEPT 9-4 Sustaining biodiversity will require more effective protection of existing parks and nature reserves, as well as the protection of much more of the earth's remaining undisturbed land area.

9-5 What is the ecosystem approach to sustaining biodiversity?

CONCEPT 9-5 We can help to sustain terrestrial biodiversity by identifying and protecting severely threatened areas (biodiversity hotspots), restoring damaged ecosystems (using restoration ecology), and sharing with other species much of the land we dominate (using reconciliation ecology).

9-6 How can we help to sustain aquatic biodiversity?

CONCEPT 9-6 We can help to sustain aquatic biodiversity by establishing protected sanctuaries, managing coastal development, reducing water pollution, and preventing overfishing.

Note: Supplements 2 (p. S3), 6 (p. S22), and 7 (p. S38) can be used with this chapter.

There is no solution, I assure you, to save Earth's biodiversity other than preservation of natural environments in reserves large enough to maintain wild populations sustainably.

EDWARD O. WILSON

9-1 What Are the Major Threats to Forest Ecosystems?

▶ **CONCEPT 9-1 Ecologically valuable forest ecosystems are being cut and burned at unsustainable rates in many parts of the world.**

Forests Vary in Their Age, Makeup, and Origins

Natural and planted forests occupy about 30% of the earth's land surface (excluding Greenland and Antarctica). Figure 7-9 (p. 128) shows the distribution of the world's northern coniferous, temperate, and tropical forests.

Forest managers and ecologists classify natural forests into two major types based on their age and structure: old-growth and second-growth forests. An **old-growth forest**, or **primary forest**, is an uncut or regenerated forest that has not been seriously disturbed by human activities or natural disasters for 200 years or more (Figure 9-2, p. 176). Old-growth forests are reservoirs of biodiversity because they provide ecological niches for a multitude of wildlife species (see Figure 7-15, p. 135).

A **second-growth forest** is a stand of trees resulting from secondary ecological succession (see Figure 5-10, p. 90). These forests develop after the trees in an area have been removed by human activities, such as clear-cutting for timber or cropland, or by natural forces such as fire, hurricanes, or volcanic eruption. Individuals also help regenerate forests by planting and tending trees (**Core Case Study**). CORE CASE STUDY

Figure 9-2 **Natural capital:** This old-growth forest is located in the U.S. state of Washington's Olympic National Forest.

A **tree plantation**, also called a **tree farm** or **commercial forest**, (Figure 9-3) is a managed forest containing only one or two species of trees that are all of the same age. They are usually harvested by clear-cutting as soon as they become commercially valuable. The land is then replanted and clear-cut again in a regular cycle. When managed carefully, such plantations can produce wood at a rapid rate and thus increase their owners' profits. Some analysts project that eventually, tree plantations could supply most of the wood used for industrial purposes such as papermaking. This would help to protect the world's remaining old-growth and secondary forests, as long as they are not cleared to plant tree plantations.

The downside of tree plantations is that, with only one or two tree species, they are much less biologically diverse and less sustainable than old-growth and second-growth forests because they violate nature's biodiversity **principle of sustainability**. In addition, repeated cycles of cutting and replanting eventually deplete the topsoil of nutrients and can lead to an irreversible ecological tipping point that hinders the regrowth of any type of forest on such land.

Forests Provide Important Economic and Ecological Services

So why should we care about forests? Answer: Because they provide highly valuable ecological and economic services (Figure 9-4 and **Concept 9-1**). For example, through photosynthesis, forests remove CO_2 from the atmosphere and store it in organic compounds (biomass). By performing this ecological service as a part of the global carbon cycle (see Figure 3-15, p. 53), forests help to stabilize average atmospheric temperatures and slow projected climate change.

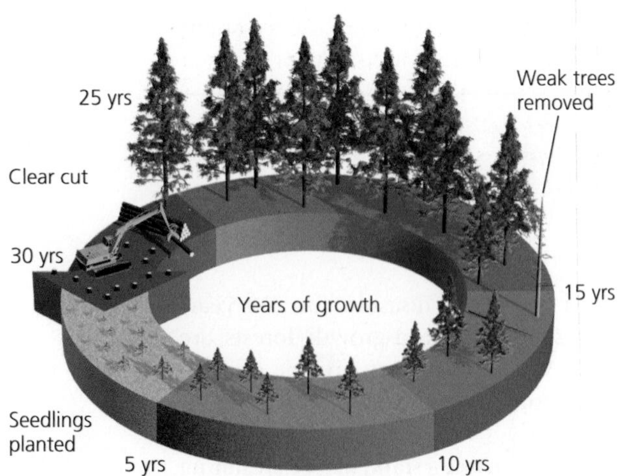

Figure 9-3 This diagram illustrates the short (25- to 30-year) rotation cycle of cutting and regrowth of a monoculture tree plantation. In tropical countries, where trees can grow more rapidly year-round, the rotation cycle can be 6–10 years. Most tree plantations (see photo, right) are grown on land that was cleared of old-growth or second-growth forests. **Question:** What are two ways in which this process can degrade an ecosystem?

Natural Capital

Forests

Ecological Services	Economic Services
Support energy flow and chemical cycling	Fuelwood
Reduce soil erosion	Lumber
Absorb and release water	Pulp to make paper
Purify water and air	Mining
Influence local and regional climate	Livestock grazing
Store atmospheric carbon	Recreation
Provide numerous wildlife habitats	Jobs

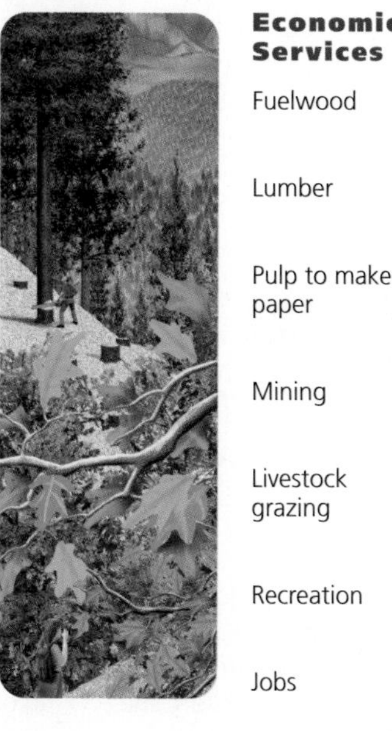

Figure 9-4 Forests provide many important ecological and economic services (**Concept 9-1**). **Question:** Which two ecological services and which two economic services do you think are the most important?

Scientists and economists have attempted to estimate the economic value of this and other ecological services provided by the world's forests and other ecosystems (Science Focus, p. 178). Regardless of such monetary values, forests provide habitats for about two-thirds of the earth's terrestrial species. Also, they are home to more than 300 million people, and one of every four people depends on a forest for making a living.

Unsustainable Logging Is a Major Threat to Forest Ecosystems

Along with highly valuable ecological services (Figure 9-4, left), forests provide us with a number of economic services (Figure 9-4, right). The first step in harvesting trees is to build roads for access and timber removal. Even carefully designed logging roads have a number of harmful effects (Figure 9-5), including increased topsoil erosion and sediment runoff into waterways, habitat fragmentation, and loss of biodiversity. (See *The Habitable Planet*, Video 9, at **www.learner.org/resources/series209.html**.) Logging roads also expose forests to invasion by nonnative pests, diseases, and wildlife species. And they open once-inaccessible forests to miners, ranchers, farmers, hunters, and off-road vehicles.

Once loggers reach a forest area, they use a variety of methods to harvest the trees. Three major methods are selective cutting, clear-cutting, and strip cutting (Figure 9-6, p. 179). With *selective cutting,* intermediate-aged or mature trees in a forest are cut singly or in small groups (Figure 9-6a). Loggers often remove all the trees

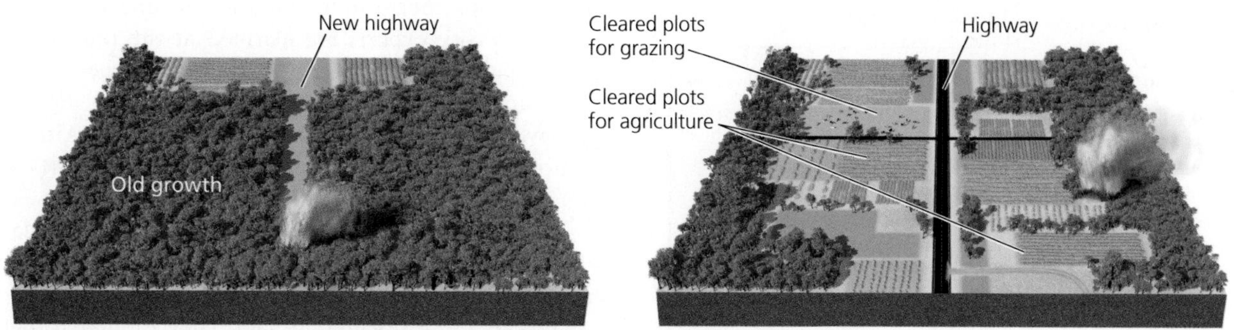

Figure 9-5 Natural capital degradation: Building roads into previously inaccessible forests is the first step in harvesting timber, but it also paves the way to fragmentation, destruction, and degradation of forest ecosystems.

SCIENCE FOCUS

Putting a Price Tag on Nature's Ecological Services

Currently, forests and other ecosystems are valued mostly for their economic services (Figure 9-4, right). However, suppose we took into account the monetary value of the ecological services provided by forests (Figure 9-4, left).

In 1997, a team of ecologists, economists, and geographers, led by ecological economist Robert Costanza of the University of Vermont, estimated the monetary worth of the earth's ecological services, which can be thought of as *ecological income*, somewhat like interest income earned from a savings account. We can think of the earth's *natural capital*—the stock of natural resources that provide the ecological services—as the savings account from which the interest income flows.

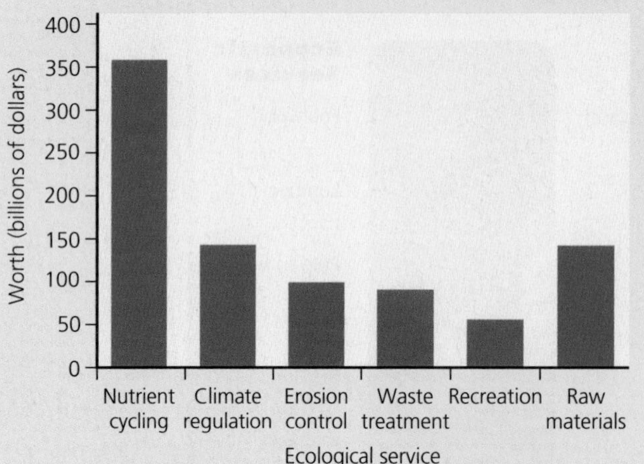

Figure 9-A This chart shows estimates of the annual global economic values of some ecological services provided by forests compared to the raw materials they produce (in billions of dollars). (Data from Robert Costanza)

Costanza's team estimated the monetary value of 17 ecological services (including, for example, pollination and regulation of atmospheric temperatures) to be at least $33.2 trillion per year—equal to about 45% of the value of all of the goods and services produced in 2010 throughout the world. The researchers also estimated that the amount of money we would have to put into a savings account in order to earn that amount of interest income would be at least $500 trillion—an average of about $71,400 for each person on earth in 2011.

According to Costanza's study, the world's forests provide us with ecological services worth at least $4.7 trillion per year—hundreds of times more than their economic value in terms of lumber, paper, and other wood products. Some of these comparative value estimates are shown in Figure 9-A. Note that the collective value of forest ecosystem services is much greater than the value of timber and other raw materials extracted from forests (**Concept 9-1**). In addition, the researchers point out that their estimates are very conservative.

Costanza's team had examined more than 100 studies and a variety of methods used to estimate the values of ecosystems. For example, one method that had been used was to estimate the costs of replacing ecosystem services such as water purification with technological services.

In 2002, Costanza and other researchers reported on a similar analysis comparing the value of preserving natural ecosystems with the values that could be obtained by using such systems to create farmland, to harvest timber, and to create aquaculture ponds, among other uses. The researchers estimated that preserving ecosystems in a global network of nature reserves occupying 15% of the earth's land surface and 30% of the ocean would provide $4.4 trillion to $5.2 trillion worth of ecological services—about 100 times the economic value of converting those systems to human uses.

If the estimates from these studies are reasonably correct, we can draw three important conclusions: **(1)** the earth's ecosystem services are essential for all humans and their economies; **(2)** the economic value of these services is huge; and **(3)** ecosystem services are an ongoing source of ecological income, as long as they are used sustainably.

Critical Thinking

Some analysts believe that we should not try to put economic values on the world's irreplaceable ecological services because their value is infinite. Do you agree with this view? Explain. What is the alternative?

from an area in what is called a *clear-cut* (Figure 9-6b and Figure 9-7). Clear-cutting is the most efficient way for a logging operation to harvest trees, but it can do considerable harm to an ecosystem.

> CONNECTIONS
> **Clear-Cutting and Loss of Biodiversity**
>
> Scientists have found that removing all the tree cover from a forest area greatly increases water runoff and loss of soil nutrients (See Chapter 2, Core Case Study, p. 25, and Figure 2-4, p. 31). This increases topsoil erosion, which in turn causes more vegetation to die, leaving barren ground that can be eroded further. More erosion also means more pollution of streams in the watershed, and all of these effects can destroy terrestrial and aquatic habitats and degrade biodiversity.

A variation of clear-cutting that allows a more sustainable timber yield without widespread destruction is *strip cutting* (Figure 9-6c). It involves clear-cutting a strip of trees along the contour of the land within a corridor narrow enough to allow natural forest regeneration within a few years. After regeneration, loggers cut another strip next to the first, and so on. GOOD NEWS

Biodiversity experts are alarmed at the practice of illegal logging taking place in 70 countries, especially in Africa and Southeast Asia (**Concept 9-1**). Such logging has ravaged 37 of the 41 national parks in the African country of Kenya (**Core Case Study**) and makes up 73–80% of all logging in Indonesia. CORE CASE STUDY

According to a 2010 British study, illegal logging has fallen by 22% since 2002, largely because of tougher enforcement of laws against it. However, it is still a serious problem, the researchers said. They estimated that every year, the global total amount of trees illegally cut, if laid end to end, would encircle the earth more than 10 times.

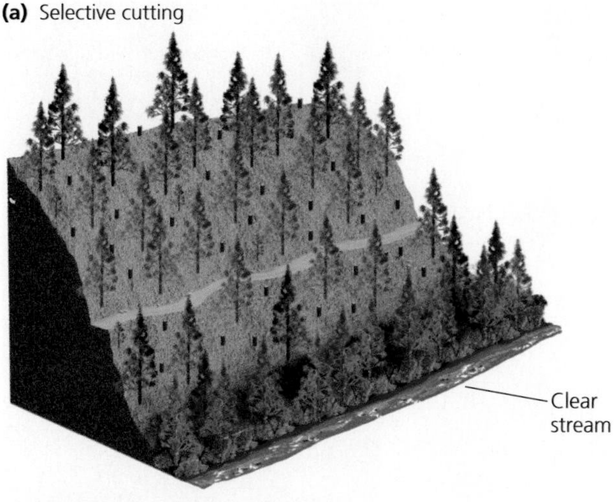

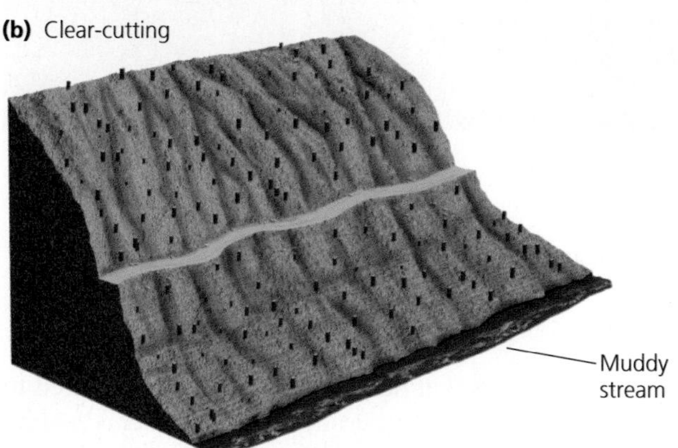

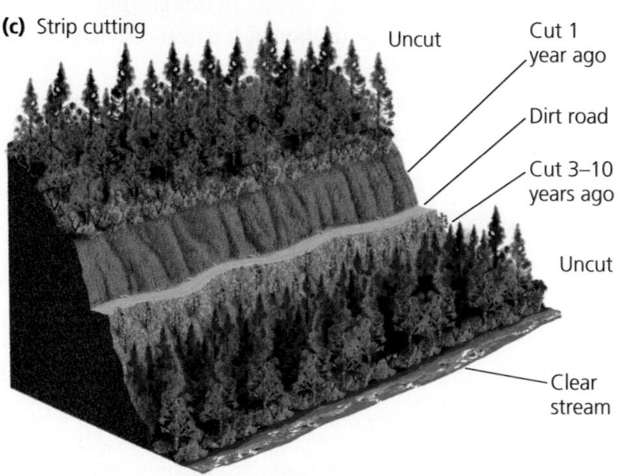

Figure 9-6 This diagram illustrates the three major tree harvesting methods. **Question:** If you were cutting trees in a forest you owned, which method would you choose and why?

Figure 9-7 This aerial photograph shows the results of clear-cut logging in the U.S. state of Washington.

Fire Can Threaten or Benefit Forest Ecosystems

Two types of fires can affect forest ecosystems. *Surface fires* (Figure 9-8, left, p. 180) usually burn only undergrowth and leaf litter on the forest floor. They may kill seedlings and small trees, but they spare most mature trees and allow most wild animals to escape.

Occasional surface fires have a number of ecological benefits. They:

- burn away flammable ground material such as dry brush and help to prevent more destructive fires;
- free valuable mineral nutrients tied up in slowly decomposing litter and undergrowth;
- release seeds from the cones of tree species such as lodgepole pines;
- stimulate the germination of certain tree seeds such as those of the giant sequoia and jack pine; and
- help to control tree diseases and insects.

Wildlife species including deer, moose, muskrat, and quail depend on occasional surface fires to maintain their habitats and to provide food in the form of the young, tender vegetation that sprouts after fires.

Another type of fire, called a *crown fire* (Figure 9-8, right), is an extremely hot fire that leaps from treetop

Figure 9-8 Surface fires (left) usually burn only undergrowth and leaf litter on a forest floor. They can help to prevent more destructive crown fires (right) by removing flammable ground material. In fact, carefully controlled surface fires sometimes are deliberately set to prevent the buildup of flammable ground material in forests. Surface fires also recycle nutrients and thus help to maintain the productivity of a variety of forest ecosystems. **Question:** What is another way in which a surface fire might benefit a forest?

to treetop, burning whole trees. Crown fires usually occur in forests that have not experienced surface fires for several decades, a situation that allows dead wood, leaves, and other flammable ground litter to accumulate. These rapidly burning fires can destroy most vegetation, kill wildlife, increase topsoil erosion, and burn or damage human structures in their paths.

CONNECTIONS

Climate Change and Forest Fires

During the next few decades, rising temperatures and increased drought from projected atmospheric warming will likely make many forest areas more suitable for insect pests, which would then multiply and kill more trees. The resulting combination of drier forests and more dead trees could increase the incidence and intensity of forest fires. This would add more of the greenhouse gas CO_2 to the atmosphere, further increasing atmospheric temperatures and causing even more forest fires in a spiraling cycle of increasingly harmful environmental changes.

Almost Half of the World's Forests Have Been Cut Down

Deforestation is the temporary or permanent removal of large expanses of forest for agriculture, settlements, or other uses. Surveys by the World Resources Institute (WRI) indicate that over the past 8,000 years, human activities have reduced the earth's original forest cover by about 46%, with most of this loss occurring in the last 60 years. According to the WRI, if current deforestation rates continue, about 40% of the world's remaining intact forests will have been logged or converted to other uses within two decades if not sooner. Clearing large areas of forests, especially old-growth forests, has important short-term economic benefits (Figure 9-4, right), but it also has a number of harmful environmental effects (Figure 9-9).

The UN Food and Agriculture Organization (FAO) reported that the net total forest cover

Natural Capital Degradation

Deforestation

- Decreased soil fertility from erosion
- Runoff of eroded soil into aquatic systems
- Premature extinction of species with specialized niches
- Loss of habitat for native species and migratory species such as birds and butterflies
- Regional climate change from extensive clearing
- Release of CO_2 into atmosphere
- Acceleration of flooding

Figure 9-9 Deforestation has some harmful environmental effects that can reduce biodiversity and degrade the ecological services provided by forests (Figure 9-4, left). **Question:** What are three products you have used recently that might have come from old-growth forests?

in several countries, including the United States (see the Case Study that follows), changed very little or even increased between 2000 and 2007. Some of the increases resulted from natural reforestation by secondary ecological succession on cleared forest areas and abandoned croplands (see Figure 5-10, p. 90). Other increases in forest cover were due to the spread of commercial tree plantations (Figure 9-3, right).

The fact that millions of trees are being planted throughout the world because of the dedication of individuals such as Wangari Maathai (**Core Case Study**) and her Green Belt Movement is very encouraging to many scientists. However, some scientists are concerned about the growing amount of land occupied by commercial tree plantations, because replacement of old-growth forests by these biologically simplified tree farms represents a loss of biodiversity, and possibly of stability, in some forest ecosystems.

■ CASE STUDY

Many Cleared Forests in the United States Have Grown Back

Forests cover about 30% of the U.S. land area, providing habitats for more than 80% of the country's wildlife species and containing about two-thirds of the nation's surface water. Today, forests in the United States (including tree plantations) cover more area than they did in 1920. The primary reason is that many of the old-growth forests that were cleared or partially cleared between 1620 and 1920 have grown back naturally through secondary ecological succession (Figure 9-10).

There are now fairly diverse second-growth (and in some cases third-growth) forests in every region of the United States except much of the West. In 1995, environmental writer Bill McKibben cited forest regrowth in the United States—especially in the East—as "the great environmental success story of the United States, and in some ways, the whole world."

Every year, more wood is grown in the United States than is cut and the total area planted with trees increases. Protected forests make up about 40% of the country's total forest area, mostly in the *National Forest System*, which consists of 155 national forests managed by the U.S. Forest Service (USFS).

On the other hand, since the mid-1960s, an increasing area of the nation's remaining old-growth (Figure 9-2) and fairly diverse second-growth forests has been cut down and replaced with biologically simplified tree plantations. According to biodiversity researchers, this reduces overall forest biodiversity and disrupts ecosystem processes such as energy flow and chemical cycling—in violation of all three **principles of sustainability**. Harvesting tree plantations too frequently also depletes forest topsoil of key nutrients. Many biodiversity researchers favor establishing tree plantations only on land that has already been degraded, instead of cutting old-growth and second-growth forests and replacing them with tree plantations.

CONNECTIONS

Toilet Paper and Old-Growth Forest Losses

Toilet paper can be made from recycled materials as well as from wood pulp. However, millions of trees in North, Central, and South American countries are cut down just to make ultra-soft toilet paper. More than half of these trees are from old- or second-growth forests, according to the Natural Resources Defense Council, including some rare old-growth forests in Canada. Where does your toilet paper come from?

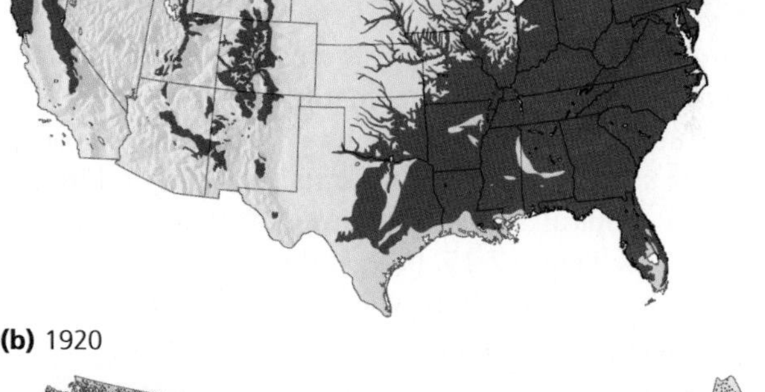

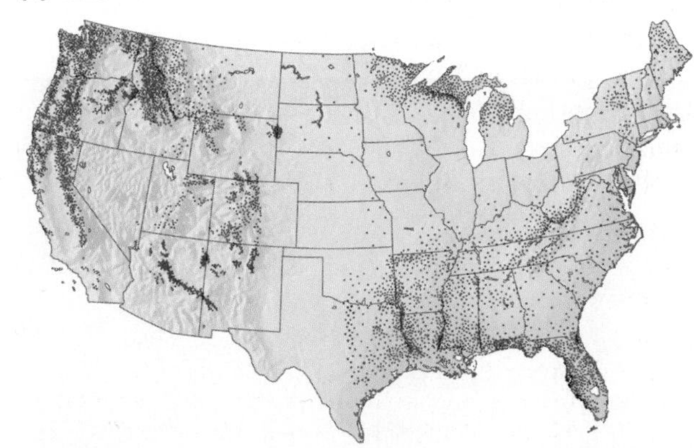

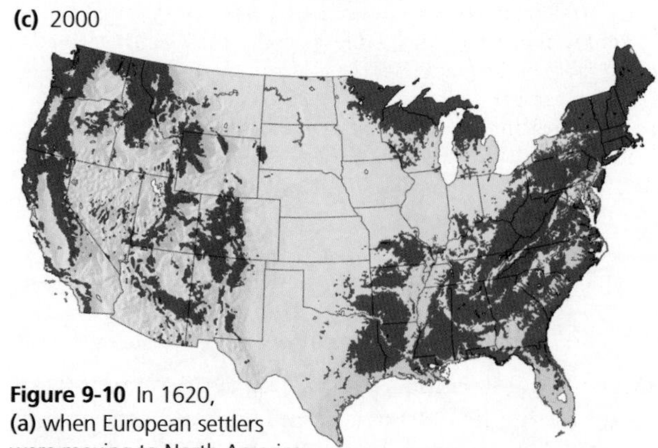

Figure 9-10 In 1620, **(a)** when European settlers were moving to North America, forests covered more than half of the current land area of the continental United States. By 1920, **(b)** most of these forests had been decimated. Since then, a combination of secondary ecological succession and the expansion of commercial forests has resulted in greatly expanded forest cover. In 2000, **(c)** secondary and commercial forests covered about a third of U.S. land in the lower 48 states.

Tropical Forests Are Disappearing Rapidly

Tropical forests (see Figure 7-14, top, p. 134) cover about 6% of the earth's land area—roughly the area of the continental United States. Climatic and biological data suggest that mature tropical forests once covered at least twice as much area as they do today. Most of this loss of half of the world's tropical forests has taken place since 1950 (see Chapter 3, Core Case Study, p. 40).

Satellite scans and ground-level surveys indicate that large areas of tropical rain forests and tropical dry forests are being cut rapidly in parts of Africa (**Core Case Study**), Southeast Asia (Figure 9-11), and South America (see Figure 3-1, p. 40)—especially in Brazil's vast Amazon Basin, which has more than 30% of the world's remaining tropical forests. About 72% of Indonesia's original forests have been lost, mostly to illegal logging. At the current rate of global deforestation, 50% of the world's remaining old-growth tropical forests will be gone or severely degraded by the end of this century.

Studies indicate that at least half of the world's known species of terrestrial plants, animals, and insects live in tropical forests. Because of their specialized niches (see Figure 7-15, p. 135) these species are highly vulnerable to extinction when their forest habitats are destroyed or degraded.

S. Chamnanrith-UNEP/Peter Arnold, Inc.

Figure 9-11 Natural capital degradation: This photo shows extreme tropical deforestation in Chiang Mai, Thailand. Such clearing of trees, which absorb carbon dioxide as they grow, helps to hasten climate change. It also dehydrates the topsoil by exposing it to sunlight. The dry topsoil can then be blown away, which can lead to an irreversible ecological tipping point, beyond which a forest cannot grow back in the area.

Causes of Tropical Deforestation Are Varied and Complex

Tropical deforestation results from a number of underlying and direct causes (Figure 9-12). Underlying causes, such as pressures from population growth and poverty, push subsistence farmers and the landless poor into tropical forests, where they try to grow enough food to survive. Government subsidies can accelerate the direct causes such as logging and ranching by reducing the costs of timber harvesting, cattle grazing, and the creation of vast plantations of crops such as soybeans.

The major direct causes of deforestation vary in different tropical areas. Tropical forests in the Amazon and other South American countries are cleared or burned (Figure 9-13) primarily for cattle grazing and large soybean plantations. In Indonesia, Malaysia, and other areas of Southeast Asia, tropical forests are being replaced with vast plantations of oil palm, which produces an oil used in cooking, cosmetics, and biodiesel fuel for motor vehicles (especially in Europe). In Africa, the primary direct cause of deforestation is individuals struggling to survive by clearing plots for small-scale farming and by harvesting wood for fuel. However, women in the Green Belt Movement (**Core Case Study**) have helped to reestablish forest areas in several African countries.

According to a 2005 study by forest scientists, widespread fires in the Amazon Basin (Figure 9-13) are changing regional weather patterns by raising temperatures and reducing rainfall. This process is converting large deforested areas of tropical forests to tropical grassland (savanna)—another example of an ecosystem reaching an irreversible ecological *tipping point.* Models project that if current burning and deforestation rates continue, 20–30% of the Amazon Basin will be turned into savanna by 2060, and most of it could become savanna by 2080.

CONNECTIONS

Burning Tropical Forests and Climate Change

The burning of tropical forests releases CO_2 into the atmosphere. The problem is that CO_2 is being added to the atmosphere from such burning and from the burning of fossil fuels faster than it is being removed by the carbon cycle (Figure 3-15, p. 53). As a result, CO_2 levels have been increasing since 1960. Rising concentrations of this gas contribute to atmospheric warming, which is projected to change the global climate during this century. Scientists estimate that tropical forest fires account for at least 17% of all human-created greenhouse gas emissions, and that each year, they emit twice as much CO_2 as all of the world's cars and trucks emit. As these forests disappear, even if savannah or second-growth forests replace them, far less CO_2 will be absorbed for photosynthesis, resulting in even more atmospheric warming.

Foreign corporations operating under government concession contracts do much of the logging in tropical countries. After the best timber has been removed,

Natural Capital Degradation

Major Causes of the Destruction and Degradation of Tropical Forests

Underlying Causes

- Not valuing ecological services
- Crop and timber exports
- Government policies
- Poverty
- Population growth

Direct Causes

- Roads
- Fires
- Settler farming
- Cash crops
- Cattle ranching
- Logging
- Tree plantations

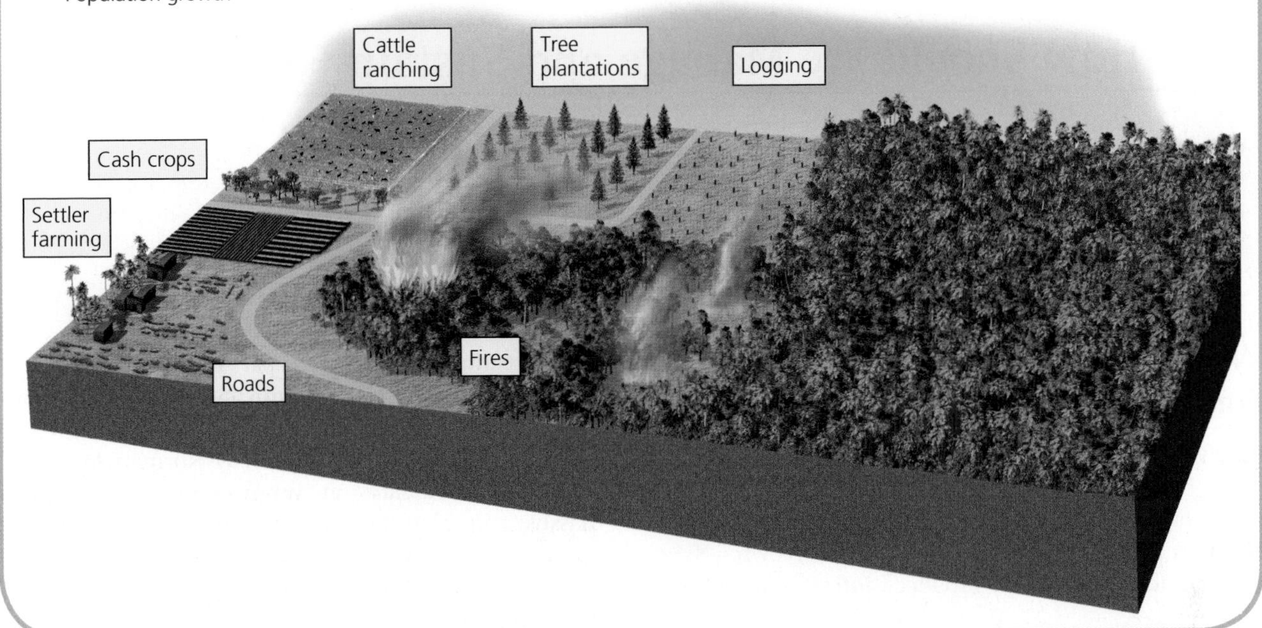

Figure 9-12 This diagram illustrates the major underlying and direct causes of the destruction and degradation of tropical forests. **Question:** If we could eliminate the underlying causes, which, if any, of the direct causes might automatically be eliminated?

Herbert Giradet/Peter Arnold, Inc.

Figure 9-13 Natural capital degradation: Large areas of tropical forest in Brazil's Amazon Basin are burned each year to make way for cattle ranches, plantation crops such as soybeans, and small-scale farms. **Questions:** What are three ways in which your lifestyle may be contributing to this process? How, in turn, might this process affect your life?

timber companies or the local government often sell the land to ranchers who burn the remaining timber to clear the land for cattle grazing. Within a few years, their cattle typically overgraze the land and the ranchers move their operations to another forest area. Then they sell the degraded land to large-scale farmers, who plow it up for plantations of crops such as soybeans, or to settlers who have migrated to tropical forests hoping to grow enough food to survive. After a few years of crop growing and erosion from rain, the nutrient-poor topsoil is depleted of nutrients. Then the farmers and settlers move on to newly cleared land to repeat this environmentally destructive process.

THINKING ABOUT
Tropical Forests

Why should you care if most of the world's remaining tropical forests are burned or cleared and converted to savanna within your lifetime? What are three ways in which this might affect your life or the lives of any children and grandchildren that you might have?

9-2 How Should We Manage and Sustain Forests?

▶ **CONCEPT 9-2 We can sustain forests by emphasizing the economic value of their ecological services, removing government subsidies that hasten their destruction, protecting old-growth forests, harvesting trees no faster than they are replenished, and planting trees.**

We Can Manage Forests More Sustainably

Biodiversity researchers and a growing number of foresters have called for more sustainable forest management. Figure 9-14 lists ways to achieve this goal (**Concept 9-2**). GOOD NEWS Certification of sustainably grown timber and of sustainably produced forest products can help consumers to play their part in reaching this goal (Science Focus, at right). Removing government subsidies and tax breaks that encourage deforestation would also help. (See the Guest Essay by Norman Myers on such *perverse subsidies* at **www.cengagebrain.com**.) Massive tree planting programs such as those run by the Green Belt Movement (**Core Case Study**) CORE CASE STUDY and the United Nations also help to restore degraded forests.

Solutions

More Sustainable Forestry

- Identify and protect forest areas high in biodiversity
- Rely more on selective cutting and strip cutting
- Stop clear-cutting on steep slopes
- Stop logging in old-growth forests
- Sharply reduce road building in uncut forest areas
- Leave most standing dead trees and fallen timber for wildlife habitat and nutrient cycling
- Put tree plantations only on deforested and degraded land
- Certify timber grown by sustainable methods
- Include ecological services of forests in estimates of their economic value

Figure 9-14 There are a number of ways to grow and harvest trees more sustainably (**Concept 9-2**). **Questions:** Which three of these methods of more sustainable forestry do you think are the most important? Why?

We Can Improve the Management of Forest Fires

In the United States, the Smokey Bear educational campaign undertaken by the Forest Service and the National Advertising Council has prevented countless forest fires. It has also saved many lives and prevented billions of dollars in losses of trees, wildlife, and human structures.

At the same time, this educational program has convinced much of the public that all forest fires are bad and should be prevented or put out. Ecologists warn that trying to prevent all forest fires can make matters worse by increasing the likelihood of destructive crown fires due to the accumulation of highly flammable underbrush and smaller trees in some forests.

Ecologists and forest fire experts have proposed several strategies for reducing fire-related harm to forests and to people who use the forests or have homes in them. One approach is to set small, contained surface fires to remove flammable small trees and underbrush in the highest-risk forest areas. Such *prescribed burns* require careful planning and monitoring to keep them from getting out of control.

SCIENCE FOCUS

Certifying Sustainably Grown Timber and Products Such as the Paper Used in This Book

Collins Pine is a forest products company that owns and manages a large area of productive timberland in the northeastern part of the U.S. state of California. Since 1940, the company has used selective cutting to help maintain the ecological and economic sustainability of its timberland.

Since 1993, Scientific Certification Systems (SCS) of Oakland, California, has evaluated the company's timber production. SCS, which is part of the nonprofit Forest Stewardship Council (FSC), was formed to develop environmentally sound and sustainable practices for use in certifying timber and timber products.

Each year, SCS evaluates Collins Pine's landholdings and has consistently found that their cutting of trees has not exceeded long-term forest regeneration; roads and harvesting systems have not caused unreasonable ecological damage; topsoil has not been damaged; and downed wood (boles) and standing dead trees (snags) are left to provide wildlife habitat. As a result, SCS judges the company to be a good employer and a good steward of its land and water resources.

The FSC reported that, by 2009, about 5% of the world's forest area in 82 countries had been certified according to FSC standards. The countries with the largest areas of FSC-certified forests are, in order, Canada, Russia, Sweden, the United States, Poland, and Brazil.

FSC also certifies 5,400 manufacturers and distributors of wood products. The paper used in this book was produced with the use of sustainably grown timber, as certified by the FSC, and contains recycled paper fibers.

Critical Thinking

Should governments provide tax breaks for sustainably grown timber to encourage this practice? Explain.

A second strategy is to allow some fires on public lands to burn, thereby removing flammable underbrush and smaller trees, as long as the fires do not threaten human structures and life. Such lands are protected from fire partly because timber companies want the commercially valuable timber there.

A third approach is to protect houses and other buildings in fire-prone areas by thinning trees and other vegetation in a zone of about 60 meters (200 feet) around them, and eliminating the use of highly flammable construction materials such as wood shingles.

A fourth approach is to thin forest areas that are vulnerable to fire by clearing away small fire-prone trees and underbrush under careful environmental controls. Many forest fire scientists warn that such thinning operations should not remove economically valuable medium-size and large trees for two reasons. *First,* these are the most fire-resistant trees. *Second,* their removal encourages dense growth of more flammable young trees and underbrush and leaves behind highly flammable slash, the debris that results from a logging operation. Many of the worst fires in U.S. history burned through cleared forest areas that contained slash. A 2006 study by Forest Service researchers found that thinning forests without using prescribed burning to remove the slash can greatly increase rather than decrease the risk of heavy fire damage.

We Can Reduce the Demand for Harvested Trees

One way to reduce the pressure on forest ecosystems is to improve the efficiency of wood use. According to the Worldwatch Institute and some forestry analysts, *as much as 60% of the wood consumed in the United States is wasted unnecessarily*. This results from inefficient use of construction materials, excess packaging, overuse of junk mail, inadequate paper recycling, and failure to reuse or find substitutes for wooden shipping containers.

One reason for cutting trees is to provide pulp for making paper, but paper can be made from fiber that does not come from trees. China uses rice straw and other agricultural residues to make much of its paper. Most of the small amount of tree-free paper produced in the United States is made from the fibers of a rapidly growing woody annual plant called *kenaf* (pronounced "kuh-NAHF"). Kenaf and other non-tree fibers such as hemp yield more paper pulp per area of land than tree farms do and require fewer pesticides and herbicides.

It is estimated that, within two to three decades, we could essentially eliminate the need to use trees to make paper. However, while timber companies successfully lobby for government subsidies to grow and harvest trees to make paper, there are no major lobbying efforts or subsidies for producing paper from kenaf and other alternative sources.

Another serious strain on forest resources is the increasing use of wood as a fuel for heating and cooking. About 50% of the wood harvested globally each year, and 75% of that amount in less-developed countries, is burned directly for fuel or converted to charcoal. More than 2 billion people in less-developed countries depend on wood for fuel. In some of these countries, where deforestation would be a problem even without the harvesting of fuelwood, the demand for fuelwood has reached crisis proportions (Figure 9-15, p. 186).

Ways to Reduce Tropical Deforestation

Analysts have pointed to various ways to protect tropical forests (see front cover photo) and use them more sustainably (Figure 9-16, p. 186).

Figure 9-15 This seaside mangrove forest in Haiti was deforested and turned into a desert by poor farmers who were harvesting fuelwood to make charcoal.

One approach is a *debt-for-nature swap*, which can make it financially attractive for countries to protect their tropical forests. In such swaps, participating countries act as custodians of protected forest reserves in return for foreign aid or debt relief. In a similar strategy, called *conservation concessions*, governments or private conservation organizations pay nations for agreeing to preserve their natural resources.

Consumers can reduce the demand for products that are supplied through illegal and unsustainable logging in tropical forests. For building projects, they can use recycled waste lumber. Substitutes for wood such as recycled plastic building materials and bamboo, which can grow up to 0.9 meters (3 feet) in a single day, are also available. We could also sharply reduce the use of throwaway paper products and replace them with reusable plates, cups, and cloth napkins.

Solutions

Sustaining Tropical Forests

Prevention

Protect the most diverse and endangered areas

Educate settlers about sustainable agriculture and forestry

Subsidize only sustainable forest use

Protect forests through debt-for-nature swaps and conservation concessions

Certify sustainably grown timber

Reduce poverty

Slow population growth

Restoration

Encourage regrowth through secondary succession

Rehabilitate degraded areas

Concentrate farming and ranching in already-cleared areas

Figure 9-16 These are some effective ways to protect tropical forests and to use them more sustainably (**Concept 9-2**). **Questions:** Which three of these solutions do you think are the most important? Why?

CONNECTIONS

Good and Bad Bamboo

Ironically, it is possible to add to an environmental problem while trying to be part of the solution. Bamboo, which is increasingly used for hardwood flooring, can be a highly sustainable building material if it is raised on degraded lands. However, some bamboo suppliers, seeing a growing market for this "green" building product, have cleared natural forests to plant rapidly growing bamboo. When buying bamboo, it is important for consumers to find out where and how it was produced. This makes it important to look for bamboo products that are certified as sustainably produced by the Forest Stewardship Council (FSC) (Science Focus, p. 185).

Individuals can plant trees—a powerful example of the idea that all sustainability is local. Wangari Maathai (**Core Case Study**) has promoted and inspired tree planting in her native country of Kenya and throughout the world.

CORE CASE STUDY

9-3 How Should We Manage and Sustain Grasslands?

▶ **CONCEPT 9-3 We can sustain the productivity of grasslands by controlling the numbers and distribution of grazing livestock, and by restoring degraded grasslands.**

Some Rangelands Are Overgrazed

Grasslands provide many important ecological services, including soil formation, erosion control, chemical cycling, storage of atmospheric carbon dioxide in biomass, and maintenance of biodiversity.

After forests, grasslands are the ecosystems most widely used and altered by human activities. **Rangelands** are unfenced grasslands in temperate and tropical climates that supply *forage,* or vegetation for grazing (grass-eating) and browsing (shrub-eating) animals. Cattle, sheep, and goats graze on about 42% of the world's grassland. The 2005 Millennium Ecosystem Assessment estimated that this could increase to 70% by 2050. Livestock also graze in **pastures**, which are managed grasslands or fenced meadows often planted with domesticated grasses or other forage crops such as alfalfa and clover. Some pastures are open areas not suited for cultivation but still able to support native grasses.

Blades of rangeland grass grow from the base, not at the tip as broadleaf plants do. Thus, as long as only the upper half of the blade is eaten and its lower half remains, rangeland grass is a renewable resource that can be grazed again and again. However, **overgrazing** occurs when too many animals graze for too long, damaging the grasses and their roots, and exceeding the carrying capacity of a rangeland area (Figure 9-17, left). Overgrazing reduces grass cover, exposes the topsoil to erosion by water and wind, and compacts the soil (which diminishes its capacity to hold water). Overgrazing also encourages the invasion of once-productive rangeland by species such as sagebrush, mesquite, cactus, and cheatgrass, which cattle will not eat.

Limited data from FAO surveys in various countries indicate that overgrazing by livestock has caused a loss in productivity in as much as 20% of the world's rangeland. About 200 years ago, grasses may have covered nearly half the land in the southwestern United States. Today, they cover only about 20%, mostly because of a combination of prolonged droughts and overgrazing, which has created footholds for invasive species that now cover many of the former grasslands.

We Can Manage Rangelands More Sustainably

The most widely used method for more sustainable management of rangelands is to control the number of grazing animals and the duration of their grazing in a given area so that the carrying capacity of the area is not exceeded (**Concept 9-3**). One way of doing this is *rotational grazing,* in which cattle are confined by portable fencing to one area for a short time (often only 1–2 days) and then moved to a new location.

USDA, Natural Resources Conservation Service

Figure 9-17 Natural capital degradation: To the left of the fence is overgrazed rangeland, while lightly grazed rangeland lies to the right of the fence.

Figure 9-18 Natural capital restoration: In the mid-1980s, cattle had degraded the vegetation and soil on this stream bank along the San Pedro River in the U.S. state of Arizona (left). Within 10 years, the area was restored through secondary ecological succession (right) after grazing and off-road vehicle use were banned (**Concept 9-3**).

Cattle tend to aggregate around natural water sources, especially along streams or rivers lined by thin strips of lush vegetation known as *riparian zones,* and around ponds created to provide water for livestock. Overgrazing by cattle can destroy the vegetation in such areas (Figure 9-18, left). Protecting overgrazed land from further grazing by moving livestock around and by fencing off these damaged areas eventually leads to their natural ecological restoration by ecological succession (Figure 9-18, right). Ranchers can also move cattle around by providing supplemental feed at selected sites and by strategically locating watering ponds and tanks and salt blocks.

GOOD NEWS

A more expensive and less widely used method of rangeland management is to suppress the growth of unwanted invader plants by the use of herbicides, mechanical removal, or controlled burning. A cheaper way to discourage unwanted vegetation in some areas is through controlled, short-term trampling by large numbers of livestock.

9-4 How Should We Manage and Sustain Parks and Nature Reserves?

CONCEPT 9-4 Sustaining biodiversity will require more effective protection of existing parks and nature reserves, as well as the protection of much more of the earth's remaining undisturbed land area.

National Parks Face Many Environmental Threats

Today, more than 1,100 major national parks are located in more than 120 countries. However, most of these national parks are too small to sustain many large animal species. And many parks suffer from invasions by nonnative species that compete with and reduce the populations of native species. Some parks are so popular that large numbers of visitors are degrading the natural features that make them attractive (see the Case Study that follows).

Parks in less-developed countries have the greatest biodiversity of all parks globally. But only about 1% of these parklands are protected. Local people in many of these countries enter the parks illegally in search of wood, cropland, game animals, and other natural products that they need for their daily survival. Loggers and miners operate illegally in many of these parks, as do wildlife poachers who kill animals to obtain and sell items such as rhino horns, elephant tusks, and furs. Park services in most of the less-developed countries have too little money and too few personnel to fight these invasions, either by force or through education.

■ CASE STUDY

Stresses on U.S. Public Parks

The U.S. national park system, established in 1912, includes 58 major national parks, sometimes called the country's crown jewels (Figure 9-19), along with 335 monuments and historic sites. States, counties, and cities also operate public parks.

Popularity is one of the biggest problems for many parks. Between 1960 and 2009, the number of visitors to U.S. national parks more than tripled, reaching about 286 million. In some parks and other public lands, noisy and polluting dirt bikes, dune buggies, jet skis, snowmobiles, and off-road vehicles degrade the aesthetic experience for many visitors, destroy or damage fragile vegetation, and disturb wildlife. Many visitors expect parks to have grocery stores, laundries, bars, and other such conveniences.

A number of parks also suffer damage from the migration or deliberate introduction of nonnative species. European wild boars, imported into the state of North Carolina in 1912 for hunting, threaten vegetation in parts of the very popular Great Smoky Mountains National Park. Nonnative mountain goats in Washington State's Olympic National Park trample and destroy the root systems of native vegetation and accelerate soil erosion.

At the same time, native species—some of them threatened or endangered—are killed in, or illegally removed from, almost half of U.S. national parks. This is what happened to the gray wolf in Yellowstone National Park until it was successfully reintroduced there after a 50-year absence (Science Focus, p. 190).

Many U.S. national parks have become threatened islands of biodiversity surrounded by a sea of commercial development. Nearby human activities that threaten wildlife and recreational values in many national parks include mining, logging, livestock grazing, coal-fired power generation, water diversion, and urban development. According to the National Park Service, air pollution, mostly from coal-fired power plants and dense vehicle traffic, degrades scenic views in U.S. national parks more than 90% of the time.

The National Park Service estimated that in 2010, the national parks had at least an $8 billion backlog for long overdue maintenance and repairs to trails, buildings, and other park facilities. Some analysts say more of these funds could come from private concessionaires who provide campgrounds, restaurants, hotels, and other services for park visitors. They pay franchise fees averaging only about 6–7% of their gross receipts, and many large concessionaires with long-term contracts pay as little as 0.75%. Analysts say these percentages could reasonably be increased to around 20%.

gray718/Shutterstock.com

Figure 9-19 Grand Teton National Park in the U.S. state of Wyoming.

CONNECTIONS

National Parks and Climate Change

According to Stephen Saunders, president of the Rocky Mountain Climate Change Organization and a former deputy assistant secretary of the U.S. Department of the Interior, projected climate change is "the greatest threat the parks have ever had." Low-lying U.S. park properties in places such as Key West, Florida, Ellis Island in New York Harbor, and large areas of Florida's Everglades National Park will likely be underwater later in this century if sea levels rise as projected. As climate zones shift in a warmer world, by 2030, Glacier National Park may not have any glaciers and the saguaro cactus (see Figure 7-10, middle, p. 129) may disappear from Saguaro National Park.

Nature Reserves Occupy Only a Small Part of the Earth's Land

Most ecologists and conservation biologists believe the best way to preserve biodiversity is to create a worldwide network of protected areas. As of 2010, less than

SCIENCE FOCUS

Reintroducing the Gray Wolf to Yellowstone National Park

Around 1800, at least 350,000 gray wolves (Figure 9-B) roamed over about three-quarters of America's lower 48 states, especially in the West. They survived mostly by preying on abundant bison, elk, caribou, and deer. However, between 1850 and 1900, most of them were shot, trapped, or poisoned by ranchers, hunters, and government employees.

When Congress passed the U.S. Endangered Species Act in 1973, only a few hundred gray wolves remained outside of Alaska, primarily in Minnesota and Michigan. In 1974, the gray wolf was listed as an endangered species in the lower 48 states.

Ecologists recognize the important role that this keystone predator species once played in parts of the West, especially in the northern Rocky Mountain states of Montana, Wyoming, and Idaho, where Yellowstone National Park is located. The wolves culled herds of bison, elk, moose, and mule deer, and kept down coyote populations. By leaving some of their kills partially uneaten, they provided meat for scavengers such as ravens, bald eagles, ermines, grizzly bears, and foxes.

When wolves declined, herds of plant-browsing elk, moose, and mule deer expanded and devastated vegetation such as willow and aspen trees growing near streams and rivers. This led to increased soil erosion. It also threatened habitats of other wildlife species and the food supplies of beaver, which eat willow and aspen. This in turn affected species that depended on wetlands created by the beavers.

In 1987, the U.S. Fish and Wildlife Service (USFWS) proposed reintroducing gray wolves into the Yellowstone National Park ecosystem to help restore and sustain biodiversity and to prevent further environmental degradation of the ecosystem. The proposal brought angry protests, some from area ranchers who feared the wolves would leave the park and attack their cattle and sheep. Other objections came from hunters who feared the wolves would kill too many big-game animals, and from mining and logging companies that feared the government would halt their operations on wolf-populated federal lands.

Tom Kitchin/Tom Stack & Associates

Figure 9-B **Natural capital restoration:** After becoming almost extinct in much of the western United States, the gray wolf was listed and protected as an endangered species in 1974. Despite intense opposition from ranchers, hunters, miners, and loggers, 31 members of this keystone species were reintroduced to their former habitat in Yellowstone National Park in 1995 and 1996.

In 1995 and 1996, federal wildlife officials caught gray wolves in Canada and relocated 31 of them to Yellowstone National Park. Scientists estimate that the long-term carrying capacity of the park is 110 to 150 gray wolves. By the end of 2010, the park had an estimated 96–98 gray wolves, a 23% decline from 124 wolves in 2008. Reasons for the decline include strife between some of the packs, food stress, and mange (a contagious skin disease caused by parasitic mites).

For over a decade, wildlife ecologist Robert Crabtree and a number of other scientists have been studying the effects of reintroducing the gray wolf into Yellowstone National Park (see *The Habitable Planet*, Video 4, at **www.learner.org/resources/series209.html**). This research has suggested that the return of this keystone predator species has sent ecological ripples through the park's ecosystem. It has contributed to a decline in populations of elk, the wolves' primary food source, which had grown too large for the carrying capacity of much of the park. In turn, more leftovers of elk killed by wolves have been an important food source for scavengers such as bald eagles and ravens.

Also, wary elk are gathering less near streams and rivers, which has helped to spur the regrowth of trees in these areas. This, in turn, has helped to stabilize and shade stream banks, lowering the water temperature and making better habitat for trout. Beavers seeking willow and aspen for food and for lodge and dam building have returned, and the dams they build establish additional wetlands and create more favorable habitat for aspens.

The wolves have also cut in half the population of coyotes—the top predators in the absence of wolves. This has reduced coyote attacks on cattle from surrounding ranches and has increased populations of smaller animals such as ground squirrels, mice, and gophers, which are hunted by coyotes, eagles, and hawks. Overall, this experiment in ecosystem restoration has helped to reestablish and sustain some of the biodiversity that the Yellowstone ecosystem once had.

Critical Thinking

Do you approve or disapprove of the reintroduction of the gray wolf into Yellowstone National Park? Explain.

13% of the earth's land area was protected, either strictly or partially in about 130,000 nature reserves, parks, wildlife refuges, wilderness tracts, and other areas. This 13% figure is misleading because no more than 5% of the earth's land is strictly protected from potentially harmful human activities. In other words, *we have reserved 95% of the earth's land for human use* (see the map in Figure 2, pp. S24–25, in Supplement 6). Conservation biologists call for full protection of at least 20% of the earth's land area in a global system of biodiversity reserves that would include multiple examples of all the earth's biomes (**Concept 9-4**).

Most developers and resource extractors oppose protecting even 13% of the earth's remaining undisturbed ecosystems. They contend that these areas might contain valuable resources that would add to current economic growth. Ecologists and conservation biologists disagree. They view protected areas as islands of biodiversity and natural capital that help to sustain all life and economies indefinitely and that serve as centers of future evolution. In other words, they serve as an "ecological insurance policy" for us and other species. (See Norman Myers' Guest Essay on this topic at **www.cengagebrain.com**.)

Whenever possible, conservation biologists call for using the *buffer zone concept* to design and manage nature reserves. This means strictly protecting an inner core of a reserve, usually by establishing two surrounding buffer zones in which local people can extract resources sustainably without harming the inner core (see the Case Study that follows). Instead of shutting people out of the protected areas and likely creating enemies, this approach enlists local people as partners in protecting a reserve from unsustainable uses such as illegal logging and poaching. By 2010, the United Nations had used this principle to create a global network of 553 *biosphere reserves* in 109 countries. So far, most biosphere reserves fall short of these ideals and receive too little funding for their protection and management. GOOD NEWS

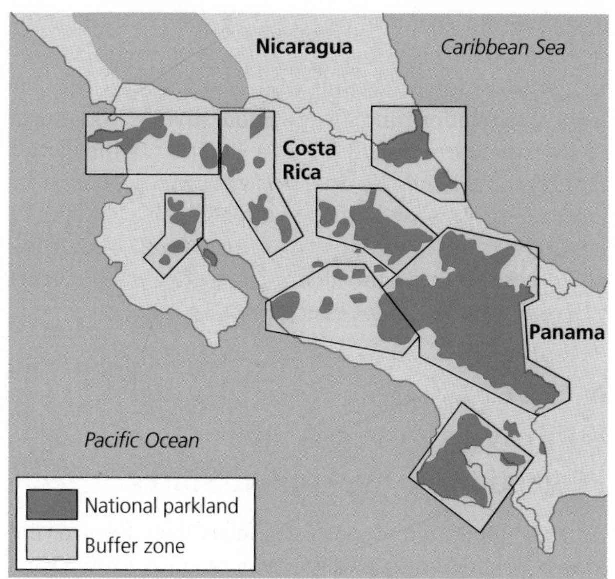

Figure 9-20 Solutions: Costa Rica has consolidated its parks and reserves into eight zoned *megareserves* designed to sustain about 80% of the country's rich biodiversity. Green areas are protected natural parklands and yellow areas are surrounding buffer zones, which can be used for sustainable forms of forestry, agriculture, hydropower, hunting, and other human activities.

■ **CASE STUDY**

Costa Rica—A Global Conservation Leader

Tropical forests once completely covered Central America's Costa Rica, which is smaller in area than the U.S. state of West Virginia and about one-tenth the size of France. Between 1963 and 1983, politically powerful ranching families cleared much of the country's forests to graze cattle.

Despite such widespread forest loss, tiny Costa Rica is a superpower of biodiversity, with an estimated 500,000 plant and animal species. A single park in Costa Rica is home to more bird species than are found in all of North America.

In the mid-1970s, Costa Rica established a system of nature reserves and national parks that, by 2010, included about a quarter of its land—6% of it reserved for indigenous peoples. Costa Rica now devotes a larger proportion of its land to biodiversity conservation than does any other country.

The country's parks and reserves are consolidated into eight zoned *megareserves* (Figure 9-20). Each reserve contains a protected inner core surrounded by two buffer zones that local and indigenous people can use for sustainable logging, crop farming, cattle grazing, hunting, fishing, and ecotourism.

Costa Rica's biodiversity conservation strategy has paid off. Today, the country's largest source of income is its $1-billion-a-year tourism industry, almost two-thirds of which involves ecotourism.

To reduce deforestation, the government has eliminated subsidies for converting forest to rangeland. It also pays landowners to maintain or restore tree cover. Between 2007 and 2008, the government planted nearly 14 million trees, which has helped to preserve the country's biodiversity, and it plans to plant 100 million trees by 2017. As they grow, the trees will remove carbon dioxide from the air and help Costa Rica to meet its goal of reducing net CO_2 emissions to zero by 2021.

The strategy has worked: Costa Rica has gone from having one of the world's highest deforestation rates to having one of the lowest. Between 1940 and 1987, forest cover in Costa Rica decreased from 75% to 21%. But by 2008, the country's forest cover had grown to more than 50%. GOOD NEWS

Protecting Wilderness Is an Important Way to Preserve Biodiversity

One way to protect undeveloped lands from human exploitation is to set them aside as **wilderness**—land officially designated as an area where natural communities have not been seriously disturbed by humans and where human activities are limited by law (**Concept 9-4**). Theodore Roosevelt, the first U.S. president to set aside protected areas, summarized what we should do with wilderness: "Leave it as it is. You cannot improve it."

Wilderness protection is not without controversy (see the Case Study that follows). Some critics oppose protecting large areas for their scenic and recreational value for a relatively small number of people. They believe this keeps some areas of the planet from being economically useful to people living today. Most conservation biologists disagree. To them the most important reasons for protecting wilderness and other areas from exploitation and degradation involve long-term needs—to *preserve biodiversity* as a vital part of the earth's natural capital and to *protect wilderness areas as centers for evolution* in response to mostly unpredictable changes in environmental conditions. In other words, protecting wilderness areas is equivalent to investing in a biodiversity insurance policy for everyone.

■ **CASE STUDY**

Controversy over Wilderness Protection in the United States

In the United States, conservationists have been trying to save wild areas from development since 1900. Overall, they have fought a losing battle. Not until 1964 did Congress pass the Wilderness Act. It allowed the government to protect undeveloped tracts of public land from development as part of the National Wilderness Preservation System.

The area of protected wilderness in the United States increased tenfold between 1970 and 2010. Even so, only about 4.7% of U.S. land is protected as wilderness—almost 75% of it in Alaska. Only about 2% of the land area of the lower 48 states is protected, most of it in the West, in 413 areas. However, only four of these areas in the lower 48 states are big enough to sustain all the species they contain.

In 2009, the U.S. government granted wilderness protection to over 800,000 hectares (2 million acres) of public land in 9 of the lower 48 states. GOOD NEWS It was the largest expansion of wilderness lands in 15 years. The new law also increased the total length of wild and scenic rivers (treated as wilderness areas) by 50%—the largest such increase ever.

9-5 What Is the Ecosystem Approach to Sustaining Biodiversity?

▶ **CONCEPT 9-5 We can help to sustain terrestrial biodiversity by identifying and protecting severely threatened areas (biodiversity hotspots), restoring damaged ecosystems (using restoration ecology), and sharing with other species much of the land we dominate (using reconciliation ecology).**

Here Are Four Ways to Protect Ecosystems

Most biologists and wildlife conservationists believe that the best way to keep from hastening the extinction of wild species through human activities is to protect threatened habitats and ecosystem services. This *ecosystems approach* would generally employ the following four-point plan:

1. Map the world's terrestrial ecosystems and create an inventory of the species contained in each of them and the natural services these ecosystems provide.
2. Locate and protect the most endangered ecosystems and species, with emphasis on protecting plant biodiversity and ecosystem services.
3. Seek to restore as many degraded ecosystems as possible.
4. Make development *biodiversity-friendly* by providing financial incentives (such as tax breaks) and technical help to private landowners who agree to help protect endangered ecosystems.

Protecting Global Biodiversity Hotspots Is an Urgent Priority

The earth's species are not evenly distributed. In fact, 17 megadiversity countries, most of them with large areas of tropical forest, contain more than two-thirds of all of the world's plant species.

To protect as much of the earth's remaining biodiversity as possible, some biodiversity scientists urge the adoption of an *emergency action* strategy to identify and quickly protect **biodiversity hotspots**—areas especially rich in plant species that are found nowhere else and are in great danger of extinction (**Concept 9-5**). These areas suffer serious ecological disruption, mostly because of rapid human population growth and the resulting pressure on natural resources. Environmental scientist Norman Myers first proposed this idea in 1988 (see his Guest Essay on this topic at **www.cengagebrain.com**).

Figure 9-21 shows 34 global, terrestrial biodiversity hotspots identified by biologists. Identified biodiversity hotspots cover only a little more than 2% of the earth's land surface, but they contain an estimated 50% of the

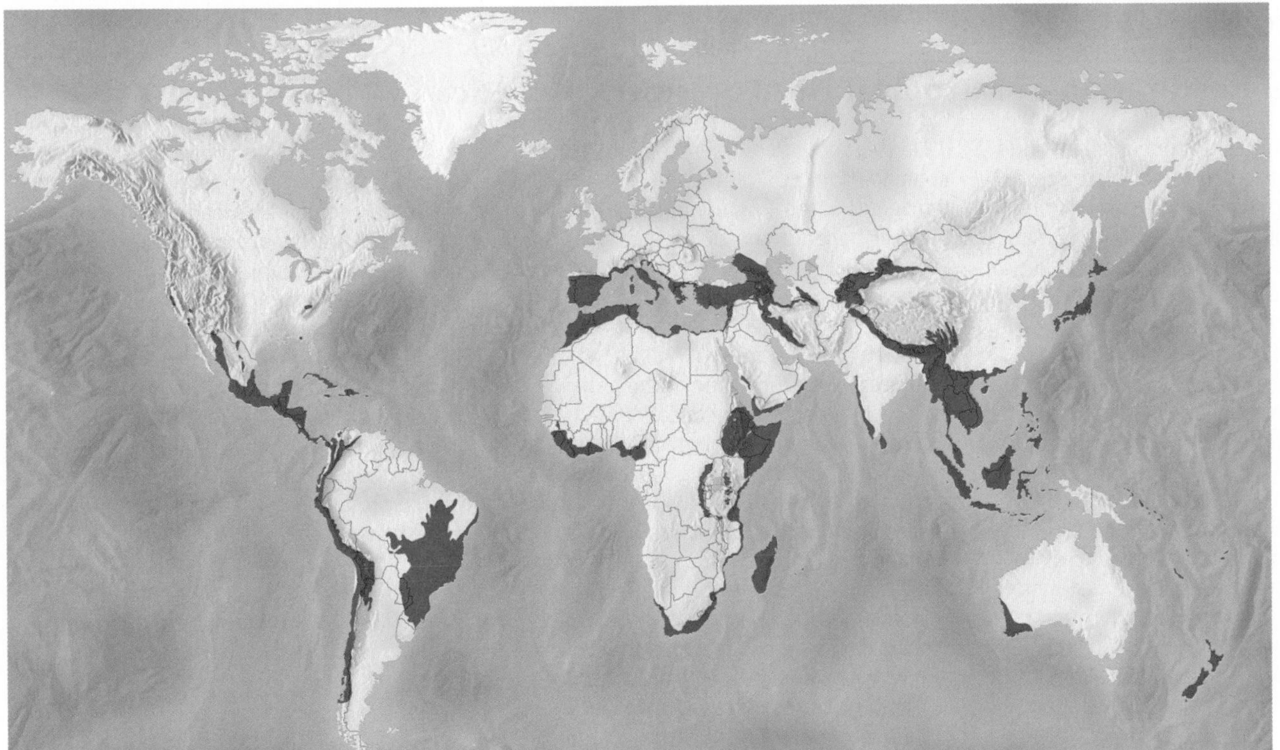

Active Figure 9-21 Endangered natural capital: This map shows 34 biodiversity hotspots identified by ecologists as important and endangered centers of terrestrial biodiversity that contain a large number of species found nowhere else. Identifying and saving these critical habitats requires a vital emergency response (**Concept 9-5**). Compare these areas with those on the global map of the human ecological footprint, as shown in Figure 2, pp. S24–25, in Supplement 6. *See an animation based on this figure at* www.cengagebrain.com. **Questions:** Are any of these hotspots near where you live? Is there a smaller, localized hotspot in the area where you live? (Data from Center for Applied Biodiversity Science at Conservation International)

world's flowering plant species and 42% of all terrestrial vertebrates (mammals, birds, reptiles, and amphibians). Yet the average proportion of these hotspots that is truly protected with government funding and law enforcement is only 5%.

Biodiversity hotspots are also home for a large majority of the world's endangered or critically endangered species and 1.2 billion people—one-fifth of the world's population. Says Norman Myers, "I can think of no other biodiversity initiative that could achieve so much at a comparatively small cost, as the hotspots strategy."

We Can Rehabilitate and Restore Ecosystems That We Have Damaged

Almost every natural place on the earth has been affected or degraded to some degree by human activities. We can at least partially reverse much of this harm through **ecological restoration**: the process of repairing damage caused by humans to the biodiversity and dynamics of natural ecosystems. Examples include replanting forests (**Core Case Study**), restoring CORE CASE STUDY grasslands, restoring coral reefs, restoring wetlands and stream banks (Figure 9-18, right), reintroducing native species (Science Focus, p. 190), removing invasive species, and freeing river flows by removing dams.

By studying how natural ecosystems recover, scientists are learning how to speed up repair operations using a variety of approaches, including the following four:

1. *Restoration:* returning a degraded habitat or ecosystem to a condition as similar as possible to its natural state.
2. *Rehabilitation:* turning a degraded ecosystem into a functional or useful ecosystem without trying to restore it to its original condition. Examples include removing pollutants and replanting to reduce soil erosion in abandoned mining sites and landfills, and in clear-cut forests.
3. *Replacement:* replacing a degraded ecosystem with another type of ecosystem. For example, a degraded forest could be replaced by a productive pasture or tree plantation.
4. *Creating artificial ecosystems:* for example, creating artificial wetlands to help reduce flooding or to treat sewage.

SCIENCE FOCUS

Ecological Restoration of a Tropical Dry Forest in Costa Rica

Costa Rica is the site of one of the world's largest *ecological restoration* projects. In the lowlands of its Guanacaste National Park, a small, tropical dry forest was burned, degraded, and fragmented for large-scale conversion of the area to cattle ranches and farms. Now it is being restored and reconnected to a rain forest on nearby mountain slopes. The goal is to eliminate damaging nonnative grasses and reestablish a tropical dry-forest ecosystem over the next 100–300 years.

Daniel Janzen, professor of biology at the University of Pennsylvania and a leader in the field of restoration ecology, helped to galvanize international support for this restoration project. He used his own MacArthur grant money to purchase this Costa Rican land for designation as a national park. He also raised more than $10 million for restoring the park.

Janzen recognizes that ecological restoration and protection of the park will fail unless the people in the surrounding area believe they will benefit from such efforts. His vision is to see that the nearly 40,000 people who live near the park play an essential role in the restoration of the degraded forest, a concept he calls *biocultural restoration*.

GOOD NEWS

By actively participating in the project, local residents reap educational, economic, and environmental benefits. Local farmers are paid to sow large areas with tree seeds and to plant tree seedlings started in Janzen's lab. Local grade school, high school, and university students and citizens' groups study the park's ecology during field trips. The park's location near the Pan American Highway makes it an ideal area for ecotourism, which stimulates the local economy.

The project also serves as a training ground in tropical forest restoration for scientists from all over the world. Research scientists working on the project give guest classroom lectures and lead field trips.

In a few decades, today's Costa Rican children will be running the park and the local political system. If they understand the ecological importance of their local environment, they will be more likely to protect and sustain its biological resources. Janzen believes that education, awareness, and involvement—not guards and fences—are the best ways to restore degraded ecosystems and to protect largely intact ecosystems from unsustainable use.

Critical Thinking

Would such an ecological restoration project be possible in the area where you live? Explain.

Researchers have suggested a science-based, four-step strategy for carrying out most forms of ecological restoration and rehabilitation.

First, identify the causes of the degradation (such as pollution, farming, overgrazing, mining, or invasive species).

Second, stop the abuse by eliminating or sharply reducing these factors. This would include removing toxic soil pollutants, improving depleted soil by adding nutrients and new topsoil, preventing fires, stopping overgrazing by moving grazing animals to another area, and controlling or eliminating disruptive nonnative species.

Third, if necessary, reintroduce key species to help restore natural ecological processes, as was done with gray wolves in the Yellowstone ecosystem (Science Focus, p. 190).

Fourth, protect the area from further degradation and allow secondary ecological succession to occur (Figure 9-18, right).

We Can Share Areas We Dominate with Other Species

Ecologist Michael L. Rosenzweig suggests that we develop a new form of conservation biology, called **reconciliation ecology**. This science focuses on inventing, establishing, and maintaining new habitats to conserve species diversity in places where people live, work, or play. In other words, we need to learn how to share with other species some of the spaces we dominate.

For example, people can learn how protecting local wildlife and ecosystems can provide economic resources for their communities by encouraging sustainable forms of ecotourism. In the Central American country of Belize, for example, conservation biologist Robert Horwich has helped to establish a local sanctuary for the black howler monkey. He convinced local farmers to set aside strips of forest to serve as habitats and corridors through which these monkeys can travel. The reserve, run by a local women's cooperative, has attracted ecotourists and biologists. The community has built a black howler museum, and local residents receive income by housing and guiding ecotourists and biological researchers. However, popular ecotourism sites can be degraded when they are overrun with visitors. In addition, developers erect hotels and other tourist facilities that can increase the degradation of natural features that attract tourists in the first place.

GOOD NEWS

In other parts of the world, people are learning how to protect vital insect pollinators such as native butterflies and bees, which are vulnerable to insecticides and habitat loss. Neighborhoods and municipal governments are doing this by agreeing to reduce or eliminate the use of pesticides on their lawns, fields, golf courses, and parks. Neighbors also work together to plant gardens of flowering plants as a source of food for pollinating insect species. Some neighborhoods and farmers have built devices out of wood and plastic that serve as hives for pollinating bees.

People have also worked together to protect bluebirds within human-dominated habitats where most of the bluebirds' nesting trees have been cut down and bluebird populations have declined. Special boxes were designed to accommodate nesting bluebirds, and the North American Bluebird Society has encouraged Canadians and Americans to use these boxes on their properties and to keep house cats away from nesting bluebirds. Now bluebird numbers are growing again.

There are many other examples of individuals and groups working together on projects to restore grasslands, wetlands, streams, and other degraded areas.

THINKING ABOUT
The Green Belt Movement and Reconciliation Ecology

List three ways in which the Green Belt Movement (**Core Case Study**) is a good example of reconciliation ecology at work.

Figure 9-22 lists some ways in which you can help to sustain the earth's terrestrial biodiversity.

What Can You Do?

Sustaining Terrestrial Biodiversity

- Adopt a forest
- Plant trees and take care of them
- Recycle paper and buy recycled paper products
- Buy sustainably produced wood and wood products
- Choose wood substitutes such as bamboo furniture and recycled plastic outdoor furniture, decking, and fencing
- Help to restore a nearby degraded forest or grassland
- Landscape your yard with a diversity of plants that are native to your area

Figure 9-22 **Individuals matter:** These are some ways in which you can help to sustain terrestrial biodiversity. **Questions:** Which two of these actions do you think are the most important? Why? Which of these things do you already do?

9-6 How Can We Help to Sustain Aquatic Biodiversity?

▶ **CONCEPT 9-6 We can help to sustain aquatic biodiversity by establishing protected sanctuaries, managing coastal development, reducing water pollution, and preventing overfishing.**

Human Activities Are Destroying and Degrading Aquatic Biodiversity

Human activities have destroyed or degraded a large portion of the world's coastal wetlands, coral reefs, mangroves, and ocean bottom, and disrupted many of the world's freshwater ecosystems. Scientists reported in 2006 that coastal habitats are disappearing at rates 2–10 times higher than the rate of tropical forest loss. During this century, rising sea levels, primarily caused by projected climate change, are likely to destroy many coral reefs (Figure 7-25, p. 143) and flood some low-lying islands along with their protective coastal mangrove forests (Figure 7-23, p. 142).

Another major threat is the loss and degradation of many sea-bottom habitats caused by dredging operations and trawler fishing boats. Trawlers drag huge nets, weighted down with heavy chains and steel plates, like submerged bulldozers over ocean bottoms to harvest a few species of bottom fish and shellfish (Figure 9-23, p. 196). Each year, thousands of trawler nets scrape and disturb an area of ocean floor many times larger than the total global area of forests that are clear-cut annually. Some areas of the ocean bottom are now partially protected by voluntary agreements among most nations to stop trawling there.

Habitat disruption is also a problem in freshwater aquatic zones. The main causes of disruption are dam building and excessive water withdrawal from rivers for irrigation and urban water supplies. These activities destroy aquatic habitats, degrade water flows, and disrupt freshwater biodiversity.

Another problem that threatens aquatic biodiversity is the deliberate or accidental introduction of hundreds of harmful invasive species (Figure 9-24, p. 196) into coastal waters, wetlands, and lakes throughout the world. According to the U.S. Fish and Wildlife Service, bioinvaders have caused about two-thirds of fish extinctions in the United States since 1900. In addition, they can disrupt and degrade whole aquatic ecosystems.

According to the International Union for the Conservation of Nature and Natural Resources (IUCN), 34% of the world's known marine fish species and 71% of the world's freshwater fish species face premature extinction within your lifetime.

Bottom line: We are putting tremendous increasing pressure on aquatic habitats and species at a time when we know very little about the great majority of such species and about how they interact with one another.

Peter J. Auster/National Undersea Research Center

Peter J. Auster/National Undersea Research Center

Figure 9-23 Natural capital degradation: These photos show an area of ocean bottom before (left) and after (right) a trawler net scraped it like a gigantic bulldozer. These ocean-floor communities could take decades or centuries to recover. According to marine scientist Elliot Norse, "Bottom trawling is probably the largest human-caused disturbance to the biosphere." Trawler fishers disagree and claim that ocean bottom life recovers after trawling. **Question:** What land activities are comparable to this?

Cigdem Sean Cooper/Shutterstock.com

Figure 9-24 The *common lionfish* has invaded the eastern coastal waters of North America, where it has few if any predators. One scientist described it as "an almost perfectly designed invasive species." It reaches sexual maturity rapidly, has large numbers of offspring, and is protected by venomous spines. Scientists believe it escaped from outdoor aquariums in Miami, Florida, that were damaged by Hurricane Andrew in 1992. Lionfish populations have exploded at the highest rate ever recorded by scientists in this part of the world. They compete with popular reef fish species like grouper and snapper, taking their food and eating their young.

Overfishing: Gone Fishing, Fish Gone

The human demand for seafood has been met historically through fisheries. A **fishery** is a concentration of a particular wild aquatic species suitable for commercial harvesting in a given ocean area or inland body of water. Today, fish are hunted throughout the world's oceans by a global fleet of about 4 million fishing boats. Modern industrial fishing (see the Case Study that follows) has caused as much as 80% depletion of some wild fish species in only 10–15 years.

Just to keep consuming seafood at our current rate, we will need 2.5 times the area of the earth's oceans, according to the *Fishprint of Nations 2006*, a study based on the concept of the human ecological footprint (Figure 1-8, p. 14). A **fishprint** is defined as the area of ocean needed to sustain the fish consumption of an average person, a nation, or the world. The study found that all fishing nations together are taking 57% more than the sustainable yield of the world's oceans. This means that we are harvesting more than half again as many wild fish as their populations can sustain in the long run.

As a result, according to the Woods Hole National Fisheries Service, 52% of the world's fisheries are fully exploited, 20% are moderately overexploited, and 28% are overexploited or depleted. Such overharvesting has led to the collapse of some of the world's major fisheries (Figure 9-25). Another effect of overfishing is that when larger predatory species dwindle, rapidly reproducing invasive species can more easily take over and disrupt ocean food webs.

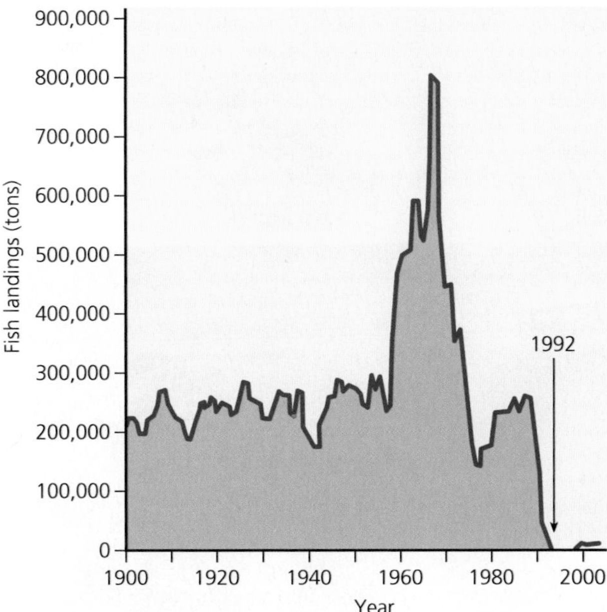

Figure 9-25 Natural capital degradation: This graph illustrates the collapse of Canada's 500-year-old Atlantic cod fishery off the coast of Newfoundland in the northwest Atlantic. Beginning in the late 1950s, fishers used bottom trawlers to capture more of the cod stock, reflected in the sharp rise in this graph. This resulted in extreme overexploitation of the fishery, which began a steady decline throughout the 1970s, followed by a slight recovery in the 1980s, and then total collapse by 1992, when the fishery was shut down. Despite a total ban on fishing, the cod population has not recovered. The fishery was reopened on a limited basis in 1998 but then closed indefinitely in 2003 and today shows no signs of recovery. (Data from Millennium Ecosystem Assessment)

CONNECTIONS

Domino Effect from the Collapse of the North Atlantic Cod Fishery

After the cod were fished out in the North Atlantic, fishers began harvesting sharks, which provide the important service of helping to control the populations of other species. Since then, overfishing of big sharks has cut Atlantic stocks of those species by 99%. With the large sharks essentially gone from the northwest Atlantic, populations of rays and skates, which the sharks once fed on, have exploded and have wiped out most of the region's bay scallops, which in turn had served as a food source for other species, including humans.

According to a 2003 study by conservation biologist Boris Worm and his colleagues, 90% or more of the large, predatory, open-ocean fishes such as tuna, swordfish, and marlin have disappeared since 1950. Worm projected that by 2048, almost all of the major commercial fish species would be driven to commercial extinction, but other scientists have objected to these claims.

As large species have become overfished, the fishing industry has begun working its way down marine food webs by shifting to smaller species. This has been referred to as "stealing the ocean's food supply," and it could unravel marine food webs and disrupt marine ecosystems and their ecosystem services.

■ CASE STUDY
Industrial Fish-Harvesting Methods

Industrial fishing fleets dominate the world's marine fishing industry. They use global satellite positioning equipment, sonar fish-finding devices, huge nets and long fishing lines, spotter planes, and gigantic refrigerated factory ships that can process and freeze their catches. These fleets help to supply the growing demand for seafood. But critics say that these highly efficient fleets are vacuuming the seas, decreasing marine biodiversity, and degrading important marine ecosystem services.

Figure 9-26 (p. 198) shows the major methods used for the commercial harvesting of various marine fishes and shellfish. Let us look more closely at a few of these methods.

Trawler fishing is used to catch fishes and shellfish—especially cod, flounder, shrimp, and scallops—that live on or near the ocean floor. It involves dragging a funnel-shaped net, held open at the neck, along the ocean bottom. The net is weighted down with chains or metal plates. This equipment scrapes up almost everything that lies on the ocean floor and often destroys bottom habitats—somewhat like clear-cutting the ocean floor (Figure 9-23, right). Newer trawling nets are large enough to swallow 12 jumbo jet planes and even larger nets are on the way.

Another method, *purse-seine fishing*, is used to catch surface-dwelling species such as tuna, mackerel, anchovies, and herring, which tend to feed in schools near the surface or in shallow areas. After a spotter plane locates a school, the fishing vessel encloses it with a large net called a purse seine. Nets that are used to capture yellowfin tuna in the eastern tropical Pacific Ocean have killed large numbers of dolphins that swim on the surface above schools of tuna.

Fishing vessels also use *long-lining*, which involves putting out lines up to 100 kilometers (60 miles) long, hung with thousands of baited hooks. The depth of the lines can be adjusted to catch open-ocean fish species such as swordfish, tuna, and sharks or ocean-bottom species such as halibut and cod. Longlines also hook and kill large numbers of endangered sea turtles, dolphins, and seabirds each year. Making simple modifications to fishing gear and fishing practices can decrease turtle, dolphin, and seabird deaths.

With *drift-net fishing*, fish are caught by huge drifting nets that can hang as deep as 15 meters (50 feet) below the surface and extend to 64 kilometers (40 miles) long. This method can lead to overfishing of the desired species and may trap and kill large quantities of unwanted fish, called *bycatch*, along with marine mammals, sea turtles, and seabirds. Almost one-third of the world's annual fish catch by weight consists of bycatch species, which are mostly thrown overboard dead or dying. This adds to the depletion of these species.

Since 1992, a U.N. ban on the use of drift nets longer than 2.5 kilometers (1.6 miles) in international waters

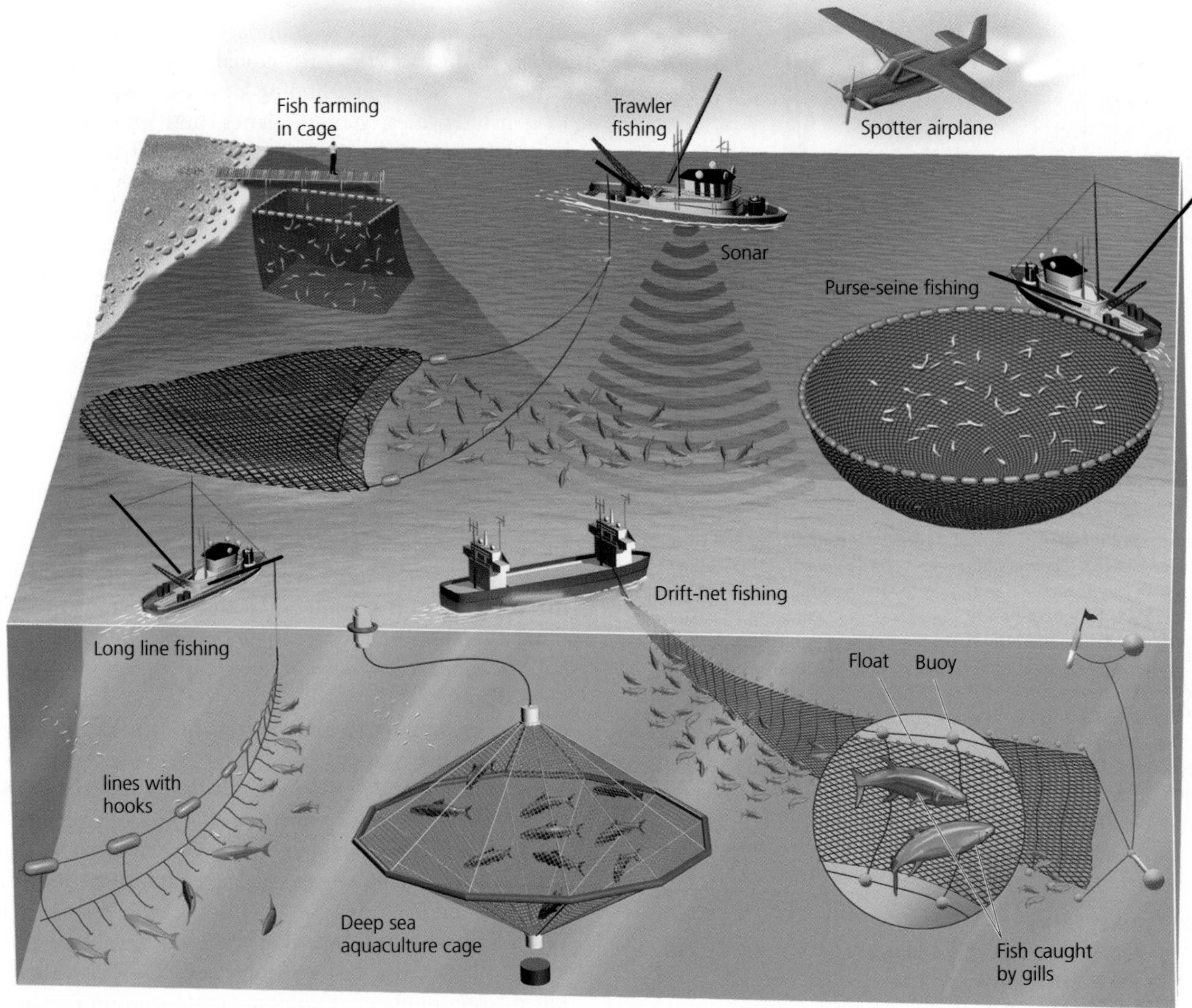

Figure 9-26 This diagram illustrates several major commercial fishing methods used to harvest various marine species (along with methods used to raise fish through aquaculture). These fishing methods have become so effective that many fish species have become commercially extinct.

has sharply reduced use of this technique. But longer nets continue to be used because compliance is voluntary and it is difficult to monitor fishing fleets over vast ocean areas. Also, the decrease in drift net use has led to increased use of longlines, which often have similar harmful effects on marine wildlife.

We Can Protect and Sustain Marine Biodiversity

Protecting marine biodiversity is difficult for several reasons. *First,* the human ecological footprint (see Figure 1-8, p. 14) and fishprint are expanding so rapidly that it is difficult to monitor their impacts. *Second,* much of the damage to the oceans and other bodies of water is not visible to most people. *Third,* many people incorrectly view the seas as an inexhaustible resource that can absorb an almost infinite amount of waste and pollution and still produce all the seafood we want. *Fourth,* most of the world's ocean area lies outside the legal jurisdiction of any country. Thus, much of it is an open-access resource, subject to overexploitation—a classic case of the tragedy of the commons (see Chapter 1, p. 13).

Nevertheless, there are several ways to protect and sustain marine biodiversity (**Concept 9-6**). For example, we can *protect endangered and threatened aquatic species,* as discussed in Chapter 8. And some individuals find economic rewards in restoring and sustaining streams, wetlands, and aquatic systems.

We can also *establish protected marine sanctuaries.* Since 1986, the IUCN has helped to establish a global system of *marine protected areas* (MPAs)—areas of ocean

INDIVIDUALS MATTER

Sylvia Earle—Champion of the Oceans

Sylvia Earle (Figure 9-C) is an oceanographer, explorer, author, and lecturer. For decades, she has been a global leader in publicizing the urgent need to increase our understanding of the global ocean that helps support all life and to protect much more of it from harmful human activities.

Currently, Earle is the Explorer-in-Residence at the National Geographic Society. She has lead more than 100 ocean research expeditions and has spent more that 7,000 hours underwater, either diving or descending in research submarines to study ocean life.

Earle's research has focused on the ecology and conservation of marine ecosystems, with an emphasis on developing deep-sea exploration technology. She is the author of more than 175 publications and has been a participant in numerous radio and television productions.

During her long career, Earle has also been the Chief Scientist of the U.S. National Oceanic and Atmospheric Administration (NOAA) and she has founded three companies devoted to developing submarines and other devices for deep-sea exploration and research. She has received more than 100 major international and national honors, including a place in the National Women's Hall of Fame.

Select Committee on Energy Independence and Global Warming

Figure 9-C Sylvia Earle is one of the world's most respected oceanographers. She has taken a leading role in helping us to understand the world's oceans and to protect them. *Time* magazine named her the first Hero for the Planet and the U.S. Library of Congress calls her a *living legend*.

These days, Earle is leading a campaign to ignite public support for a global network of *marine protected areas*, which she dubs "hope spots." Her goal is to help save and restore the oceans, which she calls "the blue heart of the planet." She says, "There is still time, but not a lot, to turn things around."

partially protected from human activities. There are more than 4,000 MPAs worldwide. However, nearly all MPAs allow dredging, trawler fishing, and other ecologically harmful resource extraction activities.

Many scientists and policymakers call for protecting whole marine ecosystems within a global network of fully protected *marine reserves*, some of which already exist. These areas are declared off-limits to destructive human activities in order to enable their ecosystems to recover and flourish. Some reserves could be made temporary or moveable to protect migrating species such as turtles.

GOOD NEWS

Marine reserves work and they work quickly. Scientific studies show that within fully protected marine reserves, commercially valuable fish populations double, fish size grows by almost a third, fish reproduction triples, and species diversity increases by almost one-fourth. Furthermore, these improvements happen within 2–4 years after strict protection begins.

Despite the importance of such protection, only about 0.8% of the world's oceans is fully protected, compared to 5% of the world's land. In other words, 99.2% of the world's oceans are not effectively protected from harmful human activities. Many marine scientists, including Sylvia Earle (Individuals Matter, above) argue that in order to sustain marine biodiversity, we need to set aside at least 30% of the world's oceans as fully protected marine reserves.

THINKING ABOUT
Marine Reserves

Do you support setting aside at least 30% of the world's oceans as fully protected marine reserves? Explain. How would this affect your life?

Figure 9-27 (p. 200) lists several ways to manage global fisheries more sustainably.

Taking an Ecosystem Approach to Sustaining Aquatic Biodiversity

Edward O. Wilson (Individuals Matter, p. 65) and other biodiversity experts have promoted an ecosystem approach to sustaining terrestrial biodiversity. Wilson has proposed the following strategies for applying this approach to aquatic biodiversity:

- Complete the mapping of the world's aquatic biodiversity, identifying and locating as many plant and animal species as possible.
- Identify and preserve the world's aquatic biodiversity hotspots and areas where deteriorating ecosystem services threaten people and other forms of life.
- Create large and fully protected marine reserves to allow damaged marine ecosystems to recover and to allow fish stocks to be replenished.

Solutions

Managing Fisheries

Fishery Regulations

Set low catch limits

Improve monitoring and enforcement

Economic Approaches

Reduce or eliminate fishing subsidies

Certify sustainable fisheries

Protect Areas

Establish no-fishing areas

Establish more marine protected areas

Consumer Information

Label sustainably harvested fish

Publicize overfished and threatened species

Bycatch

Use nets that allow escape of smaller fish

Use net escape devices for seabirds and sea turtles

Aquaculture

Restrict coastal locations of fish farms

Improve pollution control

Nonnative Invasions

Kill or filter organisms from ship ballast water

Dump ballast water at sea and replace with deep-sea water

Figure 9-27 There are a number of ways to manage fisheries more sustainably and protect marine biodiversity. **Questions:** Which four of these solutions do you think are the most important? Why?

- Protect and restore the world's lakes and river systems, which are the most threatened ecosystems of all.
- Initiate worldwide ecological restoration projects in systems such as coral reefs and inland and coastal wetlands.
- Find ways to raise the incomes of people who live in or near protected lands and waters so that they can become partners in the protection and sustainable use of ecosystems.

There is growing evidence that the current harmful effects of human activities on aquatic biodiversity and ecosystem services could be reversed over the next 2 decades. Doing this will require implementing an ecosystem approach to sustaining both terrestrial and aquatic ecosystems. According to Edward O. Wilson, such a conservation strategy would cost about $30 billion per year—an amount that could be provided by a tax of one penny per cup of coffee consumed in the world each year.

GOOD NEWS

This strategy for protecting the earth's vital biodiversity will not be implemented without bottom-up political pressure on elected officials from individual citizens and groups. It will also require cooperation among scientists, engineers, and key people in government and the private sector. And it will be important for individuals to "vote with their wallets" by trying to buy only products and services that do not have harmful impacts on terrestrial and aquatic biodiversity.

Here are this chapter's *three big ideas*:

- The economic values of the important ecological services provided by the world's ecosystems are far greater than the value of raw materials obtained from those systems.
- We can sustain terrestrial biodiversity by protecting severely threatened areas, protecting remaining undisturbed areas, restoring damaged ecosystems, and sharing with other species much of the land we dominate.
- We can sustain aquatic biodiversity by establishing protected sanctuaries, managing coastal development, reducing water pollution, and preventing overfishing.

REVISITING The Green Belt Movement and Sustainability

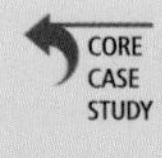

In this chapter, we looked at how humans are destroying or degrading terrestrial biodiversity in a variety of ecosystems. We also saw how we can reduce this destruction and degradation by using the earth's resources more sustainably. The **Core Case Study** showed us the importance of simply planting trees, and we learned the importance of protecting species and ecosystems in nature reserves such as parks and wilderness areas.

We also learned about the importance of preserving what remains of richly diverse and highly endangered ecosystems (biodiversity hotspots). We examined the key strategy of restoring or rehabilitating some of the ecosystems we have degraded (restoration ecology). In addition, we explored ways in which people can share with other species some of the land they occupy in order to help sustain biodiversity (reconciliation ecology).

Preserving terrestrial and aquatic biodiversity involves applying the three **principles of sustainability** (see back cover). First, it means respecting biodiversity and understanding the value of sustaining it. Then, in helping to sustain biodiversity by planting trees for example, we also help to restore and preserve the flows of energy from the sun through food webs and the cycling of nutrients within ecosystems. Also, if we rely less on fossil fuels and more on direct solar energy and its indirect forms, such as wind and flowing water, we will generate less pollution and interfere less with natural chemical cycling and other forms of natural capital that sustain biodiversity along with our own lives and societies.

We abuse land because we regard it as a commodity belonging to us. When we see land as a community to which we belong, we may begin to use it with love and respect.

ALDO LEOPOLD

REVIEW

CORE CASE STUDY

1. Describe the Green Belt Movement founded by Wangari Maathai (**Core Case Study**). CORE CASE STUDY

SECTION 9-1

2. What is the key concept for this section? Distinguish among an **old-growth (primary) forest**, a **second-growth forest**, and a **tree plantation** (**tree farm** or **commercial forest**). What major ecological and economic benefits do forests provide? Describe the efforts of scientists and economists to put a price tag on the major ecological services provided by forests and other ecosystems.

3. Describe the harm caused by building roads into previously inaccessible forests. Distinguish among selective cutting, clear-cutting, and strip cutting in the harvesting of trees. What are two types of forest fires? What are some ecological benefits of occasional surface fires? What are four ways to reduce the harmful impacts of diseases and insects on forests? What effects might projected climate change have on forests?

4. What is **deforestation** and what parts of the world are experiencing the greatest forest losses? List some major harmful environmental effects of deforestation. Summarize the encouraging news about deforestation in the United States. How serious is tropical deforestation? What are the major underlying and direct causes of tropical deforestation?

SECTION 9-2

5. What is the key concept for this section? Describe four ways to manage forests more sustainably. What is certified timber? What are four ways to reduce the harm caused by forest fires to forests and to people? What is a prescribed fire? What are three ways to reduce the need to harvest trees? What are five ways to protect tropical forests and use them more sustainably?

SECTION 9-3

6. What is the key concept for this section? Distinguish between **rangelands** and **pastures**. What is **overgrazing** and what are its harmful environmental effects? What are three ways to reduce overgrazing and use rangelands more sustainably?

SECTION 9-4

7. What is the key concept for this section? What major environmental threats affect national parks in the world and in the United States? Why are many U.S. national parks considered to be threatened islands of biodiversity? Describe some of the ecological effects of reintroducing the gray wolf to Yellowstone National Park in the United States. What percentage of the world's land has been set aside and protected as nature reserves, and what percentage do conservation biologists believe should be protected?

8. How should nature reserves be designed and managed? Describe what Costa Rica has done to establish nature reserves. What is **wilderness** and why is it important? Describe the controversy over protecting wilderness in the United States.

SECTION 9-5

9. What is the key concept for this section? Describe a four-point strategy for protecting ecosystems. What is a **biodiversity hotspot** and why is it important to protect such areas? About how much of the earth's land surface is occupied by hotspots and what percentages of the world's flowering plants and terrestrial vertebrates live in these areas? What is **ecological restoration**? Summarize a science-based, four-point strategy for carrying out ecological restoration and rehabilitation. Describe the ecological restoration of a tropical dry forest in Costa Rica. Define and give three examples of **reconciliation ecology**.

SECTION 9-6

10. What is the key concept for this section? Summarize the threats to aquatic biodiversity resulting from human activities. Define **fishery** and **fishprint** and summarize the threats to marine fisheries. Describe three industrial fish harvesting methods. Why is it difficult to protect marine biodiversity? What are three ways in which we could protect more marine biodiversity? How can the ecosystem approach be applied to protecting aquatic biodiversity? What are this chapter's *three big ideas*? Describe the relationship between preserving biodiversity as it is done by the Green Belt Movement and the three scientific **principles of sustainability**. SUSTAINABILITY

Note: Key terms are in **bold** type.

CRITICAL THINKING

1. Describe some ecological, economic, and social benefits of the Green Belt Movement (**Core Case Study**). Is there an area near where you live that could benefit from such intensive tree planting? If so, describe how it would benefit the area. CORE CASE STUDY

2. Suppose we fail to protect a much larger percentage of the world's remaining old-growth forests and tropical rain forests. Describe three harmful effects that such failure would be likely to have on any children and grandchildren you might have.

3. In the early 1990s, Miguel Sanchez, a subsistence farmer in Costa Rica, was offered $600,000 by a hotel developer for a piece of land that he and his family had been using sustainably for many years. The land, which contained an old-growth rain forest and a black sand beach, was surrounded by an area under rapid development. Sanchez refused the offer. What would you have done if you were in Miguel Sanchez's position? Explain your decision.

4. In 2009, environmental analyst Lester R. Brown estimated that reforesting the earth and restoring the earth's degraded rangelands would cost about $15 billion a year. Suppose the United States, the world's most affluent country, agreed to put up half of this money, at an average annual cost of $25 per American citizen. Would you support doing this? Explain. What other part or parts of the federal budget would you decrease to come up with these funds?

5. Are you in favor of establishing more wilderness areas in the United States, especially in the lower 48 states (or in the country where you live)? Explain. What might be some drawbacks of doing this?

6. What do you think are the three greatest threats to aquatic biodiversity and aquatic ecosystem services? Explain your selections. Imagine that you are a national official in charge of setting policy for preserving aquatic biodiversity and outline a plan for dealing specifically with these threats.

7. You are a defense attorney arguing in court for preserving a coastal wetlands area to prevent it from being developed. Give your three strongest arguments for preservation of this ecosystem. Assume that there is a coral reef offshore from this wetland, and include that fact in your arguments.

8. Congratulations! You are in charge of the world. List the three most important features of your policies for using and managing the world's **(a)** forests, **(b)** grasslands, **(c)** nature reserves such as parks and wildlife refuges, **(d)** biological hotspots, **(e)** marine aquatic systems, and **(f)** freshwater aquatic systems.

DOING ENVIRONMENTAL SCIENCE

1. Visit at least two of the following areas: **(a)** a diverse old-growth forest, **(b)** a forest area that has been recently clear-cut, and **(c)** a forest area that was clear-cut 5–10 years ago. Compare the biodiversity, soil erosion, and signs of rapid water runoff in each of the three areas. Make notes and draw some conclusions about the effects of human activities in these areas.

2. Survey the condition of a nearby wetland, coastal area, river, or stream and research its history. Has its condition improved or deteriorated during the last 10 years? What governmental or private efforts are being used to protect this aquatic system? Based on your ecological knowledge of this system, make recommendations to policy makers for protecting it.

GLOBAL ENVIRONMENT WATCH EXERCISE

Search for *Forests and Deforestation* and use the topic portal to find out the following: **(a)** whether overall tropical deforestation around the world is increasing or decreasing; **(b)** the three countries with the highest rates of deforestation; **(c)** the main causes for this deforestation, and **(d)** three countries where forests are actually growing back (where there is a net gain in forest cover when clearing of forests and regrowth of forests is considered).

ECOLOGICAL FOOTPRINT ANALYSIS

Use the table below to answer the questions that follow.

Country	Area of tropical rain forest (square kilometers)	Area of deforestation per year (square kilometers)	Annual rate of tropical forest loss
A	1,800,000	50,000	
B	55,000	3,000	
C	22,000	6,000	
D	530,000	12,000	
E	80,000	700	

1. What is the annual rate of tropical rain forest loss, as a percentage of total forest area, in each of the five countries? Answer by filling in the blank column in the table.
2. What is the annual rate of tropical deforestation collectively in all of the countries represented in the table?
3. According to the table, and assuming the rates of deforestation remain constant, which country's tropical rain forest will be completely destroyed first?
4. Assuming the rate of deforestation in country C remains constant, how many years will it take for all of its tropical rain forests to be destroyed?
5. Assuming that a hectare (1.0 hectare = 0.01 square kilometer) of tropical rain forest absorbs 0.85 metric tons (1 metric ton = 2,200 pounds) of carbon dioxide per year, what would be the total annual growth in the carbon footprint (carbon emitted but not absorbed by vegetation because of deforestation) in metric tons of carbon dioxide per year for each of the five countries in the table?

LEARNING ONLINE

The website for *Environmental Science 14e* contains helpful study tools and other resources to take the learning experience beyond the textbook. Examples include flashcards, practice quizzing, information on green careers, tips for more sustainable living, activities, and a wide range of articles and further readings. To access these materials and other companion resources for this text, please visit **www.cengagebrain.com**. For further details, see the preface, page xvi.

10 Food, Soil, and Pest Management

CORE CASE STUDY Organic Agriculture Is on the Rise

We face the critical challenges of increasing food production for the world's growing human population without causing serious environmental harm. Each day, there are about 227,000 more people at the dinner table and by 2050 there will likely be 2.6 billion more people to feed. This population increase is more than twice China's current population and more than eight times the current U.S. population. Most of these people will live in huge cities in countries where soils are eroding, irrigation wells are going dry, and good cropland is increasingly scarce.

Sustainability experts call for us to develop and phase in more sustainable ways to produce food over the next few decades. One component of more sustainable agriculture is **organic agriculture**, in which crops are grown with the use of environmentally sound and sustainable methods and without the use of synthetic pesticides, synthetic inorganic fertilizers, or genetically engineered plants or animals. To be classified as *organically grown*, animals must be raised on 100% organic feed without the use of antibiotics or growth hormones. Figure 10-1 compares organic agriculture with industrialized agriculture.

In the United States, a label of *100 percent organic* means that a product is produced only by organic methods and contains all organic ingredients. Products labeled *organic* must contain at least 95% organic ingredients and those labeled *made with organic ingredients* must contain at least 70% organic ingredients. The word *natural* is used on food labels primarily as an advertising ploy and carries no requirement for organic content.

The global market for certified organic food has been growing rapidly. Even so, certified organic farming is practiced on less than 1% of the world's cropland and only 0.6% of U.S. cropland, and organic food accounts for less than 4% of U.S. food sales.

More than two decades of research indicates that organic farming has a number of environmental advantages over conventional industrialized farming. A major advantage is that 100% certified organic food does not contain the pesticide residues that can be found in lower-grade organic and nonorganic foods. On the other hand, conventional agriculture usually can produce higher yields of crops on smaller areas of land than organic agriculture can. Another drawback is that most organically grown food costs 10–100% more than conventionally produced food (depending on specific items), primarily because organic farming is more labor intensive.

In this chapter, we look at different ways to produce food, the environmental effects of food production, and how to produce food more sustainably.

Industrialized Agriculture

Uses synthetic inorganic fertilizers and sewage sludge to supply plant nutrients

Makes use of synthetic chemical pesticides

Uses conventional and genetically modified seeds

Depends on nonrenewable fossil fuels (mostly oil and natural gas)

Produces significant air and water pollution and greenhouse gases

Is globally export-oriented

Uses antibiotics and growth hormones to produce meat and meat products

Organic Agriculture

Emphasizes prevention of soil erosion and the use of organic fertilizers such as animal manure and compost, but no sewage sludge to supply plant nutrients

Employs crop rotation and biological pest control

Uses no genetically modified seeds

Reduces fossil fuel use and increases use of renewable energy such as solar and wind power for generating electricity

Produces less air and water pollution and greenhouse gases

Is regionally and locally oriented

Uses no antibiotics or growth hormones to produce meat and meat products

Figure 10-1 There are major differences between industrialized agriculture and organic agriculture.

Key Questions and Concepts

10-1 What is food security and why is it difficult to attain?

CONCEPT 10-1A Many people in less-developed countries have health problems from not getting enough food, while many people in more-developed countries suffer health problems from eating too much.

CONCEPT 10-1B The greatest obstacles to providing enough food for everyone are poverty, corruption, political upheaval, war, bad weather, and the harmful environmental effects of industrialized food production.

10-2 How is food produced?

CONCEPT 10-2 High-input industrialized agriculture and lower-input traditional agriculture have greatly increased food supplies.

10-3 What environmental problems arise from industrialized food production?

CONCEPT 10-3 Future food production may be limited by soil erosion and degradation, desertification, irrigation water shortages, air and water pollution, climate change from greenhouse gas emissions, and loss of biodiversity.

10-4 How can we protect crops from pests more sustainably?

CONCEPT 10-4 We can sharply cut pesticide use without decreasing crop yields by using a mix of cultivation techniques, biological pest controls, and small amounts of selected chemical pesticides as a last resort (integrated pest management).

10-5 How can we improve food security?

CONCEPT 10-5 We can improve food security by reducing poverty and chronic malnutrition, relying more on locally grown food, and cutting food waste.

10-6 How can we produce food more sustainably?

CONCEPT 10-6 We can produce food more sustainably by using resources more efficiently, sharply decreasing the harmful environmental effects of industrialized food production, and eliminating government subsidies that promote such harmful impacts.

Note: Supplements 2 (p. S3), 4 (p. S10), 5 (p. S18), and 6 (p. S22) can be used with this chapter.

There are two spiritual dangers in not owning a farm.
One is the danger of supposing that breakfast comes from the grocery,
and the other that heat comes from the furnace.

ALDO LEOPOLD

10-1 What Is Food Security and Why Is It Difficult to Attain?

▶ **CONCEPT 10-1A Many people in less-developed countries have health problems from not getting enough food, while many people in more-developed countries suffer health problems from eating too much.**

▶ **CONCEPT 10-1B The greatest obstacles to providing enough food for everyone are poverty, corruption, political upheaval, war, bad weather, and the harmful environmental effects of industrialized food production.**

Many People Suffer from Chronic Hunger and Malnutrition

In a country that enjoys **food security**, all or most of the people in the country have daily access to enough nutritious food to live active and healthy lives. GOOD NEWS Today, we produce more than enough food to meet the basic nutritional needs of every person on the earth. But even with this food surplus, one of every six people in less-developed countries is not getting enough to eat. These people face **food insecurity**—living with chronic hunger and poor nutrition, which threatens their ability to lead healthy and productive lives (**Concept 10-1A**).

Most agricultural experts agree that *the root cause of food insecurity is poverty,* which prevents poor people from growing or buying enough food. This is not surprising given that about half of the world's people are

Harmut Schwartzbach/Peter Arnold, Inc.

Figure 10-2 These starving children were collecting ants to eat in famine-stricken Sudan, Africa, where a civil war lasted from 1983 until 2005. The effects of the war are still being felt, and prolonged drought has made the problem worse.

trying to survive on the equivalent of $2.25 a day and one out of six people on $1.25 a day. Other obstacles to food security are political upheaval, war (Figure 10-2), corruption, and bad weather, including prolonged drought, flooding, and heat waves (**Concept 10-1B**).

To maintain good health and resist disease, individuals need fairly large amounts of *macronutrients* (such as carbohydrates, proteins, and fats; see Table 10-1 and Figures 8, 9, and 12, pp. S14–S15, in Supplement 4), and smaller amounts of *micronutrients*—vitamins, such as A, B, C, and E, and minerals such as iron, iodine, and calcium.

People who cannot grow or buy enough food to meet their basic energy needs suffer from **chronic undernutrition**, or **hunger** (**Concept 10-1A**). Many of the world's poor can afford to live only on a low-protein, high-carbohydrate, vegetarian diet consisting mainly of grains such as wheat, rice, or corn. They often suffer from **chronic malnutrition**, a condition in which they do not get enough protein and other key nutrients. This weakens them, makes them more vulnerable to disease, and hinders the normal physical and mental development of children.

According to the UN Food and Agriculture Organization (FAO) in 2010, there were about 925 million chronically undernourished and malnourished people—nearly one of every seven people on the planet. About 40% of them live in China and India (See Figure 11, p. S30, in Supplement 6 for a map of the countries with the most undernourished people).

The FAO estimates that each year, nearly 6 million children younger than age 5—on average, 16,400 per day—die prematurely from chronic hunger and malnutrition and from increased susceptibility to normally nonfatal infectious diseases (such as measles and diarrhea) because of their weakened condition. How many children died from such poverty-related causes during your lunch hour?

Table 10-1 Key Nutrients for a Healthy Human Life

Nutrient	Food Source	Function
Proteins	Animals and some plants	Help to build and repair body tissues
Carbohydrates	Wheat, corn, and rice	Provide short-term energy
Lipids (oils and fats)	Animal fats, nuts, oils	Help to build membrane tissues and create hormones

Many People Do Not Get Enough Vitamins and Minerals

Many people, most of them in less-developed countries, suffer from a deficiency of one or more vitamins and minerals, usually *vitamin A, iron,* and *iodine* (**Concept 10-1A**). Some 250,000–500,000 children younger than age 6 go blind each year from a lack of vitamin A, and within a year, more than half of them die.

Having too little *iron* (Fe)—a component of the hemoglobin that transports oxygen in the blood—causes *anemia*. It causes fatigue, makes infection more likely, and increases a woman's chances of dying from hemorrhage in childbirth. According to the World Health Organization (WHO), one of every five people in the world—mostly women and children in less-developed countries—suffers from iron deficiency.

The chemical element *iodine* (I) is essential for proper functioning of the thyroid gland, which produces hormones that control the body's rate of metabolism. Chronic lack of iodine can cause stunted growth, mental retardation, and goiter—a swollen thyroid gland that can lead to deafness (Figure 10-3). Almost 33% of the world's people do not get enough iodine in their

John Paul Kay/Peter Arnold, Inc.

Figure 10-3 This woman in Bangladesh has a goiter, an enlargement of the thyroid gland caused by a diet containing too little iodine.

food and water. According to the United Nations, some 600 million people (almost twice the current U.S. population) suffer from goiter. And 26 million children suffer irreversible brain damage each year from lack of iodine. According to the FAO and the WHO, eliminating this serious health problem by adding traces of iodine to salt would cost the equivalent of only 2–3 cents per year for every person in the world.

Many People Have Health Problems from Eating Too Much

Overnutrition occurs when food energy intake exceeds energy use and causes excess body fat. Too many calories, too little exercise, or both can cause overnutrition. People who are underfed and underweight, as well as those who are overfed and overweight, face similar health problems: *lower life expectancy, greater susceptibility to disease and illness,* and *lower productivity and life quality* (**Concept 10-1A**).

We live in a world where about 925 million people face health problems because they do not get enough nutritious food to eat and another 1.1 billion people have health problems because they eat too much. According to a 2009 study by U.S. Centers for Disease Control and Prevention (CDC), 68% of American adults older than age 20 are overweight, and half of those people are obese (at least 20% over their ideal weight). This plays an important role in four of the top ten causes of death in the United States—heart disease, stroke, Type 2 diabetes, and some forms of cancer.

10-2 How Is Food Produced?

▶ **CONCEPT 10-2 High-input industrialized agriculture and lower-input traditional agriculture have greatly increased food supplies.**

Food Production Has Increased Dramatically

About 10,000 years ago, humans began shifting from hunting and gathering their food to growing edible plants in nutrient-rich topsoil and raising animals for food and labor. Today, three systems supply most of our food. *Croplands* produce mostly grains, primarily rice, wheat, and corn; *rangelands, pastures,* and *feedlots* produce meat and meat products; and *fisheries* and *aquaculture* (fish farming) supply fish and shellfish.

These three systems depend on a small number of plant and animal species. About 66% of the world's people survive primarily by eating three grain crops—*rice, wheat,* and *corn.* Only a few species of mammals and fish provide most of the world's meat and seafood.

Since 1960, there has been a staggering increase in global food production from all three of the major food production systems (**Concept 10-2**). GOOD NEWS This occurred because of technological advances such as increased use of tractors and other farm machinery, and high-tech fishing equipment. Another major advance has been the development of **irrigation**, a mix of methods by which water is supplied to crops by artificial means. Other technological developments include the manufacturing of inorganic chemical fertilizers and pesticides, the development of high-yield grain varieties, and industrialized production of large numbers of livestock and fish.

Today there is a growing demand for food because of population growth, rising affluence that increases the demand for more costly meat and meat products, and the increasing use of grains such as corn to produce biofuels for cars instead of food for people.

Industrialized Crop Production Relies on High-Input Monocultures

Crops are raised by means of two types of agriculture: industrialized agriculture and subsistence agriculture. **Industrialized agriculture**, or **high-input agriculture**, uses heavy equipment and large amounts of financial capital, fossil fuels, water, commercial inorganic fertilizers, and pesticides to produce single crops, or *monocultures*. The major goal of industrialized agriculture is to steadily increase each crop's *yield*—the amount of food produced per unit of land. Industrialized agriculture is practiced on 25% of all cropland, mostly in more-developed countries, and produces about 80% of the world's food (**Concept 10-2**).

Plantation agriculture is a form of industrialized agriculture used primarily in tropical less-developed countries. It involves growing *cash crops* such as bananas, soybeans (mostly to feed livestock), sugarcane (to produce sugar and ethanol fuel), coffee, palm oil (Figure 10-4, p. 208, used as a cooking oil and to produce biodiesel fuel), and vegetables on large monoculture plantations, mostly for export to more-developed countries.

Modern industrialized agriculture produces large amounts of food, but is it sustainable? A growing

age fotostock/SuperStock

Figure 10-4 These large plantations of oil palms on the island of Borneo in Malaysia were planted in an area that was once covered with tropical rain forest.

number of analysts say it is not, because it violates the three **principles of sustainability**. It relies heavily on nonrenewable fossil fuels, does not rely on a diversity of crops as a form of ecological insurance, and neglects the conservation and recycling of nutrients in topsoil. To many analysts, our economic systems promote unsustainable forms of industrialized agriculture, partly because they do not include most of the harmful environmental and health costs of such food production in the market prices of food.

Traditional Agriculture Often Relies on Low-Input Polycultures

Traditional agriculture provides about 20% of the world's food crops on about 75% of its cultivated land, mostly in less-developed countries.

There are two main types of traditional agriculture. **Traditional subsistence agriculture** supplements energy from the sun with the labor of humans and draft animals to produce enough crops for a farm family's survival, with little left over to sell or store as a reserve for hard times. In **traditional intensive agriculture**, farmers increase their inputs of human and draft-animal labor, animal manure for fertilizer, and water to obtain higher crop yields. If the weather cooperates, farmers can produce enough food to feed their families and to sell some for income.

Some traditional farmers focus on cultivating a single crop, but many grow several crops on the same plot simultaneously, a practice known as **polyculture**. Such crop diversity—an example of implementing the biodiversity **principle of sustainability** (see back cover)—reduces the chance of losing most or all of the year's food supply to pests, bad weather, and other misfortunes.

In parts of South America and Africa, some traditional farmers grow as many as 20 different crops together on small cleared plots in tropical forests. The crops rely on sunshine and natural fertilizers such as animal manure for their growth, and they mature at different times. They provide food throughout the year and keep the topsoil covered to reduce erosion from wind and water. Polyculture lessens the need for fertilizer and water, because root systems at different depths in the soil capture nutrients and moisture efficiently. Synthetic insecticides and herbicides are rarely needed because multiple habitats are created for natural predators of crop-eating insects, and weeds have trouble competing with the multitude and density of crop plants.

GOOD NEWS

Research shows that, on average, such low-input polyculture produces higher yields than does high-input monoculture. For example, ecologists Peter Reich and David Tilman found that carefully controlled polyculture plots with 16 different species of plants consistently out-produced plots with 9, 4, or only 1 type of plant species. Therefore, some analysts argue for greatly increased use of polyculture to produce food more sustainably.

All types of conventional crop production depend on having fertile topsoil (Science Focus, at right).

A Closer Look at Industrialized Crop Production

Farmers have two ways to produce more food: farming more land or getting higher yields from existing cropland. Since 1950, about 88% of the increase in global food production has come from using high-input industrialized agriculture (Figure 10-1, left) to increase crop yields in a process called the **green revolution**.

A green revolution involves three steps. *First,* develop and plant monocultures of selectively bred or genetically engineered high-yield varieties of key crops such as rice, wheat, and corn. *Second,* produce high yields by using large inputs of water and synthetic inorganic fertilizers, and pesticides. *Third,* increase the number of crops grown per year on a plot of land through *multiple cropping*. Between 1950 and 1970, in what was called the *first green revolution,* this high-input approach dramatically increased crop yields in most of the world's more-developed countries, especially the United States.

A *second green revolution* has been taking place since 1967. Fast-growing dwarf varieties of rice and wheat, specially bred for tropical and subtropical climates, have been introduced into middle-income, less-developed countries such as India, China, and Brazil. Producing more food on less land in such countries has helped to protect some biodiversity by preserving large areas of forests, grasslands, wetlands, and easily eroded mountain terrain that might otherwise be used for farming.

SCIENCE FOCUS

Soil Is the Foundation of Life on Land

Soil is a complex mixture of eroded rock, mineral nutrients, decaying organic matter, water, air, and billions of living organisms, most of them microscopic decomposers. Soil formation begins when bedrock is slowly broken down into fragments and particles by physical, chemical, and biological processes, called *weathering*. Figure 10-A shows profiles of different-aged soils. Such profiles are made up of *soil horizons*, or layers, that vary in number, composition, and thickness, depending on the type of soil.

Soil, on which all terrestrial life depends, is a key component of the earth's natural capital. It supplies most of the nutrients needed for plant growth and purifies and stores water, while organisms living in the soil help to control the earth's climate by removing carbon dioxide from the atmosphere and storing it as organic carbon compounds.

Most soils that have developed over long periods of time, called *mature soils*, contain horizontal layers, or *horizons*, (Figure 10-A), each with a distinct texture and composition that vary with different types of soils. Most mature soils have at least three of the four possible horizons. Think of them as the top three floors in the geological building of life underneath your feet.

The roots of most plants and the majority of a soil's organic matter are concentrated in the soil's two upper layers, the *O horizon* of leaf litter and the *A horizon* of topsoil. In most mature soils, these two layers teem with bacteria, fungi, earthworms, and small insects, all interacting in complex ways. Bacteria and other decomposer microorganisms, found by the billions in every handful of topsoil, break down some of the soil's complex organic compounds into a porous mixture of the partially decomposed bodies of dead plants and animals, called *humus*, and inorganic materials such as clay, silt, and sand. Soil moisture carrying these dissolved nutrients is drawn up by the roots of plants and transported through stems and into leaves as part of the earth's chemical cycling processes.

The *B horizon (subsoil)* and the *C horizon (parent material)* contain most of a soil's inorganic matter, mostly broken-down rock consisting of varying mixtures of sand, silt, clay, and gravel. Much of it is transported by water from the A horizon (Figure 10-A). The C horizon lies on a base of parent material, which is often *bedrock*.

The spaces, or *pores*, between the solid organic and inorganic particles in the upper and lower soil layers contain varying amounts of air (mostly nitrogen and oxygen gas) and water. Plant roots use the oxygen for cellular respiration. As long as the O and A horizons are anchored by vegetation, the soil layers as a whole act as a sponge, storing water and nutrients, and releasing them in a nourishing trickle.

Although topsoil is a renewable resource, it is renewed very slowly, which means it can be depleted. Just 1 centimeter (0.4 inch) of topsoil can take hundreds of years to form, but it can be washed or blown away in a matter of weeks or months when we plow grassland or clear a forest and leave its topsoil unprotected.

Critical Thinking

How does soil contribute to each of the four components of biodiversity described in Figure 4-2, p. 63?

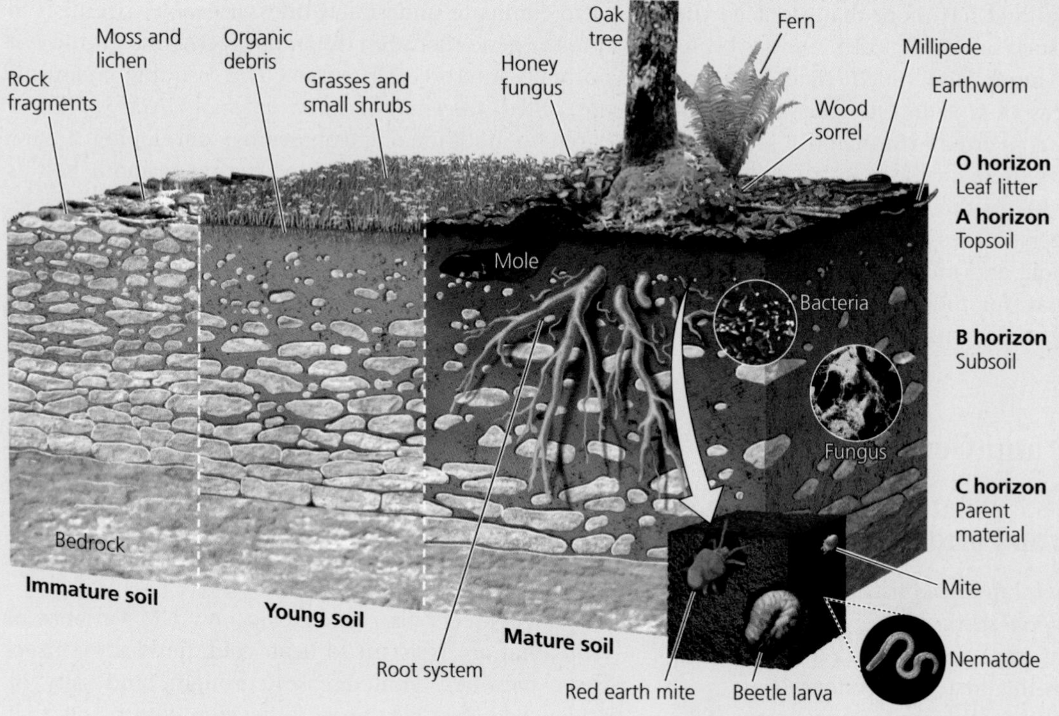

Active Figure 10-A This diagram shows a generalized soil profile and illustrates how soil is formed. *See an animation based on this figure and compare soil profiles from grassland, desert, and three types of forests at* **www.cengagebrain.com**. **Questions:** What role do you think the tree in this figure plays in soil formation? How might the soil formation process change if the tree were removed?

Largely because of the two green revolutions, GOOD NEWS world grain production tripled between 1961 and 2009 (Figure 10-5, left, p. 210). Per capita food production increased by 31% between 1961 and 1985, but since then it has declined slightly (Figure 10-5, right). The world's three largest grain-producing countries—China, India, and the United States, in that order—produce almost half of the world's grains.

People directly consume about 48% of the world's grain production. About 35% is used to feed livestock

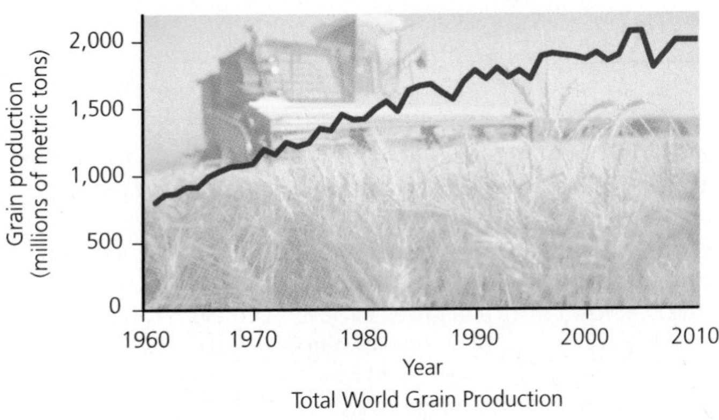

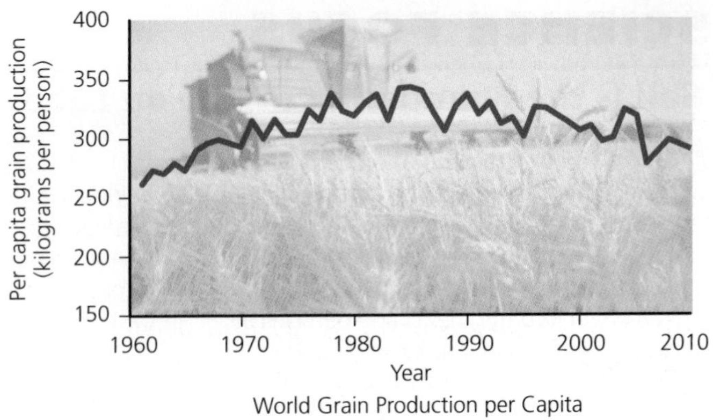

Figure 10-5 *Global outlook:* These graphs show the growth in worldwide grain production of wheat, corn, and rice (left), and per capita grain production (right) between 1961 and 2009. **Question:** Why do you think grain production per capita has grown less consistently than total grain production? (Data from U.S. Department of Agriculture, Worldwatch Institute, UN Food and Agriculture Organization, and Earth Policy Institute)

and indirectly consumed by people who eat meat and meat products. The remaining 17% (mostly corn) is used to make biofuels such as ethanol for cars and other vehicles.

In the United States, industrialized farming has evolved into *agribusiness,* as a small number of giant multinational corporations increasingly control the growing, processing, distribution, and sale of food in U.S. and global markets. One of the benefits of this is that since 1950, U.S. industrialized agriculture has become very efficient and has more than doubled the yields of key crops such as wheat, corn, and soybeans without cultivating more land. Such yield increases have kept large areas of U.S. forests, grasslands, and wetlands from being converted to farmland.

Another benefit of these increases is that Americans spend only about 13% of their disposable income on food, compared to percentages of up to 50% that people in China and India and most other less-developed countries have to pay for food. However, crop yield increases also have their downside, which we examine later in this chapter.

Crossbreeding and Genetic Engineering Produce New Varieties of Crops and Livestock

For centuries, farmers and scientists have used *crossbreeding* through *artificial selection* to develop genetically improved varieties of crops and livestock animals. Such selective breeding in this first *gene revolution* has yielded amazing results. GOOD NEWS Ancient ears of corn were about the size of your little finger, and wild tomatoes were once the size of grapes.

Traditional crossbreeding is a slow process, typically taking 15 years or more to produce a commercially valuable new crop variety, and it can combine traits only from species that are genetically similar. Typically, resulting varieties remain useful for only 5–10 years before pests and diseases reduce their effectiveness. Important advances are still being made with this method.

Today, scientists are creating a second *gene revolution* by using *genetic engineering* to develop genetically improved strains of crops and livestock animals. It involves altering an organism's genetic material through adding, deleting, or changing segments of its DNA (see Figure 11, p. S15, in Supplement 4). The goal of this process, called *gene splicing,* is to produce desirable traits or to eliminate undesirable ones. It enables scientists to transfer genes between different species that would not normally interbreed in nature. The resulting organisms are called *genetically modified organisms (GMOs).* Compared to traditional crossbreeding, developing a new crop variety through gene splicing takes about half as long, usually costs less, and allows for the insertion of genes from almost any other organism into crop cells.

Ready or not, much of the world is entering the *age of genetic engineering.* According to the U.S. Department of Agriculture (USDA), at least 70% of the food products on U.S. supermarket shelves contain some form of genetically engineered food or ingredients, and the proportion is increasing. But you won't find GM ingredient information on food labels because labeling of GM products is not required by law. However, certified organic food, which is labeled as such (**Core Case Study**), CORE CASE STUDY makes no use of genetically modified seeds or ingredients.

Bioengineers plan to develop new GM varieties of crops that are resistant to heat, cold, herbicides, insect pests, parasites, viral diseases, drought, and salty or acidic soil. They also hope to develop crop plants that can grow faster and survive with little or no irrigation and with less fertilizer and pesticides.

Many scientists believe that such innovations hold great promise for helping to improve global food security. Others warn that genetic engineering is not free of drawbacks, which we examine later in this chapter.

Meat Production Has Grown Steadily

Meat and animal products such as eggs and milk are good sources of high-quality protein and represent the world's second major food-producing system. Between 1961 and 2010, world meat production—mostly beef, pork, and poultry—increased more than fourfold and average meat consumption per person more than doubled. Global meat production is likely to more than double again by 2050 as affluence rises and more middle-income people begin consuming more meat and animal products in rapidly developing countries such as China and India.

About half of the world's meat comes from livestock grazing on grass in unfenced rangelands and enclosed pastures. The other half is produced through an industrialized factory farm system. It involves raising large numbers of animals (Figure 10-6) bred to gain weight quickly, mostly in crowded *feedlots* and *concentrated animal feeding operations* (*CAFOs*). They are fed grain, fish meal, or fish oil, which are usually doctored with growth hormones and antibiotics. Cattle in such feedlots often stand knee-deep in manure and are covered with feces when they arrive at slaughterhouses.

Most veal calves, pigs, chickens, and turkeys that are raised in more-developed countries spend their lives in very crowded pens and cages, often in huge buildings. As a result, feedlots and CAFOs, and the animal wastes and runoff associated with them, create serious environmental impacts on the air and water, which we examine later in this chapter.

Fish and Shellfish Production Have Increased Dramatically

The world's third major food-producing system consists of fisheries and aquaculture. A **fishery** is a concentration of particular aquatic species suitable for commercial harvesting in a given ocean area or inland body of water. Industrial fishing fleets harvest most of the world's marine catch of wild fish (see Chapter 9, Case Study, p. 197). Fish and shellfish are also produced through **aquaculture**—the practice of raising marine and freshwater fish in freshwater ponds and rice paddies or in underwater cages in coastal waters or in deeper ocean waters (see Figure 9-26, p. 198), instead of hunting and catching them by using fishing boats and modern industrialized fishing fleets.

Most of the world's aquaculture involves raising species that feed on algae or other plants—mainly carp in China (which raises 70% of the world's farmed fish) and India, catfish in the United States, tilapia in several countries, and shellfish in a number of coastal countries. However, the farming of meat-eating species such as shrimp and salmon is growing rapidly, especially in more-developed countries. As a result, about 37% of the global catch of wild fish is converted to fish meal and fish oil, which are fed to farmed meat-eating fish, as well as to cattle and pigs.

Global seafood production has increased dramatically since 1950 (Figure 10-7, p. 212) (**Concept 10-2**). Today, about half of the world's fish and shellfish are caught by industrial fishing fleets and the other half are raised by aquaculture.

Matt Meadows/Peter Arnold, Inc.

Figure 10-6 *Industrialized beef production:* On this cattle feedlot in Imperial Valley, California (USA) 40,000 cattle are fattened up on grain for a few months before being slaughtered.

Figure 10-7 World seafood production, including both wild catch and aquaculture, 1950 and 2008. **Question:** What are two trends that you can see in these data? (Data from UN Food and Agriculture Organization, U.S. Census Bureau, and Worldwatch Institute)

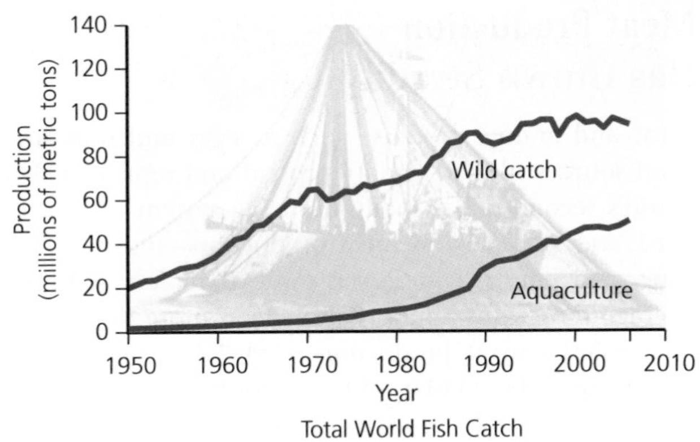

According to 2007 report by the FAO, about 50% of the world's ocean fisheries are being harvested at full capacity and around 25% are depleted or overexploited. Some fishery scientists warn that unless we reduce overfishing and ocean pollution, and slow projected climate change, most of the world's major commercial ocean fisheries could collapse by 2050.

Industrialized Food Production Requires Huge Inputs of Energy

The industrialization of food production has been made possible by the availability of energy, mostly from nonrenewable oil and natural gas. It is used to run farm machinery, irrigate crops, and produce synthetic pesticides (mostly from petrochemicals produced when oil is refined) and synthetic inorganic fertilizers. Fossil fuels are also used to process food and transport it long distances within and between countries.

When we consider the energy used to grow, store, process, package, transport, refrigerate, and cook all plant and animal food, it takes about 10 units of fossil fuel energy to put 1 unit of food energy on the table in the United States. For example, in the United States, food travels an average of 2,400 kilometers (1,300 miles) from farm to plate. According to a study by ecological economist Peter Tyedmers and his colleagues, the large-scale hunting and gathering operation by the world's fishing fleets uses about 12.5 units of energy to put 1 unit of food energy from seafood on the table. Bottom line: producing, processing, transporting, and consuming industrialized food result in a large *net energy loss*.

THINKING ABOUT
Food and Oil

What might happen to industrialized food production and to your lifestyle if oil prices rise sharply in the next two decades, as many analysts predict they will? List two ways in which you would deal with these changes.

10-3 What Environmental Problems Arise from Industrialized Food Production?

▶ CONCEPT 10-3 Future food production may be limited by soil erosion and degradation, desertification, irrigation water shortages, air and water pollution, climate change from greenhouse gas emissions, and loss of biodiversity.

Producing Food Has Major Environmental Impacts

Agricultural systems have produced spectacular increases in the world's food production since 1950. However, according to many analysts, agriculture has greater harmful environmental impacts than any other human activity and these environmental effects may limit future food production (**Concept 10-3**).

According to a 2010 study by 27 experts assembled by the United Nations Environment Programme (UNEP), one of the three chief contributors to humanity's massive ecological footprint (Figure 1-8, p. 14) is agriculture. Agriculture accounts for about 70% of global freshwater removed from aquifers and surface waters. It also uses about 38% of the world's ice-free land and emits about 25% of the world's greenhouse gas emissions. In addition, agriculture produces about 60% of the world's water pollution, mostly from synthetic nitrate and phosphate fertilizer that washes off of crop fields. Figure 10-8 summarizes the harmful effects of industrialized agriculture. We now explore such effects in greater depth, starting with the problems of erosion and degradation of soils.

Natural Capital Degradation

Food Production

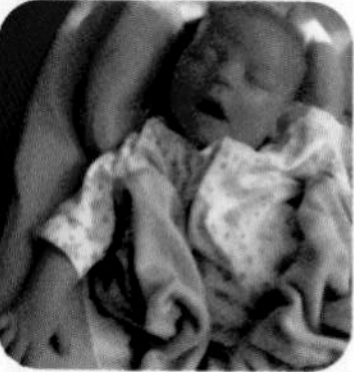

Biodiversity Loss	Soil	Water	Air Pollution	Human Health
Loss and degradation of grasslands, forests, and wetlands in cultivated areas	Erosion	Water waste	Emissions of greenhouse gas CO_2 from fossil fuel use	Nitrates in drinking water (blue baby)
Fish kills from pesticide runoff	Loss of fertility	Aquifer depletion	Emissions of greenhouse gas N_2O from use of inorganic fertilizers	Pesticide residues in drinking water, food, and air
Killing wild predators to protect livestock	Salinization	Increased runoff, sediment pollution, and flooding from cleared land	Emissions of greenhouse gas methane (CH_4) by cattle (mostly belching)	Contamination of drinking and swimming water from livestock wastes
Loss of genetic diversity of wild crop strains replaced by monoculture strains	Waterlogging	Pollution from pesticides and fertilizers	Other air pollutants from fossil fuel use and pesticide sprays	Bacterial contamination of meat
	Desertification	Algal blooms and fish kills in lakes and rivers caused by runoff of fertilizers and agricultural wastes		
	Increased acidity			

Figure 10-8 Food production has a number of harmful environmental effects (**Concept 10-3**). **Question:** Which item in each of these categories do you believe is the most harmful?

Topsoil Erosion Is a Serious Problem in Parts of the World

Soil erosion is the movement of soil components, especially surface litter and topsoil (Figure 10-A), from one place to another by the actions of wind and water. Some topsoil erosion is natural, and some is caused by human activities.

Flowing water, the largest cause of erosion, carries away particles of topsoil that have been loosened by rainfall (Figure 10-9). Severe erosion of this type leads to the formation of gullies (Figure 10-10, p. 214). Wind also loosens and blows topsoil particles away, especially in areas with a dry climate and relatively flat and exposed land. We lose natural capital in the form of fertile topsoil when we destroy soil-holding grasses through activities such as farming (see Figure 7-12, p. 105), deforestation (see Figure 9-9, p. 180), and overgrazing (see Figure 9-17, p. 187).

Erosion of topsoil has two major harmful effects. One is *loss of soil fertility* through depletion of plant nutrients in topsoil. The other is *water pollution* in nearby surface waters, where eroded topsoil ends up as sediment. This can kill fish and shellfish and clog irrigation ditches, boat channels, reservoirs, and lakes. Additional water pollution occurs when the eroded sediment contains pesticide residues. By removing vital plant nutrients from topsoil and adding excess plant nutrients to aquatic systems, we degrade the topsoil and pollute the water, and thus interfere with the carbon, nitrogen, and phosphorus cycles. In other words,

Tim McCabe/USDA Natural Resources Conservation Service

Figure 10-9 Flowing water from rainfall is the leading cause of topsoil erosion as seen on this farm in the U.S. state of Tennessee.

Figure 10-10 Natural capital degradation: Severe gully erosion has damaged this cropland in Bolivia.

we are violating the earth's chemical cycling **principle of sustainability**.

A joint survey by the UN Environment Programme (UNEP) and the World Resources Institute estimated that topsoil is eroding faster than it forms on about 38% of the world's cropland (Figure 10-11). (See the Guest Essay on soil erosion by David Pimentel at **www.cengagebrain.com**.) (For more information on topsoil erosion, see *The Habitable Planet*, Video 15, at **www.learner.org/resources/series209.html**.)

Drought and Human Activities Are Degrading Drylands

Desertification in arid and semiarid parts of the world threatens livestock and crop contributions to the world's food supply. It occurs when the productive potential of topsoil falls by 10% or more because of a combination of prolonged drought and human activities that expose topsoil to erosion.

A drop in production of more than 50% usually results in huge gullies and sand dunes (Figure 10-12). Only in extreme cases does desertification lead to what we call desert. A 2007 report by the FAO and a 2010 study by agriculture researcher Montserrat Núñez estimated that 38% of the world's land and 70% of the world's arid and semiarid lands used for agriculture are threatened by desertification (Figure 10-13). (For more information on desertification see *The Habitable Planet*, Video 22, at **www.learner.org/resources/series209.html**.)

According to a 2007 study by the Intergovernmental Panel on Climate Change (IPCC), projected climate change during this century is likely to greatly increase severe and prolonged drought, expand desertification, and threaten food supplies in various areas of the world.

Excessive Irrigation Has Serious Consequences

A major reason for the success in boosting productivity on farms is the use of irrigation, which accounts for about 70% of the water that humanity uses. Currently, the roughly 20% of the world's cropland that is irrigated produces about 45% of the world's food.

But irrigation has a downside. Most irrigation water is a dilute solution of various salts, such as sodium chloride (NaCl), that are picked up as the water flows over

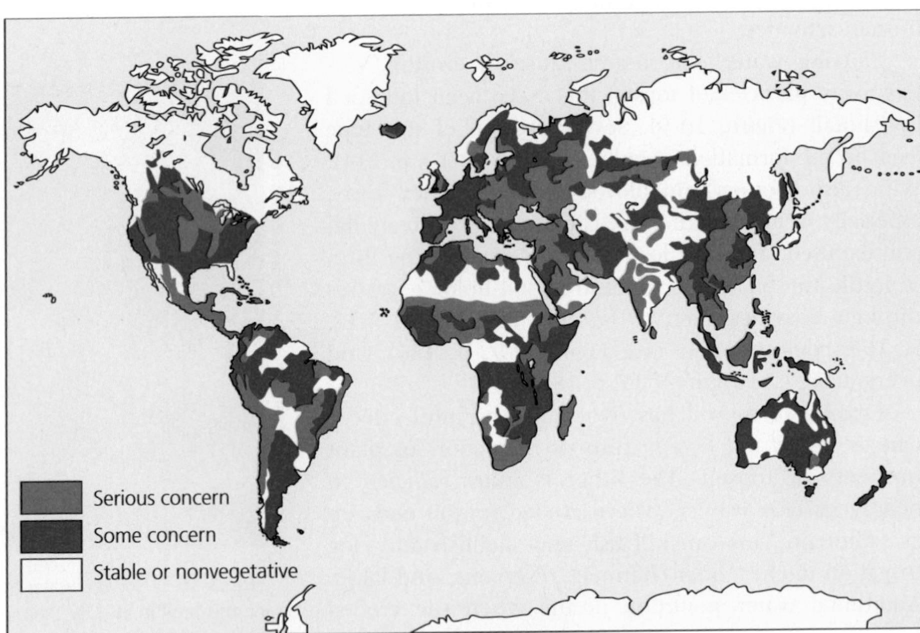

Figure 10-11 Natural capital degradation: Topsoil erosion is a serious problem in some parts of the world. **Question:** Can you see any geographical pattern associated with this problem? (Data from UN Environment Programme and the World Resources Institute)

Figure 10-12 Sand dunes threaten to take over an oasis in the Sahel region of West Africa.

or through soil and rocks. Irrigation water that has not been absorbed into the topsoil evaporates, leaving behind a thin crust of dissolved mineral salts in the topsoil. Repeated applications of irrigation water in dry climates lead to the gradual accumulation of salts in the upper soil layers—a soil degradation process called **salinization**. It stunts crop growth, lowers crop yields, and can eventually kill plants and ruin the land.

The United Nations estimates that severe salinization has reduced yields on at least 10% of the world's irrigated cropland, and almost 25% of irrigated cropland in the United States, especially in western states (Figure 10-14, p. 216).

Another problem with irrigation is **waterlogging**, in which water accumulates underground and gradually raises the water table, especially when farmers apply large amounts of irrigation water in an effort to leach salts deeper into the soil. Without adequate drainage, waterlogging lowers the productivity of crop plants and kills them after prolonged exposure. At least 10% of the world's irrigated land suffers from waterlogging, and the problem is getting worse.

Perhaps the biggest problem resulting from excessive irrigation in agriculture is that it has contributed to depletion of groundwater and surface water supplies in many areas of the world. We discuss this in Chapter 11.

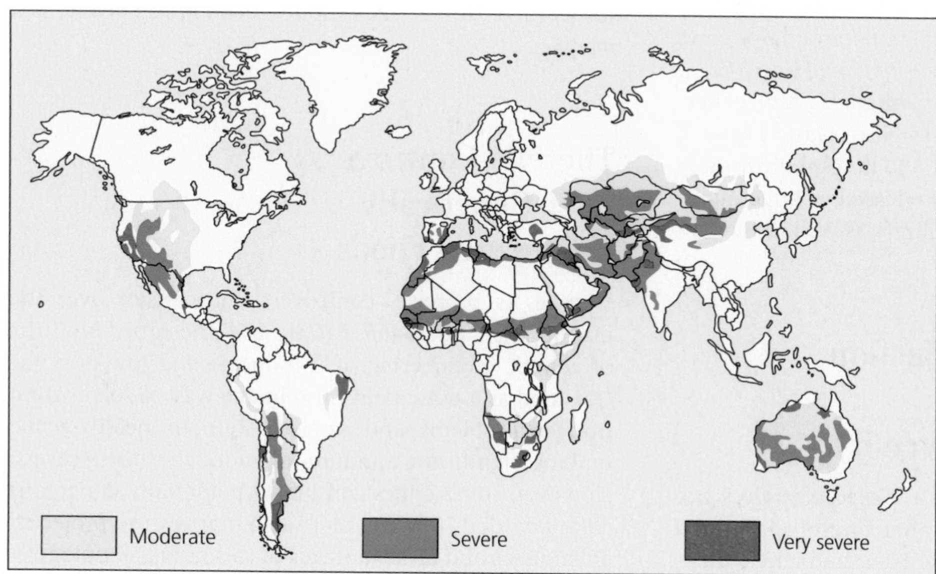

Figure 10-13 Natural capital degradation: This map shows how desertification of arid and semiarid lands varied in 2007. **Question:** Can you see any geographical pattern associated with this problem? (Data from UN Environment Programme, Harold E. Drengue, U.S. Department of Agriculture, and the Central Intelligence Agency's World Factbook, 2008)

U.S. Natural Resources Conservation Service

Figure 10-14 Natural capital degradation: White alkaline salts have displaced crops that once grew on this heavily irrigated land in the U.S. state of Colorado.

Agriculture Contributes to Air Pollution and Projected Climate Change

Agricultural activities, including the clearing (Figure 10-4) and burning of forests to raise crops or livestock, create a great deal of air pollution. They also account for more than 25% of the human-generated emissions of carbon dioxide and other greenhouse gases, which during this century are projected to sharply reduce crop productivity in some areas that currently have high crop productivity.

Industrialized livestock production alone generates about 18% of the world's greenhouse gases that can build up in the atmosphere and change the climate. This is more than all of the world's cars, trucks, buses, and planes emit, according to the 2006 FAO study, *Livestock's Long Shadow*. In particular, cattle and dairy cows release the powerful greenhouse gas methane (CH_4), mostly through belching, and methane is also generated by liquid animal manure stored in waste lagoons. Nitrous oxide (N_2O), with about 300 times the warming capacity of CO_2 per molecule, is released in huge quantities by synthetic inorganic fertilizers as well as by livestock manure.

Food and Biofuel Production Systems Have Caused Major Losses of Biodiversity

Natural biodiversity and some ecological services are threatened when tropical and other forests are cleared (see Figure 9-9, p. 180 and Figure 10-4) and when grasslands are plowed up and replaced with croplands used to produce food and biofuels for cars (**Concept 10-3**).

A related problem is the increasing loss of *agrobiodiversity*—the world's genetic variety of animal and plant species used to provide food. Scientists estimate that since 1900, we have lost 75% of the genetic diversity of agricultural crops.

For example, India once planted 30,000 varieties of rice. Now more than 75% of its rice production comes from only ten varieties and soon, almost all of its production may come from just one or two varieties. In the United States, about 97% of the food plant varieties available to farmers in the 1940s no longer exist, except perhaps in small amounts in seed banks, in the backyards of a few gardeners, and in non-profit organizations such as Seed Savers Exchange.

In other words, we are rapidly shrinking the world's genetic "library," which is critical for increasing food yields. In fact, we might soon need it more than ever to develop new plant and livestock varieties that can adapt to the effects of projected climate change on different parts of the world. This failure to preserve agrobiodiversity as an ecological insurance policy is a serious violation of the biodiversity **principle of sustainability**. SUSTAINABILITY

Wild and endangered varieties of crops important to the world's food supply are stored in about 1,400 refrigerated gene or seed banks, agricultural research centers, and botanical gardens scattered around the world. However, power failures, fires, storms, war, and unintentional disposal of seeds can cause irreversible losses of these stored plants and seeds. A new underground seed vault on a frozen Norwegian arctic island is being stocked with duplicates of much of the world's seed collections. It is not as vulnerable to power losses, fires, or storms as most other seed vaults. GOOD NEWS

However, the seeds of many plants cannot be stored successfully in gene banks. And because stored seeds do not remain alive indefinitely, they must be planted (germinated) periodically, and new seeds must be collected for storage. Unless this is done, seed banks become *seed morgues*.

There Is Controversy over Genetically Engineered Foods

Despite its promise, controversy has arisen over the use of *genetically modified (GM) food* and other products of genetic engineering. Its producers and investors see GM food as a potentially sustainable way to solve world hunger problems and improve human health while making significant amounts of money in the process. However, some critics consider it potentially dangerous "Frankenfood." Figure 10-15 summarizes the projected advantages and disadvantages of this new technology.

Trade-Offs

Genetically Modified Crops and Foods

Advantages	Disadvantages
Need less fertilizer	Unpredictable genetic and ecological effects
Need less water	Harmful toxins and new allergens in food
More resistant to insects, disease, frost, and drought	No increase in yields
Grow faster	More pesticide-resistant insects and herbicide-resistant weeds
May need less pesticides or tolerate higher levels of herbicides	Could disrupt seed market
May reduce energy needs	Lower genetic diversity

Figure 10-15 Genetically modified crops and foods have advantages and disadvantages. **Questions:** Which two advantages and which two disadvantages do you think are the most important? Why?

Critics recognize the potential benefits of GM crops (Figure 10-15, left) but they point out most of the GM crops developed so far provide very few of these benefits. In 2009, for example, biologist and genetic engineering expert Doug Gurian-Sherman pointed out that despite 20 years of research and 13 years of commercial sales, genetically engineered crops have failed to significantly increase U.S. crop yields.

Critics also warn that we know too little about the long-term potential harm to human health and ecosystems resulting from the widespread use of such crops, which have largely been assumed to be harmless until some proven harm arises. Critics point out that if GM organisms that are released into the environment cause some harmful genetic or ecological effects, as some scientists expect, those organisms cannot be recalled.

For example, genes in plant pollen from genetically engineered crops can spread among nonengineered species. The new strains can then form hybrids with wild crop varieties, which could reduce the natural genetic biodiversity of the wild strains. This could in turn reduce the gene pool needed to crossbreed new crop varieties and to develop new genetically engineered varieties—another violation of the biodiversity sustainability principle.

CONNECTIONS

GM Crops and Organic Food Prices CORE CASE STUDY

The possible unintended spread of GM crop genes threatens the production of certified organic crops (**Core Case Study**), which must be grown in the absence of such genes. Because organic farmers have to perform expensive tests to detect GMOs or take costly planting measures to prevent the spread of GMOs to their fields from nearby crop fields, they have to raise the prices of their produce. This makes it more difficult for these farmers to compete with the industrial farming operations that generate the GM genes in the first place.

Most scientists and economists who have evaluated the genetic engineering of crops believe that its potential benefits will eventually outweigh its risks. However, critics, including the Ecological Society of America, call for more controlled field experiments and long-term testing to better understand the ecological and health risks, and stricter regulation of this rapidly growing technology. Indeed, a 2008 International Assessment of Agricultural Knowledge by 400 scientists, with input from more than 50 countries, raised serious doubts about the ability of GM crops to increase food security compared to other more effective and sustainable alternative solutions (which we discuss later in this chapter).

There Are Limits to Expansion of the Green Revolutions

Several factors have limited the success of the green revolutions to date and may limit them in the future (**Concept 10-3**). Without huge inputs of water and synthetic inorganic fertilizers and pesticides, most green revolution and genetically engineered crop varieties produce yields that are no higher (and are sometimes lower) than those from traditional strains. These high inputs also cost too much for most subsistence farmers in less-developed countries.

Scientists point out that continuing to increase these inputs eventually produces no additional increase in crop yields. This helps to explain the slowdown in the rate of global grain yields per hectare from an average increase of 2.1% a year between 1950 and 1990 to 1.3% annually between 1990 and 2010.

Can we expand the green revolutions by irrigating more cropland? Since 1978, the amount of irrigated land per person has been declining, and it is projected to fall much more between 2011 and 2050. One reason for this is population growth, which is projected to add 2.6 billion more people between 2010 and 2050. Other factors are wasteful use of irrigation water, soil salinization, depletion of both underground water supplies (aquifers) and surface water, and the fact that most of the world's farmers do not have enough money to irrigate their crops.

The good news is that we can get more crops per drop of irrigation water by using known

methods and technologies to greatly improve the efficiency of irrigation. We discuss this more fully in Chapter 11.

Can we increase the food supply by cultivating more land? We have already cleared or converted about 38% of the world's ice-free land surface for use as croplands, rangelands, or pastures. Clearing tropical forests and irrigating arid land could more than double the area of the world's cropland. The problem is that much of this land has poor soil fertility, steep slopes, or both. And cultivating such land usually is expensive, is unlikely to be sustainable, and reduces biodiversity by degrading and destroying wildlife habitats.

In addition, during this century, fertile croplands in coastal areas, including many of the major rice-growing floodplains and river deltas in Asia, are likely to be flooded by rising sea levels resulting from projected climate change. Food production could also drop sharply in some major food-producing areas because of increased drought and longer and more intense heat waves, also resulting from projected climate change.

Industrialized Meat Production Has Harmful Environmental Consequences

Proponents of industrialized meat production point out that it has increased meat supplies, reduced overgrazing, kept food prices down, and yielded higher profits. But environmental scientists point out that such systems use large amounts of energy (mostly fossil fuels) and water, and produce huge amounts of animal wastes that sometimes pollute surface water and groundwater while saturating the air with their odors and emitting large quantities of climate-changing greenhouse gases into the atmosphere.

Analysts also point out that meat produced by industrialized agriculture is artificially cheap because most of its harmful environmental and health costs are not included in the market prices of meat and meat products. Figure 10-16 summarizes the advantages and disadvantages of industrialized meat production.

In 2008, the FAO reported that overgrazing and soil compaction and erosion by livestock had degraded about 20% of the world's grasslands and pastures. The same report estimated that rangeland grazing and industrialized livestock production caused about 55% of all topsoil erosion and sediment pollution, and 33% of the water pollution that had resulted from runoff of nitrogen and phosphorous from excessive inputs of synthetic fertilizers.

CONNECTIONS

Corn, Ethanol, and Ocean Dead Zones

Huge amounts of synthetic inorganic fertilizers are used in the midwestern United States to produce corn for animal feed and for conversion to ethanol fuel. Much of this fertilizer runs off cropland and eventually goes into the Mississippi River, and the added nitrate and phosphate nutrients over-fertilize coastal waters in the Gulf of Mexico, where the river flows into the ocean. Each year, this creates a "dead zone" often larger than the U.S. state of Massachusetts. This oxygen-depleted zone threatens one-fifth of the nation's seafood yield. In other words, growing corn in the Midwest, largely to fuel cars with ethanol, degrades aquatic biodiversity and seafood production in the Gulf of Mexico.

Fossil fuel energy (mostly from oil) is also an essential ingredient in industrialized meat production. Using this energy pollutes the air and water, and emits greenhouse gases that contribute to projected climate change as discussed above.

Another growing problem is the use of antibiotics in industrialized livestock production facilities. A 2009 study by the Union of Concerned Scientists (UCS) found that 70% of all antibiotics used in the United States (and 50% of those used in the world) are added to animal feed. This is done in an effort to prevent the spread of diseases in crowded feedlots and CAFOs and to make the livestock animals grow faster.

The UCS study, as well as several others, concluded that the widespread use of antibiotics in livestock production is an important factor in the rise of genetic resistance among many disease-causing microbes (see Figure 4-5, p. 67). Such resistance can reduce the effectiveness of some antibiotics used to treat infectious diseases in humans. It can also promote the development of new and aggressive disease organisms that are resistant to all but a very few antibiotics currently avail-

Trade-Offs

Animal Feedlots

Advantages	**Disadvantages**
Increased meat production	Large inputs of grain, fish meal, water, and fossil fuels
Higher profits	Greenhouse gas (CO_2 and CH_4) emissions
Less land use	Concentration of animal wastes that can pollute water
Reduced overgrazing	Use of antibiotics can increase genetic resistance to microbes in humans
Reduced soil erosion	
Protection of biodiversity	

Figure 10-16 Animal feedlots and confined animal feeding operations have advantages and disadvantages. **Questions:** Which single advantage and which single disadvantage do you think are the most important? Why?

able. In 2009, the U.S. Centers for Disease Control and Prevention reported that antibiotic-resistant infections killed about 65,000 people in the United States—more than the number of deaths from breast cancer and prostrate cancer combined and about twice the number of Americans killed by automobile accidents in 2009.

Finally, according to the USDA, animal waste produced by the American meat industry amounts to about 130 times the amount of waste produced by the country's human population. Globally, only about half of all manure is returned to the land as nutrient-rich fertilizer—a violation of the nutrient cycling **principle of sustainability**. Much of the other half of this waste ends up polluting the air, water, and soil, and producing foul odors.

Aquaculture Can Harm Aquatic Ecosystems

Figure 10-17 lists the major advantages and disadvantages of aquaculture, which in 2010 accounted for roughly half of global seafood production. Some analysts warn that the harmful environmental effects of aquaculture could limit its future production potential (**Concept 10-3**).

One environmental problem associated with aquaculture is that using fish meal and fish oil to feed farmed fish can deplete populations of wild fish. About 37% of the wild marine fish catch is used in the production of fish meal and fish oil.

Another problem is that fish such as farmed salmon raised on fish meal or fish oil can be contaminated with long-lived toxins such as PCBs and dioxins. Aquaculture producers contend that the concentrations of these chemicals are not high enough to threaten human health.

Fish farms, especially those that raise carnivorous fish such as salmon and tuna, also produce large amounts of wastes. Along with the pesticides and antibiotics used in fish farms, these wastes can pollute aquatic ecosystems and fisheries. Another problem is that farmed fish can escape their pens and mix with wild fish, changing and possibly disrupting the gene pools of wild populations.

Trade-Offs

Aquaculture

Advantages	Disadvantages
High efficiency	Large inputs of land, feed, and water
High yield	Large waste output
Reduced over-harvesting of fisheries	Loss of mangrove forests and estuaries
Low fuel use	Some species fed with grain, fish meal, or fish oil
High profits	Dense populations vulnerable to disease

Figure 10-17 Aquaculture has advantages and disadvantages. **Questions:** Which single advantage and which single disadvantage do you think are the most important? Why?

10-4 How Can We Protect Crops from Pests More Sustainably?

▶ **CONCEPT 10-4 We can sharply cut pesticide use without decreasing crop yields by using a mix of cultivation techniques, biological pest controls, and small amounts of selected chemical pesticides as a last resort (integrated pest management).**

Nature Controls the Populations of Most Pests

A **pest** is any species that interferes with human welfare by competing with us for food, invading homes, lawns, or gardens, destroying building materials, spreading disease, invading ecosystems, or simply being a nuisance. Worldwide, only about 100 species of plants (weeds), animals (mostly insects), fungi, and microbes cause most of the damage to the crops we grow.

In natural ecosystems and many polyculture agroecosystems, *natural enemies* (predators, parasites, and disease organisms) control the populations of most potential pest species. For example, the world's 30,000 known species of spiders kill far more crop-eating insects every year than humans do by using chemicals. Most spiders, including the ferocious-looking wolf spider (Figure 10-18, p. 220), do not harm humans.

When we clear forests and grasslands, plant monoculture crops, and douse fields with chemicals that kill

Peter J. Bryant/Biological Photo Service

Figure 10-18 Natural capital: Spiders are important insect predators that are killed by some pesticides.

pests, we upset many of these natural population checks and balances that help to implement the biodiversity principle of sustainability (see back cover). Then we must devise and pay for ways to protect our monoculture crops, tree plantations, lawns, and golf courses from insects and other pests that nature once largely controlled at no charge.

We Use Pesticides to Help Control Pest Populations

We have developed a variety of synthetic **pesticides**—chemicals used to kill or control populations of organisms that we consider undesirable. Common types of pesticides include *insecticides* (insect killers), *herbicides* (weed killers), *fungicides* (fungus killers), and *rodenticides* (rat and mouse killers).

We did not invent the use of chemicals to repel or kill other species. For nearly 225 million years, plants have been producing chemicals called *biopesticides* to ward off, deceive, or poison the insects and herbivores that feed on them. This battle produces a never-ending, ever-changing coevolutionary process: insects and herbivores overcome various plant defenses through natural selection and new plant defenses are favored by natural selection. This process is continuously repeated in an ongoing cycle of evolutionary punch and counterpunch.

After 1950, pesticide use increased more than 50-fold and most of today's pesticides are 10–100 times more toxic than those used in the 1950s. Use of natural biopesticides and synthetic pesticides derived from such natural chemicals is also on the rise.

Some synthetic pesticides, called *broad-spectrum agents*, are toxic not only to many pests, but also to beneficial species. Examples are chlorinated hydrocarbon compounds such as DDT and organophosphate compounds such as malathion and parathion. Others, called *selective*, or *narrow-spectrum, agents*, are effective against a narrowly defined group of organisms. Examples are algicides for algae and fungicides for fungi.

Pesticides vary in their *persistence*, the length of time they remain deadly in the environment. Some, such as DDT and related compounds, remain in the environment for years and can be biologically magnified in food chains and webs (see Figure 8-11, p. 162). Others, such as organophosphates, are active for days or weeks and are not biologically magnified but can be highly toxic to humans.

About one-fourth of the pesticides used in the United States are aimed at ridding houses, gardens, lawns, parks, playing fields, swimming pools, and golf courses of insects and other species that we view as pests. According to the U.S. Environmental Protection Agency (EPA), the average lawn in the United States is doused with ten times more synthetic pesticides per unit of land area than what is put on an equivalent amount of U.S. cropland. In 1962, biologist Rachel Carson warned against relying primarily on synthetic organic chemicals to kill insects and other species we regard as pests (see Individuals Matter, at right).

Synthetic Pesticides Have Several Advantages

Synthetic pesticides have advantages and disadvantages. Proponents contend that their benefits (Figure 10-19, left) outweigh their harmful effects (Figure 10-19, right).

Trade-Offs

Conventional Chemical Pesticides

Advantages	Disadvantages
Save lives	Promote genetic resistance
Increase food supplies	Kill natural pest enemies
Profitable	Pollute the environment
Work fast	Can harm wildlife and people
Safe if used properly	Are expensive for farmers

Figure 10-19 Synthetic pesticides have advantages and disadvantages. **Questions:** Which single advantage and which single disadvantage do you think are the most important? Why?

INDIVIDUALS MATTER

Rachel Carson

Rachel Carson (Figure 10-B) began her professional career as a biologist working for the Bureau of U.S. Fisheries (now called the U.S. Fish and Wildlife Service). In that capacity, she carried out research in oceanography and marine biology, and wrote articles and books about the oceans and topics related to the environment.

In 1958, the commonly used pesticide DDT was sprayed to control mosquitoes near the home and private bird sanctuary of one of Carson's friends. After the spraying, her friend witnessed the agonizing deaths of several birds. She begged Carson to find someone to investigate the effects of pesticides on birds and other wildlife.

Carson decided to look into the issue herself and found very little independent research on the environmental effects of pesticides. As a well-trained scientist, she reviewed the scientific literature, became convinced that pesticides could harm wildlife and humans, and gathered further information about the harmful effects of the widespread use of pesticides.

In 1962, she published her findings in popular form in *Silent Spring*, a book whose title warned of the potential silencing of "robins, catbirds, doves, jays, wrens, and scores of other bird voices" because of their exposure to synthetic pesticides. Many scientists, politicians, and policy makers read *Silent Spring*, and the public embraced it.

Figure 10-B Biologist Rachel Carson (1907–1964) alerted us to the harmful effects of the widespread use of synthetic pesticides.

Chemical manufacturers understandably saw the book as a serious threat to their booming pesticide business, and they mounted a campaign to discredit Carson. A parade of critical reviewers and industry scientists claimed that her book was full of inaccuracies, made selective use of research findings, and failed to give a balanced account of the benefits of pesticides.

Some critics even claimed that, as a woman, Carson was incapable of understanding such a highly scientific and technical subject. Others charged that she was just a hysterical woman and radical nature lover who was trying to scare the public in an effort to sell books.

During these intense attacks, Carson was a single mother and the sole caretaker of an aged parent. She was also suffering from terminal breast cancer. Yet she strongly defended her research and countered her critics. She died in 1964—about 18 months after the publication of *Silent Spring*—without knowing that many historians would consider her work to be an important contribution to the modern environmental movement then emerging in the United States.

They point to the following benefits:

- *They save human lives.* Since 1945, DDT and other insecticides probably have prevented the premature deaths of at least 7 million people (some say as many as 500 million) from insect-transmitted diseases such as malaria (carried by the *Anopheles* mosquito), bubonic plague (carried by rat fleas), and typhus (carried by body lice and fleas).
- *They have been known to increase food supplies* by reducing food losses from pests.
- *They can increase profits for farmers.* Officials of pesticide companies estimate that, under certain conditions, every dollar spent on pesticides can lead to an increase in U.S. crop yields worth as much as $4.
- *They work fast.* Pesticides control most pests quickly, have a long shelf life, and are easily shipped and applied. When genetic resistance occurs (see Figure 4-5, p. 67), farmers can use stronger doses or switch to other pesticides.
- *When used properly, the health risks of some pesticides are very low, relative to their benefits,* according to pesticide industry scientists.
- *Newer pesticides are safer and more effective than many older ones.* Greater use is being made of chemicals derived originally from plants (biopesticides), which are safer to use and less damaging to the environment than are many older pesticides. Genetic engineering is also being used to develop pest-resistant crop strains and genetically altered crops that produce natural biopesticides.

Synthetic Pesticides Have Several Disadvantages

Opponents of widespread use of synthetic pesticides contend that the harmful effects of these chemicals (Figure 10-19, right, and Case Study, p. 222) outweigh their benefits (Figure 10-19, left). They cite several serious problems.

- *They accelerate the development of genetic resistance to pesticides in pest organisms.* Insects breed rapidly, and within 5–10 years (much sooner in tropical areas), they can develop immunity to widely used pesticides through natural selection and then come back stronger than before. Since 1945, about 1,000 species of insects and rodents (mostly rats) and 550 types of weeds and plant diseases have developed genetic resistance to one or more pesticides. Since 1996, the use of glyphosate herbicide has increased on genetically modified crops designed to thrive on higher doses of this herbicide. This has led to at least 15 species of "superweeds" that are now genetically resistant to this widely used herbicide.

- *They can put farmers on a financial treadmill.* Because of genetic resistance, farmers can find themselves having to pay more and more for a chemical pest control program that can become less and less effective.
- *Some insecticides kill natural predators and parasites that help control the pest populations.* About 100 of the 300 most destructive insect pests in the United States were minor pests until widespread use of insecticides wiped out many of their natural predators.
- *Pesticides do not stay put and can pollute the environment.* According to the USDA, about 98–99.9% of the insecticides and more than 95% of the herbicides applied by aerial spraying or ground spraying do not reach the target pests. They end up in the air, surface water, groundwater, bottom sediments, food, and on nontarget organisms, including humans, livestock, and wildlife.
- *Some pesticides harm wildlife.* According to the USDA and the U.S. Fish and Wildlife Service, each year, pesticides applied to cropland wipe out about 20% of U.S. honeybee colonies and damage another 15% (see Chapter 8, Case Study, p. 162). Every year, pesticides also kill more than 67 million birds and 6–14 million fish. According to a study by the Center for Biological Diversity, pesticides also menace one of every three endangered and threatened species in the United States.
- *Some pesticides threaten human health.* The WHO and UNEP estimate conservatively that, each year, pesticides seriously poison at least 3 million agricultural workers in less-developed countries and at least 300,000 workers in the United States. They also cause 20,000–40,000 deaths per year, worldwide. Each year, more than 250,000 people in the United States become ill because of household pesticide use. In addition, such pesticides are a major source of accidental poisonings and deaths of young children. According to studies by the National Academy of Sciences, pesticide residues in food cause 4,000–20,000 cases of cancer per year in the United States. (See more on this topic in Chapter 14 and in *The Habitable Planet*, Video 7, at **www.learner.org/resources/series209.html**.)

The pesticide industry disputes these claims, arguing that the exposures are not high enough to cause serious environmental or health problems. Figure 10-20 lists some ways in which you can reduce your exposure to synthetic pesticides.

What Can You Do?

Reducing Exposure to Pesticides

- Grow some of your food using organic methods
- Buy certified organic food
- Wash and scrub all fresh fruits, vegetables, and wild foods you pick
- Eat less meat, no meat, or 100% organically produced meat
- Trim the fat from meat

Figure 10-20 Individuals matter: You can reduce your exposure to pesticides. **Questions:** Which three of these steps do you think are the most important ones to take? Why?

Pesticide Use Has Not Consistently Reduced U.S. Crop Losses to Pests

Despite some claims to the contrary, largely because of genetic resistance and the loss of many natural predators, synthetic pesticides have not always succeeded in reducing U.S. crop losses.

When David Pimentel, an expert on insect ecology, evaluated data from more than 300 agricultural scientists and economists, he reached three major conclusions. *First,* between 1942 and 1997, estimated crop losses from insects almost doubled from 7% to 13%, despite a 10-fold increase in the use of synthetic insecticides. *Second,* according to the International Food Policy Research Institute, the estimated environmental, health, and social costs of pesticide use in the United States are $5–10 in damages for every dollar spent on pesticides. *Third,* experience indicates that alternative pest management practices could cut the use of synthetic pesticides by half on 40 major U.S. crops without reducing crop yields (**Concept 10-4**).

The pesticide industry disputes these findings. However, numerous studies and experience support them. Sweden has cut pesticide use in half with almost no decrease in crop yields. Campbell Soup uses no pesticides on the tomatoes it grows in Mexico, and yields have not dropped. And after a two-thirds cut in pesticide use on rice in Indonesia, yields increased by 15%.

■ CASE STUDY
Ecological Surprises: The Law of Unintended Consequences

Malaria once infected nine of every ten people in North Borneo, now known as the eastern Malaysian state of Sabah. In 1955, the World Health Organization (WHO) sprayed the island with dieldrin (a DDT relative) to kill malaria-carrying mosquitoes. The program was so successful that the dreaded disease was nearly eliminated.

Then unexpected things began to happen. The dieldrin also killed other insects, including flies and cockroaches living in houses. The islanders applauded. Next, small insect-eating lizards living in the houses died after gorging themselves on dieldrin-contaminated insects. Then cats began dying after feeding on the lizards. In the absence of cats, rats flourished and overran the villages. When the residents became threatened by sylvatic plague carried by rat fleas, the WHO parachuted healthy cats onto the island to help control the rats. Operation Cat Drop worked.

But then the villagers' roofs began to fall in. The dieldrin had killed wasps and other insects that fed on a type of caterpillar that had either avoided or was not affected by the insecticide. With most of its predators eliminated, the caterpillar population exploded, munching its way through its favorite food: the leaves used to thatch roofs.

Ultimately, this episode ended well. Both malaria and the unexpected effects of the spraying program were brought under control. Nevertheless, this chain of unintended and unforeseen events emphasizes the unpredictability of using synthetic pesticides. It reminds us that whenever we intervene in nature, everything we do affects something else, and we need to ask, "Now what will happen?"

Laws and Treaties Can Help to Protect Us from the Harmful Effects of Pesticides

More than 25,000 different pesticide products are used in the United States. Three federal agencies, the EPA, the USDA, and the Food and Drug Administration (FDA), regulate the use of these pesticides under the Federal Insecticide, Fungicide, and Rodenticide Act (FIFRA), first passed in 1947 and amended in 1972.

Under FIFRA, the EPA was supposed to assess the health risks of the active ingredients in synthetic pesticide products already in use. However, after more than 30 years, less than 10% of the active ingredients in pesticide products have been tested for chronic health effects. Serious evaluation of the health effects of the 1,200 inactive ingredients used in pesticide products began only recently. The EPA says that the U.S. Congress has not provided them with enough funds to carry out this complex and lengthy evaluation process.

In 1996, Congress passed the Food Quality Protection Act, mostly because of growing scientific evidence and citizen pressure concerning the effects of small amounts of pesticides on children. This act requires the EPA to reduce the allowed levels of pesticide residues in food by a factor of 10 when there is inadequate information on the potentially harmful effects on children.

Between 1972 and 2010, the EPA used FIFRA to ban or severely restrict the use of 64 active pesticide ingredients, including DDT and most other chlorinated hydrocarbon insecticides. However, according to studies by the National Academy of Sciences, federal laws regulating pesticide use are inadequate and poorly enforced by the three agencies. One study found that as much as 98% of the potential risk of developing cancer from pesticide residues on food grown in the United States would be eliminated if EPA standards were as strict for pre-1972 pesticides as they are for later ones.

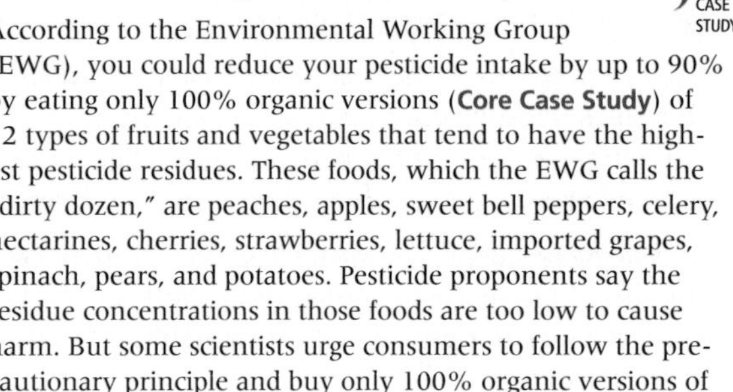
CONNECTIONS

Pesticides and Organic Foods: The Dirty Dozen CORE CASE STUDY

According to the Environmental Working Group (EWG), you could reduce your pesticide intake by up to 90% by eating only 100% organic versions (**Core Case Study**) of 12 types of fruits and vegetables that tend to have the highest pesticide residues. These foods, which the EWG calls the "dirty dozen," are peaches, apples, sweet bell peppers, celery, nectarines, cherries, strawberries, lettuce, imported grapes, spinach, pears, and potatoes. Pesticide proponents say the residue concentrations in those foods are too low to cause harm. But some scientists urge consumers to follow the precautionary principle and buy only 100% organic versions of these dirty dozen foods.

Although laws within countries protect citizens to some extent, banned or unregistered pesticides may be manufactured in one country and exported to other countries. For example, U.S. pesticide companies make and export to other countries pesticides that have been banned or severely restricted—or never even evaluated—in the United States. Other industrial countries also export banned or unapproved pesticides.

However, what goes around can come around. In what environmental scientists call a *circle of poison*, or the *boomerang effect*, residues of some banned or unapproved chemicals used in synthetic pesticides exported to other countries can return to the exporting countries on imported food. Winds can also carry persistent pesticides from one country to another.

Environmental and health scientists have urged the U.S. Congress—without success—to ban such exports. Supporters of the exports argue that such sales increase economic growth and provide jobs, and that banned pesticides are exported only with the consent of the importing countries. They also contend that if the United States did not export pesticides, other countries would.

In 1998, more than 50 countries developed an international treaty that requires exporting countries to have informed consent from importing counties for exports of 22 synthetic pesticides and 5 industrial chemicals. In 2000, more than 100 countries developed an international agreement to ban or phase out the use of 12 especially hazardous persistent organic pollutants (POPs)—9 of them persistent hydrocarbon pesticides such as DDT and other chemically similar pesticides. The United States has not signed this international agreement.

THINKING ABOUT

Exporting Pesticides

Should companies be allowed to export synthetic pesticides that have been banned, severely restricted, or not approved for use in their home countries? Explain.

There Are Alternatives to Synthetic Pesticides

Many scientists believe we should greatly increase the use of biological, ecological, and other alternative methods for controlling pests and diseases that affect crops and human health (**Concept 10-4**). Here are some of these alternatives:

- *Fool the pest.* A variety of *cultivation practices* can be used to fake out pests. Examples include rotating the types of crops planted in a field each year and adjusting planting times so that major insect pests either starve or get eaten by their natural predators.
- *Provide homes for pest enemies.* Farmers can increase the use of polyculture, which uses plant diversity to reduce losses to pests by providing habitats for the pests' predators.
- *Implant genetic resistance.* Use genetic engineering to speed up the development of pest- and disease-resistant crop strains (Figure 10-21). But controversy persists over whether the projected advantages of using GM plants outweigh their projected disadvantages (Figure 10-15, right).
- *Bring in natural enemies.* Use *biological control* by importing natural predators (Figures 10-18 and 10-22), parasites, and disease-causing bacteria and viruses to help regulate pest populations. This approach is nontoxic to other species, minimizes genetic resistance, and is usually less costly than applying pesticides. However, some biological control agents are difficult to mass-produce and are often slower acting and more difficult to apply than synthetic pesticides are. Sometimes the agents can multiply and become pests themselves.

Monsanto

Figure 10-21 **Solutions:** *Genetic engineering* can be used to reduce pest damage. Both of these tomato plants were exposed to destructive caterpillars. The normal plant's leaves are almost gone (left), whereas the genetically altered plant shows little damage (right). **Questions:** Would you have any concerns about eating the genetically engineered tomato? Why or why not?

Scott Bauer/USDA Agricultural Research Service

Figure 10-22 **Natural capital:** In this example of biological pest control, a wasp is parasitizing a gypsy moth caterpillar.

- *Use insect perfumes.* Trace amounts of *sex attractants* (called *pheromones*) can lure pests into traps or attract their natural predators into crop fields. Each of these chemicals attracts only one species. They have little chance of causing genetic resistance and are not harmful to nontarget species. However, they are costly and time-consuming to produce.
- *Bring in the hormones.* Hormones are chemicals produced by animals to control their developmental processes at different stages of life. Scientists have learned how to identify and use hormones that disrupt an insect's normal life cycle, thereby preventing it from reaching maturity and reproducing. Insect hormones have some of the same advantages and disadvantages of sex attractants. Also, they take weeks to kill an insect, are often ineffective with large infestations of insects, and sometimes break down before they can act.
- *Reduce use of synthetic herbicides to control weeds.* Organic farmers (**Core Case Study**) control weeds by methods such as crop rotation, mechanical cultivation, hand weeding, and the use of cover crops and mulches. CORE CASE STUDY

Integrated Pest Management Is a Component of More Sustainable Agriculture

Many pest control experts and farmers believe the best way to control crop pests is a carefully designed **integrated pest management (IPM)** program. In this more sustainable approach, each crop and its pests are evaluated as parts of an ecological system. Then farmers develop a carefully designed control program that uses a combination of cultivation, biological, and chemical tools and techniques, applied in a coordinated process (**Concept 10-4**).

The overall aim of IPM is to reduce crop damage to an economically tolerable level. Each year, crops are rotated, or moved from field to field, in an effort to disrupt pest infestations, and fields are monitored carefully. When an economically damaging level of pests is reached, farmers first use biological methods (natural predators, parasites, and disease organisms) and cultivation controls (such as altering planting time and using large machines to vacuum up harmful bugs). They apply small amounts of synthetic insecticides—mostly based on biopesticides—only when insect or weed populations reach a threshold where the potential cost of pest damage to crops outweighs the cost of applying the pesticide. Broad-spectrum, long-lived pesticides are not used, and different chemicals are used alternately to slow the development of genetic resistance and to help avoid killing predators of pest species.

In 1986, the Indonesian government banned 57 of the 66 synthetic pesticides used on rice and phased out pesticide subsidies over a 2-year period. It also launched a nationwide education program to help farmers switch to IPM. The results were dramatic: between 1987 and 1992, pesticide use dropped by 65% and rice production rose by 15%. In addition, nearly 1 million farmers have been trained in IPM techniques. (For more information and animations see *The Habitable Planet,* Video 7. at **www.learner.org/resources/series209.html**.) Sweden and Denmark have used IPM to cut their synthetic pesticide use by more than half. Cuba, which uses organic farming to grow its crops, makes extensive use of IPM. In Brazil, IPM has reduced pesticide use on soybeans by as much as 90%.

GOOD NEWS

According to a 2003 study by the U.S. National Academy of Sciences, these and other experiences show that a well-designed IPM program can reduce synthetic pesticide use and pest control costs by 50–65%, without reducing crop yields and food quality. IPM can also reduce inputs of fertilizer and irrigation water, and slow the development of genetic resistance, because pests are attacked less often and with lower doses of pesticides. IPM is an important form of *pollution prevention* that reduces risks to wildlife and human health and applies the biodiversity **principle of sustainability**.

SUSTAINABILITY

Despite its promise, IPM—like any other form of pest control—has some disadvantages. It requires expert knowledge about each pest situation and takes more time than does using conventional pesticides. Methods developed for a crop in one area might not apply to areas with even slightly different growing conditions. Initial costs may be higher, although long-term costs typically are lower than those of using conventional pesticides. Widespread use of IPM is hindered in the United States and a number of other countries by government subsidies for using synthetic chemical pesticides, as well as by opposition from pesticide manufacturers, and a shortage of IPM experts. **GREEN CAREER:** integrated pest management

A growing number of scientists are urging the USDA to use a three-point strategy to promote IPM in the United States. *First,* add a 2% sales tax on synthetic pesticides and use the revenue to fund IPM research and education. *Second,* set up a federally supported IPM demonstration project on at least one farm in every county in the United States. *Third,* train USDA field personnel and county farm agents in IPM so they can help farmers use this alternative. Because these measures would reduce its profits, the pesticide industry has vigorously, and successfully, opposed them.

Several UN agencies and the World Bank have joined together to establish an IPM facility. Its goal is to promote the use of IPM by disseminating information and establishing networks among researchers, farmers, and agricultural extension agents involved in IPM.

GOOD NEWS

10-5 How Can We Improve Food Security?

▶ **CONCEPT 10-5 We can improve food security by reducing poverty and chronic malnutrition, relying more on locally grown food, and cutting food waste.**

Use Government Policies to Improve Food Production and Security

Agriculture is a financially risky business. Whether farmers have a good or bad year depends on factors over which they have little control: weather, crop prices, crop pests and diseases, interest rates on loans, and global markets.

Governments use two main approaches to influence food production. First, they can *control prices* by putting a legally mandated upper limit on prices in order to keep food prices artificially low. This makes consumers happy but makes it harder for farmers to make a living.

Second, they can *provide subsidies* by giving farmers price supports, tax breaks, and other financial support to keep them in business and to encourage them to increase food production. Globally, government farm subsidies in more-developed countries average about $571,000 per minute and account for almost 33% of global farm income. However, if government subsidies are too generous and the weather is good, crop and livestock farmers may produce more food than can be sold.

Some analysts call for ending such subsidies. For example, in 1984, New Zealand ended farm subsidies. After the shock wore off, innovation took over and production of some foods such as milk quadrupled. Brazil has also ended most of its farm subsidies. Some analysts call for replacing traditional subsidies for farmers with subsidies that promote more environmentally sustainable farming practices.

Similarly, government subsidies to fishing fleets can promote overfishing and the reduction of aquatic biodiversity. For example, several governments give the highly destructive bottom-trawling industry (see Figure 9-23, right, p. 196) a total of about $150 million in subsidies a year, which is the main reason that fishers who use this practice can stay in business. Many analysts call for replacing those harmful subsidies with payments that would likely promote more sustainable fishing and aquaculture.

To improve food security and human health, some analysts urge governments to establish special programs focused on saving children from the harmful health effects of poverty. Studies by the United Nations Children's Fund (UNICEF) indicate that one-half to two-thirds of nutrition-related childhood deaths could be prevented at an average annual cost of $5–10 per child. This would involve simple measures such as immunizing more children against childhood diseases, preventing dehydration from diarrhea by giving infants a mixture of sugar and salt in their water, and preventing blindness by giving children an inexpensive vitamin A capsule twice a year.

10-6 How Can We Produce Food More Sustainably?

▶ **CONCEPT 10-6 We can produce food more sustainably by using resources more efficiently, sharply decreasing the harmful environmental effects of industrialized food production, and eliminating government subsidies that promote such harmful impacts.**

Reduce Soil Erosion

Land used for food production must have fertile topsoil (Figure 10-A), and it takes hundreds of years for fertile topsoil to form. Thus, sharply reducing topsoil erosion is the single most important component of more sustainable agriculture.

Soil conservation involves using a variety of methods to reduce topsoil erosion and restore soil fertility, mostly by keeping the land covered with vegetation. Farmers have used a number of methods to reduce topsoil erosion. For example, *terracing* is a way to grow food on steep slopes without depleting their topsoil. It is done by converting steeply sloped land into a series of broad, nearly level terraces that run across the land's contours (Figure 10-23a). This retains water for crops at each level and reduces topsoil erosion by controlling runoff.

On land with a significant slope, *contour planting* (Figure 10-23b) can be used to reduce topsoil erosion. It involves plowing and planting crops in rows across the slope of the land rather than up and down. Each row acts as a small dam to help hold topsoil and to slow runoff.

Strip cropping (Figure 10-23b) helps to reduce erosion and to restore soil fertility. It involves planting alternating strips of a row crop (such as corn or cotton) with another crop that completely covers the soil, called a *cover crop* (such as alfalfa, clover, oats, or rye). The cover crop traps topsoil that erodes from the row crop and catches and reduces water runoff. Some cover crops also add nitrogen to the soil. When one crop is harvested, the other strip is left to catch and reduce water runoff. Other ways to reduce topsoil erosion are to leave crop residues on the land after the crops are harvested or to plant cover crops immediately after harvest to help protect and hold the topsoil.

Alley cropping, or *agroforestry* (Figure 10-23c), is yet another way to slow the erosion of topsoil and to maintain soil fertility. One or more crops, usually legumes or other crops that add nitrogen to the soil, are planted together in alleys between orchard trees or fruit-bearing shrubs, which provide shade. This reduces water loss by evaporation and helps retain and slowly release soil moisture—an insurance policy during prolonged drought. The trees also can provide fruit, and tree trimmings can be used as fuelwood and as mulch for the crops.

Farmers can also establish *windbreaks,* or *shelterbelts,* of trees around crop fields to reduce wind erosion (Figure 10-23d). The trees retain soil moisture, supply wood for fuel, increase crop productivity by 5–10%, and provide habitats for birds and for insects that help with pest control and pollination.

(a) Terracing

(b) Contour planting and strip cropping

(c) Alley cropping

(d) Windbreaks

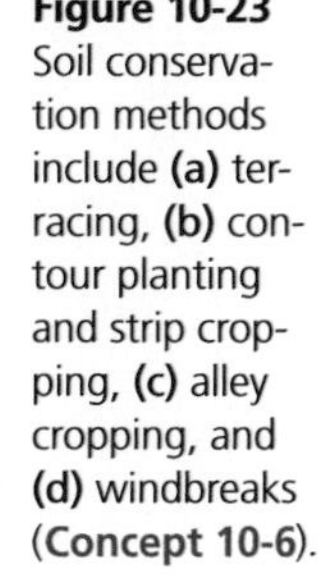

Figure 10-23 Soil conservation methods include **(a)** terracing, **(b)** contour planting and strip cropping, **(c)** alley cropping, and **(d)** windbreaks (**Concept 10-6**).

However, many farmers do not practice these effective ways to reduce soil erosion because they are in a desperate struggle to survive. Others are more interested in increasing short-term income, even if it leads to long-term environmental degradation.

The practices of eliminating or minimizing the plowing and tilling of topsoil and leaving crop residues on the ground greatly reduce topsoil erosion. Many farmers practice *conservation-tillage farming* by using special tillers and planting machines that drill seeds directly through crop residues into the undisturbed topsoil. The only disturbance to the soil is a narrow slit cut by the planter, and weeds are controlled with herbicides. Such *no-till* and *low-till* farming increases crop yields and reduces the threat of projected climate change by storing more carbon in organic compounds in the soil. It also reduces water pollution from sediment and fertilizer runoff and lessens the use of water and tractor fuel. One study showed that erosion from a conventionally tilled watershed was 700 times greater than that from a no-till watershed.

In 2008, farmers used conservation tillage on about 41% of U.S. cropland, helped by the use of herbicides. The USDA estimates that using conservation tillage on 80% of U.S. cropland would reduce topsoil erosion by at least 50%. No-till cultivation is used on less than 7% of the world's cropland, although it is widely used in some countries, including the United States, Brazil, Argentina, Canada, and Australia.

However, conservation tillage is not a cure-all. It requires costly machinery and works better in some soils than in others. It is not useful for wetland rice or for root crops such as potatoes, and it can result in increased use of herbicides.

Another way to conserve the earth's topsoil is to retire the estimated one-tenth of the world's marginal cropland that is highly erodible and accounts for the majority of the world's topsoil erosion. The goal would be to identify *erosion hotspots*, withdraw these areas from cultivation, and plant them with grasses or trees, at least until their topsoil has been renewed.

■ CASE STUDY

Soil Erosion in the United States

In the United States, a third of the country's original topsoil is gone and much of the rest is degraded. In the state of Iowa, which has the world's highest concentration of prime farmland, half of the topsoil is gone after a century of industrialized farming. According to the Natural Resources Conservation Service, 90% of American farmland is, on average, losing topsoil 17 times faster than new topsoil is being formed.

In 1935, the United States passed the *Soil Erosion Act*, which established the Soil Conservation Service (SCS) as part of the USDA. (The SCS is now called the Natural Resources Conservation Service, or NRCS.) Farmers and ranchers were given technical assistance to set up soil conservation programs.

With the help of the NRCS, U.S. farmers are sharply reducing some of their topsoil losses through a combination of conservation-tillage farming and government-sponsored soil conservation programs (**Concept 10-6**). Under the 1985 Food Security Act (Farm Act), more than 400,000 farmers participating in the Conservation Reserve Program received subsidy payments for taking highly erodible land—totaling an area larger than the U.S. state of New York—out of production and replanting it with grass or trees for 10–15 years. Since 1985, these efforts have cut topsoil losses on U.S. cropland by 40%.

There is still room for improvement, however. Effective topsoil conservation is practiced today on only half of all U.S. agricultural land. But the United States is currently the only major food-producing nation to be significantly reducing its topsoil losses.

CONNECTIONS

Corn, Ethanol, and Soil Conservation

In recent years, some U.S. farmers have been taking erodible land out of the conservation reserve in order to receive generous government subsidies for planting corn (which removes nitrogen from the soil) to make ethanol for use as a motor vehicle fuel. This has led to mounting political pressure to abandon or sharply cut back on the nation's highly successful topsoil conservation reserve program.

Restore Soil Fertility

The best way to maintain soil fertility is through topsoil conservation. The next best option is to restore some of the lost plant nutrients that have been washed, blown, or leached out of topsoil, or that have been removed by repeated crop harvesting. To do this, farmers can use **organic fertilizer** from plant and animal materials or **synthetic inorganic fertilizer** produced from various minerals.

There are several types of *organic fertilizers*. One is **animal manure**: the dung and urine of cattle, horses, poultry, and other farm animals. It improves topsoil structure, adds organic nitrogen, and stimulates the growth of beneficial soil bacteria and fungi. Another type, called **green manure**, consists of freshly cut or growing green vegetation that is plowed into the topsoil to increase the organic matter and humus available to the next crop. A third type is **compost**, produced when microorganisms in topsoil break down organic matter such as leaves, crop residues, food wastes, paper, and wood in the presence of oxygen. Farmers using 100% organic agriculture (**Core Case Study**) use only these types of organic fertilizers along with cover crops such as clover that remove nitrogen from the air and transfer it to the soil. CORE CASE STUDY

Crops such as corn and cotton can deplete nutrients in the topsoil (especially nitrogen) if they are planted on the same land several years in a row. *Crop rotation* provides one way to reduce these losses. Farmers plant areas or strips with nutrient-depleting crops one year. The next year, they plant the same areas with legumes, whose root nodules add nitrogen to the soil. This not only helps to restore topsoil nutrients but also reduces erosion by keeping the topsoil covered with vegetation.

Many farmers (especially in more developed countries) rely on *synthetic inorganic fertilizers.* The major active ingredients typically are inorganic compounds that contain *nitrogen, phosphorus,* and *potassium.* Other plant nutrients may be present in low or trace amounts. The use of synthetic inorganic fertilizers has grown more than ninefold since 1950, and it now accounts for about 25% of the world's crop yield. Unless erosion is controlled, these fertilizers can be washed off the land and can pollute nearby bodies of water and coastal estuaries. While these fertilizers can replace depleted inorganic nutrients, they do not replace organic matter. To completely restore nutrients to topsoil, both inorganic and organic fertilizers should be used.

Reduce Soil Salinization and Desertification

We know how to prevent and deal with soil salinization. One important way will be to reduce the amount of water that is put onto crop fields through more widespread use of efficient irrigation technologies that are available today (Figure 10-24).

Drip, or *trickle irrigation,* also called *microirrigation* (Figure 10-24, right), is the most efficient way to deliver small amounts of freshwater to crops precisely. It consists of a network of perforated plastic tubing installed at or below the ground level. Small pinholes in the tubing deliver drops of freshwater at a slow and steady rate, close to the roots of individual plants. GOOD NEWS

These systems drastically reduce freshwater waste because 90–95% of the water input reaches the crops. By using less freshwater, they also reduce the amount of harmful salt that irrigation water leaves in the soil. A new type of drip irrigation system developed by the nonprofit group International Development Enterprises (IDE) drastically cuts the costs of this system. (IDE is a team of specialists dedicated to providing income opportunities to poor rural households.) Other ways to prevent and reduce soil salinization are summarized in Figure 10-25.

Reducing desertification is not easy. We cannot control the timing and location of prolonged droughts caused by changes in weather patterns. But we can reduce population growth, overgrazing, deforestation, and destructive forms of planting, irrigation, and mining, which have left much land vulnerable to topsoil erosion and thus desertification. We can also work to decrease the human contribution to projected climate change, which is expected to increase severe and prolonged droughts in larger areas of the world during this century. In addition, we can restore land suffering from desertification by planting trees (see Chapter 9 Core Case Study, p. 174) and other plants that anchor topsoil and hold water. We can also grow trees and crops together and establish windbreaks.

Practice More Sustainable Aquaculture

Figure 10-26 lists some ways to make aquaculture more sustainable and to reduce its harmful environmental effects. One such approach is open-ocean aquaculture, which involves raising large carnivorous fish in under-

Figure 10-24 These three types of systems are commonly used to irrigate crops. The most efficient system is the low-energy, precision application (LEPA) center-pivot system.

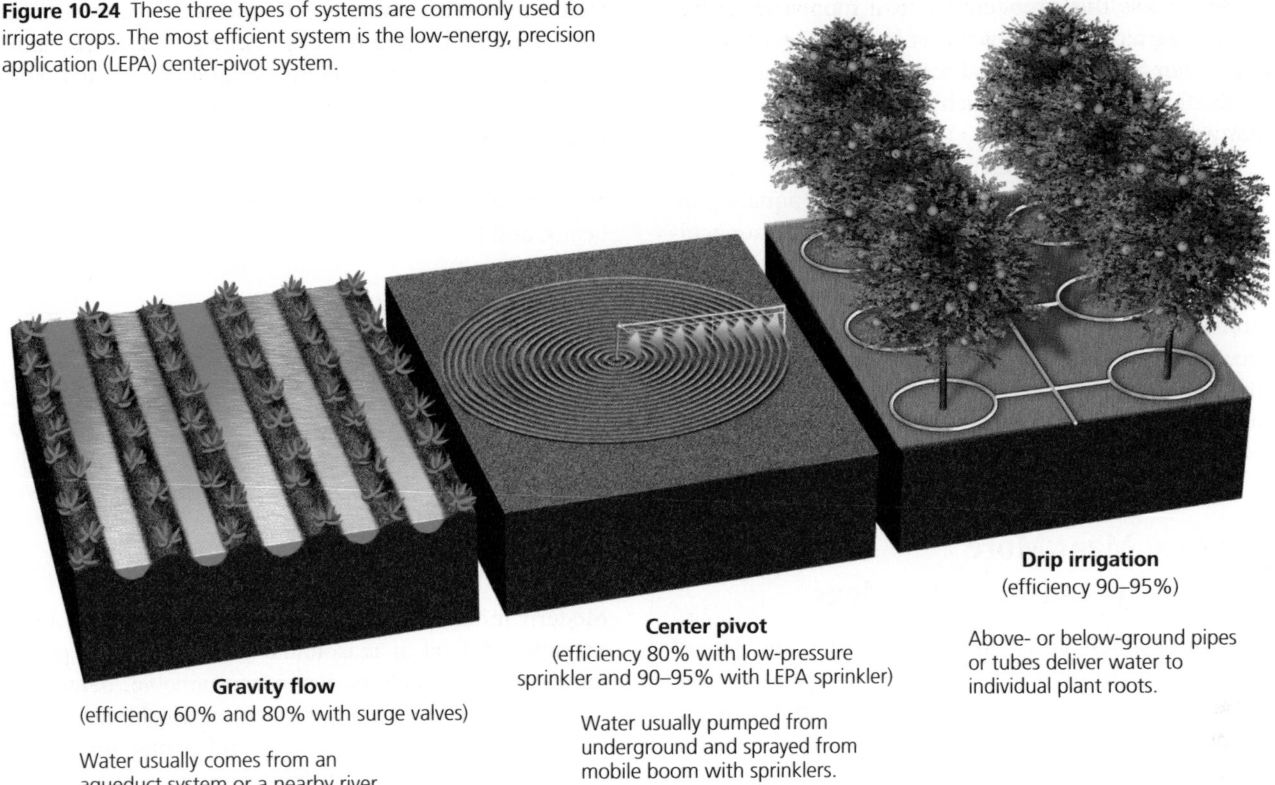

water pens located up to 300 kilometers (190 miles) offshore (see Figure 9-26, p. 198). The fish are fattened with fish meal supplied by automated buoys, and their wastes are diluted by surrounding ocean water, winds, and currents. Some farmed fish can escape and breed with wild fish; however, the environmental impact of raising fish far offshore is smaller than that of raising fish near shore and much smaller than that of industrialized commercial fishing. **GREEN CAREER:** sustainable aquaculture

In the long run, making aquaculture more sustainable will require some fundamental changes. One such change would be for more consumers to choose fish species that eat algae and other vegetation rather than other fish. Raising carnivorous fishes such as salmon, trout, tuna, grouper, and cod contributes to overfishing and population crashes within species used to feed these carnivores, and will eventually be unsustainable. Raising plant-eating fishes such as carp, tilapia, and

Solutions

Soil Salinization

Prevention	Cleanup
Reduce irrigation	Flush soil (expensive and wastes water)
Use more efficient irrigation methods	Stop growing crops for 2–5 years
Switch to salt-tolerant crops	Install underground drainage systems (expensive)

Figure 10-25 There are ways to prevent soil salinization and ways to clean it up (**Concept 10-6**). **Questions:** Which two of these solutions do you think are the most important? Why?

Solutions

More Sustainable Aquaculture

- Protect mangrove forests and estuaries
- Improve management of wastes
- Reduce escape of aquaculture species into the wild
- Raise some species in deeply submerged cages
- Set up self-sustaining aquaculture systems that combine aquatic plants, fish, and shellfish
- Certify and label sustainable forms of aquaculture

Figure 10-26 We can make aquaculture more sustainable and reduce its harmful effects. **Questions:** Which two of these solutions do you think are the most important? Why?

catfish avoids this problem and is a more sustainable form of aquaculture. However, it becomes less sustainable when aquaculture producers try to increase yields by feeding fish meal to such plant-eating species, as many of them are.

Another change would be for fish farmers to emphasize *polyaquaculture*, which has been part of aquaculture for centuries, especially in Southeast Asia. Polyaquaculture operations raise fish or shrimp along with algae, seaweeds, and shellfish in coastal lagoons, ponds, and tanks. The wastes of the fish or shrimp feed the other species, and in the best of these operations, there are just enough wastes from the first group to feed the second group. This applies the recycling and biodiversity **principles of sustainability**.

Produce Meat More Efficiently and Eat Less Meat

Meat production and consumption make the largest contributions to the ecological footprints of most individuals in affluent nations. If everyone in the world today consumed the average U.S. meat-based diet, the world's current annual grain harvest could feed only about one-third of the world's people.

A more sustainable form of meat production and consumption would involve shifting from less grain-efficient forms of animal protein, such as beef, pork, and carnivorous fish produced by aquaculture, to more grain-efficient forms, such as poultry and plant-eating farmed fish (Figure 10-27).

A number of people are eating less meat (especially red meat) and some of them regularly have one or two meatless days per week. According to agricultural science writer Michael Pollan, if all Americans picked one day per week to have no meat, the reduction in greenhouse gas emissions would be equivalent to taking 30 to 40 million cars off the road for a year. Research also indicates that people who live on a Mediterranean-style diet that includes mainly poultry, seafood, cheese, raw fruits, vegetables, and olive oil tend to be healthier and live longer than those who eat a lot of red meat.

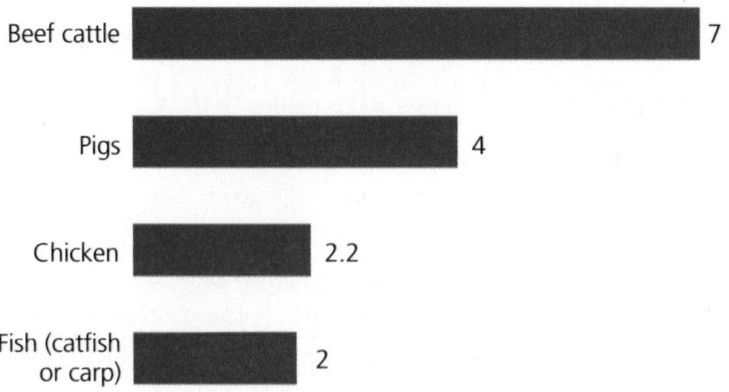

Figure 10-27 The efficiency of converting grain into animal protein varies with different types of meat. This bar graph show the kilograms of grain required for each kilogram of body weight added. **Question:** If you eat meat, what changes could you make to your diet that would reduce your environmental impact? (Data from U.S. Department of Agriculture)

Other people are going further and eliminating most or all meat from their diets. They are replacing it with a balanced vegetarian diet that includes a healthy combination of organically grown fruits and vegetables (**Core Case Study**) and protein-rich foods such as peas, beans, and lentils.

THINKING ABOUT
Meat Consumption

Would you be willing to live lower on the food chain by eating much less meat, or even no meat at all? Explain.

Shift to More Sustainable Food Production

Modern industrialized food production produces large amounts of food at reasonable prices. But to a growing number of analysts, it is unsustainable, because it violates the three **principles of sustainability**, which are based on observing and acting on what nature tells us about truly long-term sustainability. It relies heavily on the use of fossil fuels, which adds greenhouse gases to the atmosphere and thus contributes to climate change. It also reduces biodiversity and agrobiodiversity, and it reduces the cycling of plant nutrients back to topsoil. These facts are hidden from consumers because most of the harmful environmental costs of food production (Figure 10-8) are not included in the market prices of food.

Figure 10-28 lists the major components of more sustainable food production. Compared to high-input farming, low-input agriculture produces similar yields with less energy input per unit of yield and lower carbon dioxide emissions. It improves topsoil fertility and reduces topsoil erosion, can often be more profitable for farmers, and can help poor families to feed themselves (**Concept 10-6**).

One component of more sustainable food production is organic agriculture (**Core Case Study**). There is mixed evidence on whether organic foods are nutritionally healthier to eat than conventional foods. The main reason for shifting to organic farming is that it sharply reduces the harmful environmental effects of industrialized farming (Figure 10-8) as well as our exposure to pesticide residues. Also, research by the Cornucopia Institute indicates that some people buy 100% organic food because it encourages more humane treatment of animals used for food and is a more economically just system for farm workers and farmers. Figure 10-29 summarizes the major conclusions of a 22-year research study at the Rodale Institute in Kutztown, Pennsylvania (USA) comparing organic and conventional farming.

One drawback of organic farming, compared with industrialized agriculture, is that it requires more

Solutions

More Sustainable Food Production

More	Less
High-yield polyculture	Soil erosion
Organic fertilizers	Soil salinization
Biological pest control	Water pollution
Integrated pest management	Aquifer depletion
Efficient irrigation	Overgrazing
Perennial crops	Overfishing
Crop rotation	Loss of biodiversity and agrobiodiversity
Water-efficient crops	Fossil fuel use
Soil conservation	Greenhouse gas emissions
Subsidies for sustainable farming	Subsidies for unsustainable farming

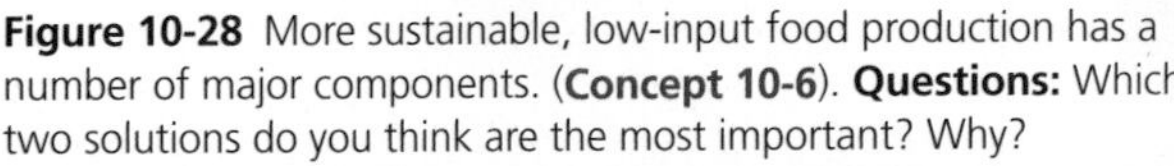

Figure 10-28 More sustainable, low-input food production has a number of major components. (**Concept 10-6**). **Questions:** Which two solutions do you think are the most important? Why?

Solutions

Organic Farming

- Improves soil fertility
- Reduces soil erosion
- Retains more water in soil during drought years
- Uses about 30% less energy per unit of yield
- Lowers CO_2 emissions
- Reduces water pollution by recycling livestock wastes
- Eliminates pollution from pesticides
- Increases biodiversity above and below ground
- Benefits wildlife such as birds and bats

Figure 10-29 More than two decades of research have revealed some of the major advantages of organic farming over conventional industrialized farming. (Data from Paul Mader, David Dubois, and David Pimentel)

human labor to use methods such as integrated pest management, crop rotation, low-till cultivation, and multi-cropping. However, it will be an important provider of jobs and income, especially for some farmers in less-developed countries. Organic farming is also a source of income for a growing number of younger farmers in more-developed countries who want to apply ecological principles to agriculture and help reduce its harmful environmental impacts.

While yields from organic farming are sometimes lower then conventional yields, farmers often make up for this by not having to use expensive synthetic pesticides, herbicides, and fertilizers, and they usually receive higher prices for their crops. As a result, the net economic return per unit of land from organic crop production is often equal to or higher than that from conventional crop production. However, there is a need for more research to evaluate the benefits of organic farming and to carry it out as sustainably as possible.

Another important component of more sustainable agriculture could be to rely less on single-crop organic agriculture and more on *organic polyculture*—in which a diversity of organic crops are grown on the same plot. For example, a diversified organic vegetable farm may grow forty or more different crops on one piece of land. Of particular interest to some scientists is the idea of using polyculture to grow *perennial crops*—crops that grow back year after year on their own (Science Focus, p. 232).

Well-designed organic polyculture, like organic monoculture, helps to conserve and replenish topsoil, requires less water, cuts water waste, and reduces the need for fertilizers and pesticides. It also reduces the air and water pollution associated with conventional industrialized agriculture.

Most proponents of more sustainable agriculture are not opposed to high-yield agriculture. Instead, they see it as vital because it protects the earth's biodiversity and it reduces the need to cultivate new and often marginal land. They call for using more environmentally sustainable forms of both high-yield polyculture and high-yield monoculture, with increasing emphasis on employing organic methods for growing crops (**Concept 10-6**) and relying more on 100% organic foods.

Another key to developing more sustainable agriculture is to shift from using fossil fuels to relying more on solar energy for food production—an important application of the solar energy **principle of sustainability**. Farmers can make greater use of renewable energy sources such as wind, flowing water, natural gas (produced from farm wastes in tanks called *biogas digesters*), and solar energy to produce electricity and fuels needed for food production. Today some farmers make money by selling any excess electricity they generate from wind and other renewable forms of energy to power companies. Agricultural science writer Michael Pollan refers to this shift toward using more renewable energy for farming as "re-solarizing our food chain."

SCIENCE FOCUS

The Land Institute and Perennial Polyculture

Some scientists call for greater reliance on conventional and organic polycultures of perennial crops as a component of more sustainable agriculture. Such crops can live for many years without having to be replanted and are better adapted to regional soil and climate conditions than most annual crops.

Over 3 decades ago, plant geneticist Wes Jackson co-founded The Land Institute in the U.S. state of Kansas. One of the institute's goals has been to grow a diverse mixture (polyculture) of edible perennial plants to supplement traditional annual monoculture crops and to help reduce the latter's harmful environmental effects. Examples in this polyculture mix include perennial grasses, plants that add nitrogen to the soil, sunflowers, grain crops, and plants that provide natural insecticides. Some of these plants could also be used as a source of renewable biofuel for motor vehicles.

This approach, called *natural systems agriculture*, copies nature by growing a diversity of perennial crops, ideally by using organic methods. It has a number of environmental benefits. Because the plants are perennials, there is no need to till the soil and replant seeds each year. This reduces topsoil erosion and water pollution from eroded sediment, because the unplowed topsoil is not exposed to wind and rain. In addition, it reduces the need for irrigation because the deep roots of such perennials retain more water than do the shorter roots of annuals (Figure 10-C). Also, there is little or no need for chemical fertilizers and pesticides, and thus little or no pollution from these sources. Perennial polycultures also remove and store more carbon from the atmosphere, and growing them requires less energy than does growing crops in conventional monocultures.

Wes Jackson calls for governments to promote this and other forms of more sustainable agriculture that would help to reduce topsoil erosion, sustain nitrogen nutrients in topsoil, cut the wasteful use of irrigation water, reduce dependence on fossil fuels, and reduce dead zones in coastal areas (Connections, p. 218). He reminds us that "if our agriculture is not sustainable, then our food supply is not sustainable."

Critical Thinking

Why do you think large seed companies generally oppose this form of more sustainable agriculture?

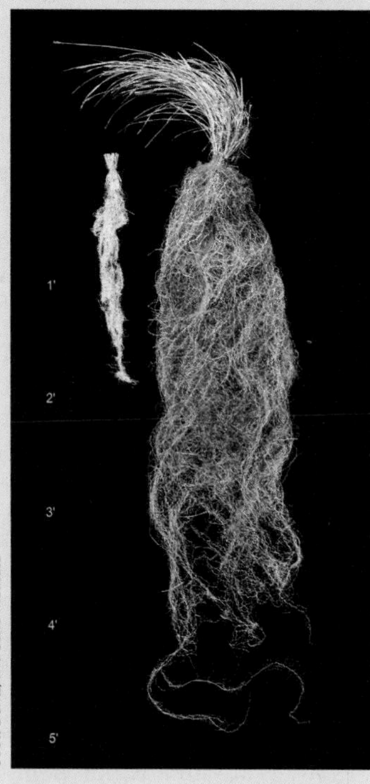

Figure 10-C The roots of an annual wheat crop plant (left) are much shorter than those of big bluestem (right), a tallgrass prairie perennial plant. The perennial plant is in the ground year-round and is much better at using water and nutrients, and at helping to maintain healthy topsoil. It also needs less fertilizer.

Proponents of more sustainable agricultural systems point out that we have all of the components needed for making a shift to such systems. GOOD NEWS They say that over the next 5 decades, a combination of education and economic policies that reward more sustainable agriculture could lead to such a shift. A key is to inform people, especially young consumers about where their food really comes from, how it is produced, and what the environmentally harmful effects of industrialized food production are, as we have tried to do in this chapter.

Critics say that this approach will not produce enough food and will lead to food shortages. However, experience and research show that well-designed forms of more sustainable polyculture can produce much higher crop yields than can conventional industrialized farming of single crops (monocultures). It can also greatly reduce the huge harmful environmental impact of conventional industrialized agriculture. Shifting to such agriculture could increase the world's food supply by up to 50%, according to a recent study by University of Michigan scientists.

> **CONNECTIONS**
>
> **Corn, Ethanol, and Food Riots**
>
> A number of scientists and energy experts call for ending U.S. government subsidies for growing corn to make ethanol fuel for cars and for returning about one-fourth of all U.S. cropland to food production. They point out that corn ethanol production has to be subsidized because it takes about as much energy to grow the corn and convert it to ethanol as we get by burning the ethanol. The recent shift to growing more corn to feed cars instead of people has also led to higher corn prices and to food riots in some low-income, corn-importing countries such as Mexico, Indonesia, and Egypt.

Analysts suggest five major strategies to help farmers and consumers to make the transition to more sustainable agriculture over the next 50 years (**Concept 10-6**). *First,* greatly increase research on more sustainable organic farming (**Core Case Study**) and perennial polyculture (Science Focus, above), and on CORE CASE STUDY improving human nutrition. *Second,* establish education and training programs in more sustainable agriculture for students, farmers, and government agricultural officials. *Third,* set up an international fund to give farmers in poor countries access to various types of more sustainable agriculture. *Fourth,* replace government subsidies for environmentally harmful forms of industrialized agriculture with subsidies that encourage more sustainable agriculture. *Fifth,* mount a massive program to educate consumers about the true environmental and health costs of the food they buy. This would help people to understand why the current system is unsustainable, and it would help to build political support for including the harmful costs of food production in the market prices of food.

Buy Locally Grown Food, Grow More Food Locally, and Cut Food Waste

Figure 10-30 lists ways in which you can promote more sustainable food production. GOOD NEWS

According to some experts, one component of more sustainable agriculture is to grow more of our food locally or regionally, ideally with certified organic farming practices. A growing number of consumers are becoming "locavores" and buying more of their food from local and regional producers in farmers' markets, which provide access to fresher seasonal foods. In addition, some people are participating in *community-supported agriculture (CSA)* programs in which they buy shares of a local farmer's crop and receive a box of fruits or vegetables each week during the summer and fall. Some individuals have started businesses where they buy produce and meat from local farmers and then sell and deliver them to consumers. This allows the consumers to get the food products they want when they want them.

Yale University is committed to growing food locally and more sustainably. Students use a plot that takes up less than half the area of a typical city block to produce more than 300 varieties of vegetables, fruits, and flowers that are used on campus and sold locally. According to the school's dining director, up to 49% of the food served is locally grown, seasonal, or organic.

What Can You Do?

More Sustainable Food Production

- Eat less meat, no meat, or organically certified meat
- Choose sustainably produced herbivorous fish
- Use organic farming to grow some of your food
- Buy certified organic food
- Eat locally grown food
- Compost food wastes
- Cut food waste

Figure 10-30 Individuals matter: There are a number of ways to promote more sustainable food production (**Concept 10-6**). **Questions:** Which three of these actions do you think are the most important? Why?

An increase in the demand for locally grown food could result in more small, diversified farms that produce organic, minimally processed food from plants and animals. Such eco-farming could be one of this century's challenging new careers for many young people. But there is a long way to go. According to Michael Pollan, locally grown food currently accounts for less than 1% of the food consumed in the United States. **GREEN CAREER:** small-scale sustainable agriculture

People who live in urban areas could grow more of their own food. According to the USDA, around 15% of the world's food is grown in urban areas, and this percentage could easily be doubled. People plant gardens and raise chickens in suburban backyards. In cities, they grow food on rooftops and balconies, on patios, in raised beds in unused or partially used parking lots (a growing practice known as *asphalt gardening*), and on floating barges near urban areas. Dwarf fruit trees can be planted in large containers of soil. One visionary scientist sees the day when much of our food will be grown in city centers within specially designed high-rise buildings.

Finally, people can sharply cut food waste as an important component of improving food security (**Concept 10-5**). In 2008, environmental scientist Vaclav Smil and other researchers estimated that Americans waste 25–50% of their food supply. This wasted food is worth at least $43 billion a year, almost twice as much as the estimated $24 billion needed to eliminate undernutrition and malnutrition in the world. Many of the world's more-developed countries also have high rates of food waste. In addition, in some less-developed countries, food waste can be high due to spoilage from heat and pests and lack of refrigeration and food storage systems.

We are at a fork in the road. We can keep on using our current food production systems, which some experts argue are unsustainable and could collapse if fossil fuel prices rise sharply or if we disrupt climate patterns by continuing to overload the atmosphere with carbon dioxide and other greenhouse gases. Or we can try a new, more sustainable approach to producing food that will likely provide large numbers of jobs and sharply reduce the harmful environmental and health impacts of industrialized food production.

Here are this chapter's *three big ideas*:

- About 925 million people have health problems because they do not get enough to eat and 1.1 billion people face health problems from eating too much.
- Modern industrialized agriculture has a greater harmful impact on the environment than any other human activity.
- More sustainable forms of food production will greatly reduce the harmful environmental impacts of industrialized food production systems while likely increasing food security.

REVISITING Organic Agriculture and Sustainability

This chapter began with a look at how we can produce food by organic agriculture, a rapidly growing component of more sustainable agriculture (**Core Case Study**). Putting more emphasis on organic farming of diverse crops and perennial polyculture could help to reduce the enormous harmful environmental impacts of agriculture as a whole (Figure 10-8).

Making the transition to more sustainable agriculture involves applying the three **principles of sustainability** (see back cover). All of these principles are violated by modern industrialized agriculture, because it depends heavily on nonrenewable fossil fuels and includes too little recycling of crop and animal wastes. It accelerates topsoil erosion and water pollution, and does too little to preserve agrobiodiversity. Further, it can destroy or degrade wildlife habitats and disrupt natural species interactions that help to control pest population sizes. Conventional commercial fishing methods and aquaculture, along with other forms of industrialized food production, violate these principles of sustainability in similar ways.

Thus, making this transition means relying more on solar and other forms of renewable energy and less on fossil fuels. It also means sustaining chemical cycling through topsoil conservation and by returning crop residues and animal wastes to the soil. It involves helping to sustain natural, agricultural, and aquatic biodiversity by relying on a greater variety of crop and animal strains and seafoods. Controlling pest populations through broader use of conventional and perennial polyculture and integrated pest management, will also help to sustain biodiversity.

Such efforts will be enhanced if we can control the growth of the human population and our wasteful use of food and resources, and stop growing more corn and other crops to fuel cars instead of to feed people. Governments could help these efforts by replacing environmentally harmful agricultural subsidies and tax breaks with more environmentally beneficial ones. The transition to more sustainable food production would be accelerated if we could find ways to include the harmful environmental and health costs of food production in the market prices of food.

Making the transition to more sustainable forms of food production will not be easy. But it can be done if we mimic nature by using the three **principles of sustainability** to meet our food needs.

While there are alternatives to oil, there are no alternatives to food.

MICHAEL POLLAN

REVIEW

CORE CASE STUDY

1. Define **organic agriculture** and compare its main components with those of conventional industrialized agriculture (**Core Case Study**). CORE CASE STUDY

SECTION 10-1

2. What are the two key concepts for this section? Define **food security** and **food insecurity**. What is the root cause of food insecurity? Distinguish between **chronic undernutrition (hunger)** and **chronic malnutrition** and describe their harmful effects. Describe the effects of diet deficiencies in vitamin A, iron, and iodine. What is **overnutrition** and what are its harmful effects?

SECTION 10-2

3. What is the key concept for this section? What three systems supply most of the world's food? Define **irrigation**. Distinguish among **industrialized agriculture (high-input agriculture)**, **plantation agriculture**, **traditional subsistence agriculture**, **traditional intensive agriculture**, and **polyculture**. How does industrialized agriculture violate the three principles of sustainability? Define **soil** and describe its formation and the major layers in mature soils. Why is topsoil one of our most important resources? What is a **green revolution**? Describe industrialized food production in the United States.

4. Distinguish between crossbreeding through artificial selection and genetic engineering. Describe the second green revolution based on genetic engineering. Describe the growth of industrialized meat production. What is a fishery? What is aquaculture? Summarize the use of energy in industrialized food production. Why does it result in a net energy loss?

SECTION 10-3

5. What is the key concept for this section? What are the major harmful environmental impacts of agriculture? What is **soil erosion** and what are its two major harmful environmental effects? What is **desertification** and what are its harmful environmental effects? Distinguish between the **salinization** and **waterlogging** of soil and describe their harmful environmental effects. What is the biggest problem resulting from excessive use of water for irrigation in agriculture?

6. Summarize industrialized agriculture's contribution to projected climate change. Explain how industrialized food production systems reduce biodiversity. What is agrobiodiversity and how is it being affected by industrialized food

production? Describe the advantages and disadvantages of using genetic engineering in food production. What factors can limit green revolutions? Compare the advantages and disadvantages of industrialized meat production. Compare the advantages and disadvantages of aquaculture.

SECTION 10-4

7. What is the key concept for this section? What is a **pest**? Define and give two examples of a **pesticide**. Summarize Rachel Carson's story. Summarize the advantages and disadvantages of modern pesticides. Describe the use of laws and treaties to help protect U.S. citizens from the harmful effects of pesticides. List seven alternatives to conventional pesticides. Define **integrated pest management (IPM)** and discuss its advantages.

SECTION 10-5

8. What is the key concept for this section? What are the two main approaches used by governments to influence food production? How have governments used subsidies to influence food production and what have been some of their effects? What are two other ways in which organizations are improving food security?

SECTION 10-6

9. What is the key concept for this section? What is **soil conservation**? Describe six ways to reduce topsoil erosion. Summarize the history of soil erosion and soil conservation in the United States. Distinguish among the use of **organic fertilizer**, **synthetic inorganic fertilizer**, **animal manure**, **green manure**, and **compost** as ways to help restore topsoil fertility. Describe ways to prevent and ways to clean up soil salinization. Describe three major methods of irrigating crops. Which ones can help reduce soil salinization? How can we reduce desertification? How can we make aquaculture more sustainable? Describe ways to produce meat more sustainably. Summarize three important aspects of making a shift to more sustainable food production. Describe the advantages of organic farming and its role in making this shift. What five strategies could help farmers and consumers shift to more sustainable agriculture? List the advantages of relying more on organic polyculture and perennial crops. What are three important ways in which individual consumers can help to promote more sustainable agriculture?

10. What are the three big ideas of this chapter? Describe the relationship between industrialized agriculture and the three **principles of sustainability**. How can these principles be applied toward making a shift to more sustainable food production systems? SUSTAINABILITY

Note: Key terms are in **bold** type.

CRITICAL THINKING

1. Do you think that the advantages of organic agriculture (**Core Case Study**) outweigh its disadvantages? Explain. Do you eat or grow organic foods? If so, or if not, explain your reasoning. CORE CASE STUDY
2. Explain why you support or oppose greatly increased use of **(a)** genetically modified food, and **(b)** organic polyculture.
3. Suppose you live near a coastal area and a company wants to use a fairly large area of coastal marshland for an aquaculture operation. If you were an elected local official, would you support or oppose such a project? Explain.
4. Explain how the widespread use of a pesticide can **(a)** increase the damage done by a particular pest and **(b)** create new pest organisms.
5. If increased mosquito populations threatened you with malaria or West Nile virus, would you want to spray DDT in your yard and inside your home to reduce the risk? Explain. What are the alternatives?
6. List three ways in which your lifestyle directly or indirectly contributes to topsoil erosion.
7. According to physicist Albert Einstein, "Nothing will benefit human health and increase the chances of survival of life on Earth as much as the evolution to a vegetarian diet." Are you willing to eat less meat or no meat? Explain.
8. Congratulations! You are in charge of the world. List the three most important features of your **(a)** agricultural policy, **(b)** policy to reduce topsoil erosion, **(c)** policy for more sustainable harvesting and farming of fish and shellfish, and **(d)** global pest management strategy.

DOING ENVIRONMENTAL SCIENCE

For a week, weigh the food that is purchased in your home and the food that is thrown out. Record and compare these numbers from day to day. Develop a plan for cutting your household food waste in half. Consider making a similar study for your school cafeteria and reporting the results and your recommendations to school officials.

GLOBAL ENVIRONMENT WATCH EXERCISE

Under the *Environment and Ecology* portal, go to the *Soil Erosion* subtopic. Look for information on causes of soil erosion and how it affects soil fertility. Write a report on your findings. If you were to overhear a group of farmers complaining about how much money they must spend on fertilizers, what suggestions would you give them for saving money? Include your answer to this question, along with your reasoning, in your report.

DATA ANALYSIS

The following table gives the world's fish harvest and population data.

	World Fish Harvest				
Years	**Fish Catch (million metric tons)**	**Aquaculture (million metric tons)**	**Total (million metric tons)**	**World Population (in billions)**	**Per Capita Fish Consumption (kilograms/person)**
1990	84.8	13.1	97.9	5.27	
1991	83.7	13.7	97.4	5.36	
1992	85.2	15.4	100.6	5.44	
1993	86.6	17.8	104.4	5.52	
1994	92.1	20.8	112.9	5.60	
1995	92.4	24.4	116.8	5.68	
1996	93.8	26.6	120.4	5.76	
1997	94.3	28.6	122.9	5.84	
1998	87.6	30.5	118.1	5.92	
1999	93.7	33.4	127.1	5.995	
2000	95.5	35.5	131.0	6.07	
2001	92.8	37.8	130.6	6.15	
2002	93.0	40.0	133.0	6.22	
2003	90.2	42.3	132.5	6.20	

1. Use the world fish harvest and population data in the table to calculate the per capita fish consumption from 1990–2003 in kilograms/person. (Hints: 1 million metric tons equals 1 billion kilograms; the human population data is measured in billions; and per capita consumption can be calculated directly by dividing the total amount consumed by a population figure for any year.)
2. Has per capita fish consumption generally increased or generally decreased between 1990 and 2003?
3. In what years has per capita fish consumption decreased?

LEARNING ONLINE

The website for *Environmental Science 14e* contains helpful study tools and other resources to take the learning experience beyond the textbook. Examples include flashcards, practice quizzing, information on green careers, tips for more sustainable living, activities, and a wide range of articles and further readings. To access these materials and other companion resources for this text, please visit **www.cengagebrain.com**. For further details, see the preface, p. xvi.

Water Resources and Water Pollution

11

The Colorado River Story

CORE CASE STUDY

The Colorado River, the major river of the arid southwestern United States, flows 2,300 kilometers (1,400 miles) through seven states to the Gulf of California (Figure 11-1). Most of its water comes from snowmelt in the Rocky Mountains. During the past 50 years, this once free-flowing river has been tamed by a gigantic plumbing system consisting of 14 major dams and reservoirs (Figure 11-2), and canals that help control flooding and supply water and electricity to farmers, ranchers, industries, and cities.

This system of dams and reservoirs provides water and electricity from hydroelectric plants at the major dams for roughly 30 million people in seven states—about one of every ten people in the United States. The river's water is used to produce about 15% of the nation's crops and livestock. It also supplies water to some of the nation's driest and hottest cities. Take away this tamed river and Las Vegas, Nevada, would be a mostly uninhabited desert area; San Diego and Los Angeles, California, could not support their present populations; and California's Imperial Valley, which grows much of the nation's vegetables, would consist mostly of cactus and mesquite plants.

So much water is withdrawn from this river to grow crops and support cities in this dry, desertlike climate, that very little of it reaches the sea. To make matters worse, the system has experienced severe *drought*, or prolonged dry weather, since 1999. This overuse of the Colorado River illustrates the challenges faced by governments and people living in arid and semiarid regions with shared river systems, as population growth and economic growth place increasing demands on limited or decreasing supplies of surface water.

To many analysts, emerging shortages of water for drinking and irrigation in many parts of the world and the water pollution problems discussed in this chapter—along with the related problems of biodiversity loss and projected climate change—are the three most serious environmental challenges the world faces during this century.

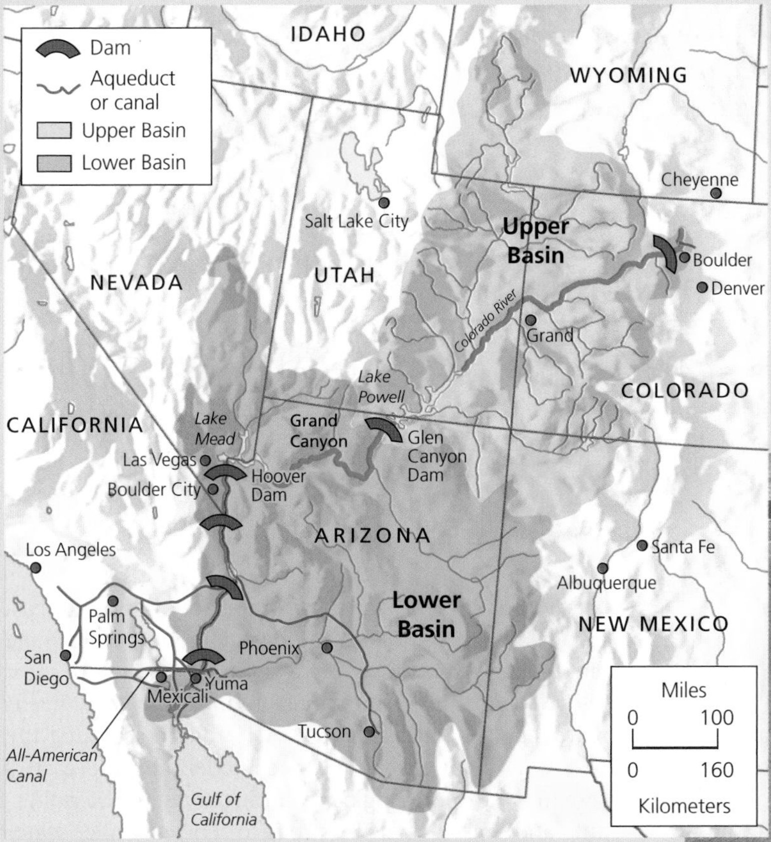

Figure 11-1 The *Colorado River basin*: The area drained by this basin is equal to more than one-twelfth of the land area of the lower 48 states. Two large reservoirs—Lake Mead behind the Hoover Dam and Lake Powell behind the Glen Canyon Dam (Figure 11-2)—store about 80% of the water in this basin. This map shows 6 of the river's 14 dams.

Figure 11-2 The Glen Canyon Dam across the Colorado River was completed in 1963. Lake Powell, behind the dam, is the second largest reservoir in the United States.

Key Questions and Concepts

11-1 Will we have enough usable water?

CONCEPT 11-1A We are using available freshwater unsustainably by wasting it, polluting it, and underpricing this irreplaceable natural resource.

CONCEPT 11-1B One of every six people does not have adequate access to clean water, and this situation will almost certainly get worse.

11-2 How can we increase freshwater supplies?

CONCEPT 11-2A Groundwater used to supply cities and grow food is being pumped from aquifers in some areas faster than it is renewed by precipitation.

CONCEPT 11-2B Using dams, reservoirs, and water transfer projects to provide water to arid regions has increased freshwater supplies in some areas but has disrupted ecosystems and displaced people.

CONCEPT 11-2C We can convert salty ocean water to freshwater, but the cost is high and the resulting large volume of salty brine must be disposed of without harming aquatic or terrestrial ecosystems.

11-3 How can we use freshwater more sustainably?

CONCEPT 11-3 We can use freshwater more sustainably by cutting water waste, raising water prices, slowing population growth, and protecting aquifers, forests, and other ecosystems that store and release freshwater.

11-4 How can we reduce the threat of flooding?

CONCEPT 11-4 We can lessen the threat of flooding by protecting more wetlands and natural vegetation in watersheds, and by not building in areas subject to frequent flooding.

11-5 How can we deal with water pollution?

CONCEPT 11-5 Reducing water pollution requires that we prevent it, work with nature to treat sewage, cut resource use and waste, reduce poverty, and slow population growth.

Note: Supplements 2 (p. S3), 3 (p. S6), 4 (p. S10), and 6 (p. S22) can be used with this chapter.

Our liquid planet glows like a soft blue sapphire in the hard-edged darkness of space. There is nothing else like it in the solar system. It is because of water.

JOHN TODD

11-1 Will We Have Enough Usable Water?

▶ **CONCEPT 11-1A We are using available freshwater unsustainably by wasting it, polluting it, and underpricing this irreplaceable natural resource.**

▶ **CONCEPT 11-1B One of every six people does not have adequate access to clean water, and this situation will almost certainly get worse.**

Freshwater Is an Irreplaceable Resource That We Are Managing Poorly

Freshwater, or water that is relatively pure and contains few dissolved salts, is vital for us and most other forms of life. We live on a watery planet with a precious layer of water—most of it saltwater—covering about 71% of the earth's surface (see Figure 7-19, p. 139). Look in the mirror. What you see is about 60% water, most of it inside your cells.

Water is an amazing and irreplaceable chemical with unique properties that help to keep us and other forms of life alive (see Science Focus, p. 52). We could survive for several weeks without food, but for only a few days without freshwater. In addition, it takes huge amounts of water to supply us with food and with most of the other things that we use to meet our daily needs and wants. Water also plays a key role in sculpting the earth's surface, controlling and moderating the earth's climate, and removing and diluting some of the pollutants and wastes that we produce. For example, water flowing in the Colorado River (**Core Case Study**) CORE CASE STUDY over a period of nearly 2 billion years, created the spectacular Grand Canyon in the United States.

Despite its importance, freshwater is one of our most poorly managed resources. We waste it and pollute it, and it is available at too low a cost to billions of

Links: CORE CASE STUDY refers to the Core Case Study.  refers to the book's sustainability theme. GOOD NEWS refers to good news about the environmental challenges we face.

consumers. This encourages still greater waste and pollution of this resource, for which we have no substitute (**Concept 11-1A**).

Access to freshwater is a *global health issue*. The World Health Organization (WHO) estimates that every day, an average of 3,900 children younger than age 5 die from waterborne infectious diseases because they do not have access to safe drinking water. Access to water is also an *economic issue* because water is vital for reducing poverty and producing food and energy. In less-developed countries, it is an *issue for women and children* because they often are responsible for finding and carrying daily supplies of freshwater (Figure 11-3) from distant wells and other sources to their homes. Water is also a *national and global security issue* because of increasing tensions both within and between nations over access to limited freshwater resources that they share.

Water is an *environmental issue* because excessive withdrawal of freshwater from rivers and aquifers results in falling water tables, decreasing river flows (**Core Case Study**), shrinking lakes, and disappearing wetlands. This loss of freshwater in combination with water pollution can decrease water quality, reduce fish populations, hasten the extinction of some species, and degrade aquatic ecosystem services (see Figure 7-20, p. 140, and Figure 7-27, p. 145).

CORE CASE STUDY

A. Ishokon-UNEP/Peter Arnold, Inc.

Figure 11-3 This young girl is carrying water to her home in a very dry area of India. According to the United Nations, over 1 billion people—more than 3 times the entire U.S. population—do not have access to clean water where they live. Every day, girls and women in this group typically spend an average of 3 hours collecting and hauling freshwater from distant sources.

Most of the Earth's Freshwater Is Not Available to Us

The earth is often called the blue planet because of its abundance of water in all forms: liquid, solid (ice), and gas (water vapor). However, only a tiny fraction of the planet's enormous water supply—*about 0.024%*—is readily available to us as liquid freshwater in accessible groundwater deposits and in lakes, rivers, and streams. The rest is in the salty oceans (about 97% of the earth's volume of liquid water), in frozen polar ice caps and glaciers, in deep underground aquifers, or in other inaccessible locations.

Fortunately, the world's freshwater supply is continually collected, purified, recycled, and distributed in the earth's *hydrologic cycle*—the movement of water in aquatic systems, in the air, and on land, which is driven by solar energy and gravity (see Figure 3-14, p. 51). This irreplaceable water recycling and purification system works well, unless we overload it with pollutants or withdraw freshwater from underground and surface water supplies faster than it is replenished. We can also alter long-term precipitation rates and distribution patterns of freshwater through our influence on projected climate change.

In some parts of the world, we are doing all of these things, mostly because we have thought of the earth's freshwater as essentially a free and infinite resource. As a result, we have placed little or no economic value on the irreplaceable ecological services it provides. (See *The Habitable Planet*, Video 8, at **www.learner.org/resources/series209.html**.)

On a global basis we have plenty of freshwater but it is not distributed evenly. Differences in average annual precipitation and economic resources divide the world's continents, countries, and people into water *haves* and *have-nots*. For example, Canada, with only 0.5% of the world's population, has 20% of the world's liquid freshwater, while China, with 19% of the world's people, has only 7% of the supply.

Groundwater and Surface Water Are Critical Resources

Some precipitation infiltrates the ground and percolates downward through spaces in soil, gravel, and rock until an impenetrable layer of rock stops it. The freshwater in these spaces is called **groundwater**—one of our most important sources of freshwater and a key component of the earth's natural capital.

The spaces in soil and rock close to the earth's surface hold little moisture. However, below a certain depth, in the **zone of saturation**, these spaces are completely filled with freshwater. The top of this groundwater zone is the **water table**. It falls in dry weather, or when we remove groundwater faster than nature can replenish it, and it rises in wet weather.

Deeper down are geological layers called **aquifers**, underground caverns and porous layers of sand, gravel, or rock through which groundwater flows. Mostly because of gravity, groundwater normally moves from points of high elevation and pressure to points of lower elevation and pressure. Some caverns have rivers of groundwater flowing through them. But the porous layers of sand, gravel, or rock in most aquifers are like large, elongated sponges through which groundwater seeps—typically moving only a meter or so (about 3 feet) per year and rarely more than 0.3 meter (1 foot) per day. Watertight layers of rock or clay below such aquifers keep the freshwater from escaping deeper into the earth. We use pumps to bring large quantities of this groundwater to the surface for drinking, irrigating crops, and supplying industries.

Most aquifers are replenished naturally by precipitation that percolates downward through exposed soil and rock, a process called *natural recharge*. Others are recharged from the side by *lateral recharge* from nearby lakes, rivers, and streams. Most aquifers recharge extremely slowly. And because so much of the earth's landscape has been built on or paved over, freshwater can no longer penetrate the ground to recharge aquifers below many urban areas. In addition, in dry areas of the world, there is little precipitation available to recharge aquifers.

Nonrenewable aquifers get very little, if any, recharge. They are found deep underground and were formed tens of thousands of years ago. Withdrawing freshwater from these aquifers amounts to *mining* a nonrenewable resource.

Another crucial resource is **surface water**, the freshwater from precipitation and melted snow that flows across the earth's land surface and into lakes, wetlands, streams, rivers, estuaries, and ultimately into the oceans. Precipitation that does not infiltrate the ground or return to the atmosphere by evaporation is called **surface runoff**. The land from which surface water drains into a particular river, lake, wetland, or other body of water is called its **watershed**, or **drainage basin**. For example, the drainage basin for the Colorado River is shown in yellow and green on the map in Figure 11-1 (**Core Case Study**).

CORE CASE STUDY

We Are Using Increasing Amounts of the World's Reliable Runoff

According to hydrologists (scientists who study water and its movements above, on, and below the earth's surface), two-thirds of the annual surface runoff of freshwater into rivers and streams is lost in seasonal floods and is not available for human use. The remaining one-third is **reliable surface runoff**, which we can generally count on as a source of freshwater from year to year.

During the last century, the human population tripled, global water withdrawals increased sevenfold, and per capita withdrawals quadrupled. As a result, we now withdraw about 34% of the world's reliable runoff of freshwater. Some water experts project that because of a combination of population growth, rising rates of water use per person, increased drought, and failure to reduce unnecessary water waste, we may be withdrawing up to 90% of the reliable freshwater runoff by 2025. (See the Case Study that follows.)

Worldwide, we use 70% of the freshwater we withdraw each year from rivers, lakes, and aquifers to irrigate cropland and raise livestock; that figure rises to 90% in arid regions. Industry uses roughly another 20% of the water withdrawn each year, and cities and residences use the remaining 10%. Affluent lifestyles require large amounts of water (Science Focus, at right), much of it unnecessarily wasted.

■ CASE STUDY
Freshwater Resources in the United States

The United States has more than enough renewable freshwater to meet its needs. But it is unevenly distributed and much of it is contaminated by agricultural and industrial practices. The eastern states usually have ample precipitation, whereas many western and southwestern states have little (Figure 11-4).

According to a 2009 study by the U.S. Geological Survey (USGS), groundwater and surface freshwater withdrawn in the United States is used for removing heat from electric power plants (41%), irrigation (37%), public water supplies (13%), industry (5%), and raising livestock (4%). In the eastern United States, most water is used for manufacturing and for cooling power plants. In many parts of this area, the most

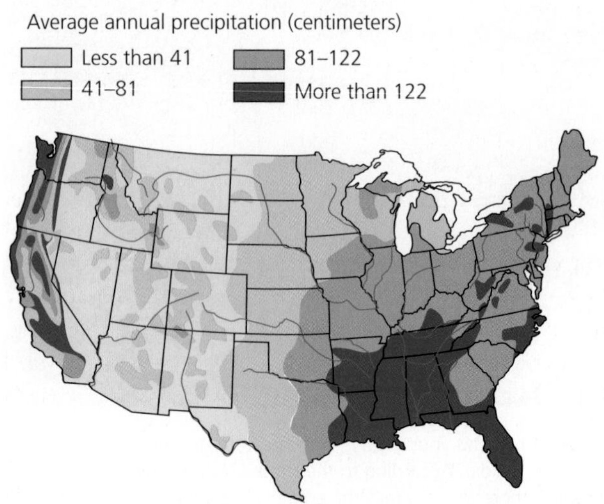

Figure 11-4 This map shows the long-term average annual precipitation and major rivers in the continental United States. (Data from U.S. Water Resources Council and U.S. Geological Survey)

SCIENCE FOCUS

Water Footprints and Virtual Water

Each of us has a **water footprint**, which is a rough measure of the volume of freshwater that we use directly and indirectly to keep us alive and to support our lifestyles.

According to the American Waterworks Association, the average American each day directly uses about 260 liters (69 gallons) of freshwater—enough water to fill about 1.7 typical bathtubs of water (each containing about 151 liters or 40 gallons of water). In the United States, this water is used mostly for flushing toilets (27%), washing clothes (22%), taking showers (17%), and running faucets (16%), or is wasted by water leaks (14%).

We use much larger amounts of freshwater indirectly to provide ourselves with food and other consumer products. Producing and delivering a typical hamburger, for example, takes about 2,400 liters (630 gallons) of freshwater—most of which is used to grow grain that is fed to cattle. This water would fill about 16 bathtubs. Similarly, according to the Coca Cola Company, it takes about 500 liters (132 gallons)—roughly 3 bathtubs of freshwater—to make a 2-liter (0.5-gallon) bottle of soda, if you include the freshwater used to grow and harvest ingredients such as sugar cane and high-fructose corn syrup.

Freshwater that is not directly consumed but is used to produce food and other products is called **virtual water**, and it makes up a large part of our water footprints, especially in more-developed nations. Figure 11-A shows one way to measure the amounts of virtual water used for producing and delivering products. These values can vary depending on how much of the supply chain is included, but they give us a rough estimate of the size of our water footprints.

Because of global trade, the virtual water used to produce and transport the wheat in a loaf of bread or the coffee beans used to make a cup of coffee (Figure 11-A) is often withdrawn as groundwater or surface water in another part of the world. For some water-short countries, it makes sense to save real freshwater by importing virtual water through food imports, instead of producing food domestically. Such countries include Egypt and other Middle Eastern nations in dry climates with little freshwater.

Large exporters of virtual water—mostly in the form of wheat, corn, soybeans, and other foods—are the European Union, the United States, Brazil, and Australia. However, a fourth of the U.S. corn crop is now being converted to ethanol to "feed" cars instead of people and livestock. If climate change threatens freshwater supplies as is projected in some important crop-growing and food-exporting regions, freshwater scarcity will turn into food scarcity as virtual water supplies drop—another example of connections in nature.

Critical Thinking

About how many bathtubs of water were used to provide all of the jeans and cotton shirts that you own? About how many tubs of water were used to provide you with the bread that you typically eat each week?

Figure 11-A Producing and delivering a single one of each of the products shown here requires the equivalent of at least one and usually many bathtubs full of freshwater, called *virtual water*. A typical bathtub contains about 151 liters (40 gallons) of water. (Data from UNESCO-IHE Institute for Water Education, UN Food and Agriculture Organization, World Water Council, Water Footprint Network, and Coca Cola Company) [All photos from Shutterstock.com unless otherwise indicated; Credits (top to bottom): Kirsty Pargeter, Aleksandra Nadeina, Alexander Kalina, Joe Belanger, Wolfgang Amri, Skip Odonnell/iStockphoto, Eky Studio, Rafal Olechowski]

serious water problems are flooding, occasional water shortages as a result of pollution, and **drought**, a prolonged period in which precipitation is at least 70% lower and evaporation is higher than normal in an area that is normally not dry.

In the arid and semiarid areas of the western half of the United States, irrigation counts for as much as 85% of freshwater use. Much of it is unnecessarily wasted and much of it is used to grow thirsty crops on arid and semiarid land. The major water problem in much of the western United States is a shortage of freshwater runoff caused by low precipitation (Figure 11-4), high evaporation, and recurring prolonged drought.

According to the USGS, about 33% of the freshwater withdrawn in the United States comes from groundwater sources, and the other 67% comes from rivers, lakes, and reservoirs. Water tables in many water-short areas, especially in the arid and semiarid western half of the lower 48 states, are dropping quickly as farmers and rapidly growing urban areas deplete many aquifers faster than they can be recharged. In 2007, the USGS projected that areas of at least 36 states are likely to face freshwater shortages by 2025 because of a combination of drought, rising temperatures, population growth, urban sprawl, and increased use and waste of water.

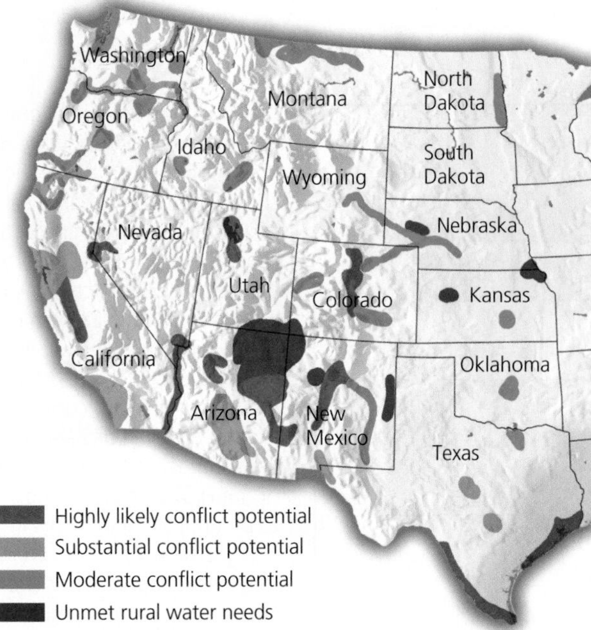

Figure 11-5 This map shows freshwater scarcity *hotspots* in 17 western states that, by 2025, could face intense conflicts over scarce freshwater needed for urban growth, irrigation, recreation, and wildlife. Some analysts suggest that this is a map of places not to live in the foreseeable future. **Question:** Which, if any, of these areas are found in the Colorado River basin (Figure 11-1)? (Data from U.S. Department of the Interior)

The U.S. Department of the Interior has mapped out *water hotspots* in 17 western states (Figure 11-5). In these areas, competition for scarce freshwater to support growing urban areas, irrigation, recreation, and wildlife could trigger intense political and legal conflicts between states and between rural and urban areas within states during the next 20 years. Based on 19 global climate models used by the UN Intergovernmental Panel on Climate Change, climate researcher Richard Seager and his colleagues at Columbia University have projected that the southwestern United States and parts of northern Mexico are very likely to have long periods of extreme drought throughout most of the rest of this century.

The Colorado River system (Figure 11-1) will be directly affected by such drought. There are five major problems associated with the use of freshwater from the river (**Core Case Study**). CORE CASE STUDY *First,* the Colorado River basin includes some of the driest lands in the United States and Mexico. *Second,* for its size, the river has only a modest flow of water. *Third,* legal pacts signed in 1922 and 1944 between the United States and Mexico allocated more freshwater for human use than the river can supply—even in rare years when there is no drought. In addition, the pacts allocated no water for protecting aquatic and terrestrial wildlife. *Fourth,* since 1960, the river has rarely flowed all the way to the Gulf of California because of its reduced water flow (due to many dams), increased freshwater withdrawals, and a prolonged drought in the American Southwest. *Fifth,* the river receives enormous amounts of pollutants from urban areas, farms, animal feedlots, and industries, as it makes its way toward the sea.

GOOD NEWS According to 2009 study by the U.S. Geological Survey, the amount of freshwater used in the United States between 1975 and 2005 has remained fairly stable despite a 30% rise in population and a significant increase in average resource use per person. The main reasons for this are increased use of more efficient irrigation systems and reduced use of wasteful types of cooling systems for many electric power plants. This is an encouraging trend but the United States still unnecessarily wastes large amounts of freshwater.

Freshwater Shortages Will Grow

The main factors that cause water scarcity in any particular area are a dry climate, drought, too many people using a freshwater supply more quickly than it can be replenished, and wasteful use of freshwater. Figure 11-6 shows the current degree of stress faced by the world's major river systems, based on a comparison of the amount of available surface freshwater with the amount used per person.

More than 30 countries—most of them in the Middle East and Africa—now face stress from freshwater scarcity. By 2050, some 60 countries, many of them in Asia, with three-fourths of the world's population, are likely to be suffering from such freshwater stress. In 2006, the Chinese government reported that two-thirds of China's 600 major cities faced freshwater shortages.

Currently, about 30% of the earth's land area—a total area roughly 5 times the size of the United States—experiences severe drought. By 2059, as much as 45% of the earth's land surface—about 7 times the area of the United States—could experience extreme drought, mostly as a result of projected climate change caused by a warmer atmosphere, according to a 2007 study by climate researcher David Rind and his colleagues.

In each of 263 of the world's water basins, two or more countries share the available freshwater supplies. Together, these areas cover nearly half of the earth's surface and are home to about 40% of the world's people. However, countries in only 158 of those basins have freshwater-sharing agreements. This explains why conflicts among nations over shared freshwater resources are likely to increase as populations grow, as demand for water increases, and as supplies shrink in many parts of the world.

In 2009, the United Nations reported that about 1 billion people—one of every seven in the world—lacked regular access to enough clean water for drinking, cooking, and washing. This is mostly due to poverty. The report also projected that by 2025, at least 3 billion people will likely lack such access to clean water. This amounts to about 3 times the current population of China and nearly 10 times the current population of the United States.

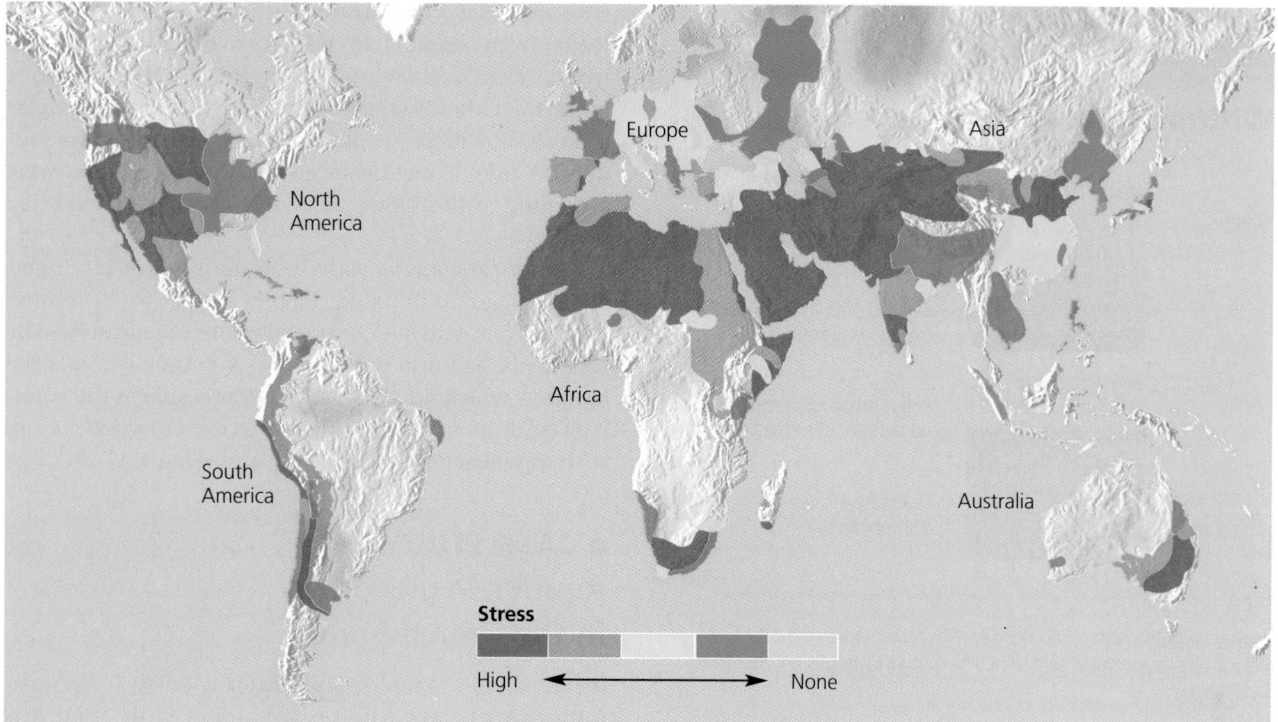

Figure 11-6 Natural capital degradation: The world's major river basins differ in their degree of freshwater-scarcity stress, based on a comparison of the amount of freshwater available with the amount used by humans (**Concept 11-1B**). **Questions:** If you live in a freshwater-stressed area, what signs of stress have you noticed? In what ways, if any, has it affected your life? (Data from World Commission on Water Use in the 21st Century)

We can increase freshwater supplies in various parts of the world by withdrawing groundwater; building dams and reservoirs to store runoff in rivers for release as needed (**Core Case Study**); transporting surface water from one area to another; and converting saltwater to freshwater (desalination). We discuss these options in the chapter section that follows. Reducing unnecessary waste of freshwater is another important way to provide more of this precious resource, as discussed in Section 11-3.

CORE CASE STUDY

11-2 How Can We Increase Freshwater Supplies?

▶ **CONCEPT 11-2A Groundwater used to supply cities and grow food is being pumped from aquifers in some areas faster than it is renewed by precipitation.**

▶ **CONCEPT 11-2B Using dams, reservoirs, and water transfer projects to provide water to arid regions has increased freshwater supplies in some areas but has disrupted ecosystems and displaced people.**

▶ **CONCEPT 11-2C We can convert salty ocean water to freshwater, but the cost is high and the resulting large volume of salty brine must be disposed of without harming aquatic or terrestrial ecosystems.**

Groundwater Is Being Withdrawn Faster Than It Is Replenished in Some Areas

Aquifers provide drinking water for nearly half of the world's people. Most aquifers are renewable resources unless the groundwater they contain becomes contaminated or is removed faster than it is replenished by rainfall, as is occurring in many parts of the world. Relying more on groundwater has advantages and disadvantages (Figure 11-7, p. 244).

Currently, water tables are falling in many areas of the world because the rate of pumping freshwater from most of the world's aquifers (mostly to irrigate crops)

Trade-Offs

Withdrawing Groundwater

Advantages	Disadvantages
Useful for drinking and irrigation	Aquifer depletion from overpumping
Exists almost everywhere	Sinking of land (subsidence) from overpumping
Renewable if not overpumped or contaminated	Pollution of aquifers lasts decades or centuries
Cheaper to extract than most surface waters	Deeper wells are nonrenewable

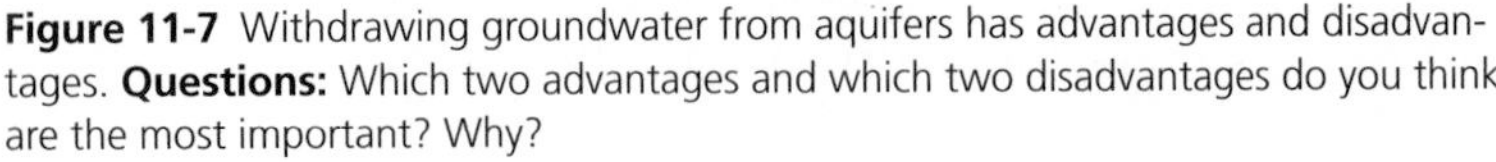

Figure 11-7 Withdrawing groundwater from aquifers has advantages and disadvantages. **Questions:** Which two advantages and which two disadvantages do you think are the most important? Why?

is greater than the rate of natural recharge from rainfall and snowmelt (**Concept 11-2A**). The world's three largest grain producers—China, India, and the United States—as well as Mexico, Saudi Arabia, Iran, Yemen, Israel, and Pakistan are overpumping many of their aquifers (see the Case Study that follows).

Every day, the world withdraws enough freshwater from aquifers to fill a convoy of large tanker trucks stretching 480,000 kilometers (300,000 miles)—more than the distance to the moon. Currently, more than 400 million people are being fed by grain produced through this unsustainable use of groundwater, according to the World Bank. This number is expected to grow.

Saudi Arabia is as water-poor as it is oil-rich. It gets about 70% of its drinking water at a high cost by removing salt from seawater, a process called *desalination*. The rest of the country's freshwater is pumped from deep aquifers, which are about as nonrenewable as the country's oil. Much of this freshwater is used to irrigate crops such as wheat grown on desert land (Figure 11-8).

■ CASE STUDY
Aquifer Depletion in the United States

In the United States, groundwater is being withdrawn from aquifers, on average, four times faster than it is replenished, according to the U.S. Geological Survey (**Concept 11-2A**). Figure 11-9 shows the areas of greatest aquifer depletion. One of the most serious overdrafts of groundwater is in the lower half of the Ogallala, the world's largest known aquifer, which lies under eight Midwestern states from southern South Dakota to Texas (most of the large red area in the center of Figure 11-9).

UN Environment Programme and U.S. Geological Survey

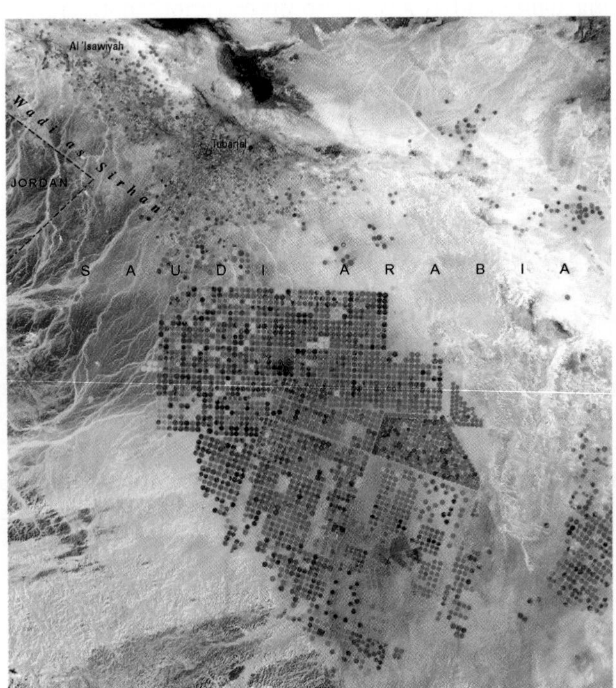

UN Environment Programme and U.S. Geological Survey

Figure 11-8 Natural capital degradation: These satellite photos show farmland irrigated by groundwater pumped from an ancient and nonrenewable aquifer in a vast desert region of Saudi Arabia between 1986 (left) and 2004 (right). Irrigated areas appear as green dots (each representing a circular spray system) and brown dots show areas where wells have gone dry and the land has returned to desert. In 2008, Saudi Arabia announced that irrigated wheat production had largely depleted its major deep aquifer and said that it would stop producing wheat by 2016 and would import grain (virtual water) to help feed its 29 million people.

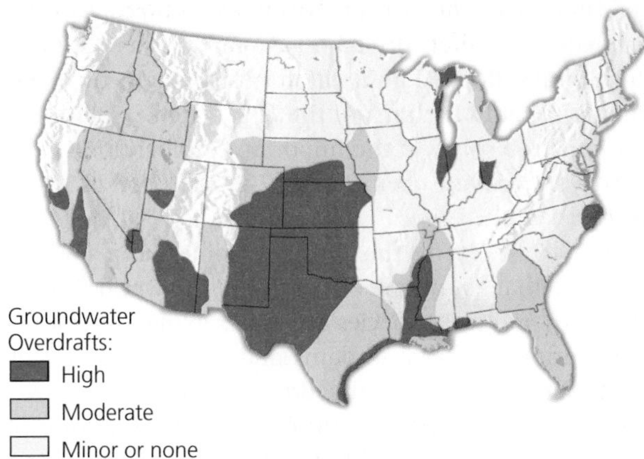

Active Figure 11-9 Natural capital degradation: This map shows areas of greatest aquifer depletion from groundwater overdraft in the continental United States (**Concept 11-2A**). Aquifer depletion is also high in Hawaii and Puerto Rico (not shown on map). *See an animation based on this figure at* www.cengagebrain.com. **Questions:** Do you depend on any of these aquifers for your drinking water? If so, what is the level of severity of overdraft where you live? (Data from U.S. Water Resources Council and U.S. Geological Survey)

The gigantic Ogallala aquifer supplies about one-third of all the groundwater used in the United States and has helped to turn the Great Plains into one of world's most productive irrigated agricultural regions. The problem is that the Ogallala is essentially a one-time deposit of liquid natural capital with a very slow rate of recharge.

In parts of the southern half of the Ogallala, groundwater is being pumped out 10–40 times faster than the slow natural recharge rate. This has lowered water tables and increased pumping costs, especially in northern Texas. Government subsidies designed to increase crop production have encouraged farmers to grow water-thirsty crops in dry areas, which has accelerated depletion of the Ogallala. Serious aquifer depletion is also taking place in California's semiarid Central Valley, which supplies half of the country's fruits and vegetables (the long red area in the California portion of Figure 11-9).

THINKING ABOUT
The Ogallala Aquifer

What are three things you would do to promote more sustainable use of the Ogallala aquifer?

Overpumping of Aquifers Has Several Harmful Effects

Overpumping of aquifers not only limits future food production, but also increases the gap between the rich and poor in some areas. As water tables drop, farmers must drill deeper wells, buy larger pumps, and use more electricity to run those pumps. Poor farmers cannot afford to do this and end up losing their land and working for richer farmers, or migrating to cities that are already crowded with poor people struggling to survive.

Withdrawing large amounts of groundwater sometimes allows the sand and rock in aquifers to collapse. This causes the land above the aquifer to *subside* or sink, a phenomenon known as *land subsidence*. Extreme, sudden subsidence is sometimes referred to as a *sinkhole*. Once an aquifer becomes compressed by subsidence, recharge is impossible. In addition, land subsidence can damage roadways, water and sewer lines, and building foundations. Areas in the United States suffering from such subsidence include California's San Joaquin Valley (Figure 11-10), Baton Rouge, Louisiana, the Phoenix area of Arizona, and parts of Florida.

Groundwater overdrafts near coastal areas, where many of the world's largest cities and industrial areas are found, can pull saltwater into freshwater aquifers. The resulting contaminated groundwater is undrinkable and unusable for irrigation. This problem is especially serious in coastal areas of the U.S. states of California,

Figure 11-10 This pole shows subsidence from overpumping of an aquifer for irrigation in California's San Joaquin Central Valley between 1925 and 1977. In 1925, this area's land surface was near the top of the pole. Since 1977 this problem has gotten worse.

Texas, Florida, Georgia, South Carolina, and New Jersey, as well as in coastal areas of Turkey, Thailand, and the Philippines.

With global shortages of freshwater looming, scientists are evaluating *deep aquifers* as future sources of freshwater. Preliminary results suggest that some of these aquifers hold enough freshwater to support billions of people for centuries. In addition, the quality of freshwater in these aquifers may be much higher than the quality of the freshwater in most rivers and lakes. GOOD NEWS

There are four major concerns about tapping these ancient deposits of freshwater. *First,* they are nonrenewable and cannot be replenished on a human timescale. *Second,* little is known about the geological and ecological impacts of pumping large amounts of freshwater from deep aquifers. *Third,* some deep aquifers flow beneath more than one country and there are no international treaties that govern rights to them. Without such treaties, wars could break out over this resource. *Fourth,* the costs of tapping deep aquifers are unknown and could be high.

Figure 11-11 lists ways to prevent or slow the problem of aquifer depletion by using this largely renewable resource more sustainably. **GREEN CAREER:** hydrogeologist

Large Dams and Reservoirs Have Advantages and Disadvantages

A **dam** is a structure built across a river to control its flow. Usually, dammed water creates an artificial lake, or **reservoir**, behind the dam (**Core Case Study** and Figure 11-2). The main goals of a dam-and-reservoir system are to capture and store the surface runoff from a river's watershed, and release it as needed to control floods; to generate electricity (hydroelectricity); and to supply freshwater for irrigation and for towns and cities. Reservoirs also provide recreational activities such as swimming, fishing, and boating. Large dams and reservoirs have both benefits and drawbacks (Figure 11-12). CORE CASE STUDY

The world's 45,000 large dams have increased the annual reliable runoff available for human use by nearly 33%. As a result, reservoirs now hold 3 to 6 times more freshwater than the total amount flowing in all of the world's natural rivers. On the down side, this engineering approach to river management has displaced 40–80 million people from their homes, flooded an area of mostly productive land totaling roughly the area of the U.S. state of California, and impaired some of the important ecological services that rivers provide (see Figure 7-27, left, p. 145 and **Concept 11-2B**).

A 2007 study by the World Wildlife Fund (WWF) estimated that about one out of five of the world's freshwater fish and plant species are either extinct or endangered, primarily because dams and water withdrawals have sharply decreased the flow of many rivers such as the Colorado. The study also found that, because of dams, excessive water withdrawals, and in some areas, prolonged severe drought, only 21 of the planet's 177 longest rivers run freely from their sources to the sea (see Figure 7-30, p. 147). Projected climate change will worsen this situation in areas that become hotter and receive less precipitation.

The reservoirs behind dams also eventually fill up with sediments such as mud and silt, typically within 50 years, which eventually makes them useless for storing water or producing electricity. About 85% of U.S. dam-and-reservoir systems will be 50 years old or more by 2025. Some aging dams are being removed in the United States because they need costly repairs and some of their reservoirs have filled with silt. In addition, a number of people have pushed for dam removal to let rivers flow free, to restore damaged river ecosystems and valuable fisheries, and to provide more river recreation such as rafting.

In recent years, some 500 dams—most of them small—have been removed from U.S. rivers. During the next decade, the licenses for more than 500 U.S. dams will be up for renewal. Many of the smaller dams may be removed. Some large dams may also be removed but this is controversial because of the flood control, irrigation water, and electricity some of them provide. In

Solutions

Groundwater Depletion

Prevention	Control
Waste less water	Raise price of water to discourage waste
Subsidize water conservation	Tax water pumped from wells near surface waters
Limit number of wells	Set and enforce minimum stream flow levels
Do not grow water-intensive crops in dry areas	Divert surface water in wet years to recharge aquifers

Figure 11-11 There are a number of ways to prevent or slow groundwater depletion by using freshwater more sustainably. **Questions:** Which two of these solutions do you think are the most important? Why?

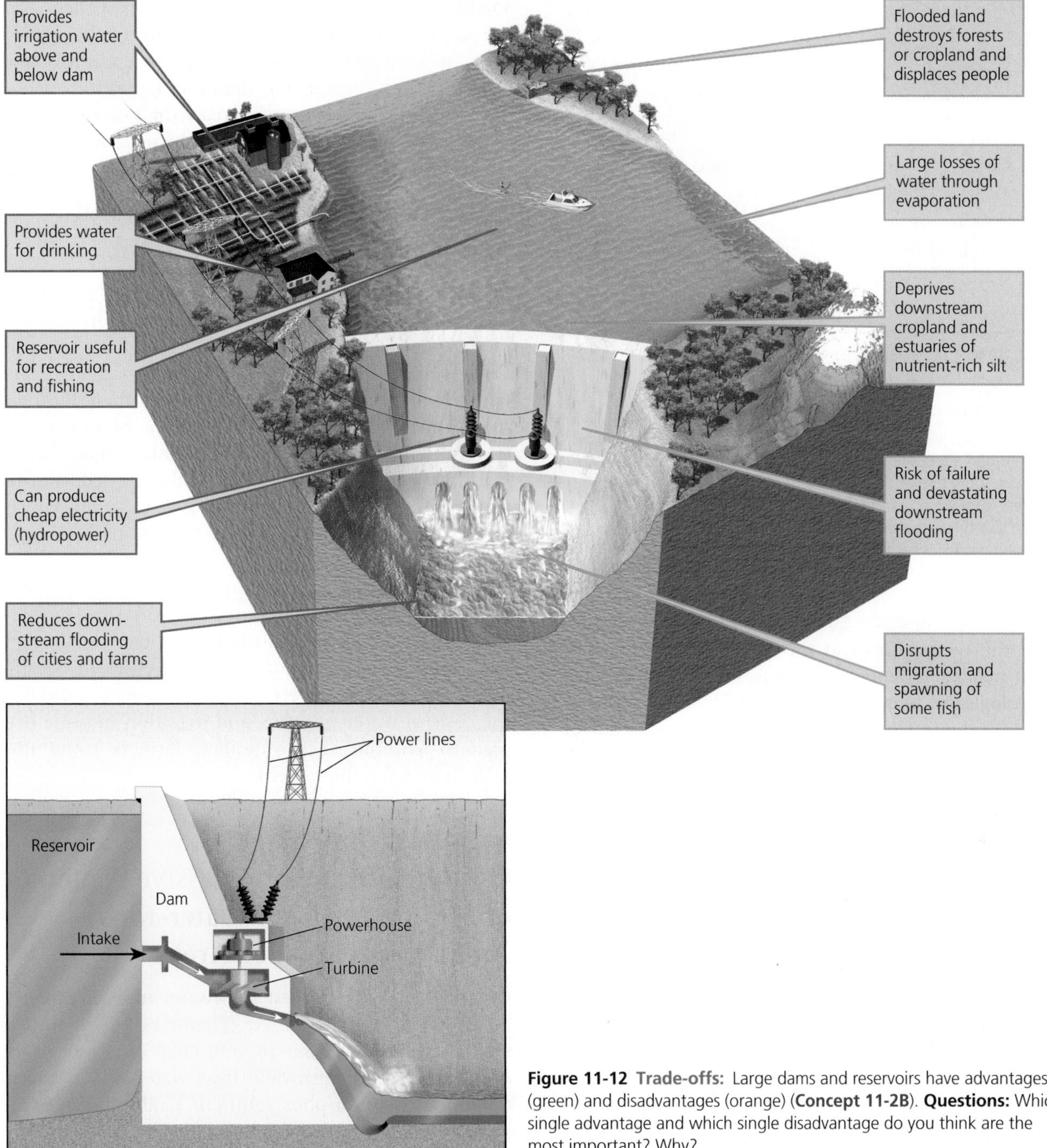

Figure 11-12 Trade-offs: Large dams and reservoirs have advantages (green) and disadvantages (orange) (**Concept 11-2B**). **Questions:** Which single advantage and which single disadvantage do you think are the most important? Why?

addition, removing a large dam is quite expensive and there is disagreement over who should bear the costs and what to do with the resulting massive amounts of concrete and other materials.

A Closer Look at the Overtapped Colorado River Basin

Since 1905, the amount of water flowing to the mouth of the heavily dammed Colorado River (**Core Case Study**) has dropped dramatically (Figure 11-13, p. 248). In most years since 1960, the river has dwindled to a small, sluggish stream by the time it reaches the Gulf of California. This threatens the survival of aquatic species that spawn in the river as well as those that live in its coastal estuary.

In 2008, a number of hydrologists warned that the current withdrawal of freshwater from the Colorado River is not sustainable. Water experts also project that if the climate continues to warm as projected, mountain snows that feed the river will melt faster and earlier, making less freshwater available to the river system when it is needed during hot summer months.

In addition, as the flow of the Colorado River slows in large reservoirs behind dams, it drops much of its

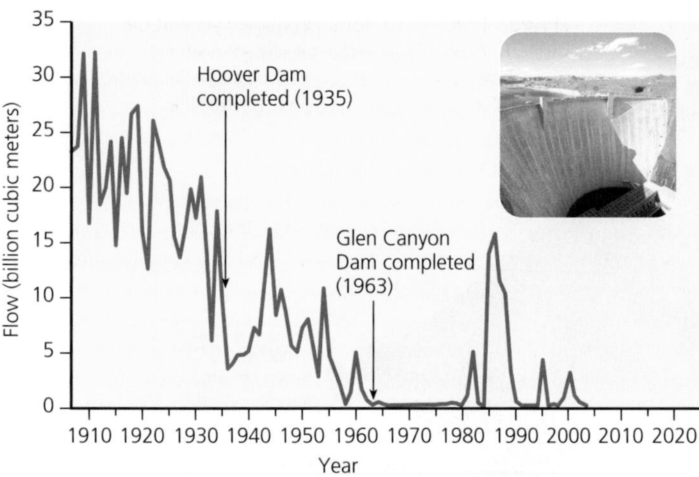

Figure 11-13 The measured water flow of the Colorado River at its mouth has dropped sharply since 1905 as a result of multiple dams, water withdrawals for agriculture and urban areas, and prolonged drought. Historical records and tree-ring analysis show that about once every century, the southwestern United States suffers from a mega-drought—a decades-long dry period. **Question:** Why do you think the flow peaks and then drops periodically? (Data from U.S. Geological Survey)

load of suspended silt. This deprives the river's coastal *delta*, the wetland area at the mouth of the river, of much-needed sediment and causes flooding and loss of ecologically important coastal wetlands. Each day, the equivalent of roughly 20,000 dump-truck loads of silt are deposited on the bottoms of the Lake Powell and Lake Mead reservoirs. Sometime during this century, these reservoirs will probably be too full of silt to control floods or to store enough water for generating hydroelectric power.

If some of the Southwest's largest reservoirs such as Lake Powell (Figure 11-2) become filled with silt during this century, the region could experience an economic and ecological catastrophe with fierce political and legal battles over who will get how much of the region's greatly diminished freshwater supply. Agricultural production would drop sharply and many people in the region's major desert cities such as Phoenix, Arizona, and Las Vegas, Nevada, likely would have to migrate to other areas. (Las Vegas gets very little rain and relies on the Hoover Dam's Lake Mead to meet more than 90% of its freshwater needs.)

Withdrawing more groundwater from aquifers is not a solution to this problem, because water tables under much of the area served by the Colorado River have been dropping, sometimes drastically, due to aquifer overpumping. Such withdrawals also threaten the survival of aquatic species that spawn in the river, and they destroy estuaries that serve as breeding grounds for numerous other aquatic species.

To deal with the water supply problems in the dry Colorado River basin, water experts call for the seven states using the river to enact and enforce strict water conservation measures and to slow population growth and urban development. They also call for phasing out state and federal government subsidies for agriculture in this region, shifting water-thirsty crops to less arid areas, and banning or severely restricting the use of water to keep golf courses and lawns green in this desert area. They suggest that the best way to implement such solutions is to sharply increase the historically low price of the river's freshwater over the next decade.

Water Transfers Can Be Wasteful and Environmentally Harmful

In many cases, water has been transferred into various dry regions of the world for growing crops and for other uses. For example, lettuce growers in the arid Central Valley of California (USA) have for many years depended on water that was transported there through huge aqueducts from the High Sierra Mountains of eastern California. Such water transfers have benefited many people, but they have also wasted a lot of water and they have degraded ecosystems from which the water was taken (**Concept 11-2B**) (see the two Case Studies that follow).

Such water waste is part of the reason why many products include large amounts of virtual water (Figure 11-A). One factor contributing to inefficient water use, such as the irrigation of lettuce crops in desert-like areas, is that governments (taxpayers) subsidize the costs of water transfers and irrigation in some dry regions. Without such subsidies, farmers could not make a living in these areas.

■ CASE STUDY

California Transfers Massive Amounts of Freshwater from Water-Rich Areas to Water-Poor Areas

One of the world's largest freshwater transfer projects is the *California Water Project* (Figure 11-14). It uses a maze of giant dams, pumps, and lined canals, or *aqueducts*, to transport freshwater from water-rich northern California to water-poor southern California's heavily populated cities and agricultural regions. This project supplies massive amounts of freshwater to areas that, without such water transfers, would be mostly desert. Agriculture consumes three-fourths of the freshwater withdrawn in California, much of it used and wasted by inefficient irrigation systems to grow crops in the desert-like conditions of the southern half of the state.

According to a 2002 study, projected climate change will make matters worse by sharply reducing surface water availability in California. Many Californians depend on *snowpacks*, bodies of densely packed, slow-melting snow in the High Sierras for their freshwater. Projected atmospheric warming will hasten the melting of these snowpacks and will also reduce precipitation during the drought years projected for the future. If these projections are correct, there will be less freshwater available for transferring to the south. Some ana-

Figure 11-14 The California Water Project and the Central Arizona Project transfer huge volumes of freshwater from one watershed to another. The red arrows show the general direction of water flow.

lysts project that during this century, many people living in arid southern California cities, as well as farmers in this area, may have to move elsewhere because of freshwater shortages.

Pumping more groundwater is not the answer, because groundwater is already being withdrawn faster than it is replenished in much of central and southern California (Figure 11-9). And desalinating seawater with current methods is too expensive to provide large amounts of freshwater for irrigation. According to many analysts, it would be quicker and less costly to reduce freshwater waste by improving irrigation efficiency, by not growing water-thirsty crops in arid areas, and by raising the historically low price of freshwater to encourage more efficient use of water.

■ CASE STUDY

The Aral Sea Disaster: A Glaring Example of Unintended Consequences

The shrinking of the Aral Sea (Figure 11-15, p. 250) is the result of a large-scale freshwater transfer project in an area of the former Soviet Union with the driest climate in central Asia. Since 1960, enormous amounts of irrigation water have been diverted from the two rivers that supply water to the Aral Sea. The goal was to create one of the world's largest irrigated areas, mostly for raising cotton and rice. The irrigation canal, the world's longest, stretches more than 1,300 kilometers (800 miles)—roughly the distance between the two U.S. cities of Boston, Massachusetts, and Chicago, Illinois.

This large-scale freshwater diversion project, coupled with drought and high evaporation rates due to the area's hot and dry climate, has caused a regional ecological, economic, and health disaster. Since 1961, the sea's salinity has risen sevenfold and the average level of its water has dropped by an amount roughly equal to the height of a six-story building. The Southern Aral Sea has lost 90% of its volume of water and is a remnant of the original sea lying in a salt-covered desert (Figure 11-15, bottom). Water withdrawal for agriculture has reduced the two rivers feeding the sea to mere trickles.

About 85% of the area's wetlands have been eliminated and about half the local bird and mammal species have disappeared. A huge area of the former Southern Aral Sea's lake bottom is now a human-made desert covered with glistening white salt. The sea's greatly increased salt concentration—3 times saltier than ocean water—has caused the presumed local extinction of 26 of the area's 32 native fish species. This has devastated the area's fishing industry, which once provided work for more than 60,000 people. Fishing villages and boats once located on the sea's coastline now sit abandoned in a salt desert.

Winds pick up the sand and salty dust and blow it onto fields as far as 500 kilometers (310 miles) away. As the salt spreads, it pollutes water and kills wildlife, crops, and other vegetation. Aral Sea dust settling on glaciers in the Himalayas is causing them to melt at a faster than normal rate—a prime example of unexpected connections and unintended consequences.

The shrinkage of the Aral Sea has also altered the area's climate. The once-huge sea acted as a thermal buffer that moderated the heat of summer and the extreme cold of winter. Now there is less rain, summers are hotter and drier, winters are colder, and the growing season is shorter. The combination of such climate change and severe salinization has reduced crop yields

1976

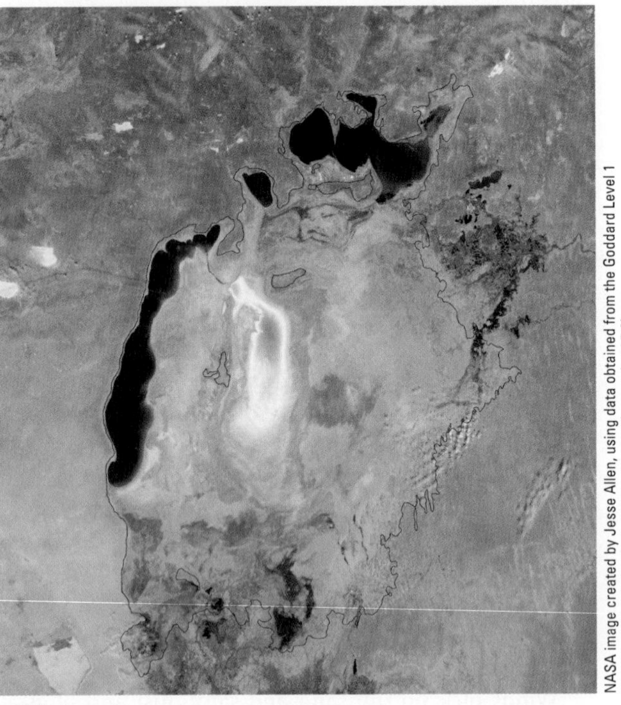

2009

Figure 11-15 Natural capital degradation: The *Aral Sea* was one of the world's largest saline lakes, straddling the borders of Kazakhstan and Uzbekistan. These satellite photos show the sea in 1976 and in 2009. **Question:** What do you think should be done to help prevent further shrinkage of the Aral Sea?

by 20–50% on almost one-third of the area's cropland—the opposite of the project's intended consequences.

To raise yields, farmers have used more herbicides, insecticides, and fertilizers, which have percolated downward and accumulated to dangerous levels in the groundwater—the source of most of the region's drinking water. Many of the 45 million people living in the Aral Sea's watershed have experienced increasing health problems—including anemia, respiratory illnesses, liver and kidney diseases, eye problems, and various cancers—from a combination of toxic dust, salt, and contaminated water.

GOOD NEWS

Since 1999, the United Nations and the World Bank have spent about $600 million to purify drinking water and upgrade irrigation and drainage systems in the area. This has improved irrigation efficiency and helped flush salts from croplands. The five countries surrounding the lake and its two feeder rivers have worked to improve irrigation efficiency. They have also partially replaced water-thirsty crops with other crops that require less irrigation water.

Removing Salt from Seawater Is Costly, Kills Marine Organisms, and Produces Briny Wastewater

Desalination is the process of removing dissolved salts from ocean water or from brackish (slightly salty) water in aquifers or lakes. It is another way to increase supplies of freshwater (**Concept 11-2C**).

The two most widely used methods for desalinating water are distillation and reverse osmosis. *Distillation* involves heating saltwater until it evaporates (leaving behind salts in solid form) and condenses as freshwater. *Reverse osmosis* (or *microfiltration*) uses high pressure to force saltwater through a membrane filter with pores small enough to remove the salt and other impurities.

Today, about 13,000 desalination plants (250 if them in the United States, half of them in Florida) operate in more than 125 countries, especially in the arid nations of the Middle East, North Africa, the Caribbean Sea, and the Mediterranean Sea. Desalination supplies less than 0.3% of the world's demand and only about 0.4% of the U.S. demand for freshwater.

There are three major problems with the widespread use of desalination. *First* is the high cost, because it takes a lot of increasingly expensive energy to remove salt from seawater. A *second* problem is that pumping large volumes of seawater through pipes and using chemicals to sterilize the water and keep down algae growth kills many marine organisms and also requires large inputs of energy (and thus money) to run the pumps. A *third* problem is that desalination produces huge quantities of salty wastewater that must go somewhere. Dumping it into nearby coastal ocean waters increases the salinity of the ocean water, which threatens food resources and aquatic life in the vicinity. Disposing of it on land could contaminate groundwater and surface water (**Concept 11-2C**).

Bottom line: Currently, significant desalination is practical only for water-short, wealthy countries and cities that can afford its high cost. However, scientists and engineers are working to develop better and more affordable desalination technologies.

11-3 How Can We Use Freshwater More Sustainably?

▶ **CONCEPT 11-3 We can use freshwater more sustainably by cutting water waste, raising water prices, slowing population growth, and protecting aquifers, forests, and other ecosystems that store and release freshwater.**

Reducing Freshwater Waste Has Many Benefits

According to water resource expert Mohamed El-Ashry of the World Resources Institute, about 66% of the freshwater used in the world and about 50% of the freshwater used in the United States is unnecessarily wasted. El-Ashry estimates that it is economically and technically feasible to reduce freshwater waste to 15%, thereby meeting most of the world's freshwater needs for the foreseeable future. GOOD NEWS

So why do we waste so much freshwater? According to water resource experts, there are two major reasons. First, the *cost of freshwater to users is low*. Such underpricing is mostly the result of government subsidies that provide irrigation water, or the electricity and diesel fuel used by farmers to pump freshwater from rivers and aquifers, at below-market prices. Because these subsidies keep freshwater prices artificially low, users have little or no financial incentive to invest in water-saving technologies. However, farmers, industries, and others benefiting from government freshwater subsidies argue that the subsidies promote the settlement and farming of arid, unproductive land, as well as stimulating local economies, and helping to keep the prices of food, manufactured goods, and electricity low.

THINKING ABOUT
Government Freshwater Subsidies

Should governments provide subsidies to farmers and cities to help keep the price of freshwater low? Explain.

Higher prices for freshwater encourage water conservation but make it difficult for low-income farmers and city dwellers to buy enough freshwater to meet their needs. When South Africa raised freshwater prices, it dealt with this problem by establishing *lifeline* rates, which give each household a set amount of free or low-priced freshwater to meet basic needs. When users exceed this amount, they pay higher prices as their freshwater use increases. This is a *user-pays* approach.

The second major cause of freshwater waste is a *lack of government subsidies for improving the efficiency of freshwater use*. Withdrawing environmentally harmful subsidies that encourage freshwater waste and replacing them with environmentally beneficial subsidies for more efficient freshwater use would sharply reduce freshwater waste and help to reduce shortages of freshwater. Understandably, farmers and industries receiving the subsidies that encourage freshwater waste have vigorously and successfully opposed eliminating or reducing their subsidies.

We Can Cut Freshwater Waste in Irrigation

Only about 60% of the world's irrigation water reaches crops. The reason for this is that in most irrigation systems, water is pumped from a groundwater or surface water source through unlined ditches where it flows by gravity to the crops being watered (see Figure 10-24, left, p. 229). About 40% of this water is lost through evaporation, seepage, and runoff. However, with existing irrigation technologies (see Figure 10-24, middle and right), this loss could be reduced to 5–10%. GOOD NEWS

Drip irrigation (see Figure 10-24, right) is used on just over 1% of the world's irrigated crop fields and about 4% of those in the United States. This percentage rises to 13% in the U.S. state of California, 66% in Israel, and 90% in Cyprus. Suppose that freshwater were priced closer to the value of the ecological services it provides and that government subsidies that encourage freshwater waste were reduced or eliminated. Then drip irrigation would probably be used to irrigate most of the world's crops.

Figure 11-16 lists other ways to reduce freshwater waste in crop irrigation. Since 1950, Israel has used

Solutions

Reducing Irrigation Water Waste

- Line canals bringing water to irrigation ditches
- Irrigate at night to reduce evaporation
- Monitor soil moisture to add water only when necessary
- Grow several crops on each plot of land (polyculture)
- Encourage organic farming
- Avoid growing water-thirsty crops in dry areas
- Irrigate with treated waste water
- Import water-intensive crops and meat

Figure 11-16 There are a number of ways to reduce freshwater waste in irrigation. **Questions:** Which two of these solutions do you think are the best ones? Why?

many of these techniques to slash irrigation water waste by 84% while irrigating 44% more land. Israel now treats and reuses 30% of its municipal sewage water for crop production and plans to increase this to 80% by 2025. The government also gradually eliminated most freshwater subsidies to raise Israel's price of irrigation water, which is now one of the highest in the world. GOOD NEWS

We Can Cut Freshwater Waste in Industries and Homes

Producers of chemicals, paper, oil, coal, primary metals, and processed foods consume almost 90% of the freshwater used by industries in the United States. Some of these industries recapture, purify, and recycle water to reduce their water use and water treatment costs. For example, more than 95% of the freshwater used to make steel can be recycled. Even so, most industrial processes could be redesigned to use much less freshwater. Figure 11-17 lists ways to use freshwater more efficiently in industries, homes, and businesses (**Concept 11-3**).

Flushing toilets with freshwater (most of it clean enough to drink) is the single largest use of domestic freshwater in the United States and accounts for about one-fourth of home water use. Since 1992, U.S. government standards have required that new toilets use no more than 6.1 liters (1.6 gallons) of freshwater per flush. Even so, just two flushings of such a toilet use more than the daily amount of freshwater available for all uses to many of the world's poor living in arid regions.

Low-flow showerheads can also save large amounts of freshwater by cutting the flow of a shower in half. And a British company recently developed a washing machine that uses as little as a cup of water for each washing cycle, thereby reducing the freshwater used to wash clothes by 98%.

According to UN studies, 30–60% of the freshwater supplied in nearly all of the world's major cities in less-developed countries is lost, primarily through leakage from water mains, pipes, pumps, and valves. Water experts say that fixing these leaks should be a high priority for governments, as it would cost less than building dams or importing freshwater. Even in advanced industrialized countries such as the United States, these losses to leakage average 10–30%. However, leakage losses have been reduced to about 3% in Copenhagen, Denmark, and to 5% in Fukuoka, Japan. GOOD NEWS

CONNECTIONS

Water Leaks and Water Bills

Any water leak unnecessarily wastes freshwater and raises water bills. You can detect a silent toilet water leak by adding a few drops of food coloring to the toilet tank and waiting 5 minutes. If the color shows up in the bowl, you have a leak. Also, a faucet leaking one drop per second wastes up to 8,200 liters (3,000 gallons) of water a year—enough to fill about 75 bathtubs. This is money going down the drain.

Solutions

Reducing Water Waste

- Redesign manufacturing processes to use less water
- Recycle water in industry
- Landscape yards with plants that require little water
- Use drip irrigation
- Fix water leaks
- Use water meters
- Raise water prices
- Use waterless composting toilets
- Require water conservation in water-short cities
- Use water-saving toilets, showerheads, and front-loading clothes washers
- Collect and reuse household water to irrigate lawns and nonedible plants
- Purify and reuse water for houses, apartments, and office buildings

Figure 11-17 There are a number of ways to reduce freshwater waste in industries, homes, and businesses (**Concept 11-3**). **Questions:** Which three of these solutions do you think are the best ones? Why?

Many homeowners and businesses in water-short areas are using drip irrigation on their properties and copying nature by replacing green lawns with a diversity of native plants that need little freshwater. Such water-thrifty landscaping saves money by reducing freshwater use by 30–85%, and by sharply reducing labor, fertilizer, and fuel requirements. This example of reconciliation ecology (see Chapter 9, p. 194) also helps to preserve biodiversity, and reduces polluted runoff, air pollution, and yard wastes. Every year, Americans spend a total of about 1.6 billion hours mowing their lawns. In the process, they use the equivalent of 20 million bathtubs full of gasoline to power noisy and highly polluting lawn mowers.

About 50–75% of the used water from bathtubs, showers, sinks, dishwashers, and clothes washers in a typical house (*gray water*) could be stored in a holding tank and then reused to irrigate lawns and nonedible plants, to flush toilets, and to wash cars. Israel reuses 70% of its wastewater to irrigate nonfood crops. In Singapore, all sewage water is treated at reclamation plants for reuse by industry. U.S. cities such as Las Vegas, Nevada, and Los Angeles, California, are also beginning to recycle some of their wastewater. These efforts mimic the way nature purifies water by recycling it, and thus they follow the chemical cycling **principle of sustainability** (see back cover). SUSTAINABILITY

The relatively low cost of water in most communities is another major cause of excessive water use and waste in homes and industries. Many water utility and irrigation authorities charge a flat fee for water use and some charge less to the largest users of water. About one-fifth of all U.S. public water systems do not have water meters and charge a single low rate for almost unlimited use of high-quality freshwater. Many apartment dwellers have little incentive to conserve water, because water-use charges are included in their rent. When the U.S. city of Boulder, Colorado, introduced water meters, water use per person dropped by 40%. **GREEN CAREER:** water conservation specialist

CONNECTIONS

Using Smart Cards to Reduce Water Use and Save Money

In parts of Brazil, an electronic device called a *water manager* allows customers to obtain freshwater on a pay-as-you-go basis. People buy *smart cards* (like long-distance phone cards), each of which contains a certain number of water credits. When they punch in the card's code on their water manager device, the water company automatically supplies them with a specified amount of freshwater. Brazilian officials say this approach saves water and electrical power, and typically reduces household water bills by 40%.

We Can Use Less Water to Remove Wastes

Currently, we use large amounts of freshwater clean enough to drink to flush away industrial, animal, and household wastes. According to the FAO, if current growth trends in population and resource use continue, within 40 years we will need the world's entire reliable flow of river water just to dilute and transport the wastes we produce each year. We could save much of this water by using systems that mimic the way nature deals with wastes by recycling them.

For example, sewage treatment plants take human waste, which contains valuable plant nutrients, and dump most of it into rivers, lakes, and oceans, which overloads aquatic systems with those nutrients. Instead, we could use the chemical cycling **principle of sustainability** by returning the nutrient-rich sludge produced by conventional waste treatment plants to the soil as a fertilizer. To make this feasible, we would have to ban the discharge of toxic industrial chemicals into sewage treatment plants; otherwise, the nutrient-rich sludge from these plants would be too toxic to apply to cropland soils.

Another way to recycle our wastes is to rely more on waterless composting toilets. These devices convert human fecal matter to a small amount of dry and odorless soil-like humus material that can be removed from a composting chamber every year or so and returned to the soil as fertilizer. One of the authors (Miller) used a composting toilet for over a decade, while living and working deep in the woods in an experimental home and office.

GOOD NEWS

We Need to Use Water More Sustainably

Figure 11-18 lists some strategies that scientists have suggested for using freshwater more sustainably (**Concept 11-3**).

Each of us can help to bring about a "blue revolution" in which we could reduce our *water footprints* by using less freshwater and cutting water waste (Figure 11-19). As with other problems, the solution starts with thinking globally and acting locally.

GOOD NEWS

Solutions

Sustainable Water Use

- Waste less water and subsidize water conservation
- Do not deplete aquifers
- Preserve water quality
- Protect forests, wetlands, mountain glaciers, watersheds, and other natural systems that store and release water
- Get agreements among regions and countries sharing surface water resources
- Raise water prices
- Slow population growth

Figure 11-18 We have a variety of ways to use the earth's freshwater resources more sustainably (**Concept 11-3**). **Questions:** Which two of these solutions do you think are the best ones? Why?

What Can You Do?

Water Use and Waste

- Use water-saving toilets, showerheads, and faucet aerators
- Shower instead of taking baths, and take short showers
- Repair water leaks
- Turn off sink faucets while brushing teeth, shaving, or washing
- Wash only full loads of clothes or use the lowest possible water-level setting for smaller loads
- Use recycled (gray) water for watering lawns and houseplants and for washing cars
- Wash a car from a bucket of soapy water, and use the hose for rinsing only
- If you use a commercial car wash, try to find one that recycles its water
- Replace your lawn with native plants that need little if any watering
- Water lawns and yards only in the early morning or evening
- Use drip irrigation and mulch for gardens and flowerbeds

Figure 11-19 Individuals matter: You can reduce your use and waste of freshwater. **Questions:** Which of these steps have you taken? Which would you like to take?

11-4 How Can We Reduce the Threat of Flooding?

▶ **CONCEPT 11-4 We can lessen the threat of flooding by protecting more wetlands and natural vegetation in watersheds, and by not building in areas subject to frequent flooding.**

Some Areas Get Too Much Water from Flooding

Some areas have too little freshwater, but others sometimes have too much because of natural flooding by streams, caused mostly by heavy rain or rapidly melting snow. Other areas flood because they receive most of their freshwater as heavy downpours during certain times of the year. A flood happens when freshwater in a stream overflows its normal channel and spills into an adjacent area, called a **floodplain**. These areas, which usually include highly productive wetlands, help to provide natural flood and erosion control, maintain high water quality, and recharge groundwater.

People settle on floodplains to take advantage of their many assets, including fertile soil, ample freshwater for irrigation, availability of nearby rivers for transportation and recreation, and flat land suitable for crops, buildings, highways, and railroads. In efforts to reduce the threat of flooding on floodplains, rivers have been narrowed and straightened (or *channelized*), surrounded by protective dikes and *levees* (long mounds of earth along their banks), and dammed to create reservoirs that store and release water as needed (Figure 11-2). However, in the long run, such measures can lead to greatly increased flood damage when heavy snowmelt or prolonged rains overwhelm them.

CORE CASE STUDY

Floods provide several benefits. They have created some of the world's most productive farmland by depositing nutrient-rich silt on floodplains. They also help recharge groundwater and refill wetlands, thereby supporting biodiversity and aquatic ecological services.

But floods also kill thousands of people each year and cost tens of billions of dollars in property damage (see the Case Study that follows). Floods are usually considered natural disasters, but since the 1960s, human activities have contributed to a sharp rise in flood deaths and damages, meaning that such disasters are partly human-made.

One such human activity is the *removal of water-absorbing vegetation*, especially on hillsides (Figure 11-20). Once a hillside has been deforested for timber, fuelwood, livestock grazing, or unsustainable farming, freshwater from precipitation rushes down the denuded slopes, erodes precious topsoil, and can increase flooding and pollution in local streams. Such deforestation can also increase landslides and mudflows. A 3,000-year-old Chinese proverb says, "To protect your rivers, protect your mountains."

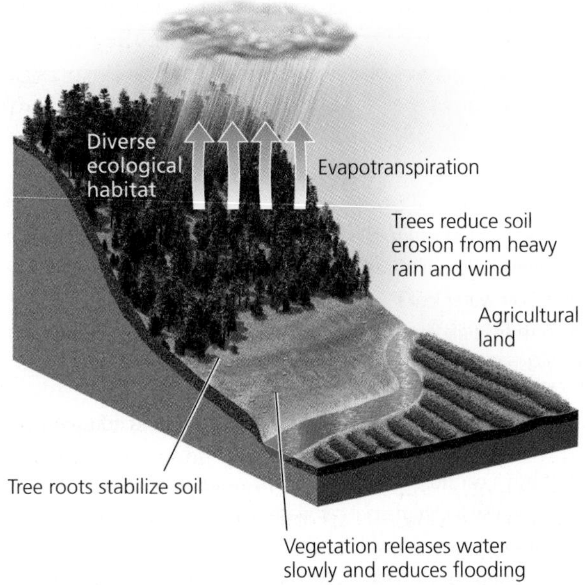

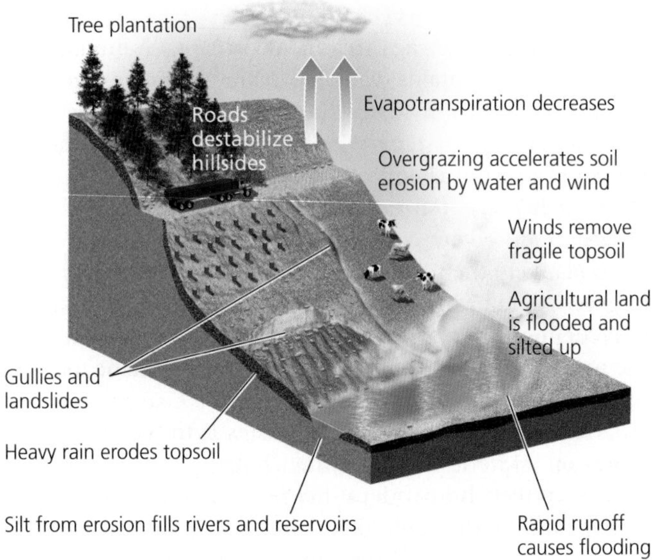

Active Figure 11-20 Natural capital degradation: These diagrams show a hillside before and after deforestation. *See an animation based on this figure at* www.cengagebrain.com. **Question:** How might a drought in this area make these effects even worse?

CONNECTIONS

Deforestation and Flooding in China

In 1998, severe flooding in China's Yangtze River watershed, home to 400 million people, left at least 15 million people homeless and caused massive economic losses. Scientists identified the causes as heavy rainfall, rapid snowmelt, and deforestation that had removed 85% of the watershed's tree cover. Chinese officials banned tree cutting in the watershed and started a tree-replanting program with the long-term goal of restoring some of the area's ecological flood-control services.

Draining and building on wetlands, which naturally absorb floodwaters, is a second human activity that increases the severity of flooding. When Hurricane Katrina struck the Gulf Coast of the United States in August 2005 and contributed to the flooding of the city of New Orleans, Louisiana, and surrounding areas, the damage was intensified because of the degradation or removal of coastal wetlands that had historically helped to absorb water and buffer this low-lying land from storm surges.

Another human-related factor that will increase flooding is a rise in sea levels during this century from projected climate change caused primarily by human activities that have led to the warming of the atmosphere (as discussed in Chapter 15). Climate change models project that, by 2075, as many as 150 million people in the world's largest cities—a number roughly equal to half of the current U.S. population—could be flooded out by rising sea levels.

■ CASE STUDY

Living Dangerously on Floodplains in Bangladesh

Bangladesh is one of the world's most densely populated countries. In 2011, its 151 million people were packed into an area roughly the size of the U.S. state of Wisconsin (which has a population of less than 6 million). And the country's population is projected to increase to 226 million by 2050. Bangladesh is a very flat country, only slightly above sea level, and it is one of the world's poorest countries.

The people of Bangladesh depend on moderate annual flooding during the summer monsoon season to grow rice and help maintain soil fertility in its delta basin region that is fed by numerous river systems. The annual floods also deposit eroded Himalayan soil on the country's crop fields. Bangladeshis have adapted to moderate flooding. Most of the houses have flat thatch roofs on which families can take refuge with their belongings in case of rising waters. The roofs can be detached from the walls, if necessary, and floated like rafts. After the waters have subsided, the roof can be reattached to the walls of the house. However, great floods can overwhelm such defenses.

In the past, great floods occurred every 50 years or so. But between 1987 and 2007 there were five severe floods, each of which covered a third or more of the country with water. Bangladesh's flooding problems begin in the Himalayan watershed, where rapid population growth, deforestation, overgrazing, and unsustainable farming on steep and easily erodible slopes have increased flows of water during the summer monsoon season. Monsoon rains now run more quickly off the denuded Himalayan foothills, carrying vital topsoil with them (Figure 11-20, right).

This increased runoff of topsoil, combined with heavier-than-normal monsoon rains, has led to more severe flooding along Himalayan rivers as well as downstream in Bangladesh's delta areas. In 1998, a disastrous flood covered two-thirds of Bangladesh's land area, in some places for 2 months, drowning at least 2,000 people and leaving 30 million people—an amount almost equal to the entire population of Tokyo, Japan—homeless. It also destroyed more than one-fourth of the country's crops, which caused thousands of people to die of starvation. In 2002, another flood left 5 million people homeless and flooded large areas of rice fields. Yet another major flood occurred in 2004.

Living at sea level on Bangladesh's coastal floodplain also means coping with storm surges, cyclones, and tsunamis. In 1970, a tropical cyclone killed as many as 1 million people. Another cyclone in 2003 killed more than a million people and left tens of millions homeless.

Many of the coastal mangrove forests in Bangladesh (and elsewhere; see Figure 7-23, p. 142) have been cleared for fuelwood, farming, and aquaculture ponds created for raising shrimp. The result: more severe flooding because these coastal wetlands had sheltered Bangladesh's low-lying coastal areas from storm surges, cyclones, and tsunamis. In areas of Bangladesh still protected by mangrove forests, damages and death tolls from cyclones have been much lower than they were in areas where the forests have been cleared.

Projected increases in sea level and storm intensity during this century, primarily because of climate change caused by atmospheric warming, will likely be a major threat to Bangladeshis who live on the largely flat delta adjacent to the Bay of Bengal. This would create millions of environmental refugees with no place to go in this already densely populated country.

Bangladesh is one of the few less-developed nations that is preparing and implementing plans to adapt to projected rising sea levels resulting from climate change. This includes using varieties of rice and other crops that can better tolerate flooding, saltwater, and drought; shifting to new crops such as corn; developing small vegetable gardens in bare patches between houses to help reduce Bangladeshi dependence on rice; building small ponds that will collect monsoon rainwater to use for irrigating vegetable gardens during dry periods; and creating a network of earthen embankments that might protect against high tides and moderate storm surges when cyclones strike.

GOOD NEWS

Solutions

Reducing Flood Damage

Prevention	Control
Preserve forests on watersheds	Straighten and deepen streams (channelization)
Preserve and restore wetlands in floodplains	Build levees or floodwalls along streams
Tax development on floodplains	Build dams
Use floodplains primarily for recharging aquifers, sustainable agriculture and forestry	

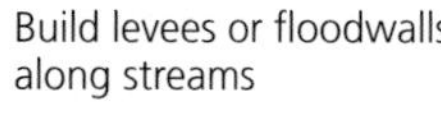

Figure 11-21 These are some methods for reducing the harmful effects of flooding (**Concept 11-4**). **Questions:** Which two of these solutions do you think are the most important? Why?

We Can Reduce Flood Risks

GOOD NEWS

Figure 11-21 lists some ways to reduce flooding risks (**Concept 11-4**). To improve flood control, we can rely less on engineered devices such as dams and levees and more on nature's systems such as wetlands and water- and topsoil-retaining vegetation in watersheds.

Channelizing streams reduces upstream flooding. But it also eliminates aquatic habitats, reduces groundwater discharge, and results in a faster flow, which can increase downstream flooding and sediment deposition. In addition, channelization encourages human settlement in floodplains, which increases the risk of damages and deaths from major floods.

Levees or floodwalls along the banks of rivers contain and speed up stream flow, but they also increase the water's capacity for doing damage downstream. They also do not protect against unusually high and powerful floodwaters such as those that occurred in 1993 when two-thirds of the levees built along the Mississippi River in the United States were damaged or destroyed. Similar flooding occurred along the Mississippi in 2011. A dam can reduce the threat of flooding by storing water in a reservoir and releasing it gradually, but dams also have a number of drawbacks (Figure 11-12).

An important way to reduce flooding is to *preserve existing wetlands* and *restore degraded wetlands* to take advantage of the natural flood control they provide in floodplains. We can also sharply reduce emissions of greenhouse gases that contribute to projected atmospheric warming and the resulting climate change, which will likely raise sea levels and flood many of the world's coastal areas during this century.

On a personal level, we can use the precautionary principle (see Chapter 8, p. 170) and *think carefully about where we choose to live*. Many poor people live in flood-prone areas because they have nowhere else to go. Most people, however, can choose not to live in areas especially subject to flooding or to water shortages.

11-5 How Can We Deal with Water Pollution?

▶ **CONCEPT 11-5 Reducing water pollution requires that we prevent it, work with nature to treat sewage, cut resource use and waste, reduce poverty, and slow population growth.**

Water Pollution Comes from Point and Nonpoint Sources

Water pollution is any change in water quality that can harm living organisms or make the water unfit for human uses such as drinking, irrigation, and recreation. It can come from single (point) sources or from larger and dispersed (nonpoint) sources. **Point sources** discharge pollutants into bodies of surface water at specific locations through drain pipes (Figure 11-22), ditches, or sewer lines. Examples include factories, sewage treatment plants (which remove some, but not all, pollutants), underground mines, oil wells, and oil tankers.

Because point sources are located at specific places, they are fairly easy to identify, monitor, and regulate. Most of the world's more-developed countries have laws that help control point-source discharges of harmful chemicals into aquatic systems. In most of the less-developed countries, there is little control of such discharges.

Nonpoint sources are broad and diffuse areas, rather than points, from which pollutants enter bodies of surface water or air. Examples include runoff of chemicals and sediments from cropland, logged forests, urban streets, parking lots, lawns, and golf courses. We have made little progress in controlling water pollu-

age fotostock/SuperStock

Figure 11-22 This is a point source of water pollution in Gargas, France.

tion from nonpoint sources because of the difficulty and expense of identifying and controlling discharges from so many diffuse sources.

Agricultural activities are by far the leading cause of water pollution. Sediment eroded from agricultural lands (see Figure 10-9, p. 213) is the most common pollutant. Other major agricultural pollutants include fertilizers and pesticides, bacteria from livestock and food-processing wastes, and excess salts from soils of irrigated cropland. *Industrial facilities,* which emit a variety of harmful inorganic and organic chemicals, are a second major source of water pollution. *Mining* is the third biggest source. Surface mining disturbs the land, creating major erosion of sediments and runoff of toxic chemicals. (See *The Habitable Planet,* Video 6 at **www.learner.org/resources/series209.html** for a discussion of how scientists measure water pollution from toxic heavy metals found in mining wastes and abandoned underground mines.)

CONNECTIONS

Atmospheric Warming and Water Pollution

Projected climate change from atmospheric warming will likely contribute to water pollution in some areas of the globe during this century. In a warmer world, some regions will get more precipitation and other areas will get less. More intense downpours will flush more harmful chemicals, plant nutrients, and microorganisms into waterways. Prolonged drought will reduce river flows that dilute wastes.

Major Water Pollutants Have Harmful Effects

Table 11-1 lists the major types of water pollutants along with examples of each and their harmful effects and sources. According to the World Health Organization (WHO), 4,400 people die each day from preventable infectious diseases that they get from drinking contaminated water. Every 18 seconds, a young child dies from diarrhea caused mostly by drinking such water.

Table 11-1 Major Water Pollutants and Their Sources

Type/*Effects*	Examples	Major Sources
Infectious agents (pathogens) *Cause diseases*	Bacteria, viruses, protozoa, parasites	Human and animal wastes
Oxygen-demanding wastes *Deplete dissolved oxygen needed by aquatic species*	Biodegradable animal wastes and plant debris	Sewage, animal feedlots, food-processing facilities, paper mills
Plant nutrients *Cause excessive growth of algae and other species*	Nitrates (NO_3^-) and phosphates (PO_4^{3-})	Sewage, animal wastes, inorganic fertilizers
Organic chemicals *Add toxins to aquatic systems*	Oil, gasoline, plastics, pesticides, fertilizers, cleaning solvents	Industry, farms, households, mining sites, runoff from streets and parking lots
Inorganic chemicals *Add toxins to aquatic systems*	Acids, bases, salts, metal compounds	Industry, households, mining sites, runoff from streets and parking lots
Sediments *Disrupt photosynthesis, food webs, other processes*	Soil, silt	Land erosion from farms and construction and mining sites
Heavy metals *Cause cancer, disrupt immune and endocrine systems*	Lead, mercury, arsenic	Unlined landfills, household chemicals, mining refuse, industrial discharges
Thermal *Make some species vulnerable to disease*	Heat	Electric power and industrial plants

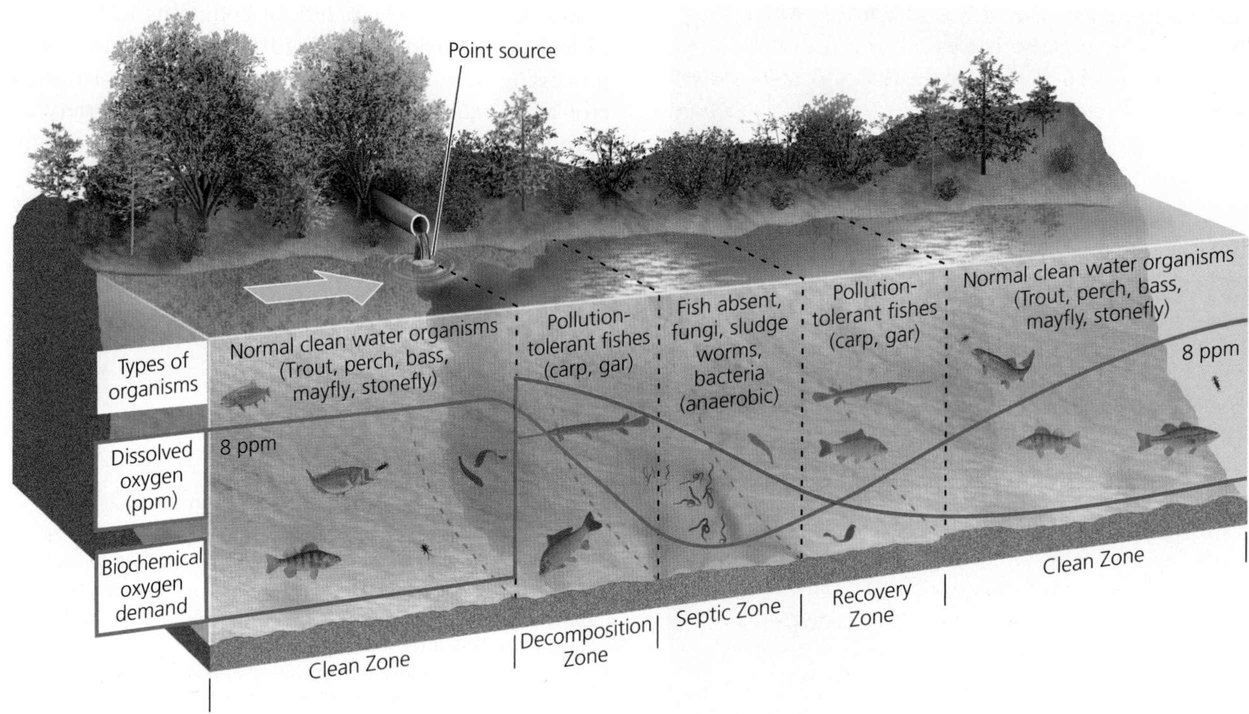

Active Figure 11-23 **Natural capital:** This figure shows the oxygen sag curve (blue) and the oxygen demand curve (red). *See an animation based on this figure at* www.cengagebrain.com. **Question:** What would be the effect of putting another discharge pipe emitting biodegradable waste to the right of the one in this picture?

Streams Can Cleanse Themselves, If We Do Not Overload Them

Flowing rivers and streams can recover rapidly from moderate levels of degradable, oxygen-demanding wastes through a combination of dilution and bacterial biodegradation of such wastes. But this natural recovery process does not work when streams become overloaded with such pollutants or when drought, damming, or water diversion reduces their flows. Also, while this process can remove biodegradable wastes, it does not eliminate slowly degradable and nondegradable pollutants.

In a flowing stream, the breakdown of biodegradable wastes by bacteria depletes dissolved oxygen and creates an *oxygen sag curve* (Figure 11-23). This reduces or eliminates populations of organisms with high oxygen requirements until the stream is cleansed of oxygen-demanding wastes. While streams can cleanse themselves, people can help by allowing these natural processes to occur (see Individuals Matter, below).

INDIVIDUALS MATTER

John Beal Planted Trees to Restore a Stream

In 1980, John Beal, an engineer with the U.S.-based Boeing Company, was told that he had only a few months to live because of severe heart problems. To help prolong his life, he began taking daily walks. His strolls took him by Hamm Creek, a small stream that flows from the hills southwest of Seattle, Washington (USA), into the Duwamish River, which empties into Puget Sound. He remembered when the stream was a spawning ground for salmon and when evergreen trees lined its banks. By 1980, the polluted stream had no fish and the trees were gone.

Beal decided to spend his last days doing something good by helping to clean up Hamm Creek. He persuaded some companies to stop polluting the creek, and he hauled out many truckloads of trash and items such as discarded washing machines and truck tires. Then he began a 15-year project of planting thousands of trees along the stream's banks. He also restored natural waterfalls and ponds that had served as salmon spawning beds.

At first, he worked alone and many people thought he was crazy, but word spread and volunteers began to join him. Television news reports and newspaper articles about the restoration project eventually brought hundreds of volunteers and schoolchildren.

The creek's water now runs clear, its vegetation has been restored, and salmon have returned to spawn. Until his death in 2006—27 years after doctors gave him only months to live—his reward was the personal satisfaction he felt about making a difference for Hamm Creek and his community. He also won more than 40 awards for his restoration work. His dedication to making the world a better place is an inspiring example of *stewardship* based on the idea that *all sustainability is local* and any individual can begin the process of restoring a polluted stream or some other degraded natural area.

Laws enacted in the 1970s to control water pollution have greatly increased the number and quality of plants that treat **wastewater**—water that contains sewage and other wastes from homes and industries—in the United States and in most other more-developed countries. Such laws also require industries to reduce or eliminate their point-source discharges of harmful chemicals into surface waters. This has enabled the United States to hold the line against increased pollution by disease-causing agents and oxygen-demanding wastes in most of its streams. This is an impressive accomplishment given the country's increased economic activity, resource consumption, and population growth since the passage of these laws.

GOOD NEWS

In most less-developed countries, stream pollution from discharges of untreated sewage, industrial wastes, and discarded trash is a serious and growing problem. A majority of these countries cannot afford to build sewage treatment plants and do not have, or do not enforce, laws for controlling water pollution.

According to the World Commission on Water in the 21st Century, half of the world's 500 major rivers are heavily polluted, and most of these polluted waterways run through less-developed countries. The Global Water Policy Project estimates that most cities in less-developed countries discharge 80–90% of their untreated sewage directly into rivers, streams, and lakes whose waters are then used for drinking, bathing, and washing clothes.

For example, in China, 54 of the 78 rivers and streams that are monitored are polluted by industrial wastes and sewage (Figure 11-24). In some parts of China, river water is too toxic to touch, much less drink. In order to clean up cities such as Beijing and Shanghai, the country has moved some of its oil refineries and factories to rural areas where two-thirds of China's population lives. Liver and stomach cancer, linked in some cases to water pollution, are among the leading causes of death in the countryside.

In 2010, Chinese officials reported that huge islands of garbage are threatening to jam the flow of water over the country's massive Three Gorges Dam. Some of these trash islands are as deep as 0.6 meters (2 feet) and in some areas, they are so compacted that people can walk on them. As it decays, this garbage will pollute the water.

In India, industrial wastes and sewage pollute more than 66% of all water resources. Also, in Latin America and Africa, most streams passing through urban or industrial areas suffer from severe pollution.

Zhao Weiming/UNEP/Peter Arnold, Inc.

Figure 11-24 Natural capital degradation: This highly polluted river in central China is greenish-black due to uncontrolled pollution from thousands of factories.

Too Little Mixing and Low Water Flow Make Lakes and Reservoirs Vulnerable to Water Pollution

Lakes and reservoirs are generally less effective at diluting pollutants than streams are, for two reasons. *First,* lakes and reservoirs often contain stratified layers (see Figure 7-30 p. 147) that undergo little vertical mixing. *Second,* they have little or no flow. The flushing and changing of water in lakes and large artificial reservoirs can take from 1 to 100 years, compared with several days to several weeks for streams.

As a result, lakes and reservoirs are more vulnerable than streams are to contamination by runoff or discharges of plant nutrients, oil, pesticides, and nondegradable toxic substances, such as lead, mercury, and arsenic. Many toxic chemicals and acids also enter lakes and reservoirs from the atmosphere.

Eutrophication is the name given to the natural nutrient enrichment of a shallow lake, estuary, or slow-moving stream. It is caused mostly by runoff of plant nutrients such as nitrates and phosphates from surrounding land. An *oligotrophic lake* is low in nutrients and its water is clear (see Figure 7-29, left, p. 146). Over time, some lakes become more eutrophic (see Figure 7-29, right) as nutrients are added from natural and human sources in the surrounding watersheds.

Near urban or agricultural areas, human activities can greatly accelerate the input of plant nutrients to a lake—a process called **cultural eutrophication**. Such inputs involve mostly nitrate- and phosphate-containing effluents from various sources, including farmland, animal feedlots, urban streets and parking lots, chemically fertilized suburban yards, mining sites, and municipal sewage treatment plants. Some nitrogen also reaches lakes by deposition from the atmosphere.

During hot weather or drought, this nutrient overload produces dense growths, or "blooms," of organisms such as algae and cyanobacteria (see Figure 7-29, right, p. 146), and thick growths of water hyacinth, duckweed, and other aquatic plants. These dense colonies of plant life can reduce lake productivity and fish growth by decreasing the input of solar energy needed for photosynthesis by phytoplankton that support fish.

When the algae die, they are decomposed by swelling populations of aerobic bacteria that deplete dissolved oxygen in the surface layer of water near the shore as well as in the bottom layer of a lake. This can kill fish and other aquatic animals. If excess nutrients continue to flow into a lake, anaerobic bacteria take over and produce gaseous products such as smelly, highly toxic hydrogen sulfide and flammable methane.

According to the U.S. Environmental Protection Agency (EPA), about one-third of the 100,000 medium to large lakes and 85% of the large lakes near major U. S. population centers have some degree of cultural eutrophication. The International Water Association also estimates that more than half of the lakes in China suffer from cultural eutrophication (Figure 11-25).

There are several ways to *prevent* or *reduce* cultural eutrophication. We can use advanced (but expensive) waste treatment to remove nitrates and phosphates before wastewater enters lakes. We can also use a preventive approach by banning or limiting the use of phosphates in household detergents and other cleaning agents, and by employing soil conservation (p. 226) and land-use control to reduce nutrient runoff.

GOOD NEWS

There are several ways to *clean up* lakes suffering from cultural eutrophication. They include mechanically removing excess weeds, controlling undesirable plant growth with herbicides and algaecides, and pumping air into lakes and reservoirs to prevent oxygen depletion, all of which are expensive and energy-intensive methods. Most lakes can recover from cultural eutrophication if excessive inputs of plant nutrients are stopped.

Groundwater Cannot Cleanse Itself Very Well

According to many scientists, groundwater pollution is a serious threat to human health. Common pollutants such as fertilizers, pesticides, gasoline, and organic solvents can seep into groundwater from numerous sources (Figure 11-26). People who dump or spill gasoline, oil, and paint thinners and other organic solvents onto the ground also contaminate groundwater.

When groundwater becomes contaminated, it cannot cleanse itself of *degradable wastes* as quickly as flowing surface water can. Groundwater flows so slowly—usually less than 0.3 meters (1 foot) per day—that contaminants are not diluted and dispersed effectively. In addition, groundwater usually has much lower concentrations of dissolved oxygen (which helps decompose many contaminants) and smaller populations of decomposing bacteria. And the usually cold temperatures of groundwater slow down chemical reactions that decompose wastes.

Thus, it can take decades to thousands of years for contaminated groundwater to cleanse itself of *slowly degradable wastes* (such as DDT). On a human time scale, *nondegradable wastes* (such as toxic lead and arsenic) remain in the water permanently.

Peter Arnold, Inc.

Figure 11-25 Severe cultural eutrophication has covered this lake near the Chinese city of Haozhou with algae.

Figure 11-26 **Natural capital degradation:** These are the principal sources of groundwater contamination in the United States. Another source in coastal areas is saltwater intrusion from excessive groundwater withdrawal. (Figure is not drawn to scale.) **Question:** What are three sources shown in this diagram that might be affecting groundwater in your area?

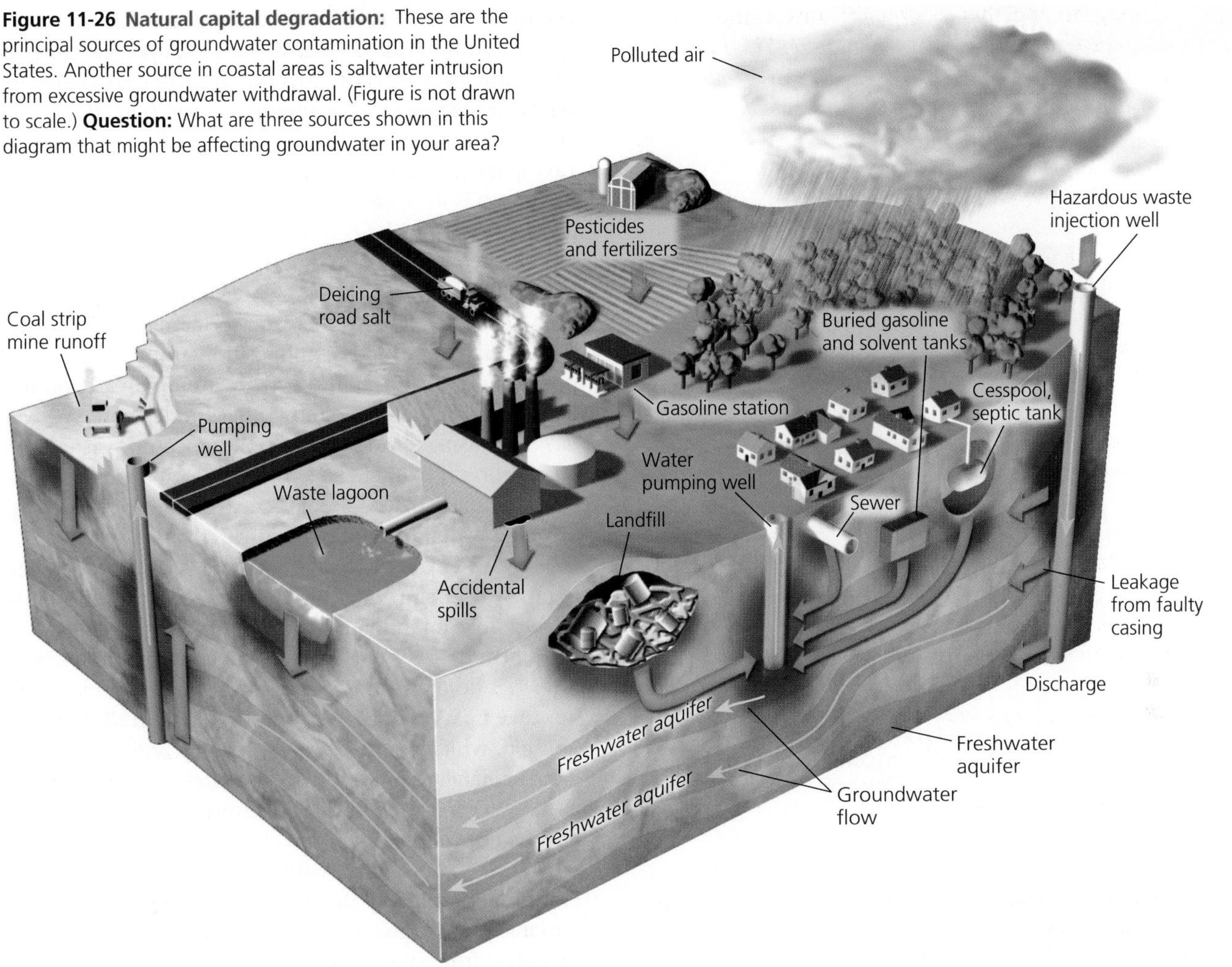

Groundwater Pollution Is a Serious Hidden Threat in Some Areas

On a global scale, we do not know much about groundwater pollution because few countries go to the great expense of locating, tracking, and testing aquifers. But the results of scientific studies in scattered parts of the world are alarming.

Groundwater provides about 70% of China's drinking water. In 2006, the Chinese government reported that aquifers in about nine of every ten Chinese cities were polluted or overexploited, and could take hundreds of years to recover.

In the United States, an EPA survey of 26,000 industrial waste ponds and lagoons found that one-third of them had no liners to prevent toxic liquid wastes from seeping into aquifers. One-third of these sites are within 1.6 kilometers (1 mile) of a drinking water well. In addition, almost two-thirds of America's liquid hazardous wastes are injected into the ground in disposal wells (Figure 11-26), some of which leak water into aquifers used as sources of drinking water.

By 2008, the EPA had completed the cleanup of about 357,000 of the more than 479,000 underground tanks in the United States that were leaking gasoline, diesel fuel, home heating oil, or toxic solvents into groundwater. During this century, scientists expect many of the millions of such tanks, which have been installed around the world, to become corroded and leaky, possibly contaminating groundwater and becoming a major global health problem. Determining the extent of a leak from a single underground tank can cost $25,000–250,000, and cleanup costs range from $10,000 to more than $250,000. If the chemical reaches an aquifer, effective cleanup is often not possible or is too costly.

Pollution Prevention Is the Only Effective Way to Protect Groundwater

Figure 11-27 (p. 262) lists ways to prevent and clean up groundwater contamination. Because of the difficulty and expense of cleaning up a contaminated aquifer, preventing groundwater contamination (Figure 11-27, left) is the only effective way to deal with this serious water pollution problem.

Solutions

Groundwater Pollution

Prevention	Cleanup
Find substitutes for toxic chemicals	Pump to surface, clean, and return to aquifer (very expensive)
Keep toxic chemicals out of the environment	
Install monitoring wells near landfills and underground tanks	Inject microorganisms to clean up contamination (less expensive but still costly)
Require leak detectors on underground tanks	
Ban hazardous waste disposal in landfills and injection wells	
Store harmful liquids in aboveground tanks with leak detection and collection systems	Pump nanoparticles of inorganic compounds to remove pollutants (still being developed)

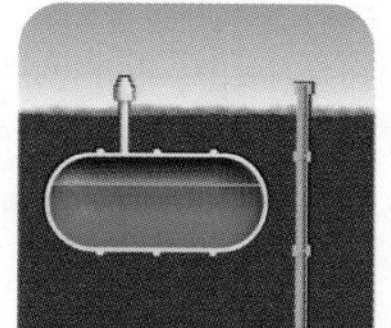

Figure 11-27 There are ways to prevent and ways to clean up contamination of groundwater but prevention is the only effective approach (**Concept 11-5**). **Questions:** Which two of these preventive solutions do you think are the most important? Why?

There Are Many Ways to Purify Drinking Water

Most of the more-developed countries have laws establishing drinking water standards. But most of the less-developed countries do not have such laws or, if they do have them, they do not enforce them.

In more-developed countries, wherever people depend on surface water sources, drinking water is usually stored in a reservoir for several days. This improves clarity and taste by increasing dissolved oxygen content and allowing suspended matter to settle. The water is then pumped to a purification plant and treated to meet government drinking water standards. In areas with very pure groundwater or surface water sources, little treatment is necessary. Some cities have found that protecting the watersheds that supply their drinking water is a lot cheaper than building water purification plants.

We have the technology to convert sewer water into pure drinking water. But reclaiming wastewater is expensive and it faces opposition from citizens and from some health officials who are unaware of the advances in this technology. In a world where we will face increasing shortages of drinking water, wastewater purification is likely to become a major growth business.

GOOD NEWS

We can also use simpler measures to purify drinking water. In tropical countries that lack centralized water treatment systems, the WHO urges people to purify drinking water by exposing a clear plastic bottle filled with contaminated water to intense sunlight. The sun's heat and UV rays can kill infectious microbes in as little as 3 hours. Painting one side of the bottle black can improve heat absorption in this simple solar disinfection method, which applies the solar energy **principle of sustainability**. Where this measure has been used, the incidence of dangerous childhood diarrhea has decreased by 30–40%.

SUSTAINABILITY

In 2007, the Danish inventor Torben Vestergaard Frandsen developed the LifeStraw™, an inexpensive, portable water filter that eliminates many viruses and parasites from water that is drawn through it (Figure 11-28). This filter has been particularly useful in Africa, where aid agencies are distributing it.

Increasingly, people all over the world are relying on bottled water, not all of which is as pure as advertised. The growing use of bottled water presents some environmental problems (see the Case Study that follows).

Vestergaard Frandsen

Figure 11-28 Four young men in Uganda demonstrate how to use the *LifeStraw*™. **Question:** Do you think the development of such devices should make prevention of water pollution less of a priority? Explain.

■ CASE STUDY

Is Bottled Water a Good Option?

GOOD NEWS

Despite some problems, experts say the United States has some of the world's cleanest drinking water. Municipal water systems in the United States are required to test their water regularly for a number of pollutants and to make the results available to citizens.

Yet about half of all Americans worry about getting sick from tap water contaminants, and many drink high-priced bottled water or install expensive water purification systems. Studies by the Natural Resources Defense Council (NRDC) reveal that in the United States, a bottle of water costs between 240 and 10,000 times as much as the same volume of tap water.

Americans are the world's largest consumers of bottled water, some of it shipped from as far away as Fiji (8,800 kilometers, or 5,500 miles). This amount of bottled water consumed by Americans is enough to meet the annual drinking water needs of the roughly 1 billion people in the world who routinely lack access to safe and clean drinking water. According to water expert Peter Gleick and the NRDC, at least 40% of the bottled water used in United States is tap water and 30% of the bottled water tested contained bacteria and synthetic organic chemicals.

Use of bottled water also causes environmental problems. According studies by the Worldwatch Institute and by water experts Peter Gleick and Heather Cooley, each year, the number of plastic water bottles thrown away, if lined up end-to-end, would circle the earth's equator eight times. And 86% of these bottles are not recycled. According to the Pacific Institute, the oil used to pump, process, bottle, transport, and refrigerate the bottled water used annually in the United States would be enough to run 3 million cars for a year. In addition, withdrawing water for bottling is helping to deplete some aquifers. There is also concern about health risks from chemicals such as bisphenol A (BPA) that can leach into the water from the plastic in some water bottles, especially if they are exposed to the hot sun.

Because of these harmful environmental impacts and the high cost of bottled water, there is a growing *back-to-the-tap* movement based on boycotting bottled water. Its motto is *"Think globally, drink locally."* From San Francisco to New York to Paris, city governments, restaurants, schools, religious groups, and many consumers are refusing to buy bottled water as this trend picks up steam. Individuals are also refilling portable bottles with tap water and using simple filters to improve the taste and color of water where necessary.

Health officials suggest that before drinking expensive bottled water or buying costly home water purifiers, consumers should have their water tested by local health departments or private labs (but not by companies trying to sell water purification equipment). Independent experts contend that unless tests show otherwise, home water-treatment systems are not worth their expense and maintenance hassles.

Ocean Pollution Is a Growing and Poorly Understood Problem

Why should we care about the oceans? Short answer: Because they keep us alive. The oceans hold 97% of the earth's water, make up 97% of the biosphere where life is found, and contain the planet's greatest diversity and abundance of life.

Oceans help to provide and recycle the planet's freshwater through the water cycle (see Figure 3-14, p. 51). They also strongly affect weather and climate, help to regulate the earth's temperature, and absorb some of the massive amounts of carbon dioxide that we emit into the atmosphere. As the famous oceanographer and explorer Sylvia A. Earle (see Chapter 9, Individuals Matter, p. 199) reminds us: "Even if you never have the chance to see or touch the ocean, the ocean touches you with every breath you take, every drop of water you drink, every bite you consume. Everyone, everywhere is inextricably connected to and utterly dependent upon the existence of the sea."

Despite its importance, we treat the ocean as the world's largest depository for the massive and growing amount of wastes and pollutants that we produce (see the Case Study that follows). And we know less about the oceans, on which our lives and well-being depend, than we do about the moon.

Coastal areas—especially wetlands, estuaries, coral reefs, and mangrove swamps—bear the brunt of our enormous inputs of pollutants and wastes into the ocean (Figure 11-29, p. 264). Roughly 40% of the world's people (53% in the United States) live on or near coastlines, which helps explain why 80% of marine pollution originates on land. (See *The Habitable Planet,* Video 5 at **www.learner.org/resources/series209.html** to learn how scientists are studying the effects of population growth and development on coastal aquatic systems.)

According to a 2006 *State of the Marine Environment* study by the UN Environment Programme (UNEP), 80–90% of the municipal sewage from most coastal areas of less-developed countries, and in some coastal areas of more-developed countries, is dumped into oceans without treatment. This often overwhelms the ability of the coastal waters to biodegrade these wastes. The coastline of China, for example, is so choked with algae growing on the nutrients provided by sewage that some scientists believe large areas of China's coastal waters can no longer sustain marine ecosystems.

Recent studies of some U.S. coastal waters have found vast colonies of viruses thriving in raw sewage and in effluents from sewage treatment plants (which do not remove viruses) and leaking septic tanks. According to one study, one-fourth of the people using coastal beaches in the United States develop ear infections, sore throats, eye irritations, respiratory disease, or gastrointestinal disease from swimming in seawater containing infectious viruses and bacteria.

Scientists also point to the underreported problem of pollution from cruise ships. A cruise liner can carry

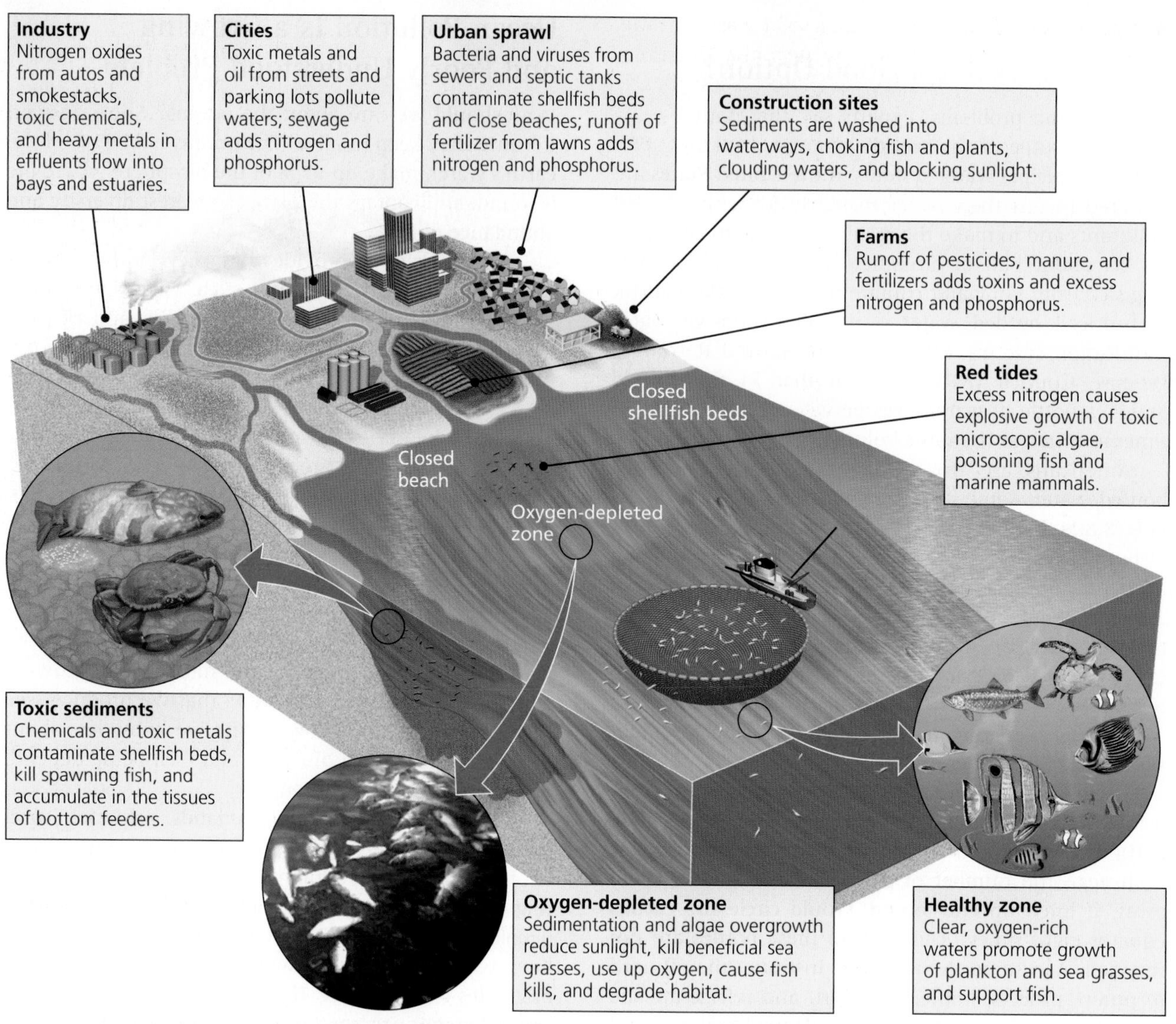

Figure 11-29 Natural capital degradation: Residential areas, factories, and farms all contribute to the pollution of coastal waters. **Questions:** What do you think are the three worst pollution problems shown here? For each one, how does it affect two or more of the ecological and economic services listed in Figure 7-20 (p. 140)?

as many as 6,300 passengers and 2,300 crewmembers, and it can generate as much waste (including toxic chemicals, garbage, sewage, and waste oil) as a small city. Many cruise ships dump these wastes at sea. In U.S. waters, such dumping is illegal, but some ships continue dumping secretively, usually at night. Some environmentally aware vacationers are refusing to go on cruise ships that do not have sophisticated systems for dealing with the wastes they produce.

Runoffs of sewage and agricultural wastes into coastal waters introduce large quantities of nitrate (NO_3^-) and phosphate (PO_4^{3-}) plant nutrients, which can cause explosive growths of harmful algae. These *harmful algal blooms* are called red (Figure 11-29), brown, or green toxic tides. They can release waterborne and airborne toxins that poison seafood, damage fisheries, kill some fish-eating birds, and reduce tourism. Annually, harmful algal blooms lead to the poisoning of about 60,000 Americans who eat shellfish contaminated by the algae.

Every year, because of harmful algal blooms, at least 400 *oxygen-depleted zones* form in coastal waters around the world, according to a 2008 study by marine scientists Robert Diaz and Rutger Rosenberg. They occur mostly in temperate coastal waters and in large bodies of water with restricted outflows, such as the Baltic and Black seas. About 43 of these zones occur in U.S. waters (Science Focus, p. 266). A 2008 study by Luan Weixin, of China's Dalain Maritime University, found that water pollutants such as nitrates and phosphates seriously contaminate about half of China's shallow coastal waters.

■ CASE STUDY
Ocean Garbage Patches: There Is No Away

In 1997, ocean researcher and sea captain Charles Moore accidentally discovered two gigantic, slowly rotating masses of plastic and other solid wastes in the middle of the North Pacific Ocean near the Hawaiian Islands. These wastes, floating on or just beneath the ocean's surface and known as the Great Pacific Garbage Patch, are trapped there by a vortex where four rotating ocean currents called *gyres* meet (Figure 11-30). Since then, four other huge garbage patches have been discovered in the world's other major oceans.

About 80–90% of this litter is long-lasting plastic. Also, roughly 80% of this trash comes from the land—washed or blown off of beaches, pouring out of storm drains, and floating down streams and rivers that empty into the sea. Most of the rest is dumped into the ocean from cargo and cruise ships, oil-drilling platforms, and cargo containers that have broken open or fallen off of ships in stormy seas, disgorging their contents.

While these garbage patches contain tons of trash, they typically are not visible from above. Much of the plastic has been partially decomposed to particles about the size of rice grains. When a slowly whirling garbage patch encounters an island, it leaves the beach littered with these particles, which mix with beach sand.

Research shows that the tiny plastic particles can be harmful to marine mammals, to some seabirds such as albatrosses, and to some fishes and other aquatic species that mistake them for food and swallow them. Because these animals cannot digest the plastic, it can clog up their digestive systems and prevent them from getting enough nutrients to stay alive. For example, a dead turtle found in Hawaii had more than a thousand pieces of plastic in its stomach and intestines.

Fish that feed on plankton ingest these tiny plastic particles, which can contain PCBs, DDT, hormone-mimicking bisphenol A (BPA), and other long-lasting harmful chemicals. These toxins can build up to high concentrations in food chains and webs (Figure 8-12, p. 163), and they can end up in our fish sandwiches and seafood dinners. Thus, toxic chemicals from a discarded plastic grocery bag, water bottle, or chips bag could end up in your stomach. Everything is connected.

Who is to blame for these patches? Each of us risks making a contribution every time we use and discard a plastic item. We may think we have simply thrown it away, but there is no away.

What can be done? There is no practical or affordable way to clean up this mess that our throwaway culture has created. The only effective solution is prevention, which means practicing the three R's of resource use: Reduce, Reuse, and Recycle.

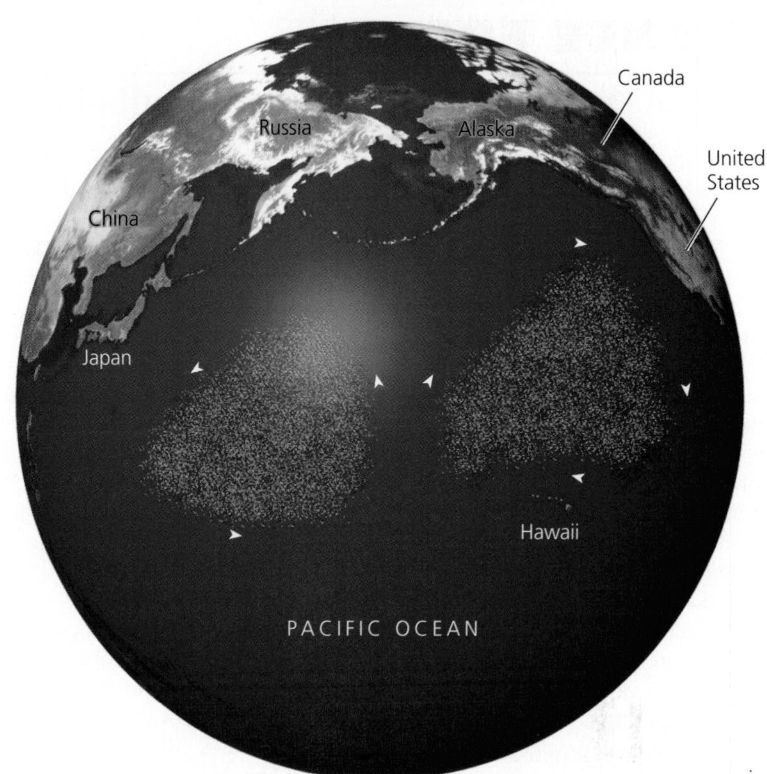

Figure 11-30 The Great Pacific Garbage Patch, by some scientists' estimates, occupies an area twice the size of the U.S. state of Texas. Other scientists say it might be as large as the continental United States. Its size is difficult to estimate because most of this plastic-laden soup is suspended slightly underwater and is difficult to see and measure.

Ocean Pollution from Oil

Crude petroleum (oil as it comes out of the ground) and *refined petroleum* (fuel oil, diesel, gasoline, and other processed petroleum products) reach the ocean from a number of sources and become highly disruptive pollutants. The most visible sources are tanker accidents, such as the huge *Exxon Valdez* oil spill in the U.S. state of Alaska in 1989, and blowouts at offshore oil drilling rigs, such as that of the BP Deep Horizon rig in the Gulf of Mexico in 2010 (discussed in Chapter 13). However, studies show that the largest source of ocean oil pollution is urban and industrial runoff from land, much of it from leaks in pipelines and oil-handling facilities.

At least 37%—and perhaps even 50%—of the oil reaching the oceans is waste oil, dumped, spilled, or leaked onto the land or into sewers from city streets and industrial sites, as well as by people changing their own motor oil. According to a 2006 UNEP study, GOOD NEWS since the mid-1980s, the amount of oil entering the marine environment from oil tanker accidents has decreased by 75% and oil discharges from industries and cities have dropped by nearly 90%. The reductions in oil spills from tankers were mostly due to increased safety requirements imposed on shippers after the 1989 Exxon Valdez tanker accident. The 2010 BP oil well blowout will probably result in tighter regulations on the design, safety features, and daily practices related to offshore drilling rigs.

Volatile organic hydrocarbons in oil and other petroleum products kill many aquatic organisms immediately upon contact, especially if these animals are in their vulnerable larval forms. Other chemicals in oil form tar-like globs that float on the surface and coat the feathers of seabirds (especially diving birds) and the fur of marine mammals. This oil coating destroys their natural heat insulation and buoyancy, causing many of them to drown or die of exposure from loss of body heat.

SCIENCE FOCUS

Oxygen Depletion in the Northern Gulf of Mexico

The world's third largest oxygen-depleted zone (after those in the Baltic Sea and the northwestern Black Sea) forms every spring and summer in a narrow stretch of the northern Gulf of Mexico off the mouth of the Mississippi River. This area includes the coastal waters of the U.S. states of Texas, Louisiana, and Mississippi (Figure 11-B). The low oxygen levels suffocate fish, crabs, and shrimp that cannot move to less polluted areas. Thus, these oxygen-depleted zones threaten aquatic biodiversity and whole ecosystems.

Scientists believe the gulf's oxygen-depleted zone forms as a result of oxygen-depleting algal blooms. Evidence indicates that it is created mostly by huge inputs of nitrate (NO_3^-) plant nutrients from farms, cities, factories, and sewage treatment plants located in the vast Mississippi River basin drain all or part of 31 U.S. states (Figure 11-B) and two Canadian provinces. This watershed contains more than half of all U.S. croplands and is one of the world's most productive agricultural regions.

Some scientists who have studied this problem fear that it could reach a tipping point beyond which the ecosystem could collapse. This would be the point at which most of organisms living in this part of the Gulf of Mexico simply can no longer move far enough or fast enough to find oxygen-rich waters. Their populations might then decline to the point where they could not recover.

In recent years, this oxygen-depleted zone has covered an area almost as large as the U.S. state of New Jersey. Because of the size and agricultural importance of the Mississippi River basin, there are no easy solutions to the problem of severe cultural eutrophication of this and other overfertilized coastal zones around the world. Preventive measures include applying less fertilizer on farms upstream, injecting fertilizer below the soil surface, and using controlled-release fertilizers that have water-insoluble coatings. Other measures include planting strips of forests and grasslands along waterways to soak up excess nitrogen, restoring Gulf Coast wetlands that once filtered some of the plant nutrients, and creating wetlands between crop fields and streams emptying into the Mississippi. Another measure would be to reduce government subsidies for growing corn to make ethanol (as discussed in Chapter 13).

Still other preventive measures involve improving flood control to prevent the release of nitrogen from floodplains within the basin during major floods and upgrading sewage treatment systems in the region to reduce discharges of nitrates into its waterways. In addition, deposition of nitrogen compounds from the atmosphere could be reduced by requiring lower emissions of nitrogen oxides from motor vehicles and by phasing in forms of renewable energy to replace the burning of fossil fuels.

Critical Thinking

How do you think each of the preventive measures described above would help to prevent pollution in the Gulf of Mexico? Can you think of other possible preventive solutions?

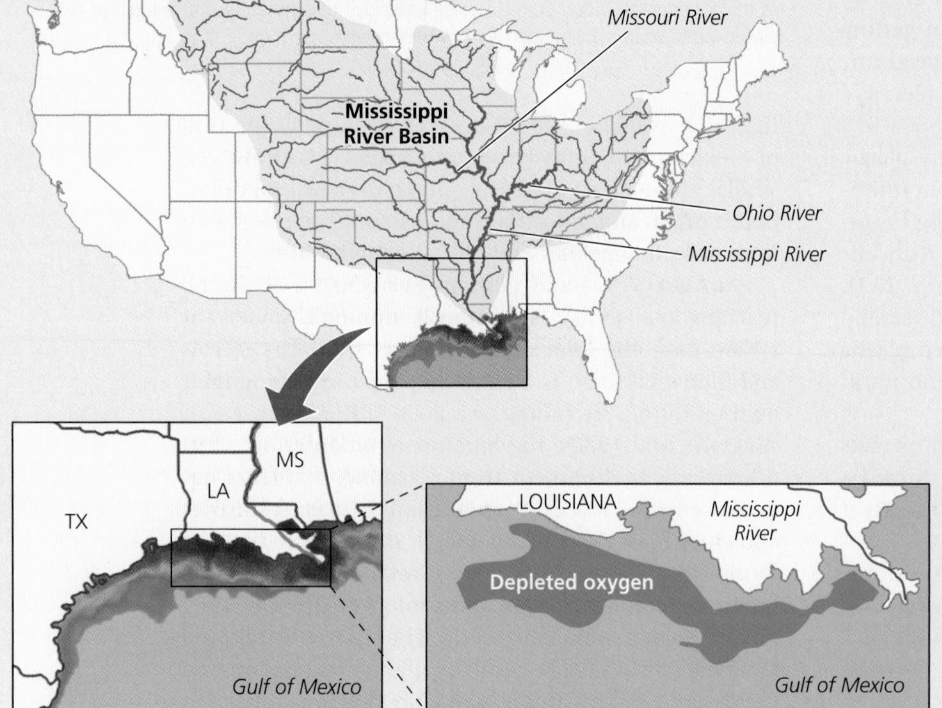

Figure 11-B Natural capital degradation: A large zone of oxygen-depleted water (containing less than 2 ppm dissolved oxygen) forms each year in the Gulf of Mexico (lower right drawing) as a result of oxygen-depleting algal blooms caused primarily by high inputs of nutrients from the Mississippi River basin (top). The bottom left inset map, based on a satellite image, shows how these inputs spread along the coast during the summer of 2006. In the image, reds and greens represent high concentrations of phytoplankton and river sediment. **Question:** Can you think of a product you used today that was directly connected to this sort of pollution? (Data from NASA)

Heavy oil components that sink to the ocean floor or wash into estuaries and coastal wetlands can smother bottom-dwelling organisms such as crabs, oysters, mussels, and clams, or make them unfit for human consumption. Some oil spills have killed coral reefs. (See *The Habitable Planet,* Video 9 at **www.learner.org/resources/series209.html** for a discussion of how scientists measure the effects of oil spills on coral reefs and fish populations.)

Research shows that populations of many forms of marine life can recover from exposure to large amounts of *crude oil* in warm waters with fairly rapid currents within about 3 years. But in cold and calm waters, full recovery can take decades. In addition, recovery from exposure to *refined oil*, especially in estuaries and salt marshes, can take 10–20 years or longer. Oil slicks that wash onto beaches can have a serious economic impact on coastal residents, who lose income normally gained from fishing and tourist activities. In 2009, some 20 years after the *Exxon Valdez* spill, researchers found patches of oil remaining on some parts of the shoreline of the U.S. state of Alaska's Prince William Sound.

If they are not too large, oil spills can be partially cleaned up by mechanical means including floating booms, skimmer boats, and absorbent devices such as large pillows filled with feathers or hair. But scientists estimate that current cleanup methods can recover no more than 15% of the oil from a major spill.

Thus, *preventing* oil pollution is the most effective and, in the long run, the least costly approach. One of the best ways to prevent tanker spills is to use oil tankers with double hulls. More stringent safety standards and inspections could help to reduce oil well blowouts at sea. And most importantly, businesses, institutions, and citizens living in coastal areas must take care to prevent leaks and spillage of even the smallest amounts of oil, as these small events altogether account for a third to half of all ocean oil pollution.

THINKING ABOUT
Ocean Oil Pollution

What are three ways in which you might be contributing to ocean oil pollution? How could you reduce your contribution to this environmental problem?

Solutions

Coastal Water Pollution

Prevention

- Reduce input of toxic pollutants
- Separate sewage and storm water lines
- Ban dumping of wastes and sewage by ships in coastal waters
- Ban dumping of hazardous material
- Strictly regulate coastal development, oil drilling, and oil shipping
- Require double hulls for oil tankers

Cleanup

- Improve oil-spill cleanup capabilities
- Use nanoparticles on sewage and oil spills to dissolve the oil or sewage (still under development)
- Require secondary treatment of coastal sewage
- Use wetlands, solar-aquatic, or other methods to treat sewage

Figure 11-31 There are methods for preventing and methods for cleaning up excessive pollution of coastal waters. **Questions:** Which two of these solutions do you think are the most important? Why?

Reducing Ocean Water Pollution

Most ocean pollution occurs in coastal waters and comes from human activities on land. Figure 11-31 lists ways to prevent and reduce the pollution of coastal waters. GOOD NEWS

The key to protecting the oceans is to reduce the flow of pollution from land and air and from streams emptying into these waters (**Concept 11-5**). Thus, ocean pollution control must be linked with land-use and air pollution control policies, which in turn are linked to energy policies (discussed in Chapter 13) and climate policies (discussed in Chapter 15).

Reducing Surface Water Pollution from Nonpoint Sources

There are a number of ways to reduce nonpoint sources of water pollution, most of which come from agricultural practices. Farmers can reduce soil erosion by keeping cropland covered with vegetation GOOD NEWS and using other soil conservation methods (see Chapter 10, pp. 226–227). They can also reduce the amount of fertilizer that runs off into surface waters by using slow-release fertilizer, using no fertilizer on steeply sloped land, and planting buffer zones of vegetation between cultivated fields and nearby surface waters. (See *The Habitable Planet*, Video 7 at **www.learner.org/resources/series209.html** to learn how scientists have reduced excessive nitrogen runoff from fertilizer.)

Organic farming (see Chapter 10, Core Case Study, p. 204) can also help to prevent water pollution caused by nutrient overload because it does not use synthetic inorganic fertilizers and pesticides. And by applying pesticides only when needed and relying more on integrated pest management (see p. 224), farmers can reduce pesticide runoff. In addition, they can control runoff and infiltration of manure from animal feedlots by planting buffer zones and by locating feedlots, pastures, and animal waste storage sites away from steeply sloped land, surface water, and flood zones.

Laws Can Help to Reduce Water Pollution from Point Sources

The Federal Water Pollution Control Act of 1972 (renamed the Clean Water Act when it was amended in 1977) and the 1987 Water Quality Act form the basis of U.S. efforts to control pollution of the country's surface waters. The Clean Water Act sets standards for allowed levels of 100 key water pollutants and requires polluters to get permits that limit how much of these various pollutants they can discharge into aquatic systems.

Also, the EPA has been experimenting with a *discharge trading policy,* which uses market forces to reduce water pollution in the United States. Under this program, a permit holder can pollute at higher levels than allowed in its permit if it buys credits from permit holders who are polluting below their allowed levels.

Environmental scientists warn that the effectiveness of such a system depends on how low the cap on total pollution levels in any given area is set and on how regularly the cap is lowered. They also warn that discharge trading could allow water pollutants to build up to dangerous levels in areas where credits are bought. They call for careful scrutiny of the cap levels and for the gradual lowering of the caps to encourage prevention of water pollution and development of better pollution-control technology. Neither adequate scrutiny of the cap levels nor gradual lowering of caps is a part of the current EPA discharge trading system.

According to the EPA, the Clean Water Act of 1972 led to numerous improvements in U.S. water quality. These are some of the encouraging developments that took place between 1992 and 2002 (the latest years for which figures available):

GOOD NEWS

- The percentage of Americans served by community water systems that met federal health standards increased from 79% to 94%.
- The percentage of U.S. stream lengths found to be fishable and swimmable increased from 36% to 60% of those tested.
- The proportion of the U.S. population served by sewage treatment plants increased from 32% to 74%.
- Annual wetland losses decreased by 80%.

These are impressive achievements given the growth of the U.S. population and its per capita consumption of water and other resources since 1972. But there is more work to be done. In 2006, the EPA found that 45% of the country's lakes and 40% of the streams surveyed were still too polluted for swimming or fishing. Also, a 2007 government study found that tens of thousands of gasoline storage tanks in 43 states were leaking. In 2009, a *New York Times* investigation found that one in five U.S. water treatment systems violated the Safe Drinking Water Act between 2003 and 2008. In addition, a 2010 *New York Times* study found that thousands of the country's largest water polluters (about 45% of them) have declared that the Clean Water Act no longer applies to waters that they are polluting. More than one of every three Americans gets their drinking water from such sources.

Some environmental scientists call for strengthening the Clean Water Act. Suggested improvements include shifting the focus of the law to water pollution prevention instead of focusing mostly on end-of-pipe removal of specific pollutants; greatly increased monitoring for violations of the law and much larger mandatory fines for violators; and regulating irrigation water quality (for which there is no federal regulation). Another suggestion is to expand the rights of citizens to bring lawsuits to ensure that water pollution laws are enforced. Still another suggestion is to rewrite the Clean Water Act to clarify that it covers all waterways (as Congress originally intended). This would eliminate confusion about which waterways are covered and stop major polluters from using this confusion to keep polluting in many areas.

Many people oppose these proposals, contending that the Clean Water Act's regulations are already too restrictive and costly. Some state and local officials argue that in many communities, it is unnecessary and too expensive to test all the water for pollutants as required by federal law.

Sewage Treatment Reduces Water Pollution

In rural and suburban areas with suitable soils, sewage from each house usually is discharged into a **septic tank** with a large drainage field. In this system, household sewage and wastewater is pumped into a settling tank, where grease and oil rise to the top, and solids fall to the bottom where they are decomposed by bacteria. The resulting partially treated wastewater is discharged into a large drainage (absorption) field through small holes in perforated pipes embedded in porous gravel or crushed stone just below the soil's surface. As these wastes drain from the pipes and percolate downward, the soil filters out some potential pollutants and soil bacteria decompose biodegradable materials.

About one-fourth of all homes in the United States are served by septic tanks. They work well as long as they are not overloaded and if their solid wastes are regularly pumped out.

In urban areas in the United States and other more-developed countries, most waterborne wastes from homes, businesses, and storm runoff flow through a network of sewer pipes to *wastewater* or *sewage treatment plants.* Raw sewage reaching a treatment plant typically undergoes one or two levels of wastewater treatment. The first is **primary sewage treatment**—a *physical* process that uses screens and a grit tank to remove large floating objects and to allow solids such as sand and rock to settle out. Then the waste stream flows into a primary settling tank where suspended solids settle out as sludge (Figure 11-32, left).

The second level is **secondary sewage treatment**—a *biological* process in which aerobic bacteria remove as much as 90% of dissolved and biodegradable, oxygen-demanding organic wastes (Figure 11-32, right). A combination of primary and secondary treatment removes 95–97% of the suspended solids and oxygen-demanding organic wastes, 70% of most toxic metal compounds and nonpersistent synthetic organic chemicals, 70% of the phosphorus, and 50% of the

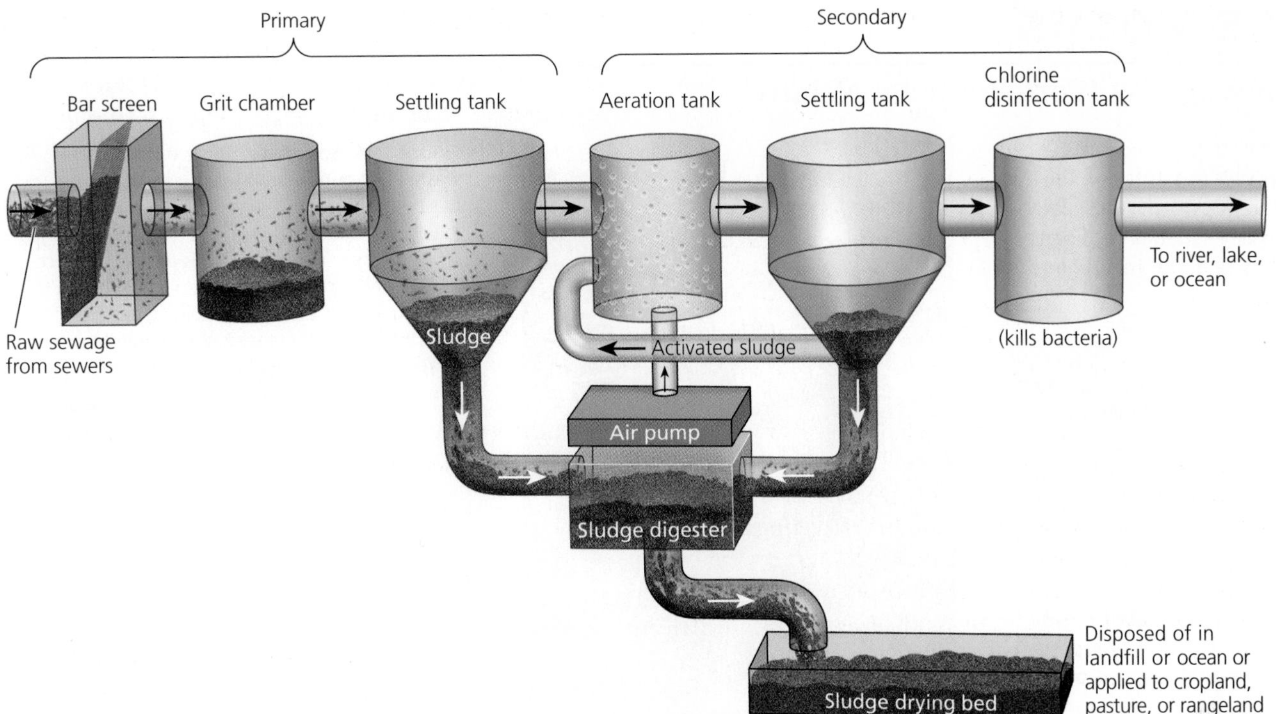

Figure 11-32 Solutions: Primary and secondary sewage treatment systems help to reduce water pollution. **Question:** What do you think should be done with the sludge produced by sewage treatment plants?

nitrogen. However, this process removes only a tiny fraction of persistent and potentially toxic organic substances found in some pesticides and in discarded medicines that people put into sewage systems, and it does not kill many disease-causing organisms.

Before discharge, water from sewage treatment plants usually undergoes *bleaching*, to remove water coloration, and *disinfection* to kill disease-carrying bacteria and some (but not all) viruses. The usual method for accomplishing this is *chlorination*. But chlorine can react with organic materials in water to form small amounts of chlorinated hydrocarbons. Some of these chemicals cause cancers in test animals, can increase the risk of miscarriages, and can damage the human nervous, immune, and endocrine systems. The use of other disinfectants, such as ozone and ultraviolet light, is increasing, but they cost more and their effects do not last as long as those of chlorination.

We Can Improve Conventional Sewage Treatment

Environmental scientist Peter Montague calls for redesigning the conventional sewage treatment system shown in Figure 11-32. The idea is to prevent toxic and hazardous chemicals from reaching sewage treatment plants and thus from getting into sludge and water discharged from such plants.

Montague suggests several ways to do this. One is to require industries and businesses to remove toxic and hazardous wastes from water sent to municipal sewage treatment plants. Another is to encourage industries to reduce or eliminate their use and waste of toxic chemicals.

Another suggestion is to require or encourage more households, apartment buildings, and offices to eliminate sewage outputs by switching to waterless, odorless *composting toilet systems*, to be installed, maintained, and managed by professionals. This process returns plant nutrients in human waste to the soil and thus mimics the natural chemical cycling **principle of sustainability**. It also reduces the need for energy-intensive and water-polluting commercial fertilizers.

On a larger scale, such systems would be cheaper to install and maintain than current sewage systems are, because they do not require vast systems of underground pipes connected to centralized sewage treatment plants. They also save large amounts of water, reduce water bills, and decrease the amount of energy used to pump and purify water.

In 2009, a Swedish entrepreneur developed a biodegradable single-use plastic bag that can be used as a toilet in urban slums and in other areas where 40% of the world's people do not have access to toilets. After it is used, the bag is knotted and buried. A thin layer of urea in this bag kills the disease-producing pathogens in the feces and helps break down the waste into plant nutrients that are then recycled—a simple and inexpensive, low-tech application of the chemical cycling **principle of sustainability**.

SCIENCE FOCUS

Treating Sewage by Working with Nature

Some communities and individuals are seeking better ways to purify sewage by working with nature (**Concept 11-5**). Biologist John Todd has developed an ecological approach to treating sewage, which he calls *living machines* (Figure 11-C).

This purification process begins when sewage flows into a passive solar greenhouse or outdoors site containing rows of large open tanks populated by an increasingly complex series of organisms. In the first set of tanks, algae and microorganisms decompose organic wastes, with sunlight speeding up the process. Water hyacinths, cattails, bulrushes, and other aquatic plants growing in the tanks take up the resulting nutrients.

After flowing though several of these natural purification tanks, the water passes through an artificial marsh made of sand, gravel, and bulrushes, which filters out algae and remaining organic waste. Some of the plants also absorb, or *sequester*, toxic metals such as lead and mercury and secrete natural antibiotic compounds that kill pathogens.

Next, the water flows into aquarium tanks, where snails and zooplankton consume microorganisms and are in turn consumed by crayfish, tilapia, and other fish that can be eaten or sold as bait. After 10 days, the clear water flows into a second artificial marsh for final filtering and cleansing. The water can be made pure enough to drink by treating it with ultraviolet light or by passing the water through an ozone generator, usually immersed out of sight in an attractive pond or wetland habitat. Operating costs are about the same as those of a conventional sewage treatment plant.

Figure 11-C Solutions: This is an ecological wastewater purification system called a *living machine*. This Solar Sewage Treatment Plant is in the U.S. city of Providence, Rhode Island. Biologist John Todd is demonstrating this ecological process he invented for purifying wastewater by using the sun and a series of tanks containing living organisms. Todd and others are conducting research to perfect such solar–aquatic sewage treatment systems based on working with nature.

More than 800 cities and towns around the world (150 in the United States) use natural or artificially created wetlands to treat sewage as a lower-cost alternative to expensive waste treatment plants. For example, in Arcata, California—a coastal town of 16,000 people—scientists and othe workers created some 65 hectares (160 acres) of wetlands between the town and the adjacent Humboldt Bay. The marshes and ponds, developed on land that was once a dump, act as a natural waste treatment plant. The project cost was less than half the estimated price of a conventional treatment plant.

This system returns purified water to Humboldt Bay, and the sludge that is removed is processed for use as fertilizer. The marshes and ponds also serve as an Audubon Society bird sanctuary, which provides habitats for thousands of seabirds, otters, and other marine animals. The town even celebrates its natural sewage treatment system with an annual "Flush with Pride" festival.

This approach and the living machine system developed by John Todd apply all three **principles of sustainability:** using solar energy, employing natural processes to remove and recycle nutrients and other chemicals, and relying on a diversity of organisms and natural processes to purify wastewater.

SUSTAINABILITY

Critical Thinking

Can you think of any disadvantages of using such a nature-based system instead of a conventional sewage treatment plant? Do you think any such disadvantages outweigh the advantages? Why or why not?

Many communities are using unconventional, but highly effective, *wetland-based sewage treatment systems*, which work with nature (Science Focus, above).

There Are Sustainable Ways to Reduce and Prevent Water Pollution

It is encouraging that since 1970, most of the world's more-developed countries have enacted laws and regulations that have significantly reduced point-source water pollution. These improvements were largely the result of *bottom-up* political pressure on elected officials from individuals and groups.

GOOD NEWS

This is a good beginning. However, to environmental and health scientists, the next step is to ramp up efforts to reduce and prevent water pollution in both more- and less-developed countries. They would begin by asking the question: *How can we avoid producing water pollutants in the first place?* (**Concept 11-5**). Figure 11-33 lists ways to achieve this goal over the next several decades.

This shift to pollution prevention will not take place unless citizens put political pressure on elected officials to promote it while also taking action to reduce their own daily contributions to water pollution. Figure 11-34 lists some steps you can take to help reduce water pollution.

GOOD NEWS

Solutions

Water Pollution

- Prevent groundwater contamination
- Reduce nonpoint runoff
- Reuse treated wastewater for drinking and irrigation
- Find substitutes for toxic pollutants
- Work with nature to treat sewage
- Practice the three R's of resource use (reduce, reuse, recycle)
- Reduce air pollution
- Reduce poverty
- Slow population growth

Figure 11-33 There are a number of ways to prevent or reduce water pollution (**Concept 11-5**). **Questions:** Which two of these solutions do you think are the best ones? Why?

What Can You Do?

Reducing Water Pollution

- Fertilize garden and yard plants with manure or compost instead of commercial inorganic fertilizer
- Minimize your use of pesticides, especially near bodies of water
- Prevent yard wastes from entering storm drains
- Do not use water fresheners in toilets
- Do not flush unwanted medicines down the toilet
- Do not pour pesticides, paints, solvents, oil, antifreeze, or other products containing harmful chemicals down the drain or onto the ground

Figure 11-34 Individuals matter: You can help reduce water pollution. **Questions:** Which three of these actions do you think are the most important ones to take? Why?

Here are this chapter's *three big ideas*:

- One of the major global environmental problems is the growing shortage of freshwater in many parts of the world.
- We can use water more sustainably by cutting water waste, raising water prices, and protecting aquifers, forests, and other ecosystems that store and release water.
- Reducing water pollution requires preventing it, working with nature to treat sewage, cutting resource use and waste, reducing poverty, and slowing population growth.

REVISITING The Colorado River and Sustainability

The **Core Case Study** that opens this chapter discusses the problems and tensions that can occur when a large number of U.S. states share a limited river water resource in a water-short region.

Generally, the water resource strategies of the 20th century have worked against natural chemical cycles and processes. Large dams, river diversions, levees, and other big engineering schemes have helped to provide much of the world with electricity, food from irrigated crops, drinking water, and flood control. But they have also degraded the aquatic natural capital necessary for long-term economic and ecological sustainability by seriously disrupting rivers, streams, wetlands, aquifers, and other aquatic systems.

This chapter also discussed water pollution in more-developed and less-developed countries. Pollution control for the benefit of the world's surface and underground water supplies is within our reach. But even more hopeful is the possibility of shifting our emphasis from cleaning up water pollution to preventing it.

The three **principles of sustainability** (see back cover) can guide us in reducing and preventing water pollution and in using water more sustainably during this century. Scientists hope to use solar energy to desalinate water, thereby increasing freshwater supplies, and solar energy is already used to purify water for drinking in some areas. Recycling more water will help us to reduce water waste, and recycling much more of the chemicals and other products that are now discarded will prevent them from becoming water pollutants. Preserving biodiversity by avoiding disruption of aquatic systems and their bordering terrestrial systems is a key factor in maintaining water supplies and water quality.

This *blue revolution*, built mostly around cutting water waste and preventing water pollution, will provide numerous economic and ecological benefits. Many environmental scientists argue that there is no time to lose in implementing it.

Thousands have lived without love—not one without water.

W. H. AUDEN

REVIEW

CORE CASE STUDY

1. Discuss the importance of the Colorado River basin in the United States and how human activities are stressing this system (**Core Case Study**). CORE CASE STUDY

SECTION 11-1

2. What are the two key concepts for this section? Define **freshwater**. Describe how water is recycled by the hydrologic cycle and how human activities can overload and alter this cycle. What percentage of the earth's freshwater is available to us? Define **groundwater**, **zone of saturation**, **water table**, **aquifer**, **surface water**, **surface runoff**, **watershed** (**drainage basin**), **reliable surface runoff**, and **drought**. What percentage of the world's reliable runoff are we using and what percentage are we likely to be using by 2025? How is most of the world's water used? Describe the availability and use of freshwater resources in the United States. Define **water footprint** and **virtual water**. What are five major problems resulting from the way people are using water from the Colorado River basin? What percentage of the earth's land suffers from severe drought today and how might this change by 2059? How many people in the world lack regular access to clean water today and how high might this number grow by 2025?

SECTION 11-2

3. What are the three key concepts for this section? What are the advantages and disadvantages of withdrawing groundwater? Describe the problem of groundwater depletion in the world and in the United States, especially in the Ogallala aquifer. Describe the problems of land subsidence and contamination of freshwater aquifers near coastal areas resulting from the overpumping of water from aquifers. List four ways to prevent or slow groundwater depletion, including the possible use of *deep aquifers*.

4. What is a **dam**? What is a **reservoir**? What are the advantages and disadvantages of using large dams and reservoirs? Describe what has happened to water flows in the Colorado River since 1960. List three possible solutions to this problem. Describe the California Water Project. Describe the environmental and health disaster caused by the Aral Sea water transfer project. Define **desalination** and distinguish between distillation and reverse osmosis as methods for desalinating water. What are three limitations of desalination?

SECTION 11-3

5. What is the key concept for this section? What percentage of available freshwater is unnecessarily wasted in the world and in the United States, and what are two causes of water waste? Describe three major irrigation methods and list ways to reduce water waste in irrigation in more- and less-developed countries. List four ways to reduce water waste in industries and homes, and three ways to use less water to remove wastes. List four ways to use water more sustainably and four ways in which you can reduce your use and waste of water.

SECTION 11-4

6. What is the key concept for this section? What is a **floodplain** and why do people like to live on floodplains? What are the benefits and drawbacks of floods? List three human activities that increase the risk of flooding. Describe the increased flooding risks that many people in Bangladesh face and what they are doing about it. List three ways to reduce the risks of flooding.

SECTION 11-5

7. What is the key concept for this section? What is **water pollution**? Distinguish between **point sources** and **nonpoint sources** of water pollution and give an example of each. List nine major types of water pollutants and give an example of each.

8. Describe how streams can cleanse themselves and how these cleansing processes can be overwhelmed. What is **wastewater**? Describe the state of stream pollution in more-developed and less-developed countries. Give two reasons why lakes cannot cleanse themselves as readily as streams can. Distinguish between **eutrophication** and **cultural eutrophication**. List ways to prevent or reduce cultural eutrophication. Explain why groundwater cannot cleanse itself very well. What are the major sources of groundwater contamination in the United States? List three ways to prevent and three ways to clean up groundwater contamination. Describe the U.S. laws passed to protect drinking water quality. Describe the environmental problems caused by the widespread use of bottled water.

9. How are coastal waters and deeper ocean waters polluted? What causes harmful algal blooms and what are their negative effects? Describe the annual oxygen depletion in the northern Gulf of Mexico. Summarize the nature, origins, and effects of the Great Pacific Garbage Patch. What are the effects of oil pollution of the oceans and what can be done to reduce such pollution?

10. List ways to reduce water pollution from **(a)** nonpoint sources and **(b)** point sources. Describe the U.S. experience with reducing point-source water pollution. What is a **septic tank** and how does it work? Explain how **primary sewage treatment** and **secondary sewage treatment** are used to help purify water. How would Peter Montague improve conventional sewage treatment? What is a composting toilet system? Describe John Todd's

use of living machines to treat sewage. Explain how wetlands can be used to treat sewage. List three ways to prevent and three ways to reduce water pollution. List five things you can do to reduce water pollution. What are this chapter's *three big ideas*? Explain how the three **principles of sustainability** can guide us in reducing and preventing water pollution and in using water more sustainably during this century.

Note: Key terms are in **bold** type.

CRITICAL THINKING

1. What do you believe are the three most important priorities for dealing with the water resource problems of the Colorado River basin, as discussed in the **Core Case Study** that opens this chapter? Explain your choices. CORE CASE STUDY

2. What role does population growth play in water supply problems? Relate this to water supply problems of the Colorado River basin (**Core Case Study**). CORE CASE STUDY

3. Explain why you are for or against **(a)** raising the price of water while providing lower lifeline rates for the poor and lower-middle class, **(b)** withdrawing government subsidies that provide farmers with water at low cost, and **(c)** providing government subsidies to farmers for improving irrigation efficiency.

4. Calculate how many liters (and gallons) of water are wasted in 1 month by a toilet that leaks 2 drops of water per second. (1 liter of water equals about 3,500 drops and 1 liter equals 0.265 gallon.) How many bathtubs (each containing about 151 liters or 40 gallons) could this wasted water fill?

5. List three ways in which human activities increase the harmful effects of flooding. What is the best way to prevent each of these human impacts? Do you think they should be prevented? Why or why not?

6. You are a regulator charged with drawing up plans for controlling water pollution. Briefly describe one idea for controlling water pollution from each of the following sources: **(a)** an effluent pipe from a factory going into a stream, **(b)** a parking lot at a shopping mall bordered by a stream, and **(c)** a farmer's field on a slope next to a stream.

7. When you flush your toilet, where does the wastewater go? Trace the actual flow of this water in your community from your toilet through sewers to a wastewater treatment plant and from there to the environment. Try to visit a local sewage treatment plant to see what it does with your wastewater. Compare the processes it uses with those shown in Figure 11-32. What happens to the sludge produced by this plant? What improvements, if any, would you suggest for this plant? List the three best ways you can think of for reducing your unnecessary waste of water. Which, if any, of these things do you already do?

8. List three ways in which you could apply **Concept 11-3** and three ways in which you could apply **Concept 11-5** to making your lifestyle more environmentally sustainable.

DOING ENVIRONMENTAL SCIENCE

1. Investigate water use at your school. Try to determine all specific sources of any water waste, taking careful notes and measurements for each of them, and try to estimate how much water is wasted per hour, per day, and per year. Develop a water conservation plan for your school and submit it to school officials.

2. Use library research, the Internet, and user interviews to evaluate the relative effectiveness of home water-purification devices. Try to determine the costs of such systems, including purchase, installation, and maintenance costs, and record these costs for each system. Determine the type or types of water pollutants that each device removes and the effectiveness of each such device.

GLOBAL ENVIRONMENT WATCH EXERCISE

Visit the *Gulf of Mexico Oil Spill 2010* topic portal and research the ecological effects of the oil well blowout and spill. Write a report that answers some or all of the following questions. Is oil pollution in the gulf still considered to be a problem? How has it affected the coastal wetlands along the gulf? How has it affected species living under the water? How has it affected bird life in the gulf? How has it affected the fishing industry there?

ECOLOGICAL FOOTPRINT ANALYSIS

In 2005, the population of the U.S. state of Florida consumed 24.5 billion liters (6.5 billion gallons) of freshwater daily. It is projected that in 2025, the daily consumption will increase to 32.1 billion liters (8.5 billion gallons) per day. Between 2005 and 2025, the population of Florida is projected to increase from 17.5 million to 25.9 million.

1. Based on total freshwater use:
   a. Calculate the per capita consumption of water per day in Florida in 2005 and the projected per capita consumption per day for 2025.
   b. Calculate the per capita consumption of water for the year in Florida in 2005 and the projected per capita consumption per year for 2025.
2. In 2005, how did the Florida *average water footprint* (consumption per person per year), based only on water used within the state, compare with the average U.S. water footprint of approximately 249,000 liters (66,000 gallons) per person per year, and with the global average water footprint of 123,770 liters (32,800 gallons) per person per year?

LEARNING ONLINE

The website for *Environmental Science 14e* contains helpful study tools and other resources to take the learning experience beyond the textbook. Examples include flashcards, practice quizzing, information on green careers, tips for more sustainable living, activities, and a wide range of articles and further readings. To access these materials and other companion resources for this text, please visit **www.cengagebrain.com**. For further details, see the preface, p. xvi.

Geology and Nonrenewable Minerals

12

The Real Cost of Gold

CORE CASE STUDY

Mineral resources extracted from the earth's crust are processed into an amazing variety of products that can make our lives easier and provide us with economic benefits and jobs. But extracting minerals from the ground and using them to manufacture products has a number of harmful environmental effects.

For example, many newlyweds would be surprised to know that mining enough gold to make their wedding rings produced roughly enough mining waste to equal the total weight of more than three mid-size cars. This waste is usually left piled near the mine site and can pollute the air and nearby surface water.

In 2010, the world's top five gold producing countries were, in order, China, Australia, South Africa, the United States, and Russia. In Australia and North America, mining companies level entire mountains of rock containing only small concentrations of gold. To extract the gold, miners spray a solution of highly toxic cyanide salts (which react with gold) onto huge piles of crushed rock. The solution then drains off the rocks, pulling some gold with it, into settling ponds (Figure 12-1). After the solution is recirculated in this process a number of times, the gold is removed from the ponds.

This cyanide is extremely toxic to birds and mammals drawn to these ponds in search of water. The ponds can also leak or overflow, which poses threats to underground drinking water supplies and to fish and other forms of life in nearby lakes and streams. Special liners in the settling ponds can help prevent leaks, but some have failed. According to the U.S. Environmental Protection Agency, all such liners will eventually leak.

In 2000, snow and heavy rains washed out an earthen dam on one end of a cyanide leach pond at a gold mine in Romania. The dam's collapse released large amounts of water laced with cyanide and toxic metals into the Tisza and Danube Rivers, which flow through parts of Romania and Hungary. Several hundred thousand people living along these rivers were told not to fish or to drink or withdraw water from the rivers or from wells adjacent to them. Businesses located near the rivers were shut down. Thousands of fish and other aquatic animals and plants were killed. This accident and a similar one that occurred in January 2005 in Romania could have been prevented if the mining company had installed a stronger containment dam and a backup settling pond to prevent leakage into nearby surface water.

In this chapter, we will look at the earth's dynamic geologic processes, the valuable minerals such as gold that some of these processes produce, and the potential supplies of these resources. We will also consider the environmental effects of using them and how we might use them more sustainably.

Creatas/SuperStock

Figure 12-1 This gold mine site in the Black Hills of the U.S. state of South Dakota has cyanide leach piles and settling ponds.

Key Questions and Concepts

12-1 What are the earth's major geological processes and hazards?

CONCEPT 12-1 Dynamic processes move matter within the earth and on its surface, and can cause volcanic eruptions, earthquakes, tsunamis, erosion, and landslides.

12-2 How are the earth's rocks recycled?

CONCEPT 12-2 The three major types of rock found in the earth's crust—sedimentary, igneous, and metamorphic—are recycled very slowly by the processes of erosion, melting, and metamorphism.

12-3 What are mineral resources and what are the environmental effects of using them?

CONCEPT 12-3 We can make some minerals in the earth's crust into useful products, but extracting and using these resources can disturb the land, erode soils, produce large amounts of solid waste, and pollute the air, water, and soil.

12-4 How long will supplies of nonrenewable mineral resources last?

CONCEPT 12-4A All nonrenewable mineral resources exist in finite amounts, and as we get closer to depleting any mineral resource, the environmental impacts of extracting it generally become more harmful.

CONCEPT 12-4B Raising the price of a scarce mineral resource can lead to an increase in its supply, but there are environmental limits to this effect.

12-5 How can we use mineral resources more sustainably?

CONCEPT 12-5 We can try to find substitutes for scarce resources, reduce resource waste, and recycle and reuse minerals.

Note: Supplements 4 (p. S10) and 6 (p. S22) can be used with this chapter.

Civilization exists by geological consent, subject to change without notice.

WILL DURANT

12-1 What Are the Earth's Major Geological Processes and Hazards?

▶ **CONCEPT 12-1 Dynamic processes move matter within the earth and on its surface, and can cause volcanic eruptions, earthquakes, tsunamis, erosion, and landslides.**

The Earth Is a Dynamic Planet

Geology, one of the subjects of this chapter, is the science devoted to the study of dynamic processes taking place on the earth's surface and in its interior. As the primitive earth cooled over eons, its interior separated into three major concentric zones: the *core*, the *mantle*, and the *crust* (see Figure 3-2, p. 42).

The **core** is the earth's innermost zone. It is extremely hot and has a solid inner part, surrounded by a liquid core of molten or semisolid material. Surrounding the core is a thick zone called the **mantle**. Most of the mantle is solid rock, but under its rigid outermost part is the **asthenosphere**—a zone of hot, partly melted rock that flows and can be deformed like pizza dough.

The outermost and thinnest zone of the earth is the **crust**. It consists of the *continental crust*, which underlies the continents (including the continental shelves extending into the oceans), and the *oceanic crust*, which underlies the ocean basins and makes up 71% of the earth's crust (Figure 12-2). The combination of the crust and the rigid, outermost part of the mantle (above the asthenosphere) is called the **lithosphere**.

The Earth Beneath Your Feet Is Moving

We tend to think of the earth's crust, mantle, and core as fairly static. In reality, *convection cells* or *currents* move large volumes of rock and heat in loops within the mantle like gigantic conveyer belts (Figure 12-3 and **Concept 12-1**). Geologists believe that the flows of energy and heated material in these convection cells caused the lithosphere to break up into a dozen or so

Abyssal hills
Oceanic crust (lithosphere)
Abyssal floor
Oceanic ridge
Abyssal floor
Trench
Volcanoes
Folded mountain belt
Craton
Abyssal plain
Abyssal plain
Continental shelf
Continental slope
Continental rise
Continental crust (lithosphere)
Mantle (lithosphere)
Mantle (lithosphere)
Mantle (asthenosphere)

Figure 12-2 The earth's crust and upper mantle have certain major features. The *lithosphere*, composed of the crust and outermost mantle, is rigid and brittle. The *asthenosphere*, a zone in the mantle, can be deformed by heat and pressure.

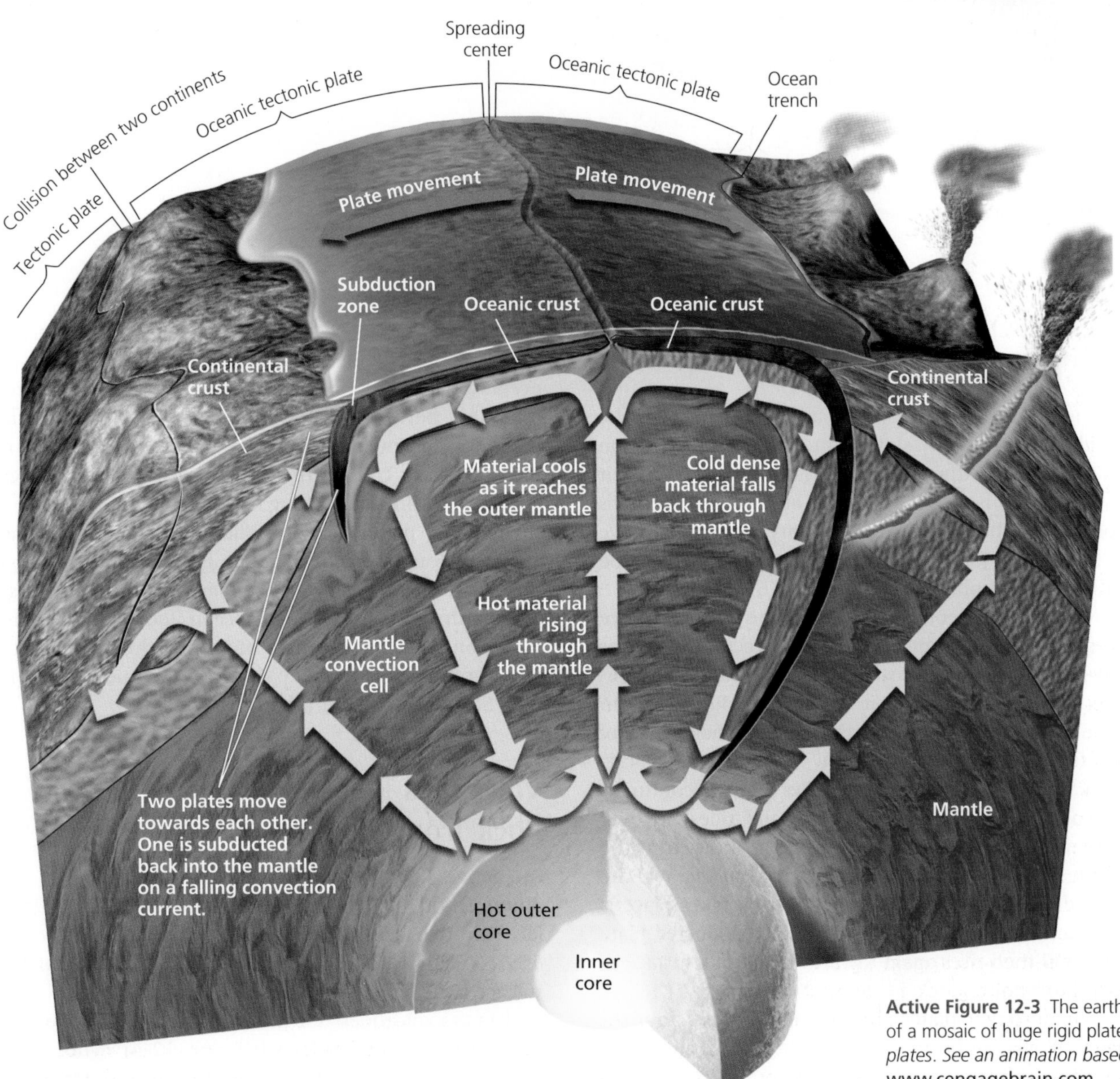

Active Figure 12-3 The earth's crust is made up of a mosaic of huge rigid plates, called *tectonic plates. See an animation based on this figure at* www.cengagebrain.com.

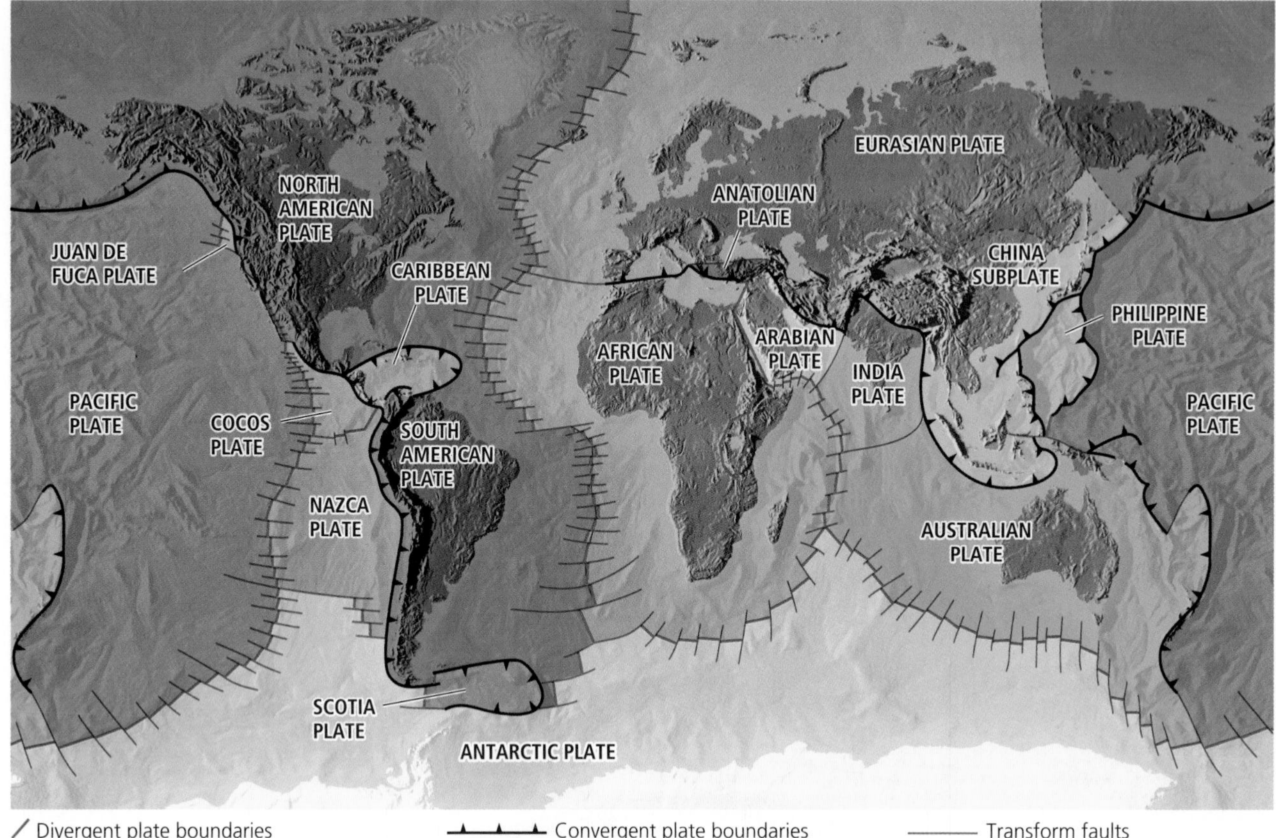

Active Figure 12-4 This map shows the earth's major tectonic plates. *See an animation based on this figure at* www.cengagebrain.com. **Question:** Which plate are you riding on?

huge rigid plates, called **tectonic plates**, which move extremely slowly atop the asthenosphere (Figures 12-3 and 12-4).

These gigantic, thick plates are somewhat like the world's largest and slowest-moving surfboards on which we ride without noticing their movement. Their typical speed is about the rate at which fingernails grow. Throughout the earth's history, continents have split apart and joined as tectonic plates drifted atop the earth's asthenosphere (Figures 4-7, p. 69, and 12-2).

Much of the geologic activity at the earth's surface takes place at the boundaries between tectonic plates as they separate, collide, or slide past one another. The tremendous forces produced at these plate boundaries can cause mountains to form, earthquakes to shake parts of the crust, and volcanoes to erupt.

When an oceanic plate collides with a continental plate, the continental plate usually rides up over the denser oceanic plate and pushes it down into the mantle (Figure 12-3) in a process called *subduction*. The area where this collision and subduction takes place is called a *subduction zone*. Over time, the subducted plate melts and then rises again toward the earth's surface as *magma*, or molten rock.

When oceanic plates move apart from one another, magma flows up through the resulting cracks and solidifies. This creates *oceanic ridges* (Figure 12-2), some of which have higher peaks and deeper canyons than the earth's continents have. On the other hand, when two oceanic plates collide, a *trench* ordinarily forms at the boundary between the two plates as the denser plate is subducted under the less dense one. When two continental plates collide, they push up mountain ranges such as the Himalayas along the collision boundary.

Tectonic plates can also slide and grind past one another along a fracture (fault) in the lithosphere—a type of boundary called a *transform fault*. Most transform faults are located on the ocean floor but a few are found on land. For example, the North American Plate and the Pacific Plate slide past each other along California's San Andreas fault (Figure 12-5).

Volcanoes Release Molten Rock from the Earth's Interior

An active **volcano** occurs where magma rising in a plume through the lithosphere reaches the earth's surface through a central vent or a long crack, called a *fissure* (Figure 12-6). Magma that reaches the earth's surface is called *lava* and often builds into a cone.

Many volcanoes form along the boundaries of the earth's tectonic plates (Figure 12-4) when one plate slides under or moves away from another plate.

Figure 12-5 This is the San Andreas Fault as it crosses part of the Carrizo plain between San Francisco and Los Angeles, California (USA). This fault, which runs almost the full length of California, is responsible for earthquakes of various magnitudes. **Question:** Is there a transform fault near where you live or go to school?

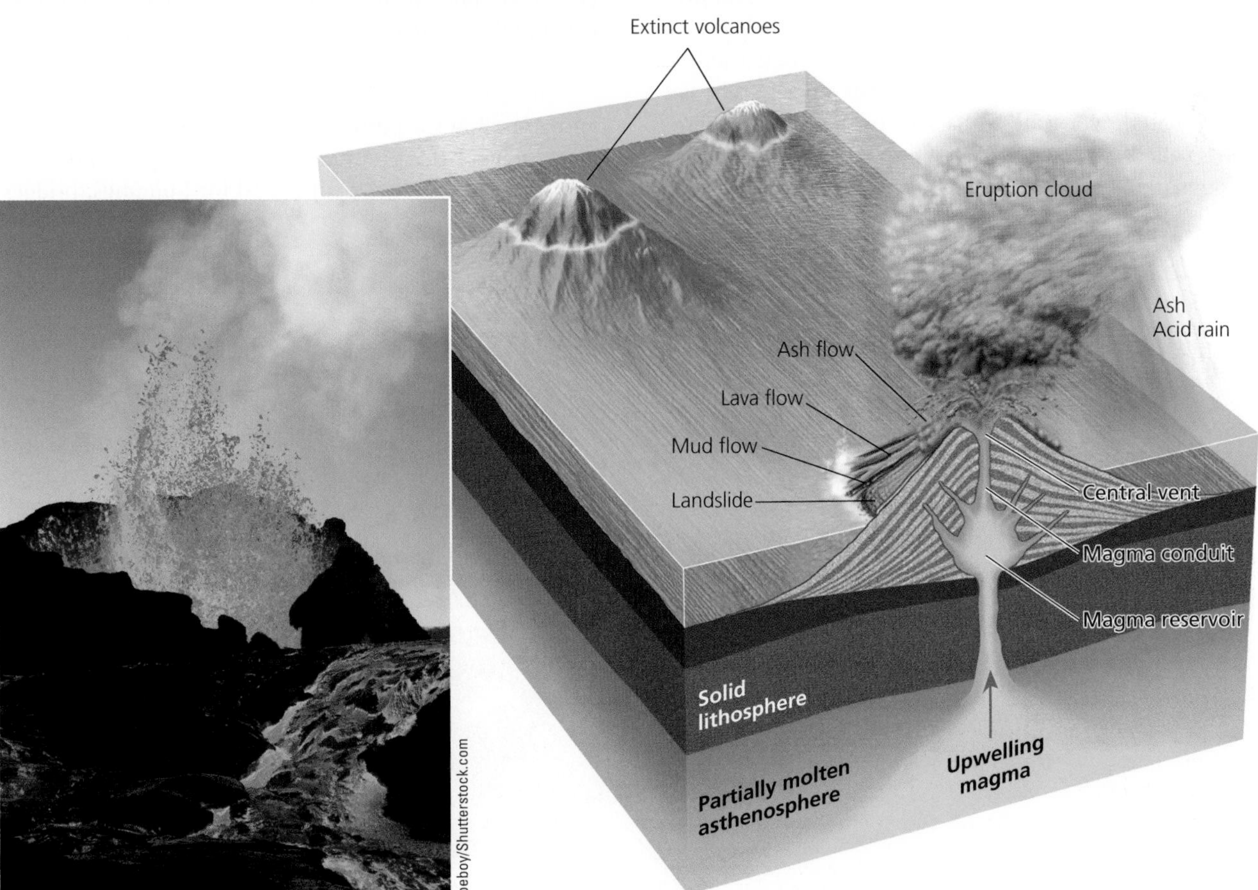

Figure 12-6 Sometimes, the internal pressure in a volcano is high enough to cause lava, ash, and gases to be ejected into the atmosphere (photo inset) or to flow over land, causing considerable damage.

Volcanoes can *erupt*, releasing large chunks of lava rock, glowing hot ash, liquid lava, and gases (including water vapor, carbon dioxide, and sulfur dioxide) into the environment (**Concept 12-1**). Eruptions can be extremely destructive, causing loss of life and obliterating ecosystems and human communities. Volcanoes that have erupted and then become inactive have created chains of islands.

While volcanic eruptions can be destructive, they do provide some benefits. They can result in the formation of majestic mountains and lakes such as Crater Lake (Figure 7-29, left, p. 146), and the weathering of lava contributes to fertile soils.

GOOD NEWS

We can reduce the loss of human life and some of the property damage caused by volcanic eruptions in several ways. We use historical records and geologic measurements to identify high-risk areas, so that some people can avoid living in those areas. Scientists use monitoring devices that warn us when volcanoes are likely to erupt, and evacuation plans have been developed for areas prone to volcanic activity.

Earthquakes Are Geological Rock-and-Roll Events

Forces inside the earth's mantle and near its surface push, stress, and deform rocks. At some point the stress can cause the rocks to suddenly shift or break and produce a transform fault, or fracture in the earth's crust (Figure 12-5). When a fault forms, or when there is abrupt movement on an existing fault, energy that has accumulated over time is released in the form of vibrations, called *seismic waves*, which move in all directions through the surrounding rock. This internal geological process is called an **earthquake** (Figure 12-7 and **Concept 12-1**). Most earthquakes occur at the boundaries of tectonic plates (Figure 12-4) when colliding plates create tremendous pressures in the earth's crust or when plates slide past one another at transform faults.

The place where an earthquake begins, often far below the earth's surface is called the *focus* (Figure 12-7). The earthquake's *epicenter* is located on the earth's surface directly above the focus. The energy of the earth's tremendous internal stress is released in the form of seismic waves that move upward and outward from the earthquake's focus like ripples in a pool of water, shaking parts of the crust.

Scientists measure the severity of an earthquake by the *magnitude* of its seismic waves. The magnitude is a measure of ground motion (shaking) caused by the earthquake, as indicated by the *amplitude*, or size of the seismic waves when they reach a recording instrument, called a *seismograph*.

Scientists use the *Richter scale*, on which each unit has amplitude 10 times greater than the next smaller unit. Thus, a magnitude 5.0 earthquake would result in 10 times more ground shaking than a magnitude 4.0 earthquake and the amount of ground movement from a magnitude 7.0 quake is 100 times greater than that of a magnitude 5.0 quake. Seismologists rate earthquakes as *insignificant* (less than 4.0 on the Richter scale), *minor* (4.0–4.9), *damaging* (5.0–5.9), *destructive* (6.0–6.9), *major* (7.0–7.9), and *great* (over 8.0). The largest recorded earthquake occurred in Chile on May 22, 1960 and measured 9.5 on the Richter scale.

The *primary effects of earthquakes* include shaking and sometimes a permanent vertical or horizontal displacement of the ground. These effects may have serious consequences for people and for buildings, bridges,

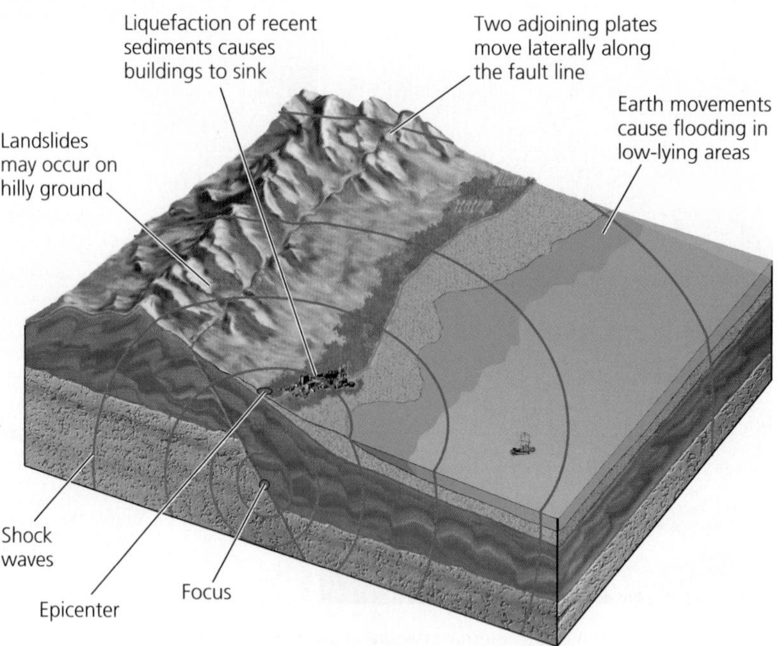

Figure 12-7 An *earthquake*, one of nature's most powerful events, has certain major features and effects.

freeway overpasses, dams, and pipelines. A major earthquake is a very large rock-and-roll geological event.

One way to reduce the loss of life and property damage from earthquakes is to examine historical records and make geologic measurements to locate active fault zones. We can then map high-risk areas and establish building codes that regulate the placement and design of buildings in such areas. Then people can evaluate the risk and factor it into their decisions about where to live. Also, engineers know how to make homes, large buildings, bridges, and freeways more earthquake resistant. (See Figure 13, p. S31, in Supplement 6 for a map of earthquake risk in various areas of the United States and Figure 14, p. S31, in Supplement 6 for a map of such areas throughout the world.)

GOOD NEWS

Earthquakes on the Ocean Floor Can Cause Huge Waves Called Tsunamis

A **tsunami** is a series of large waves generated when part of the ocean floor suddenly rises or drops (Figure 12-8). Most large tsunamis are caused when certain types of faults in the ocean floor move up or down as a result of a large underwater earthquake, a landslide caused by such an earthquake, or in some cases, a volcanic eruption (**Concept 12-1**). Such earthquakes often occur offshore in subduction zones where a tectonic plate slips under a continental plate (Figure 12-3).

Tsunamis are often called *tidal waves*, although they have nothing to do with tides. They can travel far across the ocean at the speed of a jet plane. In deep water, the waves are very far apart—sometimes hundreds of kilometers—and their crests are not very high. As a tsunami approaches a coast with its shallower waters, it slows down, its wave crests squeeze closer together, and their heights grow rapidly. It can hit a coast as a series of towering walls of water that can level buildings.

We can detect tsunamis through a network of ocean buoys or pressure recorders located on the ocean floor that provide some degree of early warning. These data are sent to tsunami emergency warning centers. But there are far too few of these recorders and emergency warning centers to adequately monitor earthquake activity that might generate tsunamis.

Between 1900 and 2011, tsunamis killed an estimated 300,000 people in regions of the Pacific Ocean. The largest loss of life occurred in December 2004

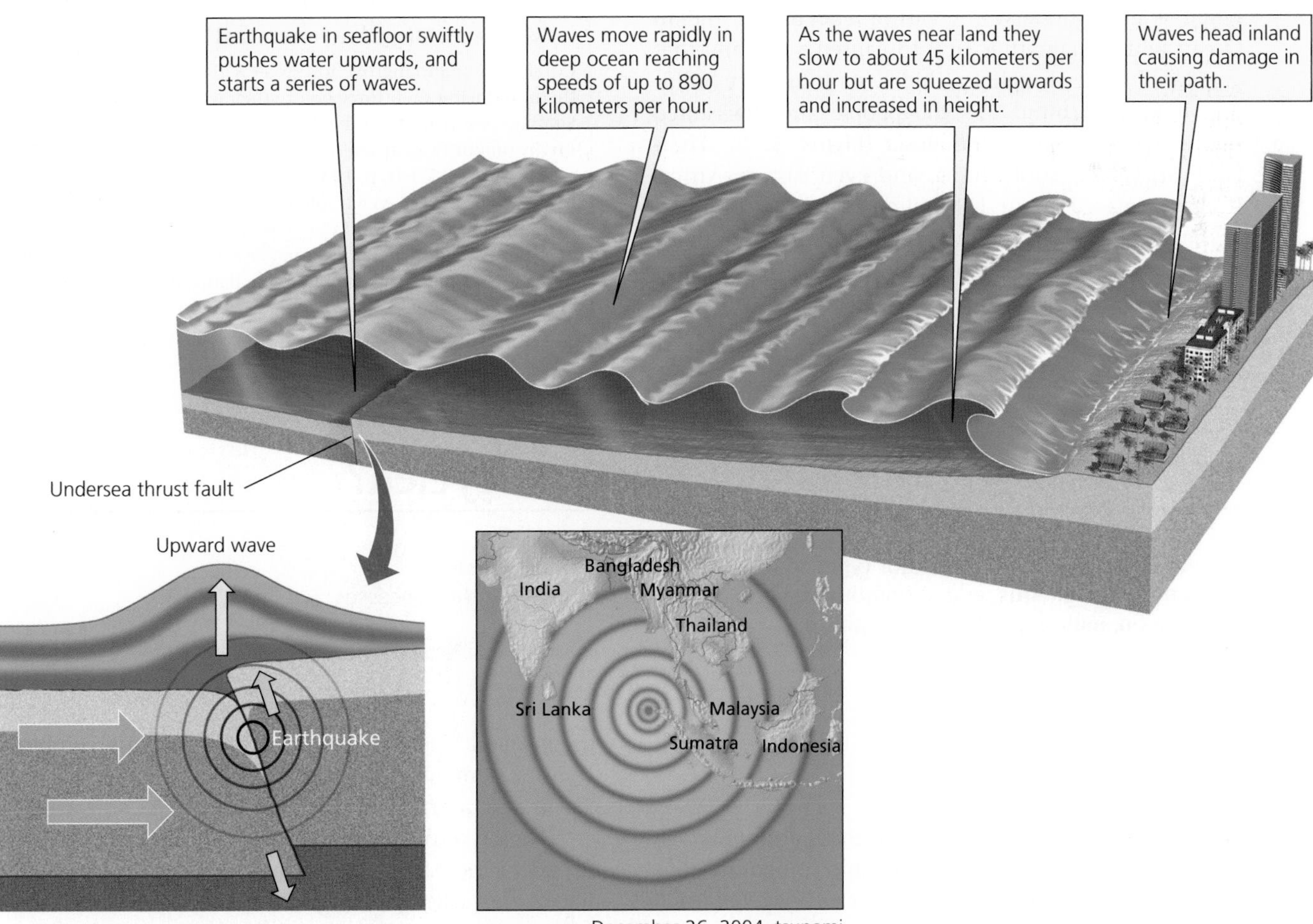

Figure 12-8 This diagram illustrates how a tsunami forms. The map shows the area affected by a large tsunami in December 2004.

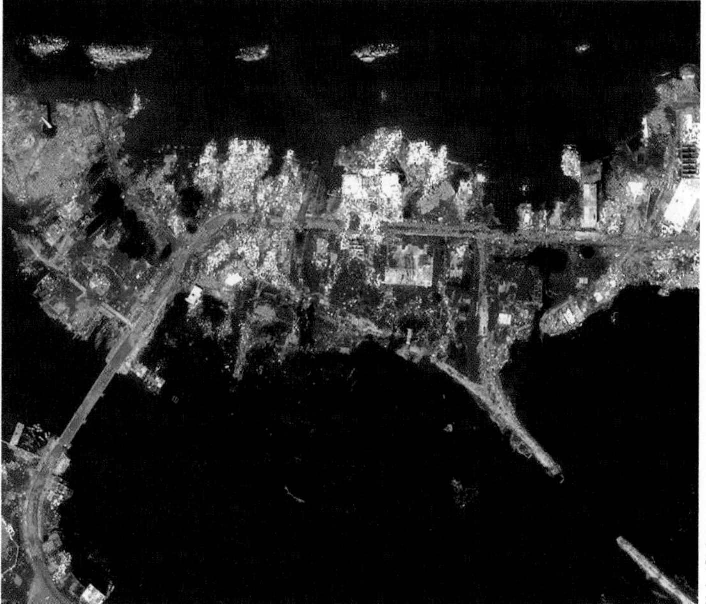

Figure 12-9 The tsunami of December 2004 killed 168,000 people in Indonesia, as well as tens of thousands more in other countries bordering the Indian Ocean. These photos show the Banda Aceh Shore near Gleebruk in Indonesia on June 23, 2004 before the tsunami (left), and on December 28, 2004 after it was stuck by the tsunami (right) (**Concept 12-1**).

when a great underwater earthquake in the Indian Ocean with a magnitude of 9.15 caused a tsunami that generated waves roughly as high as a four-story building. It killed around 279,900 people and devastated many coastal areas of Indonesia (Figure 12-9), Thailand, Sri Lanka, South India, and even eastern Africa. It also displaced about 1.8 million people (1.3 million of them in India and Indonesia), and destroyed or damaged about 470,000 buildings and houses. No buoys or gauges were in place to provide an early warning for this tsunami.

CONNECTIONS

Coral Reefs, Mangrove Forests, and Tsunami Damage

Coral reefs and mangrove forests slow the waves that roll over them, reducing their force before they hit nearby shorelines. Satellite observations and ground studies done by the UN Environment Programme showed that healthy coral reefs (see Figure 7-25, left, p. 143) and mangrove forests (see Figure 7-23, p. 142) played a role in reducing the force of the 2004 Indonesian tsunami's huge waves and the resulting death toll and destruction in some areas. In areas where mangrove forests had been removed, the damage and death toll were much higher.

12-2 How Are the Earth's Rocks Recycled?

▶ **CONCEPT 12-2 The three major types of rock found in the earth's crust—sedimentary, igneous, and metamorphic—are recycled very slowly by the processes of erosion, melting, and metamorphism.**

There Are Three Major Types of Rocks

The earth's crust consists mostly of rocks and minerals. A **mineral** is an element or inorganic compound that occurs naturally in the earth's crust as a *crystalline solid,* or one that has a regularly repeating internal arrangement of its atoms. A few minerals consist of a single element such as gold or mercury (see Figure 2-3, p. 9). But most of the more than 2,000 identified minerals occur as inorganic compounds formed by various combinations of elements. Examples are salt (sodium chloride or NaCl) and quartz (silicon dioxide or SiO_2).

Rock is a solid combination of one or more minerals found in the earth's crust. Some kinds of rock such as limestone (calcium carbonate, or $CaCO_3$) and quartzite (silicon dioxide, or SiO_2) contain only one mineral. But most rocks consist of two or more minerals. For example, granite is a mixture of mica, feldspar, and quartz crystals.

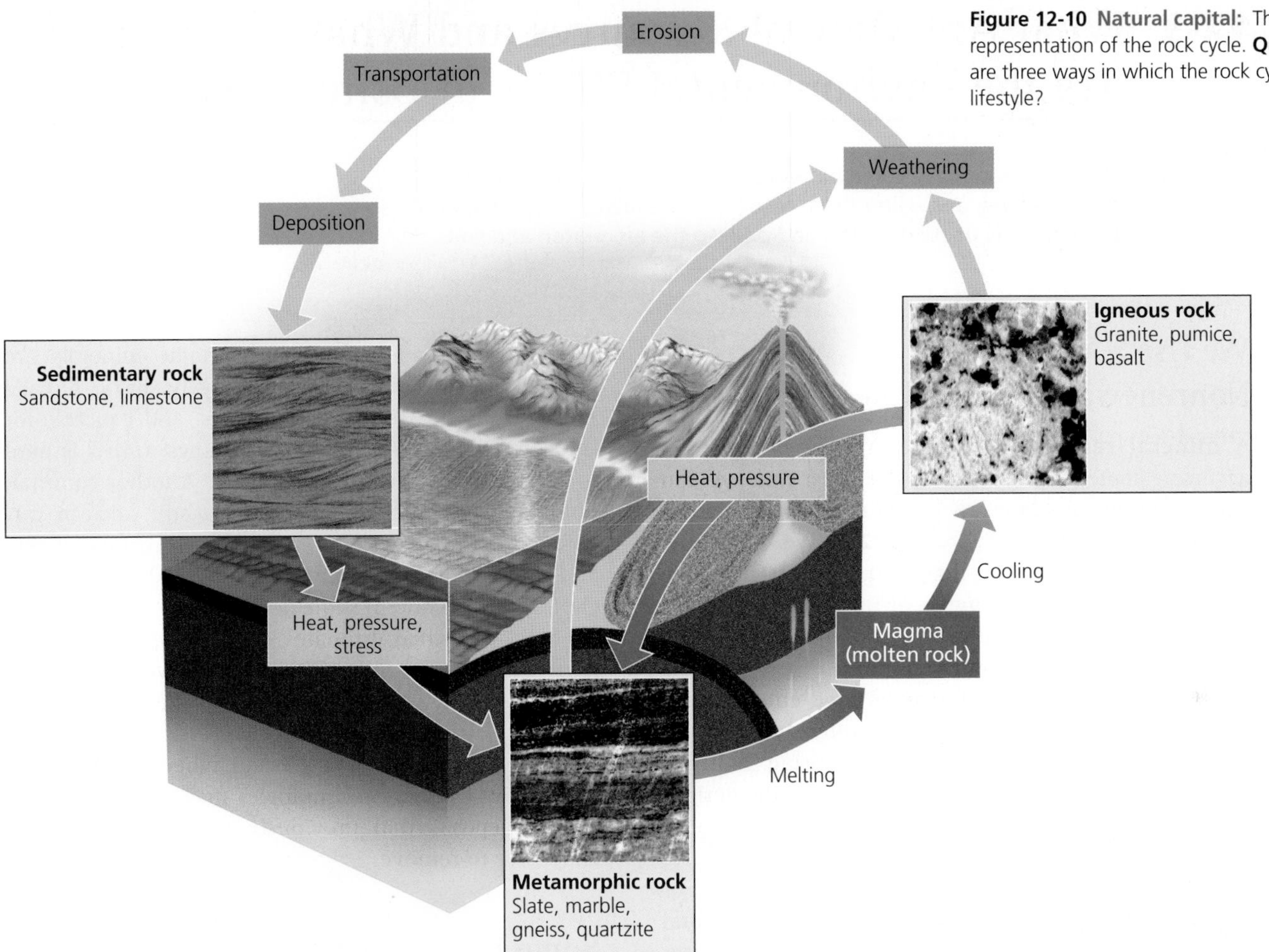

Figure 12-10 **Natural capital:** This is a simplified representation of the rock cycle. **Question:** What are three ways in which the rock cycle benefits your lifestyle?

Based on the way it forms, rock is placed in three broad classes: sedimentary, igneous, or metamorphic (**Concept 12-2**). **Sedimentary rock** is made of *sediments*—dead plant and animal remains and tiny particles of weathered and eroded rocks. These sediments are transported by water, wind, or gravity to downstream, downwind, downhill, or underwater sites. There they are deposited in layers that accumulate over time. Eventually, the increasing weight and pressure on the underlying layers convert the sedimentary layers into rock. Examples include *sandstone* and *shale* (formed from pressure created by deposited layers made primarily of sand), *dolomite* and *limestone* (formed from the compacted shells, skeletons, and other remains of dead aquatic organisms), and *lignite* and *bituminous coal* (derived from compacted plant remains).

Igneous rock forms below or on the earth's surface when magma wells up from the earth's upper mantle or deep crust and then cools and hardens. Examples include *granite* (formed underground) and *lava rock* (formed aboveground). Igneous rocks form the bulk of the earth's crust but are usually covered by sedimentary rocks.

Metamorphic rock forms when a preexisting rock is subjected to high temperatures (which may cause it to melt partially), high pressures, chemically active fluids, or a combination of these agents. These forces may transform a rock by reshaping its internal crystalline structure and its physical properties and appearance. Examples include *slate* (formed when shale and mudstone are heated) and *marble* (produced when limestone is exposed to heat and pressure).

Earth's Rocks Are Recycled Very Slowly

The interaction of physical and chemical processes that change rocks from one type to another is called the **rock cycle** (Figure 12-10). Rocks are recycled over millions of years by three processes: *erosion, melting,* and *metamorphism,* which produce *sedimentary, igneous,* and *metamorphic* rocks, respectively.

In these processes, rocks are broken down, melted, fused together into new forms by heat and pressure, cooled, and sometimes recrystallized within the earth's mantle and crust. Rock from any of these classes can be converted to rock of either of the other two classes or can be recycled within its own class (**Concept 12-2**). The *rock cycle* is the slowest of the earth's cyclic processes.

The rock cycle also plays a major role in forming concentrated deposits of mineral resources, and life on the earth depends on these minerals. In fact, without the rock cycle, you would not exist.

12-3 What Are Mineral Resources and What Are the Environmental Effects of Using Them?

▶ **CONCEPT 12-3 We can make some minerals in the earth's crust into useful products, but extracting and using these resources can disturb the land, erode soils, produce large amounts of solid waste, and pollute the air, water, and soil.**

We Use a Variety of Nonrenewable Mineral Resources

A **mineral resource** is a concentration of naturally occurring material from the earth's crust that we can extract and process into raw materials and useful products at an affordable cost (**Concept 12-3**). We know how to find and extract more than 100 minerals from the earth's crust. Two major types of minerals are *metallic minerals* (such as aluminum and gold), and *nonmetallic minerals* (such as sand and limestone).

An **ore** is rock that contains a large enough concentration of a particular mineral—often a metal—to make it profitable for mining and processing. A **high-grade ore** contains a large concentration of the desired mineral, whereas a **low-grade ore** contains a smaller concentration.

Mineral resources are used for many purposes. For example, about 60 elements—two-thirds of the total on the periodic table—are used for making computer chips. *Aluminum* (Al) is used for packaging and beverage cans, and as a structural material in motor vehicles, aircraft, and buildings. Steel, an essential material used in buildings, machinery, and motor vehicles, is a mixture (alloy) of *iron* (Fe) and other elements that are added to give it certain physical properties. *Manganese* (Mn), *cobalt* (Co), and *chromium* (Cr) are widely used in important steel alloys. *Copper* (Cu), a good conductor of electricity, is used for electrical and communications wiring as well as for plumbing pipes. *Gold* (Au) (**Core Case Study**) is used in electrical equipment, tooth fillings, jewelry, coins, and some medical implants. CORE CASE STUDY

> CONNECTIONS
> **Lithium and U. S. Energy Dependence**
>
> *Lithium* (Li) is becoming increasingly important as a vital component of lithium-ion batteries used in cell phones, iPods, laptop computers, and electric cars. The South American countries of Bolivia, Chile, and Argentina have about 80% of the global reserves of lithium. Bolivia alone has about 50%. The United States holds only about 3% of the global reserves. Japan, China, South Korea, and the United Arab Emirates are buying up access to global lithium reserves to ensure their ability to sell batteries to the rest of the world for the rapidly growing fleet of electric cars. Within a few decades, the United States may be as dependent on expensive imports of lithium and lithium batteries as the country is now dependent on imported oil.

The most widely used nonmetallic minerals are sand and gravel. *Sand*, which is mostly silicon dioxide (SiO_2), is used to make glass, bricks, and concrete for the construction of roads and buildings. *Gravel* is used for roadbeds and to make concrete. Another common nonmetallic mineral is *limestone* (mostly calcium carbonate, or $CaCO_3$), which is crushed to make concrete and cement. *Phosphate salts* are mined and used in inorganic fertilizers and in some detergents.

Most published estimates of the supply of a given mineral resource refer to its **reserves**: identified resources from which we can extract the mineral profitably at current prices. Reserves increase when we find new, profitable deposits and when higher prices or improved mining technologies make it profitable to extract deposits that previously were considered too expensive to remove.

Some Environmental Impacts of Mineral Use

We can use metals to produce many useful products. But mining, processing, using, and disposing of or recycling metals—what is known as the *life cycle of a metal* (Figure 12-11)—takes enormous amounts of energy and water, and can disturb the land, erode soil, produce solid waste, and pollute the air, water, and soil (Figure 12-12) (**Concept 12-3**). Some environmental scientists and resource experts warn that the greatest danger from continually increasing our consumption of nonrenewable mineral resources may be the environmental damage caused by extracting, processing, and converting them to products.

The environmental impacts from mining an ore are affected by its percentage of metal content, or *grade*. The more accessible and higher-grade ores are usually exploited first. As they are depleted, mining lower-grade ores takes more money, energy, water, and other materials, and increases land disruption, mining waste, and pollution.

> THINKING ABOUT
> **Low-Grade Ores**
>
> Use the second law of thermodynamics (see Chapter 2, p. 36) to explain why mining lower-grade ores requires more energy and materials, and increases land disruption, mining waste, and pollution.

Figure 12-11 Each metal resource that we use has a *life cycle*. Each step in this process uses large amounts of energy and water, and produces some pollution and waste.

There Are Several Ways to Remove Mineral Deposits

After mineral deposits are located, several different mining techniques can be used to remove them. The location and type of the mineral resource determines the technique used to remove it.

Shallow mineral deposits are removed by **surface mining**, in which materials lying over a deposit are removed to expose the resource for processing. There are different types of surface mining, but they generally begin with removal of all vegetation, including forests, from a site. Then the **overburden**, or soil and rock overlying a useful mineral deposit, is removed. It is usually deposited in piles of waste material called **spoils**. Sometimes ore deposits that contain metals such as gold (**Core Case Study**) are dredged from streams, and the unused materials are usually left in spoils piles on the land. Surface mining is used to extract about 90% of the nonfuel mineral and rock resources and 60% of the coal used in the United States.

CORE CASE STUDY

The type of surface mining that is used depends on two factors: the resource being sought and the nature of the land lying over the resource. In **open-pit mining**

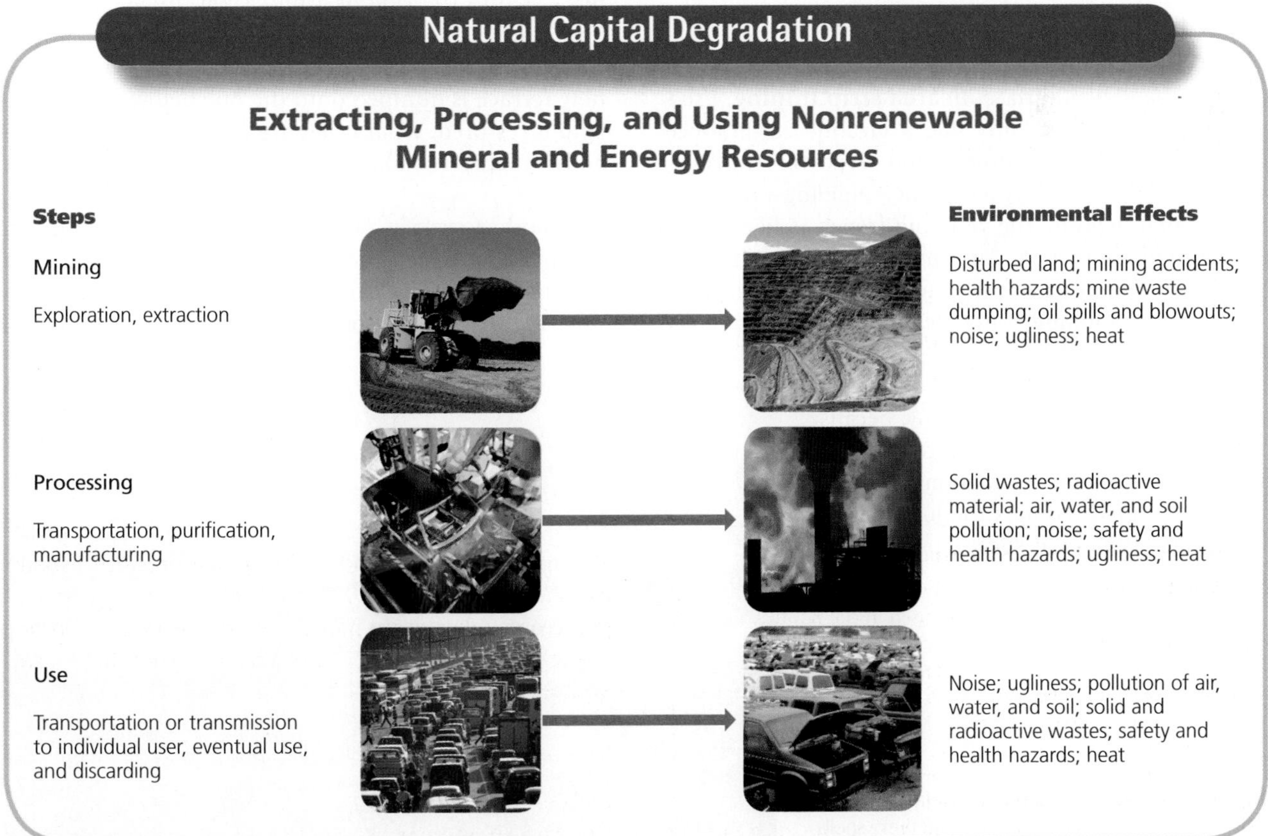

Figure 12-12 Harmful environmental effects result from the extraction, processing, and use of nonrenewable mineral or energy resources. **Questions:** What are three mineral resources that you used today? Which of these harmful environmental effects might have resulted from obtaining and using these resources?

Figure 12-13 **Natural capital degradation:** This *open-pit* copper mine, which is located near the U.S. city of Bisbee, Arizona, has been abandoned since the mid-1970s. **Question:** Should governments require mining companies to fill in and restore such sites once their ore is depleted? Explain.

BenC/Shutterstock.com

(Figure 12-13), machines dig very large holes and remove metal ores (such as iron, copper, and gold ores), as well as sand, gravel, and stone (such as limestone and marble).

Strip mining is useful and economical for extracting mineral deposits that lie in large horizontal beds close to the earth's surface. In **area strip mining**, used where the terrain is fairly flat, a gigantic earthmover strips away the overburden, and a power shovel—which can be as tall as a 20-story building—removes the mineral deposit. The resulting trench is filled with overburden, and a new cut is made parallel to the previous one. This process is repeated over the entire site.

Area strip mining often leaves a series of spoils banks (Figure 12-14) that can be eroded by water and wind. Regrowth of vegetation on these banks is quite slow, because they have no topsoil; thus, returning a site to its previous condition before it was strip mined requires the long process of primary ecological succession (see Figure 5-9, p. 89).

Contour strip mining (Figure 12-15) is used mostly to mine coal on hilly or mountainous terrain. Gigantic power shovels and bulldozers cut a series of terraces into the side of a hill. Then, huge earthmovers remove the overburden, An excavator or power shovel extracts the coal, and the overburden from each new terrace is dumped onto the one below. Unless the land is restored, what is left are a series of spoils banks and a highly erodible bank of soil and rock called a *highwall.*

Another surface mining method is **mountaintop removal**, in which the top of a mountain is removed to expose seams of coal, which are then extracted (Figure 12-16). To accomplish this, miners use explosives,

pmphoto/Shutterstock.com

Figure 12-14 **Natural capital degradation:** Area strip mining of coal in Germany created these spoils banks. In the background is a coal-burning power plant. **Question:** Should governments require mining companies to restore such sites as fully as possible? Explain.

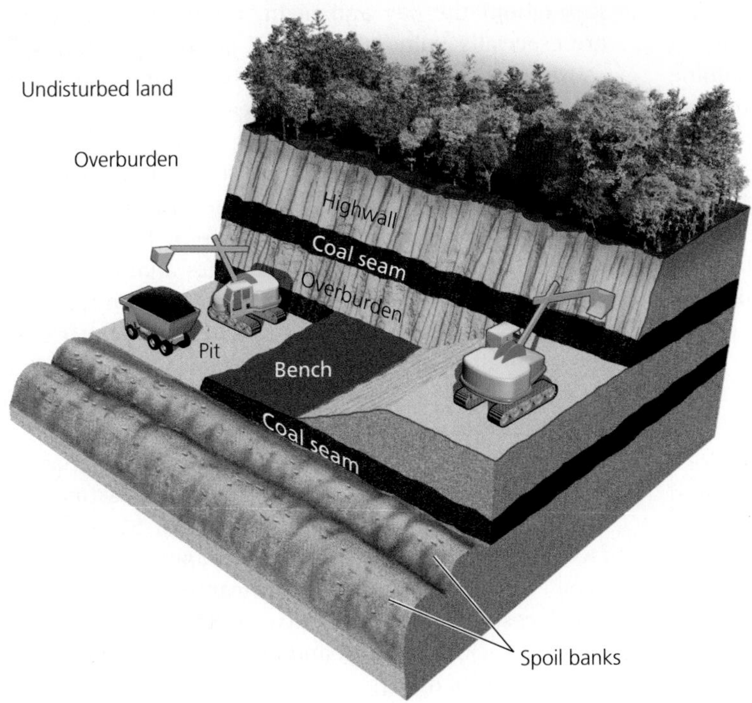

Figure 12-15 Natural capital degradation: *Contour strip mining* is used in hilly or mountainous terrain to extract minerals such as coal. Coal can also be removed from the coal seam in the highwall shown here by highwall mining. This involves drilling into the seam with an automated machine that inserts a cutterhead to loosen the coal. This method is very efficient in recovering coal without contaminating it with rock debris. However, the equipment is quite costly and expensive to operate.

earth movers, large power shovels, and other machines with huge buckets, called draglines. This method is commonly used in the Appalachian Mountains of the United States.

Deep deposits of minerals are removed by **subsurface mining**, in which underground mineral resources are removed through tunnels and shafts. This method is used to remove coal and metal ores that are too deep to be extracted by surface mining. Miners dig a deep, vertical shaft, blast open subsurface tunnels and chambers to reach the deposit, and use machinery to remove the resource and transport it to the surface.

Mining Has Harmful Environmental Effects

Mining can do long-term harm to the environment in a number of ways. One type of damage is *scarring and disruption of the land surface* (Figures 12-1 and 12-13 through 12-16). CORE CASE STUDY

In mountaintop removal (Figure 12-16), enormous machines plow great volumes of waste rock and dirt into valleys below the mountaintops. This destroys forests, buries mountain streams, and increases flood hazards. Wastewater and toxic sludge, produced when the coal is processed, are often stored behind dams in these valleys, which can overflow or collapse and release toxic substances such as arsenic and mercury.

In the United States, more than 500 mountaintops in West Virginia and other Appalachian states have been removed to extract coal. According to the U.S. Environmental Protection Agency (EPA), the resulting spoils have buried more than 1,100 kilometers (700 miles) of streams—a total length roughly equal to the distance between the two U.S. cities of New York and Chicago.

Figure 12-16 Natural capital degradation: This mountaintop coal mining operation took place in the U.S. state of West Virginia, where coal is the state rock. Mountaintop removal for coal is also occurring in the U.S. states of Tennessee, Kentucky, Virginia, and Pennsylvania. **Question:** Do you think mountaintop removal is an acceptable method for extracting a mineral resource? Explain.

Surface mining in tropical forests and other tropical areas destroys or degrades vital biodiversity when forests are cleared and rivers are polluted with mining wastes. Since 1980, millions of miners have streamed into these areas in search of gold (**Core Case Study**). These small-scale miners use destructive techniques to dig large pits and dredge sediments from rivers. Some use hydraulic mining—a technique that was outlawed in the United States—in which water cannons wash entire hillsides into collection boxes for gold removal.

CONNECTIONS

Mercury Poisoning and Tropical Gold Mining

Mercury is a highly toxic chemical that interferes with the human nervous system and brain functions, and it can build up to high levels in the human body. Mercury is used illegally and rampantly in Asia, Africa, and Latin America by tens of thousands of small-scale gold miners who use it to separate gold from stream sediments. Many miners and villagers in these illegal mining areas suffer from deadly mercury poisoning. Fish populations in one area of Borneo have dropped by 70%, due to this form of pollution. As many as 3 grams of mercury escape to the environment for every 1 gram of gold produced in this way. It is the second-biggest human-related source of mercury pollution in the world after the burning of coal.

In parts of Nigeria, the discovery of new gold deposits has led many poor people to carry out a primitive form of surface mining. They dig up gold-bearing rocks and drag them to their villages where they break and grind the rocks to extract the gold. But in the process they release a fine dust of toxic lead that is also present in the gold ore. As a result, many of these villagers' homes are contaminated with lead dust that they inhale throughout the day and night. Children under age 5 are especially vulnerable to lead poisoning, which can cause irreversible brain damage, behavioral and learning problems, convulsions, and death.

In 2010, international health and environmental experts discovered that exposure to this toxic lead had caused at least 163 deaths among the Nigerian villagers, most of them in children under age 5. In two of the villages, 82% of soil and dust samples analyzed by the U.S. Centers for Disease Control and Prevention contained lead concentrations up to 250 times higher than the level allowed in the United States. This story is a stark reminder of the connection between poverty and health hazards.

Surface mining sites can be cleaned up and restored (Figure 12-17), but it is costly. The U.S. Department of the Interior (DOI) estimates that at least 500,000 surface-mined sites dot the U.S. landscape, mostly in the West. DOI also estimates that cleaning up these sites could cost taxpayers as much as $70 billion. Worldwide, the cleanup of abandoned mining sites would cost trillions of dollars.

Subsurface mining disturbs less than one-tenth as much land as surface mining disturbs, and it usually produces less waste material. However, it creates hazards such as cave-ins, explosions, and fires. Miners often get diseases such as black lung, caused by prolonged inhalation of coal dust in subsurface mines. Another problem is *subsidence*—the collapse of land above some underground mines. It can damage houses, crack sewer lines, break gas mains, and disrupt groundwater systems.

Mining operations also produce large amounts of solid waste—three-fourths of all U.S. solid waste—and cause major water and air pollution. For example, *acid*

A. S. Zain/Shutterstock.com

Figure 12-17 This mining site in Indonesia has been ecologically restored, but such restoration is rare.

mine drainage occurs when rainwater that seeps through a mine or a spoils pile carries sulfuric acid (H_2SO_4, produced when aerobic bacteria act on iron sulfide minerals in spoils) to nearby streams and groundwater. (See *The Habitable Planet,* Video 6 at **www.learner.org/resources/series209.html**.)

According to the EPA, mining has polluted about 40% of western watersheds in the United States, and it accounts for 50% of all the country's emissions of toxic chemicals into the atmosphere. In fact, the mining industry produces more of such toxic emissions than any other U.S. industry. In the United States and in many other countries, much of this degradation comes from leaking storage ponds built to hold a toxic sludge that is produced from the mining and processing of metal ores.

How would you like to have a 2-meter (6-foot) high tsunami of red sludge as caustic as lye flowing through your neighborhood? This is what happened to several thousand residents of three villages in western Hungary on October 6, 2010, when a sludge storage pond at an aluminum production plant ruptured. Within a few hours, the pond released almost as much of the toxic red sludge as the amount of oil that gushed out of the BP oil well off the U.S. Gulf Coast over several months during the summer of 2010.

At least 120 people were injured—many from chemical burns that they got when the caustic material seeped through their clothing—and 9 people were killed. More than 270 homes and many cars were engulfed by the red muck. A large area of soil was ruined for growing crops and will have to be replaced at great expense, and most aquatic life in nearby streams was wiped out. Some of the sludge reached the Danube River, one of Europe's most important waterways. But the large volume of the river's water apparently diluted the toxic mud to levels that caused little damage to the river's ecosystem.

The red sludge was the by-product of converting aluminum ore into alumina, a compound used to make aluminum. Hungary is a big producer of aluminum, so there are many other poorly maintained red-sludge storage ponds scattered throughout the country. This story's lesson: we may wish to store away the harmful wastes we produce, but there is no away.

Removing Metals from Ores Has Harmful Environmental Effects

Ore extracted by mining typically has two components: the *ore mineral,* containing the desired metal, and waste material. The waste material that is removed from ores is called *tailings.* Particles of toxic metals in the tailings can be blown by the wind or leached by rainfall and can contaminate surface waters and groundwater.

After removing the waste material, processors use heat or chemical solvents to extract metals from the ores. Heating ores to release metals is called **smelting** (Figure 12-11). Without effective pollution control equipment, smelters emit enormous quantities of air pollutants, including sulfur dioxide and suspended toxic particles, that damage vegetation and acidify soils in the surrounding area. Smelters also cause water pollution and produce liquid and solid hazardous wastes that require safe disposal.

An example of using chemicals to remove metals from their ores is the use of highly toxic solutions of cyanide salts to extract gold from its ore (**Core Case Study**). After extracting the gold from a mine, some mining companies have deliberately declared bankruptcy. This has allowed them to walk away from cleaning up their mining operations, leaving behind large amounts of cyanide-laden water in leaking holding ponds.

CORE CASE STUDY

12-4 How Long Will Supplies of Nonrenewable Mineral Resources Last?

▶ **CONCEPT 12-4A All nonrenewable mineral resources exist in finite amounts, and as we get closer to depleting any mineral resource, the environmental impacts of extracting it generally become more harmful.**

▶ **CONCEPT 12-4B Raising the price of a scarce mineral resource can lead to an increase in its supply, but there are environmental limits to this effect.**

Mineral Resources Are Distributed Unevenly

The earth's crust contains fairly abundant deposits of nonrenewable mineral resources such as iron and aluminum. But deposits of important mineral resources such as manganese, chromium, cobalt, and platinum are relatively scarce. The earth's geologic processes have not distributed deposits of nonrenewable mineral resources evenly among countries.

Massive exports can deplete the supply of a country's nonrenewable minerals. During the 1950s, for example, South Korea exported large amounts of its iron and copper. Since the 1960s, the country has not had enough of these metals to support its rapid economic growth and now must import them.

Five nations—the United States, Canada, Russia, South Africa, and Australia—supply most of the nonrenewable mineral resources used by modern societies. South Africa, for example, is the world's largest producer of gold, chromium, and platinum.

Since 1900, and especially since 1950, there has been a sharp rise in the total and per capita use of nonrenewable mineral resources in the United States. As a result, the United States has depleted some of its once-rich deposits of metals such as lead, aluminum, and iron. Currently, the United States imports all of its supplies of 20 key nonrenewable mineral resources and more than 90% of its supplies of 4 other key minerals.

Most U.S. imports of metal resources come from reliable and politically stable countries. But experts are concerned about four *strategic metal resources*—manganese, cobalt, chromium, and platinum—which are essential for U.S. economic and military strength. The country has little or no reserves of these metals.

Other strategic minerals are metals such as tungsten, titanium, and niobium. Titanium is very important for making armor plating, aircraft, and high-end weapons, while tungsten is widely used to make filaments for light bulbs, electrodes for electronic devices, rocket engine nozzles, and steel alloys used to make turbine blades. According to the U.S. Geological Survey, China has 57% of the world's estimated tungsten reserves, Canada has 12%, and the United States only 4%. America used to produce its own tungsten. Now it must import the tungsten it uses, about half of it from China. There is also growing concern over access to a group of economically and strategically important rare earth elements (Science Focus, at right).

Supplies of Nonrenewable Mineral Resources Can Be Economically Depleted

The future supply of any nonrenewable mineral depends on two factors: the actual or potential supply of the mineral, and the rate at which we use it. We have never completely run out of any mineral, but a mineral becomes *economically depleted* when it costs more than it is worth to find, extract, transport, and process the remaining deposits (**Concept 12-4A**). At that point, there are five choices: *recycle or reuse existing supplies*, *waste less*, *use less*, *find a substitute*, or *do without*.

According to a 2006 study by Thomas Graedel of Yale University, if all nations were to extract metal resources from the earth's crust at the rate at which more-developed nations do today, within a few decades, there would likely not be enough metal resources to meet the demand, even with extensive recycling.

Market Prices Affect Supplies of Nonrenewable Minerals

Geologic processes determine the quantity and location of a mineral resource in the earth's crust. Economics determines what part of the known supply is extracted and used. An increase in the price of a scarce mineral resource can lead to increased supplies and can encourage more efficient use, but there are limits to this effect (**Concept 12-4B**).

According to standard economic theory, in a competitive market system, the price of a plentiful mineral resource is low when its supply exceeds demand. When a resource becomes scarce, its price rises. This can encourage exploration for new deposits, stimulate development of better mining technology, and make it profitable to mine lower-grade ores. It can also encourage a search for substitutes and promote resource conservation.

According to some economists, however, this price effect may no longer apply very well in most of the more-developed countries. Governments in such countries often use subsidies, taxes, regulations, and import tariffs to control the supply, demand, and price of key minerals to such an extent that a truly competitive market does not exist.

Most mineral prices are kept artificially low because governments subsidize the development of their domestic mineral resources to help promote economic growth and national security. In the United States, for instance, mining companies get various types of government subsidies, including *depletion allowances*—permission to deduct from their taxable incomes the costs of developing and extracting mineral resources. These allowances amount to 5–22% of their gross income gained from selling the minerals (see the Case Study that follows).

Most consumers are unaware that the real costs of consumer products made from mineral resources are higher than their market prices. This is because consumers are paying the taxes that provide government subsidies and tax breaks for mining companies, and that help to pay for the control of harmful environmental and health effects caused by mineral extraction, processing, and use (Figure 12-12). If these hidden extra costs were included in the market prices of such goods, some economists say these harmful effects would be sharply reduced, recycling and reuse would increase dramatically, and many minerals would be replaced with less harmful substitutes.

Mining company representatives insist that they need taxpayer subsidies and tax breaks to keep the prices of minerals low for consumers. They also claim that, without the subsidies, their companies might move their operations to other countries where they would not have to pay taxes or comply with strict mining and pollution control regulations.

Other economic factors that affect supplies of mineral resources are scarce investment capital and high

SCIENCE FOCUS

The Importance of Rare Earth Metals

The 17 *rare earth elements* or *rare earth metals* include scandium and yttrium, and 15 lanthanides found near the bottom of the periodic table (see Figure 2, p. S11, in Supplement 4). Mostly because of their magnetic strength and other unique properties, these elements and their minerals (mostly oxides that can be converted to rare earth metals) are extremely important for widely used modern technologies.

The products of such technologies include liquid crystal displays for computer and TV screens and other electronic devices, energy-efficient compact fluorescent and LED light bulbs, solar cells, computer chips, rechargeable batteries, fiber-optic cables, cell phones, digital cameras, batteries and motors for electric cars, jet engines, and generators in wind turbines. Rare earth metals are also vital for military applications such as missile guidance systems, smart bombs, lasers, radar, aircraft electronics, and satellites.

Without affordable supplies of these metals, industrialized nations could not develop the current versions of cleaner energy technology and other high-tech green products that will be major sources of economic growth and profits during this century. Nations also need these metals and their oxides to maintain their military strength.

China has roughly 37% of the world's known rare earth metal reserves, the U.S. has about 13%, Russia and Australia each have 5–6%, and Japan has none (and imports most of its supplies from China). In 2010, China produced 100% of the world's rare earth metals and 94% of the world's rare earth oxides. The United States and Japan, which are heavily dependent on these metals and their oxides, produced none. Since 2010, China has been reducing its exports of rare earth metals and oxides to other countries.

Countries such as the United States, Australia, Russia, Canada, and South Africa hold some undeveloped reserves of these metals. A large rare earth metals mine in California closed down in 2002, because of the expense of meeting pollution regulations for this very dirty mining process and because China had driven the prices of rare earth metals down to a point where the mine was too costly to operate. The mining and processing of rare earth metals and oxides cause considerable environmental damage in the form of acidified streams and toxic and radioactive wastes. Because these processes are poorly regulated in China, rare earth metals can be produced more cheaply there than in other countries. Thus, China has driven most foreign mining companies out of this market and now dominates the global supply of these immensely important metal resources.

The California mine could be retooled with more modern mining techniques. But this will take 6 or more years and will require significant private and perhaps government financial support because of the higher costs involved in meeting environmental standards. Another alternative is to extract and recycle rare earth metals from the massive amounts of electronic wastes that are being produced (as we discuss in Chapter 16). Companies making batteries for electric cars can also switch from nickel-metal-hydride batteries, which require the rare earth metal lanthanum, to lithium-ion batteries.

Also, scientists are searching for substitutes for rare earth metals that could be used to make products such as high-strength magnets. In 2010, for example, Japanese researchers announced that they had developed a motor for hybrid electric vehicles that uses readily available and less costly iron-based magnets instead of rare earth metal magnets.

Critical Thinking

Would you favor giving the owners of the California rare earth metals mine significant government tax breaks and subsidies to put the mine back into production? Explain. Would you favor reducing the environmental regulations for the mining of these metals? Explain.

financial risk. Typically, if geologists identify 10,000 possible deposits of a given resource, only 1,000 sites are worth exploring; only 100 justify drilling, trenching, or tunneling; and only 1 becomes a productive mine. When investment capital is short, mining companies are less inclined to invest, given this slim chance of recovering their investments.

■ CASE STUDY

An Outdated Mining Subsidy: The U.S. General Mining Law of 1872

Some people have become wealthy by using the little-known U.S. General Mining Law of 1872. It was designed to encourage mineral exploration and the mining of *hard rock minerals*—such as gold (**Core Case Study**), silver, copper, and uranium—on public lands and to help develop the western territories.

CORE CASE STUDY

Under this law, a person or corporation can file a mining claim or assume legal ownership of parcels of land on essentially all U.S. public land except national parks and wilderness. To file a claim, you state that you believe the land contains valuable hard rock minerals and you promise to spend at least $500 to improve it for mineral development. You must then pay $120 per year for each 8-hectare (20-acre) parcel of land used to maintain the claim, whether or not a mine is in operation.

Until 1995, when a freeze on such land transfers was declared by Congress, one could pay the federal government $6–12 per hectare ($2.50–5.00 an acre) for such land owned jointly by all U.S. citizens. One could then lease the land to someone else, build on it, sell it, or use it for essentially any purpose. People have constructed golf courses, hunting lodges, hotels, and housing subdivisions on public land that they bought from taxpayers at 1872 prices. According to a 2004 study by the Environmental Working Group, public lands containing an estimated $285 billion worth of publicly owned mineral

U.S. Geological Survey

Figure 12-18 Natural capital degradation: The Summitville gold mining site near Alamosa, Colorado, became a toxic waste site.

resources have been transferred to private companies under this law.

According to the Bureau of Land Management, mining companies remove at least $4 billion worth of hard rock minerals per year from U.S. public land—an average of $183,000 an hour. These companies pay taxpayers royalties amounting to only 2.3% of the value of the minerals, compared to royalties of 13.2% paid for oil, natural gas, and coal, and 14% for grazing rights on public lands.

After removing valuable minerals, some mining companies have walked away from their mining operations, leaving behind a toxic mess. A glaring example is the Summitville gold mine site near Alamosa, Colorado (Figure 12-18). A Canadian company used the 1872 mining law to buy the land from the federal government in 1984 at a bargain price, spent $1 million developing the site, and removed $98 million worth of gold. The company then declared bankruptcy and abandoned the site, which contained acids and toxic metals that leaked into the nearby Alamosa River. Cleanup by the EPA will eventually cost U.S. taxpayers about $120 million.

In 1992, the mining law was modified to require mining companies to post bonds to cover 100% of the estimated cleanup costs in case they go bankrupt. However, because such bonds were not required in the past, the U.S. Department of the Interior estimates that cleaning up degraded land and streams on more than 500,000 abandoned hard rock mining sites will cost U.S. taxpayers $32–72 billion. In 2009, the EPA was ordered by a federal court to develop a new rule guaranteeing that hard rock miners will pay all cleanup costs.

Mining companies point out that they must invest large sums (often $100 million or more) to locate and develop an ore site before they make any profits from mining hard rock minerals. They argue that government-subsidized land costs allow them to provide high-paying jobs to miners, to supply vital resources for industry, and to keep mineral-based products affordable. But critics argue that the money taxpayers give up as subsidies to mining companies offsets the lower prices they pay for these products.

Critics of this very old law call for permanently banning such sales of public lands, although some do support 20-year leases of designated public land for hard rock mining. Critics also call for much stricter environmental controls and cleanup restrictions on hard rock mining. They want the government to set up a fund to clean up abandoned mining sites that would be paid for by higher royalties from hard rock mining companies. They would require mining companies to pay a royalty of 8–12% on the *gross income* they earn from a given site, based on the wholesale value of all minerals removed from public land—similar to the rates paid by oil, natural gas, and coal companies.

Is Mining Lower-Grade Ores the Answer?

Some analysts contend that we can increase supplies of a mineral by extracting lower grades of ore. They point to the development of new earth-moving equipment, improved techniques for removing impurities from ores, and other technological advances in mineral extraction and processing. Such advancements have made it possible to extract some lower-grade ores and even to reduce their costs. For example, in 1900, the average copper ore mined in the United States was about 5% copper by weight. Today, that ratio is 0.5%, yet copper costs less (when adjusted for inflation).

CONNECTIONS
Metal Prices and Thievery

Because of the increased demand for copper, copper prices have risen sharply in recent years. As a result, in several U.S. communities, people have been stealing copper to sell it—stripping abandoned houses of copper pipe and wiring and stealing electrical wiring from underneath city streets and from public sports facilities. In one Oklahoma town, someone cut down several utility poles and stole the copper electrical wiring, causing a blackout. Also, because the price of the rare earth metal palladium has skyrocketed, some thieves have been stealing catalytic converters that contain palladium from cars in shopping mall parking lots. They pull up next to a vehicle, slip under it, and within minutes, remove its valuable catalytic converter.

However, several factors can limit the mining of lower-grade ores (**Concept 12-4B**). One is the increased cost of mining and processing larger volumes of ore, as the second law of thermodynamics would dictate (see Chapter 2, p. 36). Another is the increasing shortage of freshwater needed to mine and process some minerals, especially in arid and semiarid areas. A third limiting factor is the environmental impacts of the increased land disruption, along with waste material and pollution produced during mining and processing (Figure 12-12).

GOOD NEWS

One way to improve mining technology and reduce its environmental impact is to use microorganisms that can break down rock material and extract minerals in a process called *in-place*, or *in situ*, (pronounced "in SY-too") *mining*. If naturally occurring bacteria cannot be found to extract a particular metal, genetic engineering techniques could be used to produce such bacteria. Using this biological approach, sometimes called *biomining*, miners remove desired metals from ores through wells bored into the deposits. It leaves the surrounding environment undisturbed and reduces the air pollution associated with the smelting of metal ores. It also reduces hazardous chemical water pollution such as that resulting from the use of cyanide (Figure 12-1) and mercury in gold mining (**Core Case Study**).

CORE CASE STUDY

On the down side, microbiological ore processing is slow. It can take decades to remove the same amount of material that conventional methods can remove within months or years. Genetic engineers are trying to modify bacteria to speed up the process, but the possible environmental and health effects of using such bacteria are unknown. So far, biomining methods are feasible only with some low-grade ores for which conventional techniques are too expensive.

Can We Get More Minerals from the Oceans?

Some ocean mineral resources are dissolved in seawater. However, most of the chemical elements found in seawater occur in such low concentrations that recovering these mineral resources takes more energy and money than they are worth. Currently, only magnesium, bromine, and sodium chloride are abundant enough to be extracted profitably from seawater. On the other hand, in sediments along the shallow continental shelf and adjacent shorelines, there are significant deposits of sand, gravel, phosphates, sulfur, tin, copper, iron, tungsten, silver, titanium, platinum, and diamonds.

THINKING ABOUT
Extracting Minerals from Seawater

Use the second law of thermodynamics (see Chapter 2, p. 36) to explain why it costs too much to extract most dissolved minerals from seawater.

Another potential ocean source of some minerals is hydrothermal ore deposits that form when superheated, mineral-rich water shoots out of vents in volcanic regions of the ocean floor. As the hot water comes into contact with cold seawater, black particles of various metal sulfides precipitate out and accumulate as chimney-like structures, called *black smokers*, near the hot water vents. These deposits are especially rich in minerals such as copper, lead, zinc, silver, gold, and some of the rare earth metals.

Because of the rapidly rising prices of many of these metals, there is growing interest in deep-sea mining. In 2010, the Chinese government drew up plans to develop remote-controlled underwater equipment that could be used to evaluate and mine the mineral deposits around black smokers, beginning in the Indian Ocean near Madagascar. In 2011, the International Seabed Authority, a UN agency set up to manage seafloor mining in international waters, began issuing mining permits. A Canadian mining firm, Nautilus Minerals, got a permit to begin deep-sea mining for gold, copper, and other valuable metals at a site off Papua New Guinea.

Some of these deep-sea ore beds are highly concentrated. For example, Nautilus Minerals expects to be mining deposits that contain 6.7% copper, compared to 0.46% for typical deposits on land. Some analysts say that this will make the mining and processing of these ores less environmentally harmful than mining on land. However, marine biologists are concerned that the sediment stirred up by such mining could harm or kill organisms that feed by filtering seawater. Proponents of the mining say that the number of potential mining sites is quite small and that many of these organisms can live elsewhere.

Still another possible source of metals from the ocean is the potato-size *manganese nodules* that cover large areas of the Pacific Ocean floor and smaller areas of the Atlantic and Indian Ocean floors. In the future, they could be sucked up by giant vacuum pipes or scooped up by underwater mining machines. They also contain some rare earth metals (Science Focus, p. 291).

12-5 How Can We Use Mineral Resources More Sustainably?

> **CONCEPT 12-5 We can try to find substitutes for scarce resources, reduce resource waste, and recycle and reuse minerals.**

We Can Find Substitutes for Some Scarce Mineral Resources

Some analysts believe that even if supplies of key minerals become too expensive or too scarce due to unsustainable use, human ingenuity will find substitutes (**Concept 12-5**). They point to the current *materials revolution* in which silicon and other new materials, particularly ceramics and plastics, are being used as replacements for metals. They also point out the possibilities of finding substitutes for scarce minerals through nanotechnology (Science Focus, below).

For example, fiber-optic glass cables that transmit pulses of light are replacing copper and aluminum wires in telephone cables. In the future, nanowires may eventually replace the fiber-optic glass cables. High-strength plastics and composite materials, strengthened by lightweight carbon, hemp, and glass fibers, could transform the automobile and aerospace industries. They cost less to produce than metals, do not need painting (which reduces pollution and costs), can be molded into any shape, and can increase fuel efficiency by reducing the weight of motor vehicles. As nanotechnology develops, there will be more substitutes for conventional resources. However, each new application will require careful testing to avoid harmful environmental effects.

Substitution is not a cure-all. For example, platinum is currently unrivaled as a catalyst and is used in industrial processes to speed up chemical reactions, and chromium is an essential ingredient of stainless steel. We can try to find substitutes for such scarce resources, but this may not always be possible.

SCIENCE FOCUS

The Nanotechnology Revolution

Nanotechnology, or *tiny tech*, uses science and engineering to manipulate and create materials out of atoms and molecules at the ultra-small scale of less than 100 nanometers. A nanometer equals one billionth of a meter. It is 1 one-hundred-thousandth the width of a human hair. The period at the end of this sentence is about 1 million nanometers in diameter. At the nanoscale level, conventional materials have unconventional and unexpected properties.

Scientists plan to use atoms of abundant substances such as carbon, silicon, silver, titanium, and boron as building blocks to create everything from medicines and solar cells to automobile bodies. By 2010, nanomaterials were used in more than 1,100 consumer products, and the number is growing rapidly. Such products include stain-resistant and wrinkle-free coatings on clothes, odor-eating socks, self-cleaning coatings on sunglasses and windshields, sunscreens, deep-penetrating skin care products, and food containers that release nanosilver ions to kill bacteria, molds, and fungi. **GREEN CAREER:** environmental nanotechnology

Nanotechnologists envision technological innovations such as a supercomputer the size of a sugar cube that could store all the information now found in the U.S. Library of Congress; biocomposite materials smaller than a human cell that would make our bones and tendons super strong; nanovessels filled with medicines that could be delivered to cells anywhere in the body; and nanomolecules specifically designed to seek out and kill cancer cells.

We could also use nanoparticles to remove industrial pollutants in contaminated air, soil, and groundwater, and we might be able to purify water and desalinate it at an affordable cost with nanofilters. We could use the technology to turn garbage into breakfast by mimicking how nature turns wastes into plant nutrients, thus following the chemical cycling **principle of sustainability**. The list could go on.

So what's the catch? Ideally, this bottom-up manufacturing process would occur with little environmental harm, with no depletion of nonrenewable resources, and with many potential environmental benefits. But there are concerns over some possible unintended and harmful health effects on humans, because a few studies have raised red flags.

As particles get smaller, they become more reactive and potentially more toxic to humans and other animals. Laboratory studies show that nanoparticles can move across the placenta from a mother to her fetus and that they can move from the nasal passage to the brain. They might also penetrate deeply into the lungs, be absorbed into the bloodstream, and penetrate cell membranes.

Many analysts say we need to take two steps before unleashing nanotechnology more broadly. *First*, carefully investigate its potential risks. *Second*, develop guidelines and regulations for controlling its growing applications until we know more about the potentially harmful effects of this new technology. So far, governments have done little to evaluate and regulate such risks. In 2009, an expert panel of the U.S. National Academy of Sciences said that the federal government was not doing enough to evaluate the potential health and environmental risks from engineered nanomaterials.

Critical Thinking

How might the development of nanotechnology affect gold mining (**Core Case Study**)?

We Can Recycle and Reuse Valuable Metals

A more sustainable way to use nonrenewable mineral resources (especially valuable or scarce metals such as gold, iron, copper, aluminum, and platinum) is to recycle or reuse them. Recycling has a much lower environmental impact than that of mining and processing metals from ores. For example, recycling aluminum beverage cans and scrap aluminum produces 95% less air pollution and 97% less water pollution, and uses 95% less energy than mining and processing aluminum ore. Cleaning up and reusing items instead of melting and reprocessing them has an even lower environmental impact.

GOOD NEWS

Japan is investing heavily in the extraction and recycling of gold, platinum, and various rare earth metals found in its massive amounts of electronic waste, which it views as urban mines. In 2010, scientists estimated that used electronic equipment in Japan contained an amount of gold that was equivalent to about 16% of the world's gold reserves. Recovering this gold, as well as other valuable materials, is an example of following the recycling **principle of sustainability**.

SUSTAINABILITY

THINKING ABOUT
Metal Recycling and Nanotechnology

How might the development of a nanotechnology revolution (Science Focus, at left) over the next 20 years affect the recycling of metallic mineral resources?

We Can Use Mineral Resources More Sustainably

Some analysts say we have been asking the wrong question. Instead of asking how we can increase supplies of nonrenewable minerals, we should be asking, *how can we decrease our use and waste of such resources?* Posing this second question could lead us to important technologies that would help us to use mineral resources more sustainably (**Concept 12-5**). Figure 12-19 and the Case Study that follows describe some of these strategies.

THINKING ABOUT
Gold Mining

CORE CASE STUDY

How would you apply the solutions in Figure 12-19 to decreasing the need to mine gold (**Core Case Study**) and to reducing the harmful environmental effects of gold mining?

Solutions

Sustainable Use of Nonrenewable Minerals

- Do not waste mineral resources.
- Recycle and reuse 60–80% of mineral resources.
- Include the harmful environmental costs of mining and processing minerals in the prices of items.
- Reduce mining subsidies.
- Increase subsidies for recycling, reuse, and finding substitutes.
- Redesign manufacturing processes to use less mineral resources and to produce less pollution and waste (cleaner production).
- Use mineral resource wastes of one manufacturing process as raw materials for other processes.
- Slow population growth.

Figure 12-19 We can use nonrenewable mineral resources more sustainably (**Concept 12-5**). **Questions:** Which two of these solutions do you think are the best ones? Why?

■ CASE STUDY
Pollution Prevention Pays

In 1975, the U.S.-based Minnesota Mining and Manufacturing Company (3M), which makes 60,000 different products in 100 manufacturing plants, began a Pollution Prevention Pays (3P) program. The program led the company to redesign much of its equipment and processes, use fewer hazardous raw materials, identify toxic chemical outputs and recycle or sell them as raw materials to other companies, and make more nonpolluting products.

GOOD NEWS

Between 1975 and 2008, the 3M Pollution Prevention Pays program prevented more than 1.4 million metric tons (1.5 million tons) of pollutants from reaching the environment—an amount roughly equal to the weight of more than 100 empty jumbo jet airliners. The company has also saved more than $1.2 billion in waste disposal and material costs. This is an excellent example of why pollution prevention pays.

Since 1990, a growing number of companies have adopted similar pollution and waste prevention programs that have led to cleaner industrial production. (See the Guest Essay by Peter Montague on cleaner production on the website for this chapter.)

Here are this chapter's *three big ideas*:

- Dynamic forces that move matter within the earth and on its surface recycle the earth's rocks, form deposits of mineral resources, and cause volcanic eruptions, earthquakes, and tsunamis.
- The available supply of a mineral resource depends on how much of it is in the earth's crust, how fast we use it, the mining technology used to obtain it, its market prices, and the harmful environmental effects of removing and using it.
- We can use mineral resources more sustainably by trying to find substitutes for scarce resources, reducing resource waste, and reusing and recycling nonrenewable minerals.

REVISITING Gold Mining and Sustainability

In this chapter, we surveyed the harmful effects of gold mining (**Core Case Study**). We also considered a number of possibilities for extracting and using gold and other nonrenewable mineral resources in less harmful, more sustainable ways.

Technological developments could help us to expand supplies of mineral resources and to use them more sustainably. For example, if we develop it safely, we could use nanotechnology (Science Focus, p. 294) to make new materials that could replace scarce mineral resources. Another promising technology is biomining—the use of microbes to extract mineral resources without disturbing the land or polluting air and water as much as conventional mining operations do. We could make increasing use of solar energy to generate the electricity we need for powering such processes—an application of the solar energy **principle of sustainability**.

We can also use mineral resources more sustainably by reusing and recycling them, and by reducing unnecessary resource use and waste—applying the recycling **principle of sustainability**. Industries can mimic nature by using a diversity of ways to reduce the unnecessary waste of resources and to prevent pollution, thus applying the diversity **principle of sustainability**. By applying these three principles to the use of mineral resources, we would reduce the harmful environmental and health effects of mining and processing minerals, thereby sustaining nature's ability to rely on the same principles of sustainability and helping to extend mineral resource supplies into the future.

Mineral resources are the building blocks on which modern society depends. Knowledge of their physical nature and origins, the web they weave between all aspects of human society and the physical earth, can lay the foundations for a sustainable society.

ANN DORR

REVIEW

CORE CASE STUDY

1. Describe some harmful environmental effects of gold mining (**Core Case Study**).

SECTION 12-1

2. What is the key concept for this section? Define **geology**, **core**, **mantle**, **crust**, **asthenosphere**, and **lithosphere**. Define **tectonic plates**. What typically happens when tectonic plates collide, move apart, or slide by one another? Define **volcano** and describe the nature and major effects of a volcanic eruption. Define **earthquake** and describe its nature and major effects. What is a **tsunami** and what are its major effects? Explain how coral reefs and mangrove forests can help to protect against tsunamis.

SECTION 12-2

3. What is the key concept for this section? Define **mineral**, **rock**, **sedimentary rock**, **igneous rock**, and **metamorphic rock** and give an example of each. Describe the nature and importance of the **rock cycle**.

SECTION 12-3

4. What is the key concept for this section? Define **mineral resource** and list two major types of such resources. Define **ore** and distinguish between a **high-grade ore** and a **low-grade ore**. What are **reserves**? Explain the importance of lithium. Summarize the life cycle of a metal resource. Describe three major harmful environmental effects of extracting, processing, and using nonrenewable mineral resources.

5. What is **surface mining**? Define **overburden**, **spoils** and **open-pit mining**. Define **strip mining** and distinguish among **area strip mining**, **contour strip mining**, and **mountaintop removal mining**. Describe three harmful environmental effects of surface mining. What is **subsurface mining**? What is the relationship between gold mining and mercury poisoning? What caused the devastating release of caustic red mud from a sludge pond containing aluminum processing wastes at a Hungarian plant in 2000? What is **smelting** and what are its major harmful environmental effects?

SECTION 12-4

6. What are the two key concepts for this section? What five nations supply most of the world's nonrenewable mineral resources? How dependent is the United States on other countries for important nonrenewable mineral resources? Explain why rare earth minerals are so important to countries such as China, Japan, and the United States.

7. State the conventional view of the relationship between the supply of a mineral resource and its market price. Discuss the pros and cons of the U.S. General Mining Law of 1872.

8. List the opportunities and limitations of increasing mineral supplies by mining lower-grade ores. Why are thieves stealing copper from buildings and catalytic converters from cars? What are the advantages and disadvantages of biomining? Describe the opportunities and possible problems that could result from deep-sea mining.

SECTION 12-5

9. What is the key concept for this section? Discuss the opportunities and problems involved with finding substitutes for key mineral resources. What is **nanotechnology**? How could nanotechnology affect the search for substitutes, and how might it affect the mining industry? What are some problems that could arise from the widespread use of nanotechnology? Explain the benefits of recycling and reusing valuable metals. List five ways to use nonrenewable mineral resources more sustainably. Summarize 3M Company's Pollution Prevention Pays program.

10. What are the three big ideas of this chapter? Explain how we can apply the three **principles of sustainability** to obtain and use nonrenewable mineral resources in a more sustainable way.

SUSTAINABILITY

Note: Key terms are in **bold** type.

CRITICAL THINKING

1. List three ways in which lessening the need to mine gold and reducing its harmful environmental effects (**Core Case Study**) could benefit you. CORE CASE STUDY
2. You are an igneous rock. Write a report on what you experience as you move through the rock cycle (Figure 12-10). Repeat this exercise, assuming you are a sedimentary rock, and again assuming you are a metamorphic rock.
3. Use the second law of thermodynamics (see Chapter 2, p. 36) to analyze the scientific and economic feasibility of each of the following processes:
   a. extracting most of the minerals that are dissolved in seawater
   b. mining increasingly lower-grade deposits of minerals
   c. using inexhaustible solar energy to mine minerals
   d. continuing to mine, use, and recycle minerals at increasing rates, using conventional methods
4. Explain why you support or oppose each of the following proposals concerning extraction of hard rock minerals on public land in the United States (Case Study, p. 291): **(a)** halting the practice of granting title to public land for actual or claimed hard rock mineral deposits; **(b)** requiring mining companies to pay a royalty of 8–12% on the gross income they earn from hard rock minerals that they extract from public lands; and **(c)** making hard rock mining companies legally responsible for restoring degraded land and cleaning up environmental damage caused by their activities.
5. List three ways in which a nanotechnology revolution (Science Focus, p. 294) could benefit you and three ways in which it could harm you.
6. Explain why access to affordable supplies of the chemical element lithium (Li) is so important environmentally and economically to the United States. Explain the importance of rare earth metals and discuss access to these metals by the United States, Japan, and European nations.
7. List three ways in which you could apply **Concept 12-5** to making your lifestyle more environmentally sustainable.
8. Congratulations! You are in charge of the world. What are the three most important features of your policy for developing and sustaining the world's nonrenewable mineral resources?

DOING ENVIRONMENTAL SCIENCE

Do research to determine what materials each of the following items is made of and how much of those materials are required to make each item: **(a)** a class ring, **(b)** a wide-screen plasma television screen, and **(c)** a large pickup truck. Also, try to determine how much water is used to make each of these items and how much waste material is produced.

GLOBAL ENVIRONMENT WATCH EXERCISE

Search for *rare earth metals* and try to find the latest information on possible shortages of key rare earth minerals. Which countries hold the largest reserves? What is the likelihood that China will eventually control the supplies and prices of these essential minerals? What is the U.S. strategy for obtaining supplies of the key rare earth minerals?

DATA ANALYSIS

Uranium (U) is a mineral resource obtained from uranium ore in various types of rock at different concentrations. It is used as a fuel in the reactors of nuclear power plants. A high-grade ore has 2% U and a low-grade ore has 0.1% U. The estimated worldwide recoverable resources of uranium weigh 4,743,000 metric tons. The United States has about 7% of the world's uranium resources, which amounts to about 332,000 metric tons.

1. Given that current worldwide usage of uranium is about 66,500 metric tons per year, how long will the world's present recoverable uranium resources last?
2. Assume U.S. usage is about 25% of world usage. If the United States were to rely only on its domestic uranium resources, how long would they last, assuming a 100% recovery rate (meaning that 100% of the resource can be used)?
3. Assume that most U.S. ore deposits contain high-grade ore (2% U) and that the recovery rates of uranium from the ore (accounting for losses in mining, extraction, and refining) average 65%. How many metric tons of ore will have to be mined to meet U.S. needs?

LEARNING ONLINE

The website for *Environmental Science 14e* contains helpful study tools and other resources to take the learning experience beyond the textbook. Examples include flashcards, practice quizzing, information on green careers, tips for more sustainable living, activities, and a wide range of articles and further readings. To access these materials and other companion resources for this text, please visit **www.cengagebrain.com**. For further details, see the preface, p. xvi.

Energy

13

The Astounding Potential for Wind Power in the United States

CORE CASE STUDY

Between the earth's equator and its polar regions, the sun's rays strike the earth at different angles, resulting in different amounts of solar heating from north to south. Together with the earth's rotation, these differences create flows of air called *wind* (see Figure 7-3, p. 123). We can capture this indirect form of solar energy with groups of wind turbines called *wind farms* that convert it into electrical energy (Figure 13-1).

The U.S. Department of Energy (DOE) estimates that wind farms at favorable sites in the four U.S. states of North Dakota, South Dakota, Kansas, and Texas could more than meet the electricity needs of the lower 48 states. In addition, the National Renewable Energy Laboratory has estimated that U.S. offshore winds—off the Atlantic and Pacific coasts, and off the shores of the Great Lakes—could generate four times the electricity currently generated in the country. (See Figure 20, p. S36, in Supplement 6 for a map the potential supply of land- and ocean-based wind energy in the United States.)

Wind power proponents call for a crash program to develop land-based and offshore wind farms in the United States. Over time, this would help reduce dependence on coal, which when burned, produces air pollutants that kill at least 24,000 Americans a year and adds huge quantities of climate-changing carbon dioxide to the atmosphere. In 2009, for example, U.S. wind farms generated enough electricity to replace the burning of an amount of coal that would fill a coal train stretching 3,200 kilometers (2,000 miles)—more than the distance from Los Angeles to Detroit.

Between 2004 and 2010, the number of jobs in the U.S. wind industry increased from almost none to 85,000. Greatly expanding the U.S. wind turbine industry would create tens of thousands of much-needed new jobs and boost the American economy.

In this chapter, we explore and compare the benefits and drawbacks of the key energy resources on which we now depend, including nonrenewable resources such as oil, coal, and nuclear power, and renewable resources such as wind, solar energy, and flowing water.

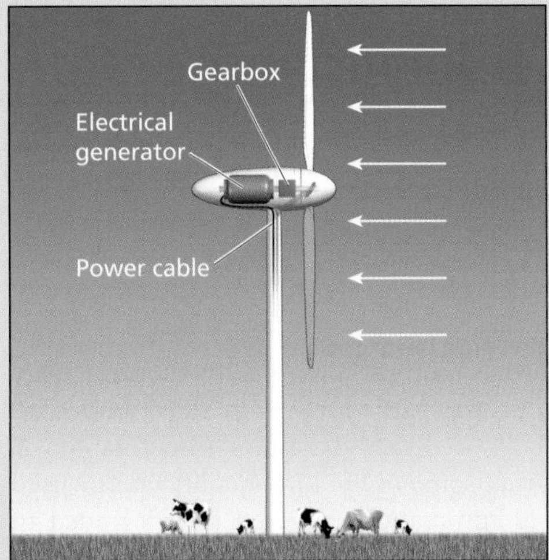

Wind turbine

Wind farm

Yegor Korzh/Shutterstock.com

Figure 13-1 A *wind turbine* (left) converts the kinetic energy of wind to electricity, another form of kinetic energy (moving electrons). Because today's wind turbines can be as tall as 22 stories and have blades as long as the height of a 7-story building, they can tap into the stronger, more reliable, and less turbulent winds found at higher altitudes. Groups of wind turbines called *wind farms* erected on the land (right and Figure 1-15, p. 21) and at sea can produce large amounts of electricity that can be fed into electrical grids.

Key Questions and Concepts

13-1 What is net energy and why is it important?

CONCEPT 13-1 Energy resources vary greatly in their net energy yields, the amount of high-quality energy available from each resource minus the amount of energy needed to make it available.

13-2 What are the advantages and disadvantages of using fossil fuels?

CONCEPT 13-2 Oil, natural gas, and coal are currently abundant and relatively inexpensive, but using them causes air and water pollution, degrades large areas of land, and releases greenhouse gases to the atmosphere.

13-3 What are the advantages and disadvantages of using nuclear power?

CONCEPT 13-3 Nuclear power has a low environmental impact and a very low accident risk, but its use has been limited by a low net energy yield, high costs, fear of accidents, long-lived radioactive wastes, and the potential for spreading nuclear weapons technology.

13-4 Why is energy efficiency an important energy resource?

CONCEPT 13-4 Improving energy efficiency could save the world at least a third of the energy it uses, and it could save the United States up to 43% of the energy it uses.

13-5 What are the advantages and disadvantages of using renewable energy resources?

CONCEPT 13-5 Using a mix of renewable energy resources—especially sunlight, wind, flowing water, sustainable biomass, and geothermal energy—can drastically reduce pollution, greenhouse gas emissions, and biodiversity losses.

13-6 How can we make the transition to a more sustainable energy future?

CONCEPT 13-6 We can make the transition to a more sustainable energy future by greatly improving energy efficiency, depending more on a mix of renewable energy resources, and including the environmental costs of energy resources in their market prices.

Note: Supplements 2 (p. S3), 4 (p. S10), 6 (p. S22), and 7 (p. S38) can be used with this chapter.

Just as the 19th century belonged to coal and the 20th century to oil,
the 21st century will belong to the sun, the wind, and energy from within the earth.

LESTER R. BROWN

13-1 What Is Net Energy and Why Is It Important?

▶ **CONCEPT 13-1 Energy resources vary greatly in their net energy yields, the amount of high-quality energy available from each resource minus the amount of energy needed to make it available.**

Basic Science: Net Energy Is the Only Energy That Really Counts

We start our discussion of energy alternatives with some basic science based on the two laws of thermodynamics (see Chapter 2, p. 36). These laws govern all physical and chemical changes involved in the use of fossil fuels and other energy resources.

Because of the first law of thermodynamics, it takes high-quality energy (see Chapter 2, p. 36) to get high-quality energy. For example, before oil becomes useful to us, it must be found, pumped up from beneath the ground or ocean floor, transferred to a refinery, converted to useful fuels, and delivered to consumers. Each of these steps requires high-quality energy, mostly obtained by burning fossil fuels such as oil and coal. The second law of thermodynamics (see Chapter 2, p. 36) tells us that some of the high-quality energy used in each step is automatically wasted and degraded to lower-quality energy, primarily waste heat that ends up in the environment. No matter how hard we try or how clever we are, we cannot violate the two scientific laws of thermodynamics.

The usable amount of *high-quality, useful energy* available from a given quantity of an energy resource is its **net energy yield**: the total amount of useful energy available from an energy resource minus the energy needed to make it available to consumers

Links: CORE CASE STUDY refers to the Core Case Study. 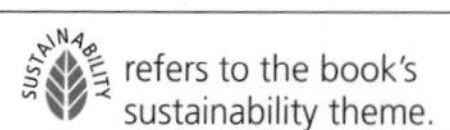 refers to the book's sustainability theme. GOOD NEWS refers to good news about the environmental challenges we face.

(**Concept 13-1**). The net energy yield for any source of energy is calculated by estimating the total amount of energy available from the resource over its projected lifetime and then subtracting the estimated amount of energy *used, automatically wasted* because of the second law of thermodynamics, and *unnecessarily wasted* in finding, extracting, processing, and transporting the useful energy to consumers.

Net energy is like the net profit earned by a business after it deducts its expenses. If a business has $1 million in sales and $900,000 in expenses, its net profit is $100,000. This is the only number that really counts to anyone who owns or plans to invest in such a business on a long-term basis.

Similarly, suppose that it takes about 9 units of energy to produce 10 units of energy by growing and processing corn to produce ethanol fuel for cars. Then the net useful energy yield is only 1 unit of energy. We can express net energy as the ratio of energy produced to the energy used to produce it. In this example, the *net energy ratio* would be 10/9, or approximately 1.1. As the ratio increases, the net energy gain also rises. When the ratio is less than 1, there is a net energy loss.

Figure 13-2 shows estimated net energy ratios for various systems that generate electricity, heat homes and buildings, produce high-temperature heat for industrial processes, and provide transportation. For example, energy and sustainability experts Ida Kubiszewski and Cutler Cleveland, in a 2007 article in the *Encyclopedia of Earth*, estimated the net energy ratio for wind energy (**Core Case Study**) to be 25.2, based on a study in which they reviewed 119 wind turbines and 50 different analyses of the net energy for wind power.

CORE CASE STUDY

Energy Resources with Low or Negative Net Energy Need Help to Compete in the Marketplace

Here is a general rule that we can use to help evaluate the economic usefulness of an energy resource based on its net energy ratio. *Any energy resource with a low or negative net energy ratio cannot compete in the open marketplace with other energy alternatives that have higher net energy ratios unless it receives financial support from the government*

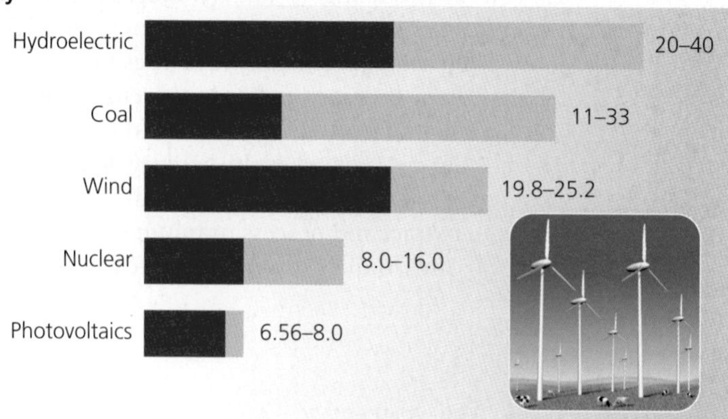

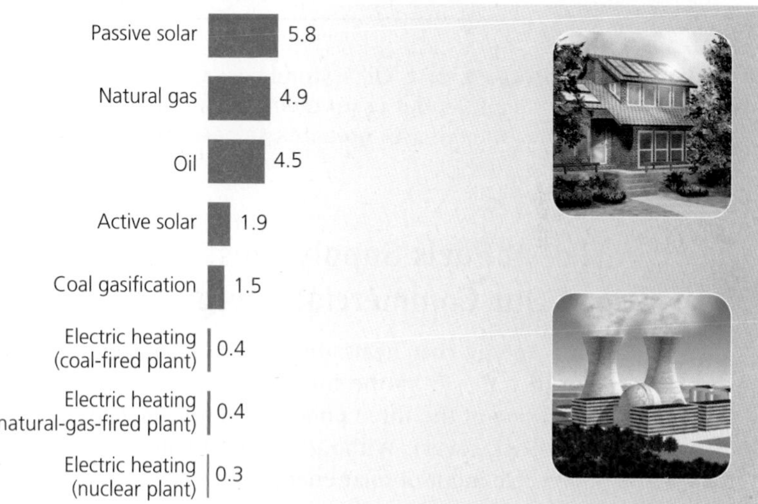

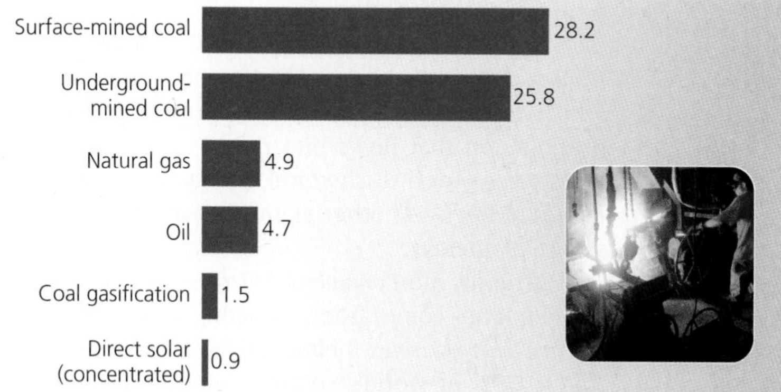

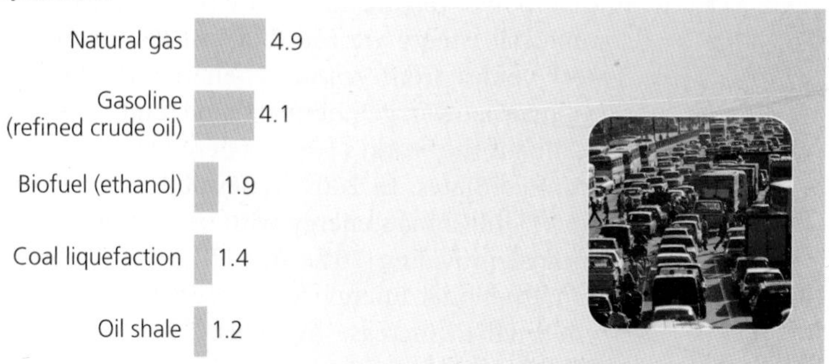

Figure 13-2 Science: *Net energy ratios* for various energy systems over their estimated lifetimes differ widely: the higher the net energy ratio, the greater the net energy available (**Concept 13-1**). **Question:** Based on these data alone, which two resources in each category should we be using? (Data from U.S. Department of Energy; U.S. Department of Agriculture; Colorado Energy Research Institute, *Net Energy Analysis*, 1976; *Encyclopedia of Earth*, 2007; and Howard T. Odum and Elisabeth C. Odum, *Energy Basis for Man and Nature*, 3rd ed., New York: McGraw-Hill, 1981)

or other outside sources of funding. Relying on such energy resources is like investing in a business that has a very low profit margin or that always loses money unless the government or some outside benefactor props it up.

For example, electricity produced by nuclear power has a low net energy yield because large amounts of energy are needed for each step in the *nuclear power fuel cycle*: to extract and process uranium ore, convert it into nuclear fuel, build and operate nuclear power plants, safely store the resulting highly radioactive wastes for thousands of years, dismantle each highly radioactive plant after its useful life (typically 40–60 years), and safely store its high-level radioactive materials for thousands of years. The low net energy yield for the whole fuel cycle is one reason why many governments throughout the world must heavily support nuclear power financially to make it available to consumers at an affordable price.

13-2 What Are the Advantages and Disadvantages of Using Fossil Fuels?

▶ **CONCEPT 13-2 Oil, natural gas, and coal are currently abundant and relatively inexpensive, but using them causes air and water pollution, degrades large areas of land, and releases greenhouse gases to the atmosphere.**

Fossil Fuels Supply Most of Our Commercial Energy

The energy that heats the earth and makes it livable for us comes from the sun at no cost to us—in keeping with one of the three **principles of sustainability** (see back cover). Without this essentially inexhaustible input of solar energy, the earth's average temperature would be –240° C (–400° F).

This direct input of solar energy produces several other forms of renewable energy resources that can be thought of as indirect solar energy: *wind* (moving air masses heated by the sun, see **Core Case Study**), *flowing water* (made possible by solar energy, which evaporates water that returns to the earth as precipitation that flows into rivers), and *biomass* (solar energy converted to chemical energy and stored in the tissues of trees and other plants that can be burned as a source of energy).

Currently, most *commercial energy*—energy sold in the marketplace—comes from extracting and burning *nonrenewable energy resources* obtained from the earth's crust. About 87% of world's commercial energy comes from nonrenewable resources—81% from carbon-containing **fossil fuels** (oil, natural gas, and coal) and 6% from nuclear power (Figure 13-3). The remaining 13% of the commercial energy we use (but just 8% in the United States) comes from *renewable* energy resources—biomass, hydropower, geothermal, wind, and solar energy.

China is the world's largest energy user, followed by the United States. In 2009, renewable energy supplied about 7% of China's energy with nonrenewable energy from coal providing 70% and oil 20%. According to the International Energy Agency, energy use in China is projected to increase by about 75% between 2009 and 2035.

We Depend Heavily on Oil

Crude oil, or **petroleum** (oil as it comes out of the ground), is a black, gooey liquid consisting of hundreds of different combustible hydrocarbons along with small amounts of sulfur, oxygen, and nitrogen impurities. It is also known as *conventional oil* and as *light* or *sweet crude oil*. Oil, coal, and natural gas are called *fossil fuels* because they were formed from the decaying remains (fossils) of organisms that lived millions of years ago.

Deposits of conventional crude oil and natural gas often are trapped together under domes deep within the earth's crust on land or under the seafloor. The crude oil is dispersed in pores and cracks in underground rock formations, somewhat like water saturating a sponge. To extract oil, developers drill a well into a deposit. Then oil, drawn by gravity out of the rock pores, flows into the bottom of the well and is pumped to the surface.

After years of pumping, usually a decade or so, the pressure in a well drops and its rate of conventional crude oil production starts to decline. This point in time is referred to as *peak production* for the well. The same thing can happen to a large oil field when the overall rate of production from its numerous wells begins to decline. *Global peak production* is the point in time when we reach the maximum overall rate of conventional crude oil production for the whole world. Then, if we continue using conventional oil faster than we can produce it, its price rises. There is disagreement over whether we have reached or will soon reach the global peak production of conventional crude oil.

After it is extracted, conventional crude oil is transported to a *refinery* by pipeline, truck, or ship (oil tanker). There it is heated to separate it into components with different boiling points (Figure 13-4) in a complex process called *refining*.

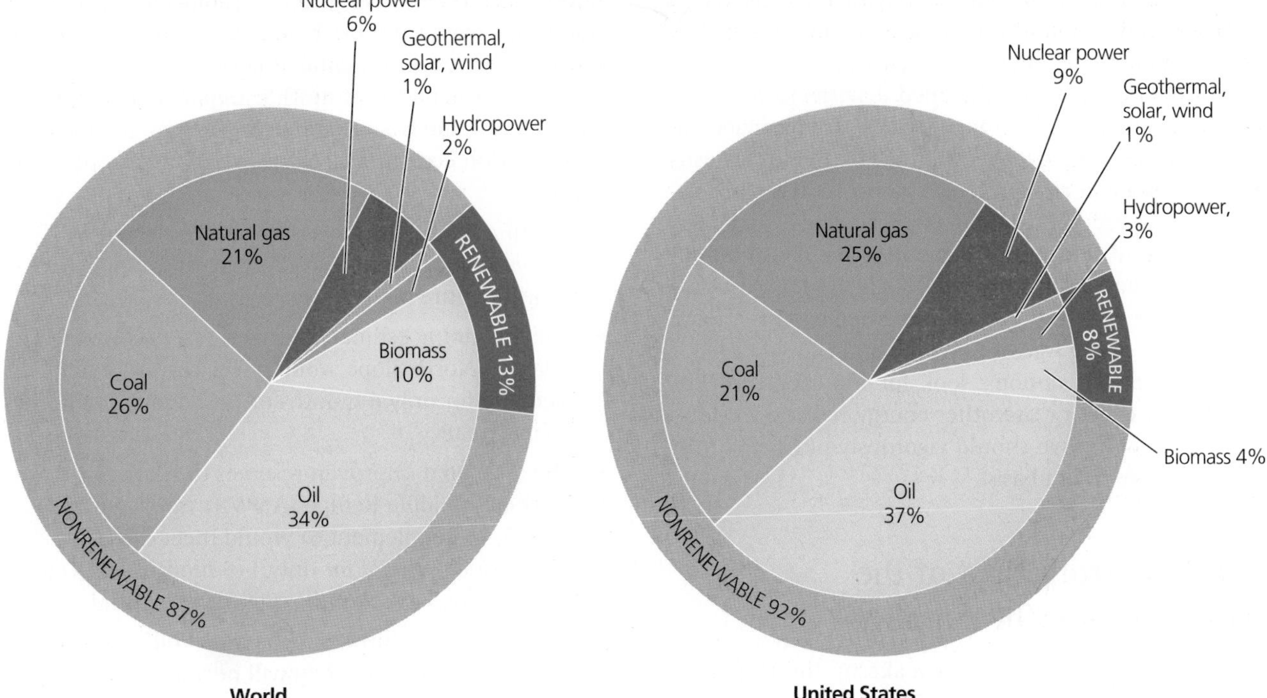

Active Figure 13-3 This figure shows energy use by source throughout the world (left) and in the United States (right) in 2009. Note that oil is the most widely use form of commercial energy and that about 81% of the energy used in the world (83% of the energy used in the United States) comes from burning nonrenewable fossil fuels. **Question:** Why do you think the world as a whole relies more on renewable energy than the United States does? *See an animation that examines and compares energy sources used in more-developed and less-developed countries at* www.cengagebrain.com. (Data from U.S. Department of Energy, British Petroleum, Worldwatch Institute, and International Energy Agency)

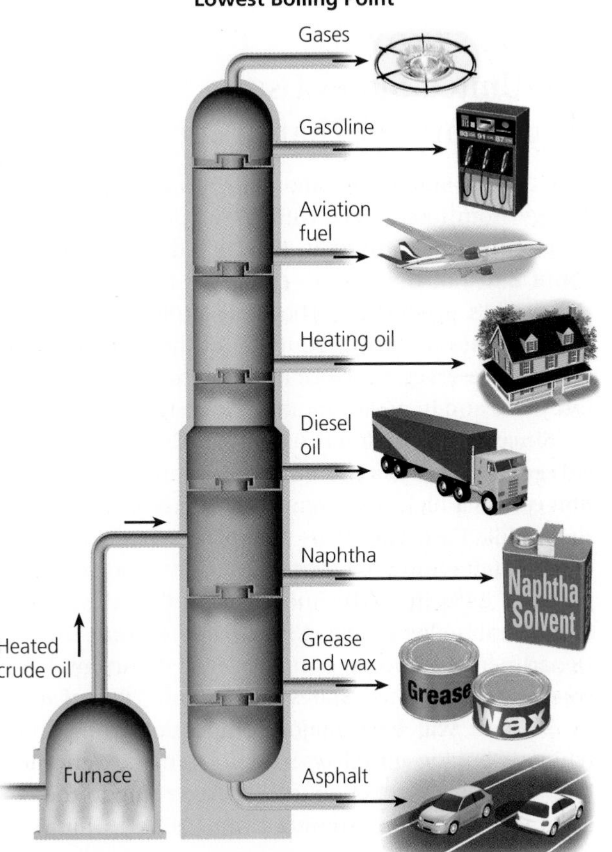

Figure 13-4 Science: When crude oil is refined, many of its components are removed at various levels, depending on their boiling points, of a giant distillation column that can be as tall as a nine-story building. The most volatile components with the lowest boiling points are removed at the top of the column.

We use some of the products of crude oil distillation, called **petrochemicals**, as raw materials in industrial organic chemicals, cleaning fluids, pesticides, plastics, synthetic fibers, paints, medicines, and many other products. Producing a desktop computer, for example, typically requires about ten times its weight in fossil fuels, mostly oil.

THINKING ABOUT
Petrochemicals

Look at your clothing and the room you are sitting in, and try to identify the items that were made from petrochemicals. What are three important ways in which your lifestyle would be different without oil?

How Long Might Supplies of Conventional Crude Oil Last?

World oil consumption has been growing rapidly since 1950. Laid end to end, the 34 billion barrels of oil used in 2010 would stretch to about 31 million kilometers

(19 million miles)—far enough to reach to the moon and back about 40 times. (One barrel of oil contains 159 liters (42 gallons) of oil.)

Proven oil reserves are identified deposits from which conventional crude oil can be extracted profitably at current prices with current technology. Geologists estimate that known and projected global reserves of conventional crude oil will be 80% depleted sometime between 2050 and 2100, depending on consumption rates. The remaining 20% will likely be too costly to remove (see Figure 5, p. S40, in Supplement 7 for a brief history of the Age of Conventional Oil).

We have four options: look for more oil, use less oil, waste less oil, or use other energy resources. Many analysts urge that we should vigorously pursue all four options on an urgent basis.

OPEC Controls Most of the World's Crude Oil Supplies

In 2010, the 12 countries that make up the Organization of Petroleum Exporting Countries (OPEC) had about 77% of the world's proven crude oil reserves and thus are likely to control most of the world's oil supplies for many decades. Today, OPEC's members are Algeria, Angola, Ecuador, Iran, Iraq, Kuwait, Libya, Nigeria, Qatar, Saudi Arabia, the United Arab Emirates, and Venezuela.

Saudi Arabia has the largest portion of the world's conventional proven crude oil reserves (20%). Other countries with large proven reserves of light crude oil are, in order, Venezuela (13%), Iran (10%), Iraq (9%), Kuwait (8%), the United Arab Emirates (7%), and Russia (6%). The United States has only about 2% of the world's proven oil reserves. China has only 1.1%, India has 0.4%, and Japan has no oil reserves. About 85% of all proven conventional crude oil reserves are in the hands of government-owned companies. Large users such as the United States and China are dependent on OPEC nations for much of the oil that supports their economies (see the Case Study that follows). Currently, the world's largest producers of oil are, in order, Russia, Saudi Arabia, and the United States. Energy experts project that by about 2020, Iraq will become the world's third largest oil producer.

One serious problem for oil supplies worldwide is that since 1984, production of conventional crude oil from proven reserves has exceeded new oil discoveries. Since 2005, global crude oil production has generally leveled off. Of the world's 64 major oil fields, 54 are now in decline.

Some analysts say that there is a lot of oil still to be found. But they are talking mostly about small, dispersed, and harder-to-extract deposits of conventional crude oil, deposits of heavy oil that must be extracted from existing oil wells, and unconventional types of heavy oils produced from tar sands and oil shales. These resources are more expensive to develop than most of today's conventional crude oil deposits and have much lower net energy ratios. Using them also results in a much higher environmental impact.

Based on data from the U.S. Department of Energy (DOE) and the U.S. Geological Survey, if global oil consumption continues to grow at about 2% per year, then:

- Saudi Arabia, with the world's largest proven conventional crude oil reserves, could supply the world's entire oil needs for only about 7 years;
- the remaining estimated proven reserves under Alaska's North Slope would meet current world demand for only 6 months or U.S. demand for less than 3 years;
- the estimated unproven reserves in Alaska's Arctic National Wildlife Refuge (ANWR) (see Figure 15, p. S32, in Supplement 6) would meet the current world demand for only 1–5 months and U.S. demand for 7–24 months (Figure 13-5); and
- if one day the Chinese use as much oil per person as Americans do today, they will need the equivalent of seven more Saudi Arabian supplies to meet their demand.

According to some analysts, in order to keep using conventional oil at the projected increasing rate of consumption, we must discover proven reserves of conventional oil equivalent to the current Saudi Arabian supply every 5 years. Most oil geologists say this is highly unlikely.

■ CASE STUDY
The United States Uses Much More Oil Than It Produces

The United States gets about 83% of its commercial energy from fossil fuels, with 37% coming from crude oil (Figure 13-3). In 2009, the United States produced about 8.5% of the world's crude oil, but used 22% of the world's production. The basic problem is that the United States has only about 2% of the world's proven crude oil reserves, much of it in environmentally sensitive areas and in areas that make it costly to produce.

Since 1984, crude oil use in the United States has exceeded new domestic discoveries, and U.S. production carries a high cost, compared to production costs in the Middle East. This helps to explain why in 2010, the United States imported about 57% of its crude oil (compared to 24% in 1970), mostly from Canada, Mexico, Saudi Arabia, Venezuela, Nigeria, and Russia. The U.S. Department of Energy estimates that if current trends continue, the United States will import 70% of its oil by 2025. It will have major competition from China, which is buying up oil reserves in Canada, Venezuela, Iraq, and several other countries. According to the U.S. Energy Information Administration (EIA), China's oil consumption is projected to increase by 80% between 2010 and 2030.

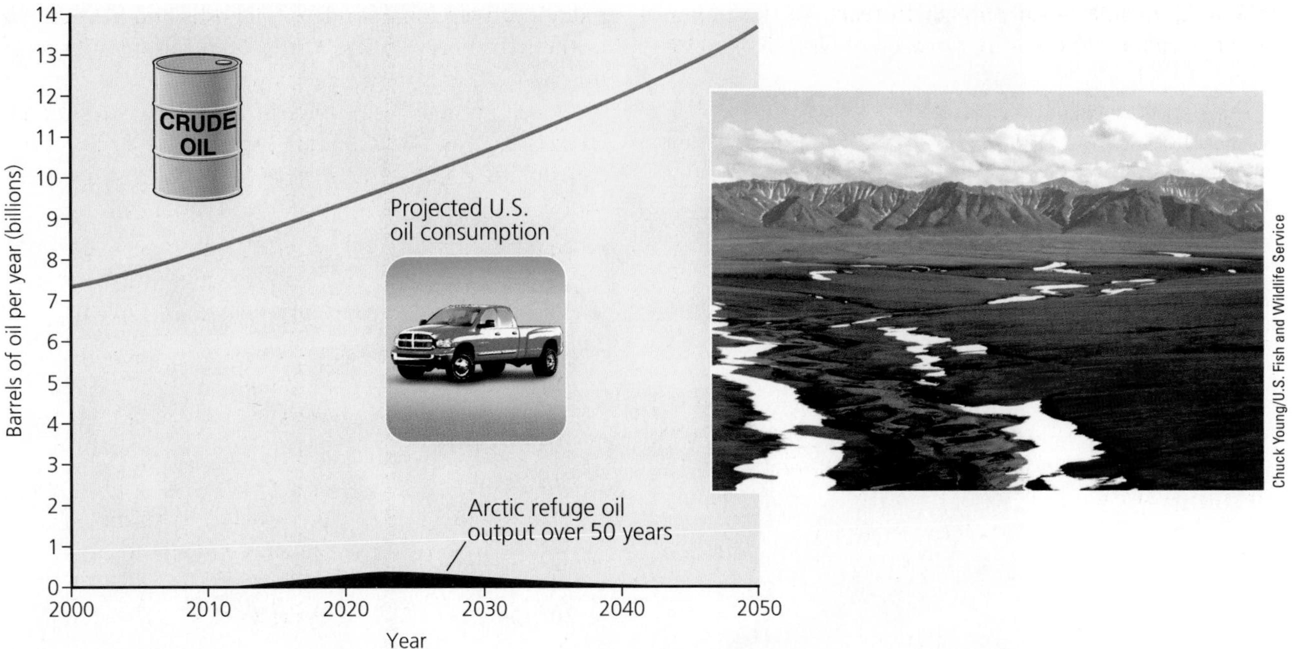

Figure 13-5 The amount of crude oil that *might* be found in the Arctic National Wildlife Refuge (right), if developed and extracted over 50 years, is only a tiny fraction of projected U.S. oil consumption. In 2008, the DOE projected that developing this oil supply would take 10–20 years and would lower gasoline prices at the pump by 6 cents per gallon at most. (Data from U.S. Department of Energy, U.S. Geological Survey, and Natural Resources Defense Council)

The EIA, as well as a number of independent geologists, estimate that if the United States opens up all of its public lands and coastal regions to oil exploration, it will find an amount of crude oil that would probably meet no more than about 1% of the country's current annual need. And this oil would be developed only at very high production costs, low net energy, and high environmental impacts. In other words, according to these energy analysts, the United States cannot even come close to meeting its huge and growing demand for crude oil and gasoline by increasing domestic supplies.

Using Crude Oil Has Advantages and Disadvantages

Figure 13-6 lists the advantages and disadvantages of using crude oil as an energy resource. The extraction, processing, and burning of nonrenewable oil and other fossil fuels have severe environmental impacts (see Figure 12-12, p. 285, and the Case Study that follows), including land disruption, greenhouse gas emissions and other forms of air pollution, water pollution, and threats to biodiversity.

In the United States, the most recent lesson about the environmental impact of oil occurred in April of 2010 in Gulf Coast waters when the BP Company's *Deepwater Horizon* oil-drilling rig exploded (Figure 13-7, p. 306), causing a rupture of the wellhead on the ocean bottom in deep water. The accident sank the rig and killed 11 of its crewmembers. During the next several weeks, an estimated 679 million liters (180 million gallons) of crude oil gushed from the ruptured wellhead. The oil contaminated ecologically vital coastal marshes, mangroves, sea-grass beds, some deep coral reefs, and deep-ocean aquatic life, although the extent of the ecological damage will not be known for years. The accident also disrupted the livelihoods of people depending on the Gulf Coast's fisheries and caused large economic losses for the area's tourism business. Before the BP

Figure 13-6 Using crude oil as an energy resource has advantages and disadvantages (**Concept 13-2**). **Questions:** Which single advantage and which single disadvantage do you think are the most important? Why?

U.S. Coast Guard

Figure 13-7 On April 20, 2010, oil and natural gas escaping from an oil-well borehole ignited and caused an explosion on the British Petroleum (BP) *Deepwater Horizon* drilling platform in the Gulf of Mexico. The ruptured wellhead eventually released a huge volume of oil into the Gulf waters.

accident, the biggest and most destructive oil spill in U.S. waters happened in 1989 when the oil tanker *Exxon Valdez* ran aground and spilled 42 million liters (11 million gallons) of oil into Alaska's Prince William Sound.

While the 2010 BP spill was an economic and ecological catastrophe, worse oil spills have received much less attention. For example, scientists estimate that since the mid-1960s, off the coast of Nigeria, more than 2.5 times the estimated amount of crude oil spilled in the 2010 Gulf Coast disaster have been spilled from various drilling sites with little media attention and no global outcry.

A critical and growing problem is that the burning of oil or any carbon-containing fossil fuel releases the greenhouse gas CO_2 into the atmosphere. According to most of the world's top climate scientists, this will warm the atmosphere and contribute to projected climate change during this century.

Will Heavy Oil Be a Useful Resource?

Heavy oil is an alternative to light crude oil that is now getting more attention. Oily shale rocks are a potential supply of heavy oil, as are tar sands (see the Case Study that follows). A certain type of rock, called *oil shale* (Figure 13-8, left), contains a solid combustible mixture of hydrocarbons called *kerogen*. After being mined, oil shale can be crushed and heated to produce the kerogen in a process that yields a distillate called shale oil (Figure 13-8, right). Before the thick shale oil is sent by pipeline to a refinery, it, too, must be heated to increase its flow rate and processed to remove sulfur, nitrogen, and other impurities.

About 72% of the world's estimated oil shale reserves are buried deep in rock formations located primarily on government-owned land in the U.S. states of Colorado, Wyoming, and Utah in an area known as the Green River formation and in the nearby Bakken oil formation (see Figure 15, p. S32, in Supplement 6). The U.S. Bureau of Land Management estimates that these deposits contain an amount of potentially recoverable heavy oil equal to almost 4 times the size of Saudi Arabia's proven reserves of conventional oil. Estimated potential global supplies of unconventional **shale oil** are about 240 times larger than estimated global supplies of conventional crude oil.

So why should we ever worry about running out of oil? The problem is that most of these oil shale deposits are locked up in rock and ore of such low grade that it

U.S. Department of Energy

Figure 13-8 Shale oil (right) can be extracted from oil shale rock (left). However, producing shale oil requires large amounts of water and has a low net energy and a very high environmental impact.

takes considerable energy and money to mine and convert the kerogen to shale oil and the shale oil to useful products. Thus, its net energy yield is very low, which means that this energy resource will not be able to compete in the open marketplace unless U.S. taxpayers provide energy companies with large government *subsidies* (payments intended to help businesses survive and thrive) and tax breaks.

Extracting shale oil would have a huge environmental impact. Large tracts of land, much of it in pristine and beautiful wilderness areas, would be torn apart. It takes about 5 barrels of water to produce one barrel of shale oil, and the big U.S. deposits are mostly in arid areas where water is already in short supply, partly because of a prolonged drought. Furthermore, digging up and processing shale oil releases 27–52% more CO_2 into the atmosphere per unit of energy produced than does producing crude oil.

■ CASE STUDY
Heavy Oil from Tar Sand

Tar sand, or **oil sand**, is a mixture of clay, sand, water, and a combustible organic material called *bitumen*—a thick, sticky, tarlike heavy oil with a high sulfur content.

Northeastern Alberta in Canada has three-fourths of the world's tar sand resources in sandy soil under a huge area of remote boreal forest (see Figure 7-14, bottom photo, p. 134), roughly equal to the area of the U.S. state of North Carolina. Other deposits are in Venezuela, Colombia, Nigeria, Russia, and the U.S. state of Utah.

The amount of unconventional heavy oil potentially available from bitumen in Canada's tar sands is roughly equal to seven times the total conventional oil reserves of Saudi Arabia. Canada has only about 2.5% of the world's proven conventional oil reserves. However, if we include known unconventional heavy oil reserves that can be extracted from tar sands and converted into synthetic crude oil, Canada has the world's second largest proven oil reserves (13%) after Saudi Arabia with 20%.

The big drawback is that developing this resource results in major harmful impacts on the land (Figure 13-9), air, water, wildlife, and climate. About 20% of Alberta's tar sand is close enough to the surface to be strip-mined. Before the mining takes place, the overlying boreal forest is clear-cut, wetlands are drained, and some rivers and streams are diverted. Next, the overburden of sandy soil, rocks, peat, and clay is stripped away to expose the tar-sand deposits. Then, five-story-high electric power shovels dig up the tar sand and load it into three-story-high trucks, which carry it to an upgrading plant. There, the oil sand is mixed with hot water and steam to extract the bitumen, which is heated by natural gas in huge cookers and converted into a low-sulfur, synthetic, heavy crude oil suitable for refining.

To extract this heavy oil, more earth is being removed in Canada's Athabasca Valley than anywhere else in the world. The project also produces huge amounts of air pollution that fill the mining region's air with dust, smoke, gas fumes, and a tarry stench. According to a 2009 study by CERA, an energy consulting group, the process also releases 3 to 5 times more greenhouse gases per barrel of the heavy synthetic crude oil than is released in the extraction and production of a barrel of conventional crude oil.

In addition, the process uses large amounts of water and creates lake-size tailing ponds of polluted and essentially indestructible toxic sludge and wastewater.

AP Photo/Jeff McIntosh

Figure 13-9 Producing heavy oil from Canada's Alberta tar sands project has involved strip-mining areas large enough to be seen from outer space, along with draining wetlands and diverting rivers. It also produces huge amounts of air and water pollutants.

Every year, many migrating birds die trying to get water and food from the ponds. Also, the dikes of compacted sand surrounding the tailings ponds have the potential to leak and release large volumes of toxic sludge onto nearby land and into the Athabasca River.

This method of producing oil takes a great deal of energy—mostly through burning natural gas and using diesel fuel to run the massive machinery—and therefore has a low net energy yield. In other words, producing heavy oil from tar sands is one of the world's least efficient, dirtiest, and most environmentally harmful ways to provide an energy resource.

Despite its severe environmental impact, oil produced from tar sands in Canada accounted for about 12% of the oil imported by the United States in 2009. By 2030, the United States may get as much as 36% of its imported oil from Canadian tar sands. Meanwhile, China is buying up long-term access to this resource.

Figure 13-10 lists the advantages and disadvantages of using tar sands and oil shale as energy resources.

THINKING ABOUT
Heavy Oils

Do the advantages of relying on heavy oil from oil shale and tar sands outweigh its disadvantages? Explain.

Natural Gas Is a Versatile, Widely Used Fossil Fuel

Natural gas is a mixture of gases of which 50–90% is methane (CH_4). It also contains smaller amounts of heavier gaseous hydrocarbons such as propane and butane. This versatile fuel has a high net energy yield (Figure 13-2) and can be burned to heat indoor space and water and to propel vehicles. Natural gas turbines, which look somewhat like jet engines, are used to produce electricity in some power plants.

Conventional natural gas lies above most reservoirs of crude oil. When such a natural gas field is tapped, propane and butane gases are liquefied under high pressure and removed as **liquefied petroleum gas (LPG)**. LPG is stored in pressurized tanks for use mostly in rural areas not served by natural gas pipelines. The rest of the gas (mostly methane) is purified and pumped into pressurized pipelines for distribution across land areas. When natural gas is found along with oil in both deep-sea and remote land areas where natural gas pipelines have not been built, it is usually burned off in the process of tapping the oil deposit because it costs too much to build pipelines in such locations. This practice adds climate-changing CO_2 and other pollutants to the air and represents a waste of a valuable resource.

Once natural gas is brought to the surface, most of it is transported across land in pipelines, but it can also be transported across oceans as **liquefied natural gas (LNG)**—gas converted to liquid at a high pressure and at a very low temperature. This highly flammable liquid is transported in refrigerated tanker ships. At its destination port, it is heated and converted back to the gaseous state to be distributed by pipeline. However, LNG has a low net energy yield because processing and delivering it uses more than a third of its energy content.

GOOD NEWS

The long-term global outlook for conventional natural gas supplies is better than that for crude oil. At current consumption rates, proven reserves of conventional natural gas should last the world 62–125 years and the United States 82–118 years. Figure 13-11 lists the advantages and disadvantages of using conventional natural gas as an energy resource.

Figure 13-10 Using heavy oil from oil shale and tar sands as an energy resource has advantages and disadvantages (**Concept 13-2**). **Questions:** Which single advantage and which single disadvantage do you think are the most important? Why?

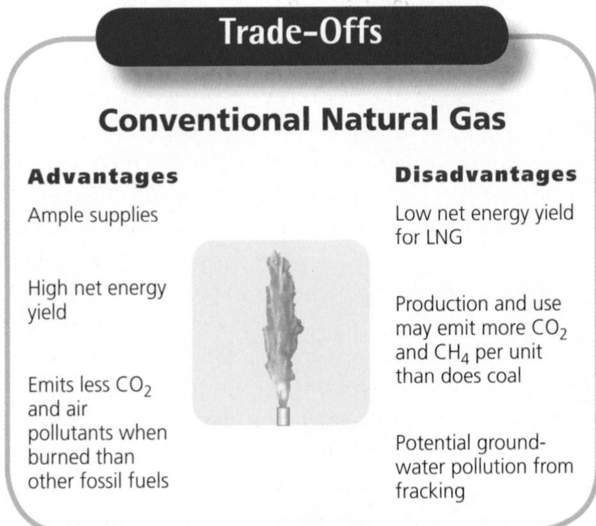

Figure 13-11 Using conventional natural gas as an energy resource has advantages and disadvantages (**Concept 13-2**). **Questions:** Which single advantage and which single disadvantage do you think are the most important? Why? Do you think that the advantages of using conventional natural gas outweigh its disadvantages? Why?

Russia has about 27% of the world's proven natural gas reserves, followed by Iran (15%) and Qatar (15%). In 2009, the United States had only 4.3% of the world's proven natural gas reserves but used about 20% of the world's annual production.

There are also several sources of *unconventional natural gas*. One is *coal bed methane gas* found in coal beds near the earth's surface across parts of the United States and Canada (see Figure 16, p. S33, in Supplement 6). However, the environmental impacts of producing natural gas from coal beds—scarring of land, large water use, a growing number of toxic waste ponds, and potential pollution of drinking water from aquifers—may limit its use.

The burning of natural gas releases CO_2 and several air pollutants into the atmosphere. However, it releases much less CO_2 per unit of energy when burned than do coal, crude oil, and synthetic crude oil from tar sands and oil shale. This makes natural gas the cleanest-burning alternative among the fossil fuels. However, other aspects of using natural gas put its clean reputation in question.

Is Natural Gas a Cleaner Fossil Fuel?

The way natural gas is extracted in some areas is causing a growing environmental problem. By a certain method called hydraulic fracturing, or *fracking*, huge amounts of water mixed with sand and some toxic chemicals are pumped underground through horizontal natural gas wells. Explosives and high pressure are used to fracture the deep rock and free up the natural gas stored there. The gas flows out of the well along with much of the water and a mix of compounds pulled from the rocks, including salts, toxic heavy metals, and naturally occurring radioactive materials. This potentially toxic slurry is stored in tanks and holding ponds.

Fracking is not new, but its use is growing dramatically, particularly in areas such as the northeastern Unites States, which lies over a deposit of natural gas large enough to supply the entire East Coast population for at least 50 years. Drillers maintain that fracking is necessary for exploiting this reserve at a reasonably low cost, and they argue that no groundwater contamination directly due to fracking has ever been recorded. However, many scientists and citizens point out that there is no guarantee that sharply increasing use of the process will not contaminate groundwater at some point. Nor can anyone say that some of the growing number of holding ponds and tanks used to store the toxic slurry will not leak and pollute rivers and streams.

People who rely on aquifers and streams in these areas for their drinking water have little protection from pollution of their water supplies that might result from natural gas drilling. This is because, under political pressure from natural gas suppliers, lawmakers writing the 2005 Energy Policy Act excluded natural gas companies from regulation under U.S. water pollution control laws. Many environmental scientists say this exclusion must be overturned soon because of the rapid growth of the use of fracking. It is used in about 25% of all U.S. natural gas production, and this percentage is expected to be 45% by 2035.

Local landowners and communities also face a dilemma. Landowners receive large amounts of money for access to natural gas deposits under their land but have to weigh possible threats to the area's water supplies. This has led to intense controversy in many once closely knit communities that have benefited from the jobs and revenues from natural gas production.

In 2011, one study by ecologist Robert Howarth and another by geoscientist David Hughes questioned the widely held assumption that natural gas is a clean energy source. Natural gas has long been thought of as a possible bridge fuel that could help us make the transition from heavy use of coal and oil to greater dependence on energy efficiency and low-carbon renewable energy sources. Natural gas does burn cleaner than oil and much cleaner than coal, and when burned completely, it emits about 30% less CO_2 than oil and about 50% less than coal.

However, these new studies estimated the total emissions of CO_2 and methane (CH_4, a much more potent greenhouse gas) throughout the processes of producing, distributing, and using natural gas. They noted that when CH_4 is found lying over oil deposits, it is vented into the atmosphere or burned (Figure 13-11, center image), which produces CO_2. They also pointed to large and numerous leaks of CH_4 from wells, pipelines, and valves. Such leaks due to fracking operations are likely to increase significantly as dependence on fracking increases.

These researchers estimated that the overall emissions of greenhouse gases from the processes of natural gas production and use could be at least 20% higher per unit of energy produced than such emissions from production and use of coal. If more research verifies these preliminary estimates, shifting from coal to natural gas could accelerate projected climate change unless the industry is required to monitor and sharply reduce such emissions.

Coal Is a Plentiful but Dirty Fuel

Coal is a solid fossil fuel formed from the remains of land plants that were buried 300–400 million years ago and exposed to intense heat and pressure over those millions of years (Figure 13-12, p. 310).

Coal is burned in power plants (Figure 13-13, p. 310) to generate about 42% of the world's electricity, 44% of the electricity used in the United States, and 73% of that used in China. It is also burned in industrial plants to make steel, cement, and other products. In order, the three largest users of coal are China, the United States, and India.

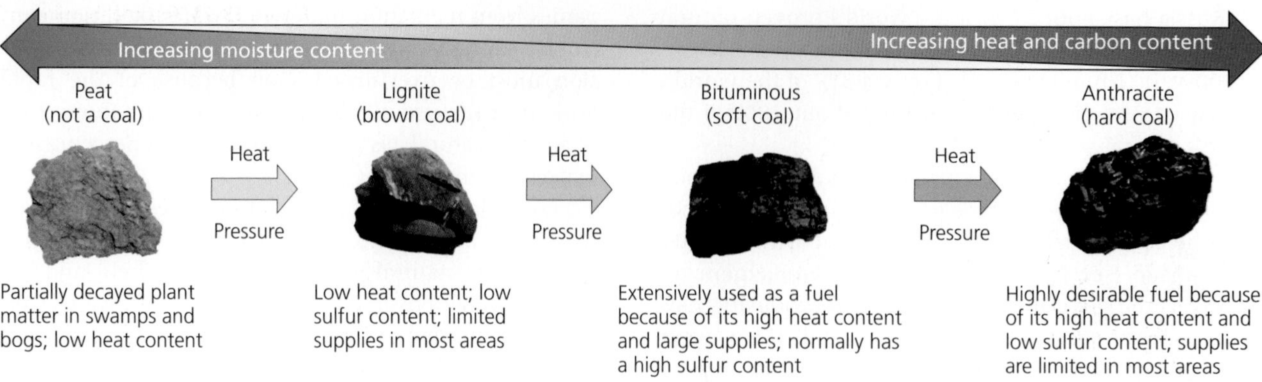

Figure 13-12 Over millions of years, different types of coal have formed. Peat is a soil material made of moist, partially decomposed organic matter and is not classified as a coal, although it, too, is used as a fuel. The different major types of coal vary in the amounts of heat, carbon dioxide, and sulfur dioxide released per unit of mass when they are burned.

Coal is the world's most abundant fossil fuel. According to the U.S. Geological Survey, identified and unidentified global supplies of coal could last for 214–1,125 years, depending on how rapidly they are used. Five countries have 76% of the world's proven coal reserves. The United States has 29% of the world's proven coal reserves (see Figure 16, p. S33, in Supplement 6). Russia has 19%, followed by China (14%), Australia (9%), and India (7%). The U.S. Geological Survey and the U.S. National Academy of Sciences estimate that identified U.S. proven coal reserves should last about 100–250 years at the current consumption rate.

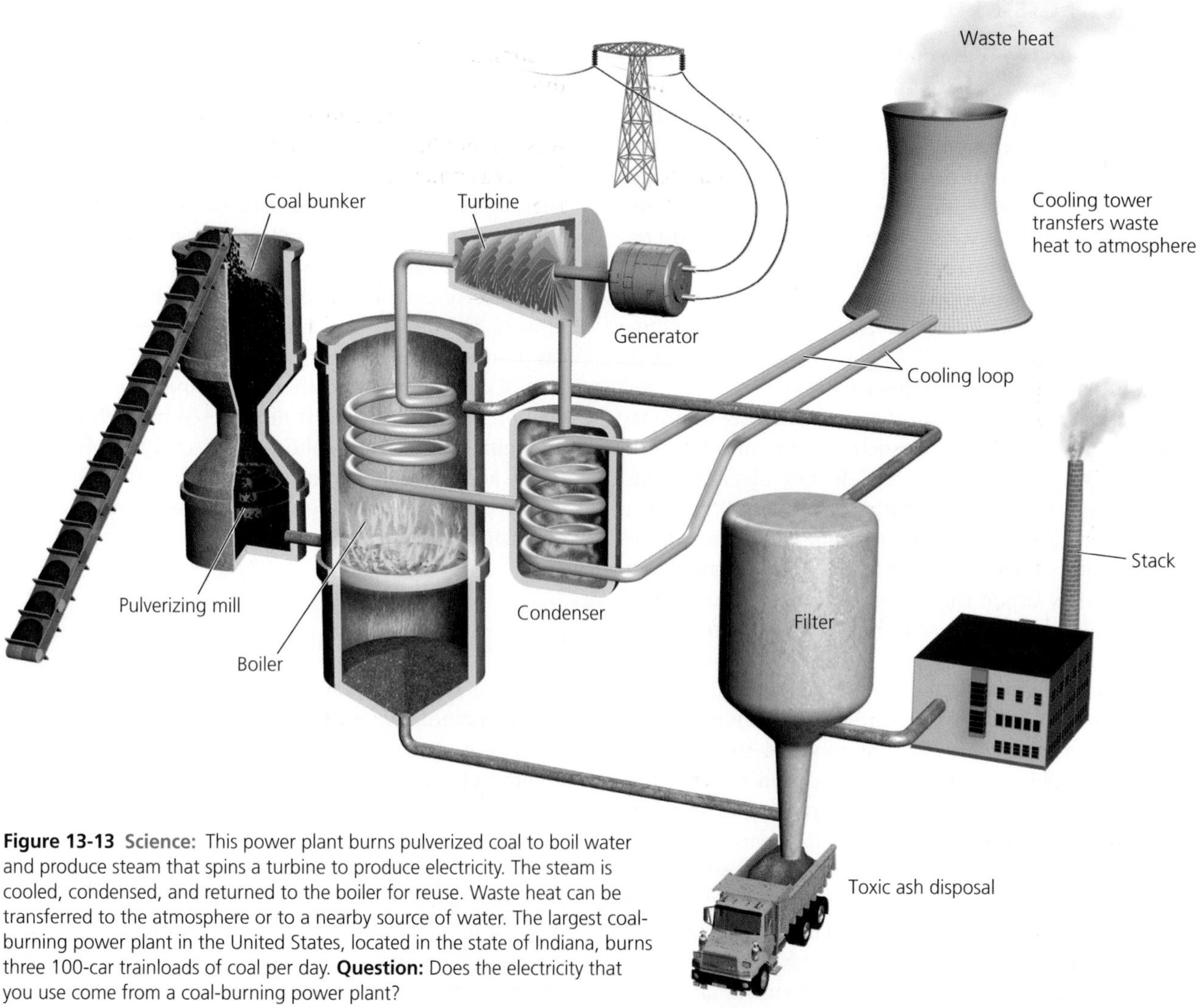

Figure 13-13 Science: This power plant burns pulverized coal to boil water and produce steam that spins a turbine to produce electricity. The steam is cooled, condensed, and returned to the boiler for reuse. Waste heat can be transferred to the atmosphere or to a nearby source of water. The largest coal-burning power plant in the United States, located in the state of Indiana, burns three 100-car trainloads of coal per day. **Question:** Does the electricity that you use come from a coal-burning power plant?

Figure 13-14 This coal-burning industrial plant in India produces large amounts of air pollution because is has inadequate air pollution controls.

Deb Kushal/Peter Arnold, Inc.

The problem is that coal is by far the dirtiest of all fossil fuels. Before it is even burned, the processes of making it available severely degrade land (see Figures 12-14 through 12-16, pp. 286–287) and pollute water and air. When coal is burned without expensive pollution control devices, it severely pollutes the air (Figure 13-14). According to a 2010 study by the Clean Air Task Force, fine-particle pollution in the United States, mostly from older coal-burning power plants without the latest air pollution control technology, prematurely kills at least 13,000 people per year. A 2010 study by Harvard Medical School's Center for Health and the Global Environment puts the figure at 24,000 people per year.

Coal-burning power and industrial plants are among the largest emitters of the greenhouse gas CO_2 (Figure 13-15). According to a 2007 study by the Center for Global Development, coal-burning power plants account for 25% of all human-generated CO_2 emissions in the world, and 40% of such emissions in the United States. Coal also contains small amounts of sulfur, which is released into the air as sulfur dioxide (SO_2) when the coal burns, and this contributes to the problem of acid precipitation and to serious human health problems (see Chapter 15).

Another problem with burning coal is that it emits trace amounts of radioactive materials as well as toxic and indestructible mercury into the atmosphere. Also, burning coal produces a highly toxic ash that must be safely stored, essentially forever.

Currently, China uses three times as much coal as the United States uses, and it has become the world's leading emitter of CO_2 and of sulfur dioxide. According to a World Bank report, China has 20 of the world's 30 most polluted cities, and outdoor and indoor air pollution—most of it from burning coal—causes 650,000 to 700,000 deaths a year there. Because of its dependence on coal, China is buying coal from the United States, Australia, Indonesia, Russia, and South Africa, even though it has a large supply of its own.

The primary reason that coal is a relatively cheap way to produce electricity is that most of its harmful environmental and health costs are not included in the market price of electricity from coal-burning power plants. In 2008, the U.S. Energy Administration and the Worldwatch Institute determined that, when the estimated harmful environmental and health costs of using coal are included, burning coal becomes the second

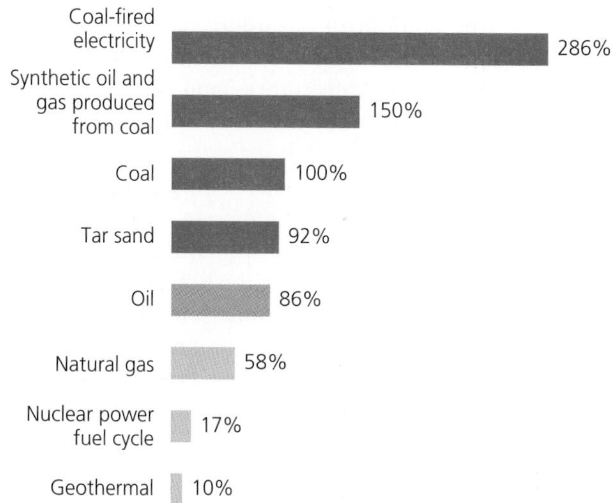

Figure 13-15 CO_2 emissions, expressed as percentages of emissions released by burning coal directly, vary with different energy resources. **Question:** Which produces more CO_2 emissions per kilogram, burning coal to heat a house or heating with electricity generated by coal? (Data from U.S. Department of Energy)

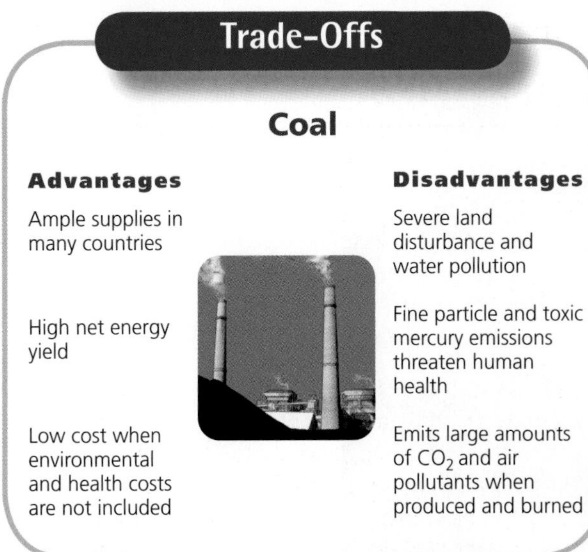

Figure 13-16 Using coal as an energy resource has advantages and disadvantages (**Concept 13-2**). **Questions:** Which single advantage and which single disadvantage do you think are the most important? Why? Do you think that the advantages of using coal as an energy resource outweigh its disadvantages?

most expensive way to produce electricity after solar cells. And this estimate does not include the additional costs of regulating coal ash (see the Case Study that follows). Figure 13-16 lists the advantages and disadvantages of using coal as an energy resource.

■ **CASE STUDY**

The Growing Problem of Coal Ash

The processes of burning coal and removing pollutants from the resulting emissions produce an ash that contains highly toxic and indestructible chemical elements such as arsenic, lead, mercury, and radioactive radium. Each year, the amount of hazardous ash produced by U.S. coal-fired power plants would fill a train with enough rail cars to stretch three-and-one-half times the distance between New York City and Los Angeles, California.

Some of the ash from coal-burning plants is sold and blended into cement and concrete, used as a base for paving roads, or converted into wallboard for use in homes and offices. In the United States, about 57% of the ash is either buried (sometimes in active or abandoned mines where it can contaminate groundwater) or made into a wet slurry that is stored in holding ponds. There the slurry can slowly leach into groundwater or break through the ponds' earthen dams and severely pollute nearby land, surface waters, and groundwater.

The hazards of poorly regulated wet-slurry storage of coal ash became clear on December 22, 2008, when a rupture occurred in a wall of a coal ash storage pond near Knoxville, Tennessee. The resulting spill of toxic sludge flooded an area larger than the combined areas of 300 football fields and destroyed or damaged 40 homes and other buildings. It also tainted waterways and soil with arsenic and other toxic chemicals, killed fish, and disrupted the lives of the people in a nearby rural community.

In 2007, the U.S. Environmental Protection Agency (EPA) determined that toxic metals and other harmful chemicals in coal ash waste ponds have contaminated groundwater used by 63 communities in 26 U.S states. In 2009, the EPA also estimated that there are 44 coal ash waste ponds in several states that are highly hazardous (see Figure 23, p. S37, in Supplement 6).

For nearly three decades, coal and electric utility companies have successfully opposed the classification and regulation of such wastes as hazardous. They argue correctly that forcing them to include these harmful environmental and health costs in the price of coal-generated electricity would make it considerably more expensive. This would make coal less competitive with other cleaner and increasingly less costly energy alternatives such as natural gas and wind power (**Core Case Study**). Cement, concrete, and wallboard producers also oppose classifying coal ash as a hazardous waste because it could prevent them from using the ash as filler in their products.

The situation in China is much worse. A 2010 Greenpeace study stated that toxic coal ash dumped in open landfills is China's largest category of solid industrial waste. From these landfills, toxic chemicals are easily dispersed into the environment by wind and rain.

The Clean Coal Campaign

For decades, economically and politically powerful U.S. coal companies and coal-burning utilities have understandably fought to preserve their profits by opposing measures such as stricter air pollution standards for coal-burning plants and classification of coal ash as a hazardous waste. For more than 30 years, these companies have also successfully led the fight against efforts to classify climate-changing CO_2 as a pollutant, which would raise their costs and threaten their long-term economic survival.

Since 2008, the U.S. coal and electric utility industries have mounted a highly effective, well-financed publicity campaign built around the misleading notion of *clean coal*. We can burn coal more cleanly by adding costly air pollution control devices. But critics point out that there is no such thing as clean coal.

Air pollution resulting from inadequate regulation of coal-burning power and industrial plants kills large numbers of people every year. Mining coal usually involves disrupting the land and polluting water and air. Even with stricter air pollution controls, burning coal will always involve some emissions of CO_2 and various air pollutants, and it will always create indestructible and hazardous coal ash.

13-3 What Are the Advantages and Disadvantages of Using Nuclear Power?

▶ **CONCEPT 13-3 Nuclear power has a low environmental impact and a very low accident risk, but its use has been limited by a low net energy yield, high costs, fear of accidents, long-lived radioactive wastes, and the potential for spreading nuclear weapons technology.**

How Does a Nuclear Fission Reactor Work?

To evaluate the advantages and disadvantages of nuclear power, we must know how a nuclear power plant and its accompanying nuclear fuel cycle work. A nuclear power plant is a highly complex and costly system designed to perform a relatively simple task: to boil water to produce steam that spins a turbine and generates electricity.

What makes nuclear power complex and costly is the use of a controlled nuclear fission reaction (see Figure 2-7, center, p. 34) to provide the heat. The fission reaction takes place in a *reactor*. The most common reactors, called *light-water reactors* (LWRs, see Figure 13-17, p. 314), produce 85% of the world's nuclear-generated electricity (100% in the United States).

The fuel for a reactor is made from uranium ore mined from the earth's crust. After it is mined, the ore must be enriched to increase the concentration of its fissionable uranium-235 by 1% to 5%. The enriched uranium-235 is processed into small pellets of uranium dioxide. Each pellet, about the size of a pencil eraser contains about the same amount of energy as that contained in a ton of coal. Large numbers of these pellets are packed into closed pipes, called *fuel rods*, which are then grouped together in *fuel assemblies*, to be placed in the core of a reactor.

Control rods are moved in and out of the reactor core to absorb neutrons, thereby regulating the rate of fission and amount of power produced. A *coolant*, usually water, circulates through the reactor's core to remove heat, and this keeps the fuel rods and other reactor components from melting and releasing massive amounts of radioactivity into the environment. A modern LWR includes an emergency core cooling system as a backup to help prevent such meltdowns.

A *containment shell* with thick, steel-reinforced concrete walls surrounds the reactor core. It is designed to keep radioactive materials from escaping into the environment in case there is an internal explosion or a melting of the reactor's core. It also protects the core from some external threats such as tornadoes and plane crashes. This is partly why these plants cost so much to build—up to $10 billion for a typical plant today.

In the process of generating electricity, LWRs are highly inefficient, losing about 65% of the high-quality energy available in their nuclear fuel as waste heat to the environment. But even before that point, an equivalent of 9% of the energy content of the fuel has already been lost as heat when the uranium fuel is mined, upgraded, and transported to the plant. At least another 8% is lost in dealing with the radioactive wastes produced by a plant, bringing the net energy loss to about 82%. If we add the enormous amount of energy needed to dismantle a plant at the end of its life and transport and store its high- and moderate-level radioactive materials for thousands of years, some scientists estimate that using nuclear power will eventually require more energy than it will ever produce.

What Is the Nuclear Fuel Cycle?

Running a nuclear power plant is only one part of the **nuclear fuel cycle** (Figure 13-18, p. 315), which also includes the mining of uranium, processing and enriching the uranium to make fuel, using it in the reactor, safely storing the resulting highly radioactive wastes for thousands of years until their radioactivity falls to safe levels, and retiring the highly radioactive plant by taking it apart and storing its high- and moderate-level radioactive material safely for thousands of years.

The final step in the cycle occurs when, after 15–60 years, a reactor comes to the end of its useful life, mostly because of corrosion and radiation damage to its metal parts, and it must be *decommissioned*, or retired. It cannot simply be shut down and abandoned, because its structure contains large quantities of high- and intermediate-level radioactive materials that must be kept out of the environment.

Each step in the nuclear fuel cycle adds to the cost of nuclear power and reduces its net energy (**Concept 13-1**). Proponents of nuclear power tend to focus on the low CO_2 emissions and multiple safety features of the reactors. But in evaluating the safety, economic feasibility, net energy yield, and overall environmental impact of nuclear power, energy experts and economists caution us to look at the entire nuclear fuel cycle, not just the power plant operations. Figure 13-19 (p. 315) lists the major advantages and disadvantages of producing electricity by using the nuclear power fuel cycle (**Concept 13-3**).

Let's look more closely at some of the challenges involved in using nuclear power.

Figure 13-17 Science: In this water-cooled nuclear power plant, water is pumped under high pressure into the reactor core where nuclear fission takes place. The fission reaction produces huge quantities of heat that is used to convert the water to steam, which spins a turbine that generates electricity. Some nuclear plants withdraw the water used for cooling the reactor from a nearby source such as a river and return the heated water to that source, as shown here. Other nuclear plants transfer the waste heat from the intensely hot water to the atmosphere by using one or more gigantic cooling towers, as shown in the inset photo of the Three Mile Island nuclear power plant near Harrisburg, Pennsylvania (USA). There, a serious accident in 1979 almost caused a meltdown of the plant's reactor. **Question:** How do you think the heated water that is returned to its source affects that aquatic ecosystem?

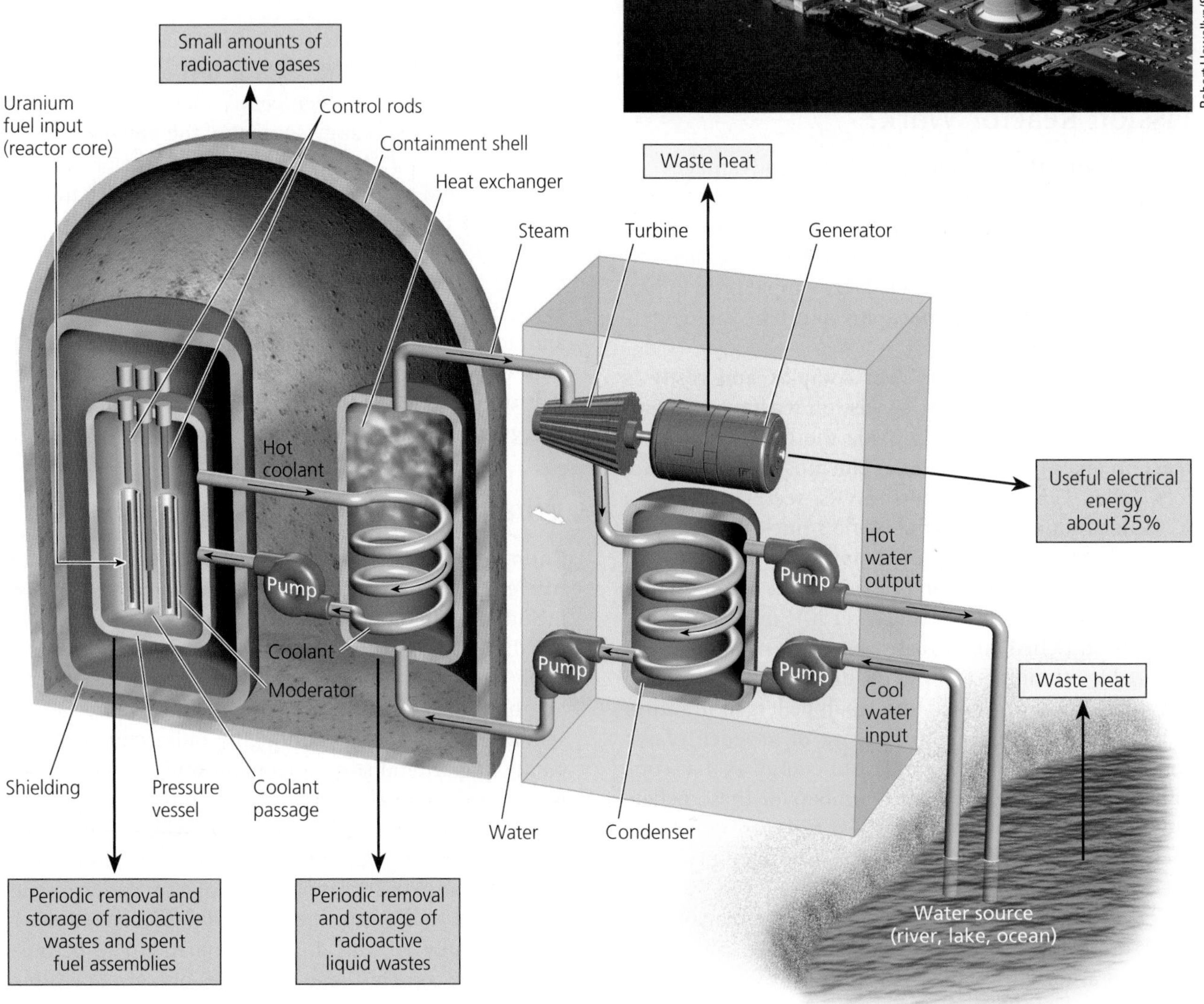

Storing Spent Radioactive Fuel Rods Presents Risks

The high-grade uranium fuel in a nuclear reactor lasts for 3–4 years, after which it becomes *spent*, or useless, and must be replaced. On a regular basis, reactors are shut down for refueling, which usually involves replacing about a third of the reactor's fuel rods that contain the spent fuel.

However, the spent fuel is intensely hot and highly radioactive and cannot simply be thrown away. Researchers have found that even 10 years after being removed from a reactor, a spent-fuel rod assembly can still emit enough radiation to kill a person standing 1 meter (39 inches) away in less than 3 minutes. Thus, after they are removed from reactors, spent fuel-rod assemblies are stored in *water-filled pools* (Figure 13-20, left, p. 316). After about 5 years of cooling, they can be transferred to *dry casks* made of heat-resistant metal alloys and concrete (Figure 13-20, right).

Currently, about 75% of all spent fuel rods in the United States are stored in pools and 25% are in dry casks. A 2005 study by the U.S. National Academy of Sciences warned that the intensely radioactive waste storage pools and dry casks at 68 nuclear power plants in 31 U.S. states are especially vulnerable to sabotage or terrorist attack. Generally, these pools and casks are located outside of reactor core buildings and thus are

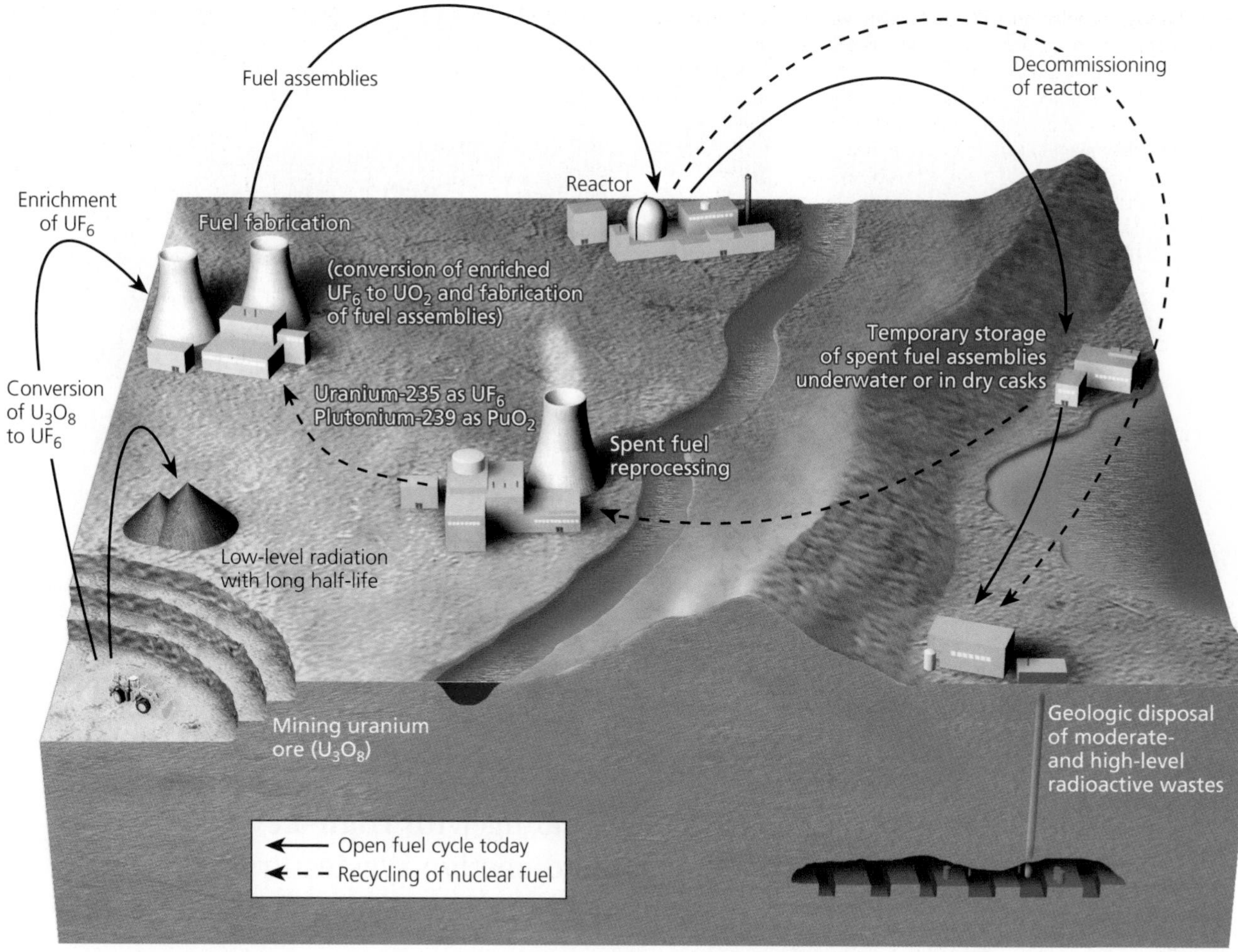

Fuel assemblies

Decommissioning of reactor

Reactor

Enrichment of UF_6

Fuel fabrication

(conversion of enriched UF_6 to UO_2 and fabrication of fuel assemblies)

Temporary storage of spent fuel assemblies underwater or in dry casks

Conversion of U_3O_8 to UF_6

Uranium-235 as UF_6
Plutonium-239 as PuO_2

Spent fuel reprocessing

Low-level radiation with long half-life

Mining uranium ore (U_3O_8)

Geologic disposal of moderate- and high-level radioactive wastes

← Open fuel cycle today
← – – Recycling of nuclear fuel

Figure 13-18 Science: Using nuclear power to produce electricity involves a sequence of steps and technologies that together are called the *nuclear fuel cycle*. As long as a reactor is operating safely, the power plant itself has a fairly low environmental impact and a very low risk of an accident. But considering the entire nuclear fuel cycle, the financial costs are high and the environmental impact and other risks increase. High-level radioactive wastes must be stored safely for thousands of years, several points in the cycle are vulnerable to terrorist attack, and the uranium-enrichment technology used in the cycle can also be used to produce nuclear weapons-grade uranium (**Concept 13-3**). All in all, an amount of energy equal to about 82% of the energy content of the nuclear fuel is wasted in the nuclear fuel cycle. **Question:** Do you think the market price of nuclear-generated electricity should include all the costs of the nuclear fuel cycle or should governments (taxpayers) continue to heavily subsidize nuclear power? Explain.

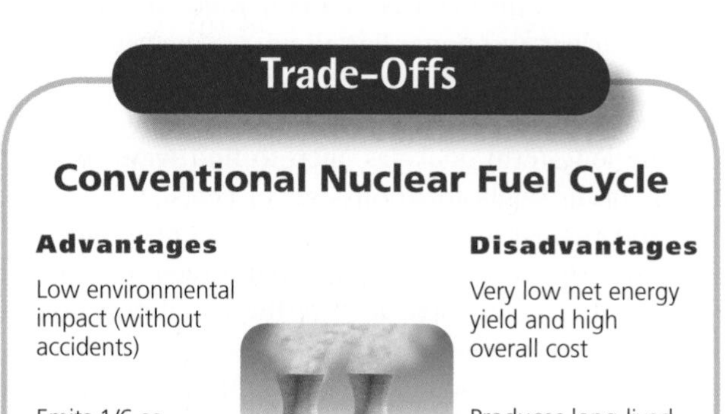

Trade-Offs

Conventional Nuclear Fuel Cycle

Advantages	Disadvantages
Low environmental impact (without accidents)	Very low net energy yield and high overall cost
Emits 1/6 as much CO_2 as coal	Produces long-lived, harmful radioactive wastes
Low risk of accidents in modern plants	Promotes spread of nuclear weapons

Figure 13-19 Using the nuclear power fuel cycle (Figure 13-18) to produce electricity has advantages and disadvantages. **Questions:** Which single advantage and which single disadvantage do you think are the most important? Why? Do you think that the advantages of using the conventional nuclear power fuel cycle to produce electricity outweigh its disadvantages? Explain.

not protected nearly as well as the reactor core is from accidents or acts of terrorism that could release their deadly radioactivity into the environment.

To assess this risk, government security agencies have sent teams of mock terrorists to nuclear plant sites to test the security of the plants. Government records reveal that, between 2005 and 2010, eight of the roughly 100 attempts to breach security at U.S. nuclear plants were successful.

U.S. Department of Energy/Nuclear Regulatory Commission

U.S. Department of Energy/Nuclear Regulatory Commission

Figure 13-20 Science: After 3–4 years in a reactor, spent fuel rods are removed and stored in a deep pool of water contained in a steel-lined concrete basin (left) for cooling. After about 5 years of cooling, the fuel rods can be stored in dry-storage casks (right). **Questions:** Would you be willing to live within a block or two of these casks or have them transported through the area where you live in the event that they were transferred to a long-term storage site? Explain. What are the alternatives?

A 2002 study by the Institute for Resource and Security Studies and the Federation of American Scientists found that in the United States, about 161 million people—53% of the population—live within 121 kilometers (75 miles) of an aboveground spent-fuel storage site. Since that study, the total amount of spent fuel has grown. According to a 2011 study by the U.S. Institute for Policy Studies, the largest concentration of lethal, high-level radioactive materials on the planet is found in spent fuel rods stored at U.S. reactor sites.

The danger is that when the water drains from a storage pool, the rods catch fire and release highly radioactive materials that could escape to the environment. In 2011, this is apparently what happened at the major nuclear accident in Japan (see Case Study, p. 319). This accident showed that storage pools are the most vulnerable and potentially dangerous component of the nuclear fuel cycle.

The Institute for Policy Studies warned in 2011 that the risk from spent nuclear fuel in the United States is much greater than in Japan. Critics have called for requiring that fuel rod storage pools be as well protected as reactor cores, because many of these pools contain more radioactive materials than their reactor cores contain. Critics also call for strictly requiring backup power for emergency core cooling systems and measures to insure that water levels will be maintained in storage pools. Such requirements are not in place in the United States, critics say, because they would add to the already high cost of electricity produced by the nuclear fuel cycle.

Dealing with High-Level Radioactive Wastes Produced by Nuclear Power Is a Difficult Problem

Each part of the nuclear power fuel cycle produces radioactive wastes. *High-level radioactive wastes* consist mainly of spent fuel rods and assemblies from commercial nuclear power plants and dismantled plants, and assorted wastes from the production of nuclear weapons.

The nuclear waste problem begins with spent fuel rods. They can be processed to remove radioactive plutonium, as is done with some of the other radioactive wastes we produce. This reduces the storage time for these wastes from 240,000 to about 10,000 years. But such processing is costly and produces plutonium that terrorists or countries that support them could use to make nuclear weapons. For these reasons the United States abandoned fuel recycling in 1977 after spending billions of dollars.

The long-term goal has been to find ways to store these wastes safely for thousands of years. Most scientists and engineers agree in principle that deep burial in a geologically acceptable underground repository is the safest and cheapest way to store these and other high-level radioactive wastes. However, after almost 60 years of research and evaluation, no country has built and tested such a repository (see the Case Study that follows). Meanwhile these deadly wastes are building up.

Another radioactive waste problem arises when a nuclear power plant reaches the end of its useful life and must be closed. Around the world, 285 of the 441 commercial nuclear reactors now operating will need to be decommissioned by 2025. A worn-out nuclear plant contains highly radioactive fuel rods and huge quantities of moderately radioactive concrete and other materials that must be stored safely. A few plants have been decommissioned and their parts removed to temporary storage facilities.

Eventually, however, all plants will have to be dismantled and their high-level radioactive materials will have to be stored safely for thousands of years. Scientists have proposed three ways to do this. One strategy is to store the highly radioactive parts in a permanent, secure repository, which so far, no country has built and tested. A second approach is to install a physical barrier around the plant and set up full-time security for 30–100 years, until the plant can be dismantled after its radioactivity has reached safer levels. These levels would still be high enough to require long-term safe storage of leftover materials.

A third option is to enclose the entire plant in a concrete and steel-reinforced tomb, called a containment structure. This is what was done with a reactor at Chernobyl, Ukraine that exploded and nearly melted down in 1986, due to a combination of poor reactor design and human operator error (see Case Study, p. 319). The explosion and the radiation released over large areas killed a number of people and contaminated a vast area of land with long-lasting radioactive fallout. However, within a few years, the containment structure began to crumble, due to the corrosive nature of the radiation inside the damaged reactor, and to leak radioactive wastes. The structure is being rebuilt at great cost and is unlikely to last even several hundred years.

Regardless of the method chosen, the high costs of retiring nuclear plants adds to the total costs of the nuclear power fuel cycle and reduces its already low net energy yield. Even if all the nuclear power plants in the world were shut down tomorrow, we would still have to find a way to protect ourselves and thousands of generations to come from the high-level radioactive wastes that have already been produced. This underscores the principle that we should be very careful about committing ourselves to any technology that produces harmful wastes that must be safely stored, essentially forever. However, with nuclear power, we no longer have this option.

■ CASE STUDY
High-Level Radioactive Wastes in the United States

In 1987, the DOE announced plans to build a repository for underground storage of high-level radioactive wastes from U.S. commercial nuclear reactors on federal land in the Yucca Mountain desert region, about 160 kilometers (100 miles) northwest of Las Vegas, Nevada. By 2008, more than $10.4 billion had been spent on preliminary development of the site with $2 billion of these costs paid by the nuclear industry and $8.4 billion paid by taxpayers. In 2008, the total cost of the repository was projected to be $96 billion.

Some scientists argued that the Yucca Mountain site should never be allowed to open, because rock fractures and tiny cracks might allow water to flow through the site. A build-up of hydrogen gas produced from the breakdown of this water by the heat of the stored waste materials could cause an immense explosion.

Proponents of the plan, argue that such worries about water make the area a good site because it is located in an arid region that typically gets less than 20 centimeters (8 inches) of rain per year. Critics argue that this could change during the next 10,000 to 240,000 years, especially with projected climate change. But proponents point to scientific studies indicating that the area's water table has not changed significantly during the past 100,000 years.

Another concern of critics was possible leakage of radioactive material into groundwater. No one has been able to establish that the storage containers planned for the site will last more than 1,000 years, and they would need to be leak-free for at least 10,000 years. Leaks could lead to radioactive contamination of drinking water wells, groundwater, and surface streams and wetlands.

Finally, critics have been concerned about the fact that nuclear waste would have to be transported across the country from at least 120 different storage sites. This would amount to thousands of shipments of dry waste containers through parts of almost every one of the lower 48 states during a period of many decades. Critics argue that this would create countless opportunities for accidents, sabotage, or theft of radioactive materials by terrorists.

There would also be political protests about such wastes passing through various areas of the country. However, the federal government has the authority to ship such wastes without getting the approval of state or local governments. In 2009, the president of the United States requested that the U.S. Congress cut off funding for the Yucca Mountain project while other, shorter-term alternatives are evaluated.

THINKING ABOUT
Shipment of Nuclear Waste

Would you oppose having high-level radioactive wastes shipped through an area near your community? Explain.

Can Nuclear Power Lessen Dependence on Imported Oil and Help Reduce Projected Global Warming?

Some proponents of nuclear power in the United States claim it will help reduce the U.S. dependence on imported crude oil. Other analysts argue that it will not do so because oil-burning power plants provide only about 1% of the electricity produced in the United States.

Critics also point out that increased use of nuclear power in the United States will make the country dependent on imports of uranium needed to fuel the plants. Currently, about 95% of the uranium used to fuel U.S. nuclear plants is imported, mostly from Russia. And China is buying up access to many of the world's uranium suppliers to support its projected expansion of nuclear power.

Nuclear power advocates also contend that increased use of nuclear power will greatly reduce or eliminate CO_2 emissions and, in turn, reduce the projected threat of climate change caused by atmospheric warming. Critics argue that the nuclear power industry has mounted a misleading but effective public relations campaign to convince the public that nuclear power does not involve emissions of CO_2 and other greenhouse gases.

Scientists point out that this argument is only partially correct. While nuclear plants are operating, they do not emit CO_2. However, during the 10 years that it typically takes to build a plant, especially in the manufacturing of many tons of construction cement, large amounts of CO_2 are emitted. In addition, every other step in the nuclear power fuel cycle (Figure 13-18) involves CO_2 emissions. Such emissions are much lower than those from coal-burning power plants (Figure 13-15) but they still contribute to projected atmospheric warming and climate change. In other words, nuclear power is not a carbon-free source of energy.

In 2009, Michael Mariotte, Executive Director of the Nuclear Information and Resource Service, estimated that in order for nuclear power to play an effective role in slowing projected climate change over the next 50 years, the world would need to build some 2,000 nuclear reactors. This would require building about 40 nuclear reactors each year during this period, or about one every 9 days. Typically, it takes a decade or more to build a nuclear power plant at a cost of up to $10 billion. Energy expert Joseph Romm estimates that we would need to build ten new long-term storage repositories, such as the one that was planned for Yucca Mountain (Case Study, p. 317) to store the resulting wastes.

Nuclear Power Is Not Expanding Very Rapidly

In the 1950s, researchers predicted that by the year 2000, at least 1,800 nuclear power plants would supply 21% of the world's commercial energy (25% in the United States) and most of the world's electricity. After almost 60 years of development, a huge financial investment, and enormous government subsidies, some 441 commercial nuclear reactors in 31 countries produced only 6% of the world's commercial energy and 14% of its electricity. In the United States, 104 licensed commercial nuclear power reactors in 31 states generate about 8% of the country's overall energy and 20% of its electricity.

In 2010, 63 new reactors were under construction globally, far from the number needed just to replace the reactors that will have to be decommissioned in coming years. Another 143 reactors are planned but even if they are completed after a decade or two, they will not replace the 285 aging reactors that must be retired around the world. This helps explain why nuclear power is now the world's slowest-growing form of commercial energy (see Figure 10, p. S42, in Supplement 7).

Experts Disagree about the Future of Nuclear Power

The future of nuclear power is a subject of debate. Critics argue that the nuclear power industry could not exist without support from governments and taxpayers, because of the extraordinarily high costs and the low net energy yield of the nuclear fuel cycle. For example, the U.S. government has provided huge subsidies, tax breaks, and loan guarantees to the industry. It also provides accident insurance guarantees, because insurance companies have refused to fully insure any nuclear reactor. Without these large taxpayer-subsidized payments, this industry would not exist in the United States or anywhere else.

Another obstacle to the growth of nuclear power has been public concerns about the safety of nuclear reactors. Because of the multiple built-in safety features, the risk of exposure to radioactivity from nuclear power plants in the United States and most other more-developed countries is extremely low. However, several explosions and partial or complete meltdowns have occurred (see the Case Study that follows). Still another serious safety concern related to commercial nuclear power is the spread of nuclear weapons technology around the world.

CONNECTIONS

Nuclear Power Plants and the Spread of Nuclear Weapons

The United States and 14 other countries have been selling commercial and experimental nuclear reactors and uranium fuel-enrichment and purification technology in the international marketplace for decades. Much of this information and equipment can be used to produce bomb-grade material for use in nuclear weapons. Energy expert John Holdren, pointed out that, with the exception of the United States, Great Britain, and the former Soviet Union, the 60 countries that have nuclear weapons or the knowledge to develop them have gained most of such information by using civilian nuclear power technology. Some critics see that as the single most important reason for not building more nuclear power plants throughout the world.

Proponents of nuclear power argue that governments should continue funding research, development, and pilot-plant testing of potentially safer and less expensive second-generation reactors. Between 2000 and 2010, the U.S. nuclear industry spent $645 million lobbying Congress and the White House to get them to support further development of nuclear power and their efforts paid off in the form of government financial support and loan guarantees. Even so, most utility companies and money lenders are unlikely to take on the huge financial risk of building new nuclear plants as long as natural gas and coal provide much cheaper ways to produce electricity.

The nuclear industry claims that hundreds of new *advanced light-water reactors (ALWRs)* could be built in just a few years. ALWRs have built-in safety features designed to make explosions and releases of radioactive emissions almost impossible. Also, some scientists call for replacing today's uranium-based reactors with new ones based on the element thorium. They argue that such reactors would be much less costly and safer, and would cut the amount of nuclear waste generated in half.

China, which is building more nuclear power plants than the rest of the world combined, is building some reactors with advanced designs along with one thorium-based reactor. However, in 2009, Great Britain's chief of Health and Safety said that he could not recommend plans for building nuclear plants using the new designs because of significant safety issues that need to be resolved.

Implementing new designs could require decades of research, according to some scientists. To be environmentally and economically acceptable, these analysts believe that any new-generation nuclear technology should meet the five criteria listed in Figure 13-21.

Solutions

- Reactors must be built so that a runaway chain reaction is impossible.
- The reactor fuel and methods of fuel enrichment and fuel reprocessing must be such that they cannot be used to make nuclear weapons.
- Spent fuel and dismantled structures must be easy to dispose of without burdening future generations with harmful radioactive waste.
- Taking its entire fuel cycle into account, nuclear power must generate a net energy yield high enough so that it does not need government subsidies, tax breaks, or loan guarantees to compete in the open marketplace.
- Its entire fuel cycle must generate fewer greenhouse gas emissions than other energy alternatives.

Figure 13-21 Some critics of nuclear power say that any new generation of nuclear power plants should meet all of these criteria. So far, no existing or proposed reactor even comes close to doing so. **Question:** Do you agree that any new nuclear power plant proposals should meet these five criteria? Explain.

According to World Bank economists, conventional and proposed new-generation nuclear fission power plants will not compete in today's energy market unless they are shielded from open-market competition by government subsidies and tax breaks. This is because the technology is extremely expensive and the net energy yield for the nuclear fuel cycle is very low. The nuclear power industry downplays such claims by focusing on the costs of operating a nuclear plant instead of on the much higher costs of the entire nuclear fuel cycle. To critics such as energy expert Amory Lovins, electricity from nuclear power, touted in the 1950s as being "too cheap to meter," has now become "too expensive to matter."

Other proponents of nuclear power hope to develop **nuclear fusion**—a nuclear change at the atomic level in which the nuclei of two isotopes of a light element such as hydrogen are forced together at extremely high temperatures until they fuse to form a heavier nucleus, releasing energy in the process (see Figure 2-7, bottom, p. 34). Some scientists hope that controlled nuclear fusion will provide an almost limitless source of energy.

With nuclear fusion, there would be no risk of a meltdown or of a release of large amounts of radioactive materials, and little risk of the additional spread of nuclear weapons. In addition to generating electricity, fusion power could be used to destroy hazardous wastes, and it could have many other uses.

However, in the United States, after more than 50 years of research and a $25 billion investment, controlled nuclear fusion is still in the laboratory stage. None of the approaches tested so far has produced more energy than they use. Unless there is some unexpected scientific breakthrough, some skeptics will continue to quip that "nuclear fusion is the power of the future and always will be."

■ **CASE STUDY**

The Three Worst Nuclear Power Plant Accidents

The world has experienced three major nuclear power plant accidents—one in the United States, one in the former Soviet Union, and one in Japan.

The first of these accidents involved the partial meltdown of a reactor core at the Three Mile Island nuclear plant in the U.S. state of Pennsylvania in 1979 (see photo in Figure 13-17). No mandatory evacuation was ordered, but within a few days, 140,000 people had voluntarily left the area. The accident caused no known deaths among plant workers or the general public. The cleanup cost the plant owners about $1 billion—about 2.5 times the cost of constructing the reactor.

Several years of investigation revealed that the accident involved a loss of coolant water due to equipment failure stemming from a violation of a U.S. Nuclear Regulatory Commission (NRC) operating rule. Another problem was that the plant operators were not trained well enough to identify and understand what was happening. The accident led to a tightening of NRC regulations. However, critics say that some of the regulations and their enforcement are still inadequate, especially for spent fuel rod storage. They also criticize the almost automatic 20-year extensions of the operating licenses that by 2010, had been granted for 63 aging U.S. nuclear reactors whose designs are now outdated, including 23 reactors like the ones involved in Japan's 2011 accident.

The world's worst nuclear power plant accident occurred in 1986 at the Chernobyl nuclear power plant in Ukraine. Two explosions in one of the reactors blew the roof off of the reactor building and partially melted the reactor core. The resulting fires burned for 10 days and released large amounts of radiation into the atmosphere that spread to several countries. Some 350,000 people living within a 19-mile radius of the plant were eventually evacuated from the area.

The Chernobyl death toll may never be known, but investigators have estimated that at least 6,000 workers and cleanup personnel died from exposure to radiation released by the accident. Projections of long-term deaths, mostly from cancers and other health effects caused by radiation exposure, range from 9,000 to 212,000. Investigations revealed that the accident was caused by a combination of poor reactor design, inadequate operating regulations, and human error. Fortunately, the reactor design is not used in most other countries.

The third major accident occurred on March 11, 2011 at the Fukushima Daiichi Nuclear Power Plant in northeast Japan. The accident, which damaged all six of the plant's reactors, was triggered by a major 8.9 offshore earthquake that caused a severe tsunami (see Figure 12-8, p. 281). A huge wave of seawater washed over the nuclear plant's protective seawalls and knocked out the circuits and backup generators of its emergency core cooling system. Then, explosions (presumably from the build up of hydrogen gas) blew the roofs off of three of the reactor buildings (Figure 13-22) and released radioactivity into the atmosphere and nearby coastal waters. Evidence indicates that these three reactors suffered partial meltdowns of their cores.

After some initial confusion and conflicting statements about the severity of the accident, the Japanese government implemented mandatory evacuation of all residents within a 10-kilometer (6-mile) radius of the plant. Later, as the severity of the accident became more apparent, this zone was extended to 31 kilometers (19 miles) and U.S. nuclear experts urged extending it to 80 kilometers (50 miles).

Experts have estimated that as many as 50 of the workers involved in dealing with the accident will die from excessive exposure to radioactivity. However, the reasonably quick evacuation of a fairly large zone around the plant should reduce numbers of deaths, cancers, and other health effects of radiation exposure among the public. In 2011, it was not known how long residents would be prevented from returning to their homes. Officials were projecting that it would take at least a year to bring the reactors under control and to clean up the radioactivity in and around the plant, but some experts believed that to be overly optimistic.

After preliminary investigations in 2011, it seemed that one of the most important lessons to be learned from this accident was about the danger of storing intensely radioactive spent fuel rods. The Fukushima plant rods were stored in pools (Figure 13-20, left) above the reactors, and the most serious radiation emissions probably resulted from the loss of water in these pools. Exposed to the air, the spent fuel rods caught fire and released clouds of radioactive materials.

AP Photo/AIR PHOTO SERVICE, file

Figure 13-22 Damage from an explosion in Unit 4 of the crippled Fukushima Daiichi nuclear power plant. This photo was taken on March 24, 2011, by a small unmanned aircraft.

It will take years to sort out the details of this accident and to evaluate the factors that contributed to it. Preliminary explanations pointed to **(1)** the failure of the utility company to develop worst-case scenarios that would have helped speed up their reaction to the crisis, **(2)** the fact that the plant's protective seawalls were not built high enough to withstand the tsunami waves, **(3)** design flaws that exposed the emergency core cooling system controls and backup generators to flooding and that failed to protect the spent fuel rod storage pools from the damages they suffered, and **(4)** a too cozy relationship between nuclear plant owners and the government's nuclear regulatory officials.

THINKING ABOUT
Government Subsidies for Nuclear Power

Do you think the benefits of nuclear power justify high government subsidies for the nuclear industry? Explain.

13-4 Why Is Energy Efficiency an Important Energy Resource?

CONCEPT 13-4 Improving energy efficiency could save the world at least a third of the energy it uses, and it could save the United States up to 43% of the energy it uses.

We Waste Huge Amounts of Energy

Many analysts urge us to increase our supply of energy and to save money by reducing our unnecessary waste of energy. The best way to do this is to improve **energy efficiency**: the measure of how much work we can get from each unit of energy we use. Every unit of energy we save eliminates the need to produce that energy, and thus saves us money.

You may be surprised to learn that roughly 84% of all commercial energy used in the United States is wasted (Figure 13-23). About 41% of this energy is unavoidably lost because of the degradation of energy quality imposed by the second law of thermodynamics (see Chapter 2, p. 36). The other 43% is wasted unnecessarily, mostly due to the inefficiency of incandescent lightbulbs, industrial motors, most motor vehicles, coal and nuclear power plants, and numerous other energy-consuming devices.

Another reason for this waste is that many people live and work in leaky, poorly insulated, and badly designed buildings that require excessive heating in the winter and cooling in the summer. Unnecessary energy waste costs the United States an average of about $570,000 per minute, according to energy analyst Amory Lovins (see his Guest Essay at **www.cengagebrain.com**). Lovins estimates that in the United States, "we could save at least half the oil and gas and three-fourths of the electricity we use at a cost of only about an eighth of what we're now paying for these forms of energy."

For years, many Americans have been buying larger, gas-guzzling vehicles and building larger houses that require more energy and cost more to heat and cool. Many live in ever-expanding suburban areas that surround most cities, and they must depend on their cars for getting around. Roughly three of every four Americans commute to work, mostly in energy-inefficient vehicles, and only 5% rely on more energy-efficient mass transit, which it is not available in most spread-out suburban areas.

Reducing energy waste has numerous economic and environmental advantages (Figure 13-24, p. 322). To most energy analysts, *reducing energy* GOOD NEWS

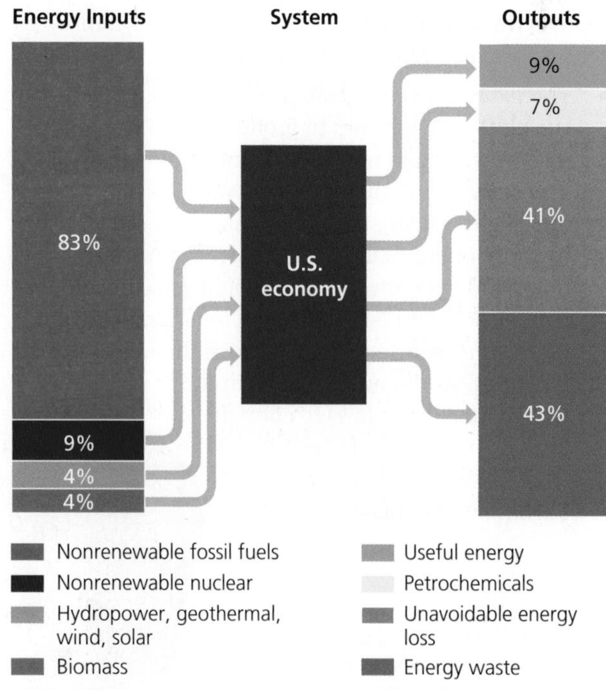

Figure 13-23 This diagram shows how commercial energy flows through the U.S. economy. Only 16% of all commercial energy used in the United States ends up performing useful tasks. **Question:** What are two examples of unnecessary energy waste? (Data from U.S. Department of Energy)

Solutions

Reducing Energy Waste

- Prolongs fossil fuel supplies
- Reduces oil imports and improves energy security

- Has very high net energy yield
- Saves a lot of money

- Reduces pollution and environmental degradation
- Buys time to phase in renewable energy

- Creates local jobs

Figure 13-24 Reducing unnecessary energy waste and thereby improving energy efficiency has several benefits. **Questions:** Which two of these benefits do you think are the most important? Why?

waste is the quickest, cleanest, and usually the cheapest way to provide more energy, reduce pollution and environmental degradation, and slow projected climate change.

We waste large amounts of energy and money by relying heavily on four devices:

- The *incandescent light bulb* uses only about 5% of the electricity it draws to produce light. The other 95% is wasted as heat, which means it is really a *heat bulb*. This highly wasteful way to produce light is being phased out and replaced by more energy-efficient compact fluorescent bulbs and even more efficient light-emitting diodes (LEDs) (Figure 13-25). Shifting to energy-efficient fluorescent lighting and LEDs in all homes, office buildings, stores, and factories in the United States would save enough energy to close all of the country's 600 large coal-burning power plants.

Figure 13-25 LEDs last so long (at least 30,000 hours) that users can install them and forget about them for up to 20 years. They contain no mercury, turn on instantly, and many models are dimmable. LED bulbs are expensive but prices are dropping because of newer designs and mass production.

Dennis Steen/Shutterstock.com

- The *internal combustion engine*, which propels most motor vehicles and wastes about 80% of the energy in its fuel.
- A nuclear *power plant* (Figure 13-17), just in generating electricity, wastes about 65% of the energy in its nuclear fuel and probably closer to 82% when we include the additional energy used in the nuclear fuel cycle (Figure 13-18).
- A *coal-fired power plant* (Figure 13-13) wastes about 65% of the energy that is released by burning coal to produce electricity, and probably 75–80% if we include the energy used to dig up the coal and transport it to the plant, as well as to transport and store the toxic ash byproduct.

We Can Save Energy and Money in Industry

Industry accounts for about 30% of the world's energy consumption and 33% of U.S. energy consumption, mostly for the production of metals, chemicals, petrochemicals, cement, and paper. There are many ways for industries to cut energy waste.

Some industries save energy and money by using *cogeneration*, which involves using a *combined heat and power (CHP)* system. In such a system, two useful forms of energy (such as steam and electricity) are produced from the same fuel source. For example, the steam produced in generating electricity in a CHP system can be used to heat the power plant or other nearby buildings, rather than released into the environment and wasted. The energy efficiency of these systems is 75–90% compared to 30–40% for coal-fired boilers and nuclear power plants.

Another way to save energy and money in industry is to *replace energy-wasting electric motors*, which use one-fourth of the electricity produced in the United States and 65% of the electricity used in U.S. industry. Most of these motors are inefficient because they run only at full speed with their output throttled to match the task—somewhat like keeping one foot on the gas pedal of your car and the other on the brake pedal to control its speed. Replacing them with variable speed motors, which run at the minimum rate needed for each job, saves energy and reduces the environmental impact of electric motor use.

Recycling materials such as steel and other metals is a third way for industry to save energy and money. For example, it takes 75% less energy to produce steel from recycled scrap iron than it takes to produce it from virgin iron ore and emits 40% less CO_2. Likewise, making aluminum products from recycled aluminum cans and other items requires much less energy than using virgin aluminum.

A fourth way for industry (and homeowners) to save energy and money is to *switch from low-efficiency incandescent lighting* to higher-efficiency fluorescent lighting and LEDs (Figure 13-25).

A growing number of major corporations are saving a lot of money by cutting their energy waste. For example, the CEO of Dow Chemical Company, which operates 165 manufacturing plants in 37 countries, estimates that between 1996 and 2006, energy efficiency improvements cost the company about $1 billion, but resulted in savings of about $8.6 billion.

There is also a great deal of energy waste in the generation and transmission of electricity to industries and communities (see the Case Study that follows). Some large U.S. energy utility companies are realizing this and taking steps to save money and to increase their profits. For example, Excelon—the largest owner of nuclear power plants in the United States—plans to invest in high-voltage transmission lines, smart meter systems, and improvements in the efficiency and capacity of its nuclear power plants. Edison International, another major utility, plans to replace some of its older coal-burning power plants with cleaner natural gas plants and renewable sources of electricity.

Another problem is that utility companies have historically made their profits by promoting electricity use instead of efficiency. However, several U.S. state utility commissions now reward utilities for the kilowatts of electricity their customers save through efficiency improvements, rather than rewarding the utilities for selling more kilowatts. In 2009, the American Council for an Energy Efficient Economy estimated that by making energy efficiency the focus of national policy, the United States could eliminate the need for building 450 new coal-fired or nuclear power plants.

GOOD NEWS

■ **CASE STUDY**

Saving Energy and Money with a Smarter Electrical Grid

Grid systems of high-voltage transmission lines carry electricity from power plants, wind farms (**Core Case Study**), and other electricity producers to users. In the United States, many energy experts place top priority on converting and expanding the outdated electrical grid system into what they call a *smart grid*. This digitally controlled, ultra-high-voltage (UHV) grid with superefficient transmission lines would be responsive to local and regional changes in demand and supply.

CORE CASE STUDY

Such a system would use smart meters to monitor the amount of electricity used and the patterns of use for each customer. It would then use this information to deliver electricity as efficiently as possible.

Smart meters would also show consumers how much energy they are using by the minute and for each appliance. This information would help them to reduce their power consumption and electricity bills. Smart appliances such as clothes washers and dryers could be programmed to perform their tasks during off-peak hours when electricity is cheaper. A smart grid could also allow people to run their air conditioners remotely so that they could turn them off when they leave home and turn them on before they return.

China plans to build an efficient and reliable UHV grid by 2020 and to become the global leader in manufacturing and selling such technology and equipment. According to the U.S. Department of Energy, building such a grid would cost the United States from $200 billion to $800 billion, but would pay for itself in a few years by saving the U.S. economy more than $100 billion a year.

THINKING ABOUT
A Smart Grid

Can you think of any drawbacks to building a smart grid in the United States or in the country where you live? Do you think the drawbacks outweigh the benefits of such a grid? Explain.

We Can Save Energy and Money in Transportation

There is a lot of room for reducing energy waste in transportation. During parts of 1973 and 1974, OPEC banned oil exports to the United States because of its support for Israel in that country's 18-day war with Egypt and Syria. The resulting sharp rise in oil prices led to gasoline shortages and to double-digit inflation in the United States. This led the U.S. government to impose higher fuel efficiency standards for new vehicles sold in the United States beginning in 1978.

By 1986, these standards had greatly increased the average fuel efficiency of new cars and light trucks (including sports utility vehicles or SUVs) sold in he United States. However, between 1986 and 2008, government fuel efficiency standards for new cars and light trucks (including SUVs) were not raised. This, along with greatly increased sales of light trucks and SUVs, lead to a decline in overall fuel efficiency in the United States between 1985 and 2005. In 2008, the U.S. Congress passed a law requiring new motor vehicles to have an average combined fuel efficiency of 15 kilometers per liter, or kpl (35 miles per gallon, or mpg) by 2016. This has led to a slight increase in the fuel efficiency of new vehicles produced since 2005.

Fuel economy standards for new vehicles in Europe, Japan, China, and Canada are much higher than are those in the United States (Figure 13-26, p. 324). The 2008 law raising U.S. fuel economy standards to 15 kpl (35 mpg), to be attained by 2016, will result in standards that are lower than current standards in China and in many other countries. China plans to achieve an average fuel economy of 17 kpl (40 mpg) by 2015 and 22 kpl (52 mpg) by 2020. Energy experts such as Joseph Romm call for the U.S. government to require all new cars sold in the United States to get at least 43 kpl (100 mpg) by 2040.

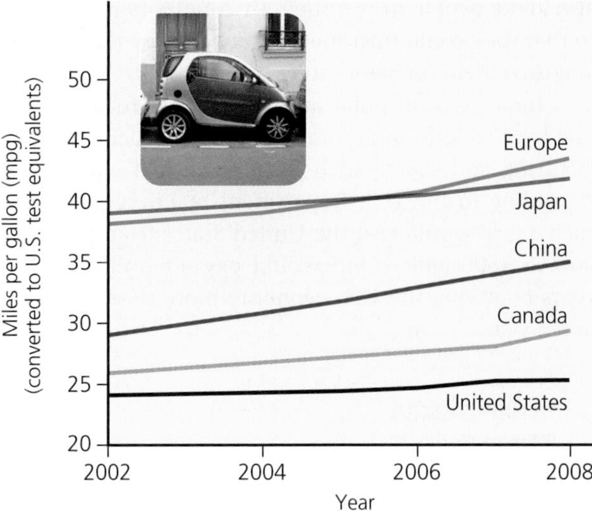

Figure 13-26 This graph shows a comparison of the changes in fuel economy standards for new motor vehicles (cars and light trucks) in various countries between 2002 and 2008. (Data from U.S. Environmental Protection Agency, National Highway Traffic Safety Administration, and International Council on Clean Transportation)

One reason that many Americans buy large and inefficient motor vehicles is that most U.S. consumers do not realize that gasoline costs them much more than the price they pay at the pump. In 2005, according to a study by the International Center for Technology Assessment, the hidden costs of gasoline for U.S. consumers were about $3.18 per liter ($12 per gallon).

These hidden costs include government subsidies and tax breaks for oil companies, car manufacturers, and road builders; costs of pollution control and cleanup; costs of military protection of oil supplies in the Middle East; time wasted idling in traffic jams; and costs of illness from air and water pollution in the form of higher medical bills and health insurance premiums. Consumers pay these hidden costs, but not at the gas pump.

One way to include more of the real cost of gasoline in its market price is through gasoline taxes, which are widely used in Europe but are politically unpopular in the United States. Some analysts call for raising U.S. gasoline taxes and reducing payroll and income taxes to balance such increases, thereby relieving consumers of any additional financial burden. So far, oil and car companies have been able to prevent such a solution in the United States by persuading elected representatives to keep gasoline taxes low, thereby keeping the true costs of gasoline hidden from consumers and providing no incentive to improve fuel efficiency.

THINKING ABOUT
The Real Cost of Gasoline

Do you think that the estimated hidden costs of gasoline should be included in its price at the pump? Explain. Would you favor much higher gasoline taxes if payroll taxes were eliminated or sharply reduced? Explain.

Another way for governments to encourage higher efficiency in transportation is to give consumers tax breaks or other economic incentives to encourage them to buy more fuel-efficient vehicles. Energy expert Amory Lovins has proposed a *fee-bate* program in which buyers of inefficient vehicles would pay a high fee, and the resulting revenues would be given to buyers of fuel-efficient vehicles as rebates. For example, the fee on a gas-guzzling vehicle averaging about 5 kpl (12 mpg) might be $10,000. The government would then give that amount as a rebate to the buyer of a fuel-efficient car that averages 20 kpl (46 mpg) or more.

Within a short time, such a program—endorsed by the U.S. National Academy of Sciences—would greatly increase sales of gas-sipping vehicles. It would also focus carmakers on producing and making their profits from such energy-efficient vehicles, and it would cost the government (taxpayers) very little. So far, the U.S. Congress has not implemented such a program.

More Energy-Efficient Vehicles Are on the Way

GOOD NEWS

There is growing interest in developing modern, superefficient, ultralight, and ultrastrong cars that could get up to 130 kpl (300 mpg) using existing technology (see Amory Lovins's Guest Essay at **www.cengagebrain.com**).

One of these vehicles is the energy-efficient, gasoline–electric *hybrid car* (Figure 13-27, left). It has a small gasoline-powered engine and a battery-powered electric motor that provides the energy needed for acceleration and hill climbing. The most efficient models of these cars, such as the 2011 Toyota Prius, get a combined city/highway mileage of up to 22 kpl (51 mpg) and emit about 65% less CO_2 per kilometer driven than a comparable conventional car emits.

A newer option is the *plug-in hybrid electric vehicle*—a hybrid with a second and more powerful battery that can be plugged into an electrical outlet and recharged (Figure 13-27, right). By running primarily on electricity, plug-in hybrids can easily get the equivalent of at least 43 kpl (100 mpg) for ordinary driving and up to 430 kpl (1,000 mpg), if used only for trips of less than 64 kilometers (40 miles) before recharging. The number of plug-in hybrid vehicles in the marketplace is growing rapidly. China plans to be producing 3 million gas-electric hybrid and plug-in hybrid vehicles per year by 2020 and to lead in global sales of these vehicles.

GOOD NEWS

According to a 2006 DOE study, replacing most of the current U.S. vehicle fleet with plug-in hybrid vehicles over 2 decades, would cut U.S. oil consumption by 70–90%, eliminate the need for costly oil imports, save consumers money, and reduce CO_2 emissions by 27%. If the batteries in these cars were recharged mostly by electricity generated by renewable resources such as wind (**Core Case Study**), U.S. emissions of CO_2 would drop by 80–90%,

CORE CASE STUDY

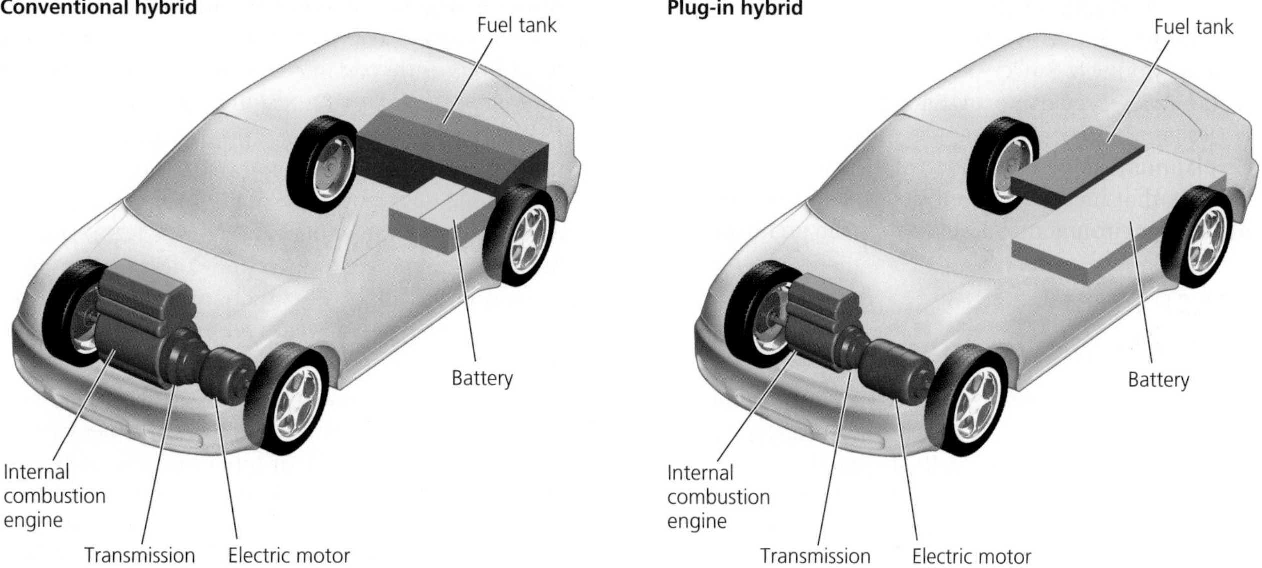

Figure 13-27 Solutions: A *conventional gasoline–electric hybrid* vehicle (left) has a small internal combustion engine and a battery. A *plug-in hybrid* vehicle (right) has a smaller internal combustion engine with a second and more powerful battery that can be plugged into a 110-volt or 220-volt outlet and recharged. This allows it to run farther on electricity alone and the gasoline engine can recharge the battery as needed on the road.

which would help to slow projected climate change. **GREEN CAREER:** plug-in hybrid car and bus technology

The next stage in the development of superefficient cars may be an electric vehicle that uses a *fuel cell*—a device that uses hydrogen gas (H_2) as a fuel to produce electricity. Fuel cells are at least twice as efficient as internal combustion engines, have no moving parts, and require little maintenance. This would essentially eliminate emissions of CO_2 and most air pollutants if the hydrogen were produced with the use of electricity from low-carbon renewable sources such as wind and solar energy. But such cars are unlikely to be widely available until 2020 or later and will probably be very expensive, in part because they have negative net energy ratios. **GREEN CAREER:** fuel-cell technology

Reducing a car's weight can increase its gas mileage. The fuel efficiency for all types of cars could nearly double if car bodies were to be made of ultralight and ultrastrong composite materials such as fiberglass and the carbon-fiber and hemp-fiber composites used in bicycle helmets and in some racing cars. Actually, in 1941, industrialist Henry Ford unveiled a light car body made of hemp fibers, and he believed that the future of the car industry was in all-electric vehicles. Now, 70 years, later we are returning to his vision.

Other ways to save energy in transportation include shifting from diesel-powered to electrified rail systems, building accessible mass transit systems within cities, and constructing high-speed rail lines between cities as is done in Japan, much of Europe (see Chapter 6, p. 114), and more recently in China. Another approach is to encourage bicycle use by designating bike lanes on highways and city streets. Also, companies are now saving money by using video conferencing as an alternative to flying their employees to meetings.

We Can Design Buildings That Save Energy and Money

According to a 2007 UN study, changes in the design and construction of buildings could save 30–40% of the energy used globally. GOOD NEWS For example, orienting a building to face the sun so it can get more of its heat from solar energy can save up to 20% of heating costs and as much as 75% of such costs when the building is well insulated and airtight (see the Case Study that follows). Orienting and designing buildings to get much of their heat from the sun is a simple application of the solar energy **principle of sustainability** that people have been using for centuries. SUSTAINABILITY

Green architecture, based on energy-efficient and money-saving designs, makes use of natural lighting, passive solar heating, solar cells, solar hot water heaters, recycled wastewater, and energy-efficient appliances and lighting. Some designs also include *living roofs,* or *green roofs,* covered with soil and vegetation. Others use white or light-colored roofs that help reduce cooling costs by reflecting incoming solar radiation, especially in hotter climates—another strategy for working with nature that people have used for centuries.

Superinsulation is very important in energy-efficient design (see the Case Study that follows). Superinsulated houses in Sweden use 90% less energy for heating and cooling than that used by typical American homes of the same size. Another example of a superinsulated house is one with thick walls of straw bales that are covered on the inside and outside with adobe (see Photos 9 and 10 in the Detailed Contents). (See the Guest Essay about straw-bale construction and solar energy houses by Nancy Wicks at **www.cengagebrain.com**.)

Green building certification standards now exist in 21 countries, thanks to the efforts of the World Green Building Council. Since 1999, the U.S. Green Building Council's Leadership in Energy and Environmental Design (LEED) program has awarded silver, gold, and platinum standard certificates to nearly 9,000 U.S. buildings that meet certain efficiency standards. **GREEN CAREERS:** environmental design and green architecture

■ CASE STUDY
The Rocky Mountain Institute

In 1984, energy analyst Amory B. Lovins completed construction of a large, solar-heated, solar-powered, superinsulated, partially earth-sheltered home and office in Snowmass, Colorado (USA), where winters are extremely cold. Lovins is co-founder and head of Rocky Mountain Institute (RMI), a nonpartisan group of scientists and analysts who do research and consulting on energy efficiency and renewable energy alternatives.

GOOD NEWS

This home and office has no conventional heating system. Instead, it makes use of energy from the sun, heavy roof insulation, thick stonewalls, energy-efficient windows, and a waste-heat recovery system. Solar energy provides this building with 99% of its heat and hot water, 95% of its daytime lighting, and 90% of its household electricity. The building's heating bill is less than $50 a year. This house is so heavily insulated and airtight that heat from direct sunlight, appliances, and human bodies warm it with little or no need for a backup heating system, even though winters in Colorado are very cold.

Excluding power for office equipment, the structure draws a little more electricity than a single 100-watt incandescent light bulb uses. This is accomplished through the use of energy-efficient lights, appliances, computers, and other electrical devices, and solar cells that generate electricity. The savings from using these devices repaid their costs in only 10 months.

The RMI building is designed to work with nature. It is oriented to collect as much sunlight as possible. It contains a central greenhouse with a variety of plants, which humidifies the building and helps to heat it and purify its air.

THINKING ABOUT
Efficient Building Design

Can you think of ways in which the building in which you live or work could have been designed to be more energy-efficient?

We Can Save Money and Energy in Existing Buildings

GOOD NEWS

There are many ways to save energy and money in existing buildings. A good first step is to have an expert make an *energy audit* of a house or other building to suggest ways to improve energy efficiency and save money. Some utility companies provide such audits at no or little cost. Such a survey might result in some or all of the following recommendations:

- *Insulate the building and plug air leaks.* About one-third of the heated air in typical U.S. homes and other buildings escapes through holes, cracks, and closed, single-pane windows (Figure 13-28). During hot weather, these windows and cracks let heat in, increasing the use of air conditioning. Adding insulation and plugging air leaks are two of the quickest and best ways to improve energy efficiency.
- *Use energy-efficient windows.* Replacing leaky windows with energy-efficient windows can cut expensive heat losses from a house or other building by two-thirds, while lowering cooling costs in the summer and reducing heating system CO_2 emissions. Window manufacturers now offer triple-pane windows that can pay for themselves within several years.
- *Heat houses more efficiently.* In order, the most energy-efficient ways to heat indoor space are: superinsulation (which would include plugging leaks); a geothermal heat pump that transfers heat stored in the earth to a home (discussed later in this chapter); passive solar heating; a high-efficiency, conventional heat pump (in warm climates only); small, cogenerating micro-turbines fueled by natural gas; and a high-efficiency natural gas furnace.
- *Heat water more efficiently.* One approach is to use a roof-mounted solar hot water heater. These are widely used in China, Israel, and a number of other countries. Another option is a *tankless instant water heater* fired by natural gas or LPG. (Using electricity for heating water is not efficient.) These devices, widely used in many parts of Europe, heat water instantly, providing hot water only when, and for as long as, it is needed.
- *Use energy-efficient appliances.* According to the EPA, if all U.S. households were to use the most efficient frost-free refrigerators available, 18 large coal or nuclear power plants could close. Microwave ovens use 25–50% less electricity than electric stoves do for cooking and 20% less than convection ovens use. Clothes dryers with moisture sensors cut energy use by 15%. Refrigerators with a freezer on the bottom cut electricity use and operating costs in half, compared to models with the freezer on the top or side where more cold air flows out when the door is opened. Front-loading clothes washers use 55% less energy and 30% less water than top-loading models use and cut operating costs in half.
- *Use energy-efficient lighting.* As energy-wasting incandescent bulbs are being phased out, homeowners, colleges and universities, and businesses are switching to more energy efficient compact fluorescent bulbs and LED bulbs (Figure 13-25). Halogen light bulbs are also available and are cheaper than CFL and LED bulbs. But they are only slightly more effi-

Figure 13-28 This *thermogram*, or infrared photo, shows heat losses (red, white, and orange) around the windows, doors, roofs, and foundations of houses and stores in Plymouth, Michigan (USA). Many homes and buildings in the United States and other countries are so full of leaks that their heat loss in cold weather and heat gain in hot weather are equivalent to what would be lost through a large, window-sized hole in a wall. **Question:** How do you think the place where you live or work would compare to these buildings in terms of heat loss and the resulting waste of money spent on heating and cooling bills?

cient than incandescent bulbs and run so hot that they can start a fire when they come into contact with flammable materials.

CONNECTIONS

Using Compact Fluorescent Bulbs Reduces Mercury Pollution

The typical compact fluorescent lightbulb (CFL) contains a small amount of toxic mercury—roughly the amount that would fit on the tip of a ballpoint pen—and newer bulbs will have only half this amount. However, burning enough coal to light just one incandescent bulb adds much more mercury to the atmosphere than the breaking of a CFL bulb. While the mercury in burned-out CFLs can be recycled, the mercury continuously spewed into the atmosphere by coal-burning power plants cannot be retrieved. Thus, shifting to CFLs helps to reduce the amount of toxic mercury released into the atmosphere.

Figure 13-29 (p. 328) lists ways to save energy and money where you live.

Why Are We Still Wasting So Much Energy and Money?

Cutting energy waste is an important step in solving our energy problems and would provide many benefits. Why is there so little emphasis on improving energy efficiency?

One reason is that fossil fuels and nuclear power are artificially cheap, primarily because of the government subsidies and tax breaks they receive and because their market prices do not include the harmful environmental and health costs of their production and use. As a result, people are more likely to waste energy and less likely to invest in energy efficiency.

Another reason is that there are few long-lasting government tax breaks, rebates, low-interest loans, and other economic incentives for consumers and businesses to invest in improving energy efficiency. In addition, the U.S. federal government has done a poor job of encouraging fuel efficiency in motor vehicles and educating the public about the environmental and money-saving advantages of cutting energy waste.

Recent studies indicate that many people begin saving energy when they have devices that continuously display household or vehicle energy consumption, and when their utility bills show comparisons of their energy use with that of similar households. With such feedback, people in some neighborhoods (and in college dorms) are competing to see who can achieve the lowest heating and electricity bills.

According to the EPA, about 30% of the energy used in commercial buildings, on average, is wasted. In 2010, the EPA sponsored its first National Building Competition to see which participating commercial buildings, including those owned by major corporations, could cut their energy use the most over 12 months.

The winner was the Morrison Residence Hall at the University of North Carolina (UNC) at Chapel Hill. Within a year, it cut its energy consumption by 36% and reduced its energy costs by $250,000. UNC accomplished this mostly by tweaking the building's heating and cooling equipment, replacing worn-out or missing weather stripping, adding more efficient lighting, expanding its solar-powered hot-water heating system, and getting students to reduce hot-water usage in the laundry room. A computer touch-screen monitor in the lobby helped students keep track of their energy consumption. Competitions between dorm floors encouraged students to turn off lights and computers when not in use and to save hot water.

Attic

- Hang reflective foil near roof to reflect heat.
- Use house fan.
- Be sure attic insulation is at least 30 centimeters (12 inches).

Outside

Plant deciduous trees to block summer sun and let in winter sunlight.

Bathroom

- Install water-saving toilets, faucets, and shower heads.
- Repair water leaks promptly.

Kitchen

- Use microwave rather than stove or oven as much as possible.
- Run only full loads in dishwasher and use low- or no-heat drying.
- Clean refrigerator coils regularly.

Other rooms

- Use compact fluorescent lightbulbs or LEDs and avoid using incandescent bulbs.
- Turn off lights, computers, TV, and other electronic devices when they are not in use.
- Use high-efficiency windows; use insulating window covers and close them at night and on sunny, hot days.
- Set thermostat as low as you can in winter and as high as you can in summer.
- Weather-strip and caulk doors, windows, light fixtures, and wall sockets.
- Keep heating and cooling vents free of obstructions.
- Keep fireplace damper closed when not in use.
- Use fans instead of, or along with, air conditioning.

Basement or utility room

- Use front-loading clothes washer. If possible run only full loads with warm or cold water.
- If possible, hang clothes on racks for drying.
- Run only full loads in clothes dryer and use lower heat setting.
- Set water heater at 140°F if dishwasher is used and 120°F or lower if no dishwasher is used.
- Use water heater thermal blanket.
- Insulate exposed hot water pipes.
- Regularly clean or replace furnace filters.

Figure 13-29 **Individuals matter:** You can take steps to save energy and money where you live. **Questions:** Which of these steps do you take? Are there other steps that you might start taking?

13-5 What Are the Advantages and Disadvantages of Using Renewable Energy Resources?

▶ CONCEPT 13-5 Using a mix of renewable energy resources—especially sunlight, wind, flowing water, sustainable biomass, and geothermal energy—can drastically reduce pollution, greenhouse gas emissions, and biodiversity losses.

We Can Use Renewable Energy for Many Purposes

One of nature's three **principles of sustainability** is to rely mostly on solar energy. The energy in the sunlight striking the earth for just 1 hour

is enough to run the world's economy for 1 year. The challenge is to find affordable ways to collect, concentrate, and distribute this endless source of energy.

We can get renewable solar energy directly from the sun or indirectly from wind (**Core Case Study**), moving water, and biomass, none of which would exist without direct solar energy. Another form of renewable energy is geothermal energy from the earth's interior.

Studies show that with increased and consistent government subsidies, tax breaks, and funding for research and development, renewable energy could provide 20% of the world's electricity by 2025 and 50% by 2050. Denmark already gets 20% of its electricity from wind and has plans to increase this to 50% by 2030. Brazil gets 45% of its automotive fuel from ethanol made from a sugarcane residue, and could phase out its use of gasoline within a decade. Studies show that with a crash program, the United States could get 20% of its energy and at least 25% of its electricity from renewable resources by 2020.

GOOD NEWS

China plans to lead the world in relying on renewable energy resources by selling the electric cars, wind turbines, solar cells, and other products it produces throughout the world. In 2010, it invested far more money ($51 billion) in developing such energy resources than any other country in the world.

A number of energy analysts call for shifting from the current economy fueled primarily by nonrenewable oil, coal, natural gas, and nuclear power to one powered by a variety of locally available renewable wind, solar, and geothermal energy resources over the next few decades. They argue that this would result in more decentralized and energy-efficient national economies that are less vulnerable to supply cutoffs and natural disasters. It would also improve economic and national security for the United States, China, Japan, and many other countries by reducing their dependence on imported crude oil. And it would greatly reduce air and water pollution, as well as slowing projected climate disruption, creating large numbers of jobs, and saving consumers money (**Concept 13-5**).

If renewable energy is so great, why does it provide only 13% of the world's energy and 8% of the energy used in the United States? There are three major reasons. *First,* since 1950, government tax breaks, subsidies, and funding for research and development of renewable energy resources have been much lower than those for fossil fuels (especially oil) and nuclear power, although subsidies and tax breaks for renewables have increased in recent years.

Second, although subsidies and tax breaks for fossil fuels and nuclear power have essentially been guaranteed in the United States for many decades, those for renewable energy have to be renewed by Congress every few years. The resulting financial uncertainty makes it risky for companies to invest in renewable energy.

Third, the prices we pay for nonrenewable fossil fuels and nuclear power do not include the harmful environmental and human health costs of producing and using them. This helps to shield them from free-market competition with renewable sources of energy.

Energy analysts such as Amory Lovins say that if these economic handicaps—*unbalanced and intermittent subsidies* and *inaccurate pricing*—were eliminated, many forms of renewable energy would be cheaper than fossil fuels and nuclear energy, and would quickly take over the energy marketplace.

Throughout the rest of this chapter, we evaluate these renewable energy options.

We Can Heat Buildings and Water with Solar Energy

We can provide some homes and other buildings with most of the heat they need by using **passive solar heating systems** (Figure 13-30, left). Such a system absorbs and stores heat from the sun directly within a well-insulated structure. Water tanks and walls and floors of concrete, adobe, brick, or stone can store much of the collected solar energy as heat and release it slowly throughout the day and night. A small backup heating system such as a vented natural gas or propane heater can be used, if necessary (see Guest Essay by Nancy Wicks at **www.cengagebrain.com**).

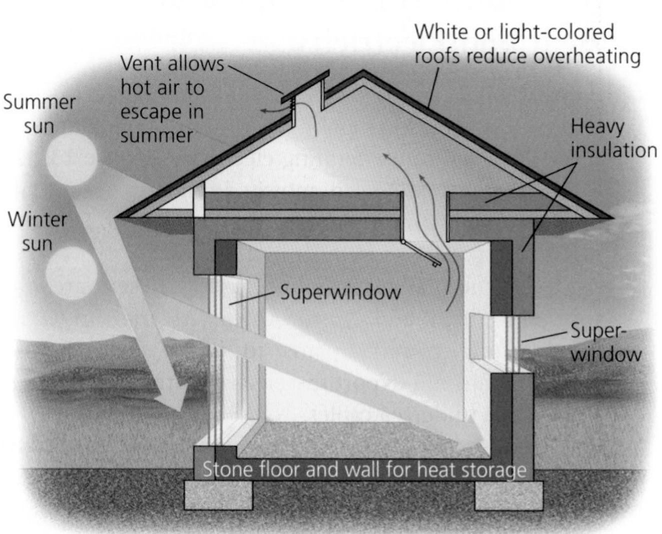

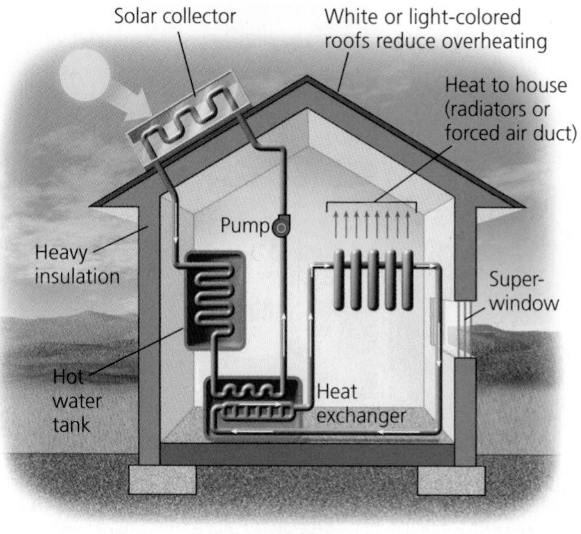

Figure 13-30 Solutions: Homes can be heated with passive or active solar systems.

An **active solar heating system** (Figure 13-30, right) captures energy from the sun by pumping a heat-absorbing fluid (such as water or an antifreeze solution) through special collectors, usually mounted on a roof or on special racks to face the sun. Some of the collected heat can be used directly. The rest can be stored in a large insulated container filled with gravel, water, clay, or a heat-absorbing chemical, for release as needed. (These are not shown in Figure 13-30.)

Rooftop active solar collectors are also used to heat water in many homes and apartment buildings. With systems that cost the equivalent of as little as $200, about one in ten houses and apartment buildings in China currently use the sun to provide hot water. Such systems are also widely used in Germany, Japan, Greece, Austria, and Turkey. In Spain and Israel, all new buildings must have rooftop systems for heating water and space. Soon millions of households in less-developed countries such as India and Brazil could be using this simple and inexpensive way to heat water.

Figure 13-31 lists the major advantages and disadvantages of using passive or active solar heating systems. They can be used to heat new homes in areas with adequate sunlight. (See the maps in Figure 18, p. S34, and Figure 19, p. S35, in Supplement 6.) But existing homes and buildings cannot use solar energy for heat, nor can they use roof-mounted solar cells to provide electricity, if they are not oriented to face the sun in order to receive the maximum amount of sunlight, or if they are blocked from sunlight by other buildings or trees.

However, there are ways around this limitation. For example, sunny Davis, California, has a national reputation as a U.S. city that promotes the use of solar energy. It also encourages planting trees for shade and using bicycles for travel. Because the city wants to avoid cutting trees, it dedicated an area of treeless land on its outskirts for establishing a *solar cell farm* with small plots of solar cells. Apartment dwellers and homeowners without enough access to sunlight can buy rights to electricity produced by this solar farm.

Trade-Offs

Passive or Active Solar Heating

Advantages	Disadvantages
Net energy is moderate (active) to high (passive)	Need access to sun 60% of time during daylight
Very low emissions of CO_2 and other air pollutants	Sun can be blocked by trees and other structures
Very low land disturbance	High installation and maintenance costs for active systems
Moderate cost (passive)	Need backup system for cloudy days

Figure 13-31 Heating a house with a passive or active solar energy system has advantages and disadvantages. **Questions:** Which single advantage and which single disadvantage do you think are the most important? Why?

THINKING ABOUT
Passive and Active Solar Systems

Do the advantages of using passive or active solar systems outweigh the disadvantages? Explain.

We Can Cool Buildings Naturally

Direct solar energy actually works against us when we want to keep a building cool, but we can use indirect solar energy (mainly wind) and other natural services to help cool buildings. We can open windows to take advantage of breezes and use fans to keep the air moving. A green roof covered with plants can also make a huge difference in keeping a building cool. When there is no breeze, superinsulation and high-efficiency windows help to keep hot air outside. Here are some other ways to keep cool:

- Block the high summer sun with shade trees, deep overhanging eaves, window awnings, or shades (Figure 13-30, left).
- In warm climates, use a light-colored roof to reflect as much as 90% of the sun's heat (compared to only 10–15% for a dark-colored roof).
- Use geothermal heat pumps for cooling (and for heating in winter).

We Can Concentrate Sunlight to Produce High-Temperature Heat and Electricity

Solar thermal systems use different methods to collect and concentrate solar energy in order to boil water and produce steam for generating electricity (Figure 13-32). These systems are used mostly in desert areas with ample sunlight (see the maps in Figure 18, p. S34, and Figure 19, p. S35, in Supplement 6).

Solar thermal plants also exist in desert areas of southern Spain and North Africa. Algeria, an OPEC oil-exporting country plans to build solar thermal power plants and export the electricity they generate to Europe via undersea cables.

Scientists estimate that we could meet all of the world's electricity needs by using a global, high-voltage smart electrical grid (see Case Study, p. 323), in combination with solar thermal power plants, that together would cover less than 1% of the world's GOOD NEWS

Sandia National Laboratories/National Renewable Energy Laboratory

Figure 13-32 *Solar thermal power:* Various systems are used to collect and concentrate solar energy in order to produce electricity. In the system shown here, an array of computer-controlled mirrors tracks the sun and focuses reflected sunlight on a central receiver, sometimes called a *power tower*. This tower near Daggett, California (USA), can collect enough heat to boil water and produce steam for generating electricity. The heat can also be used to melt a certain kind of salt. This molten salt holds the heat that was used to melt it, and this heat can then be released as needed to produce electricity at night.

deserts. One drawback is that the net energy yield for solar thermal systems is only about 3%, which means that they need large government subsidies or tax breaks in order to compete in the marketplace with alternatives that have higher net energy yields. Researchers are working to raise this net energy yield to at least 20% to make solar thermal systems more affordable.

Figure 13-33 summarizes the major advantages and disadvantages of concentrating solar energy to produce high-temperature heat or electricity.

Trade-Offs

Solar Energy for High-Temperature Heat and Electricity

Advantages	Disadvantages
Moderate environmental impact	Low net energy and high costs
No direct emissions of CO_2 and other air pollutants	Needs backup or storage system on cloudy days
Lower costs with natural gas turbine backup	High water use for cooling

Figure 13-33 Using solar energy to generate high-temperature heat and electricity has advantages and disadvantages. **Questions:** Which single advantage and which single disadvantage do you think are the most important? Why? Do you think that the advantages of using these technologies outweigh their disadvantages? Why?

THINKING ABOUT
Solar Thermal Power Systems

Do the advantages of using solar thermal power systems outweigh the disadvantages? Explain.

We can use concentrated solar energy on a smaller scale, as well. In some sunny rural areas, people use inexpensive *solar cookers* to focus and concentrate sunlight for cooking food and sterilizing water (Figure 13-34). Solar cookers can replace wood and charcoal fires, which would cut deadly indoor air pollution and help to reduce deforestation by decreasing the need for firewood.

Mark Edwards/Peter Arnold, Inc.

Figure 13-34 **Solutions:** This woman in India is using a solar cooker to prepare a meal for her family. Solar cookers costing about $10 can be made from cardboard boxes.

Sunpower (previously known as PowerLight Corporation)/National Renewable Energy Laboratories

Figure 13-35 Solutions: Photovoltaic (PV) or solar cells can provide electricity for a house or building using conventional solar panels such as those shown here on the roof of a U.S. Postal Service processing and distribution center in Inglewood, California. They can also be incorporated into metal roofing materials that can simulate the look of roof tiles of all types. A new, lightweight, modular rooftop solar cell system can be snapped together and installed in a few hours and thus reduces installation costs.

We Can Use Sunlight Directly to Produce Electricity

We can convert solar energy directly into electrical energy using **photovoltaic (PV) cells**, commonly called **solar cells** (Figure 13-35). Most solar cells are thin wafers of purified silicon (Si) or polycrystalline silicon with trace amounts of metals that allow them to produce electricity. A typical solar cell has a thickness ranging from less than that of a human hair to that of a sheet of paper. When sunlight strikes these transparent cells, they emit electrons, and many cells wired together in a panel can produce electrical power. The cells can be connected to existing electrical grid systems or to batteries that store the electrical energy until it is needed.

Solar cells have no moving parts, are safe and quiet, and produce no pollution or greenhouse gases during operation. The material used in solar cells can be made into paper-thin rigid or flexible sheets that can be printed like newspapers and incorporated into traditional-looking roofing materials and attached to a variety of surfaces such as walls, windows, and clothing.

Since 2008, Google and more than 75 public schools and community colleges in California have been covering their parking lots with canopies made of large arrays of solar cells. These solar parking lot systems provide shade, act as large carports, and help to inform students and workers about the use of solar cells to produce electricity.

Nearly 1.6 billion people in less-developed countries, or one of every four people in the world, live in rural villages that are not connected to an electrical grid. With easily expandable banks of solar cells, these people can now get electrical service (Figure 13-36). As these small, off-grid systems spread among rural villages, they will help hundreds of millions of people to lift themselves out of poverty. Eventually, new thin-film solar cells will drastically lower the cost of providing electricity to less-developed areas.

GOOD NEWS

In 2010, the three largest producers of solar-cell electricity were Japan, China, and Germany. Figure 13-37 lists the major advantages and disadvantages of using solar cells.

Jorgen Schytte/Peter Arnold, Inc.

Figure 13-36 Solutions: This system of solar cells provides electricity for a remote village in Niger, Africa. **Question:** Do you think wealthy nations should provide aid to poor countries to help them obtain solar-cell systems? Explain.

Trade-Offs

Solar Cells

Advantages	Disadvantages
Moderate net energy yield	Need access to sun
Little or no direct emissions of CO_2 and air pollutants	Need electricity storage system, backup, or connection to a grid system
Easy to install, move around, and expand as needed	High costs for older systems but decreasing rapidly
Competitive cost for newer cells	Solar-cell power plants could disrupt desert ecosystems

Figure 13-37 Using solar cells to produce electricity has advantages and disadvantages. **Questions:** Which single advantage and which single disadvantage do you think are the most important? Why?

Solar cells emit no greenhouse gases, although they are not carbon-free, because fossil fuels are usually used to produce and transport the panels. But these emissions are small compared to those of using fossil fuels and the nuclear power fuel cycle to generate electricity. Conventional solar cells also contain toxic materials that must be recovered when the cells wear out after 20–25 years of use, or when they are replaced with new systems. Also, solar cells must be cleaned because they collect dust that can reduce their efficiency. However, in 2010, research professor Malay Mazumder found a way to coat solar cells with electrically charged material that repels dust particles.

One problem with solar cells is that they are not highly energy-efficient. Even so, their overall net energy ratio is not too low because using them does not involve a series of energy-wasting steps such as those involved in the nuclear fuel cycle (Figure 13-18). Scientists are rapidly improving the efficiency of solar cells, having developed models in some laboratories that will have an efficiency of around 30%. Thus, generating electricity with such solar cells could be nearly as efficient as using coal-burning power plants but would not produce the air pollutants and climate-changing CO_2 emitted by those plants.

Until recently, the main problem with using solar cells to produce electricity has been their high cost. Despite this drawback, their production has soared in recent years and solar cells have become the world's fastest growing way to produce electricity (see the graph in Figure 11, p. S43, in Supplement 7.) This is because of their important advantages (Figure 13-37, left) and because of increased government subsidies and tax breaks for solar cell producers and users. Over the next 1-2 decades, production is likely to increase sharply as new, thin-film solar cells become inexpensive and efficient enough to compete with fossil fuels. **GREEN CAREER:** solar-cell technology

Energy analysts say that with increased research and development, plus larger and more consistent government tax breaks and other subsidies, solar cells could provide 16% of the world's electricity by 2040. In 2007, Jim Lyons, chief engineer for General Electric, projected that solar cells will be the world's number-one source of electricity by the end of this century. If that happens, it will represent a huge global application of nature's solar energy **principle of sustainability**.

We Can Produce Electricity from Falling and Flowing Water

Hydropower uses the kinetic energy of flowing and falling water to produce electricity. It is an indirect form of solar energy because it is based on the evaporation of water, which is then deposited at higher elevations where it can flow to lower elevations in rivers as part of the earth's solar-powered water cycle (see Figure 3-14, p. 51).

The most common approach to harnessing hydropower is to build a high dam across a river to create a reservoir. Some of the water stored in the reservoir is allowed to flow through large pipes at controlled rates to spin turbines that produce electricity (see Figure 11-12, p. 247).

Hydropower is the world's most widely used renewable energy source for the production of electricity. In order, the world's top six producers of hydropower are China, Canada, Brazil, the United States, Russia, and Norway.

According to the United Nations, only about 13% of the world's potential for hydropower has been developed. Much of this untapped potential is in China, India, South America, Central Africa, and parts of the former Soviet Union. China has plans to more than double its hydropower output by 2020 and is also building or funding more than 200 dams around the world.

However, some analysts expect that use of large-scale hydropower plants will fall slowly over the next several decades as many existing reservoirs fill with silt and become useless faster than new systems are built. Also, there is concern over emissions of methane, a potent greenhouse gas, from the anaerobic decomposition of submerged vegetation in hydropower plant reservoirs, especially in warm climates. In addition, if atmospheric temperatures continue to rise as projected, the electrical output of many of the world's large dams is likely to drop as mountain glaciers, a primary source of reservoir water, continue to melt.

Figure 13-38 (p. 334) lists the major advantages and disadvantages of using large-scale hydropower plants to produce electricity.

Trade-Offs

Large-Scale Hydropower

Advantages	Disadvantages
Moderate to high net energy	Large land disturbance and displacement of people
Large untapped potential	High CH_4 emissions from rapid biomass decay in shallow tropical reservoirs
Low-cost electricity	Disrupts downstream aquatic ecosystems
Low emissions of CO_2 and other air pollutants in temperate areas	

Figure 13-38 Using large dams and reservoirs to produce electricity has advantages and disadvantages. **Questions:** Which single advantage and which single disadvantage do you think are the most important? Why?

The use of *microhydropower generators* may become an increasingly important way to produce electricity. These are floating turbines, each about the size of an overnight suitcase. They use the power of flowing water to turn rotor blades, which spin a turbine to produce electric current. They can be placed in any stream or river without altering its course to provide electricity at a low cost with a very low environmental impact.

Finally, we can produce electricity from flowing water by tapping into the energy from *ocean tides* and *waves*. In some coastal bays and estuaries, water levels can rise or fall by 6 meters (20 feet) or more between daily high and low tides. Dams have been built across the mouths of some bays and estuaries to capture the energy in these large-scale flows for hydropower. In addition, several turbines, resembling underwater wind turbines, have been installed to tap the tidal flow of the East River near New York City. In addition, scientists and engineers have been trying to produce electricity by tapping wave energy along seacoasts where there are almost continuous waves.

Production of electricity from tidal and wave systems is limited because of a lack of suitable sites, citizen opposition at some sites, high costs, and equipment damage from saltwater corrosion and storms. However, improved technology could greatly increase the production of electricity from tides and waves sometime during this century.

Using Wind to Produce Electricity Is an Important Step toward Sustainability

Wind turbines (**Core Case Study**) have been erected in large numbers at favorable sites to create wind farms (Figure 13-1). So far most of the world's rapidly growing wind farms have been built on land in parts of Europe, China, and the United States. However, the new frontier for wind energy is the construction of wind farms at sea, especially in Europe (Figure 13-39) and more recently in China.

Since 1990, wind power has been the world's second fastest-growing source of energy after solar cells (see the graph in Figure 12, p. S43, in Supplement 7). In order, the largest wind-power producers in 2009 were China, the United States, Spain, Germany, and India.

Figure 13-39 Solutions: Wind turbines can be interconnected in arrays of tens to hundreds. These *wind farms* or *wind parks* can be located on land (Figure 3-1, right) or offshore. This one is located off the coast of Denmark—one of the world's leaders in wind power. **Questions:** Would you object to having a wind farm located near where you live? Why or why not?

TebNad/Shutterstock.com

In 2009, a Harvard University study led by Xi Lu estimated that wind power has the potential to produce 40 times the world's current use of electricity. The study used data on wind flows and strengths from thousands of meteorological measuring stations to determine that most electricity needs could be met by a series of large wind farms on land and at sea. Most of the land-based wind farms would be located in remote and sparsely populated areas of countries such as China, the United States, and Canada.

GOOD NEWS

Even though offshore wind farms are more costly to install, analysts expect to see increasing use of them because wind speeds over water are often stronger and steadier than those over land, and any noise made by the turbines would be muffled by the sound of waves breaking on shorelines near homes and businesses. Locating them offshore also lessens the need for negotiations among multiple landowners over the locations of turbines and electrical transmission lines.

A 2009 study published in the *Proceedings of the U.S. National Academy of Sciences* estimated that the world's top CO_2-emitting countries have ample land-based and offshore wind potential to more than meet their current electricity needs. The United States has enough wind potential to meet an estimated 16 to 22 times its current electricity needs (**Core Case Study**). Just three states—Texas, Kansas, and North Dakota—have enough usable wind power to meet all U.S. electricity needs. Texas, the nation's leading oil-producing state, is also now the nation's leading producer of electricity from wind power, followed by California. These states produce more electricity from wind than all but five of the world's countries. China has enough wind potential to meet 15 times its current electricity consumption. Canada has enough wind power to generate 39 times its electrical needs, and Russia could use wind power to meet its electrical needs 170 times over. Offshore wind power potential alone is similarly large.

CORE CASE STUDY

Unlike oil and coal, wind is widely distributed and inexhaustible, and wind power is mostly carbon-free and pollution-free. A wind farm can be built within 9 to 12 months and expanded as needed. Homeowners can also use small and quiet wind turbines that operate at low speeds to produce their own electricity.

In addition, wind power has a moderate-to-high net energy ratio. The DOE and the Worldwatch Society estimate that, when we include the harmful environmental and health costs of various energy resources in comparative cost estimates, wind energy is the least costly way to produce electricity.

Like any energy source, wind power has some drawbacks. Areas with the greatest wind power potential are often sparsely populated and located far from cities. Thus, to take advantage of the potential for using wind energy, countries such as the United States will have to invest in an upgrading and expansion of their outdated electrical grid systems (see Case Study, p. 323). The resulting increase in the number of transmission towers and lines will cause controversy. However, many of the new lines could be routed along state-owned interstate highway corridors to avoid legal conflicts.

Another problem is that winds can die down, and thus a backup source of power would be required for generating electricity. However, analysts calculate that a large number of wind farms in different areas connected to an updated electrical grid could usually take up the slack when winds die down in any one area.

Scientists are working on ways to store wind energy. Electricity produced by wind can be passed through water to produce hydrogen fuel, which could be thought of as "stored" wind power. Another option is to use wind-generated electricity to pump pressurized air deep underground into aquifers, caverns, and abandoned natural gas wells. The energy stored in the compressed air could then be released as needed to spin turbines and generate electricity when wind power is not available. This process is being used in Germany and in the U.S. state of Alabama. Also, if use of plug-in hybrid electric cars (Figure 13-27, right) grows rapidly over the next few decades, using electricity from wind farms to recharge the batteries of such cars will be another way to store wind energy. This energy would be stored at night, while millions of cars were recharging, and then used to propel the cars during the day.

CONNECTIONS

Bird Deaths and Wind Turbines

Wildlife ecologists have estimated that collisions with wind turbines kill as many as 440,000 birds each year in the United States, although other more recent estimates put the figure at 7,000 to 100,000. Compare this to much larger numbers reported by Defenders of Wildlife: domestic and feral cats kill more than 250 million birds a year; hunters, more than 100 million; cars and trucks, about 80 million; pesticide poisoning, 67 million; and collisions with tall structures and windows, at least 10 million every year. Most of the wind turbines involved in bird deaths were built with the use of outdated designs, and some were built in bird migration corridors. Wind power developers now avoid such corridors when building wind farms. Newer turbine designs reduce bird deaths considerably by using slower blade rotation speeds and by not providing places for birds to perch or nest.

Some people in populated areas oppose wind farms as being unsightly and noisy. But in windy parts of the U.S. Midwest and Canada, many farmers and ranchers welcome them and some have become wind-power producers themselves. For each wind turbine located on a farmer's or a rancher's land, the landowner typically receives $3,000 to $10,000 a year in royalties. And that farmer or rancher can still use the land for growing crops or grazing cattle.

Figure 13-40 (p. 336) lists the major advantages and disadvantages of using wind to produce electricity. According to energy analysts, wind power has more benefits and fewer serious drawbacks than any other energy resource, except for energy efficiency. **GREEN CAREER:** wind-energy engineering

Trade-Offs

Wind Power

Advantages	Disadvantages
Moderate to high net energy yield	Needs backup or storage system when winds die down
Widely available	Visual pollution for some people
Low electricity cost	Low-level noise bothers some people
Little or no direct emissions of CO_2 or air pollutants	Can kill some birds and bats if not properly designed and located
Easy to build and expand	

Figure 13-40 Using wind to produce electricity has advantages and disadvantages (**Concept 13-5**). With sufficient and consistent government incentives, wind power could supply more than 10% of the world's electricity and 20% of the electricity used in the United States by 2030. **Questions:** Which single advantage and which single disadvantage do you think are the most important? Why?

We Can Produce Energy by Burning Solid Biomass

Biomass consists of plant materials (such as wood and agricultural waste) and animal wastes that we can burn directly as a solid fuel or convert into gaseous or liquid biofuels.

Solid biomass is burned mostly for heating and cooking, but also for industrial processes and for generating electricity. Wood, wood wastes, charcoal made from wood, and other forms of biomass used for heating and cooking supply 10% of the world's energy, 35% of the energy used in less-developed countries, and 95% of the energy used in the poorest countries. In agricultural areas, *crop residues* (such as sugarcane stalks, rice husks, and corn cobs) and *animal manure* are collected and burned.

Wood is a renewable fuel only if it is harvested no faster than it is replenished. The problem is, about 2.7 billion people in 77 less-developed countries face a *fuelwood crisis* and are often forced to meet their fuel needs by harvesting wood faster than it can be replenished. One way to deal with this problem is to plant fast-growing trees, shrubs, and perennial grasses in *biomass plantations*. But repeated cycles of growing and harvesting these plantations can deplete the soil of key nutrients. Clearing forests and grasslands for such plantations destroys or degrades biodiversity. And plantation tree species such as European poplar and American mesquite are invasive species and can spread from plantations to takeover nearby natural areas.

As with any energy resource, using solid biomass has advantages and disadvantages. One problem is that clearing forests to provide the fuel reduces the amount of vegetation that would otherwise capture CO_2, and burning biomass produces CO_2. However, if the rate of use of biomass does not exceed the rate at which it is replenished by new plant growth, there is no net increase in CO_2 emissions. But monitoring and managing this balance, globally or throughout any one country, is very difficult.

Figure 13-41 lists the general advantages and disadvantages of burning solid biomass as a fuel.

We Can Convert Plants and Plant Wastes to Liquid Biofuels

Liquid biofuels such as *biodiesel* (produced from vegetable oils) and *ethanol* (ethyl alcohol produced from plants and plant wastes) are being used in place of petroleum-based diesel fuel and gasoline. The biggest producers of liquid bio-fuels are, in order, the United States, Brazil, the European Union, and China.

Biofuels have three major advantages over gasoline and diesel fuel produced from oil. *First,* while oil resources are concentrated in a small number of countries, biofuel crops can be grown almost anywhere, and thus they help countries to reduce their dependence on imported oil. *Second,* if these crops are not used faster than they are replenished by new plant growth, there is no net increase in CO_2 emissions, unless existing grasslands or forests are cleared to plant biofuel crops. *Third,* biofuels are easy to store and transport through existing

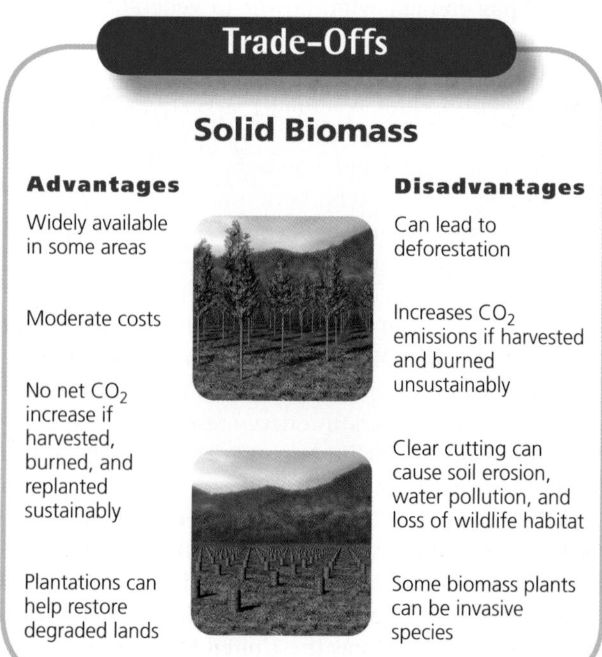

Figure 13-41 Burning solid biomass as a fuel has advantages and disadvantages. **Questions:** Which single advantage and which single disadvantage do you think are the most important? Why?

fuel networks and can be used in motor vehicles at little or no additional cost.

However, in a 2007 UN report on bioenergy, and in another study by R. Zahn and his colleagues, scientists warned that large-scale biofuel crop farming could reduce biodiversity by clearing large areas of natural forests and grasslands; increase soil degradation, erosion, and soil nutrient loss; push small farmers off their land; and raise food prices if food producers can make more money growing corn and other crops to fuel cars rather than to feed livestock and people.

CONNECTIONS

Biofuels and Climate Change

In 2007, Nobel Prize–winning chemist Paul Crutzen warned that intensive farming of biofuel crops could speed up atmospheric warming and projected climate disruption by producing more greenhouse gases than would be produced by burning fossil fuels instead of biofuels. This would happen if nitrogen fertilizers were used to grow corn and other biofuel crops. Such fertilizers, when applied to the soil, release large amounts of the potent greenhouse gas nitrous oxide. A 2008 study by Finn Danielsen and a team of other scientists concluded that keeping tropical rain forests intact is a better way to slow projected climate change than is replacing such forests with biofuel plantations.

Trade-Offs

Liquid Biofuels

Advantages	Disadvantages
Reduced CO_2 emissions for some crops	Fuel crops compete with food crops for land
High net energy yield for biodiesel from oil palms	Low net energy yield for corn ethanol and for biodiesel from soybeans
Moderate net energy yield for ethanol from sugarcane	Higher CO_2 emissions from corn ethanol

Figure 13-42 There are advantages and disadvantages to using liquid biofuels. **Questions:** Which single advantage and which single disadvantage do you think are the most important? Why?

Another problem with biofuels production is that growing corn and soybeans in climates that require irrigation could reduce water supplies in these arid regions. In fact, the two most water-intensive ways to produce a unit of energy are irrigating soybean crops to produce biodiesel fuel and irrigating corn to produce ethanol.

The challenge is to grow crops for food and biofuels by using more sustainable agriculture (see Figure 10-28, p. 231) with less irrigation, land degradation, air and water pollution, greenhouse gas emissions, and degradation of biodiversity. Also, any system for producing a biofuel should have a moderate-to-high net energy yield so that it can compete in the energy marketplace without large government subsidies.

See the Case Study that follows to learn more about ethanol production. Figure 13-42 lists the major advantages and disadvantages of using liquid biofuels.

■ CASE STUDY

Is Ethanol the Answer?

Ethanol can be made from plants such as sugarcane, corn, and switchgrass, and from agricultural, forestry, and municipal wastes. This process involves converting plant starches into simple sugars, which are processed to produce ethanol.

Brazil, by far the world's largest sugarcane producer, is the world's second largest ethanol producer after the United States. Brazil makes its ethanol from *bagasse,* a residue produced when sugarcane is crushed. GOOD NEWS This ethanol yields 8 times the amount of energy used to produce it—compared with a net energy ratio of 5 for gasoline. About 45% of Brazil's motor vehicles run on ethanol or ethanol–gasoline mixtures produced from sugarcane grown on only 1% of the country's arable land.

Within a decade, Brazil could expand its sugarcane production, eliminate all oil imports, and greatly increase ethanol exports to other countries. To do this, Brazil plans to clear and replace larger areas of its rapidly disappearing Cerrado, a wooded savanna region—one of the world's biodiversity hotspots (see Figure 9-21, p. 193)—with sugarcane plantations. This would significantly increase the harmful environmental costs of using this energy resource.

Environmental scientists David Pimentel and Tad Patzek warn that producing ethanol from sugarcane has a number of other harmful environmental effects. They include the CO_2 emissions from the burning of oil and gasoline to produce sugarcane, very high soil erosion after sugarcane plantations are harvested, and stresses on water supplies. Producing 1 liter (0.27 gallons) of ethanol from sugarcane requires the equivalent of about 174 bathtubs full of water.

In the United States, most ethanol is made from corn. Farmers profit from growing corn to produce ethanol because they receive generous government subsidies as part of the nation's energy policy. But studies indicate that using fossil-fuel-dependent industrialized agriculture to grow corn and then using more fossil fuel to convert the corn to ethanol provides a net energy ratio of only about 1.1–1.5 units of energy per unit of fossil fuel input. This explains why the U.S. government (taxpayers) must subsidize corn ethanol production to help it compete in the energy markets. It also helps to explain why Brazil, achieving a net energy ratio of 8 from bagasse, can produce ethanol at about half the cost of producing it from corn in the United States.

According to a 2007 study by environmental economist Stephen Polansky, processing all of the corn grown in the United States into ethanol each year would meet only about 30 days worth of the country's current demand for gasoline. This would leave no corn for other uses and would cause sharp increases in the prices of corn-based foods such as cereals, tortillas, poultry, beef, pork, and dairy products.

CONNECTIONS

Corn, Ethanol, and Tortilla Riots in Mexico

Traditionally, the United States has supplied approximately 75% of the world's corn. Mexico imports 80% of its corn from the United States. Since 2005, when America began using much of its corn crop to produce ethanol, the prices of food items such as corn tortillas in Mexico have risen sharply. This has drastically affected the 53 million people living in poverty in Mexico and has been a key factor in food riots and massive citizen protests.

An alternative to corn ethanol is *cellulosic ethanol,* which is produced from inedible cellulose that makes up most of the biomass of plants (see *The Habitable Planet,* Video 10, at **www.learner.org/resources/series209.html**). In this process, enzymes are used to help convert the cellulose from plant material such as leaves, stalks, and wood chips to sugars that are processed to produce ethanol. By using these widely available inedible cellulose materials to produce ethanol, producers could dodge the food vs. biofuels dilemma. A plant that could be used for cellulosic ethanol production is *switchgrass* (Figure 13-43), a tall perennial grass native to North American prairies that grows faster than corn.

One possible drawback was reported by cellulosic-ethanol expert Robert Ranier, who estimated that replacing half of the gasoline consumed in the United States with cellulosic ethanol would require about seven times the land area currently used for all corn production. Also, we need more research on how using cellulosic ethanol would affect CO_2 emissions.

Another problem is that it is difficult and costly to break down the cellulose and extract the glucose needed to make ethanol. As a result, affordable chemical processes for converting cellulosic material to ethanol are still being developed and are possibly years away.

We Can Get Energy by Tapping the Earth's Internal Heat

Geothermal energy is heat stored in soil, underground rocks, and fluids in the earth's mantle (see Figure 12-2, p. 277). We can tap into this stored energy to heat and cool buildings and to produce electricity.

National Renewable Energy Laboratory

Figure 13-43 Natural capital: The cellulose in this rapidly growing switchgrass can be converted into ethanol, but further research is needed to develop affordable production methods. This perennial plant can also help to slow projected climate change by removing carbon dioxide from the atmosphere and storing it as organic compounds in the soil.

GOOD NEWS

Scientists estimate that using just 1% of the heat stored in the uppermost 5 kilometers (8 miles) of the earth's crust would provide 250 times more energy than that stored in all the earth's crude oil and natural gas reserves.

One way to capture geothermal energy is by using a *geothermal heat pump* system (Figure 13-44). It can heat and cool a house by exploiting the temperature difference, almost anywhere in the world, between the earth's surface and underground at a depth of 3–6 meters (10–20 feet), where the earth's temperature typically is 10–20°C (50–60°F) year round.

According to the EPA, a well-designed geothermal heat pump system is the most energy-efficient, reliable, environmentally clean, and cost-effective way to heat or cool a space. It produces no air pollutants and emits no CO_2. Installation costs can be high but are generally recouped after 3–5 years; thereafter, such systems save money for their owners.

We can also tap into deeper, more concentrated *hydrothermal reservoirs* of geothermal energy. This is done

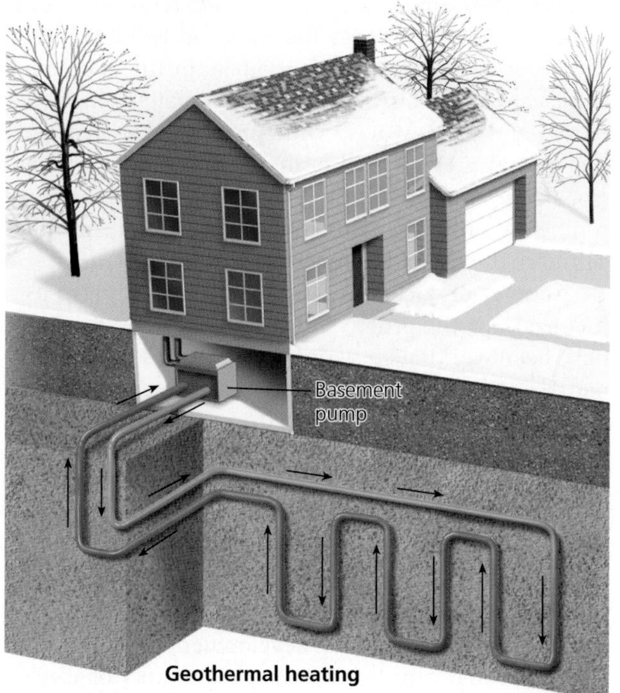

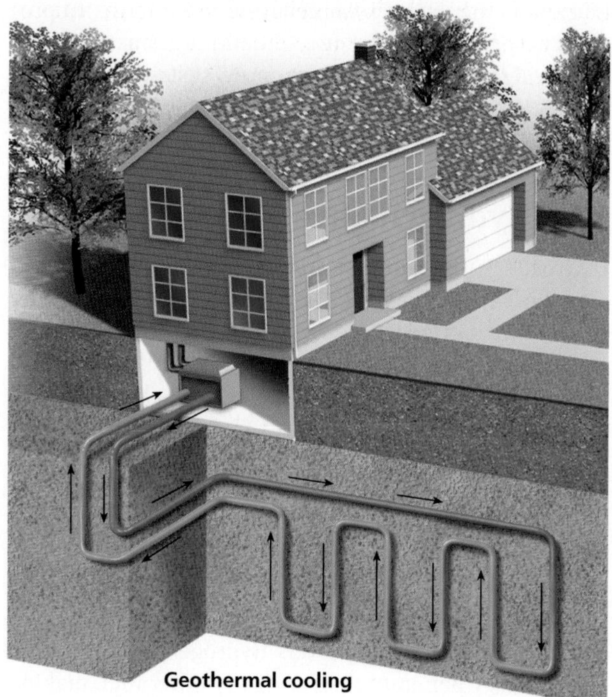

Figure 13-44 Natural capital: A geothermal heat pump system can heat or cool a house almost anywhere. In winter (left), a closed loop of buried pipes circulates a fluid, which extracts heat from the ground and carries it to a heat pump, which transfers the heat to a home's heat distribution system. In summer (right), this system works in reverse, removing heat from a home's interior and storing it in the ground.

by drilling wells into the reservoirs to extract their dry steam (with a low water content), wet steam (with a high water content), or hot water, which are then used to heat homes and buildings, provide hot water, grow vegetables in greenhouses, raise fish in aquaculture ponds, and spin turbines to produce electricity.

The United States is the world's largest producer of geothermal electricity from hydrothermal reservoirs. Most of it is produced in California, Nevada, Utah, and Hawaii (see map in Figure 22, p. S37, in Supplement 6). It meets the electricity needs of about 6 million Americans—a number roughly equal to the combined populations of Los Angeles, California, and Houston, Texas—and supplies almost 6% of California's electricity. Indonesia, with 128 active volcanoes, has abundant geothermal energy that can be tapped. In 2008, the country announced that its national oil company will develop its geothermal energy resources.

Another source of geothermal energy is *hot, dry rock* found 5 or more kilometers (3 or more miles) underground almost everywhere. Water can be injected through wells drilled into this rock. Some of this water, absorbing some of the intense heat, becomes steam that is brought to the surface and used to spin turbines to generate electricity. According to the U.S. Geological Survey, tapping just 2% of this source of geothermal energy in the United States could produce more than 2,000 times the country's current annual use of electricity. However, digging so deep into the earth's crust is costly. The high cost could be brought down by more research and improved technology. **GREEN CAREER:** geothermal engineer

Figure 13-45 lists the major advantages and disadvantages of using geothermal energy. Some analysts see this source—combined with improvements in energy

Trade-Offs

Geothermal Energy

Advantages	Disadvantages
Moderate net energy and high efficiency at accessible sites	High cost and low efficiency except at concentrated and accessible sites
Lower CO_2 emissions than fossil fuels	Scarcity of suitable sites
Low cost at favorable sites	Noise and some CO_2 emissions

Figure 13-45 Using geothermal energy for space heating and for producing electricity or high-temperature heat for industrial processes has advantages and disadvantages. **Questions:** Which single advantage and which single disadvantage do you think are the most important? Why?

efficiency, the use of solar cells and wind farms to produce electricity, and the use of natural gas as a temporary bridge fuel—as keys to a more sustainable energy future.

Will Hydrogen Save Us?

Hydrogen is the simplest and most abundant chemical element in the universe. Some scientists say that the fuel of the future is hydrogen gas (H_2). Most of their research has been focused on using fuel cells that combine H_2 and oxygen gas (O_2) to produce electricity and water vapor ($2\ H_2 + O_2 \rightarrow 2\ H_2O$ + energy), which is emitted into the atmosphere.

GOOD NEWS

Widespread use of hydrogen as a fuel would eliminate most of our outdoor air pollution problems. It would also greatly reduce the threat of projected climate disruption, because using it emits no CO_2—as long as the H_2 is not produced with the use of fossil fuels or nuclear power. Hydrogen also provides more energy per gram than does any other fuel, making it a lightweight fuel ideal for aviation.

So what is the catch? There are three challenges in turning the vision of hydrogen as a major fuel into reality. *First,* there is hardly any hydrogen gas (H_2) in the earth's atmosphere, so it must be produced from elemental hydrogen (H), which is chemically locked up in water and in organic compounds such as methane and gasoline. Producing H_2 requires using other forms of energy, and therefore, H_2 *has a negative net energy ratio.* Thus, the amount of energy it takes to make this fuel will always be more than the amount we can get by burning it.

Second, fuel cells are the best way to use H_2 to produce electricity, but current versions of fuel cells are expensive. However, progress in the development of nanotechnology (see Chapter 12, Science Focus, p. 294) could lead to less expensive and more efficient fuel cells.

Third, whether or not a hydrogen-based energy system produces less outdoor air pollution and CO_2 than a fossil fuel system depends on how the H_2 is produced. We could use electricity from coal-burning and nuclear power plants to decompose water into H_2 and O_2. But this approach does not avoid the harmful environmental effects of using coal and nuclear power.

We can also make H_2 from coal and strip it from organic compounds found in fuels such as gasoline. However, according to a 2002 study, using these methods to produce H_2 would add much more CO_2 to the atmosphere per unit of heat generated than does burning carbon-containing fuels directly. If renewable energy sources such as wind farms and solar cell power plants were used to make H_2, these CO_2 emissions would be avoided. Making hydrogen fuel by decomposing the methane in nonrenewable but cleaner burning natural gas could serve as a transition to using forms of renewable energy to produce hydrogen.

Hydrogen's negative net energy ratio is a serious limitation and means that this fuel will have to be subsidized in order for it to compete in the open marketplace with fuels that have moderate-to-high net energy yields. However, widespread use of hydrogen fuel would greatly reduce air pollution and projected global warming—two of the world's most serious environmental problems. This could make it worthwhile to subsidize the use of hydrogen.

Another obstacle to expanded use of hydrogen as a fuel has been the problem of how to store H_2. Scientists have been working on various ways to do so. For example, H_2 can be stored in a pressurized tank as liquid hydrogen. It can also be stored in solid metal hydride compounds and in sodium borohydride, which releases H_2 when heated. Another possibility is to coat the surfaces of activated charcoal or carbon nanofibers with H_2. When heated, these surfaces would release the H_2. Another possibility is to store H_2 inside nanosize glass microspheres that can easily be filled and refilled. Yet another possibility is the development of *ultracapacitors* that could quickly store large amounts of electrical energy, which could then be used to produce H_2 on demand. More research is needed to make these possibilities into usable technologies.

In the 1990s, Amory Lovins and his colleagues at the Rocky Mountain Institute designed a very light, safe, extremely efficient hydrogen-powered car. It is the basis of most prototype hydrogen fuel-cell cars now being tested by major automobile companies. Some analysts project that fuel-cell cars, running on affordable H_2 produced from natural gas, could be in widespread use by 2030 to 2050.

Larger, stationary fuel cells could provide electricity and heat for commercial and industrial users. In 2010, Bloom Energy in California began selling fuel-cell stacks, each about the size of a trash dumpster, that can be used to power buildings. They use natural gas to provide the H_2, and the company depends on significant state and federal subsidies to be competitive with producers of coal-fired electricity. In addition, Japan has built a large fuel cell that produces enough electricity to run a small town.

Canada's Toronto-based Stuart Energy is developing a fueling unit about the size of a dishwasher that will allow consumers to use electricity to produce their own H_2 from tap water. The unit could be installed in a garage and used to fuel a hydrogen-powered vehicle overnight. In sunny areas, people could install rooftop panels of solar cells to produce and store H_2 for their cars.

Another promising application is in homes, where a fuel-cell stack about the size of a refrigerator could provide heat, hot water, and electricity. Some Japanese homeowners get their electricity and hot water from such fuel cell units, which produce H_2 from the methane in natural gas. **GREEN CAREER:** hydrogen energy

While other countries have continued to explore such applications, the U.S. government, in 2009, reduced

Trade-Offs

Hydrogen

Advantages	Disadvantages
Can be produced from plentiful water at some sites	Negative net energy yield
No direct CO_2 emissions if produced from water	CO_2 emissions if produced from carbon-containing compounds
Good substitute for oil	High costs and negative net energy yield create need for subsidies
High efficiency (45–65%) in fuel cells	Needs H_2 storage and distribution system

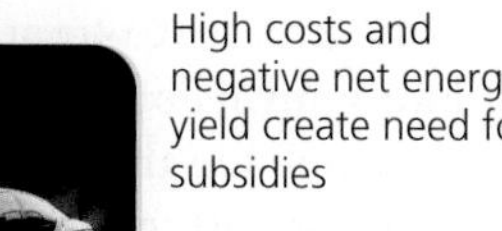

Figure 13-46 Using hydrogen as a fuel for vehicles and for providing heat and electricity has advantages and disadvantages. **Questions:** Which single advantage and which single disadvantage do you think are the most important? Why?

its research and development support for hydrogen fuel and put more emphasis on wind, direct solar energy, cleaner coal, and nuclear power. Figure 13-46 lists the major advantages and disadvantages of using hydrogen as an energy resource.

THINKING ABOUT
Using Hydrogen as a Fuel

Do the advantages of using hydrogen as a fuel outweigh the disadvantages? Explain.

13-6 How Can We Make the Transition to a More Sustainable Energy Future?

CONCEPT 13-6 We can make the transition to a more sustainable energy future by greatly improving energy efficiency, depending more on a mix of renewable energy resources, and including the environmental costs of energy resources in their market prices.

Choosing Energy Paths

We need to develop energy policies with the future in mind, because experience shows that it usually takes at least 50 years and huge financial investment to phase in new energy alternatives. Creating energy policy involves trying to answer the following questions for *each* energy alternative:

- How much of the energy resource is likely to be available in the near future (the next 25 years) and in the long term (the next 50 years)?
- What is the estimated net energy yield (p. 300) for the resource?
- What are the estimated costs for developing, phasing in, and using the resource?
- What government research and development subsidies and tax breaks, if any, will be needed to help develop the resource?
- How will dependence on the resource affect national and global economic and military security?
- How vulnerable is the resource to terrorism?
- How will extracting, transporting, and using the resource affect the environment, the earth's climate, and human health? How will these harmful costs be paid and by whom?
- Does use of the resource produce hazardous, toxic, or radioactive substances that must be safely stored for very long periods of time?

In 1977, energy analyst Amory Lovins published his pioneering book, *Soft Energy Paths*. In it, he compared what he called *hard energy paths*—based on increasing the use of nonrenewable fossil fuels and nuclear energy—to what he called *soft energy paths*—based on improving energy efficiency and increasing the use of renewable energy resources. At that time, many energy experts criticized Lovins as being unrealistic and not really understanding the energy business. Today, he is one of the world's most prominent energy experts and is helping the world make the transition to the soft energy path that he proposed over three decades ago.

In considering possible energy futures, scientists and energy experts who have evaluated energy alternatives have come to three general conclusions. First, *there will likely be a gradual shift from large, centralized power systems* such as the current coal and nuclear power plants *to smaller, decentralized power systems* (Figure 13-47) such as wind turbines, household and neighborhood solar-cell panels, rooftop solar water heaters, and small natural gas turbines. There will also be a shift from gasoline-powered motor vehicles to hybrid and plug-in electric cars. This may be followed by a shift to fuel cells for cars and to stationary fuel cells for houses and commercial buildings.

In this more electrified and decentralized system, electricity produced increasingly by renewable energy will be used to move people and goods and to heat, cool, and light buildings—all based on implementing the solar energy **principle of sustainability**. Because of the power of exponential growth, this electrification and decentralization of the global energy economy could take place in a surprisingly short time, just as the shift from relying on large mainframe computers to using mostly small personal computers happened within a decade or less.

GOOD NEWS

A shift to a more decentralized energy system would also improve national and economic security, because countries would rely on diverse, dispersed, domestic renewable energy resources instead of on a smaller number of large power plants that are vulnerable to storm damage and sabotage. Similarly, states and communities within countries could become more energy-independent and secure by making this shift.

The second general conclusion of experts about the future of energy use is that *a combination of greatly improved energy efficiency and the temporary use of nonrenewable natural gas will be the best way to make the transition to a diverse mix of renewable energy resources over the next several decades* (**Concept 13-6**). By using a variety of often locally available renewable energy resources, we would be applying the diversity **principle of sustainability** and not putting all of our "energy eggs" in only one or two baskets.

The third general conclusion is that *because of their still-abundant supplies and artificially low prices, we will continue using fossil fuels in large quantities*. This presents

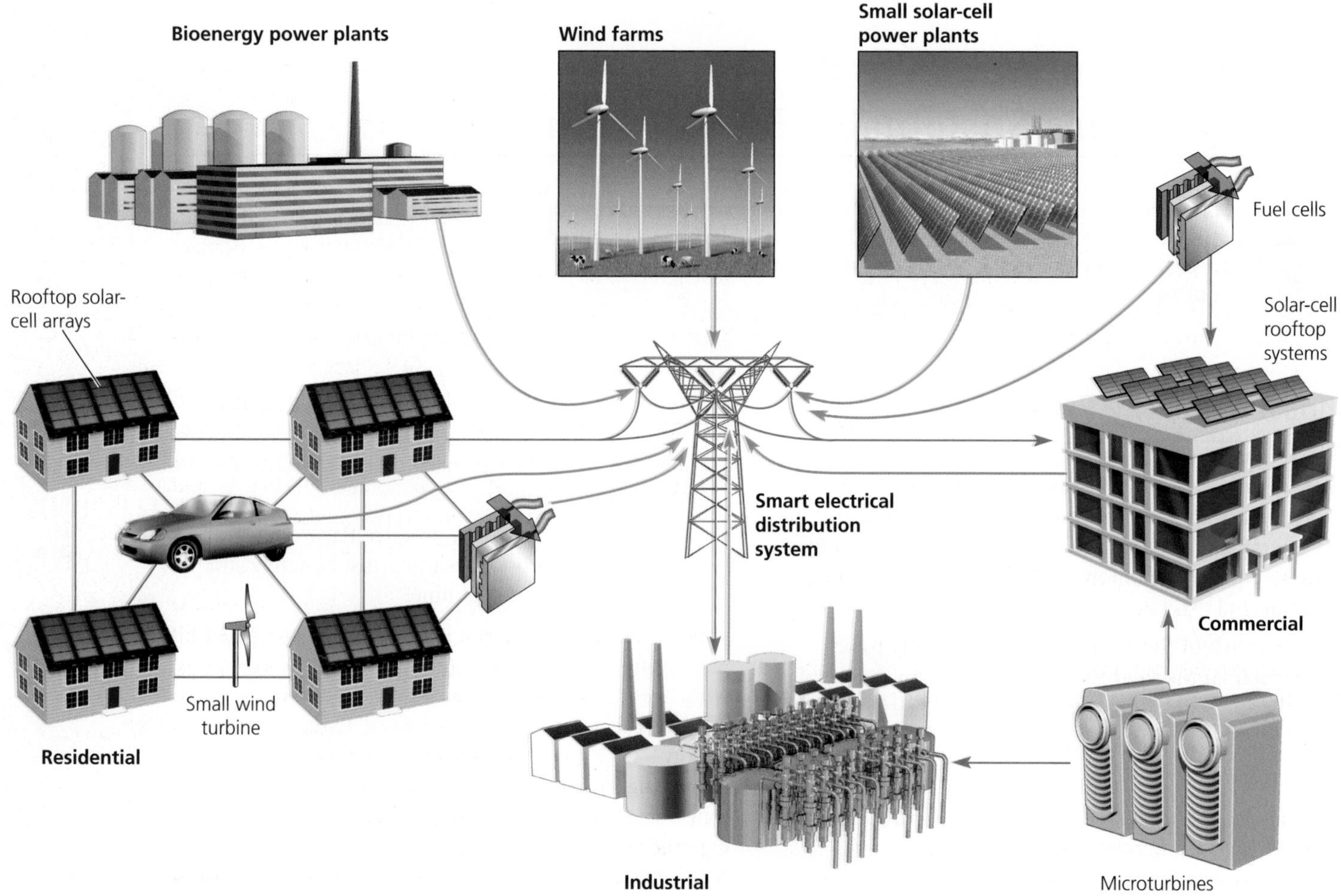

Figure 13-47 **Solutions:** During the next few decades, we will probably shift from dependence on a *centralized electric power system*, based on several hundred large-scale coal-burning and nuclear power plants, to a *decentralized power system*, in which electricity is produced by a large number of dispersed, small-scale, power generating systems that depend primarily on a variety of locally available, low-carbon renewable energy resources. **Questions:** Can you think of any disadvantages of a decentralized power system? Would such disadvantages outweigh the advantages for making such a shift? Explain.

Solutions

Making the Transition to a More Sustainable Energy Future

Improve Energy Efficiency	More Renewable Energy
Increase fuel-efficiency standards for vehicles, buildings, and appliances	Greatly increase use of renewable energy
Provide large tax credits or feebates for buying efficient cars, houses, and appliances	Provide large subsidies and tax credits for use of renewable energy
Reward utilities for reducing demand for electricity	Greatly increase renewable energy research and development
Greatly increase energy efficiency research and development	

Reduce Pollution and Health Risk

Phase out coal subsidies and tax breaks

Levy taxes on coal and oil use

Phase out nuclear power subsidies, tax breaks, and loan guarantees

Figure 13-48 Energy analysts have made a number of suggestions for helping us make the transition to a more sustainable energy future (**Concept 13-6**). **Questions:** Which five of these solutions do you think are the most important? Why?

two major challenges. One is to find ways to reduce the harmful environmental impacts of widespread fossil fuel use, with special emphasis on reducing outdoor emissions of greenhouse gases and other air pollutants. The second is to find ways to include more of the harmful environmental costs of using fossil fuels in their market prices.

Figure 13-48 summarizes these and other strategies for making the transition to a more sustainable energy future over the next 50 years (**Concept 13-6**).

Economics, Politics, and Education Can Help Us Shift to More Sustainable Energy Resources

To most analysts, economics, politics, and consumer education hold the keys to making a shift to more sustainable energy resources. Governments can use three strategies to help stimulate or reduce the short-term and long-term use of a particular energy resource.

First, they can *keep the prices of selected energy resources artificially low to encourage use of those resources.* They do this by providing research and development subsidies, tax breaks, and loan guarantees to encourage the development of those resources, and by enacting regulations that favor them. For decades, this approach has been employed to stimulate the development and use of fossil fuels and nuclear power in the United States as well as in most other more-developed countries. This has created an uneven economic playing field that *encourages* energy waste and more rapid depletion of these nonrenewable energy resources, while it *discourages* improvements in energy efficiency and the development of a variety of renewable energy resources.

Many energy analysts argue that one of the most important steps that governments can take to level the economic playing field is to phase out the $250–300 billion in annual subsidies and tax breaks now provided worldwide for fossil fuels and nuclear energy—both of which are mature industries that could be left to stand on their own, economically. For example, the U.S. oil industry—one of the country's biggest and most profitable industries—gets roughly $4 billion a year in subsidies and tax breaks, and for decades, has fought off attempts to sharply reduce or eliminate these gifts from U.S. taxpayers. Energy analysts also call for greatly increasing subsidies and tax breaks for developing and using renewable energy and energy-efficiency technologies and in developing less-costly renewable energy resources.

However, making such a shift in energy subsidies is difficult because of the immense political and financial power of the fossil fuel and nuclear power industries. They vigorously oppose any reduction of their subsidies and tax breaks, as well as any significant increase in subsidies and tax breaks for energy efficiency improvements, which would reduce the use of fossil fuels and nuclear power. They also oppose subsidies for competing renewable energy sources.

A *second* major strategy that governments can use is to *keep the prices of selected energy resources artificially high to discourage their use*. They can do this by eliminating existing tax breaks and other subsidies that favor use of a targeted resource, and by enacting restrictive regulations or taxes on its use. Such measures can increase government revenues, encourage improvements in energy efficiency, reduce dependence on imported energy, and decrease the use of energy resources that have limited supplies. To make such changes acceptable to the public, analysts suggest that governments can offset energy taxes by reducing income and payroll taxes, and providing an energy safety net for low-income users.

Third, governments can *emphasize consumer education.* Even if governments offer generous financial incentives for energy efficiency and renewable energy use, people will not make such investments if they are uninformed—or misinformed—about the availability, advantages, disadvantages, and hidden environmental costs of various energy resources.

An excellent example of what a government can do to bring about a more sustainable energy mix is the case of Germany. It is the world's most solar-powered nation, with half of the world's installed capacity. Why does cloudy Germany have more solar water heaters and solar cell panels than sunnier France and Spain have? One reason is that the German government made the public aware of the environmental benefits of these technologies. The other is that the government provided consumers with substantial economic incentives for using them.

We have the creativity, wealth, and most of the technology needed to make the transition to a more sustainable energy future within your lifetime. GOOD NEWS Making this transition depends primarily on *education, economics,* and *politics*—on how well individuals understand environmental and energy problems and their possible solutions, and on how they vote and then influence their elected officials. People can also vote with their wallets by refusing to buy energy-inefficient and environmentally harmful products and services, and by letting company executives know about their choices.

Here are this chapter's *three big ideas*:

- We should evaluate energy resources on the basis of their potential supplies, how much net energy they provide, and the environmental impacts of using them.
- Using a mix of renewable energy sources—especially solar, wind, flowing water, sustainable biofuels, and geothermal energy—can drastically reduce pollution, greenhouse gas emissions, and biodiversity losses.
- Making the transition to a more sustainable energy future will require sharply reducing energy waste, using a mix of environmentally friendly renewable energy resources, and including the harmful environmental costs of energy resources in their market prices.

REVISITING Wind Power and Sustainability

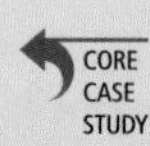

In the **Core Case Study** that opens this chapter, we considered the potential of wind as an energy resource, particularly in the United States. We saw that there is tremendous potential for this resource and that it has a very favorable net energy ratio, compared to other energy resources.

We then looked at the current energy picture in the world and in the United States, paying special attention to *net energy yield* for each resource. The guiding scientific principle underlying all energy use is that the long-term usefulness of any energy resource depends on its *net energy ratio*. Any energy resource with a low net energy ratio, such as nuclear power or corn-based ethanol, cannot compete in the open marketplace with other energy resources that have higher ratios unless it is subsidized. This principle—based on the two laws of thermodynamics, which we have never been able to violate—is very important to us for environmental, economic, and political reasons.

We also considered the environmental impacts of using each resource. Using fossil fuel resources involves high environmental impacts, although these impacts vary. For example, while burning natural gas produces relatively low greenhouse gas emissions, using coal produces very high levels of such emissions along with extensive land disturbance. And while generating electricity from a nuclear power plant involves no greenhouse gas emissions, the whole *nuclear power fuel cycle*—from mining the uranium fuel to dismantling and storing the worn-out reactor parts—involves moderately high levels of such emissions as well as other environmental impacts.

By relying mostly on nonrenewable fossil fuels, we violate the three **principles of sustainability** (see back cover), and this has become a serious long-term problem. The technologies we use to obtain energy from these resources disrupt the earth's chemical cycles by diverting huge amounts of water, degrading or destroying terrestrial and aquatic ecosystems, and emitting large quantities of greenhouse gases and other air pollutants. Using these technologies also destroys and degrades biodiversity and ecosystem services.

However, by applying the three **principles of sustainability**, we can work toward solving some of these problems. This means:

- relying much more on direct and indirect forms of solar energy for our electricity, heating and cooling, and other needs;
- recycling and reusing materials and thus reducing wasteful and excessive consumption of energy and matter; and
- mimicking nature's reliance on biodiversity by using a diverse mix of locally and regionally available renewable energy resources.

A transition to renewable energy is inevitable, not because fossil fuel supplies will run out—large reserves of oil, coal, and gas remain in the world—but because the costs and risks of using these supplies will continue to increase relative to renewable energy.

MOHAMED EL-ASHRY

REVIEW

CORE CASE STUDY

1. Explain why wind has great potential for generating electricity (**Core Case Study**). Summarize the story of the potential for wind power in the United States. CORE CASE STUDY

SECTION 13-1

2. What is the key concept for this section? What is **net energy yield** and why is it important for evaluating energy resources? What is a net energy ratio and how is it used? Explain why the nuclear fuel cycle has a low net energy yield and thus must be subsidized to compete in the open marketplace.

SECTION 13-2

3. What is the key concept for this section? What is our main source of energy? What are three forms of indirect solar energy? What is commercial energy and what major types of it are used in the world and in the United States? What are **fossil fuels**? What is **crude oil (petroleum)** and how is it extracted from the earth and refined? Define peak production. What is a **petrochemical** and why are such chemicals important? How much oil does the United States produce, compared with what it uses? What are the major advantages and disadvantages of using crude oil as an energy resource? What is **shale oil** and how is it produced? What is **tar sand**, or **oil sand**, and how is it extracted and converted to heavy oil? What are the major advantages and disadvantages of using heavy oils produced from tar sand and shale oil as energy resources?

4. Define **natural gas**, **liquefied petroleum gas (LPG)**, and **liquefied natural gas (LNG)**. What are two sources of unconventional natural gas? What are the major advantages and disadvantages of using natural gas as an energy resource? What are two problems that are affecting the reputation of natural gas as a cleaner fossil fuel? Describe each problem. What is **coal** and how is it formed? What are the major advantages and disadvantages presented by using coal as an energy resource? Describe the problems presented by coal ash. Explain why there is no such thing as clean coal.

SECTION 13-3

5. What is the key concept for this section? How does a nuclear fission reactor work and what are its major safety features? Describe the **nuclear fuel cycle** and summarize the advantages and disadvantages of relying on it. Explain how spent fuel rods are stored and what risks this presents. Why is dealing with high-level radioactive wastes a difficult problem and what has been done about it? What can we do with worn-out nuclear power plants? How has the United States dealt with the nuclear waste problem? To what extent can nuclear power lessen our dependence on imported oil? What would the nuclear industry have to do in order to play an effective role in slowing projected atmospheric warming? Summarize the stories of the world's three worst nuclear power plant accidents. Summarize the arguments of experts who disagree over the future of nuclear power. What factors have hindered the development of nuclear power? What is **nuclear fusion** and what is its potential as an energy resource?

SECTION 13-4

6. What is the key concept for this section? What is **energy efficiency**? What percentage of the energy used in the United States is unnecessarily wasted? List four widely used energy-wasting technologies. What are the major benefits of reducing energy waste? List three reasons why this source of energy has been neglected. Describe three ways to save energy and money in **(a)** industry, **(b)** transportation, **(c)** new buildings, and **(d)** existing buildings. How could a smarter electrical grid help us to save energy and money? Explain why the true cost of gasoline is much higher than what consumers pay at the pump. What is a fee-bate? Distinguish among hybrid, plug-in hybrid, and fuel-cell motor vehicles. Summarize the story of the Rocky Mountain Institute headquarters building. What are three reasons for high levels of energy waste in the world?

SECTION 13-5

7. What is the key concept for this section? Distinguish between **passive solar heating** and **active solar heating systems** and summarize the major advantages and disadvantages of such systems. What are three ways to cool houses naturally? List the major advantages and disadvantages of using solar thermal systems to generate high-temperature heat and electricity. What is a **solar cell** (**photovoltaic** or **PV cell**) and what are the major advantages and disadvantages of using such devices to produce electricity?

8. What are the major advantages and disadvantages of using hydropower to produce electricity? Summarize the benefits and drawbacks of using wind to generate electricity (**Core Case Study**). What are the major advantages and disadvantages of using **(a)** solid biomass to provide heat and electricity and **(b)** liquid biofuels to power vehicles? Summarize the arguments for and against using ethanol to power vehicles. What is **geothermal energy**, and what are the major advantages and disadvantages of using it? What are the major advantages and disadvantages of using hydrogen as a fuel?

CORE CASE STUDY

SECTION 13-6

9. What is the key concept for this section? List six questions that energy policy makers try to answer when evaluating energy resource alternatives. List three general conclusions of energy experts about possible future energy paths for the world. Describe three roles that governments play in determining which energy resources we use.

10. What are this chapter's *three big ideas*? Explain how we violate each of the three **principles of sustainability** in relying mostly on fossil fuels and nuclear power. Explain how we can use these three principles to work toward solving the problems caused by our use of fossil fuels and nuclear energy.

SUSTAINABILITY

Note: Key terms are in **bold** type.

CRITICAL THINKING

1. Suppose that a wind power developer has proposed building a wind farm near where you live (**Core Case Study**). Would you be in favor of the project or opposed to it? Write a letter to your local newspaper or a blog for a website explaining your position and your reasoning. Include the concept of *net energy* in your arguments. As part of your research, determine how the electricity you use now is generated and where the power plant is located, and include this information in your arguments.

CORE CASE STUDY

2. Should governments give a high priority to considering net energy yields when deciding what energy resources to support? What are other factors that should be considered? Which factor or factors should get the most weight in decision-making? Explain your thinking.

3. To continue using oil at the current rate, some analysts warn that we must discover and add to global oil reserves the equivalent of two new Saudi Arabian supplies every 10 years. Do you think this is possible? If not, what effects might the failure to find such supplies have on your life and on the lives of any children and grandchildren that you might have?

4. Some people in China point out that the United States and European nations fueled their economic growth since the beginning of the Industrial Revolution by burning coal, with little effort to control the resulting air pollution, and then sought cleaner energy sources later when they became more affluent. China says it is being asked to clean up before it becomes affluent enough to do so. Do you think this is a fair expectation? Explain. Because China's outdoor air pollution and greenhouse gas emissions have implications for the entire world, what role, if any, should the more-developed nations play in helping China to reduce its dependence on coal and to rely on more environmentally sustainable energy sources?

5. Explain why each of the following widely believed statements is not accurate: **(a)** coal can be a clean fuel; **(b)** natural gas is a low-carbon fuel; **(c)** nuclear power is a low-carbon energy resource.

6. Should buyers of energy-efficient motor vehicles receive large rebates funded by fees levied on gas guzzlers? Explain.

7. Explain why you agree or disagree with the following proposals made by various energy analysts:
   a. Government subsidies for all energy alternatives should be eliminated so that all energy choices can compete in a truly free-market system.
   b. All government tax breaks and other subsidies for conventional fossil fuels, synthetic natural gas and oil, and nuclear power (fission and fusion) should be phased out. They should be replaced with subsidies and tax breaks for improving energy efficiency and developing renewable energy resources.
   c. Development of solar, wind, and hydrogen energy should be left to private enterprise and should receive little or no help from the federal government, but nuclear energy and fossil fuels should continue to receive large federal government subsidies and tax breaks.

8. Congratulations! You are in charge of the world. List the five most important features of your energy policy and explain why each of them is important and how they relate to each other.

DOING ENVIRONMENTAL SCIENCE

Do a study of energy use at your school. Try to determine how the electricity used by the school is generated. If it comes from more than one source (such as coal-fired, gas-fired, or nuclear power plants, wind farms, hydroelectric dams, solar cells, and geothermal power), determine or estimate the percentage of the total electricity provided by each of the various sources. Similarly, determine how buildings are heated, how water is heated, and how vehicles are powered. You might also choose to do an energy efficiency audit—an analysis of how and where energy is wasted on all or a part of the campus (see **http://www.nwf.org/campusecology/BusinessCase/Index.cfm**). Use your findings to develop a plan for an energy-efficiency improvement program and present your plan to school officials.

GLOBAL ENVIRONMENT WATCH EXERCISE

Using the *Energy* topic portal, find *energy efficiency*, go to the *statistics* and click on the *Average Energy Consumption of New Appliances 1990 to 2005 models*. Based on this data, how do you think the large-scale replacement of 1990 appliances with 2005 appliances would impact overall energy consumption? Do further research using the GEW to learn about improvements in lighting efficiency and answer the same question concerning overall energy consumption.

ECOLOGICAL FOOTPRINT ANALYSIS

In 2008, the average fuel economy of new cars, light trucks, and SUVs in the United States was 11.4 kilometers per liter (kpl), or 26.6 miles per gallon (mpg), and the average motor vehicle in the United States was driven 19,300 kilometers (12,000 miles). There were about 250 million motor vehicles in the United States in 2008. The U.S. Environmental Protection Agency estimates that 2.3 kilograms of CO_2 are released when 1 liter of gasoline is burned (19.4 pounds of CO_2 are released when 1 gallon is burned). Use these data to calculate the *gasoline consumption* and *carbon footprints* of individual motor vehicles with different fuel efficiencies and for all of the motor vehicles in the United States by answering the following questions:

1. Suppose a car has an average fuel efficiency of 8.5 kpl (20 mpg) and is driven 19,300 kilometers (12,000 miles) a year. **(a)** How many liters (and gallons) of gasoline does this vehicle consume in a year? **(b)** If gasoline costs 80 cents per liter ($3.00 per gallon), how much will the owner spend on fuel in a year? **(c)** How many liters (and gallons) of gasoline would be consumed by a U.S. fleet of 250 million such vehicles in a year? (1 liter = 0.265 gallons and 1 kilometer = 0.621 miles)
2. Recalculate the values in Question 1, assuming that a car has an average fuel efficiency of 19.6 kpl (46 mpg).
3. Determine the number of metric tons of CO_2 emitted annually by **(a)** the car described in Question 1 with a low fuel efficiency, **(b)** a fleet of 250 million vehicles with this same fuel efficiency, **(c)** the car described in Question 2 with a high fuel efficiency, and **(d)** a fleet of 250 million vehicles with this same high fuel efficiency. These calculations provide a rough estimate of the carbon footprints for individual cars and for the entire U.S. fleet with low- and high-efficiency cars. (1 kilogram = 2.20 pounds; 1 metric ton = 1,000 kilograms = 2,200 pounds = 1.1 tons; 1 ton = 2,000 pounds).
4. If the average fuel efficiency of the U.S. fleet increased from 8.5 kpl (20 mpg) to 19.6 kpl (46 mpg), by what percentage would this reduce the CO_2 emissions from the entire fleet per year? You can think of this as the percentage reduction in the carbon footprint of the U.S. motor vehicle fleet.

LEARNING ONLINE

The website for *Environmental Science 14e* contains helpful study tools and other resources to take the learning experience beyond the textbook. Examples include flashcards, practice quizzing, information on green careers, tips for more sustainable living, activities, and a wide range of articles and further readings. To access these materials and other companion resources for this text, please visit **www.cengagebrain.com**. For further details, see the preface, p. xvi.

14 Environmental Hazards and Human Health

CORE CASE STUDY Are Baby Bottles and Food Cans Safe to Use? The BPA Controversy

In the human body, chemicals called *hormones* help to control growth and sexual development and reproduction. They also affect one's learning ability and behavior. Scientists have discovered that certain pesticides and other synthetic chemicals can act as hormone imposters that may impair reproductive systems and sexual development or cause various physical and behavioral disorders.

Some of these *hormone mimics* are chemically similar to female sex hormones called *estrogens*. In males, excess levels of female hormones, particularly estrogen, can cause feminization, smaller penises, diminished sex drive and sperm counts, and the presence of both male and female sex organs (hermaphroditism). In females, several studies have found that higher levels of BPA are associated with infertility, recurrent miscarriages, and breast cancer.

A widely used estrogen mimic is bisphenol A (BPA). It is used to harden certain plastics (especially shatter-proof polycarbonate) that are used in a variety of products including some baby bottles (Figure 14-1), sipping cups, and some pacifiers, as well as reusable water bottles, sports drink and juice bottles, microwave dishes, and food storage containers. BPA is also used to make some dental sealants as well as the plastic resins that line nearly all food and soft drink cans and cans holding baby formulas and foods. This type of liner allows the container to withstand extreme temperatures, keeps the food from interacting with the metal in the can, prevents rust, and helps to preserve the canned food. BPA is difficult to avoid, and it is rarely included on ingredient lists.

Research indicates that the BPA in plastics can leach into water or food when the plastic is heated to high temperatures, microwaved, or exposed to acidic liquids. A 2009 Harvard University Medical School study found that there was a 66% increase in BPA levels in the urine of participants who drank from polycarbonate bottles regularly for just one week. According to lead researcher Karin Michels, the study confirmed that certain containers can release BPA into the liquids stored inside them, even when they are not heated.

Three studies released in 2010 indicated that BPA was commonly found on credit card and other receipts printed by thermal imaging printers. The skin on one's fingers, especially when wet, can absorb this BPA. Gloves do not prevent such absorption because the BPA molecules can move through most glove fabrics.

A 2007 study by the Centers for Disease Control and Prevention (CDC) indicated that 93% of Americans older than age 6 had trace levels of BPA in their urine. While these levels were well below the acceptable level set by the U.S. Environmental Protection Agency (EPA), the EPA level was established in the late 1980s, long before we knew much about the potential effects of BPA on human health. The CDC study also found that children and adolescents had generally higher BPA levels than adults had.

There is scientific controversy over whether exposure to such trace amounts of BPA and other hormone imposters poses a serious threat to human health, especially for fetuses and infants. This raises several important questions: How do scientists determine the potential harm from exposure to various chemicals? How serious is the risk of harm from a particular chemical compared to other risks? What should individuals and government health and regulatory officials do if there is preliminary but not solid evidence of harm?

In this chapter, we will look at how scientists try to answer these questions about our exposure to chemicals. We also consider questions about health threats from disease-causing bacteria, viruses, and protozoa, and from other hazards that kill millions of people each year.

Anyka/Shutterstock.com

Figure 14-1 All plastic polycarbonate bottles contain bisphenol A (BPA), an estrogen mimic that can leach out of such bottles, especially when they are warmed, microwaved, or used to hold acidic juices, or are washed with alkaline soaps or detergents. Some manufacturers are no longer using polycarbonate plastic in baby bottles or in sipping cups.

Key Questions and Concepts

14-1 What major health hazards do we face?

CONCEPT 14-1 We face health hazards from biological, chemical, physical, and cultural factors, and from the lifestyle choices we make.

14-2 What types of biological hazards do we face?

CONCEPT 14-2 The most serious biological hazards we face are infectious diseases such as flu, AIDS, tuberculosis, diarrheal diseases, and malaria.

14-3 What types of chemical hazards do we face?

CONCEPT 14-3 There is growing concern about chemicals in the environment that can cause cancers and birth defects, and disrupt the human immune, nervous, and endocrine systems.

14-4 How can we evaluate chemical hazards?

CONCEPT 14-4A Scientists use live laboratory animals, case reports of poisonings, and epidemiological studies to estimate the toxicity of chemicals, but these methods have limitations.

CONCEPT 14-4B Many health scientists call for much greater emphasis on pollution prevention to reduce our exposure to potentially harmful chemicals.

14-5 How do we perceive risks and how can we avoid the worst of them?

CONCEPT 14-5 We can reduce the major risks we face by becoming informed, thinking critically about risks, and making careful choices.

Note: Supplements 2 (p. S3), 4 (p. S10), and 6 (p. S22) can be used with this chapter.

The dose makes the poison.

PARACELSUS, 1540

14-1 What Major Health Hazards Do We Face?

▶ **CONCEPT 14-1 We face health hazards from biological, chemical, physical, and cultural factors, and from the lifestyle choices we make.**

Risks Are Usually Expressed as Probabilities

A **risk** is the *probability* of suffering harm from a hazard that can cause injury, disease, death, economic loss, or damage. It is usually expressed as a mathematical statement about how likely it is that harm will be suffered from a hazard. Scientists often state probability in terms such as "The lifetime probability of developing lung cancer from smoking one pack of cigarettes per day is 1 in 250." This means that 1 of every 250 people who smoke a pack of cigarettes every day will likely develop lung cancer over a typical lifetime (usually considered to be 70 years). Probability can also be expressed as a percentage, as in a 30% chance of rain.

Risk assessment is the process of using statistical methods to estimate how much harm a particular hazard can cause to human health or to the environment. It helps us to estimate the probability of a risk, compare it with the probability of other risks, and establish priorities for avoiding or managing risks. **Risk management** involves deciding whether and how to reduce a particular risk to a certain level and at what cost. Figure 14-2 summarizes how risks are assessed and managed.

Figure 14-2 Science: Risk assessment and risk management are used to estimate the seriousness of various risks and how to reduce such risks. **Question:** What is an example of how you have applied this process in your daily living?

We Face Many Types of Hazards

All of us take risks every day. Examples include choosing to drive or ride in a car through heavy traffic; talking on a phone or texting while driving; eating foods with a high cholesterol or fat content that can contribute to potentially fatal heart attacks; consuming drinks and foods with a high sugar content, which can lead to obesity and type 2 diabetes; drinking too much alcohol; smoking or being in an enclosed space with a smoker; lying out in the sun or going to a tanning parlor, which increases the risk of getting skin cancer; and living in a tornado-, hurricane-, or flood-prone area.

At the same time, most of us would not choose to live completely risk-free. The key questions are *how serious are the risks we face* and *do the benefits of certain activities outweigh the risks?*

We can suffer harm from five major types of hazards (**Concept 14-1**):

- *Biological hazards* from more than 1,400 pathogens that can infect humans. A **pathogen** is an organism that can cause disease in another organism. Examples are bacteria, viruses, parasites, protozoa, and fungi.
- *Chemical hazards* from harmful chemicals in our air, water, soil, food, and human-made products (**Core Case Study**). CORE CASE STUDY
- *Natural hazards* such as fire, earthquakes, volcanic eruptions, floods, and storms.
- *Cultural hazards* such as unsafe working conditions, unsafe highways, criminal assault, and poverty.
- *Lifestyle choices* such as smoking, making poor food choices, drinking too much alcohol, and having unsafe sex.

14-2 What Types of Biological Hazards Do We Face?

CONCEPT 14-2 The most serious biological hazards we face are infectious diseases such as flu, AIDS, tuberculosis, diarrheal diseases, and malaria.

Some Diseases Can Spread from One Person to Another

An **infectious disease** is caused when a pathogen such as a bacterium, virus, or parasite invades the body and multiplies in its cells and tissues. Some examples are flu, malaria, tuberculosis, and measles. **Bacteria** are singe-cell organisms that are found everywhere and that can multiply very rapidly on their own. Most bacteria are harmless or beneficial but some cause infectious bacterial diseases. A bacterial disease such as tuberculosis results from an infection as the bacteria multiply and spread throughout the body. **Viruses** are smaller than bacteria and work by invading a cell and taking over its genetic machinery to copy themselves. They then multiply and spread throughout one's body, causing a viral disease such as flu or AIDS.

A **transmissible disease** (also called a *contagious* or *communicable disease*) is an infectious disease that can be transmitted from one person to another. Some are bacterial diseases such as tuberculosis, gonorrhea, and strep throat. Others are viral diseases such as the common cold, flu, and AIDS.

A **nontransmissible disease** is caused by something other than a living organism and does not spread from one person to another. Such diseases tend to develop slowly and have multiple causes. They include cardiovascular (heart and blood vessel) diseases, most cancers, asthma, diabetes, and malnutrition.

In 1900, infectious disease was the leading cause of death in the world and in the United States. Since then, and especially since 1950, the incidences of infectious diseases and the death rates from them have been greatly reduced. GOOD NEWS This has been achieved mostly by a combination of better health care, the use of antibiotics to treat infectious diseases caused by bacteria, and the development of vaccines to prevent the spread of some infectious viral diseases. As a result, average life expectancy has increased in most countries.

Infectious Diseases Are Still Major Health Threats

Despite the shift in risk levels from transmissible to nontransmissible diseases, infectious diseases remain as serious health threats, especially in less-developed countries. Figure 14-3 shows major pathways on which infectious disease organisms can enter the human body. Such diseases can then be spread through air, water, food, and body fluids such as feces, urine, blood, and droplets sprayed by sneezing and coughing.

A large-scale outbreak of an infectious disease in an area or a country is called an *epidemic.* A global epidemic such as tuberculosis or AIDS is called a *pandemic.* Figure 14-4 shows the annual death tolls from the world's seven deadliest infectious diseases (**Concept 14-2**).

Pets
Livestock
Wild animals
Insects
Food
Water
Air
Other humans
Humans
Fetus and babies

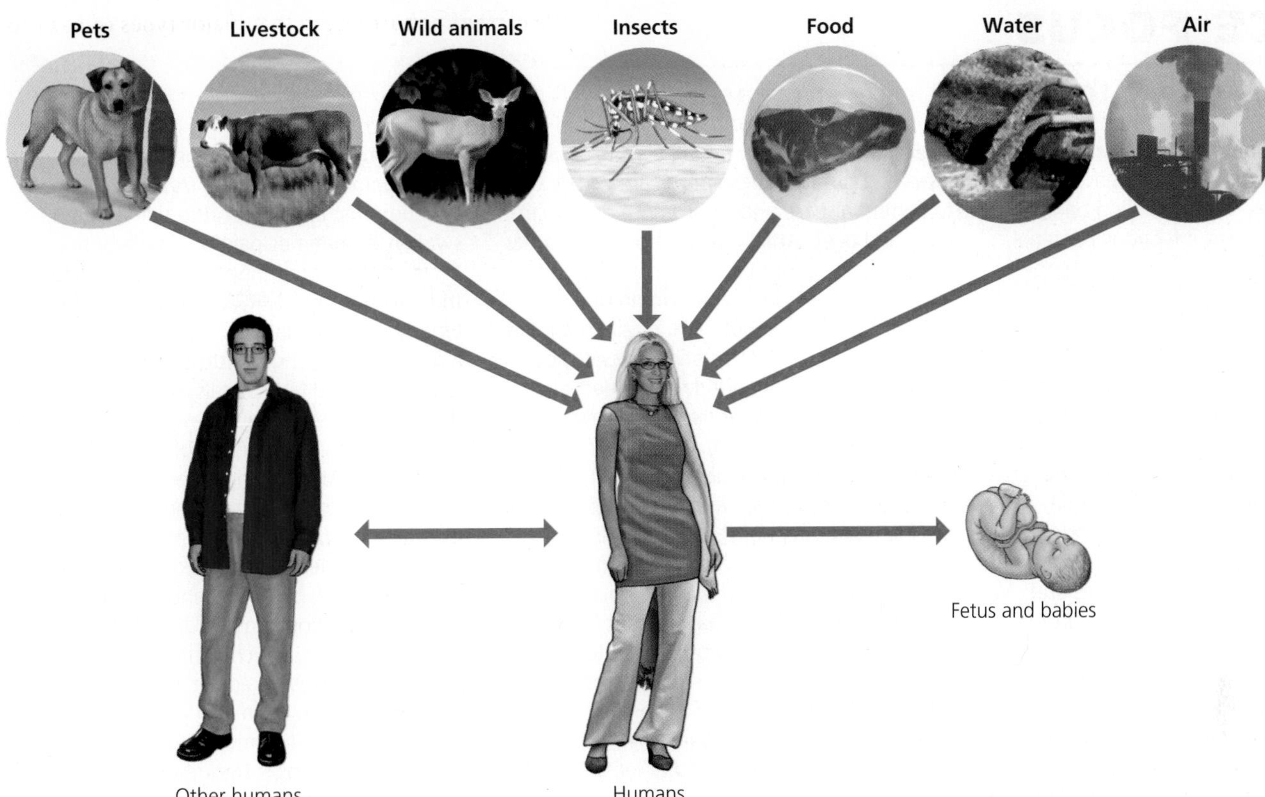

Figure 14-3 **Science:** There are a number of pathways on which infectious disease organisms can enter the human body. **Question:** Can you think of other pathways not shown here?

One reason why infectious disease is still a serious threat is that many disease-carrying bacteria have developed genetic immunity to widely used antibiotics (Science Focus, p. 352). Also, many disease-transmitting species of insects such as mosquitoes have become immune to widely used pesticides such as DDT that once helped to control their populations.

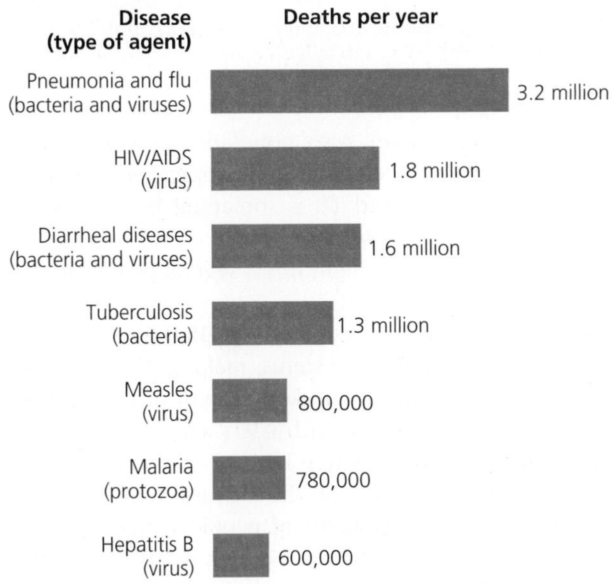

Disease
(type of agent)
Deaths per year
Pneumonia and flu
(bacteria and viruses)
3.2 million
HIV/AIDS
(virus)
1.8 million
Diarrheal diseases
(bacteria and viruses)
1.6 million
Tuberculosis
(bacteria)
1.3 million
Measles
(virus)
800,000
Malaria
(protozoa)
780,000
Hepatitis B
(virus)
600,000

Figure 14-4 *Global outlook:* In 2009, the World Health Organization estimated that the world's seven deadliest infectious diseases killed nearly 10.1 million people—most of them poor people in less-developed countries (**Concept 14-2**). This averages about 27,600 mostly preventable deaths every day. **Question:** How many people, on average, die prematurely from these diseases every hour? (Data from the World Health Organization, 2010)

■ CASE STUDY
The Growing Global Threat from Tuberculosis

Since 1990, one of the world's most underreported stories has been the rapid spread of tuberculosis (TB), an extremely contagious bacterial infection of the lungs. About 33% of the people on the planet are infected with the TB bacterium and 10% of them will eventually become sick with active TB.

Many TB-infected people do not appear to be sick, and about half of them do not know they are infected. Left untreated, each person with active TB typically infects a number of other people. Without treatment, about half of the people with active TB die from bacterial destruction of their lung tissue (Figure 14-5, p. 352).

According to the World Health Organization (WHO), TB strikes about 9.3 million people per year and kills 1.3 million—about 84% of them in less-developed countries. On average, someone dies of TB every 24 seconds.

Several factors account for the recent spread of TB. One is that there are too few TB screening and control programs, especially in less-developed countries, where 95% of the new cases occur. A second problem is that most strains of the TB bacterium have developed genetic resistance to the majority of the effective antibiotics

SCIENCE FOCUS

Genetic Resistance to Antibiotics Is Increasing

Antibiotics are chemicals that can kill bacteria. No such chemicals have been found to kill viruses. We risk falling behind in our efforts to prevent infectious bacterial diseases because of the astounding reproductive rate of bacteria, some of which can produce well over 16 million offspring in 24 hours. This allows bacteria to quickly become genetically resistant to an increasing number of antibiotics through natural selection (see Figure 4-5, p. 67).

Other factors can promote such genetic resistance. One is the spread of bacteria around the globe by human travel and international trade. Airplanes, for example, are incubators for disease-causing viruses and bacteria that are picked up in one country, circulated for hours within a closed environment, and then carried to other parts of the world as the global village shrinks. Another is the overuse of pesticides, which increases populations of pesticide-resistant insects and other carriers of bacterial diseases. In addition, some drug-resistant bacteria can quickly transfer their resistance to nonresistant bacteria by exchanging genetic material.

Yet another factor in genetic resistance is the overuse of antibiotics for colds, flu, and sore throats, most of which are viral diseases that cannot be treated with antibiotics. In many countries, antibiotics are available without a prescription, which promotes unnecessary use.

Bacterial resistance to some antibiotics has increased because these drugs are widely used to control disease and to promote growth in dairy and beef cattle, poultry, hogs, and other livestock raised for human consumption. According to a study by the Union of Concerned Scientists, about 70% of all antibiotics used in the United States and 50% of those used worldwide are added to the feed of healthy livestock. In addition, the growing use of antibacterial hand soaps and other antibacterial cleansers is probably promoting genetic resistance in bacteria, which is the opposite of the intended effect of these products.

As a result of these factors acting together, every major disease-causing bacterium has now developed strains that resist at least one of the roughly 200 antibiotics used to treat bacterial infections such as tuberculosis (see the Case Study on p. 351). Each year, about 1.7 million people pick up infections while staying in U.S. hospitals, and at least 99,000 of them die as a result, according to the U.S. Centers for Disease Control and Prevention (CDC). The genetic resistance of bacteria to antibiotics plays a role in these deaths. Major drug companies have relatively little incentive to develop new antibiotics because these drugs are less profitable for the companies than other types of drugs are.

Of particular concern is a bacterium known as *methicillin-resistant staphylococcus aureus*, commonly known as MRSA (pronounced "mersa"), which has become resistant to most common antibiotics. This type of staph infection first appears on the skin as a red, swollen, sometimes painful pimple or boil. Many victims think they have a spider bite that will not heal. MRSA can cause a vicious type of pneumonia, flesh-eating wounds, and a quick death if it gets into the bloodstream. It can be found in hospitals, nursing homes, dialysis centers, schools, playgrounds, meeting rooms, gyms, and college dormitories. It can be spread through skin contact, unsanitary use of tattoo needles, and contact with poorly laundered clothing and shared items such as towels, bed linens, athletic equipment, and razors. In 2007, the CDC reported that more than 94,000 people in the United States had MRSA infections, which contributed to the premature deaths of almost 18,600 people, many of them children.

The best way to prevent the spread of bacterial infections is for individuals to wash their hands thoroughly and frequently using plain soap (not antibacterial soap). Also, patients should ask doctors, dentists, and nurses who are treating them to make sure that they have washed their hands or put on fresh plastic gloves.

Critical Thinking

What are three things that we could do to slow the rate at which disease-causing bacteria are developing resistance to antibiotics?

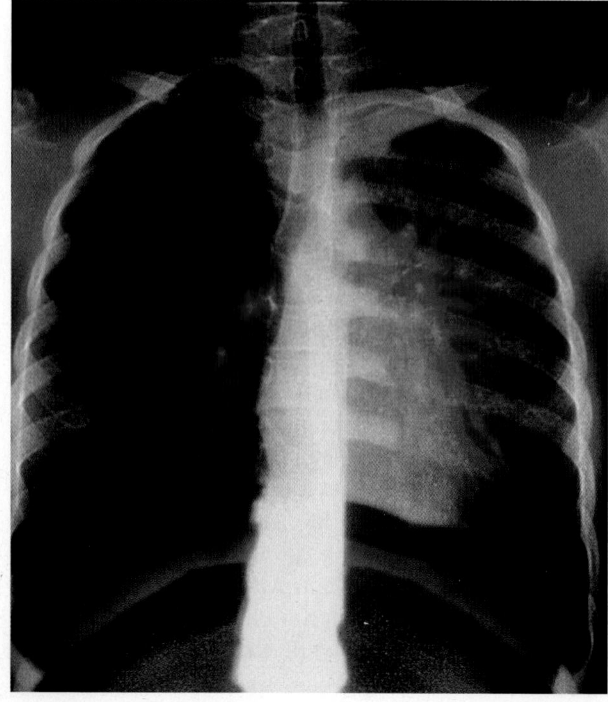

SIU/Peter Arnold, Inc.

Figure 14-5 The colorized red areas in this chest X-ray show where TB bacteria have destroyed lung tissue.

(Science Focus, above). Also, population growth, urbanization, and air travel have greatly increased person-to-person contacts, and TB is spreading faster in areas where large numbers of poor people crowd together. A person with active TB might infect several people during a single bus or plane ride.

In addition, AIDS (Case Study, p. 354) greatly weakens its victims' immune systems, making AIDS patients highly susceptible to unchecked growth of the TB bacteria. As a result, people with AIDS are 30 to 50 times more likely to develop active TB.

Slowing the spread of the disease requires early identification and treatment of people with active TB, especially those with a chronic cough, which is the primary way in which the disease is spread from person to person. Treatment with a combination of four inexpensive drugs can cure 90% of individuals with active TB. To be effective, the drugs must be taken every day for 6–9 months. Because the symptoms disappear after a few weeks, many patients, thinking they are cured, stop taking the drugs, allowing the disease to recur in drug-resistant forms and to spread to other people.

In recent years, a deadly and apparently incurable form of tuberculosis, known as *multidrug-resistant TB*, has been spreading. According to the WHO, this form of TB is now found in 49 countries and each year, there are about 490,000 new cases and 120,000 deaths. Because this disease cannot be treated effectively with antibiotics, victims must be permanently isolated from the rest of society. Victims also pose a threat to health workers. The WHO has estimated that between 2011 and 2015, more than 2 million people will contract multidrug-resistent TB.

Most people who have active TB live in the world's poorest countries and many of them cannot afford treatments. Thus, there is little financial incentive for large drug companies to invest a great deal of money on developing drugs to treat the disease. However, efforts to develop more effective antibiotics and vaccines are being undertaken by governments and private groups such as the Bill and Melinda Gates Foundation. Also, researchers are developing new and easier ways to detect TB in its victims—a key improvement that could save millions of lives (Individuals Matter, below).

GOOD NEWS

Viral Diseases and Parasites Kill Large Numbers of People

Viruses evolve quickly, are not affected by antibiotics, and can kill large numbers of people. The biggest killer is the *influenza* or *flu* virus (**Concept 14-2**), which is transmitted by the body fluids or airborne emissions of an infected person. Easily transmitted and especially potent flu viruses could spread around the world in a pandemic that could kill millions of people in only a few months.

The second biggest viral killer is the *human immunodeficiency virus* (HIV) (see the Case Study that follows). On a global scale, HIV infects about 1.8 million people every year, and the complications resulting from AIDS kill about 1.8 million people annually.

The third largest viral killer is the *hepatitis B virus* (HBV), which damages the liver and kills about 600,000 people each year. Like HIV, it is transmitted by unsafe sex, sharing of needles by drug users, infected mothers who pass the virus to their offspring before or during birth, and exposure to infected blood.

In recent years, several other viruses that cause previously unknown diseases have received widespread media coverage. They are examples of *emergent diseases*—illnesses that were previously unknown or were absent in human populations for at least 20 years. One is the *West Nile virus*, which is transmitted to humans by the bite of a common mosquito that gets infected when it feeds on birds that carry the virus. Since 1992 when this virus emerged in the United States, it has spread from coast to coast. Between 1999 and 2009, it caused severe illnesses in more than 23,500 people and killed more than 1,000 people. Such illnesses include viral encephalitis, an acute infection of the brain and spinal cord, and viral meningitis, an infection in the membranes that cover the brain and spinal cord. Fortunately, the chance of being infected and killed by West Nile virus is low (about 1 in 2,500).

Scientists believe that more than half of all infectious diseases throughout history were originally transmitted to humans from wild or domesticated animals. The West Nile virus, HIV/AIDS, and a flu strain from birds, called *avian flu*, all fall in this category. The development of such diseases has spurred the growth of the relatively new field of *ecological medicine*. **GREEN CAREER:** ecological medicine

Health officials are concerned about the spread of West Nile virus, the newest avian flu, and other emergent diseases. But in terms of annual infection rates and deaths, the three most dangerous viruses by far are still flu, HIV, and HBV (**Concept 14-2**).

INDIVIDUALS MATTER

Three College Students Have Saved Thousands of Lives

When Hersh Tapadia, Daniel Jeck, and Pavak Shah were seniors in North Carolina State University's engineering school, they recognized a problem that was contributing to the deaths of many thousands of tuberculosis (TB) victims. For their senior project, they chose to tackle this problem, and as a result, they have prevented an untold number of TB fatalities.

If TB is diagnosed early enough, a victim can receive treatment and go on to live a healthy life. However, early diagnosis requires testing a sample of a patient's feces. Many health clinics in poor countries have to send their patients' samples to labs for analysis, which can take months. Patients often die waiting for lab results, and in the meantime, they have in many cases spread the disease to other people. Even worse, international health officials estimate that perhaps 40% of TB cases are missed because of outdated, inadequate medical technology and methods of diagnosis.

Enter the three engineering students who developed a relatively simple device to detect TB. When a microscope slide with a sample from a patient is inserted into the device, any TB bacteria that are present glow white against a black background. This allows health-care workers with little training to detect TB in seconds at a cost of less than a dollar per patient.

The three inventors are working on adapting this device so that it can be used to diagnose malaria, AIDS, and other infectious diseases. They plan to create a business that will manufacture and sell the devices. In this way, they hope to make a living by helping millions of people to live longer and healthier lives.

You can greatly reduce your chances of getting infectious diseases by practicing good old-fashioned hygiene. Wash your hands thoroughly and frequently with plain soap for at least 20 seconds (the amount of time needed to sing the Happy Birthday song twice), avoid touching your face, don't share personal items such as razors or towels, keep all cuts and scrapes covered with bandages until healed, and stay away from people who have flu or other viral diseases. Some health officials warn that using antibacterial soaps, liquids, and sprays in order to avoid infectious disease may be doing more harm than good, because they can contribute to genetic resistance in infectious bacteria (Science Focus, p. 352).

Yet another growing health hazard is infectious diseases caused by parasites, especially malaria (see the second Case Study that follows).

■ CASE STUDY
The Global HIV/AIDS Epidemic

The global spread of *acquired immune deficiency syndrome (AIDS)*, caused by infection with the *human immunodeficiency virus (HIV)*, is a major global health threat. This virus cripples the immune system and leaves the body vulnerable to infections such as tuberculosis (TB) and rare forms of cancer such as *Kaposi's sarcoma*. The virus is transmitted from one person to another by unsafe sex, sharing of needles by drug users, infected mothers who pass the virus on to their children before or during birth, and exposure to infected blood. A person infected with the HIV can live a normal life but if the infection develops into AIDS, death is likely.

Since the HIV virus was identified in 1981, this viral infection has spread around the globe. According to the WHO, in 2009, a total of about 33.3 million people worldwide (about 1.1 million in the United States) were living with HIV, and there were about 2.6 million new cases of AIDS (40,000 in the United States)—half of them in people aged 15 to 24. This is down from an estimated 3.2 million new cases in 1997.

Treatment that includes combinations of expensive antiviral drugs can slow the progress of HIV and AIDS, but they are expensive. With such drugs, a person with AIDS, on average, can expect to live about 24 years after being infected at a cost of about $25,200 a year. Such drugs cost too much to be used extensively in less-developed countries where AIDS infections are widespread. About 5 million people were getting antiviral drugs in 2009, but 10 million people needed treatment and did not get it.

For people who do not take such antiviral drugs, it takes an average of 10 to 11 years for an HIV infection to progress to AIDS. This long incubation period means that infected people often spread the virus for several years without knowing they are infected. About one of every five people infected with HIV is not aware of the infection. Currently, there is no cure for AIDS. Those who get this disease will almost certainly die from it, unless something else kills them first.

Between 1981 and 2009, more than 27 million people (610,000 in the United States) died of AIDS-related diseases. Each year, AIDS kills about 1.8 million people (about 18,000 in the United States). Globally, according to the WHO, AIDS is the leading cause of death among persons 25 to 44 years old.

AIDS has reduced the life expectancy of the 750 million people living in sub-Saharan Africa from 62 to 47 years, on average, and to 40 years in the seven countries most severely affected by AIDS. The premature deaths of teachers, health-care workers, soldiers, and other young, productive adults affect the population age structure in African countries such as Botswana (see Figure 6-10, p. 104). These deaths also lead to diminished education and health care, decreased food production and economic development, and disintegrating families. Thus, prevention is the best policy for dealing with HIV/AIDS. For example, one encouraging study in South Africa showed that a gel that women applied to themselves before and after having sex cut their chances of becoming infected with HIV by 39%.

GOOD NEWS

■ CASE STUDY
Malaria—The Spread of a Deadly Parasite

Almost half of the world's people—most of them living in poor African countries—are at risk from malaria (Figure 14-6). So is anyone traveling to malaria-prone areas, because there is no vaccine that can prevent this disease.

Malaria is caused by a parasite that is spread by the bites of certain mosquito species. It infects and destroys red blood cells, causing intense fever, chills, drenching sweats, severe abdominal pain, vomiting, headaches, and increased susceptibility to other diseases. It kills an average of at least 2,100 people per day (**Concept 14-2**). About 90% of those dying are children younger than age 5. Many of the children who survive suffer brain damage or impaired learning ability.

Four species of protozoan parasites in the genus *Plasmodium* cause malaria. The cycle of infection begins when an uninfected female of any of about 60 *Anopheles* mosquito species bites a person (usually at night) who is infected with the *Plasmodium* parasite, ingests blood that contains the parasite, and later bites an uninfected person. *Plasmodium* parasites then move out of the mosquito and into the human's bloodstream and liver where they multiply. Malaria can also be transmitted by contaminated blood transfusions and by drug users sharing needles.

Over the course of human history, malarial protozoa probably have killed more people than all the wars ever fought. During the 1950s and 1960s, the spread of malaria was

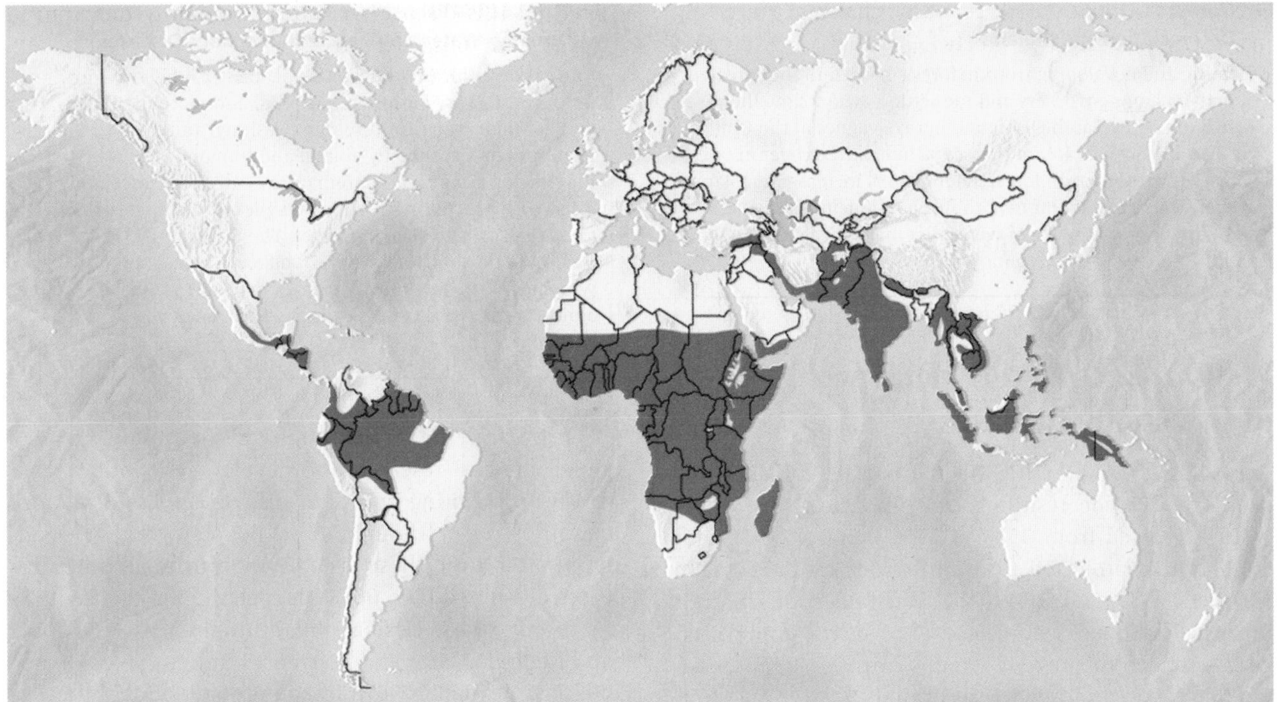

Figure 14-6 *Global outlook:* About 47% of the world's population live in areas in which malaria is prevalent. Malaria kills at least 780,000 people per year or about 3 people every 2 minutes. More than 80% of these victims live in sub-Saharan Africa and most of them are children younger than age 5. (Data from the World Health Organization and U.S. Centers for Disease Control and Prevention)

sharply slowed when swamplands and marshes where mosquitoes breed were drained or sprayed with insecticides, and drugs were used to kill the parasites in victims' bloodstreams. Since 1970, however, malaria has come roaring back. Most species of the *Anopheles* mosquito have become genetically resistant to most insecticides. Worse, the *Plasmodium* parasites have become genetically resistant to common antimalarial drugs as well. Also, people with HIV infections are more vulnerable to malaria, and people with malaria are more vulnerable to HIV.

Researchers are working to develop new antimalarial drugs and vaccines, as well as biological controls for *Anopheles* mosquitoes. But these approaches receive too little funding and have proved difficult to implement.

Another approach is to provide poor people in malarial regions with free or inexpensive, long-lasting, insecticide-treated bed nets (Figure 14-7) and window screens. Zinc and vitamin A supplements could also be given to children to boost their resistance to malaria. In addition, we can greatly reduce the incidence of malaria at a low cost by spraying the insides of homes with low concentrations of the pesticide DDT twice a year. While DDT is being phased out in most countries, the WHO supports limited use of it for malaria control.

Columbia University economist Jeffrey Sachs estimates that spending \$2–3 billion a year on preventing and treating malaria might save more than a million lives a year. Sachs notes that the economic and health benefits of such an investment would be great.

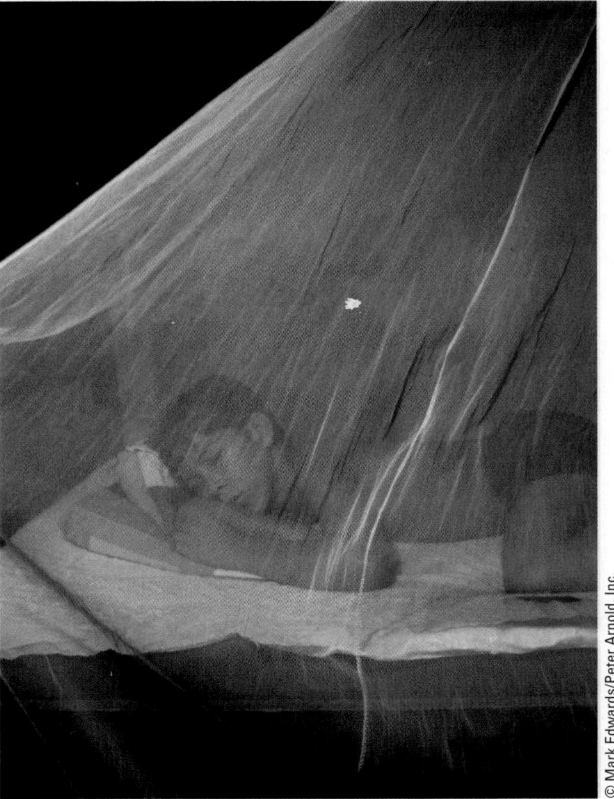

Figure 14-7 This boy, who lives in Brazil's Amazon Basin, is sleeping under an insecticide-treated mosquito net to reduce the risk of being bitten by malaria-carrying mosquitoes. Such nets cost about \$5 each and can be donated through groups such as MalariaNoMore.

CONNECTIONS

Deforestation and Malaria

Clearing and developing tropical forests has led to the spread of malaria among workers and the settlers who follow them. A 2010 study by Sarah Olson and her colleagues at the University of Wisconsin found that a 5% loss of tree cover in one part of Brazil's Amazon forest led to a 50% increase in malaria in that study area. Olson hypothesized that deforestation creates partially sunlit pools of water that make ideal breeding ponds for malaria-carrying mosquitoes.

CONNECTIONS

Drinking Water, Latrines, and Infectious Diseases

More than a third of the world's people—2.6 billion—do not have sanitary bathroom facilities, and more than 1 billion get their water for drinking, washing, and cooking from sources polluted by animal and human feces. A key to reducing sickness and premature death from infectious disease is to focus on providing people with simple latrines and access to safe drinking water. The UN estimates that this could be done for about $20 billion a year—about what people in wealthier countries, who already have almost universal access to clean water, spend each year on bottled water.

We Can Reduce the Incidence of Infectious Diseases

GOOD NEWS

According to the WHO, the percentage of all deaths worldwide resulting from infectious diseases decreased from 35% to 17% between 1970 and 2006, and is projected to drop to 16% by 2015. Also, between 1971 and 2006, the percentage of children in less-developed countries who were immunized with vaccines to prevent tetanus, measles, diphtheria, typhoid fever, and polio rose from 10% to 90%—saving about 10 million lives each year. Between 1990 and 2008, the estimated annual number of children younger than age 5 who died from preventable illnesses such as diarrhea, malaria, and pneumonia dropped from nearly 12 million to 7.7 million.

Figure 14-8 lists measures promoted by health scientists and public health officials to help prevent or reduce the incidence of infectious diseases—especially in less-developed countries. An important breakthrough has been the development of simple *oral rehydration therapy* to help prevent death from dehydration for victims of severe diarrhea, which causes about one-fourth of all deaths of children younger than age 5 (**Concept 14-2**). This therapy involves administering a simple solution of boiled water, salt, and sugar or rice at a cost of only a few cents per person. It has been the major factor in reducing the annual number of deaths from diarrhea from 4.6 million in 1980 to 1.6 million in 2008.

The WHO has estimated that implementing the solutions listed in Figure 14-8 could save the lives of as many as 4 million children younger than age 5 each year. Experts note that investing in these solutions would pay off not only in savings of young lives, but also in higher levels of security for many less-developed countries. Improving the health of people in these countries would help to stabilize their societies and would give them a boost in making the demographic transition (see Figure 6-11, p. 105) toward more sustainable economies.

However, the WHO estimates that only 10% of global medical research and development money goes toward preventing infectious diseases in less-developed countries, even though more people worldwide suffer and die from these diseases than from all other diseases combined. The problem is getting more attention. In recent years, certain philanthropists, including Bill and Melinda Gates and Warren E. Buffet, have donated billions of dollars toward improving global health, with special emphasis on infectious diseases in less-developed countries. **GREEN CAREER:** infectious disease prevention

Solutions

Infectious Diseases

- Increase research on tropical diseases and vaccines
- Reduce poverty
- Decrease malnutrition
- Improve drinking water quality
- Reduce unnecessary use of antibiotics
- Educate people to take all of an antibiotic prescription
- Reduce antibiotic use to promote livestock growth
- Require careful hand washing by all medical personnel
- Immunize children against major viral diseases
- Provide oral rehydration for diarrhea victims
- Conduct global campaign to reduce HIV/AIDS

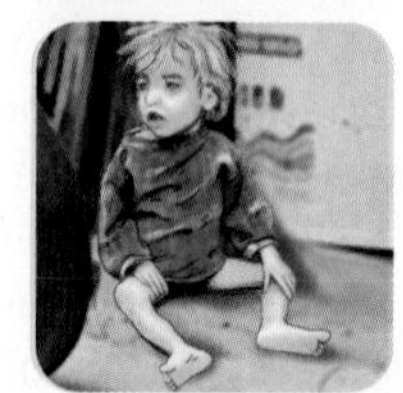

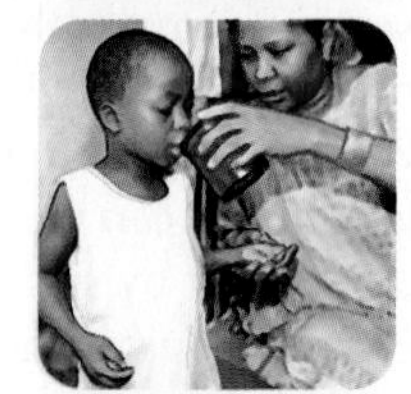

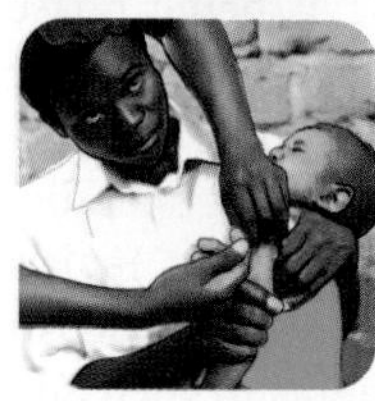

Figure 14-8 There are a number of ways to prevent or reduce the incidence of infectious diseases, especially in less-developed countries. **Questions:** Which three of these approaches do you think are the most important? Why?

14-3 What Types of Chemical Hazards Do We Face?

▶ **CONCEPT 14-3 There is growing concern about chemicals in the environment that can cause cancers and birth defects, and disrupt the human immune, nervous, and endocrine systems.**

Some Chemicals Can Cause Cancers, Mutations, and Birth Defects

A **toxic chemical** is an element or compound that can cause temporary or permanent harm or death to humans and animals. In 2004, the EPA listed arsenic, lead, mercury, vinyl chloride (used to make PVC plastics), and polychlorinated biphenyls (PCBs) as the top five toxic substances in terms of human and environmental health.

There are three major types of potentially toxic agents. **Carcinogens** are chemicals, some types of radiation, and certain viruses that can cause or promote *cancer*—a disease in which malignant cells multiply uncontrollably and create tumors that can damage the body and often lead to premature death. Examples of carcinogens are arsenic, benzene, formaldehyde, gamma radiation, PCBs, radon, ultraviolet (UV) radiation, certain chemicals in tobacco smoke, and vinyl chloride.

Typically, 10–40 years may elapse between the initial exposure to a carcinogen and the appearance of detectable cancer symptoms. Partly because of this time lag, many healthy teenagers and young adults have trouble believing that their smoking, drinking, eating, and other habits today could lead to some form of cancer before they reach age 50.

The second major type of toxic agent, **mutagens**, includes chemicals or forms of radiation that cause or increase the frequency of mutations, or changes, in the DNA molecules found in cells. Most mutations cause no harm but some can lead to cancers and other disorders. For example, nitrous acid (HNO_2), formed by the digestion of nitrite (NO_2^-) preservatives in foods, can cause mutations linked to increases in stomach cancer in people who consume large amounts of processed foods and wine containing such preservatives. Harmful mutations occurring in reproductive cells can be passed on to offspring and to future generations.

Teratogens, a third type of toxic agent, are chemicals that harm or cause birth defects in a fetus or embryo. Ethyl alcohol is a teratogen. Drinking during pregnancy can lead to offspring with low birth weight and a number of physical, developmental, behavioral, and mental problems. Other teratogens are angel dust, benzene, formaldehyde, lead, mercury (Science Focus, p. 358), PCBs, phthalates, thalidomide, and vinyl chloride. Figure 14-9 shows potential pathways on which persistent toxic chemicals such as PCBs move through the living and nonliving environment.

Some Chemicals May Affect Our Immune and Nervous Systems

Since the 1970s, research on wildlife and laboratory animals, along with some studies of humans, have yielded a growing body of evidence that suggests that

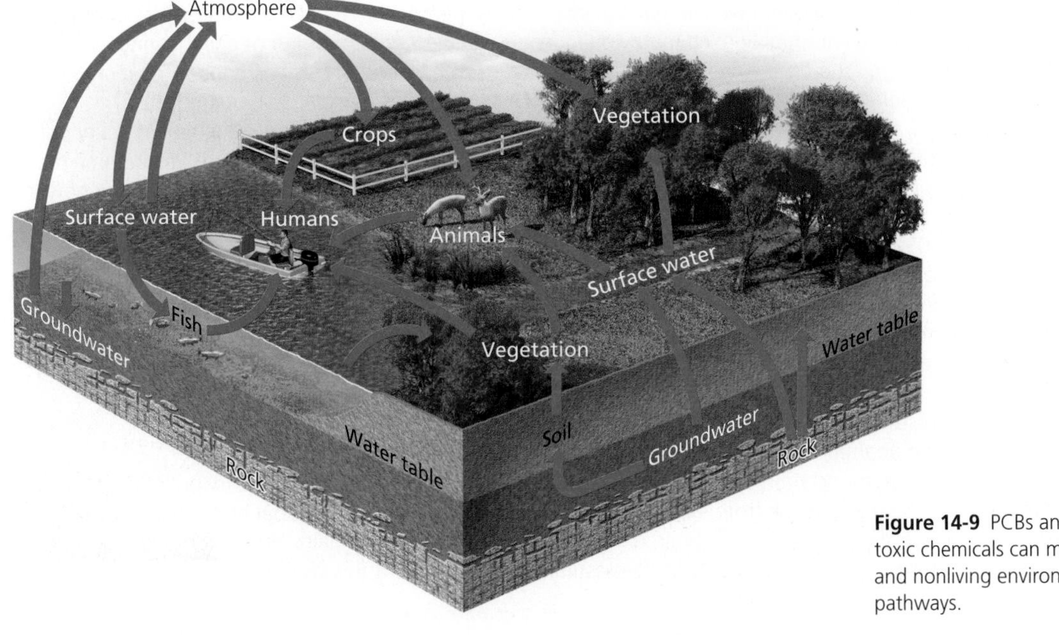

Figure 14-9 PCBs and other persistent toxic chemicals can move through the living and nonliving environment on a number of pathways.

long-term exposure to some chemicals in the environment can disrupt the body's immune, nervous, and endocrine systems (**Concept 14-3**).

The *immune system* consists of specialized cells and tissues that protect the body against disease and harmful substances by forming *antibodies,* or specialized proteins that render invading agents harmless. Some chemicals such as arsenic, methylmercury, and dioxins can weaken the human immune system and leave the body vulnerable to attacks by allergens and infectious bacteria, viruses, and protozoa.

Some natural and synthetic chemicals in the environment, called *neurotoxins,* can harm the human *nervous system* (brain, spinal cord, and peripheral nerves). Effects can include behavioral changes, learning disabilities, attention deficit disorder, retardation, paralysis, and death. Examples of neurotoxins are PCBs, arsenic, lead, and certain pesticides.

An especially dangerous neurotoxin is *methylmercury* (Science Focus, below). It has received a great deal of attention from environmental and health scientists because it is so persistent in the environment. It can be biologically magnified in food chains and webs just as DDT (see Figure 8-11, p. 162) and PCBs are.

As with all forms of pollution, prevention is the best policy. Figure 14-10 lists ways to prevent or reduce human inputs of mercury into the environment.

Some Chemicals Affect the Human Endocrine System

The *endocrine system* is a complex network of glands that release tiny amounts of *hormones* into the bloodstreams of humans and other vertebrate animals. Very low levels of these chemical messengers (often measured in

SCIENCE FOCUS

Mercury's Toxic Effects

Mercury (Hg) (see Figure 2-3, right) and its compounds are all toxic. Research indicates that long-term exposure to high levels of mercury can permanently damage the human nervous system, kidneys, and lungs. Fairly low levels of mercury can also harm fetuses and cause birth defects.

This toxic metal is released into the air from rocks, soil, and volcanoes and by vaporization from the oceans. Such natural sources account for about one-third of the mercury reaching the atmosphere each year. According to the EPA, the remaining two-thirds come from human activities—primarily from the smokestacks of coal-burning power and industrial plants, cement kilns, and solid-waste incinerators. When it rains, these emissions are washed out of the atmosphere onto the soil and into bodies of water.

Because mercury is an element, it cannot be broken down or degraded. Therefore, this indestructible global pollutant accumulates in soil, water, and the bodies of people, fish, and other animals that feed high on food chains and food webs. According to the Natural Resources Defense Council, predatory fish such as large tuna, swordfish, mackerel, and sharks can have mercury concentrations in their bodies that are as much as 10,000 times higher than the levels in the water around them.

In the atmosphere, some elemental mercury (emitted mostly by coal-burning power plants and solid waste incinerators) is converted to more toxic inorganic and organic mercury compounds that can be deposited in aquatic environments. Under certain conditions in aquatic systems, bacteria can convert inorganic mercury compounds to highly toxic methylmercury, which can be biologically magnified in food chains and webs. As a result, in such aquatic systems, high levels of methylmercury are often found in the tissues of large fishes, which feed at high trophic levels.

Humans are exposed to mercury in three ways. *First,* we may inhale vaporized elemental mercury (Hg) (see Figure 2-3, right) or particles of inorganic mercury salts such as mercury sulfide (HgS) and mercuric chloride ($HgCl_2$). *Second,* we can eat fish contaminated with highly toxic methylmercury (CH_3Hg^+). The *third* way involves consuming high-fructose corn syrup (HFCS), widely used as a sweetener in beverages and food products. A 2005 study by former FDA scientist Renee Dufault found detectable levels of mercury in 9 of 20 samples of HFCS. A 2005 study by food safety researcher David Wallinga also found detectable levels of mercury in one out of three supermarket food products that contained high levels of HFCS. There is concern that exposure to this source of mercury could pose a health threat to the fetuses of pregnant women who consume large quantities of HFCS. The U.S. Corn Refiners Association disputes these findings.

The greatest risk from exposure to low levels of methylmercury is brain damage in fetuses and young children. Studies estimate that 30,000–60,000 of the children born each year in the United States are likely to have reduced IQs and possible nervous system damage because of such exposure. Also, methylmercury has been shown to harm the hearts, kidneys, and immune systems of some adults.

According to the EPA, about 75% of all human exposure to mercury comes from eating fish. In 2004, the U.S. Food and Drug Administration (FDA) and the EPA advised nursing mothers, pregnant women, and women who may become pregnant not to eat shark, swordfish, king mackerel, or tilefish and to limit their consumption of albacore tuna to no more than 170 grams (6 ounces) per week. The EPA estimates that about one of every 12 women of childbearing age in the United States has enough mercury in her blood to harm a developing fetus.

In its 2003 report on global mercury pollution, the UN Environment Programme recommended phasing out coal-burning power plants and waste incinerators throughout the world as rapidly as possible. Other recommendations are to reduce or eliminate mercury in the production of batteries, paints, and chlorine by no later than 2020. Substitute materials and processes are available for these products.

Critical Thinking

To sharply reduce mercury pollution, should we phase out all coal burning as rapidly as possible? Explain. How might your lifestyle change if this were done?

Solutions

Mercury Pollution

Prevention	Control
Phase out waste incineration	Sharply reduce mercury emissions from coal-burning plants and incinerators
Remove mercury from coal before it is burned	Label all products containing mercury
Switch from coal to natural gas and renewable energy resources	Collect and recycle batteries and other products containing mercury

Figure 14-10 There are a number of ways to prevent or control inputs of mercury into the environment from human sources—mostly coal-burning power plants and incinerators. **Questions:** Which four of these solutions do you think are the most important? Why?

parts per billion or parts per trillion) regulate the bodily systems that control sexual reproduction, growth, development, learning ability, and behavior. Each type of hormone has a unique molecular shape that allows it to attach to certain parts of cells called *receptors,* and to transmit its chemical message. In this "lock-and-key" relationship, the receptor is the lock and the hormone is the key.

Molecules of certain pesticides and other synthetic chemicals such as bisphenol A (BPA) (**Core Case Study**) have shapes similar to those of natural hormones. This allows them to attach to molecules of natural hormones and to disrupt the endocrine systems in humans and in some other animals. These molecules are called *hormonally active agents* (HAAs).

Examples of HAAs include aluminum, Atrazine™ and several other herbicides, DDT, PCBs, mercury (Science Focus, at left), phthalates, and BPA. Some hormone imposters such as BPA are chemically similar to estrogens (female sex hormones) and can disrupt the endocrine system by attaching to estrogen receptor molecules. Others, called *hormone blockers,* disrupt the endocrine system by preventing natural hormones such as androgens (male sex hormones) from attaching to their receptors.

Estrogen mimics and hormone blockers are sometimes called *gender benders* because of their possible effects on sexual development and reproduction. Numerous studies on wild animals, laboratory animals, and humans suggest that the males of species that are exposed to hormonal disruption are generally becoming more feminine. A likely culprit is long-term exposure to low levels of gender-bending synthetic chemicals that we have added to the environment. There is also growing concern about another group of HAAs—pollutants that can act as *thyroid disrupters* and cause growth, weight, brain, and behavioral disorders.

By 2010, more than 90 published studies by independent laboratories had found a number of significant adverse effects on test animals from exposure to very low levels of BPA (**Core Case Study**). These effects include brain damage, early puberty, certain cancers, heart disease, obesity, liver damage, impaired immune function, type 2 diabetes, hyperactivity, impaired learning, impotence in males, and obesity in unborn test animals exposed to BPA.

On the other hand, 12 studies funded by the chemical industry found no evidence or only weak evidence of adverse effects from low-level exposure to BPA in test animals. In 2011, the German Society of Toxicology concluded that BPA in polycarbonate plastic containers and in the linings of food and beverage cans "represents no noteworthy risk to the health of the human population including newborns and babies."

In 2008, the U.S. Food and Drug Administration (FDA) had concluded that BPA in food and drink containers does not pose a health hazard. However, a number of environmental and health scientists have disputed this finding, including the FDA's science advisory panel. In 2009, a National Academy of Sciences report by a panel of experts strongly criticized the FDA for relying heavily on industry-sponsored studies of BPA instead of taking into account numerous BPA studies by some of the country's best independent health scientists.

It is fairly easy to use alternatives to baby bottles and other products that contain BPA. (Plastic containers having the triangular recycling code 7 on their bottoms are likely to contain BPA.) However, it is difficult to replace the plastic resins containing BPA that line food containers and that help to prevent food contamination and spoilage. There are substitutes for these liners, but they are more expensive, and the potential health effects of some chemicals they contain also need to be evaluated.

In 2010, despite intense opposition from the chemical industry, Canada classified BPA as a toxic chemical and banned its use in baby bottles. Also in 2010, the European Union voted to ban the sale of plastic baby bottles that contain BPA. Six U.S. states have banned the use of BPA in children's products, as has the city of Chicago, Illinois. In 2010, the FDA said that BPA is cause for "some concern" and needs further evaluation. The FDA also said that it was working to reduce BPA exposure for pregnant women, infants, and young children.

THINKING ABOUT
BPA

Should plastics containing BPA be banned from use in all children's products? Explain. Should plastics containing BPA be banned from use in the liners of canned food containers? Explain. What are the alternatives?

There is also growing concern over possible harmful effects from exposure to low levels of certain HAAs called *phthalates* (pronounced THALL-eights). These chemicals are used to soften polyvinyl chloride (PVC) plastic found in a variety of products, and they are used as solvents in many consumer products. Phthalates are found in many detergents, perfumes, cosmetics, baby powders, body lotions for adults and babies, hair sprays, deodorants, soaps, nail polishes, and shampoos for adults and babies. They are also found in PVC products such as soft vinyl toys, teething rings, blood storage bags, intravenous drip bags, and medical tubing.

A 2008 study by Sheela Sathyanarayana and her colleagues tested urine from the diapers of 163 infants aged 2 to 28 months. They found that 81% of the urine samples had measurable amounts of seven or more phthalates related to the use of phthalate-containing baby lotions, baby powders, and baby shampoos.

Exposing laboratory animals to high doses of various phthalates has caused birth defects and liver cancer, kidney and liver damage, premature breast development, immune system suppression, and abnormal sexual development. The European Union and at least 14 other countries have banned phthalates. But scientists, government regulators, and manufacturers in the United States are divided on its risks to human health and reproductive systems. Toy makers in the United States say that phthalates, which have been used for more than 20 years in baby items, pose no threats. In addition, they warn that substitutes could make plastic toys more brittle and subject to breaking, and thus less safe for children.

Some scientists hypothesize that more frequent occurrences of certain health problems may be related to rising levels of hormone disruptors in our bodies. Such problems include sharp drops in male sperm counts and male sperm mobility found in 20 countries on six continents, rising rates of testicular cancer and genital birth defects in men, and increased breast cancer rates in women. Other scientists disagree and point out that there are not enough scientific studies and statistical evidence to strongly link these medical problems with HAA levels in humans. They call for more research before banning or severely restricting HAAs, which would cause huge economic losses for the companies that make them.

The scientific and economic controversies over possible health risks from exposure to chemicals such as BPA and phthalates highlight the difficulty in assessing possible harmful health effects from exposure to very low levels of various chemicals widely found in the environment and in the products that we use. Resolving these uncertainties will take decades of research.

Some scientists believe that as a precaution during this period of research, governments and individual consumers should act to sharply reduce the use of potentially harmful hormone disrupters, especially in products frequently used by pregnant women, infants, young children, and teenagers. They also call for manufacturers to search for less harmful substitutes for such chemicals.

Consumers now have more choices, since most makers of baby bottles, sipping cups, and sports water bottles offer BPA-free alternatives. To avoid BPA contamination, some consumers are avoiding plastic containers with a #7 recycling code, using powdered infant formula instead of liquid formula from metal cans, choosing glass bottles and food containers instead of those made of plastic or lined with plastic resins, avoiding bottled water in plastic containers, avoiding use of plastic wrap for food, and using glass, ceramic, or stainless steel coffee mugs rather than plastic cups.

THINKING ABOUT
Hormone Disrupters

Should we ban or severely restrict the use of potential hormone disrupters? What beneficial or harmful effects might this have in your life?

14-4 How Can We Evaluate Chemical Hazards?

▶ **CONCEPT 14-4A Scientists use live laboratory animals, case reports of poisonings, and epidemiological studies to estimate the toxicity of chemicals, but these methods have limitations.**

▶ **CONCEPT 14-4B Many health scientists call for much greater emphasis on pollution prevention to reduce our exposure to potentially harmful chemicals.**

Many Factors Determine the Harmful Health Effects of Chemicals

Toxicology is the study of the harmful effects of chemicals on humans and other organisms. In effect, it is a study of poisonous substances.

Toxicity is a measure of the harmfulness of a substance—its ability to cause injury, illness, or death to a living organism. A basic principle of toxicology is that *any synthetic or natural chemical can be harmful if ingested in a large enough quantity*. But the critical question is this: *At what level of exposure to a particular toxic chemical will the* *chemical cause harm?* This is the meaning of the chapter-opening quote by the German scientist Paracelsus: *The dose makes the poison.*

This is a difficult question to answer because of the many variables involved in estimating the effects of human exposure to chemicals. A key factor is the **dose**, the amount of a harmful chemical that a person has ingested, inhaled, or absorbed through the skin at any one time.

The effects of a particular chemical can also depend upon the age of the person exposed to it. For example, toxic chemicals usually have a greater effect on fetuses, infants, and children than on adults (see the Case Study that follows). Toxicity also depends on *genetic makeup*, which determines an individual's sensitivity to a particular toxin. Some people are sensitive to a number of toxins—a condition known as *multiple chemical sensitivity* (MCS). Another factor is how well the body's detoxification systems (such as the liver, lungs, and kidneys) work.

Several other variables can affect the level of harm caused by a chemical. One is its *solubility*. Water-soluble toxins (which are often inorganic compounds) can move throughout the environment and get into water supplies and the aqueous solutions that surround the cells in our bodies. Oil- or fat-soluble toxins (which are usually organic compounds) can penetrate the membranes surrounding cells, because the membranes allow similar oil-soluble chemicals to pass through them. Thus, oil- or fat-soluble toxins can accumulate in body tissues and cells.

Another factor is a substance's *persistence*, or resistance to breakdown. Many chemicals such as DDT and PCBs have been widely used because they are not easily broken down in the environment. This means that more people and wildlife are likely to come in contact with them and they are more likely to remain in the body and have long-lasting harmful health effects.

Biological accumulation and magnification can also play a role in toxicity. Animals higher on the food chain are more susceptible to the effects of fat-soluble toxic chemicals because of the magnified concentrations of the toxins in their bodies. Examples of chemicals that can be biomagnified include DDT (see Figure 8-13, p. 163), PCBs, and methylmercury.

The damage to health resulting from exposure to a chemical is called the **response**. One type of response, an *acute effect*, is an immediate or rapid harmful reaction ranging from dizziness and nausea to death. A *chronic effect* is a permanent or long-lasting consequence (kidney or liver damage, for example) of exposure to a single dose or to repeated lower doses of a harmful substance.

Some people have the mistaken idea that all natural chemicals are safe and all synthetic chemicals are harmful. In fact, many synthetic chemicals, including many of the medicines we take, are quite safe if used as intended, while many natural chemicals such as mercury and lead are deadly.

■ CASE STUDY
Protecting Children from Toxic Chemicals

In 2005, the Environmental Working Group analyzed umbilical cord blood from 10 randomly selected newborns in U.S. hospitals. Of the 287 chemicals detected in that study, 180 have been shown to cause cancers in humans or animals, 217 have damaged the nervous systems in test animals, and 208 have caused birth defects or abnormal development in test animals. Scientists do not know what harm, if any, might be caused by the very low concentrations of these chemicals found in infants' blood. But these results are cause for concern, and they underscore the urgent need for more research.

Infants and young children are more susceptible to the effects of toxic substances than are adults, for three major reasons. *First,* they generally breathe more air, drink more water, and eat more food per unit of body weight than do adults. *Second,* they are exposed to toxins in dust or soil when they put their fingers, toys, or other objects in their mouths. *Third,* children usually have less well-developed immune systems and body detoxification processes than adults have. Fetuses are also highly vulnerable to trace amounts of toxic chemicals that they receive from their mothers.

In 2003, the EPA proposed that in determining any risk, regulators should assume children have a 10-times higher risk factor than adults have. Some health scientists suggest that to be on the safe side, we should assume that this risk for children is 100 times the risk for adults. However, a 2010 study by the U.S. Government Accountability Office found that, between 1999 and 2009, there was a lapse in the EPA's commitment to protecting children from harmful chemicals, which could indicate that such higher risk factors were not always considered.

THINKING ABOUT
Toxic Chemical Levels for Children

Should environmental regulations require that allowed exposure levels for children to toxic chemicals be 100 times lower than for adults? Explain your reasoning.

Scientists Use Live Laboratory Animals and Non-Animal Tests to Estimate Toxicity

The most widely used method for determining toxicity is to expose a population of live laboratory animals to measured doses of a specific substance under controlled conditions. Laboratory-bred mice and rats are widely used because, as mammals, their systems function, to some degree, similarly to human systems. Also, they are small and can reproduce rapidly under controlled laboratory conditions.

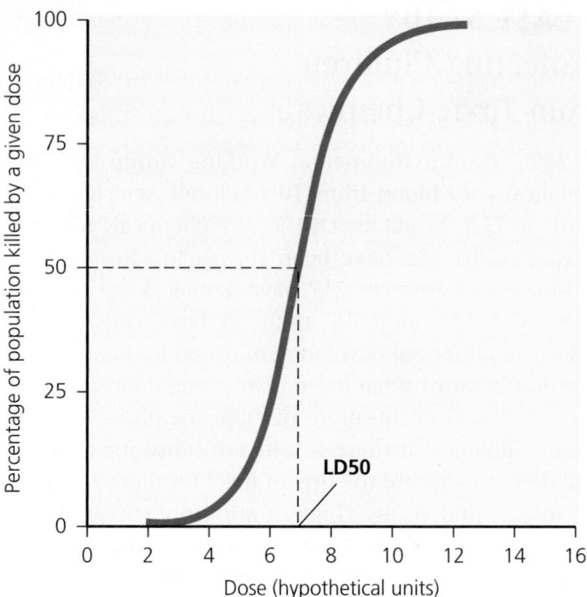

Figure 14-11 Science: This hypothetical *dose-response curve* illustrates how scientists can estimate the LD50, the dosage of a specific chemical that kills 50% of the animals in a test group. Toxicologists use this method to compare the toxicities of different chemicals.

Animal tests typically take 2–5 years, involve hundreds to thousands of test animals, and can cost as much as $2 million per substance tested. Such tests can be painful to the test animals and can kill or harm them. Animal welfare groups want to limit or ban the use of test animals and, at the very least, to ensure that they are treated in the most humane manner possible.

Scientists estimate the toxicity of a chemical by determining the effects of various doses of the chemical on test organisms and plotting the results in a **dose-response curve** (Figure 14-11). One approach is to determine the *lethal dose*—the dose that will kill an animal. A chemical's *median lethal dose (LD50)* is the dose that can kill 50% of the animals (usually rats and mice) in a test population within an 18-day period.

Chemicals vary widely in their toxicity (Table 14-1). Some poisons can cause serious harm or death after a single exposure at very low dosages. Others cause such harm only at dosages so huge that it is nearly impossible to get enough into the body to cause injury or death. Most chemicals fall between these two extremes.

Some scientists challenge the validity of extrapolating data from test animals to humans, because human physiology and metabolism often differ from those of the test animals. Other scientists say that such tests and models work fairly well (especially for revealing cancer risks) when the correct experimental animal is chosen or when a chemical is toxic to several different test-animal species.

GOOD NEWS

More humane methods for toxicity testing are available and are being used more often to replace testing on live animals. They include making computer simulations and using tissue cultures of cells and bacteria, chicken egg membranes, and individual animal cells, instead of whole, live animals. High-speed robot testing devices can now screen the biological activity of more than one million compounds a day to help determine their possible toxic effects.

The problems with estimating toxicities by using laboratory experiments get even more complicated (**Concept 14-4A**). In real life, each of us is exposed to a variety of chemicals, some of which can interact in ways that decrease or enhance their short- and long-term individual effects. Toxicologists already have great difficulty in estimating the toxicity of a single substance. Adding the problem of evaluating *mixtures of potentially toxic substances,* isolating the culprits, and determining how they can interact with one another is overwhelming from a scientific and economic standpoint. For example, just studying the interactions among 3 of the 500 most widely used industrial chemicals would take 20.7 million experiments—a physical and financial impossibility.

THINKING ABOUT
Animal Testing

Should laboratory-bred mice, rats, and other animals be used to determine toxicity and other effects of chemicals? Explain.

Table 14-1 Toxicity Ratings and Average Lethal Doses for Humans

Toxicity Rating	LD50 (milligrams per kilogram of body weight)*	Average Lethal Dose**	Examples
Supertoxic	Less than 5	Less than 7 drops	Nerve gases, botulism toxin, mushroom toxin, dioxin (TCDD)
Extremely toxic	5–50	7 drops to 1 teaspoon	Potassium cyanide, heroin, atropine, parathion, nicotine
Very toxic	50–500	1 teaspoon to 1 ounce	Mercury salts, morphine, codeine
Moderately toxic	500–5,000	1 ounce to 1 pint	Lead salts, DDT, sodium hydroxide, sodium fluoride, sulfuric acid, caffeine, carbon tetrachloride
Slightly toxic	5,000–15,000	1 pint to 1 quart	Ethyl alcohol, Lysol, soaps
Essentially nontoxic	15,000 or greater	More than 1 quart	Water, glycerin, table sugar

*Dosage that kills 50% of individuals exposed.

**Amounts of substances in liquid form at room temperature that are lethal when given to a 70-kilogram (150-pound) human.

There Are Other Ways to Estimate the Harmful Effects of Chemicals

Scientists use several other methods to get information about the harmful effects of chemicals on human health. For example, *case reports,* usually made by physicians, provide information about people suffering some adverse health effect or dying after exposure to a chemical. Such information often involves accidental or deliberate poisonings, drug overdoses, homicides, or suicide attempts.

Most case reports are not reliable sources for estimating toxicity because the actual dosage and the exposed person's health status are often unknown. But such reports can provide clues about environmental hazards and suggest the need for laboratory investigations.

Another source of information is *epidemiological studies,* which compare the health of people exposed to a particular chemical (the *experimental group*) with the health of a similar group of people not exposed to the agent (the *control group*). The goal is to determine whether the statistical association between exposure to a toxic chemical and a health problem is strong, moderate, weak, or undetectable.

Four factors can limit the usefulness of epidemiological studies. *First,* in many cases, too few people have been exposed to high enough levels of a toxic agent to detect statistically significant differences. *Second,* the studies usually take a long time. *Third,* closely linking an observed effect with exposure to a particular chemical is difficult because people are exposed to many different toxic agents throughout their lives and can vary in their sensitivity to such chemicals. *Fourth,* we cannot use epidemiological studies to evaluate hazards from new technologies or chemicals to which people have not yet been exposed.

Are Trace Levels of Toxic Chemicals Harmful?

Almost everyone who lives in a more-developed country is now exposed to potentially harmful chemicals (Figure 14-12) that have built up to trace levels in their blood and in other parts of their bodies. Also, trace

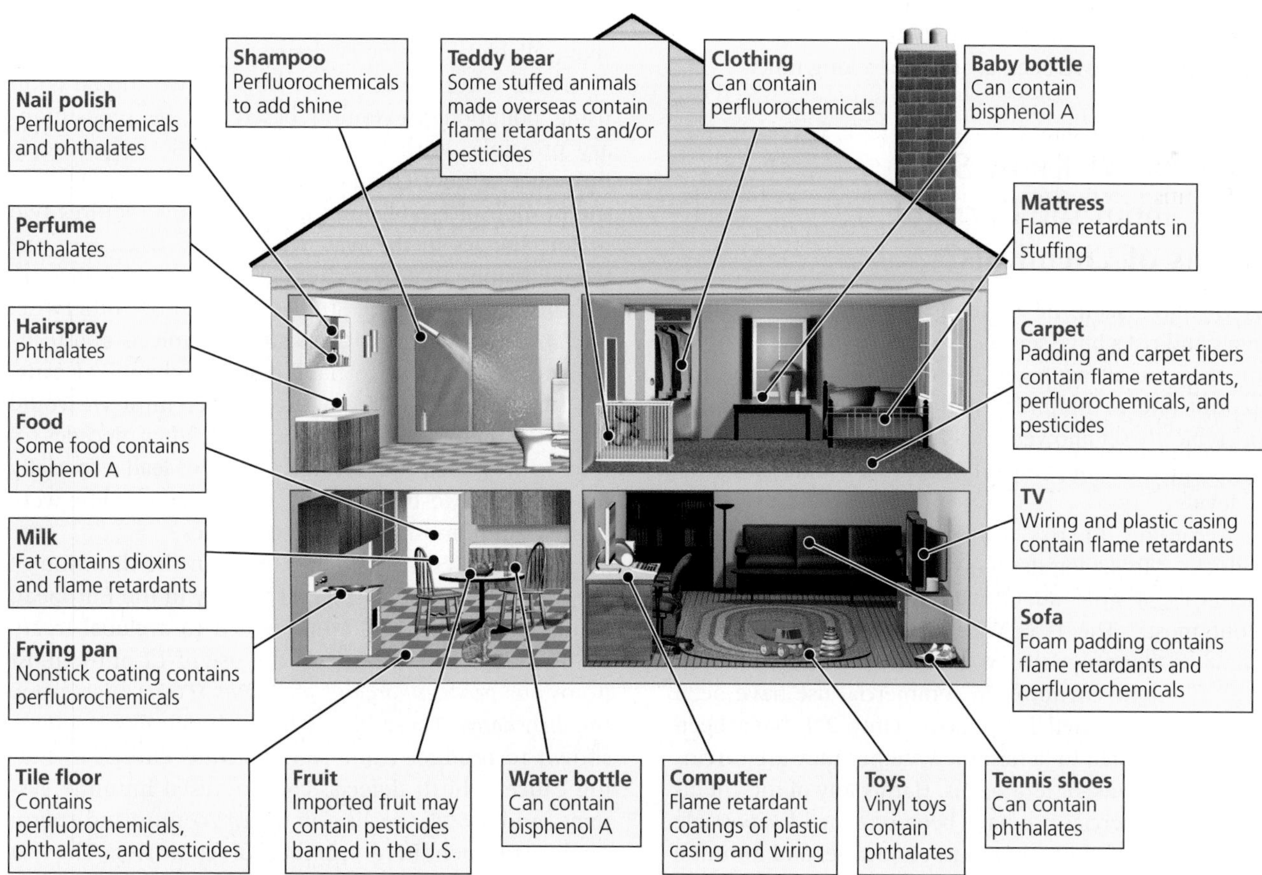

Figure 14-12 A number of potentially harmful chemicals are found in many homes. Most people living in more-developed countries have traces of these chemicals in their blood and body tissues. (Data from U.S. Environmental Protection Agency, Centers for Disease Control and Prevention, and New York State Department of Health)
Questions: Does the fact that we do not know much about the long-term harmful effects of these chemicals make you more likely or less likely to minimize your exposure to them? Why?

amounts of birth control pills, blood pressure medicines, antidepressants, and other chemicals with unknown effects on human health are being flushed down toilets and released into waterways from sewage treatment plants or are leaching into groundwater from home septic systems.

Should we be concerned about trace amounts of various synthetic chemicals in air, water, food, and our bodies? The honest answer is that, in most cases, we do not know because there is too little data and because of the difficulty in determining the effects of exposures to low levels of these chemicals (**Concept 14-4A**).

Some scientists view trace amounts of such chemicals with alarm, especially because of their potential long-term effects on the human immune, nervous, and endocrine systems (**Core Case Study**). Others view the risks from trace levels as minor. They point out that average life expectancy has been increasing in most countries, especially more-developed countries, for decades. Some scientists contend that the concentrations of such chemicals are so low that they are harmless.

CORE CASE STUDY

Chemists are now able to detect much smaller amounts of potentially toxic chemicals in air, water, and food than they had previously been able to measure. This is good news, but it can give the false impression that dangers from toxic chemicals are on the rise. In some cases, we may simply be uncovering levels of chemicals that have been around for a long time.

Why Do We Know So Little about the Harmful Effects of Chemicals?

As we have seen, all methods for estimating toxicity levels and risks have serious limitations (**Concept 14-4A**). But they are all we have. To take this uncertainty into account and to minimize harm, scientists and regulators typically set allowed levels of exposure to toxic substances at 1/100 or even 1/1,000 of the estimated harmful levels.

According to risk assessment expert Joseph V. Rodricks, "Toxicologists know a great deal about a few chemicals, a little about many, and next to nothing about most." The U.S. National Academy of Sciences estimates that only 10% of the more than 80,000 registered synthetic chemicals in commercial use have been thoroughly screened for toxicity. Only 2% have been adequately tested to determine whether they are carcinogens, mutagens, or teratogens. Hardly any of the chemicals in commercial use have been screened for possible damage to the human nervous, endocrine, and immune systems. Because of insufficient data and the high costs of regulation, federal and state governments do not supervise the use of nearly 99.5% of the commercially available chemicals in the United States, and the problem is much worse in some less-developed countries.

How Far Should We Go in Using Pollution Prevention and the Precautionary Principle?

We know little about the potentially toxic chemicals around us and inside of us, and estimating their effects is very difficult, time-consuming, and expensive. So where does this leave us?

Some scientists and health officials, especially those in European Union countries, are pushing for much greater emphasis on *pollution prevention*. They say we should not release chemicals that we know or suspect can cause significant harm into the environment. This means looking for harmless or less harmful substitutes for toxic and hazardous chemicals (Individuals Matter, at right). Another option is to recycle them within production processes to keep them from reaching the environment, as the U.S. companies 3M and DuPont have been doing.

Pollution prevention is a strategy for implementing the *precautionary principle* (see Chapter 8, p. 170). According to this principle, which can be applied to many other problems, when there is substantial preliminary evidence that an activity, technology, or chemical substance can harm humans or the environment, we should take precautionary measures to prevent or reduce such harm, rather than waiting for more conclusive (reliable) scientific evidence (**Concept 14-4B**).

There is controversy over how far we should go in using pollution prevention based on the precautionary principle. With this approach, those proposing to introduce a new chemical or technology would bear the burden of establishing its safety. This requires two major changes in the way we evaluate risks. *First,* we would assume that new chemicals and technologies are harmful until scientific studies could show otherwise. *Second,* we would remove existing chemicals and technologies that appear to have a strong chance of causing significant harm from the market until we could establish their safety. For example, after decades of research revealed the harmful effects of lead, especially on children, lead-based paints and leaded gasoline were phased out in most developed countries.

Some movement is being made in the direction of the precautionary principle, especially in the European Union. In 2000, negotiators agreed to a global treaty that would ban or phase out the use of 12 of the most notorious *persistent organic pollutants (POPs),* also called the *dirty dozen.* These highly toxic chemicals have been shown to produce numerous harmful effects, including cancers, birth defects, compromised immune systems, and a 50% decline in sperm counts and sperm quality in men in a number of countries. The list includes DDT and eight other pesticides, PCBs, and dioxins. In 2009, nine more POPs were added, some of which are widely used in pesticides and in flame-retardants added to clothing, furniture, and other consumer goods.

INDIVIDUALS MATTER 

Ray Turner and His Refrigerator

Life as we know it could not exist on land or in the upper layers of the oceans and other bodies of water without the thin layer of ozone (O_3) found in the lower stratosphere (see Figure 3-3, p. 42). It is critical for protecting us from the sun's harmful UV rays. Therefore, a basic rule of sustainability relating to pollution prevention is: *Do not mess with the ozone layer*.

However, for decades, people have done just that, violating the principle of pollution prevention by releasing large amounts of chemicals called chlorofluorocarbons (CFCs) into the troposphere. These chemicals have drifted into the stratosphere where they react with and destroy some of the ozone.

In 1974, scientists alerted the world to this threat. After further research and lengthy debate, in 1992, most of the world's nations signed a landmark international agreement to phase out the use of CFCs and other ozone-destroying chemicals. The discovery of these chemicals led scientists to use the principle of pollution prevention to search for less harmful alternatives.

Ray Turner, a manager at Hughes Aircraft in the U.S. state of California, was concerned about this problem. His company was using CFCs as cleaning agents to remove films, caused by oxidation, from the electronic circuit boards they manufactured. Turner's concern for the environment led him to search for a cheap and simple substitute for these chemicals. He found it in his refrigerator.

Turner decided to put drops of some common kitchen substances on a corroded penny to see whether any of them would remove the film caused by oxidation. Then he used his soldering gun to see whether solder would stick to the surface of the penny, which would indicate that the film had been cleaned off.

First he tried vinegar. No luck. Then he tried some ground-up lemon peel. Another failure. Next he tried a drop of lemon juice and watched as the solder took hold. The rest, as they say, is history.

Today, Hughes Aircraft uses inexpensive, CFC-free, citrus-based solvents to clean circuit boards. This new cleaning technique has reduced circuit board defects by about 75% at the company. In addition, Turner got a hefty bonus. Now, other companies clean computer boards and chips using acidic chemicals extracted from cantaloupes, peaches, and plums. Maybe you can find a solution to an environmental problem in your refrigerator.

New chemicals will be added to the list when the harm they could potentially cause is seen as outweighing their usefulness. The POPs treaty went into effect in 2004 but has not been ratified and implemented by the United States.

In 2007, the European Union enacted regulations known as REACH (for *registration, evaluation, and authorization of chemicals*). It required the registration of 30,000 untested, unregulated, and potentially harmful chemicals. In the REACH process, the most hazardous substances are not approved for use if safer alternatives exist. In addition, when there is no alternative, producers must present a research plan aimed at finding one.

REACH puts more of the burden on industry to show that chemicals are safe. Conventional regulation has put the burden on governments to show that they are dangerous. For example, the U.S. chemical regulation structure was enacted in 1976. At U.S. Congressional hearings in 2009, experts testified that the current system makes it virtually impossible for the government to limit or ban the use of toxic chemicals. The hearings found that under this system, the EPA had required testing for only 200 of the more than 80,000 chemicals used in industry—that is, less than one of every 400—and had issued regulations to control just 5 of those chemicals.

Manufacturers and businesses contend that widespread application of the much more precautionary REACH approach would make it too expensive and almost impossible to introduce any new chemical or technology. Proponents of increased reliance on using pollution prevention agree that we can go too far, but argue we have an ethical responsibility to reduce known or potentially serious risks to human health and to the earth's life-support system (**Concept 14-4B**).

Proponents also point out that using this principle focuses the efforts and creativity of scientists, engineers, and businesses on finding solutions to pollution problems based on prevention rather than on cleanup. It also reduces health risks for employees and society, frees businesses from having to deal with pollution regulations, and reduces the threat of lawsuits from injured parties. In some cases, it helps companies to increase their profits from sales of safer products and innovative technologies, and it improves the public image of businesses operating in this manner.

For almost 40 years, most laws and technologies for dealing with pollution have focused on cleaning up or diluting pollution after it has been produced. Experience shows that such pollution control is only a temporary solution that can be overwhelmed by more people consuming more things and producing more pollution. Environmental and health scientists say we can do better than this by putting much greater emphasis on pollution prevention.

THINKING ABOUT
Pollution Prevention

If you could run the government of the country where you live, would you require it to regulate potentially hazardous chemicals using measures such as the POPs treaty and the REACH regulations? If so, explain your reasoning. If you were to implement such regulations, what would be the most important benefits and the most troublesome drawbacks of such rules? Do you think the benefits would outweigh the drawbacks, or vice versa? Explain.

14-5 How Do We Perceive Risks and How Can We Avoid the Worst of Them?

▶ **CONCEPT 14-5 We can reduce the major risks we face by becoming informed, thinking critically about risks, and making careful choices.**

The Greatest Health Risks Come from Poverty, Gender, and Lifestyle Choices

Risk analysis involves identifying hazards and evaluating their associated risks (*risk assessment;* Figure 14-2, left), ranking risks (*comparative risk analysis*), determining options and making decisions about reducing or eliminating risks (*risk management;* Figure 14-2, right), and informing decision makers and the public about risks (*risk communication*).

Statistical probabilities based on past experience, animal testing, and other assessments are used to estimate risks from older technologies and chemicals. To evaluate new technologies and products, risk evaluators use more uncertain statistical probabilities, based on models rather than on actual experience and testing.

The greatest risks many people face today are rarely dramatic enough to make the daily news. In terms of the number of premature deaths per year (Figure 14-13) and reduced life span (Figure 14-14), *the greatest risk by far is poverty*. The high death toll ultimately resulting from poverty is caused by malnutrition, increased susceptibility to normally nonfatal infectious diseases, and often-fatal infectious diseases transmitted by unsafe drinking water.

After poverty and gender, the greatest risks of premature death result primarily from lifestyle choices that people make (Figures 14-13 and 14-14) (**Concept 14-1**). The best ways to reduce one's risk of premature death and serious health problems are to avoid smoking and exposure to smoke (see the Case Study that follows), lose excess weight, reduce consumption of foods containing cholesterol and saturated fats, eat a variety of fruits and vegetables, exercise regularly, drink little or no alcohol (no more than two drinks in a single day), avoid excess sunlight (which ages skin and can cause skin cancer), and practice safe sex (**Concept 14-5**). A 2005 study by Majjid Ezzati with participation by 100 scientists around the world estimated that if everyone were to follow these guidelines, one-third of the 7.6 million annual deaths from cancer could be prevented. Each year the number of lives this would save roughly equals the entire population of the U.S. city of Chicago, Illinois.

About two-thirds of Americans are classified as either overweight or obese. In 2009, the American Cancer Society reported that an 18-year study of 900,000

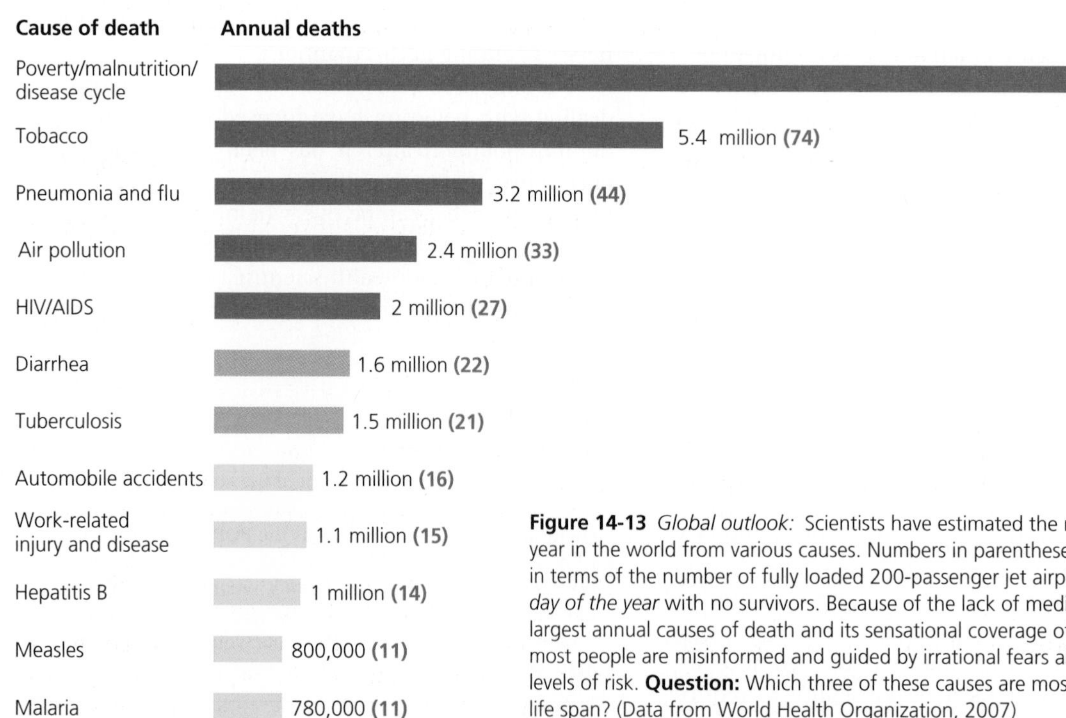

Figure 14-13 *Global outlook:* Scientists have estimated the number of deaths per year in the world from various causes. Numbers in parentheses represent death tolls in terms of the number of fully loaded 200-passenger jet airplanes crashing *every day of the year* with no survivors. Because of the lack of media coverage of the largest annual causes of death and its sensational coverage of other causes of death, most people are misinformed and guided by irrational fears about the comparative levels of risk. **Question:** Which three of these causes are most likely to shorten your life span? (Data from World Health Organization, 2007)

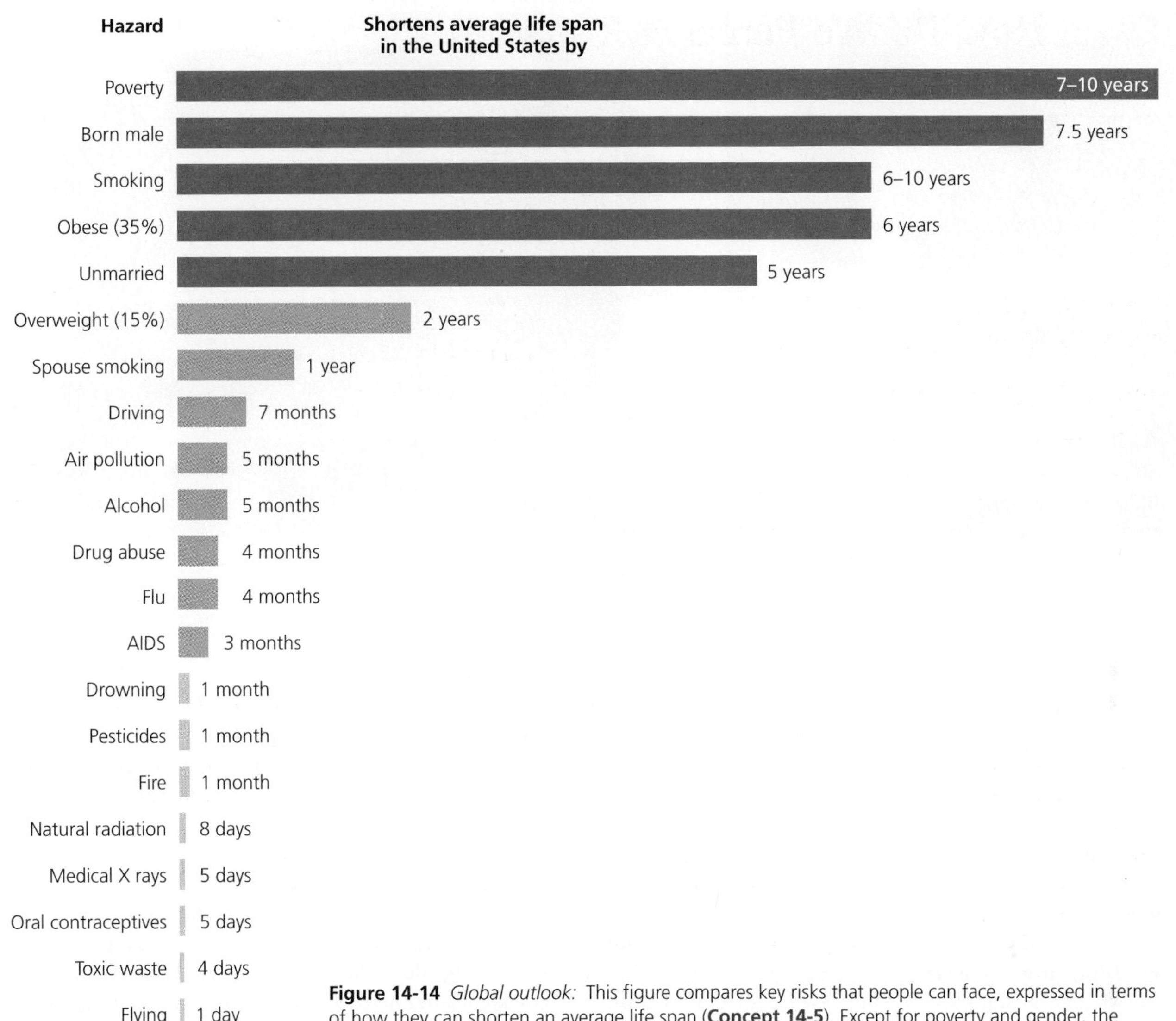

Figure 14-14 *Global outlook:* This figure compares key risks that people can face, expressed in terms of how they can shorten an average life span (**Concept 14-5**). Except for poverty and gender, the greatest risks people face come mostly from the lifestyle choices they make. These are generalized relative estimates. Individual responses to these risks differ because of factors such as genetic variation, family medical history, emotional makeup, stress, and social ties and support. **Question:** Which three of these factors are most likely to shorten your life span? (Data from Bernard L. Cohen)

cancer-free U.S. adults indicated that more than 90,000 cancer deaths could be prevented each year if Americans maintained healthy body weights.

■ CASE STUDY
Death from Smoking

What is roughly the diameter of a 30-caliber bullet, can be bought almost anywhere, is highly addictive, and kills an average of about 14,800 people every day, or about one every 6 seconds? It is a cigarette. *Cigarette smoking is the world's most preventable major cause of suffering and premature death among adults.*

In 2007, the WHO estimated that tobacco use helped to kill 100 million people during the 20th century and could kill 1 billion people during this century unless individuals and governments act now to dramatically reduce smoking. The WHO estimates that each year, tobacco contributes to the premature deaths of at least 5.4 million people (about half from more-developed and half from less-developed countries) from 25 illnesses including heart disease, stroke, lung and other cancers, and bronchitis.

In 2010, U.S. research scientist Rachel A. Whitmer and her colleagues reported on a study that had tracked the health of 21,123 individuals for 30 years. They found that people who smoked 1–2 packs of cigarettes daily between the ages of 50 and 60 had a 44% higher chance of getting Alzheimer's disease or vascular dementia (which reduces blood flow to the brain and can trigger strokes that erode memory) by age 72. Another disease related to smoking is *emphysema,* which results in irreversible lung damage and chronic shortness of breath (Figure 14-15, p. 368).

According to the WHO, lifelong smokers reduce their life spans by an average of 15 years. By 2030, the annual death toll from smoking-related diseases

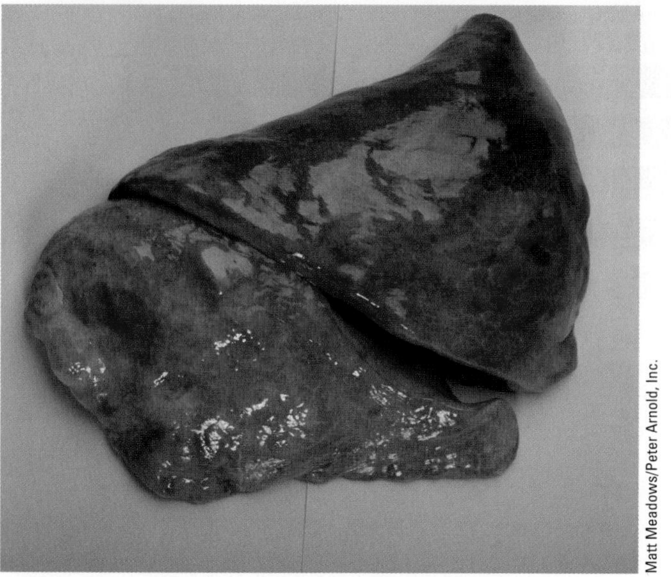

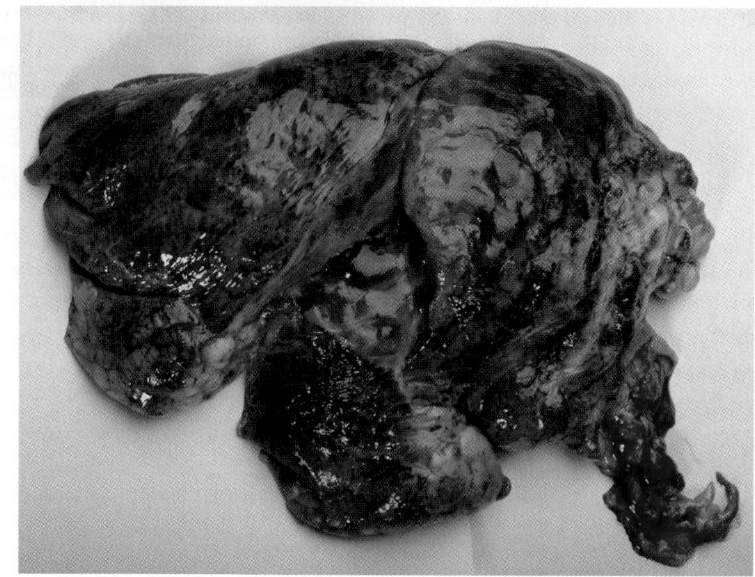

Figure 14-15 There is a startling difference between normal human lungs (left) and the lungs of a person who died of emphysema (right). The major causes are prolonged smoking and exposure to air pollutants.

is projected to reach more than 8 million—an average of 21,900 preventable deaths per day. About 80% of these deaths are expected to occur in less-developed countries, especially China, where 30% of the world's smokers live. The annual death toll from smoking in China is about 1.2 million, an average of 2,000 people a day. In India, where 11% of all smokers live, the annual death toll from smoking is about 900,000.

In 2009, the CDC estimated that smoking kills about 443,000 Americans per year—an average of 1,211 deaths per day, or nearly one every minute (Figure 14-16). This death toll is roughly equivalent to six fully loaded 200-passenger jet planes crashing *every day of the year* with no survivors! Smoking also causes about 8.6 million illnesses every year in the United States.

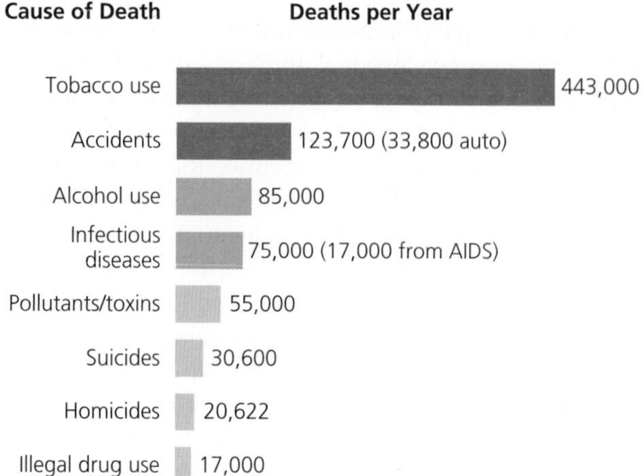

Figure 14-16 Smoking is by far the leading cause of preventable deaths in the United States, causing more premature deaths each year than all the other categories in this figure combined. (Data from U.S. Centers for Disease Control and Prevention, 2009)

According to 2010 study by David L. Ashley and other researchers at the U.S. Centers for Disease Control and Prevention (CDC), American-made cigarettes tend to contain more cancer-causing chemicals than do cigarettes made in other countries.

The overwhelming scientific consensus is that the nicotine inhaled in tobacco smoke is highly addictive. Only one in ten people who try to quit smoking succeeds. Smokers suffer about the same relapse rate as do recovering alcoholics and those addicted to heroin or crack cocaine. Also, a British government study showed that adolescents who smoke more than one cigarette have an 85% chance of becoming smokers.

Studies also show that *passive smoking*, or breathing secondhand smoke, poses health hazards for children and adults. Children who grow up living with smokers are more likely to develop allergies and asthma. Among adults, nonsmoking spouses of smokers have a 30% higher risk of both heart attack and lung cancer than spouses of nonsmokers have. In 2010, the CDC and the California Environmental Protection Agency estimated that each year, secondhand smoke causes an estimated 3,400 lung cancer deaths and 46,000 deaths from heart disease in the United States. A 2010 study by British researchers found that, globally, about 600,000 people die every year from exposure to secondhand smoke.

A study published in 2004 by Richard Doll and Richard Peto found that cigarette smokers die, on average, 10 years earlier than nonsmokers, but that kicking the habit—even at 50 years of age—can cut such a risk in half. If people quit smoking by the age of 30, they can avoid nearly all the risk of dying prematurely, but again, the longer one smokes, the harder it is to quit.

In the United States, the percentage of adults who smoke dropped from roughly 40%

in the 1960s to 21% in 2009 (24% of males and 18% of females). Such declines can be attributed to media coverage about the harmful health effects of smoking, sharp increases in cigarette taxes in many states, mandatory health warnings on cigarette packs, and the banning of smoking in workplaces, bars, restaurants, and public buildings.

Despite some encouraging trends, tobacco companies produced more than 5 trillion cigarettes in 2010—an average of about 700 for every person on the planet. In China about 350 million people—more than the entire U.S population—are hooked on tobacco. One major problem is that the price of a pack of cigarettes does not always reflect their harmful health effects. In 2006, CDC researchers calculated that the cost of treating smoking-related illnesses in the United States and the cost of lost worker productivity resulting from such illnesses amounted to about $10.50 a pack. Some states have raised taxes on cigarettes to include some of these costs, reasoning that higher prices might discourage smoking and that smokers, not the general public, should pay these costs.

Estimating Risks from Technologies Is Not Easy

The more complex a technological system, and the more people needed to design and run it, the more difficult it is to estimate the risks of using the system. The overall *reliability*—the probability (expressed as a percentage) that a person, device, or complex technological system will complete a task without failing—is the product of two factors:

System reliability (%) = Technology reliability (%) × Human reliability (%)

With careful design, quality control, maintenance, and monitoring, a highly complex system such as a nuclear power plant or space shuttle can achieve a high degree of technological reliability. But human reliability usually is much lower than technological reliability and is almost impossible to predict: *To err is human.*

Suppose the technological reliability of a nuclear power plant is 95% (0.95) and human reliability is 75% (0.75). Then the overall system reliability is 71% (0.95 × 0.75 = 71%). Even if we could make the technology 100% reliable (1.0), the overall system reliability would still be only 75% (1.0 × 0.75 = 75%).

One way to make a system more foolproof, or fail-safe, is to move more of the potentially fallible elements from the human side to the technological side. However, chance events such as a lightning strike can knock out an automatic control system, and no machine or computer program can completely replace human judgment. Also, the parts in any automated control system (such as the blowout protectors for the BP gulf oil well that ruptured in 2010) are manufactured, assembled, tested, certified, inspected, and maintained by fallible human beings. In addition, computer software programs used to monitor and control complex systems can be flawed because of human design error or can be deliberately sabotaged to cause their malfunction.

Most People Do a Poor Job of Evaluating Risks

Most of us are not good at assessing the relative risks from the hazards that surround us. Many people deny or shrug off the high-risk chances of death (or injury) from voluntary activities they enjoy such as *motorcycling* (1 death in 50 participants), *smoking* (1 in 250 by age 70 for a pack-a-day smoker), *hang gliding* (1 in 1,250), and *driving* (1 in 3,300 without a seatbelt and 1 in 6,070 with a seatbelt). Indeed, the most dangerous thing that many people around the world do each day is to drive or ride in a car.

Yet some of these same people may be terrified about their chances of being killed by *the flu* (a 1 in 130,000 chance), *a nuclear power plant accident* (1 in 200,000), *West Nile virus* (1 in 1 million), *lightning* (1 in 3 million), *a commercial airplane crash* (1 in 9 million), *snakebite* (1 in 36 million), or *shark attack* (1 in 281 million). Worldwide each year, sharks kill an average of 6 people, while elephants kill at least 200 people.

Five factors can cause people to see a technology or a product as being more or less risky than experts judge it to be. First is *fear.* Research going back 3 decades shows that fear causes people to overestimate risks and worry more about unusual risks than they do for common, everyday risks. Studies show that people tend to overestimate numbers of deaths caused by tornadoes, floods, fires, homicides, cancer, and terrorist attacks, and to underestimate numbers of deaths from flu, diabetes, asthma, heart attack, stroke, and automobile accidents. Many people also fear a new, unknown product or technology more than they do an older, more familiar one. For example, some people fear genetically modified food and trust food produced by traditional plant-breeding techniques.

The second factor in our estimation of risk is the *degree of control* we have in a given situation. Most of us have a greater fear of things over which we do not have personal control. For example, some individuals feel safer driving their own car for long distances through bad traffic than traveling the same distance on a plane. But look at the numbers. The risk of dying in a car accident in the United States while using a seatbelt is 1 in 6,070, whereas the risk of dying in a U.S. commercial airliner crash is about 1 in 9 million.

The third factor is *whether a risk is catastrophic,* not chronic. We usually are more frightened by news of catastrophic accidents such as a plane crash than we are of a cause of death such as smoking, which has a much larger death toll spread out over time.

Fourth, some people suffer from *optimism bias,* the belief that risks that apply to other people do not apply to them. While people get upset when they see others driving erratically while talking on a cell phone or text messaging, they may believe that talking on the cell phone or texting does not impair their own driving ability.

A fifth factor is that many of the risky things we do are highly pleasurable and give *instant gratification,* while the potential harm from such activities comes later. Examples are smoking cigarettes, eating lots of ice cream, and getting a tan.

We Can Get Better at Evaluating and Reducing Risk

Here are four guidelines for evaluating and reducing risk (**Concept 14-5**):

- *Compare risks.* Is there a risk of getting cancer by eating a charcoal-broiled steak once or twice a week for a lifetime? Yes, because almost any chemical can harm you if the dose is large enough. The question is whether this danger is great enough for you to worry about. In evaluating a risk, the key question is not "Is it safe?" but rather *"How risky is it compared to other risks?"*
- *Determine how much risk you are willing to accept.* For most people, a 1 in 100,000 chance of dying or suffering serious harm from exposure to an environmental hazard is a threshold for changing their behavior. However, in establishing standards and reducing risk, the U.S. EPA generally assumes that a 1 in 1 million chance of dying from an environmental hazard is acceptable. People involuntarily exposed to such risks believe that this standard is too high.
- *Evaluate the actual risk involved.* The news media usually exaggerate the daily risks we face in order to capture our interest and sell newspapers and magazines or gain television viewers. As a result, most people who are exposed to a daily diet of such exaggerated reports believe that the world is much more risk-filled than it really is.
- *Concentrate on evaluating and carefully making important lifestyle choices.* This will help you to have a much greater chance of living a longer, healthier, happier, and less fearful life. When you worry about a risk, the most important question to ask is, "Do I have any control over this?" There is no point worrying about risks over which you have no control. But you do have control over major ways to reduce risks from heart attack, stroke, and many forms of cancer, because you can decide whether to smoke, what to eat, and how much alcohol to drink. Other factors under your control are whether you practice safe sex, how much exercise you get, how safely you drive, and how often you expose yourself to ultraviolet rays from the sun or from a tanning bed.

GOOD NEWS

Here are this chapter's *three big ideas*:

- We face significant hazards from infectious diseases such as flu, AIDS, tuberculosis, diarrheal diseases, and malaria, and from exposure to chemicals that can cause cancers and birth defects, and disrupt the human immune, nervous, and endocrine systems.
- Because of the difficulty in evaluating the harm caused by exposure to chemicals, many health scientists call for much greater emphasis on pollution prevention.
- Becoming informed, thinking critically about risks, and making careful choices can reduce the major risks we face.

REVISITING Bisphenol A and Sustainability

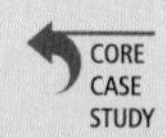

In the **Core Case Study** that opens this chapter, we saw that certain chemicals such as bisphenol A can act as hormone disruptors and may have a number of harmful health effects on humans, especially children. In this chapter, we also saw how difficult it is to evaluate the nature and severity of threats from BPA and other potentially toxic chemicals. In addition, we evaluated the risks to human health from other chemical, biological, physical, cultural, and lifestyle hazards.

One of the important facts discussed in this chapter is that on a global basis, the greatest threat to human health is the poverty–malnutrition–disease cycle, followed by the threats from smoking, pneumonia and flu, air pollution, and HIV/AIDS. There are some threats that we can do little to avoid, but we can reduce other threats, partly by applying the three **principles of sustainability**. For example, we can greatly reduce exposure to air and water pollutants by shifting from nonrenewable fossil fuels (especially coal) to a diversity of renewable energy resources, including solar, wind, flowing water, and biomass. We can reduce our exposure to harmful chemicals used in the manufacturing of various goods by cutting resource use and waste, and by reusing and recycling material resources. We can also mimic biodiversity by using diverse strategies for solving environmental and health problems, and especially for reducing poverty and controlling population growth. In so doing, we also help to preserve the earth's biodiversity.

Is this idealistic? Sure. But if creative and caring people throughout human history had not acted to improve the world by pursuing goals that others said were impossible or too idealistic, we would have accomplished very little on this marvelous planet. Each of us can make a difference.

The burden of proof imposed on individuals, companies, and institutions should be to show that pollution prevention options have been thoroughly examined, evaluated, and used before lesser options are chosen.

JOEL HIRSCHORN

REVIEW

CORE CASE STUDY

1. Describe the potential risks from exposure to trace amounts of hormone mimics such as bisphenol A (**Core Case Study**).

SECTION 14-1

2. What is the key concept for this section? Distinguish among **risk**, **risk assessment**, and **risk management**. Distinguish between possibility and probability. Give an example of a risk from each of the following: biological hazards, chemical hazards, physical hazards, cultural hazards, and lifestyle choices. What is a **pathogen**?

SECTION 14-2

3. What is the key concept for this section? Distinguish between **bacteria** and **viruses**, and how the diseases they can cause are usually treated. Distinguish among **infectious disease**, **transmissible disease**, and **nontransmissible disease**, and give an example of each. List four ways in which infectious organisms can enter the body. In terms of death rates, what are the world's four most serious infectious diseases? Distinguish between an epidemic and a pandemic of an infectious disease. List the causes and possible solutions for the increasing genetic resistance to commonly used antibiotics. What is MRSA and why is it so threatening?

4. Describe the global threat from tuberculosis. Describe the threat from flu. Summarize the story of health threats from the global HIV/AIDS pandemic and list six ways to reduce these threats. Describe and compare the threats from the hepatitis B and West Nile viruses.

5. Describe the threat from malaria for about 47% of the world's people and how we can reduce this threat. Explain how deforestation can increase the spread of malaria. List five major ways to reduce the global threat from infectious diseases.

SECTION 14-3

6. What is the key concept for this section? What is a **toxic chemical**? Distinguish among **carcinogens**, **mutagens**, and **teratogens**, and give an example of each. Describe the human immune, nervous, and endocrine systems and give an example of a chemical that can threaten each of these systems. Describe the toxic effects of the various forms of mercury and ways to reduce these threats. What are hormonally active agents, what risks do they pose, and how can we reduce these risks? Summarize concerns about exposure to bisphenol A (BPA) and the controversy over what to do about exposure to this chemical (**Core Case Study**). Summarize concerns over exposure to phthalates.

CORE CASE STUDY

SECTION 14-4

7. What are the two key concepts for this section? Define **toxicology**, **toxicity**, **dose**, and **response**. Give three reasons why children are more vulnerable to harm from toxic chemicals than are adults. Explain how the toxicity of a substance can be estimated by testing laboratory animals, and discuss the limitations of this approach. What is a **dose-response curve**? Explain how toxicities are estimated through use of case reports and epidemiological studies, and discuss the limitations of these approaches. Why do we know so little about the harmful effects of chemicals? Discuss the use of the precautionary principle and pollution prevention in dealing with health threats from chemicals.

SECTION 14-5

8. What is the key concept for this section? What is **risk analysis**? In terms of premature deaths, what are the three greatest threats that people face? Describe the health threats from smoking and how we can reduce these threats.

9. How can we reduce the threats resulting from the use of various technologies? What are five factors that can cause people to misjudge risks? List four guidelines for evaluating and reducing risk.

10. What are this chapter's *three big ideas*? Discuss how we can lessen the threats of harm from chemicals such as hormone mimics by applying the three **principles of sustainability**.

SUSTAINABILITY

Note: Key terms are in **bold** type.

CRITICAL THINKING

1. Should we ban the use of hormone mimics such as bisphenol A (**Core Case Study**) in products used by children younger than age 5? Should we ban them for use in all products? Explain.

2. What are three actions you would take to reduce the global threats to human health and life from **(a)** tuberculosis, **(b)** HIV/AIDS, and **(c)** malaria?

3. Evaluate the following statements:
   a. We should not worry much about exposure to toxic chemicals because almost any chemical, at a large enough dosage, can cause some harm.
   b. We should not worry much about exposure to toxic chemicals because, through genetic adaptation, we can develop immunities to such chemicals.
   c. We should not worry much about exposure to toxic chemicals because we can use genetic engineering to reduce our susceptibility to their effects.
   d. We should not worry about exposure to a chemical such as bisphenol A (BPA) because it has not been absolutely proven scientifically that BPA has killed anyone.

4. Workers in a number of industries are exposed to higher levels of various toxic substances than is the general public. Should we reduce workplace levels allowed for such chemicals? What economic effects might this have?

5. Explain why you agree or disagree with the proposals for reducing the death toll and other harmful effects of smoking listed in the Case Study on p. 367. Do you believe there should be a ban on smoking indoors in all public places? Explain.

6. What are the three major risks you face from **(a)** your lifestyle, **(b)** where you live, and **(c)** what you do for a living? Which of these risks are voluntary and which are involuntary? List three steps you could take to reduce these risks. Which of these steps do you already take or plan to take?

7. In deciding what to do about risks from chemicals in the area where you live, would you support legislation requiring the use of pollution prevention based on the precautionary principle? Explain.

8. Congratulations! You are in charge of the world. List the three most important features of your program to reduce the risks from exposure to **(a)** infectious disease organisms and **(b)** toxic and hazardous chemicals.

DOING ENVIRONMENTAL SCIENCE

Pick a potentially harmful chemical not discussed in detail in this chapter and use the library or Internet to evaluate **(a)** what it is used for and how widely it is used, **(b)** its potential harm, **(c)** the scientific evidence for such claims, and **(d)** solutions for dealing with this threat. Pick a study area, such as your dorm or apartment building, your block, or your city. In this area, try to determine the level of presence of the chemical you are studying. You could do this by finding 4 or 5 examples of items or locations containing the chemical and then estimating the total amount based on your sample.

GLOBAL ENVIRONMENT WATCH EXERCISE

Search for *bisphenol A* (*BPA*) and research the latest developments in studies of the potentially harmful health effects of BPA (**Core Case Study**). List the names of three or four countries that have restricted its use. Pick one of these countries and try to find some details on how that country is regulating the use of BPA. Write a short report summarizing your findings. Have any countries completely banned the use of BPA? What are they? Try to find reports of two studies that give opposing conclusions abut how BPA should be regulated. Summarize the arguments for these conclusions on both sides. Based on what you have found, do you believe the use of BPA should be regulated more strictly in the state or country where you live? Should it be banned altogether? Explain your reasoning.

DATA ANALYSIS

The graph below shows the effects of AIDS on life expectancy at birth in Botswana, 1950–2000, and projects these effects to 2050. Answer the questions below.

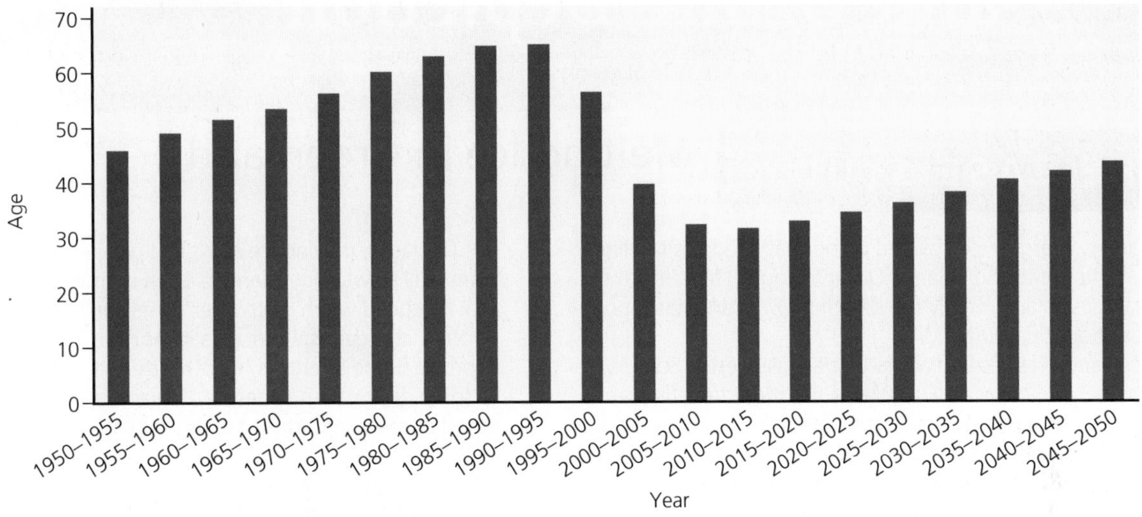

Source: Data from United Nations and U.S. Census Bureau

1. a. By what percentage did life expectancy in Botswana increase between 1950 and 1995?
   b. By what percentage was life expectancy in Botswana projected to decrease between 1995 and 2015?
2. a. By what percentage is life expectancy in Botswana projected to increase between 2015 and 2050?
   b. By what percentage was life expectancy in Botswana projected to decrease between 1995 and 2050?

LEARNING ONLINE

The website for *Environmental Science 14e* contains helpful study tools and other resources to take the learning experience beyond the textbook. Examples include flashcards, practice quizzing, information on green careers, tips for more sustainable living, activities, and a wide range of articles and further readings. To access these materials and other companion resources for this text, please visit **www.cengagebrain.com**. For further details, see the preface, p. xvi.

15 Air Pollution, Climate Disruption, and Ozone Depletion

CORE CASE STUDY Melting Ice in Greenland

Greenland is a largely ice-covered island that straddles the line between the Arctic and Atlantic Oceans (Figure 15-1). It is the world's largest island, nearly the size of Mexico, and has a population of about 56,000 people.

Greenland's ice lies in glaciers that are as deep as 3.2 kilometers (2 miles) and cover about 80% of this mountainous island. This ice formed during the last ice age and survives because of its huge mass. Scientists are now studying Greenland's ice closely because it appears to be melting at a slow but accelerating rate. Some hypothesize that this melting is one effect of *atmospheric warming*—the gradual rise of the average temperature of the atmosphere near the surface of the earth. There is concern that this warming could lead to rapid *climate disruption* during this century that could have harmful effects on most of the planet, its ecosystems, and its people.

Greenland's glaciers contain about 10% of the world's freshwater. If this ice were to melt, it would raise sea levels, just as an ice cube raises the level of water in a glass when you drop it in. Greenland's ice contains enough water to raise the global sea level by as much as 7 meters (23 feet) if all of it were to melt and drain into the sea. It is highly unlikely that all of the ice will melt, but even a moderate loss of this ice over one or several centuries would raise sea levels considerably.

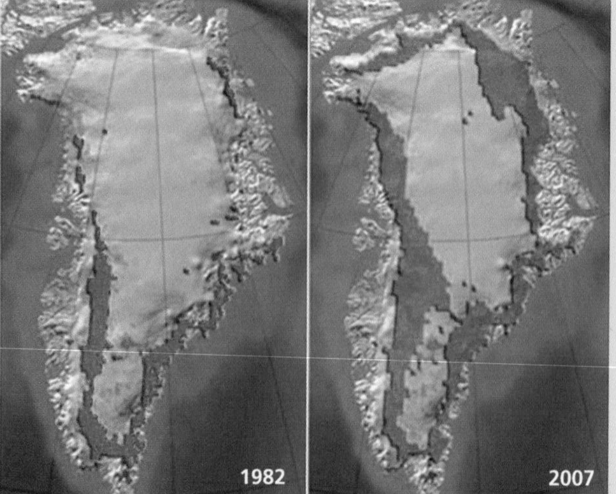

Figure 15-1 The total area of Greenland's glacial ice that melted during the summer months of 2007 (red area on right image) was much greater than the amount that melted during the summer of 1982. Some of this ice is replaced by snow during winter months. But the annual net loss of Greenland's land-based ice has increased during recent years. (Data from Konrad Steffen and Russell Huff, University of Colorado, Boulder)

The large, moving ice mass in a glacier scrapes along extremely slowly, but scientists have demonstrated that it can pick up speed when meltwater on its surface, flowing down through its crevices, lubricates its bottom layer, which sits on bedrock. Some scientists hypothesize this is what is happening to some of northern Greenland's glaciers. In 2009, scientists noted another important factor that can speed this glacial melting: they found that glaciers flowing to the sea are getting thinner at their ocean outlets. As a glacier gets thinner, at some point it can float free of the ground at the shore, which is like taking off the brakes, and it can then flow faster toward the sea.

In addition, as the overall thickness of a glacier decreases, its grip on the land weakens, further accelerating its movement toward the sea. Finally, as the weight of the glacier lessens due to melting, the land under it rebounds, somewhat like a cushion resuming its shape when a person sitting on it gets up. This further destabilizes the ice and can also cause it to slide faster.

These effects are still extremely slow—undetectable to the human eye. But according to a 2009 study by scientists led by Michiel van den Broeke, satellite measurements and computer modeling confirm that the Greenland ice sheet has been losing mass at an accelerating rate since the 1990s (Figure 15-1).

The Intergovernmental Panel on Climate Change (IPCC), created in 1988, is a group of more than 2,500 climate scientists and other researchers working in disciplines related to climate studies. For over two decades, they have volunteered without pay to study the origins of climate change and its possible effects, and to report their findings to the world. In its latest report in 2007, the IPCC made several projections. One of them was that the average global sea level is likely to rise by 18–59 centimeters (7–23 inches) by the end of this century. Since then, however, several scientists studying the melting of Greenland's ice have concluded that the rise in the average global sea level by 2100 will likely be closer to 100 centimeters (39 inches) and perhaps twice that. They estimate that Greenland's ice loss will make up half to two-thirds of that rise.

This change in projections about the rising average global sea level and Greenland's contribution to it is one example of how scientists are revising some IPCC projections upward due to the acceleration of rising temperatures at the earth's poles. There have also been findings that have caused these scientists to revise some of their projections downward. Throughout this chapter, we examine the nature of the atmosphere and air pollution and the findings of the IPCC and other climate scientists about likely causes and effects of projected climate disruption during this century. We also look at some possible ways to deal with this environmental challenge.

Key Questions and Concepts

15-1 What is the nature of the atmosphere?

CONCEPT 15-1 The two innermost layers of the atmosphere are the troposphere, which supports life, and the stratosphere, which contains the protective ozone layer.

15-2 What are the major air pollution problems?

CONCEPT 15-2A Three major outdoor air pollution problems are *industrial smog* primarily from burning coal, *photochemical smog* from motor vehicle and industrial emissions, and *acid deposition* from coal burning and motor vehicle exhaust.

CONCEPT 15-2B The most threatening indoor air pollutants are smoke and soot from wood and coal fires (mostly in less-developed countries), cigarette smoke, and chemicals used in building materials and cleaning products.

15-3 How should we deal with air pollution?

CONCEPT 15-3 Legal, economic, and technological tools can help us to clean up air pollution, but the best solution is to prevent it.

15-4 How might the earth's climate change in the future?

CONCEPT 15-4 Considerable scientific evidence indicates that the earth's atmosphere is warming because of a combination of natural effects and human activities, and that this warming is likely to lead to significant climate disruption during this century.

15-5 What are some possible effects of a warmer atmosphere?

CONCEPT 15-5 The projected rapid change in the atmosphere's temperature could have severe and long-lasting consequences, including increased drought and flooding, rising sea levels, and shifts in the locations of croplands and wildlife habitats.

15-6 What can we do to slow projected climate disruption?

CONCEPT 15-6 We can reduce greenhouse gas emissions and the threat of climate disruption while saving money and improving human health if we cut energy waste and rely more on cleaner renewable energy resources.

15-7 How have we depleted ozone in the stratosphere and what can we do about it?

CONCEPT 15-7A Widespread use of certain chemicals has reduced ozone levels in the stratosphere, which has allowed more harmful ultraviolet radiation to reach the earth's surface.

CONCEPT 15-7B To reverse ozone depletion, we must stop producing ozone-depleting chemicals and adhere to the international treaties that ban such chemicals.

Note: Supplements 2 (p. S3), 4 (p. S10), 5 (p. S18), 6 (p. S22), and 7 (p. S38) can be used with this chapter.

Civilization has evolved during a period of remarkable climate stability,
but this era is drawing to a close. We are entering a new era,
a period of rapid and often-unpredictable climate change.

LESTER R. BROWN

15-1 What Is the Nature of the Atmosphere?

▶ **CONCEPT 15-1 The two innermost layers of the atmosphere are the *troposphere*, which supports life, and the *stratosphere*, which contains the protective ozone layer.**

The Atmosphere Consists of Several Layers

We live under a thin blanket of gases surrounding the earth, called the *atmosphere*. It is divided into several spherical layers (Figure 15-2, p. 376). Our focus in this book is on the atmosphere's two innermost layers: the troposphere and the stratosphere.

About 75–80% of the earth's air mass is found in the **troposphere**, the atmospheric layer closest to the earth's surface. This layer extends only about 17 kilometers (11 miles) above sea level over the equator and 6 kilometers (4 miles) above sea level over the poles. If the earth were the size of an apple, this lower layer containing the air we breathe would be no thicker than the apple's skin.

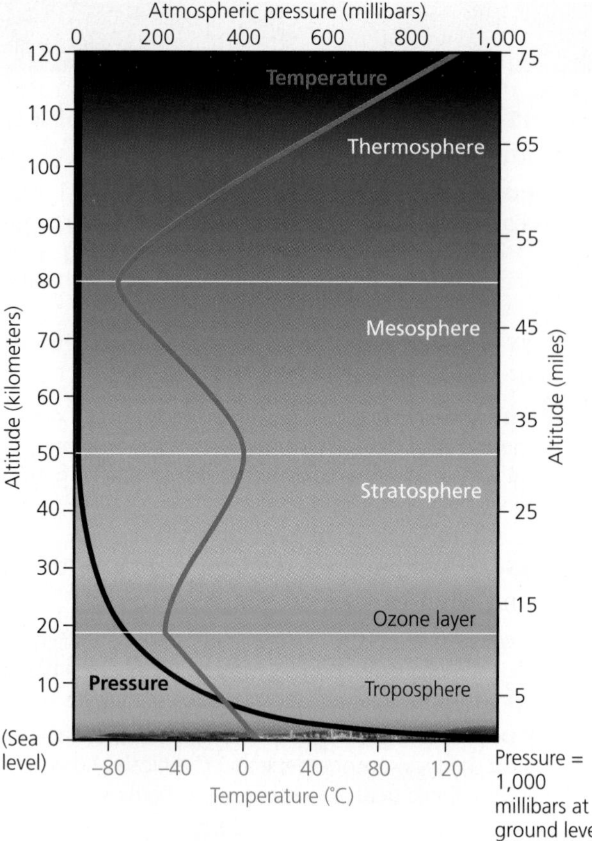

Figure 15-2 **Natural capital:** The earth's atmosphere is a dynamic system that includes four layers. The average temperature of the atmosphere varies with altitude (red line) and with differences in the absorption of incoming solar energy. **Question:** Why do you think most of the planet's air is in the troposphere?

Take a deep breath. About 99% of the volume of air you inhaled consists of two gases: nitrogen (78%) and oxygen (21%). The remainder consists of water vapor (varying from 0.01% at the frigid poles to 4% in the humid tropics, for an average of about 1%), 0.93% argon (Ar), 0.039% carbon dioxide (CO_2), and trace amounts of dust and soot particles as well as other gases including methane (CH_4), ozone (O_3), and nitrous oxide (N_2O).

Rising and falling air currents, winds, and concentrations of CO_2 and other greenhouse gases in the troposphere play a major role in the planet's short-term *weather* (see Supplement 5, p. S18) and long-term *climate*.

The atmosphere's second layer is the **stratosphere**, which extends from about 17 to about 48 kilometers (from 11 to 30 miles) above the earth's surface (Figure 15-2). Although the stratosphere contains less matter than the troposphere, its composition is similar, with two notable exceptions: its volume of water vapor is about 1/1,000 that of the troposphere and its concentration of ozone (O_3) is much higher.

Much of the atmosphere's small amount of ozone (O_3) is concentrated in a portion of the stratosphere called the **ozone layer**, found roughly 17–26 kilometers (11–16 miles) above sea level. Most stratospheric ozone is produced when some of the oxygen molecules in this layer interact with ultraviolet (UV) radiation emitted by the sun ($3\ O_2 + UV \rightleftarrows 2\ O_3$). This "global sunscreen" of ozone in the stratosphere keeps about 95% of the sun's harmful UV radiation (see Figure 3-3, p. 42) from reaching the earth's surface.

15-2 What Are the Major Air Pollution Problems?

▶ **CONCEPT 15-2A Three major outdoor air pollution problems are *industrial smog* primarily from burning coal, *photochemical smog* from motor vehicle and industrial emissions, and *acid deposition* from coal burning and motor vehicle exhaust.**

▶ **CONCEPT 15-2B The most threatening indoor air pollutants are smoke and soot from wood and coal fires (mostly in less-developed countries), cigarette smoke, and chemicals used in building materials and cleaning products.**

Air Pollution Comes from Natural and Human Sources

Air pollution is the presence of chemicals in the atmosphere in concentrations high enough to harm organisms, ecosystems, or human-made materials, or to alter climate. Air pollutants come from natural and human sources. Natural sources include wind-blown dust, pollutants from wildfires and volcanic eruptions, and volatile organic chemicals released by some plants. Most natural air pollutants are spread out over the globe or removed by chemical cycles, precipitation, and gravity. But in areas experiencing volcanic eruptions or forest fires, chemicals emitted by these events can temporarily reach harmful levels.

Most human inputs of outdoor air pollutants occur in industrialized and urban areas where people, cars, and factories are concentrated. These pollutants are generated mostly by the burning of fossil fuels in power plants and industrial facilities (*stationary sources*) and in motor vehicles (*mobile sources*).

Scientists classify outdoor air pollutants into two categories: **Primary pollutants** are chemicals or substances emitted directly into the air from natural pro-

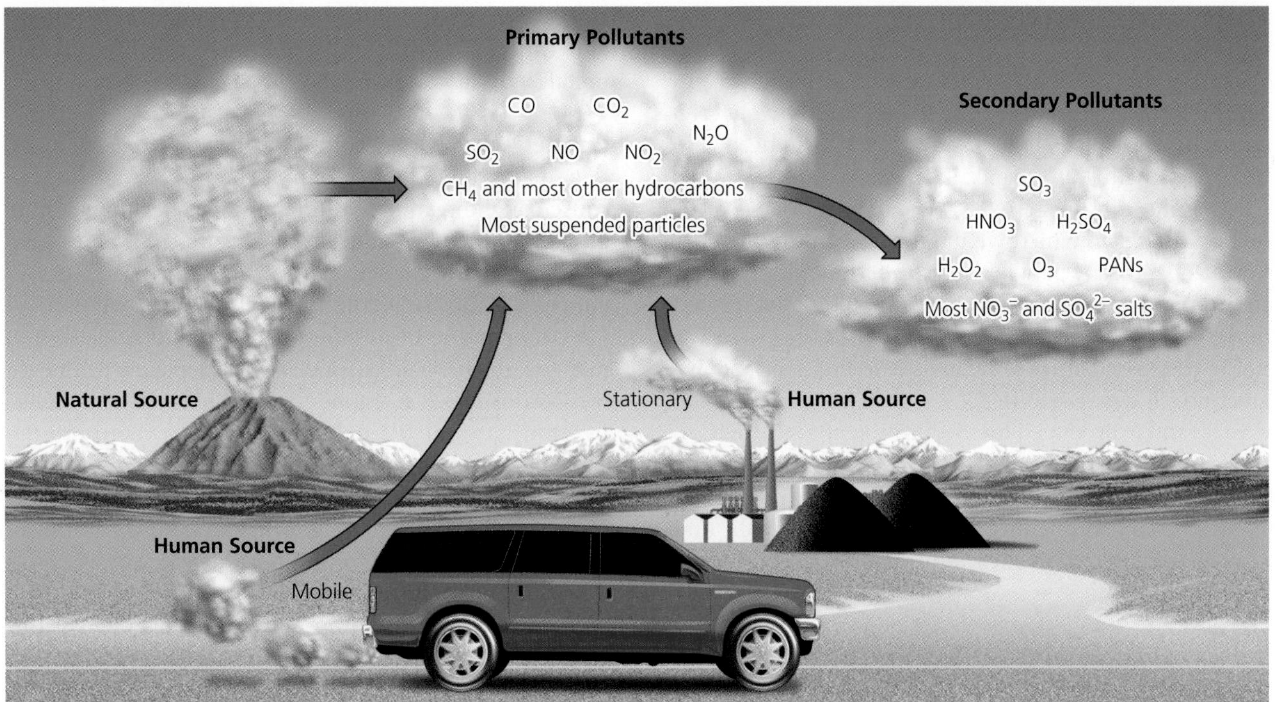

Figure 15-3 Human inputs of air pollutants come from *mobile sources* (such as cars) and *stationary sources* (such as industrial, power, and cement plants). Some *primary air pollutants* react with one another and with other chemicals in the air to form *secondary air pollutants*.

cesses and human activities (Figure 15-3) at concentrations high enough to cause harm. While in the atmosphere, some primary pollutants react with one another and with other natural components of air to form new harmful chemicals, called **secondary pollutants** (Figure 15-3, right).

Outdoor air pollution was once a regional problem limited mostly to cities. Now it is a global problem, largely due to the sheer volume of pollutants produced by human activities. Long-lived pollutants entering the atmosphere in India and China now find their way across the Pacific where they affect the west coast of North America. There is no place on the planet's surface that has not been affected by air pollution.

What Are the Major Outdoor Air Pollutants?

Carbon oxides. *Carbon monoxide* (CO) is a colorless, odorless, and highly toxic gas that forms during the incomplete combustion of carbon-containing materials. Major sources are motor vehicle exhaust, the burning of forests and grasslands, smokestacks of fossil fuel–burning power plants and industries, tobacco smoke, and open fires and inefficient stoves used for cooking.

CO can combine with hemoglobin in red blood cells, which reduces the ability of blood to transport oxygen to body cells and tissues. Long-term exposure can trigger heart attacks and aggravate lung diseases such as asthma and emphysema. At high levels, CO can cause headache, nausea, drowsiness, confusion, collapse, coma, and death.

Carbon dioxide (CO_2) is a colorless, odorless gas. About 93% of the CO_2 in the atmosphere is the result of the natural carbon cycle (see Figure 3-15, p. 53). The rest comes from human activities, primarily the burning of fossil fuels and the clearing of CO_2-absorbing forests and grasslands. Until recently it has not been classified in many countries as an air pollutant. But there is considerable and growing scientific evidence that increasing levels of CO_2 (see Figure 14, p. S44, in Supplement 7) caused by human activities are contributing to atmospheric warming and projected climate change, which can contribute to human health problems, as discussed in Sections 15-4 through 15-6 of this chapter.

Note that, at a high enough level, essentially any chemical introduced into the atmosphere can become a pollutant, including CO_2, which we add every time we exhale or burn a carbon-containing chemical. The key question is whether or not we are adding CO_2 to the atmosphere faster than it is removed by the earth's carbon cycle. Data from scientific measurements show that we are, as revealed by increases in CO_2 levels in the atmosphere since 1960 (see Figure 14, p. S44, in Supplement 7).

Nitrogen oxides and nitric acid. *Nitric oxide* (NO) is a colorless gas that forms when nitrogen and oxygen gas react under high-combustion temperatures in automobile engines and coal-burning power and industrial plants. Lightning and certain bacteria in soil and water also produce NO as part of the nitrogen cycle (see Figure 3-16, p. 54).

In the air, NO reacts with oxygen to form *nitrogen dioxide* (NO_2), a reddish-brown gas. Collectively, NO and NO_2 are called *nitrogen oxides* (NO_X). Some of the NO_2 reacts with water vapor in the air to form *nitric acid* (HNO_3) and nitrate salts (NO_3^-), components of harmful *acid deposition*, which we discuss later in this chapter. Both NO and NO_2 play a role in the formation of **photochemical smog**—a mixture of pollutants formed under the influence of sunlight in cities with heavy traffic (discussed further below). *Nitrous oxide* (N_2O), a greenhouse gas, is emitted from fertilizers and animal wastes, and is produced by the burning of fossil fuels.

At high enough levels, nitrogen oxides can irritate the eyes, nose, and throat, aggravate lung ailments such as asthma and bronchitis, suppress plant growth, and reduce visibility when they are converted to nitric acid and nitrate salts.

Sulfur dioxide and sulfuric acid. *Sulfur dioxide* (SO_2) is a colorless gas with an irritating odor. About one-third of the SO_2 in the atmosphere comes from natural sources as part of the sulfur cycle (see Figure 3-18, p. 56). The other two-thirds (and as much as 90% in some urban areas) come from human sources, mostly combustion of sulfur-containing coal in power and industrial plants, oil refining, and smelting of sulfide ores.

In the atmosphere, SO_2 can be converted to *aerosols*, which consist of microscopic suspended droplets of *sulfuric acid* (H_2SO_4) and suspended particles of sulfate (SO_4^{2-}) salts that return to the earth as a component of acid deposition. Sulfur dioxide, sulfuric acid droplets, and sulfate particles reduce visibility and aggravate breathing problems. They can also damage crops, trees, soils, and aquatic life in lakes, and they corrode metals and damage paint, paper, leather, and stone on buildings and in statues and monuments. SO_2 emissions in China have increased by more than a third in the past decade, according to a 2007 U.S. National Academy of Sciences report.

Particulates. *Suspended particulate matter* (SPM) consists of a variety of solid particles and liquid droplets that are small and light enough to remain suspended in the air for long periods. The U.S. Environmental Protection Agency (EPA) classifies particles as fine, or PM-10 (with diameters less than 10 micrometers), and ultrafine, or PM-2.5 (with diameters less than 2.5 micrometers). About 62% of the SPM in outdoor air comes from natural sources such as dust, wild fires, and sea salt. The remaining 38% comes from human sources such as coal-burning power and industrial plants, motor vehicles, and road construction.

These particles can irritate the nose and throat, damage the lungs, aggravate asthma and bronchitis, and shorten one's life span. Toxic particulates such as lead, cadmium, and polychlorinated biphenyls (PCBs), can cause genetic mutations, reproductive problems, and cancer. Particulates also reduce visibility, corrode metals, and discolor clothes and paints.

Ozone. *Ozone* (O_3), a colorless and highly reactive gas, is a major ingredient of photochemical smog. It can cause coughing and breathing problems, aggravate lung and heart diseases, reduce resistance to colds and pneumonia, and irritate the eyes, nose, and throat. It also damages plants, rubber in tires, fabrics, and paints.

Ozone in the troposphere near ground level is often referred to as "bad" ozone, while we view ozone in the stratosphere as "good" ozone because it protects us from harmful UV radiation. However, both are the same chemical. Significant evidence indicates that some human activities are *decreasing* the amount of beneficial ozone in the stratosphere and *increasing* the amount of harmful ozone in the troposphere near ground level—especially in some urban areas. We examine the issue of stratospheric ozone thinning in Section 15-7 of this chapter.

Volatile organic compounds (VOCs). Organic compounds that exist as gases in the atmosphere or that evaporate from sources on earth into the atmosphere are called *volatile organic compounds* (VOCs). Examples are hydrocarbons, emitted by the leaves of many plants, and *methane* (CH_4), a greenhouse gas that is 20 times more effective per molecule than CO_2 is at warming the atmosphere through the greenhouse effect (see Figure 3-3, p. 42).

About a third of global methane emissions comes from natural sources, mostly plants, wetlands, and termites. The rest comes from human sources, primarily rice paddies, landfills, oil and natural gas wells, and cows (mostly from their belching). Other VOCs are liquids than can evaporate into the atmosphere. Examples are benzene and other liquids used as industrial solvents, dry-cleaning fluids, and various components of gasoline, plastics, and other products.

Burning Coal Produces Industrial Smog

Sixty years ago, cities such as London, England, and the U.S. cities of Chicago, Illinois, and Pittsburgh, Pennsylvania, burned large amounts of coal in power plants and factories, and for heating homes and often for cooking food. People in such cities, especially during winter, were exposed to **industrial smog**, consisting mostly of an unhealthy mix of sulfur dioxide, suspended droplets of sulfuric acid, and a variety of suspended solid particles in outside air. Those burning coal inside their homes were exposed to dangerous levels of indoor air pollutants.

Today, urban industrial smog is rarely a problem in most of the more-developed countries where coal and heavy oil are burned only in large power and industrial plants with reasonably good pollution control or in facilities with tall smokestacks that send the pollutants high into the air, where prevailing winds disperse them. However, industrial smog remains a problem in industrialized areas of China, India, Ukraine, and some other eastern European countries, where large quantities of coal are still burned in houses, power plants, and factories with inadequate pollution controls.

Because of its heavy reliance on coal, China has some of the world's highest levels of industrial smog and 16 of the world's 20 most polluted cities. A 2007 study by the World Bank puts the annual death toll from air pollution in China at about 750,000; about 250,000 of those deaths are from outdoor air pollution and 500,000 are from indoor air pollution (mostly from burning coal for heating and cooking and from smoking cigarettes).

The history of air pollution control in Europe and the United States shows that we can reduce industrial smog fairly quickly by setting standards for coal-burning industries and utilities, and by shifting from coal to cleaner-burning natural gas in urban industries and in dwellings.

Sunlight Plus Cars Equals Photochemical Smog

A *photochemical reaction* is any chemical reaction activated by light. Photochemical smog is a mixture of primary and secondary pollutants formed under the influence of UV radiation from the sun. In greatly simplified terms,

VOCs + NO_x + heat + sunlight → ground level ozone (O_3) + other photochemical oxidants + aldehydes + other secondary pollutants

The formation of photochemical smog begins when exhaust from morning commuter traffic releases large amounts of NO and VOCs into the air over a city. The NO is converted to reddish-brown NO_2, explaining why photochemical smog is sometimes called *brown-air smog*. When exposed to ultraviolet radiation from the sun, some of the NO_2 reacts in complex ways with VOCs released by certain trees (such as some oak species, sweet gums, and poplars), motor vehicles, and businesses (such as bakeries and dry cleaners). The resulting mixture of pollutants, dominated by ground-level ozone, usually builds up to peak levels by late morning, irritating people's eyes and respiratory tracts. Some of its pollutants, known as *photochemical oxidants*, can damage lung tissue.

All modern cities have some photochemical smog, but it is much more common in cities with sunny, warm, and dry climates, and a great number of motor vehicles. Examples are Los Angeles, California, and Salt Lake City, Utah, in the United States; Sydney, Australia; São Paulo, Brazil; Bangkok, Thailand; Mexico City, Mexico; and Santiago, Chile (Figure 15-4). According to a 1999 study, if 400 million people in China were to drive conventional gasoline-powered cars by 2050, the resulting photochemical smog could regularly cover the entire western Pacific, extending to the United States. (Watch a video at **www.cengagebrain.com** to see how photochemical smog forms and how it affects us.)

CONNECTIONS

Short Driving Trips and Air Pollution

About 60% of the pollution from motor vehicle emissions occurs in the first minutes of operation before pollution control devices are working at top efficiency. Yet 40% of all car trips take place within 3 kilometers (2 miles) of drivers' homes, and half of the U.S. working population drives 8 kilometers (5 miles) or less to work.

Julio Etchart/Peter Arnold, Inc.

Figure 15-4 *Global Outlook:* Photochemical smog is a serious problem in Santiago, Chile. **Question:** How serious is photochemical smog where you live?

Several Factors Can Decrease or Increase Outdoor Air Pollution

Five natural factors help *reduce* outdoor air pollution. First, *particles heavier than air* settle out as a result of gravitational attraction to the earth. Second, *rain and snow* partially cleanse the air of pollutants. Third, *salty sea spray from the oceans* washes out many pollutants from air that flows from land over the oceans. Fourth, *winds* sweep pollutants away and mix them with cleaner air. Fifth, some pollutants are removed by *chemical reactions*. For example, SO_2 can react with O_2 in the atmosphere to form SO_3, which reacts with water vapor to form droplets of H_2SO_4 that fall out of the atmosphere as acid precipitation.

Six other factors can *increase* outdoor air pollution. First, *urban buildings* slow wind speed and reduce the dilution and removal of pollutants. Second, *hills and mountains* reduce the flow of air in valleys below them and allow pollutant levels to build up at ground level. Third, *high temperatures* promote the chemical reactions leading to formation of photochemical smog. Fourth, *emissions of volatile organic compounds (VOCs)* from certain trees and plants in heavily wooded urban areas can play a large role in the formation of photochemical smog.

A fifth factor—the so-called *grasshopper effect*—occurs when air pollutants are transported at high altitudes by evaporation and winds from tropical and temperate areas through the atmosphere to the earth's polar areas. This happens mostly during winter. It explains why, for decades, pilots have reported seeing dense layers of reddish-brown haze over the Arctic. It also explains why polar bears, sharks, and native peoples in remote arctic areas have high levels of various toxic pollutants in their bodies.

Sixth, *temperature inversions* can cause pollutants to build to high levels. During daylight, the sun warms the air near the earth's surface. Normally, this warm air and most of the pollutants it contains rise to mix with the cooler air above and are dispersed. Under certain atmospheric conditions, however, a layer of warm air can temporarily lie atop a layer of cooler air nearer the ground, creating a **temperature inversion**. Because the cooler air is denser than the warmer air above it, the air near the surface does not rise and mix with the air above. If this condition persists, pollutants can build up to harmful and even lethal concentrations in the stagnant layer of cool air near the ground. (Go to **www.cengagebrain.com** to see a video on thermal inversions and to learn what they can mean for people in some cities.)

Acid Deposition Is a Serious Regional Air Pollution Problem

Most coal-burning power plants, ore smelters, and other industrial facilities in more-developed countries use tall smokestacks to vent the exhausts from burned fuel. The exhausts contain sulfur dioxide, suspended particles, and nitrogen oxides, and the smokestacks send them high into the atmosphere where wind can dilute and disperse them.

This method reduces *local* air pollution, but it can increase *regional* air pollution downwind. Prevailing winds may transport the primary pollutants SO_2 and NO_x, emitted high into the troposphere, as far as 1,000 kilometers (600 miles). During their trip, these compounds form secondary pollutants such as droplets of sulfuric acid, nitric acid vapor, and particles of acid-forming sulfate and nitrate salts (Figure 15-3).

These acidic substances remain in the atmosphere for 2–14 days, depending mostly on prevailing winds, precipitation, and other weather patterns. During this period, they descend to the earth's surface in two forms: *wet deposition*, consisting of acidic rain, snow, fog, and cloud vapor, and *dry deposition*, consisting of acidic particles. The resulting mixture is called **acid deposition** (Figure 15-5)—sometimes called *acid rain*. Most dry deposition occurs within 2–3 days of emission, relatively close to the industrial sources, whereas most wet deposition takes place within 4–14 days in more distant downwind areas.

Acid deposition is mostly a *regional* air pollution problem (**Concept 15-2A**) in areas that lie downwind from coal-burning facilities and from urban areas with large numbers of cars, as shown in Figure 15-6. In some areas, soils contain compounds that can react with and neutralize, or *buffer*, some inputs of acids. The areas most sensitive to acid deposition are those with thin, already acidic soils that provide no such natural buffering (Figure 15-6, green and most red areas) and those where the buffering capacity of soils has been depleted by decades of acid deposition.

In the United States, older coal-burning power and industrial plants without adequate pollution controls, mostly located in the Midwest, emit the largest quantities of SO_2 and other pollutants that cause acid deposition. Because of these emissions and those of other urban industries and motor vehicles, as well as the prevailing west-to-east winds, typical precipitation in the eastern United States is at least 10 times more acidic than natural precipitation is. Some mountaintop forests in the eastern United States, as well as east of Los Angeles, California, are bathed in fog and dews that are as acidic as lemon juice—with about 1,000 times the acidity of normal precipitation.

This is also an international problem. Acid-producing chemicals generated in one country are often exported to other countries by prevailing winds.

Acid Deposition Has a Number of Harmful Effects

Acid deposition damages statues and buildings, contributes to human respiratory diseases, and can leach toxic metals (such as lead and mercury) from soils and rocks

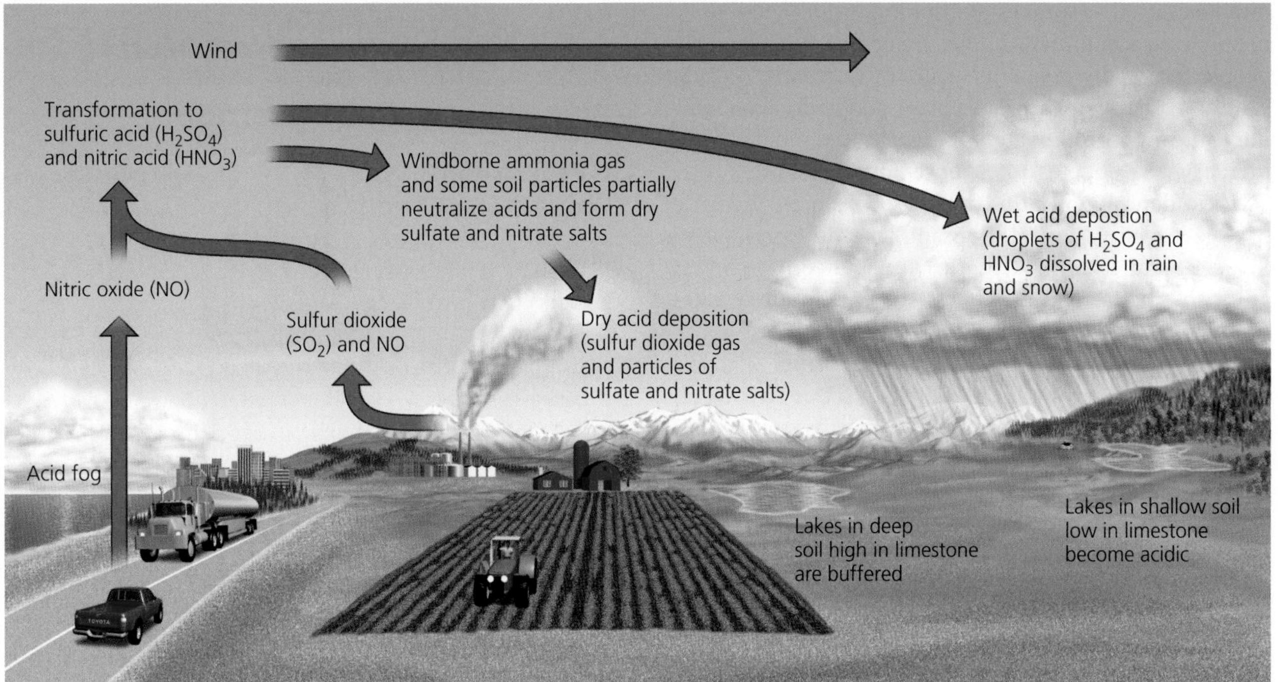

Active Figure 15-5 Natural capital degradation: *Acid deposition*, which consists of rain, snow, dust, or gas with a pH lower than 5.6, is commonly called acid rain. *See an animation based on this figure at* **www.cengagebrain.com**. **Question:** What are three ways in which your daily activities contribute to acid deposition?

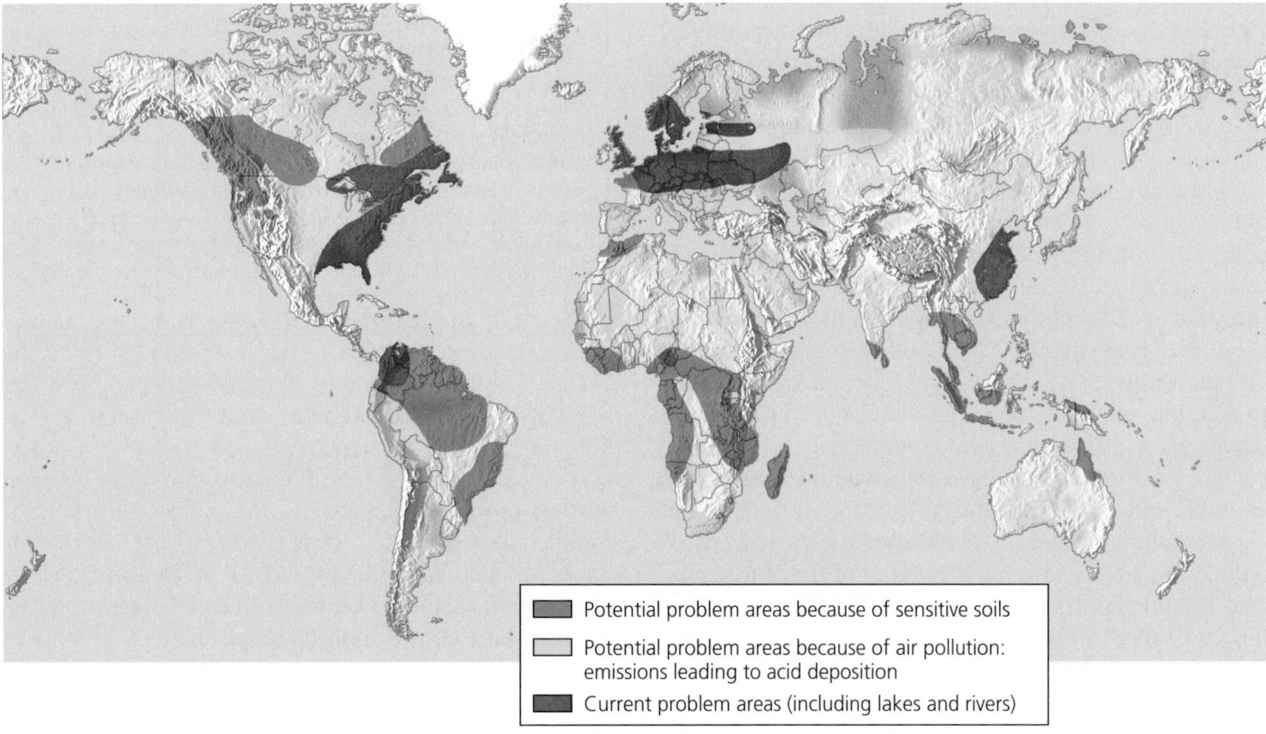

Figure 15-6 This map shows regions where acid deposition is now a problem and regions with the potential to develop this problem. Such regions have large inputs of air pollution (mostly from power plants, industrial facilities, and ore smelters) or are sensitive areas with naturally acidic soils and bedrock that cannot neutralize (buffer) inputs of acidic compounds. (Data from World Resources Institute and U.S. Environmental Protection Agency) **Question:** Do you live in or near an area that is affected by acid deposition or an area that is likely to be affected by acid deposition in the future?

into lakes used as sources of drinking water. These toxic metals can accumulate in the tissues of fish eaten by people (especially pregnant women). Currently, 45 U.S. states have issued warnings telling people to avoid eating fish caught from waters that are contaminated with mercury (see Chapter 14, Science Focus, p. 358).

Acid deposition harms aquatic ecosystems. Because of excess acidity, several thousand lakes in Norway and Sweden, and 1,200 in Ontario, Canada, contain few if any fish. In the United States, several hundred lakes (most in the Northeast) are similarly threatened.

Acid deposition (often along with other air pollutants such as ozone) can affect forests in two ways. One is by leaching essential plant nutrients such as calcium and magnesium from forest soils. The other is by releasing ions of aluminum, lead, cadmium, and mercury, which are toxic to the trees. These two effects rarely kill trees directly, but they can weaken them and leave them vulnerable to stresses such as severe cold, diseases, insect attacks, and drought. Mountaintop forests are especially vulnerable to acid deposition because they tend to have thin soils with little buffering capacity and are continuously exposed to highly acidic fog and clouds. (See a video to examine how acid deposition can harm a pine forest and what it means to surrounding land and waters at **www.cengagebrain.com**.)

We Know How to Reduce Acid Deposition

Figure 15-7 summarizes ways to reduce acid deposition. According to most scientists studying the problem, the best solutions are *preventive approaches* that reduce or eliminate emissions of sulfur dioxide, nitrogen oxides, and particulates. GOOD NEWS

Although we know how to prevent acid deposition (Figure 15-7, left), implementing these solutions is politically difficult. One problem is that the people and ecosystems it affects often are quite far downwind from the sources of the problem. Also, countries with large supplies of coal (such as China, India, Russia, Australia, and the United States) have a strong incentive to use it. In addition, politically powerful owners of coal-burning power plants argue that taking additional measures to cut their contribution to acid rain would increase the cost of electricity for consumers.

Air pollution laws in the United States helped to reduce the acidity of rainfall in parts of the Northeast, Mid-Atlantic, and Midwest regions. But there is still a long way to go in reducing emissions from older coal-burning power and industrial plants. Meanwhile, China has become the world's largest emitter of sulfur dioxide and has one of the world's most serious acid deposition problems. However, between 2006 and 2009, SO_2 emissions in China fell by more than 13%, despite the construction of many new coal-burning power plants. This is encouraging but China has a long way to go in curtailing acid deposition.

Solutions

Acid Disposition

Prevention

Reduce coal use

Burn low-sulfur coal

Increase use of natural gas and renewable energy resources

Remove SO_2 particulates and NO_X from smokestack gases and remove NO_X from motor vehicular exhaust

Tax emissions of SO_2

Cleanup

Add lime to neutralize acidified lakes

Add phosphate fertilizer to neutralize acidified lakes

Figure 15-7 There are several ways to reduce acid deposition and its damage. **Questions:** Which two of these solutions do you think are the best ones? Why?

CONNECTIONS

Low-Sulfur Coal, Atmospheric Warming, and Toxic Mercury

Some U.S. power plants have lowered SO_2 emissions by switching from high-sulfur to low-sulfur coals. However, this has increased CO_2 emissions that contribute to atmospheric warming and projected climate change, because low-sulfur coal has a lower heat value, which means that more coal must be burned to generate a given amount of electricity. Low-sulfur coal also has higher levels of toxic mercury and other trace metals, so burning it emits more of these hazardous chemicals into the atmosphere.

Indoor Air Pollution Is a Serious Problem

In less-developed countries, the indoor burning of wood, charcoal, dung, crop residues, coal, and other fuels in open fires (Figure 15-8) or in unvented or poorly vented stoves exposes people to dangerous levels of particulate air pollution (**Concept 15-2B**). According to the World Health Organization (WHO) and the World Bank, *indoor air pollution is the world's most serious air pollution problem, especially for poor people.*

Indoor air pollution is also a serious problem in more-developed areas of all countries, mostly because of chemicals used in building materials and products. Figure 15-9 shows some typical sources of indoor air pollution in a modern home.

Joerg Boethling/Peter Arnold, Inc.

Figure 15-8 Burning wood to cook food inside this dwelling in India exposes the occupants to dangerous levels of air pollution. Other fuels often burned in leaky stoves or indoor open fires in many dwellings include charcoal, dung, coal, and crop residues.

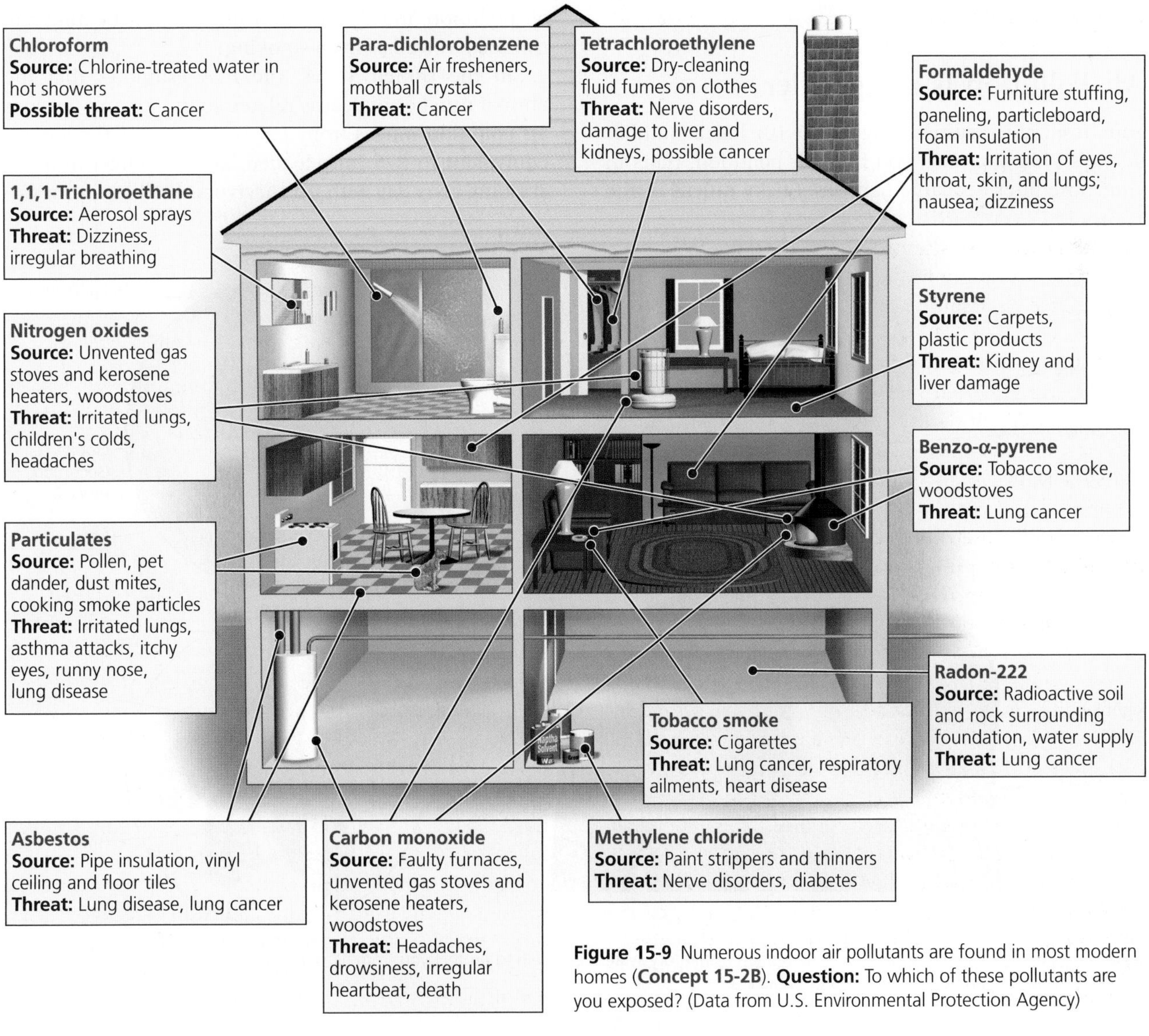

Chloroform
Source: Chlorine-treated water in hot showers
Possible threat: Cancer

Para-dichlorobenzene
Source: Air fresheners, mothball crystals
Threat: Cancer

Tetrachloroethylene
Source: Dry-cleaning fluid fumes on clothes
Threat: Nerve disorders, damage to liver and kidneys, possible cancer

Formaldehyde
Source: Furniture stuffing, paneling, particleboard, foam insulation
Threat: Irritation of eyes, throat, skin, and lungs; nausea; dizziness

1,1,1-Trichloroethane
Source: Aerosol sprays
Threat: Dizziness, irregular breathing

Styrene
Source: Carpets, plastic products
Threat: Kidney and liver damage

Nitrogen oxides
Source: Unvented gas stoves and kerosene heaters, woodstoves
Threat: Irritated lungs, children's colds, headaches

Benzo-α-pyrene
Source: Tobacco smoke, woodstoves
Threat: Lung cancer

Particulates
Source: Pollen, pet dander, dust mites, cooking smoke particles
Threat: Irritated lungs, asthma attacks, itchy eyes, runny nose, lung disease

Radon-222
Source: Radioactive soil and rock surrounding foundation, water supply
Threat: Lung cancer

Tobacco smoke
Source: Cigarettes
Threat: Lung cancer, respiratory ailments, heart disease

Asbestos
Source: Pipe insulation, vinyl ceiling and floor tiles
Threat: Lung disease, lung cancer

Carbon monoxide
Source: Faulty furnaces, unvented gas stoves and kerosene heaters, woodstoves
Threat: Headaches, drowsiness, irregular heartbeat, death

Methylene chloride
Source: Paint strippers and thinners
Threat: Nerve disorders, diabetes

Figure 15-9 Numerous indoor air pollutants are found in most modern homes (**Concept 15-2B**). **Question:** To which of these pollutants are you exposed? (Data from U.S. Environmental Protection Agency)

EPA studies have revealed some alarming facts about indoor air pollution. *First,* levels of 11 common pollutants generally are 2 to 5 times higher inside U.S. homes and commercial buildings than they are outdoors, and in some cases they are as much as 100 times higher. *Second,* pollution levels inside cars in traffic-clogged urban areas can be up to 18 times higher than outside levels. *Third,* the health risks from exposure to such chemicals are magnified because most people in developed urban areas spend 70–98% of their time indoors or inside vehicles. Smokers, children younger than age 5, the elderly, the sick, pregnant women, people with respiratory or heart problems, and factory workers are especially at risk from indoor air pollution. **GREEN CAREER:** indoor air pollution specialist

According to the EPA and public health officials, the four most dangerous indoor air pollutants in more-developed countries are *tobacco smoke* (see Chapter 14, Case Study, p. 367) *formaldehyde* emitted from many building materials and various household products; *radioactive radon-222 gas,* which can seep into houses from underground rock deposits; and *very small (ultrafine) particles* of various substances in emissions from motor vehicles, coal-burning facilities, wood burning, and forest and grass fires.

Air Pollution Is a Big Killer

Your respiratory system (Figure 15-10) has a number of ways to help protect you from air pollution. Hairs in your nose filter out large particles. Sticky mucus in the lining of your upper respiratory tract captures smaller (but not the smallest) particles and dissolves some gaseous pollutants. Sneezing and coughing expel contaminated air and mucus when pollutants irritate your respiratory system.

In addition, hundreds of thousands of tiny, mucus-coated, hairlike structures, called *cilia,* line your upper respiratory tract. They continually move back and forth and transport mucus and the pollutants it traps to your throat where they are swallowed or expelled.

Prolonged or acute exposure to air pollutants, including tobacco smoke, can overload or break down these natural defenses. Fine and ultrafine particulates get lodged deep in the lungs, contributing to lung cancer, asthma, heart attack, and stroke. Years of smoking or breathing polluted air can lead to other lung ailments such as chronic bronchitis and emphysema, which leads to acute shortness of breath.

According to the WHO, at least 2.4 million people worldwide die prematurely each year—an average of nearly 274 deaths every hour—from the effects of air pollution. According to a 2007 World Bank study, most of these deaths occur in Asia, with about 750,000 deaths per year in China alone (500,000 from indoor air pollution, mostly from burning coal for heating and cooking, as well as from smoking).

In the United States, the EPA estimates that the annual number of deaths related to indoor and outdoor air pollution ranges from 150,000 to 350,000 people—equivalent to 2–5 fully loaded, 200-passenger airliners crashing *every day* with no survivors. Millions more suf-

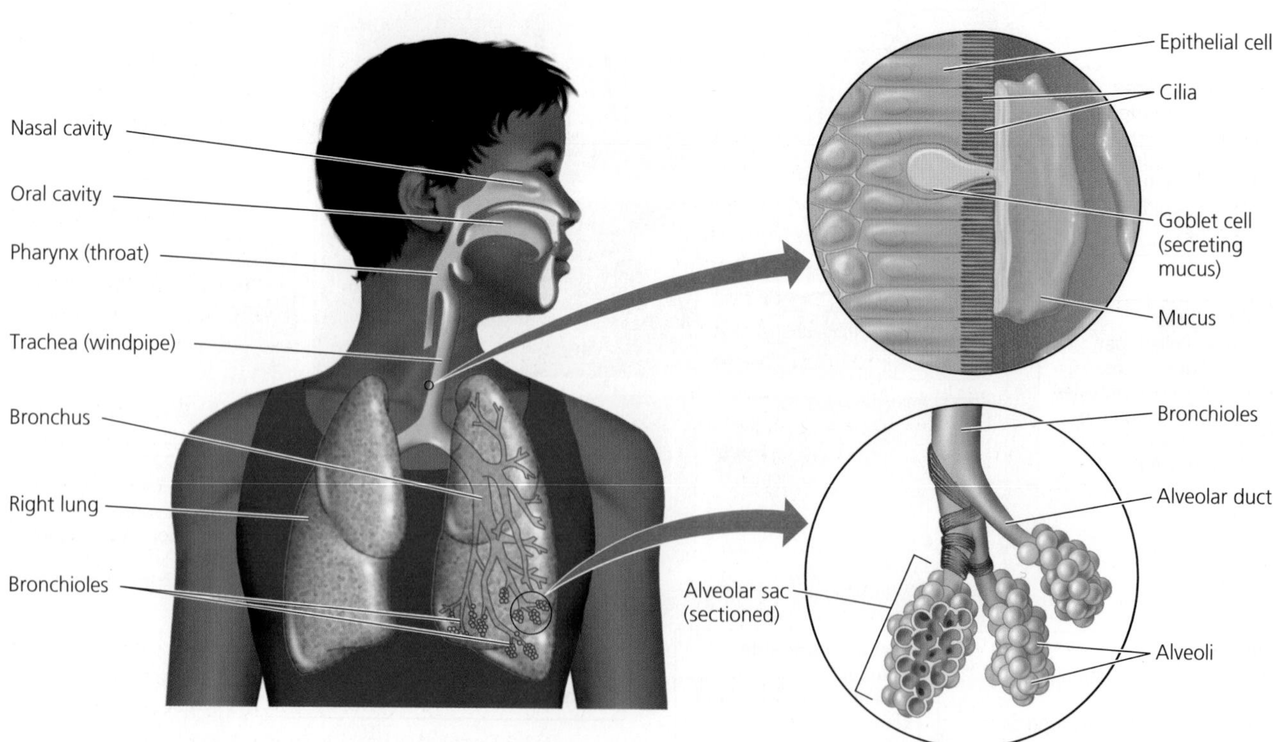

Figure 15-10 The major components of the human respiratory system can help to protect you from air pollution, but these defenses can be overwhelmed or breached.

fer from asthma attacks and other respiratory disorders. Inhalation of very small particulates, mostly from coal-burning power plants, is responsible for as many as 68,000 premature deaths a year in the United States (Figure 15-11).

According to EPA studies, every year, more than 125,000 Americans (96% of them in urban areas) get cancer from breathing soot-laden diesel fumes emitted by buses, trucks, tractors, construction equipment, and train engines. According to the EPA, a large diesel truck emits as much particulate matter as 150 cars, and particulate emissions from a diesel train engine equal those from 1,500 cars. Also, according to a 2009 study led by Daniel Lack, the world's 100,000 or more diesel-powered oceangoing ships emit almost half as much particulate pollution as do the world's 760 million cars. This is largely because many cargo ships burn low-grade oil called *bunker fuel* in which the concentration of polluting sulfur is 30 times higher than that of diesel fuel sold at the pumps of U.S. gas stations. Thus, the largely unregulated shipping industry is one of the largest polluters of the atmosphere.

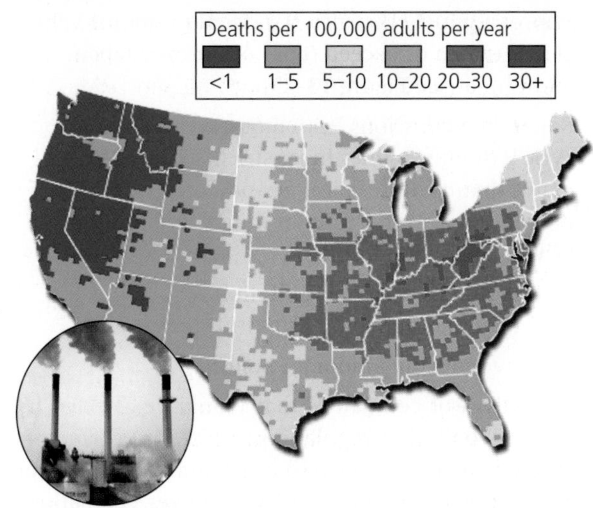

Figure 15-11 This map shows the distribution of premature deaths from air pollution in the United States, mostly from very small, fine, and ultra-fine particles added to the atmosphere by coal-burning power plants. (Data from U.S. Environmental Protection Agency) **Questions:** Why do the highest death rates occur in the eastern half of the United States? If you live in the United States, what is the risk at your home or where you go to school?

15-3 How Should We Deal with Air Pollution?

▶ **CONCEPT 15-3 Legal, economic, and technological tools can help us to clean up air pollution, but the best solution is to prevent it.**

Laws and Regulations Can Reduce Outdoor Air Pollution

The United States provides an excellent example of how a regulatory approach can reduce air pollution (**Concept 15-3**). The U.S. Congress passed the Clean Air Acts in 1970, 1977, and 1990. With these laws, the federal government established air pollution regulations for key pollutants that are enforced by states and major cities.

Congress directed the EPA to establish air quality standards for six major outdoor pollutants—carbon monoxide (CO), nitrogen dioxide (NO_2), sulfur dioxide (SO_2), suspended particulate matter (SPM, smaller than PM-10), ozone (O_3), and lead (Pb). Each standard specifies the maximum allowable level for a pollutant, averaged over a specific period.

According to a 2009 EPA report, the combined emissions of the six major pollutants decreased by about 54% between 1980 and 2008, even with significant increases during the same period in gross domestic product, vehicle miles traveled, energy consumption, and population. The decreases in emissions during this period were 78% for lead, 68% for CO, 59% for SO_2, 35% for NO_2, 31% for PM-10, and 14% for ground-level ozone.

GOOD NEWS

In 2010, the EPA announced a new rule designed to tackle regional air pollution problems, including acid deposition. The rule requires power plants to reduce their emissions of SO_2 and nitrogen oxides (NO_x). These requirements are projected cut SO_2 emissions by 71% and NO_x emissions by 52% below their 2005 levels.

The reduction of outdoor air pollution in the United States since 1970 has been a remarkable success story, mostly because of two factors. *First*, U.S. citizens insisted that laws be passed and enforced to improve air quality. *Second*, the country was affluent enough to afford such controls and improvements.

Environmental scientists applaud this success, but they call for strengthening U.S. air pollution laws by:

- Putting much greater emphasis on preventing air pollution. The power of prevention (**Concept 15-3**) was made clear by the large drop in atmospheric lead levels after lead in gasoline was banned in 1976.
- Sharply reducing emissions from older coal-burning power and industrial plants, cement plants, oil refineries, and waste incinerators. Approximately 20,000 of such facilities in the United States have been excused from having to meet the air pollution standards for new facilities.

- Improving fuel efficiency standards for motor vehicles to match or exceed those of Europe, Japan, and China (see Figure 13-25).
- Regulating emissions from motorcycles and two-cycle gasoline engines more strictly. The EPA estimates that using a typical gas-powered riding lawn mower for 1 hour creates as much air pollution as driving a car for 34 hours.
- Setting much stricter air pollution regulations for airports and oceangoing ships in U.S. waters.
- Sharply reducing indoor air pollution.

Executives of companies that would be affected by stronger air pollution regulations claim that correcting these problems would cost too much and would hinder economic growth. Proponents of stronger regulations contend that history has shown that most industry cost estimates for implementing U.S. air pollution control standards have been much higher than the costs actually proved to be.

The ultimate success of any emissions trading approach depends on two factors: how low the initial cap is set and how often it is lowered in order to promote continuing innovation in air pollution prevention and control. Without these two elements, emissions trading programs mostly shift pollution problems from one area to another without achieving any overall improvement in air quality.

GOOD NEWS

Between 1990 and 2006, the emissions trading system helped to reduce SO_2 emissions from power plants in the United States by 53%, at a cost of less than one-tenth what was projected by the industry.

An emissions trading system is also being used for NO_x. However, environmental and health scientists strongly oppose a cap-and-trade program for controlling emissions of toxic mercury by coal-burning power plants and industries. They warn that coal-burning plants that choose to buy permits instead of sharply reducing their mercury emissions will create toxic hotspots with unacceptably high levels of mercury.

We Can Use the Marketplace to Reduce Outdoor Air Pollution

One approach to reducing pollutant emissions has been to allow producers of air pollutants to buy and sell government air pollution allotments in the marketplace. With the goal of reducing SO_2 emissions, the Clean Air Act of 1990 authorized an *emissions trading,* or *cap-and-trade program,* which enables the 110 most polluting coal-fired power plants in 21 states to buy and sell SO_2 pollution rights. Each plant is annually given a number of pollution credits, which allow it to emit a certain amount of SO_2. A utility that emits less than its allotted amount has a surplus of pollution credits. That utility can use its credits to offset SO_2 emissions at its other plants, keep them for future plant expansions, or sell them to other utilities or private parties.

Proponents of this approach say it is cheaper and more efficient than government regulation of air pollution. Critics of this approach contend that it allows utilities with older, dirtier power plants to buy their way out of their environmental responsibilities and to continue polluting. In addition, without strict government oversight, this approach makes cheating possible, because cap-and-trade is based largely on self-reporting of emissions.

There Are Many Ways to Reduce Outdoor Air Pollution

Figure 15-12 summarizes several ways to reduce emissions of sulfur oxides, nitrogen oxides, and particulate matter from stationary sources such as coal-burning power plants and industrial facilities.

Figure 15-13 lists several ways to prevent or reduce emissions from motor vehicles, the primary factor in the formation of photochemical smog.

Unfortunately, the already poor air quality in urban areas of many less-developed countries is worsening as the numbers of motor vehicles in these nations rise. Many of these vehicles are 10 or more years old, have no pollution-control devices, and burn leaded gasoline.

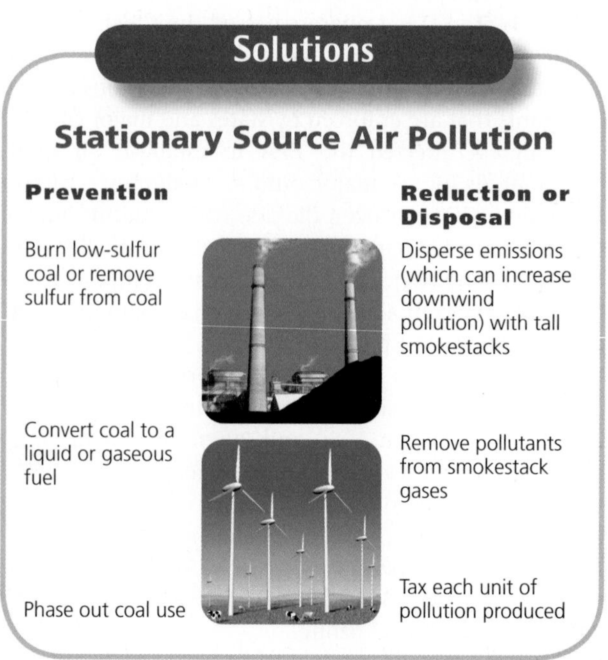

Figure 15-12 There are several ways to prevent, reduce, or disperse emissions of sulfur oxides, nitrogen oxides, and particulate matter from stationary sources such as coal-burning power plants and industrial facilities (**Concept 15-3**). **Questions:** Which two of these solutions do you think are the best ones? Why?

Solutions

Motor Vehicle Air Pollution

Prevention	Reduction
Walk, bike, or use mass transit	Require emission control devices
Improve fuel efficiency	Inspect car exhaust systems twice a year
Get older, polluting cars off the road	Set strict emission standards

Figure 15-13 There are a number of ways to prevent or reduce emissions from motor vehicles (**Concept 15-3**). **Questions:** Which two of these solutions do you think are the best ones? Why?

Solutions

Indoor Air Pollution

Prevention	Cleanup or Dilution
Ban indoor smoking	Use adjustable fresh air vents for work spaces
Set stricter formaldehyde emissions standards for carpet, furniture, and building materials	Circulate air more frequently
Prevent radon infiltration	Circulate a building's air through rooftop greenhouses
Use less polluting cleaning agents, paints, and other products	Use efficient venting systems for wood-burning stoves

Figure 15-14 There are several ways to prevent, clean up, or disperse indoor air pollution (**Concept 15-2B**). **Questions:** Which two of these solutions do you think are the best ones? Why?

Reducing Indoor Air Pollution Should Be a Priority

Little effort has been devoted to reducing indoor air pollution, even though it poses a much greater threat to human health than does outdoor air pollution (**Concept 15-2B**). Air pollution experts suggest several ways to prevent, clean up, or disperse indoor air pollution, as shown in Figure 15-14.

In less-developed countries, indoor air pollution from open fires (Figure 15-8) and inefficient stoves could be reduced. More people could use inexpensive clay or metal stoves (Figure 15-15) that burn fuels (including straw and other crop wastes) more efficiently and vent their exhausts to the outside, or they could use stoves that use solar energy to cook food (see Figure 13-34, p. 331). GOOD NEWS

Figure 15-16 (p. 388) lists some ways in which you can reduce your exposure to indoor air pollution.

We Can Emphasize Pollution Prevention

Environmental and health scientists argue that over the next few decades, we will need to put much greater emphasis on preventing air pollution (**Concept 15-3**). With this approach, the question is not *What can we do about the air pollutants we produce?* but rather *How can we avoid producing these pollutants in the first place?*

Like the shift to *controlling outdoor air pollution* between 1970 and 2010, this new shift to *preventing outdoor and indoor air pollution* will not take place unless

National Renewable Energy Laboratories

Figure 15-15 The energy-efficient Turbo Stove™ can greatly reduce indoor air pollution and the resulting premature deaths in less-developed countries such as India. (See Figure 15-8.)

What Can You Do?

Indoor Air Pollution

- Test for radon and formaldehyde inside your home and take corrective measures as needed
- Do not buy furniture and other products containing formaldehyde
- Remove your shoes before entering your house to reduce inputs of dust, lead, and pesticides
- Switch to phthalate-free detergents
- Use baked lemons as natural fragrance
- Test your house or workplace for asbestos fiber levels, and check for any crumbling asbestos materials if it was built before 1980
- Do not store gasoline, solvents, or other volatile hazardous chemicals inside a home or attached garage
- If you smoke, do it outside or in a closed room vented to the outside
- Make sure that wood-burning stoves, fireplaces, and kerosene and gas-burning heaters are properly installed, vented, and maintained
- Install carbon monoxide detectors in all sleeping areas
- Use fans to circulate indoor air
- Grow house plants, the more, the better

Figure 15-16 **Individuals matter:** You can reduce your exposure to indoor air pollution. **Questions:** Which three of these actions do you think are the most important? Why?

individual citizens and groups put political pressure on elected officials to enact and enforce appropriate regulations. In addition, citizens can, through their purchases, put economic pressure on companies to get them to manufacture and sell products and services that do not add to pollution problems.

15-4 How Might the Earth's Climate Change in the Future?

▶ **CONCEPT 15-4 Considerable scientific evidence indicates that the earth's atmosphere is warming because of a combination of natural effects and human activities, and that this warming is likely to lead to significant climate disruption during this century.**

Weather and Climate Are Not the Same

In thinking about climate change, it is very important to distinguish between weather and climate. *Weather* consists of short-term changes in atmospheric variables such as the temperature, precipitation, wind, and barometric pressure in a given area over a period of hours or days (see Supplement 5, p. S18). By contrast, *climate* is determined by the *average* weather conditions of the earth or of a particular area, especially temperature and precipitation, over periods of at least three decades to thousands of years.

During any period of 30 or more years, in a given area of the planet, there will often be hotter years and cooler years, and wetter years and drier years, as weather fluctuates widely from day to day and from year to year. There is now a growing controversy over atmospheric warming and projected climate change, largely because not everyone understands the difference between weather and climate. One or two warmer or colder years or decades can result simply from changes in the weather and do not necessarily tell us that the earth's climate is warming or cooling.

Climate scientists look at data on normally changing weather conditions to see if there has been a general rise or fall in any measurements such as average temperature or precipitation over a period of at least 30 years. Then and only then do they make projections about long-term changes in climate. Unless we have at least 30 years of data we cannot make meaningful statements about climate.

Climate Change Is Not New

Climate change is neither new nor unusual. Over the past 3.5 billion years, the planet's climate has been altered by volcanic emissions, changes in solar input, continents moving slowly atop shifting tectonic plates, impacts by large meteors, and other factors.

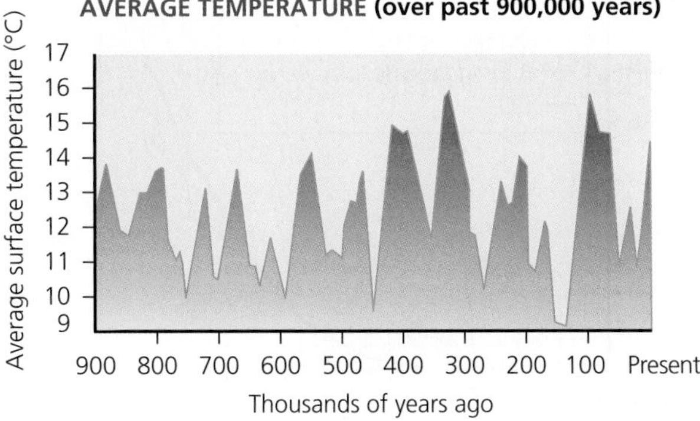

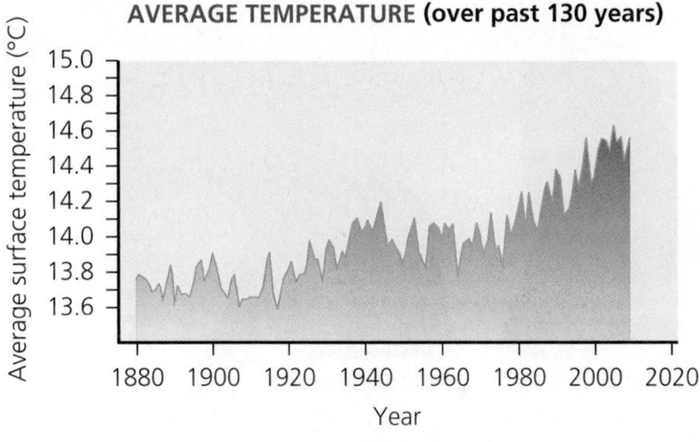

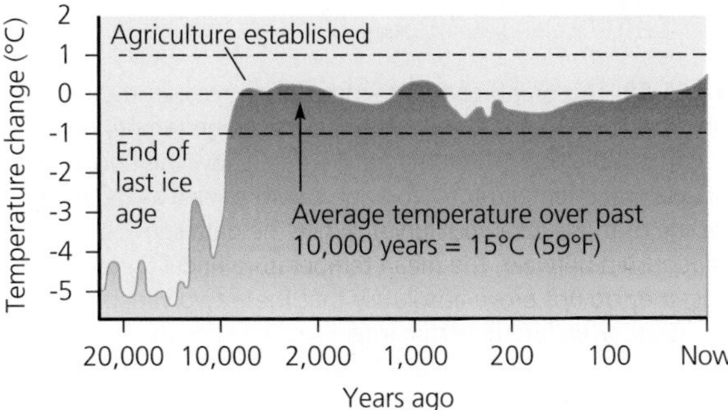

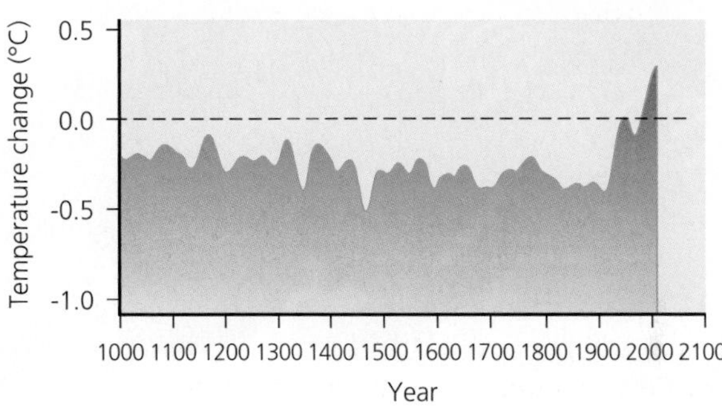

Figure 15-17 Science: The two graphs in the top half of this figure are rough estimates of long- and short-term global average temperatures, and the two graphs on the bottom are estimates of changes in the average temperature of the earth's lower atmosphere over thousands of years. They are based on scientific evidence that contains gaps, but they do indicate general trends. (Data from Goddard Institute for Space Studies, Intergovernmental Panel on Climate Change, National Academy of Sciences, National Aeronautics and Space Administration, National Center for Atmospheric Research, and National Oceanic and Atmospheric Administration) **Question:** Assuming these are good estimates, what are two conclusions you can draw from these graphs?

Over the past 900,000 years, the atmosphere has experienced prolonged periods of *global cooling* and *global warming* (Figure 15-17, top left). These alternating cycles of freezing and thawing are known as *glacial and interglacial* (between ice ages) *periods.*

For roughly 10,000 years, we have had the good fortune to live in an interglacial period characterized by a fairly stable climate based mostly on a generally steady global average surface temperature (Figure 15-17, bottom left). These favorable climate conditions allowed the human population to grow as agriculture developed, and later as cities grew. For the past 1,000 years, the average temperature of the atmosphere has remained fairly stable but began to rise during the last century (Figure 15-17, bottom right) when more people began clearing more forests and burning more fossil fuels. The top right graph in Figure 15-17 shows that most of the recent increase in temperature has taken place since 1975.

Past temperature changes such as those depicted in Figure 15-17 are estimated through analysis of a number of types of evidence, including radioisotopes in rocks and fossils; plankton and radioisotopes in ocean sediments; tiny bubbles, layers of soot, and other materials trapped in different layers of ancient air found in ice cores from glaciers (Figure 15-18, p. 390); pollen from the bottoms of lakes and bogs; tree rings; and temperature measurements taken regularly since 1861. Such measurements have limitations, but they show general changes in temperature, which in turn can affect the earth's climate. (See *The Habitable Planet,* Video 12, at **www.learner.org/resources/series209.html** for a discussion of how scientists are analyzing ice cores from mountain glaciers.)

Human Activities Emit Large Quantities of Greenhouse Gases

Along with solar energy, a natural process called the *greenhouse effect* (see Figure 3-3, p. 42) plays a key role in determining the earth's climate. It occurs when some of the solar energy absorbed by the earth radiates into the atmosphere as infrared radiation (heat). As this radiation interacts with molecules in the air—especially the four *greenhouse gases,* water vapor (H_2O), carbon dioxide (CO_2), methane (CH_4), and nitrous oxide (N_2O)—it increases their kinetic energy and warms the lower atmosphere and the earth's surface, helping to create a livable climate. (See a video on how greenhouse gases help to generate heat in the lower atmosphere and raise the earth's temperature at **www.cengagebrain.com.**)

Figure 15-18 **Science:** *Ice cores* are extracted by drilling deep holes into ancient glaciers at various sites such as this one near the South Pole in Antarctica. Analysis of ice cores yields information about the past composition of the lower atmosphere, temperature trends such as those shown in Figure 15-17, greenhouse gas concentrations, solar activity, snowfall, and forest fire frequency.

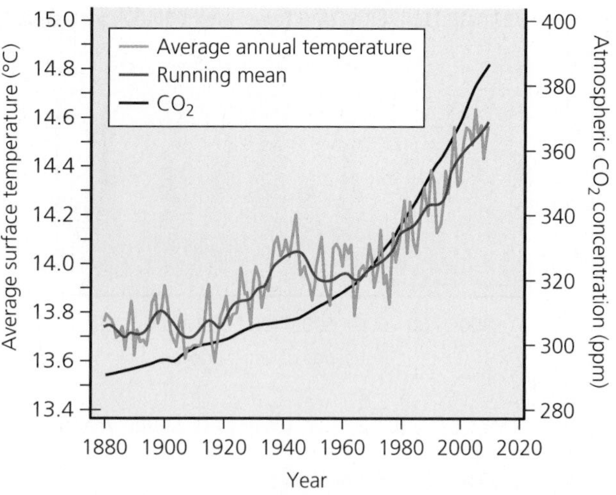

Figure 15-19 This graph compares atmospheric concentrations of carbon dioxide (CO_2) and the atmosphere's average temperature for the period 1880–2009. The temperature data show the fluctuating seasonal average temperatures (orange curve) as well as the *mean*, or average of these data (red curve). While the data show an apparent relationship between the mean temperature and CO_2 concentrations, they do not prove absolutely that these two variables are related. (Data from Intergovernmental Panel on Climate Change, NASA Goddard Institute for Space Studies, and NOAA Earth System Research Laboratory)

Since the beginning of the Industrial Revolution in the mid-1700s, human actions—mainly the burning of fossil fuels—have led to significant increases in the concentrations of several greenhouse gases in the lower atmosphere. In particular, the average atmospheric concentration of CO_2 rose dramatically during that time, along with the average temperature of the atmosphere (Figure 15-19).

Working through the World Meteorological Organization and the United Nations Environment Programme, several nations established the Intergovernmental Panel on Climate Change (IPCC) in 1988 to document past climate changes and project future changes. The IPCC network includes more than 2,500 climate scientists and scientists in disciplines related to climate studies from more than 130 countries. In 2007, the IPCC issued a report based on more than 29,000 sets of data, finding that:

1. The earth's lower atmosphere has warmed, especially since 1980, due mostly to increased levels of CO_2 and other greenhouse gases (Figure 15-19).
2. Most of these increases are due to human activities, especially the burning of fossil fuels and deforestation.
3. These human-induced changes are beginning to change the earth's climate.
4. If greenhouse gas concentrations continue to rise, the earth is likely to experience rapid atmospheric warming and climate disruption during this century (Science Focus, p. 392).

Here is some of the evidence that the IPCC and other climate scientists have used to support the major conclusions of the 2007 IPCC report:

- Between 1906 and 2005, the average global surface temperature has risen by about 0.74 C° (1.3 F°). Most of this increase has taken place since 1980 (Figure 15-17, top right, and Figure 15-19).
- Average levels of CO_2 in the atmosphere rose sharply between 1960 and 2010 (see Figure 14, p. S44, in Supplement 7).
- The first decade in this century (2000–2009) was the warmest decade since 1881 (Figure 15-17, top right), and 2010 was the warmest year on record.
- In some parts of the world, glaciers are melting (Figure 15-20 and **Core Case Study**) and floating sea ice is shrinking (Figure 15-21), both at increasing rates. CORE CASE STUDY
- In 2010, NASA scientists reported on a survey of the world's major lakes, which showed that these lakes have warmed since 1985 at rates of 0.81–1.8 F° per decade.
- During the 20th century, the world's average sea level rose by 19 centimeters (7 inches), mostly because of runoff from melting land-based ice and the expansion of ocean water as its temperature increased. By comparison, sea levels rose about 2 centimeters (3/4 of an inch) in the

Figure 15-20 Much of Alaska's Muir Glacier in the popular Glacier Bay National Park and Preserve melted between 1948 and 2004. **Question:** How might melting glaciers in Alaska and other parts of the world affect your life?

Sept. 1979

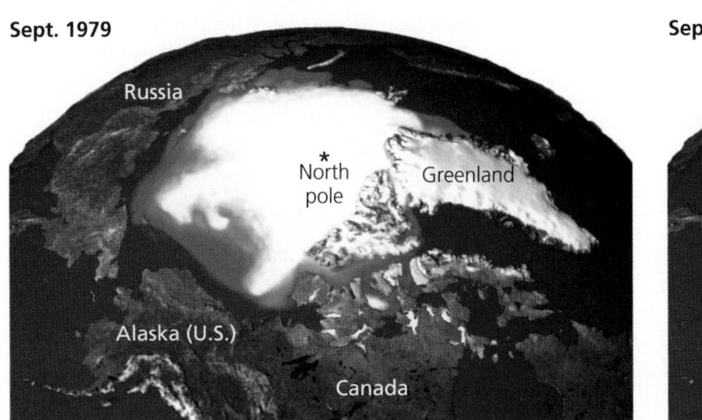

Sept. 2010

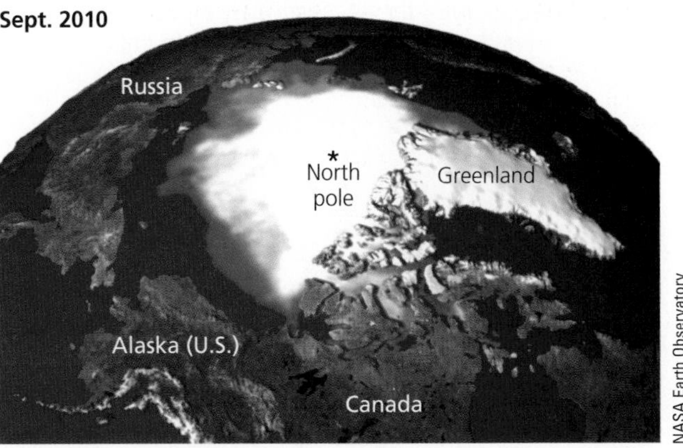

Figure 15-21 *The big melt:* Each summer, some of the floating ice in the Arctic Sea melts and then refreezes during winter. But in most recent years, rising average atmospheric and ocean temperatures have caused more and more ice to melt during the summer months. Satellite data show a 39% drop in the average cover of summer arctic sea ice between 1979 and 2010. If this trend continues, this summer ice may be gone by 2040, and perhaps earlier, according to a 2007 estimate made by NASA climate scientist Jay Zwally. (Data U.S. Goddard Space Flight Center, NASA, National Snow and Ice Data Center)

18th century and 6 centimeters (2 inches) in the 19th century.

A growing number of highly regarded scientists argue that evidence of the likelihood of severe, long-term climate disruption has strengthened since the 2007 IPCC report. For example, in 2010, the U.S. National Academy of Sciences concluded that "A strong and credible body of evidence shows that climate change is occurring, is caused largely by human activities, and poses significant risks for a broad range of human and natural systems."

Scientists have identified a number of natural and human-influenced factors that might *amplify* or *dampen* projected changes in the average temperature of the atmosphere (Science Focus, p. 392). We now examine a few of the most important factors.

CO_2 Emissions Play an Important Role

Data from the National Oceanic and Atmospheric Administration (NOAA) show that the atmospheric concentration of CO_2 rose from about 285 parts per million (ppm) in 1850, shortly after the start of the Industrial Revolution, to 390 ppm in 2010 (Figure 15-19).

SCIENCE FOCUS

Using Models to Project Future Changes in Atmospheric Temperatures

To project the effects of increasing levels of greenhouse gases on average global temperatures, scientists have developed complex *mathematical models*, which simulate interactions among the earth's sunlight, clouds, landmasses, oceans, ocean currents, concentrations of greenhouse gases and air pollutants, and other factors within the earth's complex climate system. They run these continually improving models on supercomputers and compare the results to known past climate changes, from which they project future changes in the earth's average temperature. Figure 15-A gives a greatly simplified summary of some of the interactions in the global climate system.

Such models provide *projections* of what is *likely* or *very likely* to happen to the average temperature of the lower atmosphere. How well these projections match what actually happens in the real world depends on the validity of the assumptions and variables built into the models, and on the accuracy of the data used.

Recall that no scientific research can give us absolute proof or certainty, but instead provides us with varying levels of certainty. In other words, there will always be some degree of uncertainty in the results of science. However, when most experts in a particular field, such as climate science, generally agree that the uncertainty involved in a set of measurements, a scientific theory, or a model is less than 10% (which means the level of certainty is at least 90%), we should take such results very seriously because they are very likely to be correct.

In 1990, 1995, 2001, and 2007, the IPCC published reports on how global temperatures have changed in the past (Figure 15-17) and made projections of how they are likely to change during this century and how such changes are likely to affect the earth's climate. According to the 2007 report, based on analyzing past climate data and the use of 19 climate models, it is very likely that human activities, especially the burning of fossil fuels, have played an important role in the observed atmospheric warming of the past 30 years (Figure 15-17, top right). The researchers base this on the fact that, after thousands of times running the models, the only way they can get the model results to match actual measurements is by including the human activities factor.

The 2007 IPCC report and more recent runs of the 19 climate models suggest that it is very likely that the earth's mean surface temperature will increase by 2–4.5C° (3.6–8.1F°) between 2005 and 2100 (Figure 15-B) unless the world halts deforestation and makes drastic cuts in greenhouse gas emissions from fossil fuel–burning power plants, factories, and cars. The lower temperature in this range is likely only if global greenhouse gas emissions fall 50–85% below 2000 levels by 2050, which is unlikely.

In any scientific model, there is always some degree of uncertainty, as indicated by the fairly wide range of projected temperature changes in Figure 15-B. Climate experts are working hard to improve these models

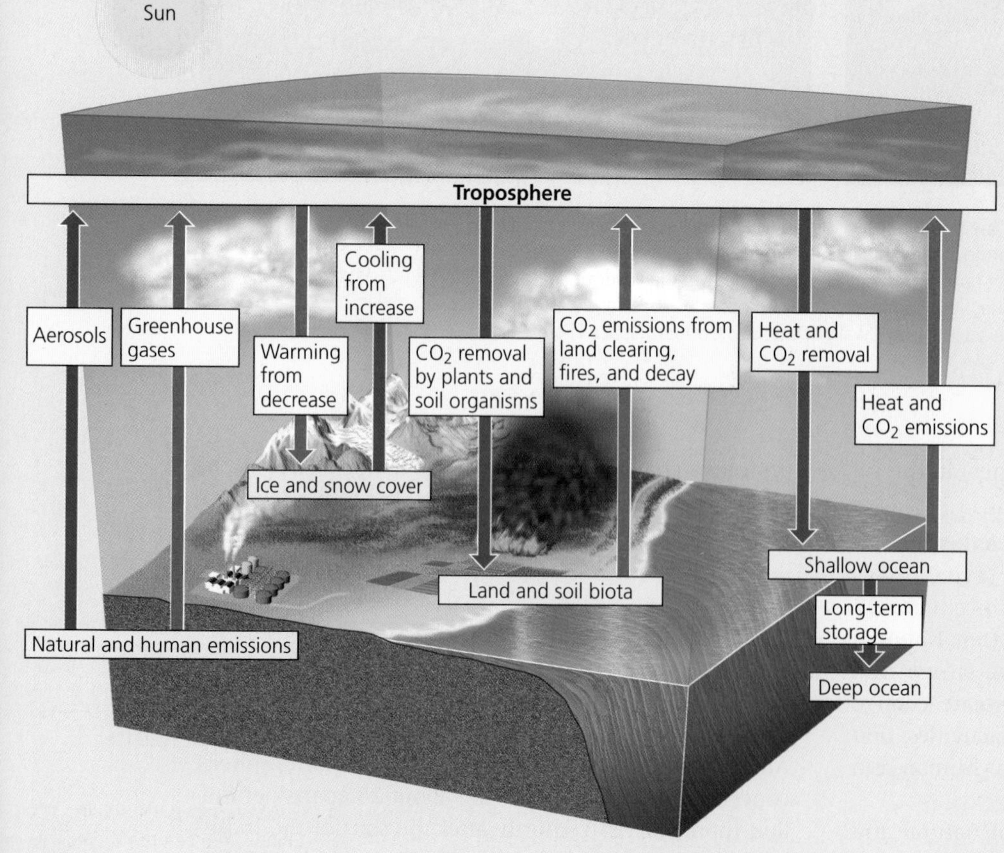

Figure 15-A Science: This is a simplified model of some major processes that interact to determine the average temperature and greenhouse gas content of the lower atmosphere and thus the earth's climate. Red arrows show processes that warm the atmosphere and blue arrows show those that cool it. **Question:** Why do you think a decrease in snow and ice cover is adding to the warming of the atmosphere?

and to narrow the projected range of temperatures. Also, current climate models are better at projecting likely general changes in climate factors at the global level than they are at making such projections for particular areas. Despite such limitations, these models are the best tools that we have for estimating likely overall climate change in coming decades.

Critical Thinking

If the projected temperature increases in Figure 15-B take place, what are three major ways in which this will affect your lifestyle and that of any children or grandchildren you might have?

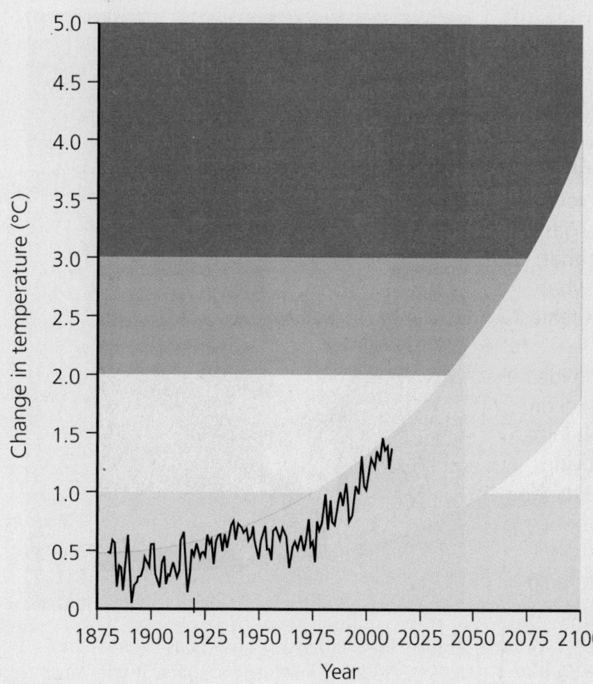

Figure 15-B Science: This figure shows estimated (yellow area) and measured changes (black curve) in the average temperature of the atmosphere at the earth's surface between 1860 and 2008, and the projected range of temperature increase during the rest of this century (**Concept 15-1**). (Data from U.S. National Academy of Sciences, National Center for Atmospheric Research, Intergovernmental Panel on Climate Change, and Hadley Center for Climate Prediction and Research)

This amounts to a 37% increase in the amount of CO_2 in the atmosphere, and this is a long-term increase, because CO_2 remains in the atmosphere for decades to hundreds of years.

According to a 2007 study by climate scientists Christopher Field and Gregg Marland, if CO_2 emissions continue to increase at their current rate, levels in the atmosphere are likely to rise to 560 ppm by 2050 and could soar to 1,390 ppm by 2100. Climate models project that such dramatic increases would bring about significant changes in the earth's climate, which would likely cause major ecological and economic disruption in the latter half of this century.

A number of scientific studies and major climate models indicate that we need to prevent CO_2 levels from exceeding 450 ppm—an estimated threshold, or irreversible *tipping point*, that could set into motion large-scale climate changes that would last for hundreds to thousands of years. According to NASA climate scientist James Hansen (Individuals Matter, p. 394), we need to bring CO_2 levels down to 350 ppm, the level we had in 1990, to maintain a climate similar to that in which our civilizations developed over the last 10,000 years.

In 2009, China and the United States together accounted for about 42% of the world's annual greenhouse gas emissions (Figure 15-22). However, considering emissions per person, the United States still leads the world, emitting about 5 times more CO_2 per person than China emits and almost 200 times more CO_2 per person than the poorest countries emit. It is somewhat encouraging that U.S. emissions per person have remained at roughly the same level since 1990, but many analysts argue that this level is still a major environmental threat.

What Role Does the Sun Play?

The energy output of the sun plays the key role in the earth's temperature and this output has varied over millions of years. However, in 2007, climate researchers Claus Froelich, Mike Lockwood, and Ben Santer all

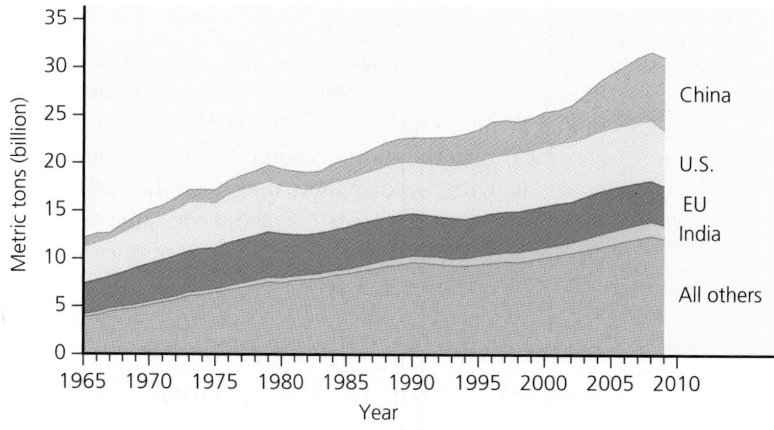

Figure 15-22 This graph shows how CO_2 emissions from the burning of fossil fuels in selected countries increased between 1965 and 2009. China surpassed the United States as the leading emitter in 2007. (Data from BP Statistical Review)

INDIVIDUALS MATTER

Sounding the Alarm—James Hansen

In 1988, American climate scientist James Hansen (Figure 15-C) appeared before a U.S. Congressional committee and stated that, based on the results of climate models that he and others had developed, atmospheric warming was a grave threat made worse every day by large and growing emissions of carbon dioxide and other greenhouse gases resulting from human activities. With that, he kicked off the public debate over what most climate scientists believe is our greatest environmental challenge.

Hansen also helped to promote the idea of creating the Intergovernmental Panel on Climate Change (IPCC) in 1988 to organize climate experts with the goals of studying this issue and reporting their findings to the world. He has used his decades of experience as a climate scientist and head of NASA's Goddard Institute for Space Studies to help develop several climate models (Science Focus, p. 392).

Hansen is well respected by his peers and is also known for his willingness to speak out. His message is that, based on what he has learned by studying climate and climate change for many decades, rising levels of greenhouse gases will likely lead to catastrophic climate disruption during this century unless those levels are drastically reduced. He has stated repeatedly that what motivates him is his concern for future generations. "The difficulty of this problem," Hansen said, "is that its main impacts will be felt by our children and by our grandchildren."

Bruce Gilbert/Courtesy of NASA Goddard Institute for Space Studies

Figure 15-C James E. Hansen has been head of NASA's Goddard Institute for Space Studies since 1981.

For his outspokenness, he has been threatened and harassed by some industry and government representatives. He has refused to yield to pressure from his powerful critics.

In 2006, Hansen received the American Association for the Advancement of Science Award for Scientific Freedom and Responsibility. In 2009, he received the American Meteorological Society's highest award for his "outstanding contributions to climate modeling, understanding climate change, and for clear communication of climate science in the public arena."

concluded in separate studies that most of the rise in global average atmospheric temperatures since 1980 (Figure 15-17, top right, and Figure 15-19) could not be the result of increased solar output. Instead, they determined that the energy output of the sun has dropped slightly during the past several decades.

These researchers noted that, according to satellite and weather balloon measurements since 1975, the troposphere has warmed while the stratosphere has cooled. This is the opposite of what a hotter sun would do. According to Santer, "If you increase output from the sun, you increase the amount of energy that arrives at the top of the Earth's atmosphere and get heating throughout the atmosphere."

Instead, the data show that the atmosphere is now heating from the bottom up, which indicates that inputs at the earth's surface (most likely from human activities) play the more important role in atmospheric warming. Other studies show that since the mid-1970s, the sun's output has remained about the same and thus cannot account for the rise in temperature since 1975.

What Role Do Oceans Play in Projected Climate Disruption?

The world's oceans absorb CO_2 from the atmosphere as part of the carbon cycle (see Figure 3-15, p. 53) and thus help to moderate the earth's average surface temperature and its climate. It is estimated that the oceans remove about 25–30% of the CO_2 pumped into the lower atmosphere by human activities. The oceans also absorb heat from the lower atmosphere. Then, partly driven by this heat, ocean currents slowly transfer some the CO_2 to the deep ocean (see Figure 7-5, p. 124), where it is buried in carbon compounds in bottom sediments for several hundred million years.

The ability of the oceans to absorb CO_2 decreases as water temperatures increase. As the oceans warm up, some of their dissolved CO_2 is released into the lower atmosphere, like CO_2 bubbling out of a warm carbonated soft drink. According to scientific measurements from a 2007 study, the upper portion of the oceans warmed by an average of 0.32–0.67C° (0.6–1.2F°) during the last century—an astounding increase considering the huge volume of water involved—most likely due to increasing atmospheric temperatures. In 2009, the world's oceans were the warmest in 130 years of record keeping.

According to a 2005 report by the United Kingdom's Royal Society and a 2010 report by the U.S. National Academy of Sciences, rising levels of CO_2 in the ocean have increased the acidity of its surface waters by 30% since about 1800. This is because much of the CO_2 absorbed by the ocean reacts with water to produce carbonic acid (H_2CO_3). The higher acidity threatens corals and other organisms with shells made of calcium carbonate, which dissolves when acidity reaches a certain

Tatiana Grozetskaya/Shutterstock.com

Cheryl Casey/Shutterstock.com

Figure 15-23 Cumulus clouds (left) are thick, relatively low-lying clouds that tend to decrease surface warming by reflecting some incoming solar radiation back into space. Cirrus clouds (right) are thin and float at high altitudes; they tend to warm the earth's surface by preventing some heat from flowing into space.

level. This further reduces the ability of the oceans to absorb CO_2 and could thus accelerate projected climate change.

There Is Uncertainty about the Effects of Cloud Cover on Atmospheric Warming

The role played by clouds in the warming of the atmosphere is uncertain. Warmer temperatures increase evaporation of surface water and create more clouds, which can either cool or warm the atmosphere.

An increase in thick and continuous *cumulus clouds* at low altitudes (Figure 15-23, left) could decrease surface warming by reflecting more sunlight back into space. But an increase in thin, wispy *cirrus clouds* at high altitudes (Figure 15-23, right) could increase the warming of the lower atmosphere by preventing more heat from escaping into space. Climate modelers have not been able to show which of these effects is the most important and are working hard to understand more about the role of clouds in their climate models.

In 2010, Axel Lauer and other researchers developed a new model for evaluating the effects of cloud cover on atmospheric warming. According to their model, by the end of this century, higher CO_2 levels and a warmer atmosphere will shrink the size of low cumulus clouds. This will allow more solar radiation to reach the lower atmosphere, which will likely accelerate atmospheric warming toward the higher estimated temperatures shown in Figure 15-B (p. 393).

> CONNECTIONS
> **Air Travel and Atmospheric Warming**
>
> Infrared satellite images indicate that wispy condensation trails (contrails) left behind by jet planes expand and turn into cirrus clouds, which tend to heat the atmosphere. If these preliminary results are confirmed, we may find that emissions from jet planes are responsible for as much as half of the warming of the lower troposphere in the northern hemisphere where much of the world's air travel is concentrated.

15-5 What Are Some Possible Effects of a Warmer Atmosphere?

▶ CONCEPT 15-5 The projected rapid change in the atmosphere's temperature could have severe and long-lasting consequences, including increased drought and flooding, rising sea levels, and shifts in the locations of croplands and wildlife habitats.

Enhanced Atmospheric Warming Could Have Serious Consequences

Most historic changes in the temperature of the lower atmosphere took place over thousands of years (Figure 15-17, top left and bottom left). What makes the current problem urgent is that we face *a rapid projected increase in the average temperature of the lower atmosphere during this century* (Figure 15-B). This, in turn, is very likely to rapidly change the fairly mild climate that we have had for the past 10,000 years. Thus, what we are facing is the threat of *climate disruption*, based on a very rapid rise in atmospheric temperature within only a century compared to such increases in the past occurring over thousands of years (Figure 15-17, top left).

Climate models indicate that such changes will determine where we can grow food and how much of it we can grow; which areas will suffer from increased drought and which will experience increased flooding; and in what areas people and many forms of wildlife can live. If the models are correct, we will have to deal with these disruptive effects within this century.

A 2003 U.S. National Academy of Sciences report laid out a nightmarish *worst-case scenario* in which human activities, alone or in combination with natural factors, trigger new and abrupt climate and ecological changes that could last for thousands of years. The report describes ecosystems collapsing, floods in low-lying coastal cities, forests consumed in vast wildfires, and grasslands, dried out from prolonged drought, turning into dust bowls. It warns of rivers drying up—and with them supplies of drinking and irrigation water—as the mountain glaciers that feed the rivers melt. It also describes premature extinction of up to half of the world's species, prolonged droughts, more intense and longer-lasting heat waves, more destructive storms and flooding, much colder weather in some parts of the world, and the rapid spread of some infectious tropical diseases.

These possibilities were supported by a 2003 analysis carried out for the U.S. Department of Defense by Peter Schwartz and Doug Randall. They concluded that projected climate disruption "must be viewed as a serious threat to global stability and should be elevated to a U.S. national security concern."

Let us look more closely at some of the current signs and projected effects of such climate disruption.

Severe Drought Is Likely to Increase

According to a 2005 study by National Center for Atmospheric Research, scientist Aiguo Dai and his colleagues, severe and prolonged drought now affects at least 30% of the earth's land (excluding Antarctica)—an area the size of Asia. According to a 2007 study by climate researchers at NASA's Goddard Institute for Space Studies, by 2059, up to 45% of the world's land area could be experiencing extreme drought. A 2010 study by Dai projected that by the 2030s, the western two-thirds of the United States, which is already quite dry, will be significantly drier and that some parts of the United States may face extreme drought.

As droughts get worse and more widespread, the growth of trees and other plants declines, which reduces the removal of CO_2 from the atmosphere. As forests and grasslands also dry out, wildfires become more frequent, and this adds CO_2 to the atmosphere. Climate scientists project that these combined effects from the projected worsening of prolonged droughts are likely to speed up atmospheric warming.

Other effects of this browning of the land include declining stream flows and less available surface water; falling water tables with more evaporation, worsened by farmers irrigating more to make up for drier conditions; shrinking lakes, reservoirs, and inland seas; dwindling rivers; water shortages for 1–3 billion people; and declining biodiversity.

More Ice and Snow Are Likely to Melt

Models project that climate change will be the most severe in the world's polar regions. Light-colored ice and snow in these regions help to cool the earth by reflecting incoming solar energy. The melting of such ice and snow exposes much darker land and sea areas, which reflect significantly less sunlight and absorb more solar energy. This warms the atmosphere above the poles more and faster than the atmosphere is warming at lower latitudes, as projected by all major climate models. The result is more melting of snow and ice, which causes further warming in an escalating spiral of change.

According to the 2007 IPCC report, arctic temperatures have risen almost twice as fast as average temperatures in the rest of the world during the past 50 years. Also, soot generated by North American, European, and Asian industries is darkening arctic ice and lessening its ability to reflect sunlight. Mostly as a result of these two factors, summer sea ice in the Arctic is disappearing faster than scientists thought only a few years ago (Figure 15-21). Because of changes in short-term weather conditions, summer sea ice coverage is likely to fluctuate. But the overall projected long-term trend is for the average summer ice coverage to shrink.

Mountain glaciers are affected by two climatic factors: average snowfall, which adds to their mass during the winter, and average warm temperatures, which spur their melting during the summer.

During the past 25 years, many of the world's mountain glaciers have been slowly melting and shrinking (Figure 15-20). Mountain glaciers, such as those of the Himalayan Mountains in Asia and the Andes Mountains in South America, are high-elevation reservoirs—sources of water for drinking, irrigation, and hydropower. Thus, water, food, and power shortages could threaten billions of people in Asia and South America as these glaciers slowly melt over the next century or two.

CONNECTIONS

Melting Permafrost and Atmospheric Warming

Ice within arctic soils and ocean bottom sediments, or *permafrost,* is also melting. According to the 2004 Arctic Climate Impact Assessment, 10–20% of the Arctic's current permafrost soils might thaw during this century. Recent research indicated that the amount of carbon stored as methane (CH_4) and CO_2 in permafrost is more than twice the amount estimated by the IPCC in 2007. Scientists are concerned about another methane source—a layer of permafrost on the Arctic Sea floor. Some scientists refer to these sources as "methane time bombs." They could eventually release amounts of CH_4 and CO_2 that would be many times the current levels in the atmosphere, resulting in even more rapid atmospheric warming and climate disruption.

Sea Levels Are Rising

Sea levels are rising faster than IPCC scientists had reported in 2007, according to more recent studies. For example, a 2008 U.S. Geological Survey report concluded that the world's average sea level will most likely rise 0.8–2 meters (3–6.5 feet) by the end of this century, or 3 to 5 times the increase estimated in the 2007 IPCC report, and will probably keep rising for centuries. This rise is due to the expansion of seawater as it warms and to the melting of land-based ice.

Such a rise in sea level would be more dramatic if the vast land-based glaciers in Greenland (**Core Case Study**) CORE CASE STUDY and western Antarctica, as well as many of the world's mountaintop glaciers, continue melting at their current, or higher, rates as the atmosphere warms. A 2009 study by British climate scientist Jonathan Bamber estimated that a loss of just 15% of Greenland's ice sheet (Figure 15-1) would cause a 1-meter (3.3-foot) rise in sea level.

According to the IPCC, if the world's average sea level rises by 1 meter (3.3 feet) by 2100, this could degrade or destroy at least one-third of the world's coastal estuaries, wetlands, and coral reefs, and disrupt many of the world's coastal fisheries. It would also cause flooding and erosion of low-lying barrier islands and gently sloping coastlines, especially in U.S. coastal states such as Texas, Louisiana, North and South Carolina, and Florida (Figure 15-24).

The projected rise in sea levels would also flood agricultural lowlands and deltas in coastal areas where much of the world's rice is grown. This is already happening on parts of the east coast of India and in Bangladesh. Another result will be saltwater contamination of freshwater coastal aquifers and decreasing amounts of groundwater needed for irrigation and drinking water supplies.

Figure 15-24 If the average sea level rises by 1 meter (3.3 feet), the areas shown here in red in the U.S. state of Florida will be flooded. (Data from Jonathan Overpeck and Jeremy Weiss based on U.S. Geological Survey Data)

Rising seas will likely submerge low-lying islands around the world, flooding out a total population greater than that of the United States. Flooding in some of the world's largest cities located on coasts would displace at least 150 million people and threaten trillions of dollars worth of buildings, roads, and other forms of infrastructure. (See *The Habitable Planet*, Video 5 at **www.learner.org/resources/series209.html**.)

Extreme Weather Is Likely to Increase in Some Areas

According to the IPCC, atmospheric warming will raise the number and intensity of extreme weather events such as severe droughts and heat waves in some areas, which could kill large numbers of people, reduce crop production, and expand deserts. At the same time, because a warmer atmosphere can hold more moisture, other areas will experience increased flooding from heavy and prolonged snow or rainfall. Also, in some areas, global atmospheric warming will likely lead to colder winter weather, according to climate models.

CONNECTIONS

Atmospheric Warming and Colder Winter Weather

As the average temperature of the atmosphere rises, some parts of earth will likely experience colder winter weather. In 2010, research by NOAA scientist James Overland indicated that continued melting of sea ice in the Arctic region will likely lead to colder and snowier winters in Europe, eastern North America, and eastern Asia. His hypothesis is that the loss of sea ice is causing the Arctic Ocean to absorb more heat during the summer. This could weaken the *jet stream*—a high, rapidly moving flow of air that affects global weather patterns (see Figure 2, p. S18, in Supplement 5)—and cause it to move southward during the winter, bringing cold air into the areas listed above. So a warmer Arctic not only threatens polar bears (see Chapter 8 Core Case Study, p. 151) but also will likely bring more severe winters to large areas of the world.

The effect of atmospheric warming on tropical storms and hurricanes is uncertain because of conflicting data and hypotheses. In 2010, a World Meteorological Organization panel of experts on hurricanes and climate change, from both sides of this scientific debate, reached the following consensus view: projected atmospheric warming is likely to lead to fewer but stronger hurricanes that could cause more damage.

Climate Disruption Is a Threat to Biodiversity

According to the 2007 IPCC report, projected climate disruption is likely to upset ecosystems and decrease biodiversity in areas of every continent. According to the 1997 IPCC study, approximately 30% of the land-based plant and animal species assessed so far could disappear if the average global temperature change exceeds 1.5–2.5°C (2.7–4.5°F). This percentage could grow to 70% if the temperature change exceeds 3.5°C (6.3°F) (**Concept 15-5**).

The hardest hit will be plant and animal species in colder climates such as the polar bear in the Arctic and penguins in Antarctica; species that live at higher elevations; species with limited ranges such as some amphibians (see Chapter 4, Case Study, p. 73); and those with limited tolerance for temperature change. On the other hand, the populations of plant and animal species that thrive in warmer climates could grow.

The ecosystems most likely to suffer disruption and species loss from climate change are coral reefs. Other vulnerable ecosystems are polar seas, coastal wetlands, high-elevation mountaintops, and alpine and arctic tundra. Primarily because of drier conditions, forest fires might become more frequent and intense in some areas such as the southeastern and western United States. A warmer climate could also greatly increase populations of insects and fungi that damage trees (Figure 15-25).

Shifts in regional climate would also threaten many existing state and national parks, wildlife reserves, wilderness areas, and wetlands, along with much of the biodiversity they contain. Thus, slowing the rate of projected climate change would help to sustain the earth's biodiversity, which in turn supports us and our economies.

Agriculture Could Face an Overall Decline

According to the 2007 IPCC report, crop productivity is projected to increase slightly at middle to high latitudes with moderate warming, but decrease if warming goes too far. For example, moderately warmer temperatures and higher precipitation in parts of mid-western Canada, as well as in Russia and Ukraine, could lead to higher crop productivity there. But this effect may be limited because soils generally are more lacking in sufficient plant nutrients in these northern regions.

Climate change models predict a decline in agricultural productivity in tropical and subtropical regions, especially in Southeast Asia and Central America, where many of the world's poorest people live. Also, flooding of river deltas due to rising sea levels could reduce crop production in these areas and fish production in nearby coastal aquaculture ponds. Food production could also decrease in farm regions that are dependent on rivers fed by snow and glacial melt, and in any arid and semiarid areas where droughts become more prolonged. In addition, these disruptions in food production could be largely unpredictable because of the projected higher frequency of extreme weather events resulting from a warmer atmosphere.

According to the IPCC, food is likely to be plentiful for a while because of the longer growing season in northern regions in a warmer world. But the scientists warn that by 2050, some 200–600 million of the world's poorest and most vulnerable people could face starvation and malnutrition due to the effects of projected climate disruption.

A Warmer World Is Likely to Threaten the Health of Many People

According to IPCC and other reports, more frequent and prolonged heat waves in some areas will raise the numbers of deaths and illnesses, especially among older people, people in poor health, and the urban poor who cannot afford air conditioning.

On the other hand, in a warmer world, fewer people will die from cold weather. But a 2007 study led by Mercedes Medina-Ramon of the Harvard University School of Public Health suggests that increased numbers of heat-related deaths will be greater than the projected drop in cold-related deaths. In addition, hunger and

Natural Resources Canada, Canadian Forest Service

Figure 15-25 With warmer winters, exploding populations of mountain pine beetles have munched their way through large areas of lodgepole pine forest in the Canadian province of British Columbia. In this photo, the orange-colored trees are those that are dead or dying—killed by these beetles.

malnutrition will increase in areas where agricultural production drops.

A warmer, CO_2-rich world will favor rapidly multiplying insects, microbes, toxic molds, and fungi that make us sick, as well as plants that produce pollens that cause allergies and asthma attacks. Microbes that cause infectious tropical diseases such as dengue fever and yellow fever are likely to expand their ranges and numbers if mosquitoes that carry them spread to warmer temperate and higher elevation areas as they have begun to do (Figure 15-26).

A 2009 study by a research institute led by former UN Secretary General Kofi Annan estimated that climate disruption already contributes to the premature deaths of more than 300,000 people—an average of 822 people a day. The report estimated that 325 million people—a number greater than the U.S. population—are now seriously affected by accelerating climate change through natural disasters and environmental degradation.

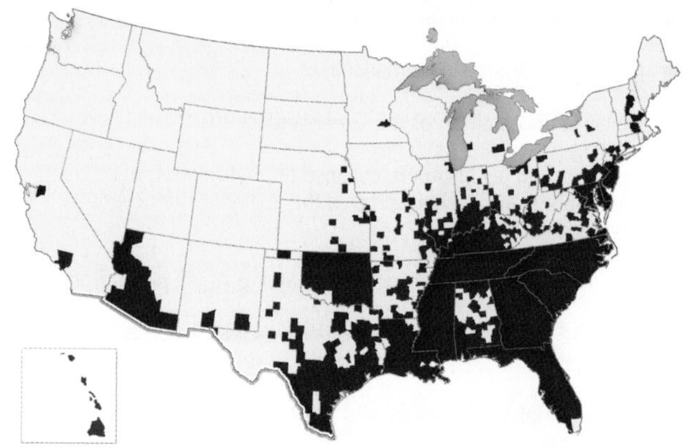

Figure 15-26 Areas in blue show counties in 28 U.S. states where one or both species of mosquitoes that transmit dengue fever have been found as of 2005. This supports the claim by some scientists that projected climate disruption will expand the range of mosquitoes that carry dengue fever, a debilitating and potentially deadly disease. (Data from Natural Resources Defense Council, 2009)

15-6 What Can We Do to Slow Projected Climate Disruption?

▶ **CONCEPT 15-6 We can reduce greenhouse gas emissions and the threat of climate disruption while saving money and improving human health if we cut energy waste and rely more on cleaner renewable energy resources.**

What Are Our Options?

For several years, many environmental analysts have been calling for urgent action at the national and international levels to curb greenhouse gas emissions, primarily from the top down, by regulating and taxing such emissions. Now a growing number of analysts are contending that this approach, by itself, will not work.

These analysts make three points. *First,* psychological research indicates that using fear and guilt, and promoting sacrifice to change behavior rarely works. *Second,* people are primarily interested in the short-term, not long-term, benefits of changing their behavior. *Third,* for elected officials whose future depends on being reelected every few years, spending their efforts on long-term problems is most often not in their best short-term interests.

So these analysts argue that, rather than emphasizing the long-term harmful effects of climate disruption, we should stress the *short-term benefits* for individuals, corporations, schools, and universities of working locally to reduce greenhouse gas emissions. This is a bottom-up approach to dealing with the threat of climate disruption. Such important short-term benefits include:

- Money saved from cutting energy use and waste; better health because of cleaner air; more jobs in the less polluting and more cost-competitive wind and solar industries; and
- Improved national and economic security for many nations due to reduced dependence on imported oil.

Regardless of how we approach climate change, most climate scientists argue that our most urgent priority is *to do all we can to avoid any and all* **climate change tipping points**—thresholds beyond which natural systems can change irreversibly. Figure 15-27 (p. 400) summarizes some of the tipping points that scientists are most concerned about.

There are two basic approaches to dealing with the projected harmful effects of global climate disruption. One, called *mitigation,* is to act to slow it and to avoid climate change tipping points. For example, climate scientist James Hansen (Individuals Matter, p. 394) and other prominent scientists have urged policy makers and business leaders to mount a crash program to cut global CO_2 emissions by 50–85% by 2050. The other approach, called *adaptation,* is to recognize that some climate change is unavoidable and to try to reduce some of its harmful effects.

Regardless of the approach we take, many scientists and business leaders see the climate change problem as an urgent challenge, but also as one that is filled with

- Atmospheric carbon level of 450 ppm
- Melting of all Arctic summer sea ice
- Collapse and melting of the Greenland ice sheet
- Severe ocean acidification, collapse of phytoplankton populations, and a sharp drop in the ability of the oceans to absorb CO_2
- Massive release of methane from thawing Arctic permafrost
- Collapse and melting of most of the western Antarctic ice sheet
- Severe shrinkage or collapse of the Amazon rain forest

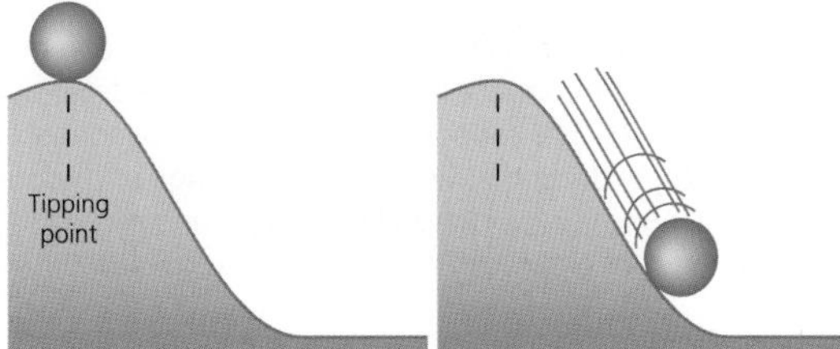

Figure 15-27 Environmental and climate scientists have come up with this list of possible climate change tipping points.

Solutions

Slowing Climate Disruption

Prevention

- Cut fossil fuel use (especially coal)
- Shift from coal to natural gas
- Improve energy efficiency
- Shift to renewable energy resources
- Transfer energy efficiency and renewable energy technologies to developing countries
- Reduce deforestation
- Use more sustainable agriculture and forestry
- Put a price on greenhouse gas emissions
- Reduce poverty
- Slow population growth

Cleanup

- Remove CO_2 from smokestack and vehicle emissions
- Store (sequester) CO_2 by planting trees
- Sequester CO_2 in soil by using no-till cultivation and taking cropland out of production
- Sequester CO_2 deep underground (with no leaks allowed)
- Sequester CO_2 in the deep ocean (with no leaks allowed)
- Repair leaky natural gas pipelines and facilities
- Use animal feeds that reduce CH_4 emissions from cows (belching)

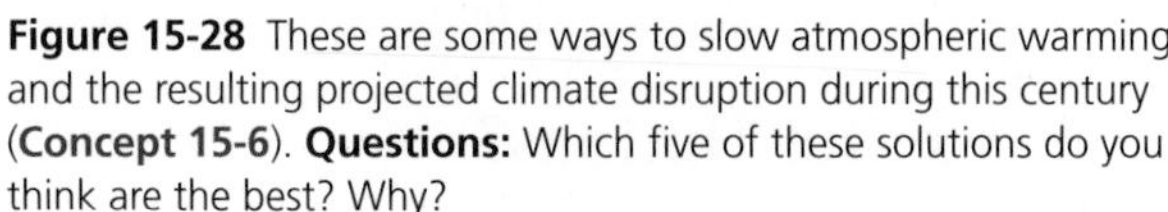

Figure 15-28 These are some ways to slow atmospheric warming and the resulting projected climate disruption during this century (**Concept 15-6**). **Questions:** Which five of these solutions do you think are the best? Why?

economic opportunity. In 2009, a group of investors who collectively manage more than $13 trillion of the world's financial assets warned that climate disruption poses a threat to the global economy, but they projected that the challenge of making the transition to a low-carbon world presents an enormous global investment opportunity.

For example, in the United States, electric cars are being made in Michigan, and wind generators are being produced in Illinois, Oregon, and Iowa. The U.S. state of Texas produces more electricity from wind power than all but five of the world's countries. In 2010, Bill Gates, one of the world's most successful business leaders, suggested that making a shift to a low-carbon economy will lead us into a new era of economic growth and prosperity.

Preventing and Reducing Greenhouse Gas Emissions

Figure 15-28 lists a number of ways to slow the rate and degree of atmospheric warming and the resulting climate disruption caused by human activities (**Concept 15-6**). Among these solutions are four major *prevention* strategies that, by 2050, could reduce human greenhouse gas emissions by 57–83%, according to a 2010 study by the U.S. National Academy of Sciences:

1. Improve energy efficiency to reduce fossil fuel use, especially the use of coal.
2. Shift from nonrenewable carbon-based fossil fuels to a mix of low-carbon renewable energy resources based on local and regional availability.
3. Stop cutting down tropical forests and plant trees to help remove more CO_2 from the atmosphere.
4. Shift to more sustainable and climate-friendly agriculture (Figure 10-30, p. 233).

THINKING ABOUT
Climate Disruption Prevention Strategies

How are each of the four strategies listed above related to one or more of the **principles of sustainability** (see back cover)?

The solutions listed on the right side of Figure 15-28 are *output*, or *cleanup*, strategies, which focus on dealing with CO_2 after it has been produced. Among these is one that is generally referred to as **carbon capture and storage**, or **CCS**, which involves removing CO_2 from the smokestacks of coal-burning power and industrial plants and storing it deep underground in abandoned coal beds and oil and gas fields or under the sea floor.

However, it is becoming clear to scientists that stored CO_2 would have to remain sealed from the atmosphere forever. Any large-scale leaks caused by earthquakes or other shocks, as well as any number of smaller continuous leaks from CO_2 storage sites around the world, could dramatically increase atmospheric warming in a very short time. Some scientists doubt that we can develop the technology to guarantee that all such stored CO_2 would remain safely sequestered without some of it leaking out. Meanwhile, use of CCS would enable increasing use of coal, considered to be the world's most harmful energy resource.

CONNECTIONS

Sea Creatures, Carbon Dioxide, and Cement

In 2010, earth science professor Brent Constantz developed a process for removing CO_2 from the smokestack emissions of a power plant in Moss Landing, California, and converting it to cement by spraying it with mineral rich seawater. He developed this CCS method by mimicking sea creatures that convert CO_2 to calcium carbonate, or limestone, to form their skeletons and shells. He envisions converting CO_2 emissions into stone and locking them away essentially forever in the concrete foundations of our cities.

Many analysts argue that, rather than spending huge amounts of time and resources on cleaning up greenhouse gas emissions, we should focus on reducing and preventing greenhouse gas emissions, as soon as possible. In addition to cutting CO_2 emissions, some scientists urge us to increase efforts to reduce emissions of other greenhouse gases. For example, methane (CH_4) is 25 times more effective in warming the atmosphere than CO_2. It is emitted from landfills, rice paddies, older natural-gas pipelines, and many other sources. In 2010, atmospheric scientists Veerabhadran Ramanathan and David Victor estimated that, just by collecting CH_4 from all landfills (which is being done in a growing number of large landfills in more-developed countries), minimizing the rotting of rice in paddies, and replacing old leaky natural-gas pipelines, we could cut global CH_4 emissions by 40%.

Another climate-changing pollutant is soot generated by inefficient coal-burning power plants, diesel engines, brick-making kilns, and cook stoves, among many other sources. Soot is accumulating on Himalayan Mountain glaciers and Arctic ice fields and contributing to the melting of this ice and to atmospheric warming. Ramanathan and Victor report that in China and India, programs to make power plants more efficient, to filter soot from diesel engines, to cut emissions from kilns, and to replace leaky stoves with efficient ones (Figure 15-15) could cut soot levels by almost two-thirds.

Many scientists and engineers want to use prevention strategies that fall under the umbrella of *geoengineering,* or trying to manipulate certain natural conditions to help counter an enhanced greenhouse effect. For example, some scientists have called for injecting sulfate particles into the stratosphere to reflect some of the incoming sunlight into space and cool the troposphere. Other scientists have called for placing a series of giant mirrors in orbit above the earth to reflect incoming sunlight for the same purpose.

According to skeptical scientists, one major problem with most of these technological fixes, and with some carbon capture and storage schemes is that they require huge investments of energy and materials, and there is no guarantee that they will work. If we rely on these systems and continue emitting greenhouse gases, and if the systems then fail, atmospheric temperatures will likely soar at a rapid rate, greatly speeding up the processes of climate disruption.

Governments Can Help to Reduce the Threat of Climate Disruption

Governments can use six major strategies to promote the solutions listed in Figure 15-28 (**Concept 15-6**). They can:

- Strictly regulate carbon dioxide (CO_2) and methane (CH_4) as climate-changing pollutants.
- Phase out the most inefficient polluting coal-burning power plants and replace them with more efficient and cleaner natural gas and renewable energy alternatives such as wind power.
- Put a price on carbon emissions by phasing in taxes on each unit of CO_2 or CH_4 emitted, or phasing in energy taxes on each unit of any fossil fuel burned (Figure 15-29) and offsetting these tax increases by reducing taxes on income, wages, and profits.
- Use a cap-and-trade system (Figure 15-30, p. 402), which uses the marketplace to help reduce emissions of CO_2 and CH_4.

Figure 15-29 Using carbon and energy taxes or fees to help reduce greenhouse gas emissions has advantages and disadvantages. **Question:** Which two advantages and which two disadvantages do you think are the most important and why?

Trade-Offs

Cap and Trade Policies

Advantages	Disadvantages
Clear legal limit on emissions	Revenues not predictable
Rewards cuts in emissions	Vulnerable to cheating
Record of success	Rich polluters can keep polluting
Low expense for consumers	Puts variable price on carbon

Figure 15-30 Using a cap-and-trade policy to help reduce greenhouse gas emissions has advantages and disadvantages. **Question:** Which two advantages and which two disadvantages do you think are the most important and why?

- Phase out government subsidies and tax breaks for fossil fuels and industrialized food production, and phase in such subsidies and breaks for energy-efficiency technologies, low-carbon renewable energy sources, and more sustainable agriculture.
- Focus research and development efforts on innovations that lower the cost of clean energy alternatives, so that they can compete more favorably with fossil fuels. This is probably quicker and more politically feasible than trying to raise the cost of energy from fossil fuels.
- Work out agreements to finance and monitor efforts to reduce deforestation—which accounts for 12% to 17% of global greenhouse gas emissions—and to promote global tree-planting efforts (see Chapter 9, Core Case Study, p. 174).
- Encourage more-developed countries to help fund the transfer of the latest energy-efficiency and cleaner energy technologies to less-developed countries so that they can bypass older, energy wasting and polluting technologies.

Environmental economists and a growing number of business leaders argue that, regardless of which of these strategies governments use, the most critical goal for governments is to find ways to put a price on carbon emissions. This will help us come closer to paying the estimated harmful environmental and health costs of our carbon emissions. Without this, carbon emissions will only continue to grow. Economists such as Gary Yohe contend that the resulting higher costs for fossil fuels will spur innovation in finding ways to reduce carbon emissions, improve energy efficiency, and phase in a mix of cleaner, low-carbon renewable energy alternatives. However, establishing laws and regulations that raise the price of fossil fuels is politically difficult because of the immense political and economic power of the fossil fuel industries.

Governments have also entered into international climate negotiations. In December 1997, delegates from 161 nations met in Kyoto, Japan, to negotiate a treaty to slow global warming and its projected climate disruption. The first phase of the resulting *Kyoto Protocol* went into effect in February 2005 with 187 of the world's 194 countries (not including the United States) ratifying the agreement by late 2009.

This agreement requires the 36 participating more-developed countries to cut their emissions of CO_2, CH_4, and N_2O to certain levels by 2012, when the treaty expires. Less-developed countries were excluded from this requirement, because such reductions would curb their economic growth. In 2005, participating countries began negotiating a second phase of the treaty, which is supposed to go into effect after 2012. But these negotiations have failed to extend the original agreement.

Some analysts see top-down international climate agreements as slow and ineffective responses to an urgent global problem.

Some Countries, States and Localities Are Leading the Way

Some nations are leading others in facing the challenges of projected climate disruption. Costa Rica aims to be the first country to become *carbon neutral* by cutting its net carbon emissions to zero by 2030. The country generates 78% of its electricity with renewable hydroelectric power and another 18% from renewable wind and geothermal energy. GOOD NEWS

While China emits more CO_2 into the atmosphere than any other country, it also has one of the world's most intensive energy efficiency programs, according to ClimateWorks, a global energy consulting firm. Chinese automobile fuel-efficiency standards are higher than U.S. standards but enforcement of the standards is still weak. China's government is working with the country's top 1,000 industries to implement tough energy efficiency goals. China is also rapidly becoming the world's leader in developing and selling solar cells, solar water heaters, wind turbines, advanced batteries, and plug-in hybrid electric cars.

By 2010, at least 30 U.S. states had set goals for reducing greenhouse gas emissions. California plans to get 33% of its electricity from low-carbon renewable energy sources by 2030.

Since 1990, local governments in more than 650 cities around the world (including more than 450 U.S. cities) have established programs to reduce their greenhouse gas emissions.

Some Companies Are Reducing Their Carbon Footprints

Leaders of some of the most prominent U.S. companies, including Alcoa, DuPont, Ford Motor Company, General Electric, Johnson & Johnson, PepsiCo, GOOD NEWS and Shell Oil, have joined with leading environmental organizations to form the U.S. Climate Action Partnership. They have called on the government to enact strong national climate change legislation, saying "in our view, the climate change challenge will create more economic opportunities than risks for the U.S. economy."

However, realizing that significantly reducing their own greenhouse gas emissions will save them money, they are not waiting for government action. For example, between 1990 and 2006, DuPont slashed its energy usage and cut its greenhouse gas emissions by 72% and saved $3 billion, while the company increased its business by almost a third.

Leaders of these and many other major companies see an enormous profit opportunity in developing or using energy-efficient and cleaner-energy technologies, such as fuel-efficient cars, wind turbines, and solar cells. They understand that there is gold in going green. Chinese leaders understand this and are moving rapidly to make China a world leader in the development of hybrid-electric cars, wind turbines, solar cell systems, high-speed trains, and other green technologies that they are selling throughout the world.

Colleges and Universities Are Going Green

Many colleges and universities are also taking action. The Tempe campus of Arizona State University (ASU) boasts the largest collection of solar panels of any U.S. university. ASU was the first U.S. university to establish an entire academic program with its School of Sustainability. GOOD NEWS

The College of the Atlantic in Bar Harbor, Maine (USA) has been carbon neutral since 2007. It gets all of its electricity from renewable hydropower, and many of its buildings are heated with the use of renewable wood pellets. Some students there built a wind turbine that powers a nearby organic farm, which offers organic produce to the campus, to local schools, and to food banks.

EARTH University in Costa Rica has a mission to promote sustainable development in tropical countries. Its sustainable agriculture degree program has attracted students from more than 20 different countries.

The number of examples such as these is large and growing. Here are just a few more: Students and faculty at Oberlin College in the U.S. state of Ohio created a Web-based system in some of the school's dorms that monitors use of energy and water, giving students real-time feedback that can help them to reduce their waste of these resources. The University of British Columbia was the first school in Canada to open a sustainability office, and it offers more than 300 sustainability-related courses. In the United Kingdom, Leeds University replaced its trash bins with recycling containers, doubling the amount of recycling on campus, and changed its purchasing practices to favor recycled goods.

THINKING ABOUT
What Your School Can Do

What are three steps that you think your school should take to help reduce its CO_2 emissions?

Individual Choices Make a Difference

Each of us plays a part in the projected acceleration of atmospheric warming and climate disruption during this century. Whenever we use energy generated by fossil fuels, for example, we add a certain amount of CO_2 to the atmosphere. Each use of energy adds to an individual's *carbon footprint,* the amount of carbon dioxide generated by one's lifestyle. Figure 15-31 lists some ways in which you can cut your CO_2 emissions. GOOD NEWS

We Can Prepare for Climate Disruption

According to global climate models, the world needs to make a 50–85% cut in emissions of greenhouse gases by 2050 to stabilize concentrations of these gases in the atmosphere. This would help to prevent the planet from heating up by more than 2°C (3.6°F) and to head off rapid changes in the world's climate and the projected harmful effects that would result, as discussed in Section 15-5. However, because of the difficulty of making

What Can You Do?

Reducing CO_2 Emissions

- Calculate your carbon footprint (there are many helpful websites)
- Walk, bike, carpool, and use mass transit or drive fuel-efficient cars
- Reduce garbage by recycling and reusing more items
- Use energy-efficient appliances and compact fluorescent or LED lightbulbs
- Wash laundry in warm or cold water
- Dry clothes on a rack or line
- Use a low-flow showerhead
- Eat less meat or no meat
- Heavily insulate your house and seal all air leaks
- Use energy-efficient windows
- Insulate your hot water heater and set it at 49°C (120°F)
- Plant trees to shade your house during summer
- Buy from businesses working to reduce their emissions

Figure 15-31 **Individuals matter:** You can reduce your annual emissions of CO_2. **Question:** Which of these steps, if any, do you take now or plan to take in the future?

such large reductions, many analysts believe that while we work to slash emissions, we should also begin to prepare for the likely harmful effects of projected climate disruption.

For example, relief organizations such as the International Red Cross are turning their attention to projects such as expanding mangrove forests as buffers against storm surges, building shelters on high ground, and planting trees on slopes to help prevent landslides in the face of projected higher levels of precipitation and rising sea levels. The U.S. state of Alaska has plans to relocate coastal villages at risk from rising sea levels and storm surges. And some coastal communities in the United States now require that new houses and other new construction be built high enough off of the ground to survive projected higher storm surges.

A No-Regrets Strategy

The threat of climate disruption pushes us to look for preventive solutions such as those listed in Figure 15-28. But suppose we had the ability to look into the future and, in doing so, we found that the climate models were wrong and that climate disruption was not a serious threat after all. Should we then return to the present and abandon the search for preventive solutions?

A number of climate and environmental scientists, economists, and business leaders say *no* to this question, and they call for us to begin implementing such changes now as a *no-regrets strategy*. They argue that actions such as those listed in Figure 15-28 should be implemented on an urgent basis because they will lead to very important environmental, health, and economic benefits.

For example, improving energy efficiency has numerous economic and environmental advantages (see Figure 13-24, p. 322). Individuals, companies, and local governments can save large amounts of money and improve their economic security by cutting their energy bills. In addition, the use of fossil fuels, especially coal, causes a great deal of air pollution, which harms or kills large numbers of people. Thus, a sharp decrease in the use of these fuels will improve the overall health of the population and save money now spent on health care. In addition, cutting coal use will greatly reduce land disruption resulting from surface mining.

Another benefit of the no-regrets strategy is that, by relying more on a mix of renewable, domestic energy resources, many countries will cut their dependence on imports of fossil fuels and strengthen their national security. The large sums of money that they now spend for fuel imports will then be available for investing in health care, education, and jobs, which will further bolster their economic security. In addition, sharply decreasing or halting the destruction of tropical forests would help to preserve the increasingly threatened biodiversity that makes up an important part of the earth's vital natural capital (see Figure 1-3, p. 9).

The no-regrets strategy seems to have worked already in a part of the U.S. state of Kansas where public opinion polls show that most residents do not believe that projected climate disruption will happen or that it is due to human activities. There, a year-long competition sponsored in 2009 by a group concerned about carbon emissions yielded some interesting results. Instead of focusing on climate change, the group challenged people to cut their use of fossil fuels for other reasons: to save money, to help strengthen national security, and to carry out religious and ethical commitments to care for the earth. In the towns where the competition was held, energy use declined by as much as 5% compared to other nearby areas. This was a substantial savings, according to local officials.

15-7 How Have We Depleted Ozone in the Stratosphere and What Can We Do about It?

▶ **CONCEPT 15-7A Widespread use of certain chemicals has reduced ozone levels in the stratosphere, which has allowed more harmful ultraviolet radiation to reach the earth's surface.**

▶ **CONCEPT 15-7B To reverse ozone depletion, we must stop producing ozone-depleting chemicals and adhere to the international treaties that ban such chemicals.**

Our Use of Certain Chemicals Threatens the Ozone Layer

A layer of ozone in the lower stratosphere keeps about 95% of the sun's harmful ultraviolet (UV-A and UV-B) radiation from reaching the earth's surface (see Figure 15-2). However, measurements taken by meteorologists show a considerable seasonal depletion (thinning) of ozone concentrations in the stratosphere above Antarctica (Figure 15-32) and the Arctic. Similar measurements reveal a lower overall ozone thinning everywhere except over the tropics. The loss of

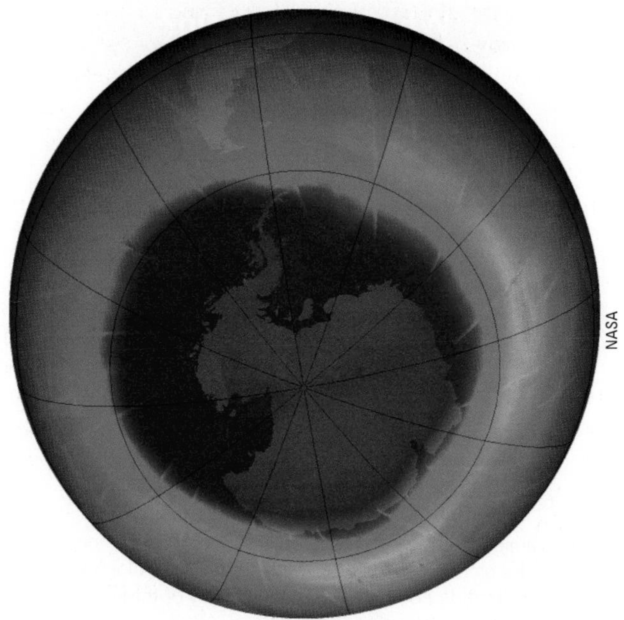

Figure 15-32 Natural capital degradation: This colorized satellite image shows massive ozone thinning over Antarctica during several months in 2009. The center of this image shows a large area where the concentration of ozone has decreased by 50% or more (100% in some places). (Data from NASA)

ozone over Antarctica has been called an *ozone hole*. A more accurate term is *ozone thinning* because the ozone depletion varies with altitude and location.

Based on these measurements as well as mathematical and chemical models, the overwhelming consensus of researchers is that ozone depletion in the stratosphere poses a serious threat to humans, other animals, and some primary producers (mostly plants) that use sunlight to support the earth's food webs (**Concept 15-7A**).

This problem began with the discovery of the first chlorofluorocarbon (CFC) in 1930. Chemists soon developed similar compounds to create a family of highly useful CFCs, known by their trade name of Freons. These chemically unreactive, odorless, nonflammable, nontoxic, and noncorrosive compounds seemed to be dream chemicals. Inexpensive to manufacture, they became popular as coolants in air conditioners and refrigerators, propellants in aerosol spray cans, cleaners for electronic parts such as computer chips, fumigants for granaries and ships' cargo holds, and as gases used to make insulation and packaging.

It turned out that CFCs were too good to be true. Starting in 1974 with the work of chemists Sherwood Rowland and Mario Molina (Individuals Matter, below), scientists demonstrated that CFCs are persistent chemicals that destroy protective ozone in the stratosphere. Measurements and models indicate that 75–85% of the observed ozone losses in the stratosphere since 1976 resulted from people releasing CFCs and other ozone-depleting chemicals into the atmosphere, beginning in the 1950s.

Why Should We Worry about Ozone Depletion?

Why should we care about ozone depletion? One reason is that more biologically damaging UV-A and UV-B radiation is reaching the earth's surface (**Concept 15-7A**).

INDIVIDUALS MATTER

Sherwood Rowland and Mario Molina—A Scientific Story of Expertise, Courage, and Persistence

In 1974, calculations by chemists Sherwood Rowland and Mario Molina of the University of California–Irvine indicated that chlorofluorocarbons (CFCs) were lowering the average concentration of ozone in the stratosphere. They shocked both the scientific community and the $28-billion-per-year CFC industry by calling for an immediate ban of CFCs in spray cans, for which substitutes were available.

The research done by these two scientists led them to four major conclusions. *First*, once injected into the atmosphere, these persistent CFCs remain there.

Second, over 11–20 years, these compounds rise into the stratosphere through convection, random drift, and the turbulent mixing of air in the lower atmosphere.

Third, once they reach the stratosphere, the CFC molecules break down under the influence of high-energy UV radiation. This releases highly reactive chlorine atoms (Cl), as well as atoms of fluorine (F) and bromine (Br), all of which accelerate the breakdown of ozone (O_3) into O_2 and O in a cyclic chain of chemical reactions. As a consequence, ozone is destroyed faster than it forms in some parts of the stratosphere.

Fourth, each CFC molecule can last in the stratosphere for 65–385 years, depending on its type. During that time, each chlorine atom released during the breakdown of CFC can convert hundreds of O_3 molecules to O_2.

The CFC industry (led by DuPont) was a powerful, well-funded adversary with a lot of profits and jobs at stake. It attacked Rowland and Molina's calculations and conclusions. But the two researchers held their ground, expanded their research, and explained their results to other scientists, elected officials, and the media. After 14 years of delaying tactics, DuPont officials acknowledged in 1988 that CFCs were depleting the ozone layer, and they agreed to stop producing them. They turned to producing and selling higher-priced alternatives that their chemists had developed.

In 1995, Rowland and Molina received the Nobel Prize in Chemistry for their work on CFCs. In awarding the prize, the Royal Swedish Academy of Sciences said that these two scientists contributed to "our salvation from a global environmental problem that could have had catastrophic consequences."

Natural Capital Degradation

Effects of Ozone Depletion

Human Health

- Worse sunburns
- More eye cataracts and skin cancers
- Immune system suppression

Food and Forests

- Reduced yields for some crops
- Reduced seafood supplies from reduced phytoplankton
- Decreased forest productivity for UV-sensitive tree species

Climate Change

- While in troposphere, CFCs act as greenhouse gases

Wildlife

- Increased eye cataracts in some species
- Decreased populations of aquatic species sensitive to UV radiation
- Reduced populations of surface phytoplankton
- Disrupted aquatic food webs from reduced phytoplankton

Air Pollution and Materials

- Increased acid deposition
- Increased photochemical smog
- Degradation of outdoor paints and plastics

Figure 15-33 Decreased levels of ozone in the stratosphere can have a number of harmful effects. (**Concept 15-4A**). **Questions:** Which three of these effects do you think are the most threatening? Why?

This is causing more eye cataracts, worse sunburns, and more skin cancers. Figure 15-33 lists these and other expected effects of decreased ozone levels in the stratosphere. Figure 15-34 lists ways in which you can protect yourself from harmful UV radiation.

Another serious threat from ozone depletion is that the resulting increase in UV radiation could impair or destroy phytoplankton, especially in Antarctic waters (see Figure 3-11, p. 48). These tiny marine plants play a key role in removing CO_2 from the atmosphere. They also form the base of ocean food webs. Destroying them would eliminate the vital ecological services they provide. It could also accelerate projected atmospheric warming by reducing the ability of the oceans to remove CO_2 from the atmosphere—another example of the disruption of the global carbon cycle (see Figure 3-15, p. 53).

What Can You Do?

Reducing Exposure to UV Radiation

- Stay out of the sun, especially between 10 A.M. and 3 P.M.
- Do not use tanning parlors or sunlamps.
- When in the sun, wear protective clothing and sunglasses that protect against UV-A and UV-B radiation.
- Be aware that overcast skies do not protect you.
- Do not expose yourself to the sun if you are taking antibiotics or birth control pills.
- When in the sun, use a sunscreen with a protection factor of at least 15.
- Examine your skin and scalp at least once a month for moles or warts that change in size, shape, or color and sores that do not heal. If you observe any of these signs, consult a doctor immediately.

Figure 15-34 **Individuals matter:** You can reduce your exposure to harmful UV radiation. **Question:** Which of these precautions do you already take?

We Can Reverse Stratospheric Ozone Depletion

According to researchers in this field, we should immediately stop producing all ozone-depleting chemicals (**Concept 15-7B**). However, models indicate that even with immediate and sustained action, it will take about 60 years for the earth's ozone layer to recover the levels of ozone it had in 1980, and it could take about 100 years to recover to pre-1950 levels.

CONNECTIONS

Atmospheric Warming and Repair of the Ozone Layer

Scientists from Johns Hopkins University reported in 2009 on their discovery that warming of the troposphere makes the stratosphere cooler, which slows down its rate of ozone repair. Thus, as the planet warms, ozone levels may take much longer to return to where they were before ozone thinning began, or they may never recover completely.

In 1987, representatives of 36 nations met in Montreal, Canada, and developed the *Montreal Protocol*. This treaty's goal was to cut emissions of CFCs (but not other ozone-depleting chemicals) by about 35% between 1989 and 2000. After hearing more bad news about seasonal ozone thinning above Antarctica in 1989, representatives of 93 countries had more meetings and in 1992 adopted the *Copenhagen Protocol*, an amendment that accelerated the phase-out of key ozone-depleting chemicals. The ozone protocols set an important precedent by using *prevention*, with governments and companies working together, to solve a serious global environmental problem.

Substitutes are available for most CFCs, and others are being developed. However, the most widely used substitutes are hydrofluorocarbons (HFCs), which act as greenhouse gases. An HFC molecule can be up to 10,000 times more potent in warming the atmosphere than a molecule of CO_2 is. Currently, HFCs account for only 2% of all greenhouse gas emissions, but the IPCC has warned that global use of HFCs is growing rapidly. Scientists, including Mario Molina (Individuals Matter, p. 405), have called for applying the Montreal protocol to HFCs to prevent a great deal of atmospheric warming. As of 2010, their proposal had not been adopted.

The landmark international agreements on stratospheric ozone, now signed by all 196 of the world's countries, are important examples of global cooperation in response to a serious environmental problem. GOOD NEWS If nations continue to follow these agreements, ozone levels should return to 1980 levels by 2068 (18 years later than originally projected) and to 1950 levels by 2108 (**Concept 15-7B**).

Here are this chapter's *three big ideas*:

- All countries need to step up efforts to control and prevent outdoor and indoor air pollution.
- Reducing the projected harmful effects of rapid climate disruption during this century requires emergency action to increase energy efficiency, sharply reduce greenhouse gas emissions, rely more on renewable energy resources, and slow population growth.
- We need to continue phasing out the use of chemicals that have reduced ozone levels in the stratosphere and allowed more harmful ultraviolet radiation to reach the earth's surface.

REVISITING Melting Ice in Greenland and Sustainability

In this chapter, we have seen that human activities such as the burning of fossil fuels and the widespread clearing and burning of forests for planting crops have contributed to higher levels of dangerous pollutants and greenhouse gases in the atmosphere. This threatens the health of millions of people and has contributed to increased atmospheric warming, which in turn is projected to result in climate disruption during this century.

The effects of climate disruption include rapid and increased melting of land-based ice (**Core Case Study**) and sea ice, worsening drought, rising sea levels, and declining biodiversity. Many of these effects further accelerate climate disruption in a worsening spiral of change. We have also seen how the widespread use of certain chemicals has thinned the stratospheric ozone layer, which has led to plant and animal health problems.

We can apply the three **principles of sustainability** (see back cover) to help reduce the harmful effects of air pollution, projected climate disruption, and stratospheric ozone depletion. We can reduce inputs of pollutants, greenhouse gases, and ozone-depleting chemicals into the atmosphere by relying more on direct and indirect forms of solar energy than on fossils fuels; reducing the waste of matter and energy resources; recycling and reusing matter resources much more widely than we do now; and mimicking biodiversity by using a variety of low-carbon renewable energy resources in place of fossil fuels and nuclear power. We can enhance these strategies by finding substitutes for ozone-depleting chemicals, emphasizing pollution prevention, and reducing human population growth.

The atmosphere is the key symbol of global interdependence.
If we can't solve some of our problems in the face of threats to this global commons,
then I can't be very optimistic about the future of the world.

MARGARET MEAD

REVIEW

CORE CASE STUDY

1. Summarize the story of Greenland's melting glaciers (**Core Case Study**) and the possible effects of this process. Explain how it fits into the IPCC projections about climate change.

SECTION 15-1

2. What is the key concept for this section? Define **troposphere**, **stratosphere**, and **ozone layer**. Describe how the troposphere and stratosphere differ. Why is the ozone layer important?

SECTION 15-2

3. What are the two key concepts for this section? What is **air pollution**? Distinguish between **primary pollutants** and **secondary pollutants** and give an example of each. List the major outdoor air pollutants and their harmful effects. Distinguish between **industrial smog** and

photochemical smog. List and briefly describe five natural factors that help to reduce outdoor air pollution and six natural factors that help to worsen it. What is a **temperature inversion** and how can it affect air pollution levels? What is **acid deposition**, how does it form, and what are its major environmental impacts on vegetation, lakes, human-built structures, and human health? List three major ways to reduce acid deposition.

4. What is the major indoor air pollutant in many less-developed countries? What are the four most dangerous indoor air pollutants in more-developed countries? Briefly describe the human body's defenses against air pollution, how they can be overwhelmed, and illnesses that can result. About how many people die from air pollution each year, worldwide and in the United States?

SECTION 15-3

5. What is the key concept for this section? Summarize the accomplishments of air pollution control laws in the United States and discuss how they can be improved. List the advantages and disadvantages of using an emissions trading program to reduce outdoor air pollution. Summarize the major ways to reduce emissions from power plants and motor vehicles. What are four ways to reduce indoor air pollution?

SECTION 15-4

6. What is the key concept for this section? Describe the warming and cooling of the atmosphere that has occurred over the past 900,000 years and during the last century. How do scientists get information about past temperatures and climates? What is the greenhouse effect and why is it so important to life on the earth? List the four conclusions of the IPCC in 2007 regarding global temperature changes during the last half of the past century and about projected temperature changes during this century. What are five pieces of evidence that support these conclusions?

7. What role do CO_2 emissions play in atmospheric warming? How are models used to project future atmospheric temperature changes? Summarize the contributions of James Hansen to climate science. Describe how each of the following might affect average atmospheric temperatures and projected climate change during this century: **(a)** the sun, **(b)** the oceans, and **(c)** cloud cover.

SECTION 15-5

8. What is the key concept for this section? Briefly explain the projections of scientists on how projected climate change is likely to affect drought, ice cover, sea levels, extreme weather events, biodiversity, crop yields, and human health during this century.

SECTION 15-6

9. What is the key concept for this section? Define **climate change tipping points** and list four examples. What are four major prevention strategies for slowing projected climate change? What is **carbon capture and storage (CCS)**? Describe five problems associated with capturing and storing carbon dioxide emissions. Describe two geoengineering strategies for dealing with climate change. List six things that governments could do to help slow projected climate change. What are the pros and cons of the Kyoto Protocol? Describe what some countries, cities, major corporations and schools have done to reduce their carbon footprints. List five ways in which you can reduce your carbon footprint. List five ways in which we can prepare for the possible long-term harmful effects of projected climate change. Explain the no-regrets strategy.

SECTION 15-7

10. What are the two key concepts for this section? How have human activities depleted ozone in the stratosphere? List five harmful effects of such depletion. Describe how scientists Sherwood Roland and Mario Molina helped to awaken the world to this threat. What has the world done to help reduce the threat of ozone depletion in the stratosphere? What are the *three big ideas* for this chapter? Explain how the three **principles of sustainability** can be applied to dealing with the problems of air pollution, climate change, and ozone depletion.

Note: Key terms are in **bold** type.

CRITICAL THINKING

1. If you had convincing evidence that at least half of Greenland's glaciers (**Core Case Study**) were sure to melt during this century, would you argue for taking serious actions now to slow projected climate disruption? If so, summarize the arguments you would use. If not, explain why you would be opposed to such actions.

2. Suppose someone tells you that carbon dioxide (CO_2) should not be classified as an air pollutant because it is a natural chemical that is part of the carbon cycle (see Figure 3-15, p. 53) and a chemical that we add to the atmosphere every time we exhale. Explain the faulty reasoning in this statement about CO_2.

3. China relies on coal for two-thirds of its commercial energy usage, partly because the country has abundant supplies of this resource. Yet China's coal burning has caused innumerable and growing problems for the country and for its neighboring nations. In addition, pollution generated in China now spreads across the Pacific Ocean to the west coast of North America. Do you think China is justified in developing this resource to the maximum as

other countries, including the United States, have done with their coal resources? Explain. What are China's alternatives?

4. Photochemical smog is largely the result of motor vehicle emissions. Considering your use of motor vehicles, now and in the future, what are three ways in which you could reduce your contribution to photochemical smog?

5. A top U.S. economic adviser to the president once gave a speech in Williamsburg, Virginia (USA), to representatives of governments from a number of countries. He told his audience not to worry about atmospheric warming because the average global temperature increases predicted by scientists were much less than the temperature increase he had experienced that day in traveling from Washington, DC, to nearby Williamsburg. What was the flaw in his reasoning? Write an argument that you could use to counter his claim.

6. Explain why you would support or oppose each of the strategies listed in Figure 15-28 for slowing projected climate disruption caused by atmospheric warming.

7. Some scientists have suggested that, in order to help cool the warming atmosphere, we could annually inject huge quantities of sulfate particles into the stratosphere. This might have the effect of reflecting some incoming sunlight back into space. Explain why you would support or oppose this geoengineering scheme.

8. What are three consumption patterns or other aspects of your lifestyle that directly add greenhouse gases to the atmosphere? Which, if any, of these habits would you be willing to give up in order to help slow projected climate disruption? Write up an argument that you could use to convince others to do the same. Compare your answers to those of your classmates.

DOING ENVIRONMENTAL SCIENCE

1. **a.** Conduct a test for radon in your room or apartment or in the main living area of the house or building where you live. (Radon testing kits are available at affordable prices in most hardware stores, drug stores, and home centers.) If radon is present, do some research and make a plan for how you will minimize your exposure to it.
   b. Find out whether or not the buildings at your school have been tested for radon. If so, what were the results? What has been done about areas with unacceptable levels of radon? If this testing has not been done, talk with school officials about having it done.

2. Gather data on the average annual temperature and average annual precipitation for the area where you live over the last 30 years. Try to find data for as many of these years as possible, and plot this data to determine whether the average temperature and precipitation during this period has increased, decreased, or stayed about the same. Write a report summarizing your search for data, your results, and your conclusions.

GLOBAL ENVIRONMENT WATCH EXERCISE

Using the World Map, select *Climate Change* and click on the pin for Greenland. Find the latest information on the melting of glaciers in Greenland (**Core Case Study**). What is the rate of melting and how might this affect sea levels during this century? Briefly summarize the evidence used by scientists to support their statements about melting ice in Greenland. Summarize any information you find about on-going and future studies of Greenland's ice.

CORE CASE STUDY

ECOLOGICAL FOOTPRINT ANALYSIS

According to the International Energy Agency, the average American adds 19.6 metric tons (21.6 tons) of CO_2 per year to the atmosphere, compared with a world average of 4.23 metric tons (4.65 tons). The table on the next page is designed to help you understand the sources of your personal inputs of CO_2 into the atmosphere. You will be making calculations to fill in the blanks in this table.

Some typical numbers are provided in the "Typical Quantity per Year" column of the table. However, your calculations will be more accurate if you can use information based on your own personal lifestyle, which you can enter in the blank "Personal Quantity per Year" column. For example, you could add up your monthly utility bills for a year and divide the total by the number of persons in your household to get a rough estimate of your own utility use.

	Units per Year	Personal Quantity per Year	Typical Quantity per Year	Multiplier	Emissions per Year (lbs. CO_2)
Residential Utilities					
Electricity	kwh		4,500	1.5	
Heating oil	gallons		37	22	
Natural gas	hundreds of cubic feet (ccf)		400	12	
Propane	gallons		8	13	
Coal	tons		–	4,200	
Transportation					
Automobiles	gallons		600	19	
Air travel	miles		2,000	0.6	
Bus, urban	miles		12	0.07	
Bus, intercity	miles		0	0.2	
Rail or subway	miles		28	0.6	
Taxi or limousine	miles		2	1	
Other motor fuel	gallons		9	22	
Household Waste					
Trash	pounds		780	0.75	
Recycled Items	pounds		337	–2	
				Total (pounds)	
				Total (tons)	
				Total (kilograms)	
				Total (metric tons)	

Source: Thomas B. Cobb at Ohio's Bowling Green State University developed this CO_2 calculator.

1. Calculate your carbon footprint. To calculate your emissions, first complete the blank "Personal Quantity per Year" column as described above. Wherever you cannot provide personal data, use that listed in the "Typical Quantity per Year" column. Then, for each activity, calculate your annual consumption (using the units specified in the "Units per Year" column), and multiply your annual consumption by the associated number in the "Multiplier" column to obtain an estimate of the pounds of CO_2 resulting from that activity, which you will enter in the "Emissions per Year" column. Finally, add the numbers in that column to find your carbon footprint, and express the final CO_2 result in both pounds and tons (1 ton = 2,000 lbs) and in kilograms and metric tons (1 kilogram = 2.2 pounds; 1 metric ton = 1.1 tons).
2. Compare your emissions with those of your classmates and with the per capita U.S. average of 19.6 metric tons (21.6 tons) of CO_2 per person per year. Actually, your answer should be considerably less—roughly about half the per capita value—because this computation only accounts for direct emissions. For instance, CO_2 resulting from driving a car is included, but the CO_2 emitted in manufacturing and disposing of the car is not. You can find more complete carbon footprint calculators at several sites on the Web.
3. Consider and list actions you could take to reduce your carbon footprint by 20%.

LEARNING ONLINE

The website for *Environmental Science 14e* contains helpful study tools and other resources to take the learning experience beyond the textbook. Examples include flashcards, practice quizzing, information on green careers, tips for more sustainable living, activities, and a wide range of articles and further readings. To access these materials and other companion resources for this text, please visit **www.cengagebrain.com**. For further details, see the preface, p. xvi.

Solid and Hazardous Waste

16

E-waste—An Exploding Problem

CORE CASE STUDY

Electronic waste, or *e-waste*, consists of discarded television sets, cell phones, computers, and other electronic devices (Figure 16-1). It is the fastest-growing solid waste problem in the United States and in the world. Each year, Americans discard an estimated 155 million cell phones, 48 million personal computers, and many more millions of television sets, iPods, Blackberries, FAX machines, printers, and other electronic products.

Most e-waste ends up in landfills and incinerators, even though 80% of the components in electronic devices contain valuable materials that can be recycled or reused. These wasted resources include high-quality plastics and valuable metals such as aluminum, copper, platinum, silver, gold, and rare earth metals that are used in most of the world's electronic devices and military weapons systems (see Science Focus, Chapter 12, p. 291). E-waste is also a source of toxic and hazardous pollutants, including polyvinylchloride (PVC), brominated flame retardants, lead, and mercury, which can contaminate air, surface water, groundwater, and soil, and cause serious and even life-threatening health problems for e-waste workers.

Figure 16-1 *Rapidly growing electronic waste* (e-waste) consumes resources and pollutes the air, water, and land with harmful compounds. **Question:** Have you disposed of an electronic device lately? If so, how did you dispose of it?

Much of the e-waste in the United States that is not buried or incinerated is shipped to China and other Asian and African countries where labor is cheap and environmental regulations are weak. Workers there—many of them children—dismantle, burn, and treat e-waste products with acids to recover valuable metals and reusable parts. As they do this, they are exposed to toxic metals and other harmful chemicals. The remaining scrap is dumped into waterways and fields or burned in open fires, exposing many people to highly toxic chemicals called dioxins. This practice is a dark and dangerous side of recycling.

Transfer of such hazardous waste from more-developed to less-developed countries was banned by the International Basel Convention, a treaty now signed by 175 countries to reduce and control the movement of hazardous waste across international borders. Despite this ban, much of the world's e-waste is not officially classified as hazardous waste or is illegally smuggled out of some EU countries. The United States can export its e-waste legally because it is the only industrialized nation that has not ratified the Basel Convention.

The European Union (EU) has led the way in dealing with e-waste, taking a *cradle-to-grave* approach, which requires manufacturers to take back electronic products at the end of their useful lives and repair, remanufacture, or recycle them. In the EU, e-waste is banned from landfills and incinerators. Japan is also taking this approach. To cover the costs of these programs, consumers pay a recycling tax on electronic products. However, because of the high costs involved in complying with these laws, much of this waste that is supposed to be recycled domestically is shipped illegally to other countries.

Recycling probably will not keep up with the explosive growth of e-waste, and there is money to be made from illegally sending such materials to other countries. According to Jim Puckett, coordinator of the Basel Action Network, an organization working to stem the flow of toxic waste, the only real long-term solution is a *prevention* approach through which electrical and electronic products are designed and manufactured without the use of toxic materials in such a way that they can be easily repaired, remanufactured, and recycled. Electronic waste is just one of many types of solid and hazardous waste discussed in this chapter.

Key Questions and Concepts

16-1 What are solid waste and hazardous waste, and why are they problems?

CONCEPT 16-1 Solid waste contributes to pollution and wastes valuable resources that could be reused or recycled; hazardous waste contributes to pollution as well as to natural capital degradation, health problems, and premature deaths.

16-2 How should we deal with solid waste?

CONCEPT 16-2 A sustainable approach to solid waste is first to reduce it, then to reuse or recycle it, and finally to safely dispose of what is left.

16-3 Why is reusing and recycling materials so important?

CONCEPT 16-3 Reusing items decreases the consumption of matter and energy resources, and reduces pollution and natural capital degradation; recycling does so to a lesser degree.

16-4 What are the advantages and disadvantages of burning or burying solid waste?

CONCEPT 16-4 Technologies for burning and burying solid wastes are well developed, but burning contributes to air and water pollution and greenhouse gas emissions, and buried wastes eventually contribute to the pollution and degradation of land and water resources.

16-5 How should we deal with hazardous waste?

CONCEPT 16-5 A more sustainable approach to hazardous waste is first to produce less of it, then to reuse or recycle it, then to convert it to less-hazardous materials, and finally to safely store what is left.

16-6 How can we make the transition to a more sustainable low-waste society?

CONCEPT 16-6 Shifting to a low-waste society requires individuals and businesses to reduce resource use and to reuse and recycle wastes at local, national, and global levels.

Note: Supplements 2 (p. S3), 4 (p. S10), and 6 (p. S22) can be used with this chapter.

Solid wastes are only raw materials we're too stupid to use.

ARTHUR C. CLARKE

16-1 What Are Solid Waste and Hazardous Waste, and Why Are They Problems?

▶ **CONCEPT 16-1 Solid waste contributes to pollution and wastes valuable resources that could be reused or recycled; hazardous waste contributes to pollution as well as to natural capital degradation, health problems, and premature deaths.**

We Throw Away Huge Amounts of Useful Things and Hazardous Materials

In the natural world, wherever humans are not dominant, there is essentially no waste because the wastes of one organism become nutrients or raw materials for others (see Figure 3-11 p. 48). This natural recycling of nutrients follows one of the three **principles of sustainability** (see Figure 1-2, p. 8, or back cover).

Modern humans violate this principle in a big way. We produce huge amounts of waste materials that go unused and that pollute the environment. Because of the law of conservation of matter (see Chapter 2, p. 34) and the nature of human lifestyles, we will always produce some waste. However, studies indicate that we could reduce this waste of potential resources and the resulting environmental harm it causes by as much as 90%.

One major category of waste is **solid waste**—any unwanted or discarded material we produce that is not a liquid or a gas. Solid waste can be divided into two types. One type is **industrial solid waste** produced by mines, farms, and industries that supply people with goods and services. The other is **municipal solid waste (MSW)**, often called *garbage* or *trash,* which consists of the combined solid waste produced by homes and workplaces other than factories. Examples include paper and cardboard, food wastes, cans, bottles, yard wastes, furniture, plastics, metals, glass, wood, and e-waste (**Core Case Study**). Much of this waste is thrown away and ends up in rivers (see Figure 1-7, p. 12), lakes, the ocean (Figure 11-30, p. 266), and open trash dumps. Some resource experts believe that we should change the name of the trash we produce from MSW of MWR—mostly wasted resources.

CORE CASE STUDY

In more-developed countries, most MSW is buried in landfills or burned in incinerators. In many less-developed countries, much of it ends up in open dumps (Figure 1-12, p. 17), where poor people eke out a living finding items they can sell for reuse or recycling (see Photo 1 in the Detailed Contents). In 2008, the Organization for Economic Cooperation and Development projected that between 2005 and 2030, the world's output of MSW will almost double.

Another major category of waste is **hazardous**, or **toxic**, **waste**, which threatens human health or the environment because it is poisonous, dangerously chemically reactive, corrosive, or flammable. Examples include industrial solvents, hospital medical waste, car batteries (containing lead and acids), household pesticide products, dry-cell batteries (containing mercury and cadmium), and ash from incinerators and coal-burning power plants (see Chapter 13, Case Study, p. 312). Figure 16-2 lists some of the harmful chemicals found in many homes. The two largest classes of hazardous wastes are *organic compounds* (such as various solvents, pesticides, PCBs, and dioxins) and nondegradable *toxic heavy metals* (such as lead, mercury, and arsenic).

Another form of extremely hazardous waste is highly radioactive waste produced by nuclear power plants and nuclear weapons facilities (see Chapter 13, p. 316). Such wastes must be stored safely for 10,000 to 240,000 years, depending on what radioactive isotopes are present in them. After 60 years of research, scientists and governments have not found a scientific and politically acceptable way to safely isolate these dangerous wastes for periods of time that, in some cases, may be longer than the roughly 200,000 years that our species has been around.

According to the UN Environment Programme (UNEP), more-developed countries produce 80–90% of the world's hazardous wastes, and the United States is the largest producer. However, as China continues to rapidly industrialize, largely without adequate pollution controls, it may soon take over the number-one spot. In order, the top three U.S. producers of hazardous waste are the military, the chemical industry, and the mining industry.

What Harmful Chemicals Are in Your Home?

Cleaning
- Disinfectants
- Drain, toilet, and window cleaners
- Spot removers
- Septic tank cleaners

Gardening
- Pesticides
- Weed killers
- Ant and rodent killers
- Flea powders

Paint Products
- Paints, stains, varnishes, and lacquers
- Paint thinners, solvents, and strippers
- Wood preservatives
- Artist paints and inks

Automotive
- Gasoline
- Used motor oil
- Antifreeze
- Battery acid
- Brake and transmission fluid

General
- Dry-cell batteries (mercury and cadmium)
- Glues and cements

Figure 16-2 Harmful chemicals are found in many homes. The U.S. Congress has exempted the disposal of these household chemicals and other items from government regulation. **Question:** Which of these chemicals are found where you live?

■ CASE STUDY
Solid Waste in the United States

The United States leads the world in total solid waste production and in solid waste per person. With only 4.5% of the world's people, the United States produces about 33% of the world's solid waste. About 98.5% of all solid waste produced in the United States is industrial solid waste from mining (76%), agriculture (13%), and industry (9.5%).

The remaining 1.5% of U.S. solid waste is MSW. In 2009, the largest categories of this MSW were paper and cardboard (28% of total), food waste (14%), yard waste (14%), plastics (12%), and metals (9%), followed by rubber, leather, and textiles (8%), wood (7%), glass (5%) and miscellaneous wastes (3%). Since 1990, the total amount of MSW generated in the United States has leveled off, and the MSW per person has declined slightly. However, the total volume of MSW produced is still huge.

Every year, the United States generates enough MSW to fill a bumper-to-bumper convoy of garbage trucks encircling the globe almost 8 times! Here are some of the solid wastes that consumers throw away in the high-waste economy of the United States:

- Enough tires each year to encircle the planet almost three times (Figure 16-3, p. 414).

Figure 16-3 Hundreds of millions of discarded tires have accumulated in this massive tire dump in Midway, Colorado (USA).

Jim Wark/Peter Arnold, Inc.

- An amount of disposable diapers each year that, if linked end to end, would reach to the moon and back seven times.
- Enough carpet each year to cover the U.S. state of Delaware.
- About 2.5 million nonreturnable plastic bottles *every hour*. If laid end-to-end, the number of these bottles produced each year would reach from the earth to the moon and back about 6 times.
- About 274 million plastic shopping bags per day, an average of about 3,200 every second.
- Enough office paper each year to build a wall 3.5 meters (11 feet) high across the country from New York City to San Francisco, California.
- Some 186 billion pieces of junk mail (an average of 660 pieces per American) each year, about 45% of which are thrown in the trash unopened.

Most of these wastes break down very slowly, if at all. Lead, mercury, glass, plastic foam, and most plastic bottles essentially take forever to break down; an aluminum can takes 500 years; plastic bags, 400 to 1,000 years; and a plastic six-pack holder, 100 years.

GOOD NEWS

Some encouraging news is that in 2008 and 2009, both the total amount of MSW and the amount of MSW per person declined slightly for the first time since 1960. Major reasons for this include the economic downturn, which led to a decrease in the purchasing of goods, and increased rates of recycling.

16-2 How Should We Deal with Solid Waste?

▶ CONCEPT 16-2 A sustainable approach to solid waste is first to reduce it, then to reuse or recycle it, and finally to safely dispose of what is left.

We Can Burn or Bury Solid Waste or Produce Less of It

We can deal with the solid wastes we create in two ways. One method is **waste management**, in which we attempt to control wastes in ways that reduce their environmental harm without seriously trying to reduce the amount of waste produced. This output approach begins with the question, "What do we do with solid waste?" It typically involves mixing wastes together and then transferring them from one part of the environment to another, usually by burying them, burning them, or shipping them to another location.

The second approach is **waste reduction**, in which we produce much less waste and pollution, and the wastes we do produce are considered to be potential

First Priority	Second Priority	Last Priority
Primary Pollution and Waste Prevention	**Secondary Pollution and Waste Prevention**	**Waste Management**
■ Change industrial process to eliminate use of harmful chemicals ■ Use less of a harmful product ■ Reduce packaging and materials in products ■ Make products that last longer and are recyclable, reusable, or easy to repair	■ Reuse ■ Repair ■ Recycle ■ Compost ■ Buy reusable and recyclable products	■ Treat waste to reduce toxicity ■ Incinerate waste ■ Bury waste in landfills ■ Release waste into environment for dispersal or dilution

Figure 16-4 *Integrated waste management:* The U.S. National Academy of Sciences suggests these priorities for dealing with solid waste. To date, these waste-reduction priorities have not been followed in the United States or in most other countries. Instead, most efforts are devoted to waste management through disposal (bury it, burn it, or send it somewhere else). **Question:** Why do you think most countries do not follow these priorities, even though they are based on reliable science? (Data from U.S. Environmental Protection Agency and U.S. National Academy of Sciences)

resources that we can reuse, recycle, or compost (**Concept 16-2**). This waste prevention approach begins with the question, "How can we avoid producing so much solid waste?"

There is no single solution to the solid waste problem. Most analysts call for using **integrated waste management**—a variety of coordinated strategies for both waste disposal and waste reduction. Environmental scientists and the U.S. Environmental Protection Agency (EPA) call for much greater emphasis on waste prevention and reduction (Figure 16-4) rather than on waste disposal. In 2009, the EPA reported that 54% of the MSW produced in the United States was buried in landfills, 12% was incinerated with some of the energy used for heating and producing electricity, and 34% was recycled (25%) or composted (9%).

Some scientists and economists estimate that 75–90% of the solid waste we produce could be eliminated by a combination of the strategies shown in Figure 16-4. Let us look more closely at these options in the order of priorities suggested by the U.S. National Academy of Scientists. GOOD NEWS

We Can Cut Solid Wastes by Reducing, Reusing, and Recycling

Waste reduction (**Concept 16-2**) is based on three Rs:

- **Reduce:** consume less and live a simpler lifestyle.
- **Reuse:** rely more on items that we can use repeatedly instead of on throwaway items, and buy necessary items secondhand or borrow or rent them.
- **Recycle:** separate and recycle paper, glass, cans, plastics, metal, and other items, and buy products made from recycled materials.

From an environmental standpoint, the first two Rs are preferred because they are *input*, or *prevention*, approaches that tackle the problem of waste production at the front end—before it occurs. Recycling is important but it deals with wastes after they have been produced. Reducing, reusing, and recycling strategies save matter and energy resources, reduce pollution (including greenhouse gas emissions), help protect biodiversity, and save money. Figure 16-5 lists some ways you can use the three Rs of waste reduction to reduce your output of solid waste. GOOD NEWS

Here are five strategies that industries and communities have used to reduce resource use, waste, and pollution.

First, *redesign manufacturing processes and products to use less material and energy*. For example, the weight of a

What Can You Do?

Solid Waste

- Follow the three Rs of resource use: Reduce, Reuse, and Recycle
- Ask yourself whether you really need a particular item, and refuse packaging where possible
- Rent, borrow, or barter goods and services when you can, buy secondhand, and donate or sell unused items
- Buy things that are reusable, recyclable, or compostable, and be sure to reuse, recycle, and compost them
- Avoid disposables and do not use throwaway paper and plastic plates, cups, eating utensils, and other disposable items when reusable or refillable versions are available
- Use e-mail or text-messaging in place of conventional paper mail
- Read newspapers and magazines online and read e-books
- Buy products in bulk or concentrated form whenever possible

Figure 16-5 **Individuals matter:** You can save resources by reducing your output of solid waste and pollution. **Questions:** Which three of these actions do you think are the most important? Why? Which of these things do you do?

typical car has been reduced by about one-fourth since the 1960s through use of lighter steel and lightweight plastics and composite materials.

Second, *develop products that are easy to repair, reuse, remanufacture, compost, or recycle*. For example, a Xerox photocopier made of reusable or recyclable parts that allow for easy remanufacturing could eventually save the company $1 billion in manufacturing costs.

Third, *eliminate or reduce unnecessary packaging*. Use the following hierarchy for product packaging: no packaging, reusable packaging, and recyclable packaging. The 37 European Union countries require the recycling of 55–80% of all packaging waste.

Fourth, *use fee-per-bag waste collection systems* that charge consumers for the amount of waste they throw away but provide free pickup of recyclable and reusable items.

Five, *establish cradle-to-grave responsibility laws* that require companies to take back various consumer products such as electronic equipment (**Core Case Study**), appliances, and motor vehicles, as Japan and many European countries do. CORE CASE STUDY

16-3 Why Is Reusing and Recycling Materials So Important?

▶ **CONCEPT 16-3 Reusing items decreases the consumption of matter and energy resources, and reduces pollution and natural capital degradation; recycling does so to a lesser degree.**

Reuse Is an Important Way to Reduce Solid Waste and Pollution, and to Save Money

In today's industrialized societies, we have increasingly substituted throwaway items for reusable ones, which has resulted in growing masses of solid waste.

Reuse involves cleaning and using materials over and over, and thus increasing the typical life span of a product. For example, soft drinks and beer can be sold in reusable bottles. This form of waste reduction decreases the use of matter and energy resources, cuts waste and pollution (including greenhouse gases), creates local jobs, and saves money (**Concept 16-3**).

Reuse is alive and well in many less-developed countries (see Photo 1 in the Detailed Contents). But reuse has a downside for some people. The poor who scavenge in open dumps (see Figure 1-12, p. 17) for food scraps and items that they can reuse or sell are often exposed to toxins and infectious diseases. Reuse strategies in more-developed countries include yard sales, flea markets, secondhand stores, and online sites such as eBay and craigslist.

In general, reuse is on the rise. Denmark, Finland, and the Canadian province of Prince Edward Island have banned all beverage containers that cannot be reused. In Finland, 95% of the soft drink, beer, wine, and spirits containers are refillable. The latest rechargeable batteries come fully charged, can hold a charge for up to 2 years when they are not used, can be recharged in about 15 minutes, and greatly reduce toxic waste when used in place of discarded conventional batteries. GOOD NEWS

Instead of using throwaway paper or plastic bags to carry home groceries and other items they purchase, many people have switched to reusable cloth or string bags. Both paper and plastic bags are environmentally harmful, and the question of which is more damaging has no clear-cut answer. In 2010, some reuseable cloth bags, mostly made in China, had traces of lead in paint used to print words and logos on their outside surfaces. Use of lead in such paint is being banned. Despite concerns raised because of this incident, scientists generally view the environmental problems created by disposable bags as more serious than those resulting from this limited lead paint problem.

There is a growing backlash against plastic shopping bags. To encourage people to carry reusable bags, the governments of Ireland, Taiwan, and the Netherlands tax plastic shopping bags. In Ireland, a tax of about 25¢ per bag helped to cut plastic bag litter by 90% as people switched to reusable bags. Kenya, Uganda, Rwanda, South Africa, Bangladesh, Bhutan, parts of India, Taiwan, China, Australia, France, Italy, and the U.S. city of San Francisco have banned the use of all or most types of plastic shopping bags. Not surprisingly, plastics industry officials have mounted a massive advertising and political campaign to prevent such bans.

There are many other ways to reuse various items, some of which are listed in Figure 16-6.

There Are Two Types of Recycling

Recycling involves reprocessing discarded solid materials into new, useful products. Households and workplaces produce five major types of materials that we

What Can You Do?

Reuse

- Buy beverages in refillable glass containers instead of in cans or throwaway bottles
- Use reusable plastic or metal lunchboxes
- Carry sandwiches and store refrigerated food in reusable containers instead of wrapping them in aluminum foil or plastic wrap
- Use rechargeable batteries and recycle them when their useful life is over
- When eating out, bring your own reusable silverware and napkin
- Bring your own reusable container for takeout food or restaurant meal leftovers
- Carry groceries and other items in a reusable basket or cloth bag
- Buy used furniture, computers, cars, and other items instead of buying new
- Give away or sell items you no longer use

Figure 16-6 Individuals matter: There are many ways to reuse the items we purchase. **Question:** Which of these suggestions have you tried and how did they work for you?

can recycle: paper products, glass, aluminum, steel, and some plastics. We can reprocess such materials in two ways. In **primary**, or **closed-loop**, **recycling**, materials such as aluminum cans are recycled into new products of the same type. In **secondary recycling**, waste materials are converted into different products. For example, we can shred used tires and turn them into rubberized road-surfacing material.

Just about anything (except energy) is recyclable, but there are two key questions that are important to the success of any recycling program. *First*, do the items that are separated for recycling actually get recycled? *Second*, do businesses, governments, and individuals complete the recycling loop by buying products that are made from recycled materials?

In 2009, the United States recycled about 25% of its MSW—up from 6.4% in 1960. This included the recycling of more than 62% of all the paper and paperboard used, 74% of all high-grade office paper, and 54% of all magazines. This increase in recycling rates is partly due to about 9,000 curbside-pickup recycling programs that serve about half of the U.S. population. Experts say that with education and proper incentives, the United States could recycle 60–80% of its MSW, in keeping with the chemical cycling **principle of sustainability**.

GOOD NEWS

CONNECTIONS

The Environmental Impact of Magazines

According to a 2009 study by *Audubon* magazine, fewer than 200 of the 17,000 magazines published in the United States contain any recycled paper. As a result, producing each issue of most magazines requires roughly the equivalent of 2,100 trees along with enough energy to power 49 homes for a year and enough water to fill 41,300 bathtubs. It also produces about the same amount of carbon dioxide emissions as 78 cars would emit in a year. (For information on the efforts of Cengage Learning to produce this and other textbooks more sustainably, please see p. xxiv.)

To promote separation of wastes for recycling, more than 4,000 communities in the United States use a pay-as-you-throw (PAUT) or fee-per-bag waste collection system. They charge households and businesses for the amount of garbage that is picked up, but do not charge for pickup of materials separated for recycling or reuse.

Composting is another form of recycling that mimics nature's recycling of nutrients. It involves using bacteria to decompose yard trimmings, vegetable food scraps, and other biodegradable organic wastes. The resulting organic material can be added to soil to supply plant nutrients, slow soil erosion, retain water, and improve crop yields. Homeowners can compost such wastes in simple backyard containers, in composting piles that must be turned over occasionally, or in small composting drums (Figure 16-7) that are rotated to mix the wastes and speed up the decomposition process.

GOOD NEWS

Courtesy of Green Culture, Inc.

Figure 16-7 When the compost in this backyard composter drum is ready to be used, the device can be wheeled to a garden and emptied into the soil.

Some cities in Canada and in many European Union countries collect and compost more than 85% of their biodegradable wastes in centralized community facilities. In the United States, about 3,000 municipal composting programs recycle about 60% of the yard wastes in the country's MSW. This is likely to rise as the number of states (now 20) that ban yard wastes from sanitary landfills increases. The resulting compost can be used as organic soil fertilizer, topsoil, or landfill cover. It can also be used to help restore eroded soil on hillsides and along highways, as well as on strip-mined land, overgrazed areas, and eroded cropland.

To be successful, a large-scale composting program must be located carefully and odors must be controlled, especially near residential areas. Sometimes, composting takes place in huge indoor buildings. In the Canadian city of Edmonton, Alberta, an indoor facility the size of eight football fields composts 50% of the city's organic solid waste. Composting programs must also exclude toxic materials that can contaminate the compost and make it unsafe for fertilizing crops and lawns.

Trade-Offs

Recycling

Advantages	Disadvantages
Reduces energy and mineral use and air and water pollution	Can cost more than burying in areas with ample landfill space
Reduces greenhouse gas emissions	Reduces profits for landfill and incinerator owners
Reduces solid waste	Source separation inconvenient for some
Can save landfill space	

Figure 16-8 Recycling solid waste has advantages and disadvantages (**Concept 16-3**). **Questions:** Which single advantage and which single disadvantage do you think are the most important? Why?

Recycling Has Advantages and Disadvantages

Figure 16-8 lists the advantages and disadvantages of recycling (**Concept 16-3**). Whether recycling makes economic sense depends on how we look at its economic and environmental benefits and costs.

Critics of recycling programs argue that recycling is costly and adds to the taxpayer burden in communities where recycling is funded through taxation. Proponents of recycling point to studies showing that the net economic, health, and environmental benefits of recycling (Figure 16-8, left) far outweigh the costs. For example, the EPA estimates that recycling and composting in the United States in 2009 saved an amount of energy roughly equal to that in 28 billion liters (10 billion gallons) of gasoline and reduced emissions of climate-changing carbon dioxide by an amount roughly equal to that emitted by 33 million passenger vehicles. Proponents also argue that the U.S. recycling industry employs about 1.1 million people and that its annual revenues are much larger than those of the waste management industry, which is also financed through taxation in most communities.

Critics say that recycling may make economic sense for valuable and easy-to-recycle materials in MSW such as paper and paperboard, steel, and aluminum, but probably not for cheap or plentiful resources such

INDIVIDUALS MATTER

Mike Biddle's Contribution to Plastics Recycling

In 1994, Mike Biddle, a former PhD engineer with Dow Chemical, and his business partner Trip Allen founded MBA Polymers, Inc. Their goal was to develop a commercial process for recycling high-value plastics from complex streams of manufactured goods such as computers, electronics, appliances, and automobiles. They succeeded by designing a 16-step automated process that separates plastics from nonplastic items in mixed waste streams, and then separates plastics from each other by type and grade. The process then converts them to pellets that can be used to make new products.

The pellets are cheaper than virgin plastics because the company's process uses 90% less energy than that needed to make a new plastic and because the raw material is discarded junk that is cheap or free. The environment also wins because greenhouse gas emissions from this process are much lower than those from the process of making virgin plastics. Also, recycling plastic wastes reduces the need to incinerate them or bury them in landfills.

The company is considered a world leader in plastics recycling. It operates a large, state-of-the-art research and recycling plant in Richmond, California (USA), and recently opened the world's two most advanced plastics recycling plants in China and Austria. MBA Polymers has won many awards, including the Thomas Alva Edison Award for Innovation, and it was selected by *Inc.* magazine as one of "America's Most Innovative Companies."

Maybe you can be an environmental entrepreneur by using your brainpower to develop an environmentally beneficial and financially profitable process or business.

SCIENCE FOCUS

Bioplastics

Most of today's plastics are made from organic polymers produced from petroleum-based chemicals. This may change as scientists shift to developing plastics made from biologically based chemicals.

Henry Ford, who developed the first Ford motorcar, supported research on the development of a bioplastic made from soybeans and another made from hemp. A 1914 photograph shows him using an ax to strike the body of a car made from soy bioplastic to demonstrate its strength and resistance to denting.

However, as oil became widely available, petrochemical plastics took over the market. Now, with projected climate change and other environmental problems associated with the use of oil, chemists are stepping up efforts to make biodegradable and more environmentally sustainable plastics. These *bioplastics* can be made from corn, soy, sugarcane, switchgrass (see Figure 13-43, p. 338), chicken feathers, and some components of garbage.

Compared to conventional oil-based plastics, with proper design and mass production, bioplastics could be lighter, stronger, and cheaper, and the process of making them would require less energy and produce less pollution per unit of weight. Instead of being sent to landfills, packaging made from bioplastics could be composted to produce a soil conditioner, in keeping with the chemical recycling **principle of sustainability**.

SUSTAINABILITY

Critical Thinking

What might be some disadvantages of more rapidly degradable bioplastics? Do you think they outweigh the advantages?

as glass made from silica. Currently, only about 7% by weight of all plastic wastes in the United States is recycled because there are many different types of plastic resins that are difficult to separate from products. But progress is being made in the recycling of plastics (Individuals Matter, at left) and in the development of more degradable bioplastics (Science Focus, above).

We Can Encourage Reuse and Recycling

Three factors hinder reuse and recycling. *First,* the market prices of almost all products do not include the harmful environmental and health costs associated with producing, using, and discarding them.

Second, the economic playing field is uneven, because in most countries, resource-extracting industries receive more government tax breaks and subsidies than reuse and recycling industries get.

Third, the demand and thus the price paid for recycled materials fluctuates, mostly because buying goods made with recycled materials is not a priority for most governments, businesses, and individuals.

How can we encourage reuse and recycling? Proponents say that leveling the economic playing field is the best way to start. Governments can *increase* subsidies and tax breaks for reusing and recycling materials, and *decrease* subsidies and tax breaks for making items from virgin resources.

Another strategy is to greatly increase use of the fee-per-bag waste collection system. When the U.S. city of Fort Worth, Texas, instituted such a program, the proportion of households recycling their trash went from 21% to 85%. The city went from losing $600,000 in its recycling program to making $1 million a year because of increased sales of recycled materials to industries.

One type of recycling program processes mixed wastes in centralized *materials-recovery facilities* where machines and workers separate the mixed waste to recover valuable materials for sale to manufacturers as raw materials. The remaining paper, plastics, and other combustible wastes are recycled or burned in waste-to-energy incinerators. Another type of program uses a *source separation approach,* in which households and businesses separate their trash into recyclable categories such as glass, paper, and plastics. This approach saves more energy and can yield cleaner and more valuable recyclables, usually at a lower cost.

Another important strategy is to encourage or require government purchases of recycled products to help increase demand for and lower prices of these products. Governments can also pass laws requiring companies to take back and recycle or reuse packaging and electronic waste discarded by consumers (**Core Case Study**), as is done in Japan and some European Union countries.

CORE CASE STUDY

Citizens can pressure governments to require product labeling that lists the recycled content of products as well as the types and amounts of any hazardous materials they contain. This would help consumers to make more informed choices about the environmental consequences of buying certain products. It would also help to expand the market for recycled materials by spurring demand for them.

One reason for the popularity of recycling is that it helps to soothe the consciences of people living in a throwaway society. Many people think that recycling their newspapers and aluminum cans is all they need do to meet their environmental responsibilities. Recycling is important, but reducing resource consumption and reusing resources are more effective *prevention* approaches to reducing the flow and waste of resources (**Concept 16-3**). Proponents call for increasing the U.S. recycling and composting rate of MSW from 34% to 80% over the next two decades, and using consumer education and financial incentives to put much greater emphasis on reuse.

16-4 What Are the Advantages and Disadvantages of Burning or Burying Solid Waste?

▶ **CONCEPT 16-4 Technologies for burning and burying solid wastes are well developed, but burning contributes to air and water pollution and greenhouse gas emissions, and buried wastes eventually contribute to the pollution and degradation of land and water resources.**

Burning Solid Waste Has Advantages and Disadvantages

Globally, MSW is burned in more than 600 large *waste-to-energy incinerators,* which use the heat they generate to boil water and make steam for heating water or interior spaces, or for producing electricity. (Trace the flow of waste materials through this process, as shown in Figure 16-9.) There are 87 of these incinerators in the United States, down from 102 in 2000. Most of them were built before 1990.

The United States incinerates only about 12% of its MSW. One reason for this fairly low percentage is that incineration has a bad reputation stemming from past use of highly polluting and poorly regulated incinerators. Also, incineration competes with an abundance of low-cost landfills in many parts of the country.

By contrast, Denmark incinerates 54% of its MSW in waste-to-energy incinerators with newer designs. The country also recycles or composts 42% of its MSW (compared to 34% in the United States) and buries only 4% of it in landfills (compared to 54% in the United States). Denmark's incinerators use dozens of filters to remove pollutants such as toxic mercury and dioxins. They run so cleanly that they release just a fraction of the toxic dioxins released by typical home fireplaces and backyard barbecues. According to a 2009 EPA study, modern waste-to-energy incinerators emit about half of the amount of greenhouse gases emitted by sanitary landfills.

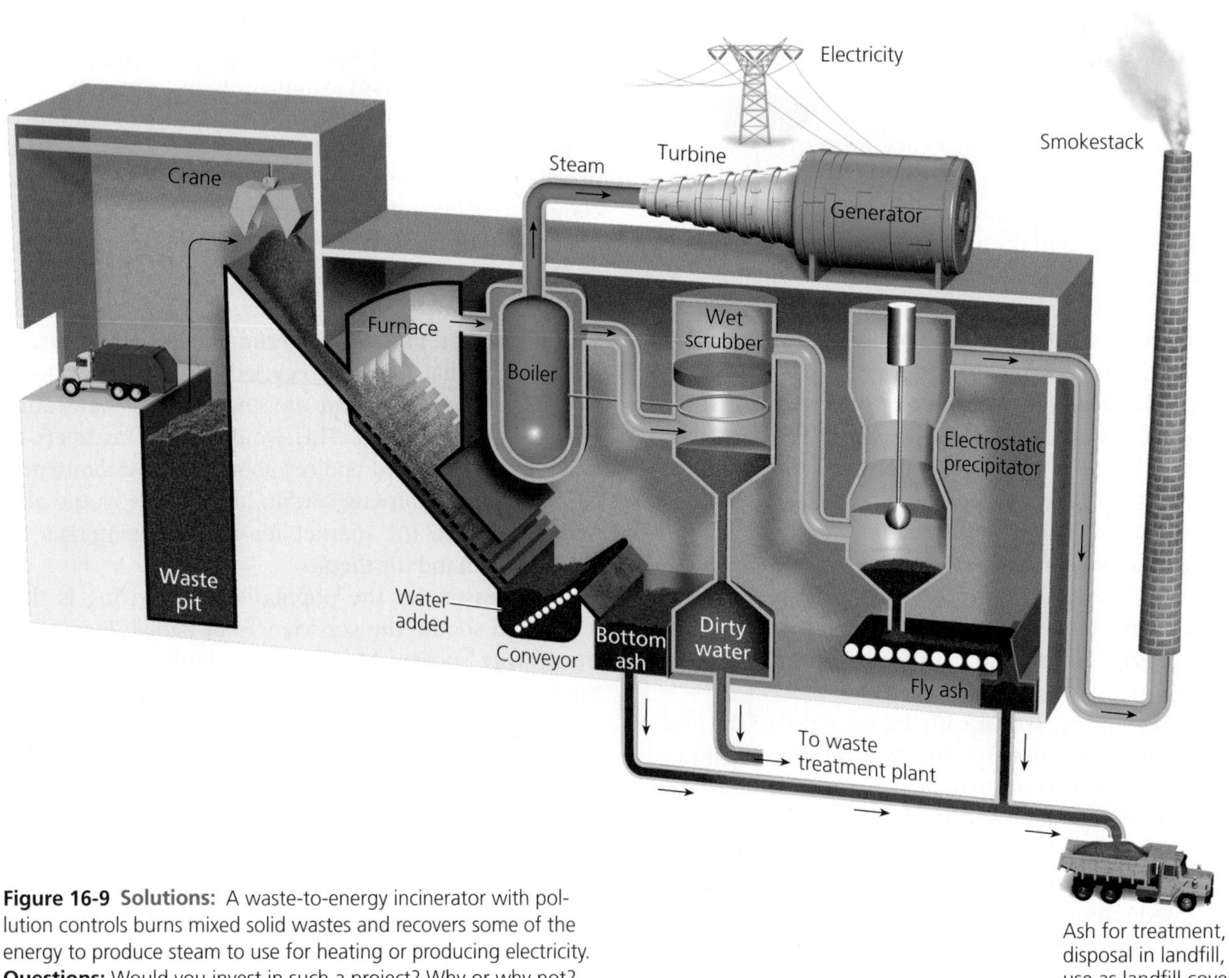

Figure 16-9 Solutions: A waste-to-energy incinerator with pollution controls burns mixed solid wastes and recovers some of the energy to produce steam to use for heating or producing electricity. **Questions:** Would you invest in such a project? Why or why not?

However, many citizens, local governments, and environmental scientists oppose any use of waste incinerators because it undermines efforts to increase recycling and reuse. Once an incinerator is built, it must be fed large volumes of trash in order to be profitable for its owners. Rather than encouraging people to reduce, reuse, and recycle, widespread incineration would tend to promote the creation of more wastes.

Figure 16-10 lists the advantages and disadvantages of using incinerators to burn solid waste.

Burying Solid Waste Has Advantages and Disadvantages

About 54%, by weight, of the MSW in the United States is buried in sanitary landfills, compared to 80% in Canada, 15% in Japan, and 4% in Denmark.

There are two types of landfills. In newer landfills, called **sanitary landfills** (Figure 16-11), solid wastes

Trade-Offs

Waste-to-Energy Incineration

Advantages	Disadvantages
Reduces trash volume	Expensive to build
Produces energy	Produces a hazardous waste
Concentrates hazardous substances into ash for burial	Emits some CO_2 and other air pollutants
Sale of energy reduces cost	Encourages waste production

Figure 16-10 Incinerating solid waste has advantages and disadvantages (**Concept 16-4**). These trade-offs also apply to the incineration of hazardous waste. **Questions:** Which single advantage and which single disadvantage do you think are the most important? Why?

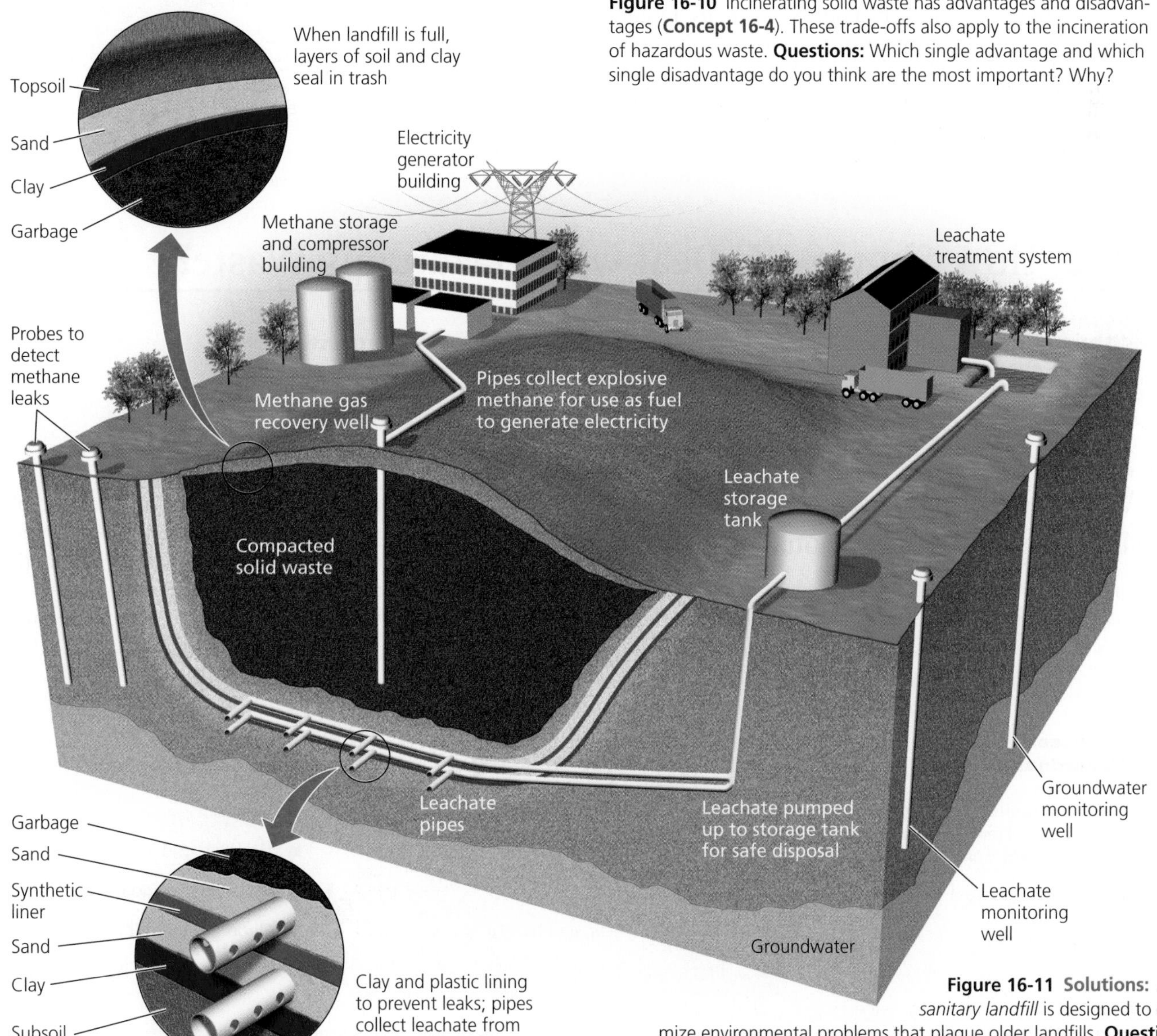

Figure 16-11 Solutions: A state-of-the-art *sanitary landfill* is designed to eliminate or minimize environmental problems that plague older landfills. **Question:** Some experts say that these landfills will eventually develop leaks and could emit toxic liquids. How do you think this could happen?

are spread out in thin layers, compacted, and covered daily with a fresh layer of clay or plastic foam. This process helps keep the material dry and reduces leakage of contaminated water (leachate) from the landfill. This covering also lessens the risk of fire, decreases odor, and reduces accessibility to vermin.

The bottoms and sides of these landfills have strong double liners and containment systems that collect the liquids leaching from them. Some landfills also have systems for collecting and burning methane, the potent greenhouse gas that is produced when the wastes decompose in the absence of oxygen. The number of sanitary landfills in the United States dropped from nearly 8,000 in 1988 to about 1,900 in 2009. This occurred because new regulations led to the closing of many small, older landfills.

The second type of landfill is an **open dump**, essentially a field or large pit where garbage is deposited and sometimes burned. They are rare in more-developed countries, but are widely used near major cities in many less-developed countries (see Figure 1-12, p. 17). China disposes of about 85% of its rapidly growing mountains of solid waste mostly in rural open dumps or in poorly designed and poorly regulated landfills that do not have most of the features of sanitary landfills. Also, some solid wastes end up in rivers, lakes, and oceans. Figure 16-12 lists the advantages and disadvantages of using sanitary landfills to dispose of solid waste.

Trade-Offs

Sanitary Landfills

Advantages	Disadvantages
Low operating costs	Noise, traffic, and dust
Can handle large amounts of waste	Release greenhouse gases (methane and CO_2) unless they are collected
Filled land can be used for other purposes	Output approach that encourages waste production
No shortage of landfill space in many areas	Eventually leak and can contaminate groundwater

Figure 16-12 Using sanitary landfills to dispose of solid waste has advantages and disadvantages (**Concept 16-4**). **Questions:** Which single advantage and which single disadvantage do you think are the most important? Why?

16-5 How Should We Deal with Hazardous Waste?

▶ **CONCEPT 16-5 A more sustainable approach to hazardous waste is first to produce less of it, then to reuse or recycle it, then to convert it to less-hazardous materials, and finally to safely store what is left.**

We Can Use Integrated Management of Hazardous Waste

Figure 16-13 shows an integrated management approach suggested by the U.S. National Academy of Sciences that establishes three priority levels for dealing with hazardous waste: produce less; convert as much of it as possible to less-hazardous substances; and put the rest in long-term, safe storage (**Concept 16-5**). Denmark follows these priorities, but most countries do not.

As with solid waste, the top priority should be pollution prevention and waste reduction. With this

Produce Less Hazardous Waste
- Change industrial processes to reduce or eliminate hazardous waste production
- Recycle and reuse hazardous waste

→

Convert to Less Hazardous or Nonhazardous Substances
- Natural decomposition
- Incineration
- Thermal treatment
- Chemical, physical, and biological treatment
- Dilution in air or water

→

Put in Perpetual Storage
- Landfill
- Underground injection wells
- Surface impoundments
- Underground salt formations

Figure 16-13 *Integrated hazardous waste management:* The U.S. National Academy of Sciences has suggested these priorities for dealing with hazardous waste (**Concept 16-5**). **Question:** Why do you think most countries do not follow these priorities? (Data from U.S. National Academy of Sciences)

approach, industries try to find substitutes for toxic or hazardous materials, reuse or recycle the hazardous materials within industrial processes, or use them as raw materials for making other products. (See the Guest Essays on this subject by Lois Gibbs and Peter Montague at **www.cengagebrain.com**.)

At least 33% of industrial hazardous wastes produced in the European Union are exchanged through clearinghouses where they are sold as raw materials for use by other industries. The producers of these wastes do not have to pay for their disposal and recipients get low-cost raw materials. About 10% of the hazardous waste in the United States is exchanged through such clearinghouses, a figure that could be raised significantly. GOOD NEWS

We can also recycle e-waste (**Core Case Study**), but most e-waste recycling efforts create further hazards, especially for workers in some less-developed countries (see the Case Study that follows). In addition, producing electronic devices and eventually throwing them away as e-waste can result in serious threats to other species, especially some that are endangered and threatened. CORE CASE STUDY

CONNECTIONS

Cell Phones and Endangered African Gorillas

Most cell phones contain coltan, a mineral extracted in the deep forests of the Democratic Republic of the Congo (DRC) in central Africa, which is also the home of the endangered lowland gorillas. Coltan is a metallic ore that is refined to produce metallic tantalum, which can hold an electrical charge. In recent years, rapid expansion of both legal and illegal coltan mining in the DRC, which holds about 80% of the world's coltan reserves, has dramatically reduced the habitat of lowland gorillas and other species. It has also contributed to the killing of lowland gorillas for bush meat to feed the miners. An international outcry has helped to shift much of the coltan mining from Africa to areas such as western Australia, but the threat to the gorillas remains. By recycling old cell phones for removal and reuse of coltan, we can reduce the need to mine it, thereby helping to save the remaining population of lowland gorillas.

■ CASE STUDY
Recycling E-Waste

In some countries, workers in e-waste recycling operations—many of them children—are often exposed to toxic chemicals as they dismantle the electronic trash to extract its valuable metals or other parts that can be sold for reuse or recycling (**Core Case Study**). CORE CASE STUDY

According to the United Nations, more than 70% of the world's e-waste ends up in China. A center for such waste is the small port city of Guiyu, where the air reeks of burning plastic and acid fumes. There, more than 5,500 small-scale e-waste businesses employ over 30,000 people (including some children) who work at very low wages in dangerous conditions to extract valuable metals like gold and copper and various rare earth metals (see Science Focus, Chapter 12, p. 291) from millions of discarded computers, television sets, and cell phones.

These workers usually wear no masks or gloves, often work in rooms with no ventilation, and are usually exposed to a cocktail of toxic chemicals. They carry out dangerous activities such as smashing TV picture tubes with large hammers to recover certain components—a method that releases large amounts of toxic lead dust into the air. They also burn computer wires to expose copper, melt circuit boards in metal pots over coal fires to extract lead and other metals, and douse the boards with strong acid to extract gold.

After the valuable metals are removed, leftover parts are burned or dumped into rivers or onto the land. Atmospheric levels of deadly dioxin in Guiyu are up to 86 times higher than World Health Organization safety standards, and an estimated 82% of the area's children younger than age 6 suffer from lead poisoning.

The United States produces roughly 50% of the world's e-waste (**Core Case Study**) and recycles only about 15% of it. This is beginning to change. Thirty-five states have banned the disposal of computers and TV sets in landfills and incinerators, and these measures are setting the stage for an emerging, highly profitable *e-cycling* industry. Also, 13 states as well as New York City make manufacturers responsible for recycling most electronic devices. CORE CASE STUDY GOOD NEWS

Some U.S. electronics manufacturers have free recycling programs. A few will arrange to pick up discarded electronic products or will pay shipping costs to have them returned. Nonprofit groups, such as Free Geek in Portland, Oregon, are motivating many people to donate, recycle, and reuse old electronic devices. Some have called for a standardized U.S. federal law that makes manufacturers responsible for taking back all electronic devices they produce and recycling them domestically.

We Can Detoxify Hazardous Wastes

The first step in dealing with hazardous wastes is to collect them. In Denmark, all hazardous and toxic waste from industries and households is delivered to any of 21 transfer stations throughout the country. From there it is taken to a large processing facility, where three-fourths of the waste is detoxified by physical, chemical, and biological methods. The rest is buried in a carefully designed and monitored landfill.

Some scientists and engineers consider *biological methods* for treatment of hazardous waste to be the wave of the future. One such approach is *bioremediation*, in which bacteria and enzymes help to destroy toxic or hazardous substances, or convert them to harmless compounds. Bioremediation takes a little longer to work than most physical and chemical methods, but it costs much less. (See the Guest Essay by John Pichtel on this topic at **www.cengagebrain.com**.)

Another approach is *phytoremediation*, which involves using natural or genetically engineered plants to absorb, filter, and remove contaminants from polluted soil and water. Various plants have been identified as "pollution sponges" that can help to clean up soil and water contaminated with chemicals such as pesticides, organic solvents, and radioactive or toxic metals. Phytoremediation is still being evaluated and is slow (it can take several growing seasons) compared to other alternatives. However, it can be used to help clean up toxic hotspots.

We can incinerate hazardous wastes to break them down and convert them to harmless or less harmful chemicals such as carbon dioxide and water. This has the same combination of advantages and disadvantages as does burning solid wastes (Figure 16-10). But incinerating hazardous waste without effective and expensive air pollution controls can release air pollutants such as highly toxic dioxins, and it produces an extremely toxic ash that must be safely and permanently stored in a landfill or vault especially designed for this hazardous end product.

We can also detoxify hazardous wastes by using a *plasma arc torch*, somewhat similar to a welding torch, to break them down at very high temperatures. This process decomposes liquid or solid hazardous organic waste into ions and atoms that can be converted into simple molecules, cleaned up, and released as a gas. Scientists are working on affordable ways to convert the gas to transportation fuels (biofuels) such as methanol, ethanol, and clean diesel. The high-temperature heat can also be used to convert hazardous inorganic ash and other matter into a molten glassy material that can be used to encapsulate toxic wastes in order to keep them from leaching into groundwater.

So why are we not making widespread use of the plasma arc torch to detoxify hazardous wastes? The main reason is its high cost. Plasma arc companies are working to bring the cost down.

We Can Store Some Forms of Hazardous Waste

Ideally, we should use burial on land or long-term storage of hazardous and toxic wastes in secure vaults only as the third and last resort after the first two priorities have been exhausted (Figure 16-13 and **Concept 16-5**). But currently, burial on land is the most widely used method in the United States and in most countries, largely because it is the least expensive of all methods.

The most common form of burial is *deep-well disposal*, in which liquid hazardous wastes are pumped under pressure through a pipe into dry, porous rock formations far beneath aquifers that are tapped for drinking and irrigation water. Theoretically, these liquids soak into the porous rock material and are isolated from overlying groundwater by essentially impermeable layers of clay and rock. The cost is low and the wastes can often be retrieved if problems develop.

However, this fairly cheap, out-of-sight and out-of-mind approach presents some problems. There are a limited number of such sites and limited space within them. Sometimes the wastes can leak into groundwater from the well shaft or migrate into groundwater in unexpected ways. Also, this is an output approach that encourages the production of hazardous wastes. In the United States, almost two-thirds of liquid hazardous wastes are injected into deep disposal wells. Many scientists believe that current regulations for deep-well disposal in the United States are inadequate and should be improved.

Surface impoundments are lined ponds, pits, or lagoons in which liquid hazardous wastes are stored (Figure 16-14). Sometimes they include liners to help contain the waste. As the water evaporates, the waste

Ken Sherman/Bruce Coleman USA, Inc.

Figure 16-14 This is a surface impoundment for storing liquid hazardous wastes located in Niagara Falls, New York (USA). Such sites can pollute the air and nearby groundwater and surface water.

Trade-Offs

Surface Impoundments

Advantages	Disadvantages
Low cost	Groundwater contamination from leaking liners (and overflow from flooding)
Wastes can often be retrieved	Air pollution from volatile organic compounds
Can store wastes indefinitely with secure double liners	Output approach that encourages waste production

Figure 16-15 Storing liquid hazardous wastes in surface impoundments has advantages and disadvantages. **Questions:** Which single advantage and which single disadvantage do you think are the most important? Why?

settles and becomes more concentrated. Figure 16-15 lists the advantages and disadvantages of this method.

EPA studies have found that 70% of the storage ponds in the United States have no liners and could threaten groundwater supplies. According to the EPA, eventually all impoundment liners are likely to leak and could contaminate groundwater.

There are some highly toxic materials (such as mercury; see Chapter 14, Science Focus, p. 358) that we cannot destroy, detoxify, or safely bury. The best way to deal with such materials is to prevent or reduce their use and to put what is produced in metal drums or other containers. These containers are placed above ground in specially designed storage buildings or underground in salt mines or bedrock caverns where they can be inspected on a regular basis and retrieved if necessary.

Sometimes both liquid and solid hazardous wastes are put into drums or other containers and buried in carefully designed and monitored *secure hazardous waste landfills* (Figure 16-16). This is the least-used method because of the expense involved.

Figure 16-17 lists some ways in which you can reduce your output of hazardous waste—the first step in dealing with it. GOOD NEWS

■ CASE STUDY

Hazardous Waste Regulation in the United States

About 5% of all hazardous waste produced in the United States is regulated under the Resource Conservation and Recovery Act (RCRA, pronounced "RICK-ra"), passed by the U.S. Congress in 1976 and amended in 1984. Under this act, the EPA sets standards for the

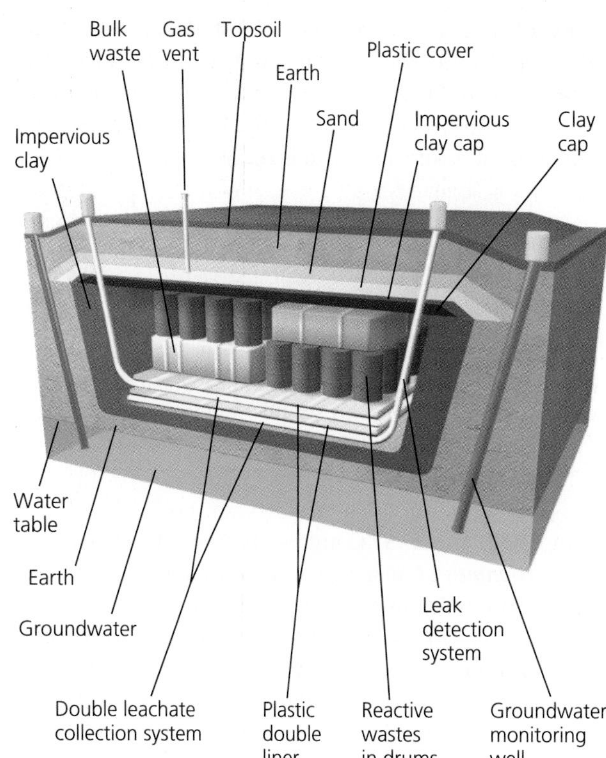

Figure 16-16 **Solutions:** This diagram shows how hazardous wastes can be isolated and stored in a secure hazardous waste landfill.

management of several types of hazardous waste and issues permits to companies that allow them to produce and dispose of a certain amount of those wastes by approved methods. Permit holders must use a *cradle-to-grave* system to keep track of waste they transfer from a point of generation (cradle) to an approved off-site disposal facility (grave), and they must submit proof of this disposal to the EPA.

What Can You Do?

Hazardous Waste

- Avoid using pesticides and other hazardous chemicals, or use them in the smallest amounts possible
- Use less harmful and usually cheaper substances instead of commercial chemicals for most household cleaners. For example, use vinegar to polish metals, clean surfaces, and remove stains and mildew; baking soda to clean utensils and to deodorize and remove stains; and borax to remove stains and mildew.
- Do not dispose of pesticides, paints, solvents, oil, antifreeze, or other hazardous chemicals by flushing them down the toilet, pouring them down the drain, burying them, throwing them into the garbage, or dumping them down storm drains. Instead, use hazardous waste disposal services available in many cities.

Figure 16-17 **Individuals matter:** You can reduce your output of hazardous wastes (**Concept 16-5**). **Questions:** Which two of these measures do you think are the most important? Why?

RCRA is a good start, but about 95% of the hazardous and toxic wastes, including e-waste, produced in the United States is not regulated. In most other countries, especially less-developed countries, an even smaller amount of this waste is regulated.

THINKING ABOUT
Hazardous Waste

CORE CASE STUDY

Why is it that 95% of the hazardous waste, including the growing amounts of e-waste (**Core Case Study**) produced in the United States, is not regulated? Do you favor regulating such wastes? What do you think would be the economic consequences of doing so? How would this change the way producers of hazardous waste deal with it?

In 1980, the U.S. Congress passed the *Comprehensive Environmental Response, Compensation, and Liability Act,* commonly known as the *CERCLA* or *Superfund* program, supervised by the EPA. Its goals are to identify sites, commonly called Superfund sites, where hazardous wastes have contaminated the environment (Figure 16-18) and to clean them up on a priority basis. The worst sites—those that represent an immediate and severe threat to human health—are put on a *National Priorities List* and scheduled for cleanup using EPA-approved methods.

In 2010, there were about almost 1,300 sites on this list, and more than 1,300 sites had been cleaned up and removed from the list. The Waste Management Research Institute estimates that at least 10,000 sites should be on the priority list and that cleanup of these sites would cost about $1.7 trillion, not including legal fees. This is a glaring example of the economic and environmental value of emphasizing waste reduction and pollution prevention over the *end-of-pipe* cleanup approach that most countries rely on.

In 1984, Congress amended the Superfund Act to give citizens the right to know what toxic chemicals are being stored or released in their communities. This required 23,800 large manufacturing facilities to report their annual releases into the environment of any of nearly 650 toxic chemicals. If you live in the United States, you can find out what toxic chemicals are being stored and released in your neighborhood by going to the EPA's *Toxic Release Inventory* website.

The Superfund Act, designed to make polluters pay for cleaning up abandoned hazardous waste sites, greatly reduced the number of illegal dumpsites (Figure 16-18) around the country. It also forced waste producers who were fearful of liability claims to reduce their production of such waste and to recycle or reuse much more of it. However, under pressure from polluters, the U.S. Congress refused to renew the tax on oil and chemical companies that had financed the Superfund legislation after it expired in 1995. The Superfund is now broke, and taxpayers, not polluters, are footing the bill for cleanups when the responsible parties cannot be found. As a result, the pace of cleanup has slowed.

The U.S. Congress and several state legislatures have also passed laws that encourage the cleanup of *brownfields*—abandoned industrial and commercial sites such as factories, junkyards, older landfills, and gas stations. In most cases, they are contaminated with hazardous wastes. Brownfields can be cleaned up and reborn as parks, nature reserves, athletic fields, ecoindustrial parks (see the Case Study at the end of this chapter), and residential neighborhoods. By 2009, more than 42,000 former brownfield sites had been redevel-

GOOD NEWS

Figure 16-18 These leaking barrels of toxic waste were found at a Superfund site in the United States that has since been cleaned up.

Phototake, Inc.

oped in the United States, according to the EPA. However, the U.S. General Accounting Office reports there are as many as 500,000 more brownfield sites throughout the United States.

Various other federal and state laws have done much to deal with hazardous waste on a prevention basis. One of the most successful was the 1976 law requiring that use of leaded gasoline be phased out completely in the United States by 1986 (see the following Case Study).

■ CASE STUDY

Lead Is a Highly Toxic Pollutant

Lead (Pb) is a pollutant that is found in air, water, soil, plants, and animals. Because it is a chemical element, lead does not break down in the environment. This indestructible and potent neurotoxin can harm the nervous system, especially in young children. Every year, 12,000–16,000 American children under age 9 are treated for acute lead poisoning, and about 200 die, according to the U.S. Centers for Disease Control and Prevention (CDC). About 30% of the survivors suffer from palsy, partial paralysis, blindness, and mental retardation.

According to the CDC, unborn fetuses and children under age 6, even with only low blood levels of lead, are especially vulnerable to nervous system impairment, lowered IQ (by an average of 7.4 points), shortened attention span, hyperactivity, hearing damage, and various behavior disorders. A 1993 study by the U.S. National Academy of Sciences, as well as numerous other studies since then, indicate that there is no safe level of lead in children's blood. As a result, health scientists call for sharply reducing the currently allowed levels for lead in the air and water.

GOOD NEWS

Data from the CDC indicates that between 1976 and 2007, the percentage of U.S. children between the ages of 1 and 5 with blood lead levels above the safety standard dropped from 85% to about 1%, which prevented at least 9 million childhood lead poisonings. The primary reason for this drop was that government regulations banned leaded gasoline in 1976 (with a complete phase-out by 1986) and greatly reduced the allowable levels of lead in paints in 1970 (even though illegal use continued until about 1978). This is an excellent example of the effectiveness of pollution prevention.

Despite this success, the CDC estimates that at least 310,000 U.S. children still have unsafe blood levels of lead caused by exposure to a number of sources. The major source is peeling lead-based paint found in about one in five (24 million) older houses built before 1960. Lead can also leach from water pipes and faucets containing lead parts or lead solder. Other sources are older coal-burning power plants that have been able to delay meeting the emission standards of new plants, as well as lead smelters and waste incinerators. In 2010, test results revealed that some reusable cloth bags had low levels of lead, mostly in their rigid bottom inserts and in the colored labels on the outsides of the bags. Most U.S. retailers no longer sell such bags.

Solutions

Lead Poisoning

Prevention

- Phase out leaded gasoline worldwide
- Phase out waste incineration
- Ban use of lead solder
- Ban use of lead in computer and TV monitors
- Ban lead glazing for ceramicware used to serve food
- Ban candles with lead cores
- Test blood for lead by age 1

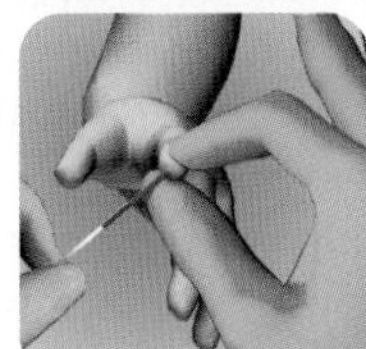

Control

- Replace lead pipes and plumbing fixtures containing lead solder
- Remove leaded paint and lead dust from older houses and apartments
- Sharply reduce lead emissions from incinerators
- Remove lead from TV sets and computer monitors before incineration or land disposal
- Test for lead in existing ceramicware used to serve food
- Test existing candles for lead
- Wash fresh fruits and vegetables

Figure 16-19 **Individuals matter:** There are several ways to help protect children from lead poisoning. **Questions:** Which two of these solutions do you think are the most important? Why?

Health scientists have proposed a number of ways to help protect children from lead poisoning, as listed in Figure 16-19. Although the threat from lead has been greatly reduced in the United States, this is not the case in many less-developed countries. About 80% of the gasoline sold in the world today is unleaded, but about 100 countries still use leaded gasoline and most countries have not banned lead-based paint.

The WHO estimates that 130–200 million children around the world are at risk of lead poisoning, and that 15–18 million children in less-developed countries have permanent brain damage because of lead poisoning. These children are acutely affected primarily because of the use of leaded gasoline and poorly regulated lead smelters and waste incinerators in their countries.

GOOD NEWS

Environmental and health scientists call for global bans on leaded gasoline and lead-based paints, and for much stricter controls on emissions from coal-burning power plants, smelters, and waste incinerators. China recently phased out leaded gasoline in less than 3 years, a process that took several decades in the United States.

16-6 How Can We Make the Transition to a More Sustainable Low-Waste Society?

▶ **CONCEPT 16-6 Shifting to a low-waste society requires individuals and businesses to reduce resource use and to reuse and recycle wastes at local, national, and global levels.**

Grassroots Action Has Led to Better Solid and Hazardous Waste Management

In the United States, individuals have organized grassroots (bottom-up) citizen movements to prevent the construction of hundreds of incinerators, landfills, treatment plants for hazardous and radioactive wastes, and polluting chemical plants in or near their communities. Health risks from incinerators and landfills, when averaged over the entire country, are quite low, but the risks for people living near such facilities are much higher.

GOOD NEWS

Manufacturers and waste industry officials point out that something must be done with the toxic and hazardous wastes created in the production of certain goods and services. They contend that even if local citizens adopt a "not in my back yard" (NIMBY) approach, the waste will always end up in someone's back yard.

Many citizens do not accept this argument. To them, the best way to deal with most toxic and hazardous waste is to produce much less of it, as suggested by the U.S. National Academy of Sciences (Figure 16-13). For such materials, they believe that the goal should be "not in anyone's back yard" (NIABY) or "not on planet Earth" (NOPE), which calls for drastically reducing production of such wastes by emphasizing pollution prevention and using the *precautionary principle* (see Chapter 8, p. 170).

Providing Environmental Justice for Everyone Is an Important Goal

Environmental justice is an ideal whereby every person is entitled to protection from environmental hazards regardless of race, gender, age, national origin, income, social class, or any political factor. (See the Guest Essay on this subject by Robert Bullard on the website for this chapter.)

Studies have shown that a lopsided share of polluting factories, hazardous waste dumps, incinerators, and landfills in the United States are located in communities populated mostly by African Americans, Asian Americans, Latinos, and Native Americans. Studies have also shown that, in general, toxic waste sites in white communities have been cleaned up faster and more completely than similar sites in African American and Latino communities have.

Such environmental discrimination in the United States and in other parts of the world has led to a growing grassroots approach to this problem that is known as the *environmental justice movement*. Supporters of this group have pressured governments, businesses, and environmental organizations to become aware of environmental injustice and to act to prevent it. They have made some progress toward their goals, but have a long way to go.

THINKING ABOUT
Environmental Injustice

Have you or anyone in your family ever been a victim of environmental injustice? If so, what happened? What would you do to help prevent such environmental injustice?

International Treaties Have Reduced Hazardous Waste

Environmental justice also applies at the international level. For decades, some more-developed countries had been shipping hazardous wastes to less-developed countries. However, since 1992, an international treaty known as the Basel Convention has been in effect. It bans participating countries from shipping hazardous waste (including e-waste; see **Core Case Study**) to or through other countries without their permission. In 1995, the treaty was amended to outlaw all transfers of hazardous wastes from industrial countries to less-developed countries. By 2010, this agreement had been signed by 175 countries and ratified (formally approved and implemented) by 172 countries. The United States, Afghanistan, and Haiti have signed but have not ratified the convention.

CORE CASE STUDY

This ban will help, but it will not wipe out the very profitable illegal shipping of hazardous wastes. Hazardous waste smugglers evade the laws by using an array of tactics, including bribes, false permits, and mislabeling of hazardous wastes as recyclable materials.

In 2000, delegates from 122 countries completed a global treaty known as the Stockholm Convention on Persistent Organic Pollutants (POPs). It regulates the use of 12 widely used persistent organic pollutants that can accumulate in the fatty tissues of humans and other animals that occupy high trophic levels in food webs. At such levels, pollutants can reach levels hundreds of thousands of times higher than their levels in the general environment (see Figure 8-13, p. 163). Because they persist in the environment, POPs can also be transported long distances by wind and water.

The original list of 12 chemicals, called the *dirty dozen*, includes DDT and 8 other chlorine-containing persistent pesticides, PCBs, dioxins, and furans. Using blood tests and statistical sampling, medical researchers at New York City's Mount Sinai School of Medicine (USA) found that it is likely that nearly every person on earth has detectable levels of POPs in their bodies. The long-term health effects of this involuntary global chemical experiment are largely unknown.

By 2009, 169 countries had signed a strengthened version of the POPs treaty that seeks to ban or phase out the use of these chemicals and to detoxify or isolate them. It allows 25 countries to continue using DDT to combat malaria until safer alternatives are found. The United States has not yet ratified this treaty.

Environmental scientists consider the POPs treaty to be an important milestone in international environmental law and pollution prevention because it uses the precautionary principle to manage and reduce the threats from toxic chemicals. The list of regulated POPs is expected to grow.

In 2000, the Swedish Parliament enacted a law that, by 2020, will ban all chemicals that are persistent in the environment and that can accumulate in living tissue. This law also requires industries to perform risk assessments on the chemicals they use and to show that these chemicals are safe to use, as opposed to requiring the government to show that they are dangerous. In other words, chemicals are assumed to be guilty until proven innocent—the reverse of the current policy in the United States and most other countries. There is strong opposition to this approach in the United States, especially from most of the industries that produce and use potentially dangerous chemicals.

We Can Make the Transition to Low-Waste Societies

According to physicist Albert Einstein, "A clever person solves a problem; a wise person avoids it." Many people are taking these words seriously. The governments of Norway, Austria, and the Netherlands have committed to reducing their resource waste by 75%. In a pilot study, residents of the U.S. city of East Hampton, New York, cut their solid waste production by 85%. Many school cafeterias, restaurants, national parks, and corporations are participating in a rapidly growing "zero waste" movement to reduce, reuse, and recycle in order to lower their waste outputs by up to 80%.

GOOD NEWS

To prevent pollution and reduce waste, many environmental scientists argue that we can make a transition to a low-waste society by understanding and following these key principles:

- Everything is connected.
- There is no *away*, as in *to throw away*, for the wastes we produce.
- Producers and polluters should pay for the wastes they produce.

We can mimic nature by reusing, recycling, composting, or exchanging most of the municipal solid wastes we produce (see the Case Study that follows).

■ CASE STUDY
Industrial Ecosystems: Copying Nature

An important goal for a more sustainable society is to make its industrial manufacturing processes cleaner and more sustainable by redesigning them to mimic how nature deals with wastes. In nature, according to the chemical cycling **principle of sustainability**, the waste outputs of one organism become the nutrient inputs of another organism, so that all of the earth's nutrients are endlessly recycled. This explains why there is essentially no waste in undisturbed ecosystems.

SUSTAINABILITY

One way for industries to mimic nature is to reuse or recycle most of the minerals and chemicals they use, instead of burying or burning them or shipping them somewhere. Another method that industries can use to mimic nature is to interact with each other through *resource exchange webs* in which the wastes of one manufacturer become the raw materials for another—similar to food webs in natural ecosystems (see Figure 3-11, p. 48).

This is exactly what is happening in Kalundborg, Denmark, where an electric power plant and nearby industries, farms, and homes are collaborating to save money and to reduce their outputs of waste and pollution within what is called an *ecoindustrial park*, or *industrial ecosystem*. They exchange waste outputs and convert them into resources, as shown in Figure 16-20 (p. 430). This cuts pollution and waste, and reduces the flow of nonrenewable mineral and energy resources through the local economy. **GREEN CAREER:** industrial ecology

These and other industrial forms of *biomimicry* provide many economic benefits for businesses. By encouraging recycling and pollution prevention, they reduce the costs of managing solid wastes, controlling pollution, and complying with pollution regulations. They also reduce a company's chances of being sued because of damages to people and the environment caused by their actions. In addition, companies improve the health and safety of workers by reducing their exposure to toxic and hazardous materials, thereby reducing company health insurance costs.

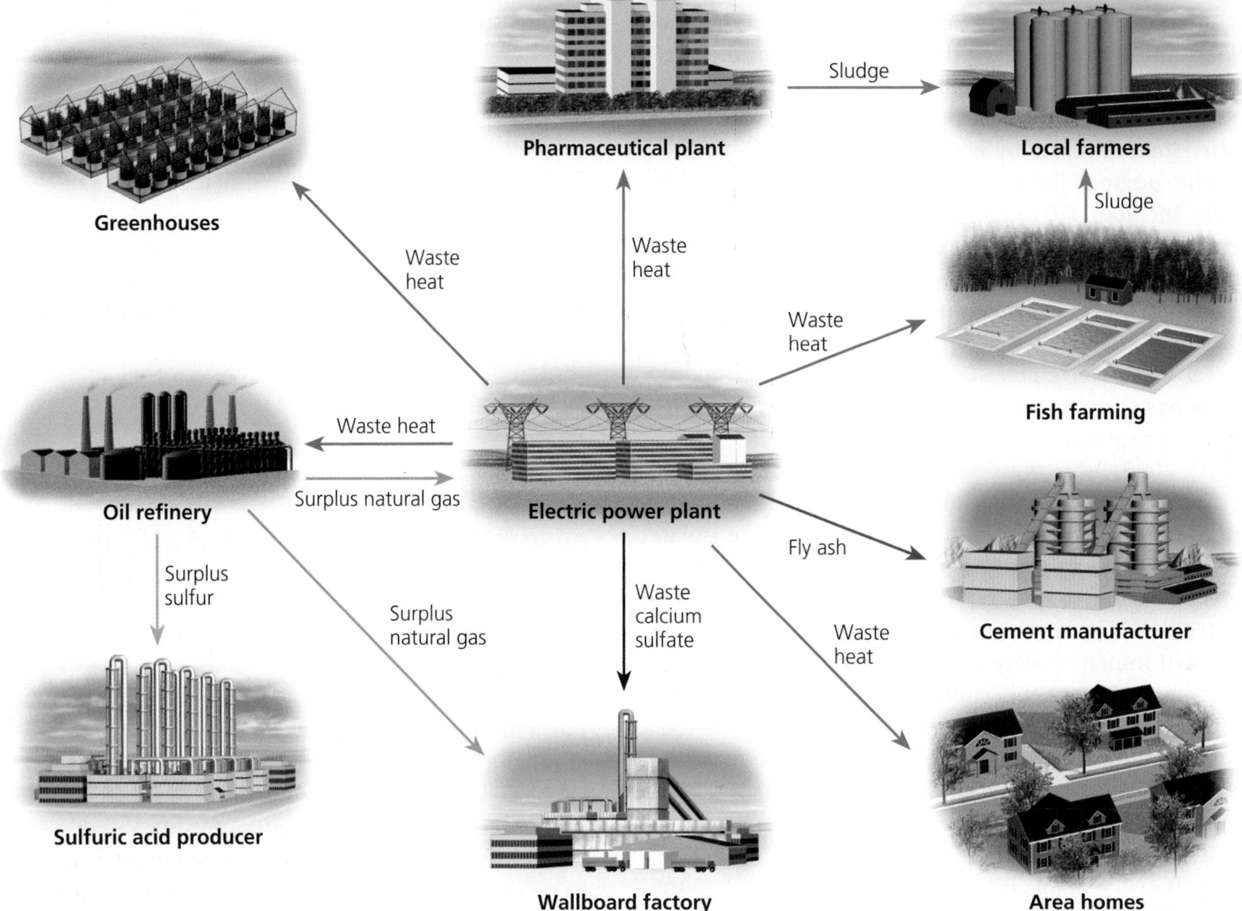

Figure 16-20 **Solutions:** This *industrial ecosystem* in Kalundborg, Denmark, reduces waste production by mimicking a natural ecosystem's food web. The wastes of one business become the raw materials for another, thus mimicking the way that nature recycles chemicals. **Question:** Is there an industrial ecosystem near where you live or go to school? If not, think about where and how such a system could be set up.

Biomimicry also encourages companies to come up with new, environmentally beneficial, and less resource-intensive chemicals, processes, and products that they can sell worldwide. In addition, these companies convey a better image to consumers based on actual results rather than public relations campaigns.

Here are this chapter's *three big ideas*:

- The order of priorities for dealing with solid waste should be to produce less of it, reuse and recycle as much of it as possible, and safely burn or bury of what is left.
- The order of priorities for dealing with hazardous waste should be to produce less of it, reuse or recycle it, convert it to less-hazardous material, and safely store what is left.
- We need to view solid wastes as wasted resources, and hazardous wastes as materials that we should not be producing in the first place.

REVISITING E-Waste and Sustainability

One of the problems of maintaining a high-waste society is the growing mass of e-waste (**Core Case Study**) and other types of solid and hazardous waste discussed in this chapter. The challenge is to make the transition from an unsustainable high-waste, throwaway economy to a more sustainable low-waste, reducing–reusing–recycling economy as soon as possible.

Such a transition will require applying the three **principles of sustainability**. We can reduce our outputs of solid and hazardous waste by relying much less on fossil fuels and nuclear power (which produces long-lived, hazardous radioactive wastes) while relying much more on renewable energy from the sun, wind, and flowing water. We can mimic nature's chemical cycling processes by reusing and recycling materials. Integrated waste management, which uses a diversity of approaches and emphasizes waste reduction and pollution prevention, is a way to mimic nature's use of biodiversity. Slowing the growth of the human population and of levels of resource use per person would also decrease the demand for materials that eventually become solid and hazardous wastes.

The key to addressing the challenge of toxics use and wastes rests on a fairly straightforward principle: harness the innovation and technical ingenuity that has characterized the chemicals industry from its beginning and channel these qualities in a new direction that seeks to detoxify our economy.

ANNE PLATT MCGINN

REVIEW

CORE CASE STUDY

1. Describe the problems associated with electronic waste (e-waste) (**Core Case Study**). CORE CASE STUDY

SECTION 16-1

2. What is the key concept for this section? Distinguish among **solid waste**, **industrial solid waste**, **municipal solid waste (MSW)**, and **hazardous (toxic) waste**, and give an example of each. Summarize the types and sources of solid waste in the United States and explain what happens to it.

SECTION 16-2

3. What is the key concept for this section? Distinguish among **waste management**, **waste reduction**, and **integrated waste management**. Summarize the priorities that prominent scientists believe we should use for dealing with solid waste. Distinguish among **reducing**, **reusing**, and **recycling** in dealing with the wastes we produce. List five ways in which industries and communities can reduce resource use, waste, and pollution.

SECTION 16-3

4. What is the key concept for this section? Explain why reusing and recycling materials are so important and give two examples of each. Explain the importance of using refillable containers and list five other ways to reuse various items. What is composting?

5. What are the major advantages and disadvantages of recycling? Distinguish between **primary (closed-loop) recycling** and **secondary recycling**. What are some benefits of composting? What are three factors that discourage recycling? What are three ways to encourage recycling and reuse? Explain how some plastics are being recycled. What are bioplastics?

SECTION 16-4

6. What is the key concept for this section? What are the major advantages and disadvantages of using incinerators to burn solid and hazardous waste? Distinguish between **sanitary landfills** and **open dumps**. What are the major advantages and disadvantages of burying solid waste in sanitary landfills?

SECTION 16-5

7. What is the key concept for this section? What are the priorities that scientists believe we should use in dealing with hazardous waste? Explain the connection between cell phones and African lowland gorillas. Summarize the problems involved in sending e-wastes to some less-developed countries for recycling. Describe three ways to detoxify hazardous wastes. What is bioremediation? What is phytoremediation? What are the major advantages and disadvantages of incinerating hazardous wastes? What are the major advantages and disadvantages of using a plasma arc torch to detoxify hazardous wastes?

8. What are the major advantages and disadvantages of disposing of liquid hazardous wastes in **(a)** deep underground wells and **(b)** surface impoundments? What is a secure hazardous waste landfill? List three ways to reduce your output of hazardous waste. Describe the regulation of hazardous waste in the United States under the Resource Conservation and Recovery Act and the Comprehensive Environmental Response, Compensation, and Liability (or Superfund) Act. What is a brownfield? Describe the effects of lead as a pollutant and how we can reduce our exposure to this chemical. Why is the reduction of lead pollution in the United States a good example of successful use of legislation to prevent pollution?

SECTION 16-6

9. What is the key concept for this section? How has grassroots action improved solid and hazardous waste management in the United States? What is **environmental justice** and how well has it been applied in locating and cleaning up hazardous waste sites in the United States? Describe regulation of hazardous wastes at the global level through the Basel Convention and the treaty to control persistent organic pollutants (POPs).

10. What are this chapter's *three big ideas*? Describe how we can deal with the growing problems of e-waste and other wastes (**Core Case Study**) by applying the three **principles of sustainability**. CORE CASE STUDY SUSTAINABILITY

Note: Key terms are in **bold** type.

CRITICAL THINKING

1. Do you think that manufacturers of computers, television sets, and other forms of e-waste (**Core Case Study**) should be required to take their products back at the end of their useful lives for repair, remanufacture, or recycling in a manner that is both environmentally responsible and that does not threaten the health of recycling workers? Explain. Would you be willing to pay more for these products to cover the costs of such a take-back program? If so, what percent more per purchase would you be willing to pay for electronic products?
2. Find three items you regularly use once and then throw away. Are there other reusable items that you could use in place of these disposable items? Compare the cost of using the disposable option for a year versus the cost of using the alternatives.
3. Use the second law of thermodynamics (see Chapter 2, p. 36) to explain why dilution is not always the solution to pollution from liquid hazardous wastes.
4. A company called Changing World Technologies has built a pilot plant to test a process it has developed for converting a mixture of computers, old tires, turkey bones and feathers, and other wastes into oil by mimicking and speeding up natural processes for converting biomass into oil. Explain why this recycling process, if it turns out to be technologically and economically feasible, could increase waste production.
5. Would you oppose having a hazardous waste landfill, waste treatment plant, deep-injection well, or incinerator in your community? For each of these facilities, explain your answer. If you oppose having such disposal facilities in your community, how do you believe the hazardous waste generated in your community should be managed?
6. How does your school dispose of its solid and hazardous wastes? Does it have a recycling program? How well does it work? Does it have a hazardous waste collection system? If so, what does it do with these wastes? List three ways to improve your school's waste reduction and management system.
7. List three ways in which you could apply **Concept 16-6** to making your lifestyle more environmentally sustainable.
8. Congratulations! You are in charge of the world. List the three most important components of your strategy for dealing with **(a)** solid waste and **(b)** hazardous waste.

DOING ENVIRONMENTAL SCIENCE

Collect the trash (excluding food waste) that you generate in a typical week. Measure its total weight and volume. Sort it into major categories such as paper, plastic, metal, and glass. Then weigh each category and calculate its percentage by weight of the total amount of trash that you have measured. What percentage by weight of this waste consists of materials that could be recycled? What percentage consists of materials for which you could have used a reusable substitute, such as a coffee mug instead of a disposable cup? What percentage by weight of the items could you have done without? Tally and compare your answers to these questions with those of your classmates. Together with your classmates, combine all the results and do the same analysis for the entire class. Use these results to estimate the same values for the entire student population at your school.

GLOBAL ENVIRONMENT WATCH EXERCISE

Use the *E-waste* topic portal to get the latest statistics and information on electronic waste (**Core Case Study**). Try to find statistics on how rapidly the world's production of e-waste is growing and how rapidly e-waste production is growing in the United States. Write a brief report on what the United States and one other country of your choice are doing to deal with this growing waste problem. Compare the two approaches in terms of how successful they are.

ECOLOGICAL FOOTPRINT ANALYSIS

The average daily municipal solid waste production per person in the United States in 2009 was 1.97 kilograms (4.34 pounds). Use the data in the figure below to get an idea of a typical annual MSW ecological footprint for each American by calculating the total weight in kilograms and pounds for each category generated during a year (1 kilogram = 2.20 pounds).

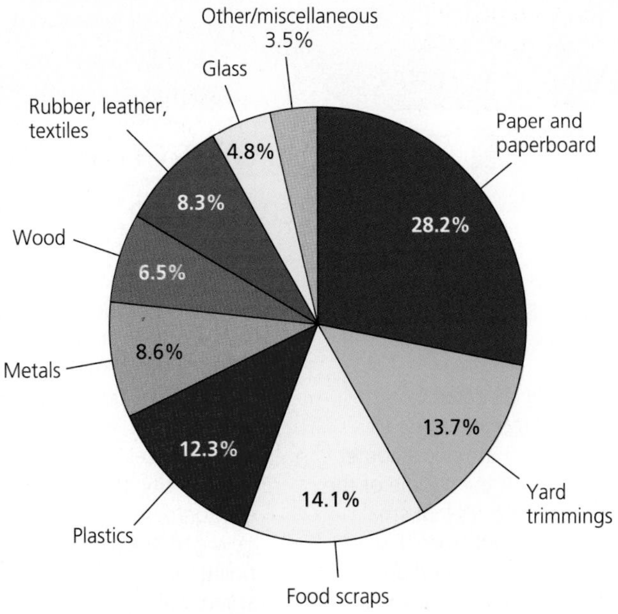

This pie chart shows a typical composition of U.S. municipal solid waste (MSW) in 2009. (Data from U.S. Environmental Protection Agency, 2010)

LEARNING ONLINE

The website for *Environmental Science 14e* contains helpful study tools and other resources to take the learning experience beyond the textbook. Examples include flashcards, practice quizzing, information on green careers, tips for more sustainable living, activities, and a wide range of articles and further readings. To access these materials and other companion resources for this text, please visit **www.cengagebrain.com**. For further details, see the preface, p. xvi.

17 Environmental Economics, Politics, and Worldviews

CORE CASE STUDY The Environmental Transformation of Chattanooga, Tennessee

Local officials, business leaders, and citizens have worked together to transform Chattanooga, Tennessee, from a highly polluted city to one of the most sustainable and livable cities in the United States (Figure 17-1).

During the 1960s, U.S. government officials rated Chattanooga as one of the dirtiest cities in the United States. Its air was so polluted by smoke from its industries that people sometimes had to turn on their vehicle headlights in the middle of the day. The Tennessee River, flowing through the city's industrial center, bubbled with toxic waste. People and industries fled the downtown area and left a wasteland of abandoned and polluting factories, boarded-up buildings, high unemployment, and crime.

In 1984, the city decided to get serious about improving its environmental quality. Civic leaders started a *Vision 2000* process with a 20-week series of community meetings in which more than 1,700 citizens from all walks of life gathered to build a consensus about what the city could be at the turn of the century. Citizens identified the city's main problems, set goals, and brainstormed thousands of ideas for solutions.

By 1995, Chattanooga had met most of its original goals. The city had encouraged zero-emission industries to locate there and replaced its diesel buses with a fleet of quiet, zero-emission electric buses, made by a new local firm.

The city also launched an innovative recycling program after environmentally concerned citizens blocked construction of a new garbage incinerator that would have emitted harmful air pollutants. These efforts paid off. Since 1989, the levels of the seven major air pollutants in Chattanooga have been lower than the levels required by federal standards.

Another project involved renovating much of the city's low-income housing and building new low-income rental units. Chattanooga also built the nation's largest freshwater aquarium, which became the centerpiece for downtown renewal. The city developed a riverfront park along both banks of the Tennessee River, which runs through downtown. The park typically draws more than a million visitors per year. As property values and living conditions have improved, people and businesses have moved back downtown.

In 1993, the community began the process again in *Revision 2000*. Goals included transforming an abandoned and blighted area in South Chattanooga into a mixed community of residences, retail stores, and zero-emission industries where employees can live near their workplaces. By 2009, most of these goals were met.

Chattanooga's environmental success story is a shining example of people working together to produce a more livable and sustainable city. It shows how people motivated by ethical and economic concerns can use economic and political tools to develop workable solutions to serious environmental problems. This chapter focuses on how we can use economics, politics, and ethics to address environmental problems.

Chattanooga Area Convention and Visitors Bureau

Figure 17-1 Since 1984, citizens have worked together to make the city of Chattanooga, Tennessee, one of the best and most sustainable places to live in the United States.

Key Questions and Concepts

17-1 How are economic systems related to the biosphere?

CONCEPT 17-1 Ecological economists and most sustainability experts regard human economic systems as subsystems of the biosphere.

17-2 How can we use economic tools to deal with environmental problems?

CONCEPT 17-2 We can use resources more sustainably by including harmful environmental and health costs in the market prices of goods and services (*full-cost pricing*), by subsidizing environmentally beneficial goods and services, and by taxing pollution and waste instead of wages and profits.

17-3 How can we implement more sustainable and just environmental policies?

CONCEPT 17-3 Individuals can work together to become part of political processes that influence how environmental policies are made and implemented.

17-4 What are some major environmental worldviews?

CONCEPT 17-4 Major environmental worldviews differ on which is more important—human needs and wants, or the overall health of ecosystems and the biosphere.

17-5 How can we live more sustainably?

CONCEPT 17-5 We can live more sustainably by becoming environmentally literate, learning from nature, living more simply and lightly on the earth, and becoming active environmental citizens.

Note: Supplements 2 (p. S3), 3 (p. S6), 6 (p. S22), and 7 (p. S38) can be used with this chapter.

We did not weave the web of life; we are merely strands within it. Whatever we do to the web, we do to ourselves.

CHIEF SEATTLE

17-1 How Are Economic Systems Related to the Biosphere?

▶ **CONCEPT 17-1 Ecological economists and most sustainability experts regard human economic systems as subsystems of the biosphere.**

Not All Market Systems are Free-Market Systems

Economics is a social science that deals with the production, distribution, and consumption of goods and services to satisfy people's needs and wants. In a market-based economic system, buyers and sellers interact competitively to make economic decisions about how goods and services are produced, distributed, and consumed. In a truly *free-market* economic system, all economic decisions are governed solely by the competitive interactions of *supply* (the amount of a good or service that is available), *demand* (the amount of a good or service that people want), and *price* (the market value of a good or service). If the demand for a good or service is greater than the supply, its price rises, and when supply exceeds demand, the price falls.

Also, in a *truly* free-market economic system **(1)** no company or small group of companies would control the prices of any goods or services; **(2)** the market prices of goods and services would include all of their direct and indirect costs (full-cost pricing); and **(3)** consumers would have full information about the beneficial and harmful environmental and health effects of the goods and services available to them.

However, this rarely happens in today's capitalist market systems. The primary goal of any business is to make as large a profit as possible for its owners or stockholders. To increase their profits, most businesses try to take business away from their competitors and

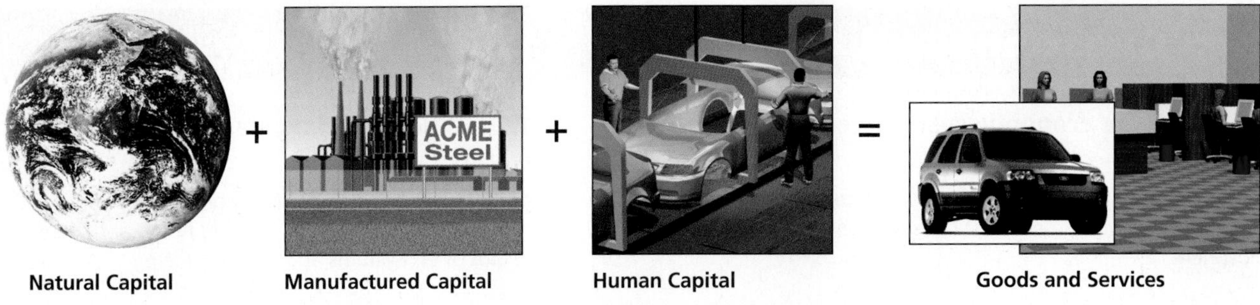

Figure 17-2 Most economic systems use three types of resources to produce goods and services.

to exert as much control as possible over the prices of the goods and services they provide. Many companies push for government support such as *subsidies*, or payments intended to help a business grow and thrive, tax breaks, trade barriers, or regulations that will give their products a market advantage over their competitors' products. When they succeed, the result is an uneven economic playing field that helps to undermine true free-market competition.

Some companies also try to withhold information from consumers about the costs and dangers their products may pose to human health or the environment, unless the government requires them to provide such information. Thus, buyers often do not get complete information about the harmful environmental impacts of the goods and services they buy. According to Paul Hawken, a successful entrepreneur who has founded several companies and a respected writer on business and the environment, "True free-market capitalism is a great idea. Maybe some day we will try it." (See the Guest Essay on natural capital by Paul Hawken at **www.cengagebrain.com**.)

Most economic systems use three types of capital, or resources, to produce goods and services (Figure 17-2). **Natural capital** (see Figure 1-3, p. 9) includes resources and services produced by the earth's natural processes, which support all economies and all life. **Human capital**, or **human resources**, include people's physical and mental talents that provide labor, organizational and management skills, and innovation. **Manufactured capital**, or **manufactured resources**, are tools and materials such as machinery, factories, and chemical compounds that people create using natural resources.

Economists Disagree over the Importance of Natural Capital and the Sustainability of Economic Growth

Economic growth for a city, state, country, or company is an increase in its capacity to provide goods and services to people. **Economic development** is the improvement of human living standards through economic growth. Today, a typical advanced industrialized country depends on a **high-throughput economy**, which attempts to boost economic growth by increasing the flow of natural matter and energy resources through the economic system to produce more goods and services (Figure 17-3).

For more than 200 years, there has been a debate over whether there are limits to economic growth. Now the debate is shifting to questions about what kinds of economic growth and development we should encourage.

Neoclassical economists, following the ideas of Alfred Marshall (1842–1924) and Milton Friedman (1912–2006), view the earth's natural capital as a subset, or part, of a human economic system. They assume that the potential for economic growth is essentially unlimited and is necessary for providing profits for businesses and jobs for workers. They also consider natural capital as important but not indispensable because they believe we can find substitutes for essentially any resource.

Ecological economists such as Herman Daly (see his Guest Essay at **www.cengagebrain.com**) and Robert Costanza disagree with this model. They point out that there are no substitutes for many vital natural resources such as clean air, clean water, fertile soil, and biodiversity, or for nature's free ecological or natural services such as climate control, air and water purification, pest control, and nutrient recycling (Figure 1-3, p. 9). In contrast to neoclassical economists, they view human economic systems as subsystems of the biosphere that

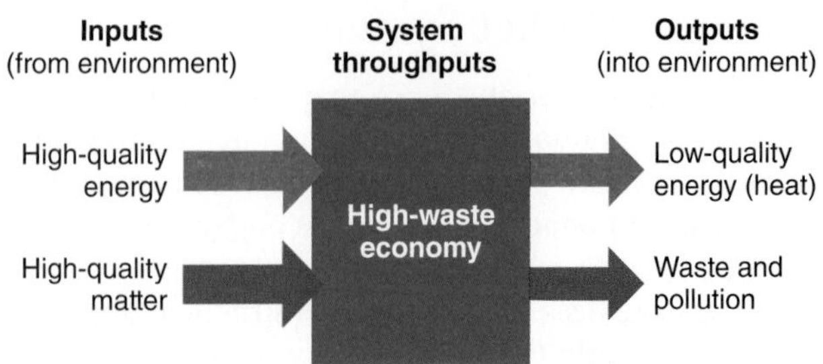

Active Figure 17-3 The *high-throughput economies* of most of the world's more-developed countries rely on continually increasing the flow of energy and matter resources to increase economic growth. *See an animation based on this figure at* www.cengagebrain.com. **Question:** What are three of your daily activities that regularly add to this throughput of matter and energy?

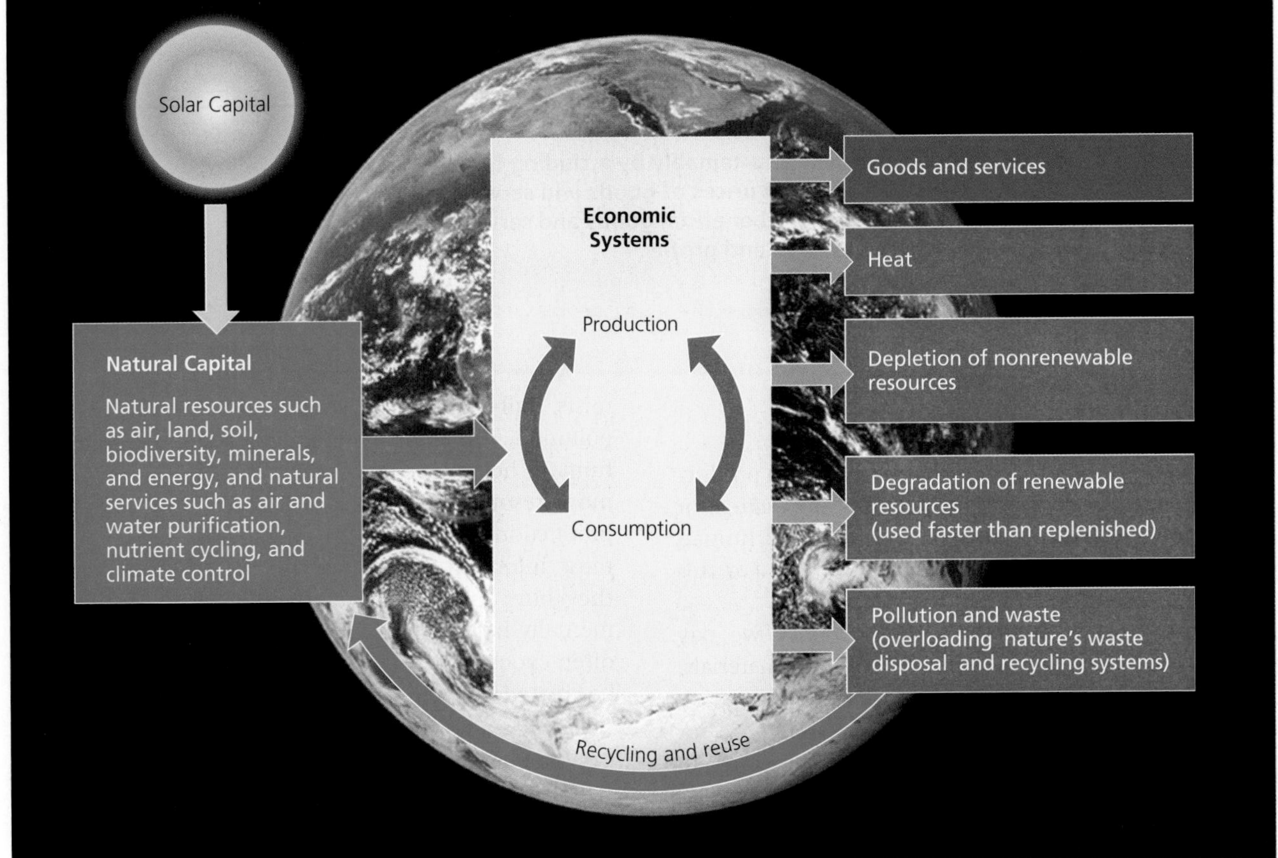

Active Figure 17-4 *Ecological economists* see all human economies as subsystems of the biosphere that depend on natural resources and services provided by the sun and earth. *See an animation based on this figure at* www.cengagebrain.com. **Question:** Do you agree or disagree with this model? Explain.

depend heavily on the earth's irreplaceable natural resources (**Concept 17-1**) (Figure 17-4).

Ecological economists also believe that conventional economic growth eventually will become unsustainable because it can deplete or degrade various irreplaceable forms of natural capital, on which all human economic systems depend, and because it will exceed the capacity of the environment to handle the pollutants and wastes we produce.

In general, the models of ecological economists are built on three major assumptions:

1. Resources are limited and we should not waste them, and there are no substitutes for most types of natural capital. In 2008, Achim Steiner, head of the UN Environment Programme (UNEP), said, "I believe the 21st century will be dominated by the concept of natural capital, just as the 20th century was dominated by financial capital."
2. We should encourage environmentally beneficial and sustainable forms of economic development, and discourage environmentally harmful and unsustainable forms of economic growth.
3. The harmful environmental and health effects of producing and using economic goods and services should be included in their market prices (*full-cost pricing*), so that consumers will have more accurate information about these effects.

Taking the middle ground between neoclassical economists and ecological economists are *environmental economists*. Many of them generally agree with the model proposed by ecological economists (Figure 17-4), and they argue that some forms of economic growth are not sustainable and should be discouraged. However, unlike ecological economists, they would accomplish this by fine-tuning existing economic systems and tools, rather than by replacing some of them with new or redesigned systems or tools.

THINKING ABOUT
Economic Growth

Do you think that the economy of the country where you live is sustainable or unsustainable? Explain.

We will now look at several strategies suggested by some ecological and environmental economists for making the transition to more sustainable eco-economies over the next several decades.

17-2 How Can We Use Economic Tools to Deal with Environmental Problems?

▶ **CONCEPT 17-2 We can use resources more sustainably by including harmful environmental and health costs in the market prices of goods and services (*full-cost pricing*), by subsidizing environmentally beneficial goods and services, and by taxing pollution and waste instead of wages and profits.**

Most Things Cost More Than We Might Think

The *market price*, or *direct price*, that we pay for a product or service usually does not include all of the *indirect*, or *external, costs* of harm to the environment and human health associated with its production and use. For this reason, such costs are often called *hidden costs*.

For example, if we buy a car, the price we pay includes the *direct*, or *internal, costs* of raw materials, labor, shipping, and a markup for dealer profit. In using the car, we pay additional direct costs for gasoline, maintenance, repairs, and insurance. However, in order to extract and process raw materials to make a car, manufacturers use energy and mineral resources, produce solid and hazardous wastes, disturb land, pollute the air and water, and release greenhouse gases into the atmosphere. These are the hidden external costs that can have short- or long-term harmful effects on us, on other people, on future generations, and on the earth's life-support systems.

Because these harmful external costs are not included in the market price of a car, most people do not connect them with car ownership. Still, the car buyer and other people in a society pay these hidden costs sooner or later in the forms of poorer health, higher costs of health care and insurance, higher taxes for pollution control, the hazards of traffic congestion, and the destruction of ecosystems that are replaced by highways and parking lots.

Many economists and environmental experts cite the failure to include the harmful environmental costs in the market prices of goods and services as one of the major causes of the environmental problems we face (Figure 1-10, p. 16). They argue for including such costs in order to make market prices reflect *full costs* as closely as possible, in keeping with one of the requirements of a true free-market economy. In other words, the prices would reflect the facts related to the harmful environmental and health effects of producing and using the goods and services that we buy. For example, the estimated costs of the harmful environmental and heath effects of using gasoline in the United States are about \$3 per liter (\$11.30 per gallon) (Chapter 13, p. 324). When we add these costs to a typical 2011 pump price of \$1.00 per liter (\$3.75 per gallon), the real cost of gasoline is about \$4 per liter (\$15 per gallon).

According to environmental and ecological economists, full-cost pricing would reduce resource waste, pollution, and environmental degradation and improve human health by encouraging producers to invent more resource-efficient and less-polluting methods of production. It would also enable consumers to make more informed decisions about the goods and services they buy. Jobs and profits would be lost in environmentally harmful businesses as consumers would more often choose green products, but jobs and profits would be created in environmentally beneficial businesses.

Such shifts in job markets and profits and changes in the structure of the marketplace are normal and important developments in creative market-based capitalism. This is why we replaced whale oil lamps with lightbulbs in the 1800s and are now gradually replacing inefficient incandescent lightbulbs with newer, more efficient bulbs.

If a shift to full-cost pricing were phased in over two decades, many environmentally harmful businesses would have time to transform themselves into profitable, environmentally beneficial businesses. Consumers would also have time to adjust their buying habits to favor more environmentally beneficial products and services.

Full-cost pricing seems to make a lot of sense. Why, then, is it not used more widely? *First*, most producers of harmful and wasteful products and services would have to charge more for them, and some would go out of business. Naturally, these producers oppose such pricing. *Second*, it is difficult to estimate many environmental and health costs, and to know how they might change in the future. But ecological and environmental economists argue that making the best possible estimates is far better than continuing with the current misleading and eventually unsustainable system, which excludes such costs.

Environmental Economic Indicators Could Help Us Reduce Our Environmental Impact

Economic growth is usually measured by the percentage of change in a country's **gross domestic product (GDP)**: the annual market value of all goods and services produced by all firms and organizations, foreign and domestic, operating within a country. Changes in a country's economic growth per person are measured by **per capita GDP**: the GDP divided by the country's total population at midyear.

GDP and per capita GDP indicators provide a standardized, useful method for measuring and comparing the economic outputs of nations. The GDP is deliberately designed to measure such outputs without distinguishing between goods and services that are environmentally or socially beneficial and those that are harmful. Nobel Prize–winning economist Simon Kuznets, who, with economist John Maynard Keynes developed the GDP indicator, warned that "GDP-based measures were never meant to be used as a measure of progress."

Environmental and ecological economists and environmental scientists call for the development and widespread use of new indicators—called *environmental indicators*—to help monitor environmental quality and human well-being. One such indicator is the **genuine progress indicator (GPI)**—the GDP plus the estimated value of beneficial transactions that meet basic needs but in which no money changes hands, minus the estimated harmful environmental, health, and social costs of all transactions. It incorporates 26 social and economic variables. The *per capita GPI* is the GPI for a country divided by that country's population at mid-year.

Redefining Progress, a nonprofit organization that develops economic and policy tools to help promote environmental sustainability developed the GPI in 1995. (This group also developed the concept of ecological footprints; see Figure 1-8, p. 14.)

Examples of beneficial transactions included in the GPI are unpaid volunteer work, health care provided by family members, childcare, and housework. Harmful costs that are subtracted to arrive at the GPI include pollution, resource depletion and degradation, and crime. The GPI is calculated as follows:

Genuine progress indicator	= GDP +	benefits not included in market transactions	−	harmful environmental and social costs

Figure 17-5 compares the per capita GDP with the per capita GPI for the United States between 1950 and 2004 (the latest data available). While the per capita GDP rose sharply over this period, the per capita GPI stayed flat, or in some cases even declined slightly. This shows that even if a nation's economy is growing, its people are not necessarily better off.

The GPI and other environmental indicators under development are far from perfect, which is true of the GDP as well. However, without environmental indicators, we would not know much about what is happening to people, to the environment, or to the planet's natural capital. And we would not have a way to evaluate policies and to determine which ones work best for improving environmental quality and life satisfaction. This is like navigating a ship through fog in iceberg-infested waters without radar.

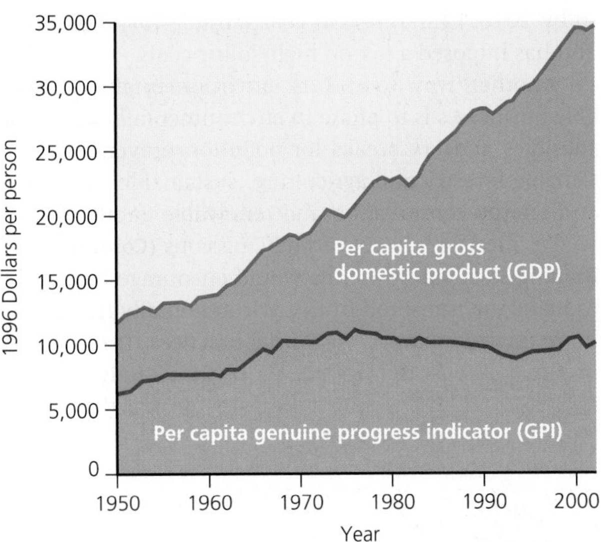

Figure 17-5 *Monitoring environmental progress:* This graph compares the per capita gross domestic product (GDP) and the per capita genuine progress indicator (GPI) in the United States between 1950 and 2004 (the latest year for which data is available). (Data from Redefining Progress, 2006) **Questions:** Would you favor making widespread use of this or similar green economic indicators? Why hasn't this been done?

We Can Reward Environmentally Sustainable Businesses

Governments can use several strategies, including subsidies, to encourage or force producers to work toward full-cost pricing. However, some subsidies enable businesses to operate in such a way that they do damage to the environment or to human health. According to environmental scientist Norman Myers, such *perverse* subsidies and tax breaks cost the world's governments (taxpayers) at least $2 trillion a year—an average $3.8 million a minute! (See Myer's Guest Essay on perverse subsidies at **www.cengagebrain.com**.)

Perverse subsidies and tax breaks can distort the economic playing field and create a huge economic incentive for unsustainable resource waste, depletion, and environmental degradation. Examples include depletion subsidies and tax breaks for extracting minerals and fossil fuels, cutting timber on public lands, irrigating with low-cost water, and overfishing commercially valuable aquatic species.

Many experts have called for phasing out such subsidies. However, the economically and politically powerful interests receiving them spend a lot of time and money *lobbying*, or trying to influence governments to continue and even increase their subsidies. They also lobby against subsidies and tax breaks for more environmentally beneficial competitors. When they are successful, their efforts undermine truly free-market competition.

GOOD NEWS

Some countries have begun reducing perverse subsidies. Japan, France, and Belgium have phased out all coal subsidies and Germany plans to do so by 2018. China has cut coal subsidies by about 73% and has imposed a tax on high-sulfur coals.

Another way to reward environmentally sustainable businesses is to phase in environmentally beneficial subsidies and tax breaks for pollution prevention, sustainable forestry and agriculture, sustainable water use, and energy conservation and renewable energy use, as well as measures to cut carbon emissions (**Concept 17-2**). Making such subsidy shifts would encourage businesses to make the transition from environmentally harmful to more environmentally beneficial practices. The U.S. city of Chattanooga, Tennessee, (**Core Case Study**) used such subsidies and tax breaks for exactly that purpose and succeeded. CORE CASE STUDY

THINKING ABOUT
Subsidies

Do you favor phasing out environmentally harmful government subsidies and tax breaks, and phasing in environmentally beneficial ones? Explain. What are three things you could you do to help bring this about? How might such subsidy shifting affect your lifestyle?

Tax Pollution and Wastes instead of Wages and Profits

Another way to discourage pollution and resource waste is to tax them. This would involve using *green taxes*, or *ecotaxes*, to help include many of the harmful environmental and health costs of production and consumption in market prices (**Concept 17-2**). Taxes could be levied on a per-unit basis on the amount of pollution and hazardous waste produced by a farm, business, or industry, and on the use of fossil fuels, nitrogen fertilizer, timber, minerals, water, and other resources. Figure 17-6 lists some of the advantages and disadvantages of using green taxes.

To many analysts, the tax system in most countries is backward. It *discourages* what we want more of—jobs, income, and profit-driven innovation—and *encourages* what we want less of—pollution, resource waste, and environmental degradation. A more environmentally sustainable economic and political system would *lower* taxes on labor, income, and wealth, and *raise* taxes on environmental activities that produce pollution, wastes, and environmental degradation. Some 2,500 economists, including eight Nobel Prize winners in economics, have endorsed this *tax-shifting* concept.

Proponents of green taxes point out three requirements for their successful implementation. *First,* green taxes would have to be phased in over 10–20 years to allow businesses to plan for the future. *Second,* income, payroll, or other taxes would have to be reduced by the same amount as the green tax so that there would be no net increase in taxes. *Third,* the poor and middle class would need a safety net to reduce the regressive nature of any new taxes on essentials such as fuel and food.

Trade-Offs

Environmental Taxes and Fees

Advantages	Disadvantages
Help bring about full-cost pricing	Low-income groups are penalized unless safety nets are provided
Encourage businesses to develop environmentally beneficial technologies and goods to save money	Hard to determine optimal level for taxes and fees
Easily administered by existing tax agencies	Governments may use money as general revenue instead of improving environmental quality and reducing taxes on income, payroll, and profits

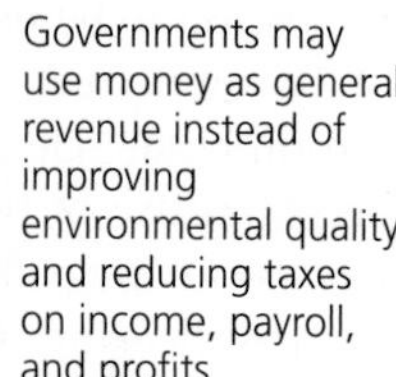

Figure 17-6 Using green taxes to help reduce pollution and resource waste has advantages and disadvantages. **Questions:** Which single advantage and which single disadvantage do you think are the most important? Why?

In Europe and the United States, polls indicate that once such tax shifting is explained to voters, 70% of them support the idea. Germany's green tax on fossil fuels, introduced in 1999, has reduced pollution and greenhouse gas emissions, helped to create up to 250,000 new jobs, lowered taxes on wages, and greatly increased use of renewable energy resources. Sweden, Denmark, Spain, and the Netherlands have raised taxes on several environmentally harmful activities while cutting taxes on income, wages, or both. GOOD NEWS

The U.S. Congress has not enacted green taxes, mostly because economically and politically powerful industries, including the automobile, fossil fuel, mining, and chemical industries, claim that such taxes will reduce their competitiveness and harm the economy by raising the prices of their goods and services, and causing a loss of jobs. In addition, most voters have been conditioned to oppose any new taxes and have not been educated about the economic and environmental benefits of tax shifting.

Environmental Laws and Regulations Can Discourage or Encourage Innovation

Environmental regulation is a form of government intervention in the marketplace that is widely used to help control or prevent pollution and to reduce resource waste and environmental degradation. It involves enacting and enforcing laws that set pollution standards, regulating harmful activities such as the release of toxic chemicals into the environment, and protecting certain irreplaceable or slowly replenished resources such as public forests from unsustainable use.

So far, most environmental regulation in the United States and in many other more-developed countries has involved passing laws that are enforced through a *command-and-control* approach. Critics say that this strategy can unnecessarily increase costs and discourage innovation, because many of these regulations concentrate on cleanup instead of prevention. Some regulations also set compliance deadlines that are too short to allow companies to find innovative solutions to pollution and excessive resource waste. They may also require the use of specific technologies where less costly but equally effective alternatives might be available or are in the process of being developed.

GOOD NEWS

A different approach favored by many economists as well as environmental and business leaders, is to use *incentive-based environmental regulations*. Rather than requiring all companies to follow the same fixed procedures or use the same technologies, this approach uses the economic forces of the marketplace to encourage businesses to be innovative in reducing pollution and resource waste. The experience of several European nations shows that *innovation-friendly environmental regulation* sets goals, frees industries to meet them in any way that works, and allows enough time for innovation. This can motivate companies to develop green products and industrial processes that create jobs, increase company profits, and make the companies more competitive in national and international markets. The U.S. city of Chattanooga, Tennessee, (**Core Case Study**) used this approach to help it become a more environmentally sustainable and livable city.

CORE CASE STUDY

Using the Marketplace to Reduce Pollution and Resource Waste

In one incentive-based regulation system, the government decides on acceptable levels of total pollution or resource use, sets limits, or *caps*, to maintain these levels, and gives or sells companies a certain number of *tradable pollution* or *resource-use permits* governed by the caps.

With this *cap-and-trade* approach, a permit holder not using its entire allocation can save credits for future expansion, use them in other parts of its operation, or sell them to other companies. The United States has used this *cap-and-trade* approach to reduce the emissions of sulfur dioxide (see Chapter 15, p. 386) and several other air pollutants. Tradable rights can also be established among countries to help them preserve biodiversity and reduce emissions of greenhouse gases and other regional and global pollutants.

GOOD NEWS

Figure 17-7 lists the advantages and disadvantages of using tradable pollution and resource-use permits.

Trade-Offs

Tradable Environmental Permits

Advantages	Disadvantages
Flexible	Big polluters and resource wasters can buy their way out
Easy to administer	May not reduce pollution at dirtiest plants
Encourage pollution prevention and waste reduction	Caps can be too high and not regularly reduced to promote progress
Permit prices determined by market transactions	Self-monitoring of emissions can allow cheating

Figure 17-7 Using tradable permits to reduce pollution and resource waste has advantages and disadvantages. **Questions:** Which two advantages and which two disadvantages do you think are the most important? Why?

The effectiveness of such programs depends on how high or low the initial cap is set and on the rate at which the cap is reduced to encourage further innovation.

Reducing Pollution and Resource Waste by Selling Services Instead of Things

In the mid-1980s, German chemist Michael Braungart and Swiss industry analyst Walter Stahel independently proposed a new economic model that would provide profits while greatly reducing resource use, pollution, and waste for a number of goods. Their idea for more sustainable economies focuses on shifting from the current *material-flow economy* (Figure 17-3) to a *service-flow economy*. Instead of buying many goods outright, customers lease or rent the *services* that such goods provide. In a service-flow economy, a manufacturer or service provider makes more money if its product uses the minimum amount of materials, lasts as long as possible, is energy efficient, produces as little pollution as possible in its production and use, and is easy to maintain, repair, reuse, or recycle.

Such an economic shift is under way in some businesses. Since 1992, Xerox has been leasing most of its copy machines as part of its mission to provide *document services* instead of selling photocopiers. When a customer's service contract expires, Xerox takes the machine back for reuse or remanufacture. It has a goal of sending no material to landfills or incinerators. To save money, Xerox designs machines to have the fewest possible parts, be energy efficient, and emit as little noise, heat, ozone, and chemical waste as possible. Canon, a Japanese manufacturer of optical imaging products, and Fiat, the Italian carmaker, are taking similar measures.

In Europe, Carrier has begun shifting from selling heating and air conditioning equipment to providing a service to customers that gives them the indoor temperatures they want. It makes higher profits by leasing and installing energy-efficient heating and air conditioning equipment that is durable and easy to rebuild or recycle. Carrier also makes money helping clients to save energy by adding insulation, eliminating heat losses, and boosting energy efficiency in their offices and homes.

Reducing Poverty Can Help Us to Deal with Environmental Problems

Poverty is defined as the inability to meet one's basic economic needs. According to the World Bank and the United Nations, 1.4 billion people—a number greater than the entire population of China and 4.5 times the size of the U.S. population—struggle to survive on an income equivalent to less than $1.25 a day (Figure 17-8).

Poverty has numerous harmful health and environmental effects (Figure 1-13, p. 18, and Figure 14-13, p. 366) and has been identified as one of the four major causes of the environmental problems we face. Reducing poverty benefits individuals, economies, and the environment, and helps to slow population growth.

To reduce poverty and its harmful effects, governments, businesses, international lending agencies, and wealthy individuals in more-developed countries could:

- Mount a massive global effort to combat malnutrition and the infectious diseases that kill millions of people prematurely (Figure 14-13, p. 366).
- Provide primary school education for all children and for the world's nearly 800 million illiterate adults (a number that is 2.5 times the size of the U.S. population).
- Provide assistance to stabilize population growth in less-developed countries as soon as possible, mostly by investing in family planning, reducing poverty, and elevating the social and economic status of women.
- Sharply reduce the total and per capita ecological footprints of developed countries and rapidly developing countries such as China and India (Figure 1-8, p. 14) because of the threat from these growing footprints to the world's environmental and economic security.
- Make large investments in small-scale infrastructure such as solar-cell power facilities for rural villages (Figure 13-36, p. 332) and sustainable agriculture projects to help less-developed nations promote more energy-efficient eco-economies.
- Encourage lending agencies to make small loans to poor people who want to increase their income (Individuals Matter, at right).

Ron Giling/Peter Arnold, Inc.

Figure 17-8 This mother and her child live in poverty on the edges of Mumbai (formerly Bombay), India, a thriving urban center.

INDIVIDUALS MATTER

Muhammad Yunus—A Pioneer in Microlending

Most of the world's poor people want to work and earn enough to climb out of poverty. But few of them have credit records or assets that they could use for collateral to secure loans. With loans, they could buy what they would need to start farming or to start small businesses.

For almost 3 decades, an innovation called *microlending*, or *microfinance*, has helped a number of people living in poverty to deal with this problem. For example, since economist Muhammad Yunus (Figure 17-A) started the Grameen (Village) Bank in Bangladesh in 1983, it has provided microloans ranging from $10 to $1,000 to more than 7.6 million impoverished people in Bangladesh who do not qualify for loans at traditional banks. Women use about 97% of the loans to start small businesses, to plant crops, to buy small irrigation pumps, to buy cows and chickens for producing and selling milk and eggs, or to buy bicycles for transportation.

To promote loan repayment, the bank puts borrowers into groups of five. If a group member fails to make a weekly payment, other members must pay it. The average repayment rate on its microloans has been 95% or higher—much greater than the average repayment rate for loans by conventional banks. Typically, about half of Grameen's microborrowers move above the poverty line within 5 years of receiving their loans. One of the bank's goals is to provide small loans at reasonable interest rates to help protect borrowers from loan sharks who charge very high interest rates that can bankrupt individuals who are trying to work their way out of poverty.

Grameen Bank microloans are also being used to develop daycare centers, health-care clinics, reforestation projects, drinking water supply projects, literacy programs, and small-scale solar- and wind-power systems in rural villages.

AP Photo/John McConnico

Figure 17-A Economist Muhammad Yunus's work has helped millions of people to work their way out of poverty. This in turn has also helped to slow the country's population growth and to empower women.

In 2006, Yunus and his colleagues at the bank jointly won the Nobel Peace Prize for their pioneering use of microcredit. Banks based on the Grameen microcredit model have spread to 58 countries, including the United States. By 2009, this approach had helped at least 133 million people worldwide to work their way out of poverty—85% of them in Asia.

Unfortunately, some other banks and microlenders have damaged the public image of microlending by making loans at very high interest rates. Thus, the Grameen Bank, founded in part to put loan sharks out of business, has helped some loan sharks in unintended ways. However, Yunus and his supporters are still working to help people escape poverty and improve their lives.

The Millennium Development Goals Present Challenges

In 2000, the world's nations set goals—called Millennium Development Goals—for sharply reducing hunger and poverty, improving health care, achieving universal primary education, empowering women, and moving toward environmental sustainability by 2015. More-developed countries pledged to donate 0.7% of their annual national income to less-developed countries to help them in achieving these goals. But by 2009, only five countries—Denmark, Luxembourg, Sweden, Norway, and the Netherlands—had donated what they had promised.

In fact, the average amount donated in most years has been 0.25% of national income. The United States—the world's richest country—gives only 0.16% of its national income to help poor countries; Japan, another wealthy country, gives only 0.18% compared with the 0.9% given by Sweden. For any country, deciding whether or not to commit 0.7% of annual national income toward the Millennium Development Goals is an ethical issue that requires individuals and nations to evaluate their priorities (Figure 17-9, p. 444).

We Can Use Lessons from Nature to Shift to More Environmentally Sustainable Economies

In this chapter, we have seen a sharp contrast between the beliefs of neoclassical economists and ecological economists. In 2001, pioneering American environmental scientist Donella Meadows (1941–2001) contrasted these two views as follows:

- The first commandment of economics is: Grow. Grow forever. The first commandment of the earth is: Enough. Just so much and no more.

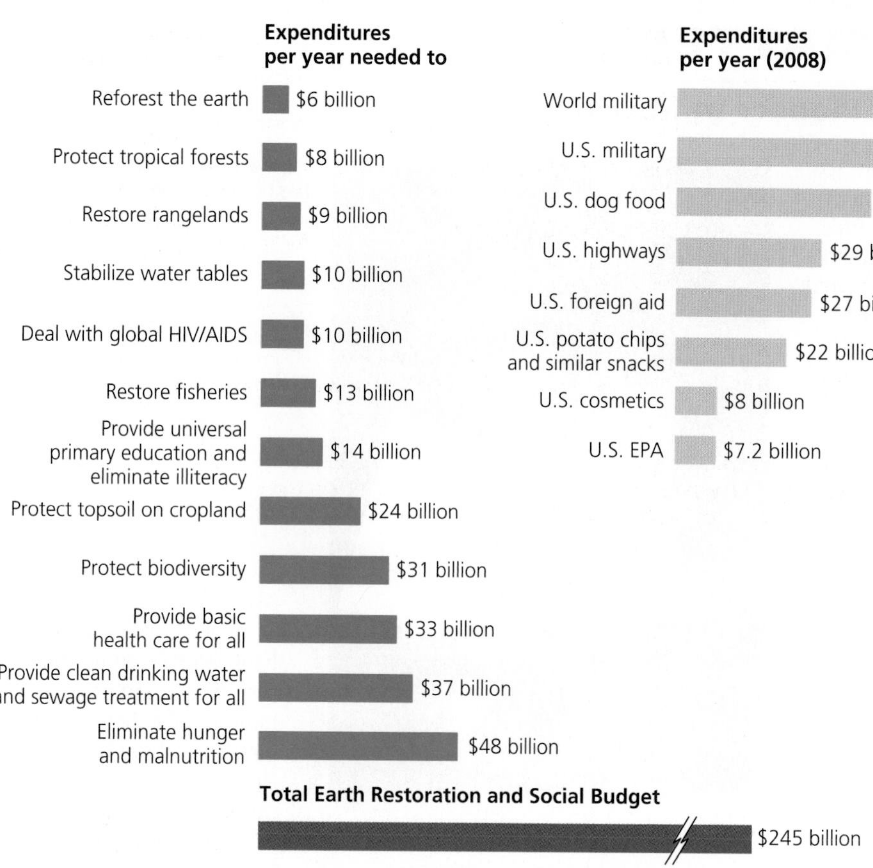

Figure 17-9 What should our priorities be? (Data from United Nations, World Health Organization, U.S. Department of Commerce, U.S. Office of Management and Budget, World Bank, Earth Policy Institute, and Stockholm International Peace Research Institute) **Question:** Looking at the right side of the figure, what three items would you reduce to help pay for solving some of the problems listed on the left side of the figure?

- Economics says: Compete. The earth says: Compete, yes, but keep your competition in bounds. Don't annihilate. Take only what you need. Leave your competitor enough to live. Wherever possible, don't compete, cooperate.
- Economics says: Use it up fast. Don't bother with repair; the sooner something wears out, the sooner you'll buy another. This makes the gross national product go round. Throw things out when you get tired of them. Get the oil out of the ground and burn it now. The earth says: What's the hurry? When something wears out, don't discard it; turn it into food for something else.
- Economics discounts the future. Take your profits from a resource such as a forest now. The earth says: Nonsense. Give to the future. Never take more in your generation than you give back to the next.
- The economic rule is: Do whatever makes sense in monetary terms. The earth says: Money measures nothing more than the relative power of some humans over other humans, and that power is puny compared with the power of the climate, the oceans, the uncounted multitudes of one-celled organisms that created the atmosphere, that recycle the waste, and that have lasted for 3 billion years. The fact that the economy, which has lasted about 200 years, puts zero values on these things means that the economy knows nothing about value—or about lasting.

The three scientific laws governing matter and energy changes (see Chapter 2, pp. 34 and 36) and the three **principles of sustainability** (see back cover) suggest that the best long-term solution to our environmental and resource problems is to shift away from a high-throughput (high-waste) economy based on ever-increasing matter and energy flow (Figure 17-3). We could then move toward a more sustainable **low-throughput (low-waste) economy**, a system based on energy flow and matter recycling, in which we could work with nature to reduce excessive throughput and the unnecessary waste of matter and energy resources (Figure 17-10). A growing number of economists and business leaders (Individuals Matter, at right) see such a transition as a great investment opportunity.

Environmental economists say we could make this transition by **(1)** reusing and recycling most nonrenewable matter resources; **(2)** using renewable resources no faster than natural processes can replenish them; **(3)** reducing resource waste by using matter and energy resources more efficiently; **(4)** reducing environmentally harmful forms of consumption; **(5)** emphasizing pollution prevention and waste reduction; and **(6)** slowing population growth to keep the number of matter and energy consumers growing slowly. Entrepreneur and environmental writer Paul Hawken suggests a simple golden rule for making such a shift: *Leave the world better than you found it, take no more than you need, try not to harm life or the environment, and make amends if you do.*

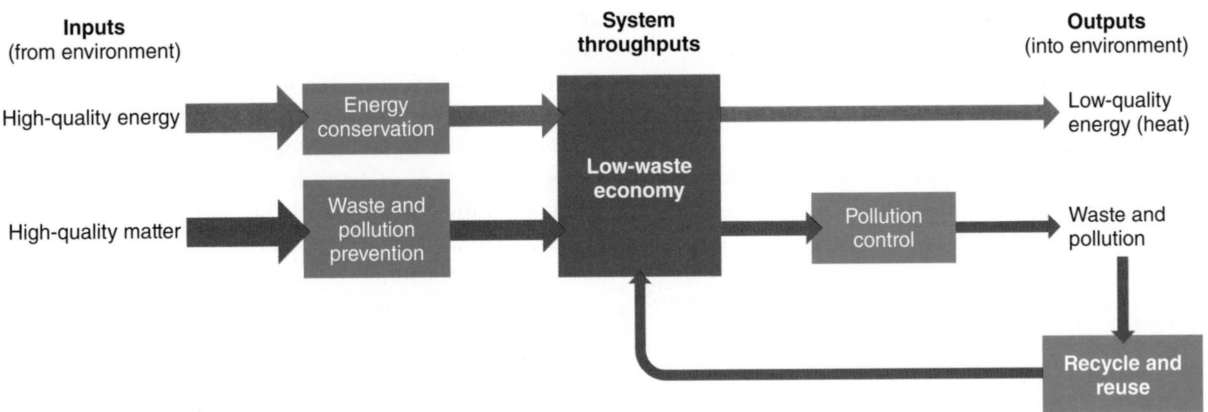

Active Figure 17-10 **Solutions:** Learning and applying lessons from nature can help us to design and manage more sustainable low-throughput economies. *See an animation based on this figure at* www.cengagebrain.com. **Question:** What are three ways in which your school could promote more sustainable economic and environmental practices?

INDIVIDUALS MATTER

Ray Anderson

Ray Anderson (Figure 17-B) was the founder of Interface, a company based in Atlanta, Georgia (USA). The company is the world's largest commercial manufacturer of carpet tiles, with 25 factories in seven countries, customers in 110 countries, and almost $900 million in annual sales in 2009.

Anderson, who died in 2011, was once called "the greenest CEO in America." He had changed the way he viewed the world—and his business—after reading Paul Hawken's book *The Ecology of Commerce*. In 1994, he announced plans to develop the nation's first totally sustainable green corporation. He then implemented hundreds of projects with goals of producing zero waste, greatly reducing energy use, reducing fossil fuel use, relying on solar energy, and copying nature. Three of his goals were to have the company put back more into nature than it took; to do good, not just to do no harm; and to influence other business leaders to adopt similar goals by providing a successful example.

Observing that there is very little waste in nature, Anderson began the transition of his business with a focus on sharply cutting waste in his company. Within 15 years of setting this and other environmental goals, Interface had decreased water usage by 74%, cut its net greenhouse gas emissions by 71%, reduced solid waste by 63%, reduced fossil fuel use by 60%, and lowered energy use by 44%. The company also gets 27% of its total energy and 88% of its electricity from renewable resources. These efforts have saved the company more than $405 million.

Figure 17-B Ray Anderson is one of the world's most respected and effective leaders in making businesses more sustainable.

Anderson also sent his carpet design team into the forest and told them to learn how nature would design floor covering. What the team discovered is nature's biodiversity sustainability principle. They observed that there were no regular repeating patterns on the forest floor. Instead of order, there was diversity and chaos. They came back and created a line of carpet tiles, no two of which had the same design. Within 18 months after it was introduced, this new product line was the company's top-selling design. Buyers like it because it seems to bring nature indoors. Similar product lines based on the diversity principle now make up 52% of the company's sales.

By implementing nature's biodiversity principle of sustainability, the company has also applied nature's chemical cycling, or no waste, sustainability principle. The tiles are made from recycled fibers and because they do not have to look alike and can be taken out of a box and laid down randomly, there is no need for large orders of one style, some of which are usually wasted. The tile factory also runs partly on solar energy, thus applying another of nature's sustainability principles. In addition, the company invented a new carpet recycling process that does not emit carbon dioxide to the atmosphere.

Anderson was one of a growing number of business leaders committed to finding more economically and ecologically sustainable, yet profitable, ways to do business. Anderson also created a consulting group as part of Interface to help other businesses start on the path toward being more sustainable. He believed that businesses can and should practice environmental stewardship and social responsibility in order to benefit their profit margins and the planet.

Several analysts have thought about specifically how to make a shift to more environmentally sustainable economies. Figure 17-11 summarizes suggestions by Hawken, Lester R. Brown, and other environmental and business leaders for using the economic tools discussed in this chapter to make this transition during the next several decades. Such strategies would apply the three **principles of sustainability** (see back cover).

Improving environmental quality and striving for environmental sustainability is now a major growth industry. It is creating profits and large numbers of new *green jobs,* which are devoted to improving environmental quality, developing cleaner and low-carbon energy resources, and promoting environmental sustainability.

Figure 17-12 might give you some ideas for a green career choice in the rapidly emerging eco-economy. Examples of such jobs include those devoted to developing more efficient batteries, retrofitting homes and other buildings with insulation and energy-efficient windows and more efficient heating and cooling systems, building a modern electric grid, and developing affordable low-carbon renewable energy resources such as solar power and wind power.

THINKING ABOUT
Green Careers

Are there any green careers listed in Figure 17-12 that interest you? For any of those that do, how could each of them help you to apply the three **principles of sustainability**?

Economics

- Reward (subsidize) environmentally sustainable economic development
- Penalize (tax and do not subsidize) environmentally harmful economic growth
- Shift taxes from wages and profits to pollution and waste
- Use full-cost pricing
- Sell more services instead of more things
- Do not deplete or degrade natural capital
- Live off income from natural capital
- Reduce poverty
- Use environmental indicators to measure progress
- Certify sustainable practices and products
- Use eco-labels on products

Resource Use and Pollution

- Cut resource use and waste by reducing, reusing, and recycling
- Improve energy efficiency
- Rely more on renewable solar, wind, and geothermal energy
- Shift from a nonrenewable carbon-based (fossil fuel) economy to a non-carbon renewable energy economy

Ecology and Population

- Mimic nature
- Preserve biodiversity
- Repair ecological damage
- Stabilize human population

Environmentally Sustainable Economy (Eco-Economy)

Figure 17-11 Solutions: We can use these strategies for shifting to more environmentally sustainable economies, or *eco-economies*, during this century. **Questions:** Which three of these solutions do you think are the most important? Why?

Environmentally Sustainable Businesses and Careers

Aquaculture	Environmental law
Biodiversity protection	Environmental nanotechnology
Biofuels	Fuel-cell technology
Climate change research	Geographic information systems (GIS)
Conservation biology	Geothermal geologist
Ecotourism management	Hydrogen energy
Energy-efficient product design	Hydrologist
Environmental chemistry	Marine science
Environmental design and architecture	Pollution prevention
Environmental economics	Recycling and reuse
Environmental education	Selling services in place of products
Environmental engineering	Solar cell technology
Environmental entrepreneur	Sustainable agriculture
Environmental health	Sustainable forestry
	Urban gardening
	Urban planning
	Waste reduction
	Water conservation
	Watershed hydrologist
	Wind energy

Figure 17-12 Green careers: These environmentally sustainable, eco-friendly businesses and careers are expected to flourish during this century, while environmentally harmful, or *sunset*, businesses are expected to decline. See the website for this book for more information on various environmental careers. **Question:** What are three of these careers and businesses that you think will be especially promising in the next 10 years? Explain.

17-3 How Can We Implement More Sustainable and Just Environmental Policies?

▶ **CONCEPT 17-3 Individuals can work together to become part of political processes that influence how environmental policies are made and implemented.**

Dealing with Environmental Problems in Democracies Is Not Easy

The roles played by a government are determined largely by its policies—the set of laws and regulations it enforces and the programs it funds. **Politics** is the process by which individuals and groups try to influence or control the policies and actions of governments at local, state, national, and international levels. One important application of this process is the development of **environmental policy**—environmental laws and regulations that are designed, implemented, and enforced, and environmental programs that are funded by one or more government agencies.

Representative democracy is government by the people through elected officials and representatives. In a *constitutional democracy*, a constitution (a document recording the rights of citizens and the laws by which a government functions) provides the basis of government authority and, in most cases, limits government power by mandating free elections and guaranteeing the right of free speech.

Political institutions in most constitutional democracies are designed to allow gradual change that ensures economic and political stability. In the United States, for example, rapid and destabilizing change is curbed by a system of checks and balances that distributes power among three branches of government—*legislative, executive*, and *judicial*—and among federal, state, and local governments.

The major function of government in democratic countries is to develop and implement policies for dealing with various issues. The first step in this complex process is to develop a policy and enact it into a law. The next step involves getting enough funds set aside by an elected legislative body to implement and enforce the new law. Developing and adopting a budget can be one of the most important and controversial activities of the executive and legislative branches of democratic governments. Once a law has been passed and funded, the appropriate government department or agency must draw up regulations or rules for implementing it.

In passing laws, developing budgets, and formulating regulations, elected and appointed government officials must deal with pressure from many competing *special-interest groups*. Each of these groups advocates passing laws, providing subsidies or tax breaks, or establishing regulations favorable to its cause, and weakening or repealing laws, subsidies, tax breaks, and regulations favorable to its opposition. Some special-interest groups such as corporations are *profit-making organizations*. Others are *nongovernmental organizations (NGOs)*, most of which are nonprofit, such as labor unions and environmental organizations.

Certain Principles Can Guide Us in Making Environmental Policy

Analysts suggest that when evaluating existing or proposed environmental policies, legislators and individuals should be guided by six principles designed to minimize environmental harm:

- *The reversibility principle:* Try not to make a decision that cannot be reversed later if the decision turns out to be wrong. For example, two essentially irreversible actions affecting the environment are the production of indestructible hazardous and toxic waste in coal-burning power plants (see Chapter 13, Case Study, p. 312), which we must try to store safely and essentially forever; and production of deadly radioactive wastes through the nuclear power fuel cycle, which must be store safely for 10,000–240,000 years (see Chapter 13, p. 316). A possible third such irreversible action in the making is the capturing and storing of carbon dioxide underground or under the ocean to help slow projected climate change, which commits us to trying to ensure that these deposits will never leak out (see Chapter 15, p. 400).
- *The net energy principle:* Do not encourage the widespread use of energy alternatives or technologies with low net energy yields (see Chapter 13, p. 300), which cannot compete in the open marketplace without government subsidies. Examples of energy alternatives with fairly low or negative net energy yields include nuclear power (including the nuclear fuel cycle), oil from tar sands, shale oil, ethanol from corn, biodiesel from soybeans, and hydrogen, as discussed in Chapter 13.
- *The precautionary principle:* When substantial evidence indicates that an activity threatens human health or the environment, we can take precautionary measures to prevent or reduce such harm, even if some of the cause-and-effect relationships are not well established scientifically.

- *The prevention principle:* Whenever possible, make decisions that help to prevent a problem from occurring or becoming worse.
- *The polluter-pays principle:* Develop regulations and use economic tools such as green taxes to ensure that polluters bear the costs of dealing with the pollutants and wastes they produce. This is an important way to include some of the harmful environmental and health effects of goods and services in their market prices (*full-cost pricing*).
- *The environmental justice principle:* Establish environmental policy so that no group of people bears an unfair share of the burden created by pollution, environmental degradation, or the execution of environmental laws. (See the Guest Essay on this subject by Robert D. Bullard at **www.cengagebrain.com**.)

These principles are useful ideals. However, implementing them always involves compromises and tradeoffs.

■ CASE STUDY
Managing Public Lands in the United States—Politics in Action

No nation has set aside as much of its land for public use, resource extraction, enjoyment, or wildlife habitat as has the United States. The federal government manages roughly 35% of the country's land, which belongs to every American. About three-fourths of this federal public land is in Alaska and another fifth is in the western states (Figure 17-13).

Some federal public lands are used for many purposes. For example, the *National Forest System* consists of 155 national forests and 22 national grasslands. These lands, managed by the U.S. Forest Service (USFS), are used for logging, mining, livestock grazing, farming, oil and gas extraction, recreation, and conservation of watershed, soil, and wildlife resources.

Figure 17-13 **Natural capital:** This map shows the national forests, parks, and wildlife refuges managed by the U.S. federal government. (Data from U.S. Geological Survey) **Questions:** Do you think U.S. citizens should jointly own more or less of the nation's land? Why or why not?

The Bureau of Land Management (BLM) oversees large areas of land—40% of all land managed by the federal government and 13% of the total U.S. land surface—mostly in the western states and Alaska. These lands are used primarily for mining, oil and gas extraction, and livestock grazing.

The U.S. Fish and Wildlife Service (USFWS) manages 553 *National Wildlife Refuges* (see Figure 8-17 p. 170). Most refuges protect habitats and breeding areas for waterfowl and big game to provide a harvestable supply for hunters. Permitted activities in most refuges include hunting, trapping, fishing, oil and gas development, mining, logging, grazing, some military activities, and farming.

The uses of some other public lands are more restricted. The *National Park System,* managed by the National Park Service (NPS), includes 58 major parks (Figure 17-14) and 335 national recreation areas, monuments, memorials, battlefields, historic sites, parkways, trails, rivers, seashores, and lakeshores. Only camping, hiking, sport fishing, and boating can take place in the national parks, whereas sport hunting, mining, and oil and gas drilling are allowed in national recreation areas.

The most restricted public lands are 756 roadless areas that make up the *National Wilderness Preservation System*. These areas lie within the other public lands and are managed by the agencies in charge of those surrounding lands. Most of these areas are open only for recreational activities such as hiking, sport fishing, camping, and non-motorized boating.

Many federal public lands contain valuable oil, natural gas, coal, geothermal, timber, and mineral resources (see Figure 15, p. S32, in Supplement 6). Since the 1800s, there has been intense controversy over how to use and manage the resources on these lands.

Most conservation biologists, environmental economists, and many free-market economists believe that four principles should govern use of public lands:

1. They should be used primarily for protecting biodiversity, wildlife habitats, and ecosystems.
2. No one should receive government subsidies or tax breaks for using or extracting resources on public lands.
3. The American people deserve fair compensation for the use of their property.
4. All users or extractors of resources on public lands should be fully responsible for any environmental damage they cause.

There is strong and effective opposition to these ideas. Developers, resource extractors, many economists, and many citizens tend to view public lands in terms of their usefulness in providing mineral, timber, and other resources and increasing short-term economic growth. They have succeeded in blocking implementation of the four principles listed above. For example, in recent years, analyses of budgets and appropriations reveal that the government has given an average of $1 billion a year—an average of $2.7 million a day—in subsidies and tax breaks to privately owned

karam/Shutterstock.com

Figure 17-14 Yosemite National Park in northern California (USA)—a vast, gorgeous valley in the High Sierra Mountains—is a symbol of the National Park System, which manages pristine areas such as this for their natural value and for use by future generations.

interests that use U.S. public lands for activities such as mining, fossil fuel extraction, logging, and grazing.

Some developers and resource extractors have sought to go further. Here are five of the proposals that such interests have made to get the U.S. Congress to open up more federal lands (owned jointly by all citizens) for development:

1. Sell public lands or their resources to corporations or individuals, usually at proposed prices that are less than market value, or turn over their management to state and local governments.
2. Slash federal funding for administration of regulations over public lands.
3. Cut diverse old-growth forests (Figure 9-2, p. 176) in the national forests for timber and for making biofuels, and replace them with simplified tree plantations to be harvested for the same purposes.
4. Open national parks, national wildlife refuges, and wilderness areas to oil drilling, mining, off-road vehicles, and commercial development.
5. Eliminate or take regulatory control away from the National Park Service and launch a 20-year construction program in the parks to build new concessions and theme parks that would be run by private firms.

Since 2002 there has been increased pressure to expand the extraction of mineral, timber, and fossil fuel resources on U.S. public lands and to weaken environmental laws and regulations protecting such lands from abuse and exploitation.

Although this Case Study has focused on the debate over the use of public lands in the United States, the same issues apply to the use of government or publicly owned lands in many other countries.

Individuals Can Influence Environmental Policy

A major theme of this book is that *individuals matter*. History shows that significant change usually comes from the *bottom up* when individuals join together to bring about change. Without previous bottom-up (grassroots) political action by millions of individual citizens and organized citizen groups (Figure 17-15), the air that millions of people breathe today and the water they drink would be much more polluted, and much more of the earth's biodiversity would have disappeared (**Concept 17-3**).

Figure 17-16 lists ways in which you can influence and change local, state, and national government policies in constitutional democracies. This is how the citizens and elected officials of the U.S. city of Chattanooga, Tennessee (**Core Case Study**), made their city more sustainable. CORE CASE STUDY

At a fundamental level, all politics is local. What we do to improve environmental quality in our own neighborhoods, schools, and work places also has national and global implications, much like the ripples spreading outward from a pebble dropped in a pond. This is the meaning of the slogan, "Think globally; act locally."

Angelo Doto-UNEP/Peter Arnold, Inc.

Figure 17-15 *Global outlook:* Children in Turin, Italy, wear gas masks as part of an organized protest against high levels of air pollution. **Question:** What environmental issue, if any, would lead you to participate in a demonstration related to this issue? Explain.

Environmental Leaders Can Make a Big Difference

Not only can we participate, but each of us can also provide environmental leadership in several different ways. First, we can *lead by example*, using our GOOD NEWS

What Can You Do?

Influencing Environmental Policy

- Become informed on issues
- Make your views known at public hearings
- Make your views known to elected representatives, and understand their positions on environmental issues
- Contribute money and time to candidates who support your views
- Vote
- Run for office (especially at local level)
- Form or join nongovernmental organizations (NGOs) seeking change
- Support reform of election campaign financing that reduces undue influence by corporations and wealthy individuals

Figure 17-16 Individuals matter: These are some ways in which you can influence environmental policy (**Concept 17-3**). **Questions:** Which three of these actions do you think are the most important? Which ones, if any, do you take?

own lifestyles and values to show others that change is possible and can be beneficial. For example, we can use fewer disposable products, eat food that has been sustainably produced, and walk, bike, or take mass transit to work or school. We can reuse and recycle many items, and we can reduce our matter and energy consumption by thinking more about whether we should acquire everything we want or limit more of our purchases to things we really need. In addition to setting a good example, we can also save money by doing many of these things.

Second, we can *work within existing economic and political systems to bring about environmental improvement* by campaigning and voting for informed and eco-friendly candidates, and by communicating with elected officials. We can also send a message to companies that we feel are harming the environment through their products or policies by *voting with our wallets*—not buying their products or services—and letting them know why. Another way to work within the system is to choose one of the many rapidly growing green careers highlighted throughout this book and described in Figure 17-12 and on the book's companion website.

Third, we can *run for some sort of local office*. Look in the mirror. Maybe you are one who can make a difference as an officeholder.

Fourth, we can *propose and work for better solutions to environmental problems*. Leadership is more than being against something. It also involves coming up with solutions to problems and persuading people to work together to achieve them. If we care enough, each of us can make a difference, as have Denis Hayes (Individuals Matter, below), Wangari Maathai (see Chapter 9 Core Case Study, p. 174), Muhammad Yunus (Individuals Matter, p. 443), and many of the citizens of Chattanooga, Tennessee (**Core Case Study**). CORE CASE STUDY

U.S. Environmental Laws and Regulations Have Been under Attack

Concerned citizens have persuaded the U.S. Congress to enact a number of important federal environmental and resource protection laws, most of them in the 1970s (Figure 17-17, p. 452, and Supplement 3, p. S6).

One of the most far-reaching of these laws, the National Environmental Policy Act of 1969 created the U.S. Environmental Protection Agency (EPA). The agency is required to establish and enforce regulations for implementing environmental laws passed by the U.S. Congress. Its critics often allege that EPA

INDIVIDUALS MATTER

Denis Hayes—A Practical Environmental Visionary

After studying ecology and political science in college, Denis Hayes (Figure 17-C) started traveling to see "what was actually going on" in the countries he had studied. After some on-the-ground experience, he decided to spend his life figuring out how human societies could benefit from organizing themselves around ecological principles.

At age 25, Hayes was enrolled in the Kennedy School of Government at Harvard University in Cambridge, Massachusetts (USA), at the same time that Senator Gaylord Nelson of the U.S. state of Wisconsin was organizing environmental teach-ins on college campuses. Hayes approached Nelson, offering to organize such an event at Harvard, and later the senator asked Hayes to organize an event for the whole country.

What started out as a plan for a number of teach-ins became Earth Day—April 22, 1970—considered by many analysts as the beginning of the modern environmental movement. That day involved teach-ins and much more—thousands of public demonstrations focused on pollution, toxic waste, nuclear power, coal mining, lead contamination, and other environmental issues. More than 20 million people took part in the first Earth Day. Later, Hayes worked on building the Earth Day Network, which now includes more than 180 nations. As a result, each year, Earth Day is now celebrated globally.

Courtesy of Denis Hayes/The Bullitt Foundation

Figure 17-C In 2008, the Audubon Society listed Denis Hayes as one of the 100 Environmental Heroes of the Twentieth Century.

Hayes has held a variety of important environmental positions. Since the first Earth Day, he has worked as an environmental lobbyist, a researcher for the Worldwatch Institute, the head of the U.S. Solar Energy Research Institute, a professor of energy and resource studies at the University of California at Santa Cruz (USA), and director of the Illinois State Energy Office.

Currently, he is president and CEO of The Bullitt Foundation of Seattle, Washington (USA). Hayes's goal is to use the resources of the foundation to help make the U.S. Pacific Northwest a global model for more sustainable development. Denis Hayes is a person who has made and is still making a difference.

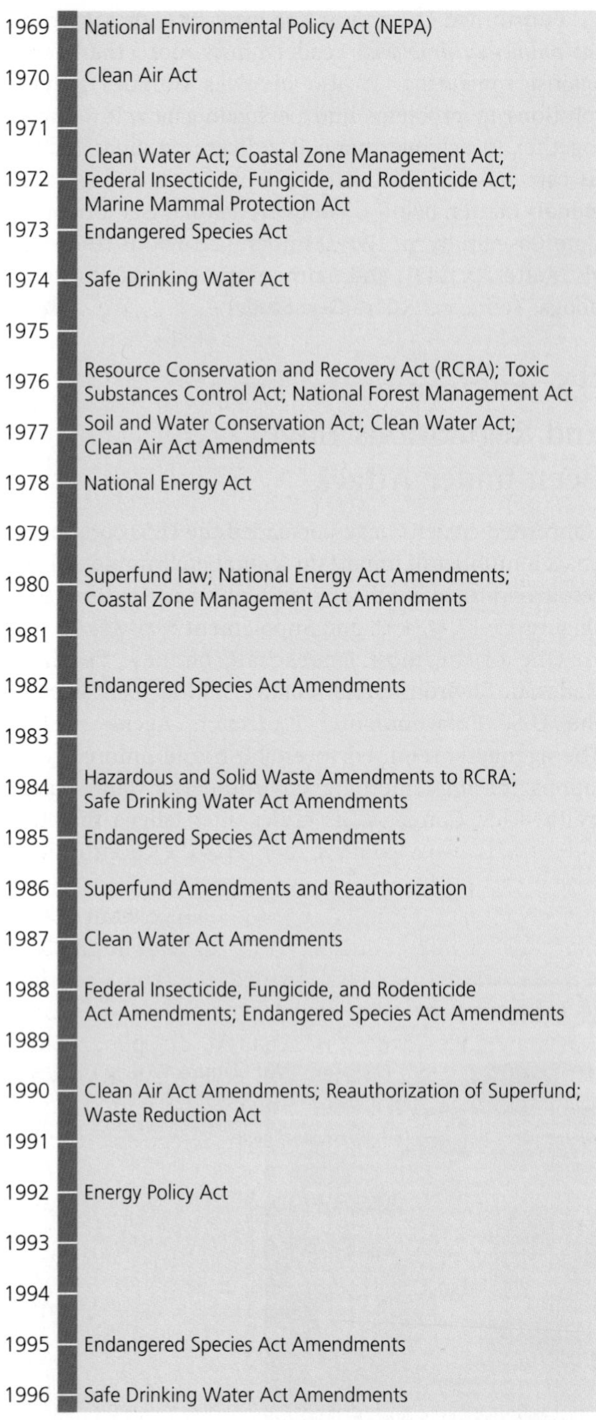

Figure 17-17 Some major environmental laws and their amended versions enacted in the United States since 1969. **Question:** Why do you think that since 1980, so few new U.S. environmental laws have been enacted and existing environmental laws have been under attack?

regulations are "job killers." But according to the U.S. Commerce Department, by 2010, EPA regulations used to implement Congressional laws had created a national industry that employed more than 1.5 million Americans, working to protect the country's environmental quality.

U.S. environmental laws have been highly effective, especially in controlling some forms of pollution. However, since 1980, a well-organized and well-funded movement has mounted a strong campaign to weaken or repeal existing U.S. environmental laws and regulations and to change the ways in which public lands are used (see Case Study, p. 448).

Three major groups are strongly opposed to various U.S. environmental laws and regulations: some corporate leaders and owners, and other powerful people who see such laws and regulations as threats to their profits, wealth, and power; citizens who see them as threats both to their private property rights and their jobs; and state and local government officials who resent having to implement federal laws and regulations with little or no federal funding, or who disagree with certain federal regulations. Since 2000, efforts to weaken most major U.S. environmental laws and regulations have escalated.

On the other hand, some concerned citizens have worked together to improve environmental quality in their local communities. The story of Chattanooga, Tennessee, (**Core Case Study**) shows how people on all sides of important environmental issues can listen to one another's concerns, try to find areas of agreement, and work together to find solutions to environmental problems.

Independent polls show that more than 80% of the U.S. public strongly support environmental laws and regulations and do not want them weakened. However, polls also show that less than 10% of the U.S. public (and in economic downturns only about 2–3%) considers the environment to be one of the nation's most pressing problems. As a result, environmental concerns often do not get transferred to the ballot box or the pocketbook.

Citizen Environmental Groups Play Important Roles

The spearheads of the global conservation, environmental, and environmental justice movements are the tens of thousands of nonprofit nongovernmental organizations (NGOs) working at the international, national, state, and local levels (**Concept 17-3**).

NGOs range in size from *grassroots* groups with just a few members to *mainline* organizations like the World Wildlife Fund (WWF), a 5-million-member global conservation organization, which operates in 100 countries. Expert lawyers, scientists, economists, lobbyists, and fund-raisers usually staff such groups. Other international groups with large memberships include Greenpeace, the Nature Conservancy, Conservation International, and the Grameen Bank (Individuals Matter, p. 443). In the United States, more than 8 million citizens belong to more than 30,000 NGOs that deal with environmental issues.

The largest environmental groups have become powerful and important forces within the U.S. political system. They have helped to persuade Congress to pass and strengthen environmental laws, and they fight attempts to weaken or repeal these laws.

The base of the environmental movement in the United States and throughout the world consists of thousands of grassroots citizens' groups organized to improve environmental quality, often at the local level. According to political analyst Konrad von Moltke, "There isn't a government in the world that would have done anything for the environment if it weren't for the citizen groups." Taken together, a loosely connected worldwide network of grassroots NGOs working for bottom-up political, social, economic, and environmental change can be viewed as an emerging citizen-based *global sustainability movement.*

GOOD NEWS

Most grassroots environmental groups use the nonviolent and nondestructive tactics of protest marches, tree sitting, lawsuits and other devices for generating publicity to help educate and encourage the public to oppose various environmentally harmful activities. Such tactics often work because they produce bad publicity for practices and businesses that threaten or degrade the environment.

Much more controversial are the few militant environmental groups that use violent means such as destroying bulldozers and SUVs and breaking into some types of research laboratories. Most environmentalists strongly oppose such tactics.

Students and Educational Institutions Can Play Important Environmental Roles

Since the mid-1980s, there has been a boom in environmental awareness on many college campuses and in public and private schools across the United States. Most student environmental groups work with members of their faculty and administration to bring about environmental improvements in their schools and local communities.

GOOD NEWS

Many of these groups make *environmental audits* of their campuses or schools. They gather data on practices affecting the environment and use them to propose changes that will make their campuses or schools more environmentally sustainable, usually while saving money in the process. Any school without a serious sustainability program is now viewed as outdated. Campus audits have resulted in numerous environmental improvements (see the Case Study that follows).

■ CASE STUDY
The Greening of American Campuses

Environmental audits performed by students at American colleges and universities have focused on implementing or improving recycling programs (now found at almost 80% of U.S. colleges and universities). They have also aimed for getting university food services to buy more food from local organic farms, shifting from fossil fuels to renewable energy, and improving energy efficiency. Students are directly helping their institutions to reduce dependence on cars, cut water waste, retrofit campus buildings to make them more energy efficient, and design more environmentally sustainable campus buildings.

In 2010, *Sierra Magazine* listed the nation's 100 greenest campuses. One such school was Oberlin College in Ohio, where students helped to design a more sustainable environmental studies building powered by solar panels, which produce 30% more electricity than the building uses. A living machine (see Chapter 11, p. 270) in the building's lobby purifies all of its wastewater, and half of the school's electricity comes from green sources. The school has a car-sharing program, and student activity fees subsidize public transportation.

At Northland College in Ashland, Wisconsin, students helped to design a green living and learning center (Figure 17-18) that houses 150 students and features a wind turbine, panels of solar cells, furniture made of recycled materials, and waterless (composting) toilets. Northland students voted to impose a *green fee* of $40 per semester on themselves to help finance the college's sustainability programs. Green Mountain College in Poultney, Vermont, gets its power from renewable biomass and biogas and, in 2011, was the country's first carbon-neutral school.

Dickinson College in Carlisle Pennsylvania integrates sustainability throughout its curriculum. In 2008, Arizona State University opened the country's first School of Sustainability, which employs more than 60 faculty members from more than 40 disciplines. Since 1990, De Anza Community College in Cupertino, California, has been integrating sustainability concepts into its curriculum. In addition, a team of students, faculty,

Courtesy of Northland College

Figure 17-18 The Environmental Living and Learning Center is an eco-friendly residence hall and meeting space at Northland College in Ashland, Wisconsin. Northland students had a major role in the design of the building.

administrators, and members of the local community worked together on developing an LEED-platinum certified building known as the Kirsch Center for Environmental Studies.

THINKING ABOUT
The Greening of Your Campus

What major steps is your school taking to increase its own environmental sustainability and to educate its students about environmental sustainability?

Environmental Security Is as Important as Military and Economic Security

Countries are legitimately concerned with *military security* and *economic security.* However, ecologists and many economists point out that all economies are supported by the earth's natural capital (see Figure 1-3, p. 9, and Figure 17-4). Today, some of the most serious new threats to global and national military and economic security are the potential for rapid climate change, increasing hunger and malnutrition, spreading water shortages, and environmental degradation. There is also an increase in the number of *failing states* where governments can no longer provide security and basic services such as education, health care, and safe supplies of water for their citizens.

According to environmental expert Norman Myers:

> *If a nation's environmental foundations are degraded or depleted, its economy may well decline, its social fabric deteriorate, and its political structure become destabilized as growing numbers of people seek to sustain themselves from declining resource stocks. Thus, national security is no longer about fighting forces and weaponry alone. It relates increasingly to watersheds, croplands, forests, genetic resources, climate, and other factors that, taken together, are as crucial to a nation's security as are military factors.*

Myers and other analysts call for all countries to make environmental security a major focus of diplomacy and government policy at all levels. (See Myers' Guest Essay on this subject at **www.cengagebrain.com**.)

A number of international environmental organizations help to shape and set global environmental policy. Perhaps the most influential is the United Nations, which supervises a large family of organizations including the U.N. Environment Programme (UNEP), the World Health Organization (WHO), the U.N. Development Programme (UNDP), and the Food and Agriculture Organization (FAO).

Other organizations that make or influence environmental decisions are the World Bank, the Global Environment Facility (GEF), and the World Conservation Union (also known as the IUCN). Despite their often limited funding, these and other international organizations have played important roles in:

- Expanding global understanding of environmental issues;
- Gathering and evaluating environmental data;
- Developing and monitoring international environmental treaties;
- Providing grants and loans for sustainable economic development and reducing poverty; and
- Helping more than 100 nations to develop environmental laws and institutions.

In 2008, environmental leader Gus Speth argued that global environmental problems are getting worse and that international efforts to solve them are inadequate. Speth and other environmental leaders propose the creation of a World Environmental Organization, on the order of the World Health Organization and the World Trade Organization, to help deal with global environmental challenges.

17-4 What Are Some Major Environmental Worldviews?

▶ **CONCEPT 17-4 Major environmental worldviews differ on which is more important —human needs and wants, or the overall health of ecosystems and the biosphere.**

There Are a Variety of Environmental Worldviews

People disagree on how serious various environmental problems are as well as on what we should do about them. These conflicts arise mostly out of differing **environmental worldviews**—ways of thinking about how the world works and beliefs that people hold about their roles in the natural world. Another factor is the widespread lack of understanding of how the earth works, keeps us alive, and supports our economies.

An environmental worldview is determined partly by a person's **environmental ethics**—what one believes about what is right and what is wrong in our behavior toward the environment. According to environmental ethicist Robert Cahn:

> *The main ingredients of an environmental ethic are caring about the planet and all of its inhabitants, allowing unselfishness to control the immediate self-interest that harms others, and living each day so as to leave the lightest possible footprints on the planet.*

People with widely differing environmental worldviews can take the same data, be logically consistent in their analysis of those data, and arrive at quite different conclusions because they start with different assumptions and values. Figure 17-19 summarizes the four major beliefs of each of three major environmental worldviews.

Most People Have Human-Centered Environmental Worldviews

It is not surprising that most environmental worldviews are human centered. One such worldview held by many people is the **planetary management worldview**. Figure 17-19 (left) summarizes the major beliefs of this worldview.

According to this view, humans are the planet's most important and dominant species, and we can and should manage the earth, mostly for our own benefit. The values of other species and parts of nature are based primarily on how useful they are to us. According to this view of nature, human well-being depends on the degree of control that we have over natural processes. And, as the world's most important and intelligent species, we can also redesign parts of the planet and its life-support systems to support us and our ever-growing economies.

Another human-centered environmental worldview is the **stewardship worldview**. It assumes that we have an ethical responsibility to be caring and responsible managers, or *stewards*, of the earth. Figure 17-19 (center) summarizes the major beliefs of this worldview.

According to the stewardship view, as we use the earth's natural capital, we are borrowing from the earth and from future generations. We have an ethical responsibility to pay this debt by leaving the earth in at least as good a condition as what we now enjoy.

Some people believe any human-centered worldview will eventually fail because it wrongly assumes we now have or can gain enough knowledge to become effective managers or stewards of the earth. Critics of human-centered worldviews point out that we do not even know how many species live on the earth, much less what their roles are, how they interact with one another and their nonliving environment, and how they support our lives and economies.

We still have much to learn about what goes on in a handful of soil, a patch of forest, the bottom of the ocean, and most other parts of the planet. At the same time, however, ecological footprints are spreading as we are using and degrading up to 80% of the earth's land ecosystems (see Figure 2, p. S24, in Supplement 6)

Figure 17-19 This is a comparison of three major environmental worldviews (**Concept 17-4**). Some environmental worldviews are *human-centered*, focusing primarily on the needs and wants of people; others are *life-* or *earth-centered*, focusing on individual species, the entire biosphere, or some level in between. **Questions:** Which of these descriptions most closely fits your worldview? Which of them most closely fits the worldviews of your parents?

SCIENCE FOCUS

Biosphere 2—A Lesson in Humility

In 1991, eight scientists (four men and four women) were sealed inside Biosphere 2, a $200 million glass and steel enclosure designed to be a self-sustaining life-support system (Figure 17-D) that would increase our understanding of Biosphere 1: the earth's life-support system.

This sealed system of interconnected domes was built in the desert near Tucson, Arizona (USA). It contained artificial ecosystems including a tropical rain forest, a savanna, a desert, a lake, streams, freshwater and saltwater wetlands, and a mini-ocean with a coral reef.

Biosphere 2 was designed to mimic the earth's natural chemical recycling systems. The facility was stocked with more than 4,000 species of plants and animals, including small primates, chickens, cats, and insects, selected to help maintain life-support functions. Human and animal excrement and other wastes were treated and recycled to help support plant growth.

Figure 17-D Biosphere 2, constructed near Tucson, Arizona, was designed to be a self-sustaining life-support system for eight people.

PRNewsFoto/Huron Valley Travel

Sunlight and external natural gas–powered generators provided energy. The Biospherians were to be isolated for 2 years and to raise their own food using intensive organic agriculture. They were to breathe air recirculated by plants and to drink water cleansed by natural recycling processes.

From the beginning, many unexpected problems cropped up and the life-support system began unraveling. The level of oxygen in the air declined when soil organisms converted it to carbon dioxide. Additional oxygen had to be pumped in from the outside to keep the Biospherians from suffocating.

Tropical birds died after the first freeze. An ant species got into the enclosure, proliferated, and killed off most of the system's original insect species. In total, 19 of the Biosphere's 25 small animal species became extinct. Before the 2-year period was up, all plant-pollinating insects became extinct, thereby dooming to extinction most of the plant species.

Despite many problems, the facility's waste and wastewater were recycled. With much hard work, the Biospherians were also able to produce 80% of their food supply. However, they suffered from persistent hunger and weight loss.

Ecologists Joel Cohen and David Tilman, who evaluated the project, concluded, "No one yet knows how to engineer systems that provide humans with life-supporting services that natural ecosystems provide for free."

Critical Thinking

Some analysts argue that the problems with Biosphere 2 resulted mostly from inadequate design and that a better team of scientists and engineers could make it work. Explain why you agree or disagree with this view.

and up to 90% of the world's oceans and other aquatic systems. As biologist David Ehrenfeld puts it, "In no important instance have we been able to demonstrate comprehensive successful management of the world, nor do we understand it well enough to manage it even in theory." This belief is illustrated by the failure of the Biosphere 2 project (Science Focus, above).

Some Environmental Worldviews Are Life-Centered and Others Are Earth-Centered

Critics of human-centered environmental worldviews argue that they should be expanded to recognize that all forms of life have value as participating members of the biosphere, regardless of their potential or actual use to humans.

Eventually all species become extinct. However, most people with a life-centered worldview believe we have an ethical responsibility to avoid hastening the extinction of any species through our activities, for two reasons. *First,* each species is a unique storehouse of genetic information that should be respected and protected simply because it exists. *Second,* every species has the potential for providing economic benefits.

Some people think we should go beyond focusing mostly on species. They believe we have an ethical responsibility to prevent the degradation of the earth's ecosystems, biodiversity, and biosphere. This *earth-centered* environmental worldview is devoted to preserving the earth's biodiversity and the functioning of its life-support systems for the benefit of humans and other forms of life, now and in the future.

People with earth-centered worldviews believe that humans are not in charge of the world and that human economies and other systems are subsystems of the biosphere (Figure 17-4). They understand that the earth's natural capital (see Figure 1-3, p. 9) keeps us and other species alive and supports our economies. They also understand that preventing the depletion and degradation of this natural capital is a key way to promote environmental sustainability. They argue that an important way to preserve the earth's natural capital is to mimic the ways in which nature has sustained itself for 3.5 billion years. One way to do this is to apply the three **principles of sustainability** (see back cover) to human economies and lifestyles.

SUSTAINABILITY

One earth-centered worldview is called the **environmental wisdom worldview**. Figure 17-19 (right) summarizes its major beliefs. According to this view, we are part of—not apart from—the community of life and the ecological processes that sustain all life. Therefore, we should work with the earth to promote environmental sustainability instead of trying to conquer and manage it mostly for our own benefit. In many respects, it is the opposite of the planetary management worldview (Figure 17-19, left).

Chief Seattle (1786–1866), leader of the Suquamish and Duwamish Native American tribes in what is now the U.S. state of Washington, summarized an ethical belief included in the environmental wisdom worldview: "The earth does not belong to us. We belong to the earth." (Another part of this statement by Chief Seattle is this chapter's opening quotation.)

The environmental wisdom worldview suggests that the earth does not need us managing it in order to go on, whereas we depend on the earth for our survival. From this perspective, it makes little sense to talk about saving the earth. It has been around for billions of years and doesn't need saving. What we need to save is the existence of our own species and cultures, which have been around for less than an eye blink of the 3.5-billion-year history of life on the earth, as well as the existence of other species that may become extinct because of our activities. (See the Guest Essay on this topic by sustainability expert Lester W. Milbrath at **www.cengagebrain.com**.)

17-5 How Can We Live More Sustainably?

▶ **CONCEPT 17-5 We can live more sustainably by becoming environmentally literate, learning from nature, living more simply and lightly on the earth, and becoming active environmental citizens.**

We Can Become More Environmentally Literate

There is widespread evidence and agreement that we are a species in the process of degrading our own life-support system and that, during this century, this behavior will threaten human civilization and the existence of up to half of the world's species. Part of the problem stems from ignorance about how the earth works, how our actions affect its life-sustaining systems, and how we can change our behavior toward the earth and thus toward ourselves. Correcting this ignorance begins by understanding three important ideas that form the foundation of environmental literacy:

1. *Natural capital matters* because it supports the earth's life and our economies.
2. *Our ecological footprints are immense and are expanding rapidly*; in fact, they already exceed the earth's ecological capacity (see Figure 1-8, p. 14).
3. *Ecological and climate change tipping points are irreversible and should never be crossed.* Once we cross such a point, neither money nor technology will save us from the harmful consequences that could last for thousands of years.

According to the environmental wisdom worldview, learning how to live more sustainably requires a foundation of environmental education aimed at producing environmentally literate citizens. Acquiring environmental literacy involves being able to answer certain key questions and having a basic understanding of certain key topics, as summarized in Figure 17-20 (p. 458).

We Can Learn from the Earth

Formal environmental education is important, but is it enough? Many analysts say no. They call for us to appreciate not only the economic value of nature, but also its ecological, aesthetic, and spiritual values. To these analysts, the problem is not just a lack of environmental literacy but also, for many people, a lack of intimate contact with nature and little understanding of how nature works and sustains us.

We face a dangerous paradox. At a time when humans have more technology and power than ever before to degrade and disrupt nature, most people know little about nature, and have little direct contact with it. Technology has lead many people to see themselves as being apart from nature instead of being part of it. This

Questions to answer

- How does life on earth sustain itself?
- How am I connected to the earth and other living things?
- Where do the things I consume come from and where do they go after I use them?
- What is environmental wisdom?
- What is my environmental worldview?
- What is my environmental responsibility as a human being?

Key Topics

- Basic concepts: sustainability, natural capital, exponential growth, carrying capacity
- Three scientific principles of sustainablility
- Environmental history
- The two laws of thermodynamics and the law of conservation of matter
- Basic principles of ecology: food webs, nutrient cycling, biodiversity, ecological succession
- Population dynamics
- Sustainable agriculture and forestry
- Soil conservation and sustainable water use
- Nonrenewable mineral resources
- Nonrenewable and renewable energy resources
- Climate disruption and ozone depletion
- Pollution prevention and waste reduction
- Environmentally sustainable economic and political systems
- Environmental worldviews and ethics
- Three social science principles of sustainability

Figure 17-20 Achieving environmental literacy involves being able to answer certain questions and having an understanding of certain key topics (**Concept 17-5**). (For an overview of U.S. environmental history, see Supplement 3, p. S6.) **Question:** After taking this course, do you feel that you can answer the questions asked here and have a basic understanding of each of the key topics listed in this figure?

can reduce our ethical ability to act responsibly toward the earth and thus toward ourselves.

A growing chorus of analysts suggest that we have much to learn from nature. Many have argued that we can kindle a sense of awe, wonder, mystery, excitement, and humility by standing under the stars, sitting in a forest, or taking in the majesty and power of the sea. We might pick up a handful of topsoil and try to sense the teeming microscopic life within it that helps keep us alive by producing most of the food we eat. We might look at a tree, a mountain, a rock, or a bee, or listen to the sound of a bird and try to sense how each of them is connected to us and we to them, through the earth's life-sustaining processes.

Such direct experiences with nature can reveal parts of the complex web of life that cannot be bought, re-created with technology (Science Focus, p. 456) or in a chemistry lab, or reproduced with genetic engineering. Understanding and directly experiencing the precious gifts we receive from nature can help us to foster within ourselves the ethical commitment that we need in order to live more sustainably on this earth and thus to preserve our own species and cultures.

This might lead us to recognize that the healing of the earth and the healing of the human spirit are one and the same. We might discover and tap into what conservationist Aldo Leopold calls "the green fire that burns in our hearts" and use this as a force for respecting and working with the earth and with one another.

CONNECTIONS

Disconnecting from Technology and Reconnecting with Nature

Many of us who venture into the natural world want to carry our GPS units, cell phones, I-pods, and other technological marvels that keep us in touch with the world we have temporarily left behind. But these technological devices can divert much of our attention from the natural world that surrounds us during such adventures. If our goal is to reconnect with and experience the natural world, then these devices can defeat that purpose.

We Can Live More Simply and Lightly on the Earth

Sustainability is not only about sustaining resources for our use. It is about sustaining the entire web of life, because all past, present, and future forms of life are connected. Here are six ethical guidelines for achieving more sustainable and compassionate societies by converting environmental concerns, literacy, and wisdom into environmentally responsible actions.

1. Use the three **principles of sustainability** (see back cover) to mimic the ways in which nature sustains itself. SUSTAINABILITY
2. Do not deplete or degrade the earth's natural capital.
3. Do not waste matter and energy resources.
4. Protect biodiversity.
5. Repair the ecological damage that we have caused.
6. Leave the earth in as good a condition as we found it, or better.

Analysts urge people who have a habit of excessive consumption to *learn how to live more simply and sustainably*. Seeking happiness through the pursuit of material things is considered folly by almost every major religion and philosophy. Yet, modern advertising persistently encourages people to buy more and more things to fill a growing list of wants as a way to achieve happiness. As American humorist and writer Mark Twain (1835–1910) observed: "Civilization is the limitless multiplication of unnecessary necessities." American comedian George Carlin (1937–2008) put it another way: "A house is just a pile of stuff with a cover on it. It is a place to keep your stuff while you go out and get more stuff."

According to research by psychologists, what a growing number of people really want, deep down, is more community, not more stuff. They want greater and more fulfilling interactions with family, friends, and neighbors, and a greater opportunity to express their creativity and to have more fun.

Some affluent people are adopting a lifestyle of *voluntary simplicity,* in which they seek to learn how to live with much less than they are accustomed to having. These people have found that a life based mostly on what one owns is not fulfilling for them. They are living with fewer material possessions and using products and services that have a smaller environmental impact (**Concept 17-5**). Instead of working longer to pay for bigger vehicles and houses, they are spending more time with their loved ones, friends, and neighbors. They are shifting from a culture of "faster, bigger, and more" to one of "slower, smaller, and less."

Practicing voluntary simplicity is a way to apply Mahatma Gandhi's *principle of enoughness*: "The earth provides enough to satisfy every person's need but not every person's greed. . . . When we take more than we need, we are simply taking from each other, borrowing from the future, or destroying the environment and other species." Most of the world's major religions have similar teachings.

Living more lightly starts with asking the question: How much is enough? Similarly, one can ask: What do I really need? These are not easy questions to answer, because people in affluent societies are conditioned to want more and more material possessions. Thus, as a result of a lifetime of exposure to commercial advertising, they often think of such wants as needs.

Throughout this text, you have encountered lists of steps we can take to live more lightly by reducing the *size* and *impact* of our ecological footprints on the earth. It would be difficult for most of us to do all or even most of these things. So which ones are the most important? To decide, consider the fact that the human activities that have the greatest harmful impacts on the environment are *food production, transportation, home energy use,* and our *overall resource use*. Based on this fact, Figure 17-21 lists the *sustainability eight*—8 key ways in which some people are choosing to live more simply.

Food

Reduce meat consumption

Buy or grow organic food and buy locally grown food

Transportation

Reduce car use by walking, biking, carpooling, car-sharing, and using mass transit

Drive an energy-efficient vehicle

Home Energy Use

Insulate your house, plug air leaks, and install energy-efficient windows

Use energy-efficient heating and cooling systems, lights, and appliances

Resource Use

Reduce, reuse, recycle, compost, replant, and share

Use renewable energy resources whenever possible

Figure 17-21 *The sustainability eight* is a list of eight ways in which people can live more lightly on the earth (**Concept 17-5**). **Questions:** Which of these things do you already do? Which, if any, do you hope to do?

THINKING ABOUT
The Sustainability Eight

Which three of the eight actions listed in Figure 17-21 do you think are the most important? Which of these things do you already do? Which of them are you thinking about doing? How do your answers to these questions relate to **(a)** the three principles of sustainability, and **(b)** your environmental worldview?

Living more sustainably is not easy, and we will not make this transition by relying primarily on technological fixes such as recycling, using energy-efficient light bulbs, and driving energy-efficient cars. These are, of course, important things to do, and they can help us to shrink our ecological footprints and to feel less guilty about our harmful impacts on our life-support system. But these efforts will not solve the underlying problems of excessive consumption and unnecessary waste of matter and energy resources.

In the end, it comes down to what each of us does to make the earth a better place to live for current and future generations, for other species, and for the ecosystems that support us. It is important to recognize that there is no single correct or best solution to any of the environmental problems we face. By being flexible and adaptable in trying a variety of cultural and technological solutions to such problems, we might find the best ways to adapt to the earth's largely unpredictable, ever-changing conditions.

Finally, we should have fun and take time to enjoy life. Laugh every day and enjoy and celebrate nature, beauty, connectedness, friendship, and love. This can empower us to become good earth citizens who practice *good earthkeeping*. As Mahatma Gandhi reminded us, "Power based on love is a thousand times more effective and permanent than power derived from fear."

We Can Bring About a Sustainability Revolution in Your Lifetime

The Industrial Revolution was a remarkable global transformation that has taken place over the past 275 years. Now, in this century, environmental leaders say it is time for another sort of global transformation—an *environmental* or *sustainability revolution* that could lead to the kind of world we envisioned in the Chapter 1 Core Case Study (see p. 5). Figure 17-22 lists some of the major cultural shifts in emphasis that we will need to make in order to bring about such an environmental revolution.

In Chapter 1, we introduced three science-based principles of sustainability (Figure 1-2). We used them throughout this book to show how life has sustained itself on the earth for billions of years and how we can mimic these principles to help save and sustain our own civilization. In this last chapter, we have explored several findings from the social sciences—economics and political science—and from the field of ethics. From those findings, we can draw three principles that, along with our three science-based principles of sustainability, can help human societies to make the shift to more sustainable living for the long-term future. We propose the following three *social science principles of sustainability* (Figure 17-23).

- **Full-cost pricing** (from economics): in working toward this goal, we would find ways to include in market prices the harmful environmental and health costs of producing and using goods and services.
- **Win-win solutions** (from political science): by focusing on solutions that will benefit the largest possible number of people, as well as the environment, we might learn to work together consistently in dealing with environmental problems, just as citizens did in Chattanooga (**Core Case Study**).
- **A responsibility to future generations** (from ethics): through this principle, we would accept our responsibility to leave the planet's life-support systems in at least as good a shape as what we now enjoy, for all future generations.

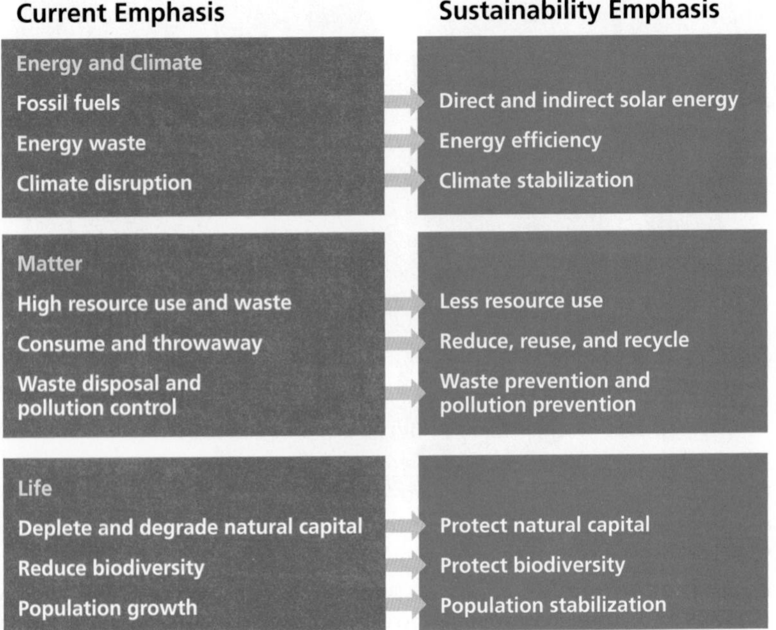

Figure 17-22 Solutions: These are some of the cultural shifts in emphasis that will be necessary to bring about the *environmental* or *sustainability revolution*. **Questions:** Which three of these shifts do you think are most important? Why?

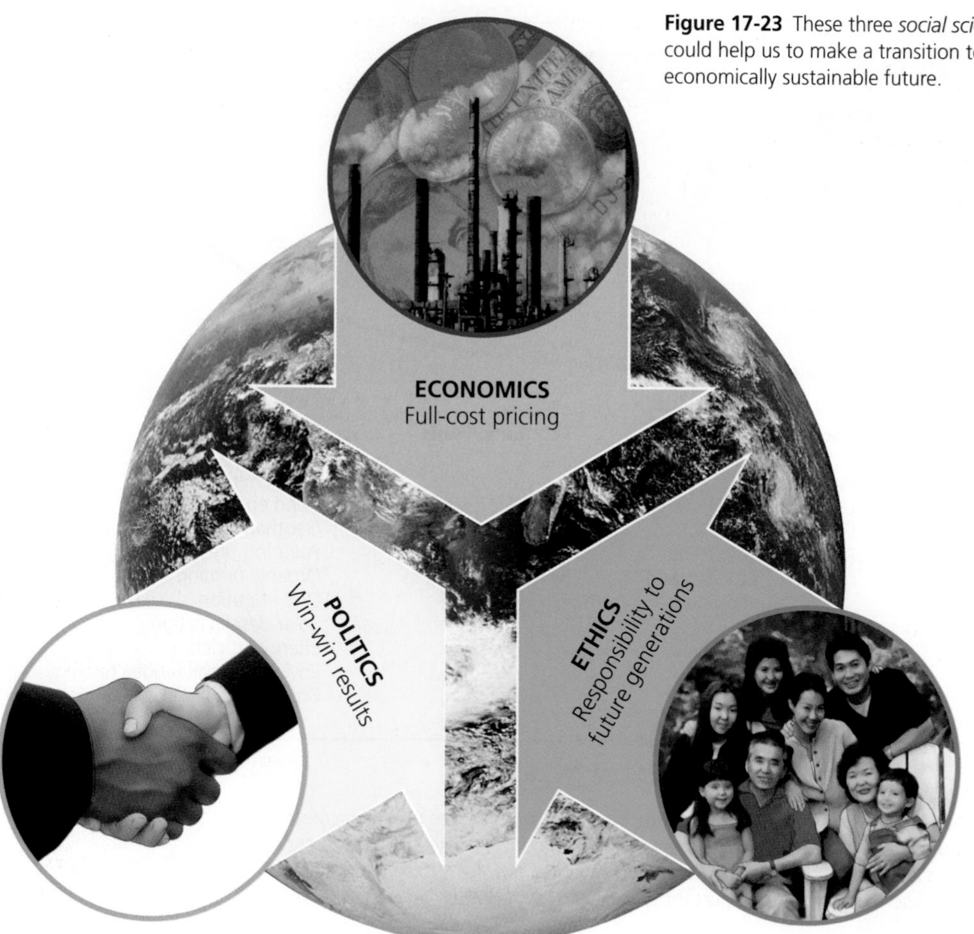

Figure 17-23 These three *social science principles of sustainability* could help us to make a transition to a more environmentally and economically sustainable future.

We can use the incredible power of exponential growth to help us bring about a sustainability revolution. Recall that if you could fold a piece of paper in half 50 times, you would have a stack of paper high enough to reach the sun—some 149 million kilometers (93 million miles) away. We can use this power of exponential growth to promote more sustainable environmental, social, economic, and technological changes that could take place rather quickly (Figure 17-24, p. 462).

GOOD NEWS

We know what needs to be done and we can change. According to social science research, in order for a major social change to occur, only 5–10% of the people in the world, or in a country or locality, must be convinced that the change must take place and then act to bring it about. We, the authors, believe that we are close to this *political and ethical tipping point,* in terms of our awareness of major environmental issues in many parts of the world. This could lead us to take individual and collective actions to prevent us from reaching a number of irreversible *ecological tipping points* (Figure 15-27, p. 400).

History also shows that we can change faster than we might think, once we have the courage to leave behind ideas and practices that no longer work, and to nurture new ideas for positive change (Figure 17-23). We can no longer afford to make big mistakes in our treatment of planet Earth, which after all, will always affect our own lives as well as those of our children and grandchildren. As biodiversity expert Edward O. Wilson points out, there is only "One Earth. One Experiment."

Some say that making this shift is idealistic and unrealistic. Others say that it is idealistic, unrealistic, and dangerous to keep assuming that our present course is sustainable, and they warn that we have precious little time to change. As Irish playwright George Bernard Shaw (1856–1950) wrote, "I dream things that never were; and I say, 'Why not?'" If certain individuals had not had the courage to forge ahead with ideas that others called idealistic and unrealistic, very few of the human achievements that we now celebrate would have come to pass.

Here are this chapter's *three big ideas*:

- A more sustainable economic system would include in market prices the harmful environmental and health costs of producing and using goods and services; subsidize environmentally beneficial goods and services; tax pollution and waste instead of wages and profits; and reduce poverty.
- Individuals can work together to become part of the political processes that influence how environmental policies are made and implemented.
- Living more sustainably means becoming environmentally literate, learning from nature, living more simply, and becoming active environmental citizens.

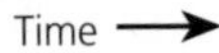

Figure 17-24 *Change can occur very rapidly.* Exponential growth starts off slowly, but at some point it increases at a very rapid rate and heads sharply upward. Exponential growth in any or all of the factors listed below this curve could help us to shift toward a more sustainable world within a very short time. **Questions:** Which two factors in each of these four categories do you believe are the most important to promote? What other factors would you add to these lists?

REVISITING Chattanooga, Tennessee, and Sustainability

Citizens and elected officials of Chattanooga, Tennessee (**Core Case Study**), have worked together for more than 25 years to make their city more sustainable. They have shown that, by using economic, political, and ethical tools, committed people can work together to improve the environmental quality of the place where they live.

As we explore different paths toward sustainability, we must first understand that our lives and societies depend on *natural capital* and that one of the biggest threats to our ways of life is our active role in *natural capital degradation*. With that understanding, we begin the search for *solutions* to difficult environmental problems. One or more of the three **principles of sustainability** will likely govern most of the solutions we devise. Competing interests working together to find the solutions must make *trade-offs* because this is the essence of realistic political change.

All of this requires our understanding that *individuals matter*. Virtually all of the environmental progress we made during the last few decades occurred because individuals banded together to insist that we can do better. This journey begins in our own communities and with our own lifestyles, because in the final analysis, *all sustainability is local*, as the citizens of Chattanooga demonstrated. This is the meaning of the motto, "Think globally; act locally." At the community level we can learn from and empower one another by communicating and working together to seek creative solutions to problems.

In working to make the earth a better and more sustainable place to live, we should be guided by historian Arnold Toynbee's observation: "If you make the world ever so little better, you will have done splendidly, and your life will have been worthwhile." Each of us has to decide whether we want to be part of the problem or part of the solution to the environmental challenges we face. This is an incredible and exciting time to be alive as we tackle the challenges of learning how to live more lightly on our planetary home.

When there is no dream, the people perish.

PROVERBS 29:18

REVIEW

CORE CASE STUDY

1. Describe the efforts of citizens in the U.S. city of Chattanooga, Tennessee (**Core Case Study**), to make their city more sustainable and livable, and summarize the major lessons learned from this successful effort. CORE CASE STUDY

SECTION 17-1

2. What is the key concept for this section? What is **economics**? Distinguish among **natural capital**, **human capital** (**human resources**), and **manufactured capital** (**manufactured resources**). Distinguish between **economic growth** and **economic development**. What is a **high-throughput economy**? Compare how neoclassical economists and ecological and environmental economists view economic systems. What are three major assumptions that ecological economists use to build their economic models?

SECTION 17-2

3. What is the key concept for this section? Why do products and services actually cost more than most people think? What is *full-cost pricing* and what are some benefits of using it to determine the market values of goods and services? Give two reasons why it is not widely used. Define and distinguish between **gross domestic product (GDP)** and **per capita GDP**. What is the **genuine progress indicator (GPI)** and how does it differ from the GDP economic indicator? Describe the benefits of shifting from environmentally unsustainable to more environmentally sustainable government subsidies and tax breaks. Give two reasons why making this subsidy shift is difficult. Discuss whether we should tax pollution and wastes instead of wages and profits. What are the major advantages and disadvantages of green taxes? What are three requirements for the successful implementation of green taxes?

4. Distinguish between command-and-control and incentive-based government regulations and describe the advantages of the second approach. What are the major advantages and disadvantages of using the cap-and-trade approach to implementing environmental regulations for controlling pollution and resource use? What are some environmental benefits of selling services instead of goods? Give two examples of this approach. What is **poverty** and how is it related to population growth and environmental degradation? List six ways in which governments and businesses can help to reduce poverty. What are the benefits of making microloans to the poor? What is a **low-throughput (low-waste) economy**? Describe Ray Anderson's attempts to develop a more environmentally sustainable carpet business. List six ways to shift to more environmentally sustainable economies. Name five new green businesses or careers that would be important in such eco-economies.

SECTION 17-3

5. What is the key concept for this section? Define **politics**, **environmental policy**, and **representative democracy**. List six principles that decision makers can use in making environmental policy. What are four major types of public lands in the United States? Describe the political controversy over managing these lands.

6. List six ways in which individuals in democracies can help to influence environmental policy. What does it mean to say that we should *think globally and act locally*? What are four ways to provide environmental leadership? Summarize the story of environmental leader Denis Hayes.

7. What are four major U.S. environmental laws? Describe the anti-environmental movement in the United States. Describe the roles of grassroots and mainstream environmental organizations and give an example of each type of organization. Give two examples of successful roles that students have played in improving environmental quality. Explain the importance of environmental security, relative to economic and military security. Describe efforts to develop international environmental policies.

SECTION 17-4

8. What is the key concept for this section? What is an **environmental worldview**? What are **environmental ethics**? Distinguish among the following environmental worldviews: **planetary management**, **stewardship**, and **environmental wisdom**. Summarize the debate over whether we can effectively manage the earth. Summarize the ecological lessons learned from the failure of the Biosphere 2 project.

SECTION 17-5

9. What is the key concept for this section? What three important ideas make up the foundation of environmental literacy? List eight goals for a person seeking environmental literacy. Describe three ways in which we can learn from the earth. List six guidelines for achieving more sustainable and compassionate societies. What is *voluntary simplicity*? List eight important steps that individuals can take to help make the transition to more sustainable societies. What are the four categories of human activities that have the highest environmental impacts?

10. List five cultural shifts involved in making the transition to a more environmentally sustainable society. Describe the power of exponential growth in making such a transition. What are three social science-based principles of sustainability? What are this chapter's *three big ideas*? Describe how the citizens of Chattanooga, Tennessee, have developed a more environmentally sustainable city by applying the three **principles of sustainability**. SUSTAINABILITY

Note: Key terms are in **bold** type.

CRITICAL THINKING

1. Use the model of the U.S. city of Chattanooga, Tennessee (**Core Case Study**) to suggest three ways to make the area where you live more sustainable. CORE CASE STUDY
2. Should we attempt to maximize economic growth by producing and consuming more and more economic goods and services? Explain. What are the alternatives?
3. Suppose that over the next 20 years, the harmful environmental and health costs of goods and services will be gradually added to the market prices of these goods and services to more closely reflect their full costs. What harmful effects and what beneficial effects might such full-cost pricing have on your lifestyle?
4. Explain why you agree or disagree with **(a)** each of the major strategies for shifting to a more environmentally sustainable economy listed in Figure 17-11 and **(b)** each of the six principles listed on pp. 447–448, which some analysts have proposed for use in making environmental policy decisions.
5. Explain why you agree or disagree with **(a)** each of the four principles that biologists and some economists have suggested for using public lands in the United States (p. 449), and **(b)** each of the five suggestions made by developers and resource extractors for managing and using U.S. public lands (p. 450).
6. This chapter summarized several different environmental worldviews. Go through these worldviews (Figure 17-19) and find the beliefs you agree with, and then describe your own environmental worldview. Which, if any, of your beliefs were changed as a result of taking this course? Compare your answer with those of your classmates.
7. Explain why you agree or disagree with the following ideas: **(a)** everyone has the right to have as many children as they want; **(b)** all people have a right to use as many resources as they want; **(c)** individuals should have the right to do whatever they want with land they own, regardless of whether such actions harm the environment, their neighbors, or the local community; **(d)** other species exist to be used by humans; **(e)** all forms of life have an intrinsic value and therefore have a right to exist. Are your responses to each of these ideas consistent with the beliefs that make up your environmental worldview, which you described in answering question 6? If not, explain.
8. If you could use television or YouTube to speak to everyone in the world today about our environmental problems, what are the three most important pieces of environmental wisdom that you would give in your speech? What beliefs from your environmental worldview influenced your selection of these three items? Compare your choices with those of your classmates.

DOING ENVIRONMENTAL SCIENCE

1. Polls have identified five categories of citizens in terms of their concern over environmental quality: **(a)** those involved in a wide range of environmental activities, **(b)** those who do not want to get involved but are willing to pay more for a cleaner environment, **(c)** those who are not involved because they disagree with many environmental laws and regulations, **(d)** those who are concerned but do not believe individual action will make much difference, and **(e)** those who strongly oppose the environmental movement. To which group do you belong? Compile your answer and those of your classmates, and determine what percentage of the total number of people in your class represents each category. As a class, conduct a similar poll of your entire school and compile the results.
2. Does your school's curriculum provide *all* graduates with the basic elements of environmental literacy? To what extent are the funds in your school's financial endowments invested in enterprises that are working to promote environmental sustainability? Are the managers of your school's facilities committed to making those facilities more environmentally sustainable, and do they have a plan for doing so? Use these questions to gather information in order to rate your school on a 1–10 scale in terms of its contributions to environmental awareness and sustainability. Develop a detailed plan illustrating how your school could become better at making such contributions, and present this information to school officials, alumni, parents, and financial backers.

GLOBAL ENVIRONMENT WATCH EXERCISE

Search the Global Environment Watch site to learn what colleges and universities are doing to try to become more sustainable. Pick two campuses and write a brief report summarizing their sustainability programs. If possible, pick one campus from the country where you live and one from another country. Compare the two programs and pick out what you consider to be the best three features of each. Then do some research to find out if your campus sustainability program includes any of these best features.

ECOLOGICAL FOOTPRINT ANALYSIS

Working with classmates, conduct an ecological footprint analysis of your campus. Work with a partner, or in small groups, to research and investigate an aspect of your school such as recycling and/or composting; water use; food service practices; energy use; building management and energy conservation; transportation for both on- and off-campus trips; grounds maintenance; and institutional environmental awareness and education. Depending on your school and its location, you may be able to add more areas to the investigation. You may decide to study the campus as a whole, or you may decide to break down the campus into smaller research areas such as dorms, administrative buildings, classrooms and classroom buildings, grounds, and other areas. To accomplish this, take the following steps:

1. After deciding on your group's research area, conduct your analysis. As part of your analysis, develop a list of questions that will help to determine the ecological impact related to your chosen topic. Each question item in this questionnaire could have a range of responses on a scale of 1 (poor) to 10 (excellent).
2. Analyze your results and share them with the class to determine what can be done to shrink the ecological footprint of your school.
3. Arrange a meeting with school officials to share your action plan with them.

LEARNING ONLINE

The website for *Environmental Science 14e* contains helpful study tools and other resources to take the learning experience beyond the textbook. Examples include flashcards, practice quizzing, information on green careers, tips for more sustainable living, activities, and a wide range of articles and further readings. To access these materials and other companion resources for this text, please visit **www.cengagebrain.com**. For further details, see the preface, page xvi.

Supplements

SUPPLEMENT

1 Measurement Units (Chapters 2, 3)

LENGTH

Metric

1 kilometer (km) = 1,000 meters (m)
1 meter (m) = 100 centimeters (cm)
1 meter (m) = 1,000 millimeters (mm)
1 centimeter (cm) = 0.01 meter (m)
1 millimeter (mm) = 0.001 meter (m)

English

1 foot (ft) = 12 inches (in)
1 yard (yd) = 3 feet (ft)
1 mile (mi) = 5,280 feet (ft)
1 nautical mile = 1.15 miles

Metric–English

1 kilometer (km) = 0.621 mile (mi)
1 meter (m) = 39.4 inches (in)
1 inch (in) = 2.54 centimeters (cm)
1 foot (ft) = 0.305 meter (m)
1 yard (yd) = 0.914 meter (m)
1 nautical mile = 1.85 kilometers (km)

AREA

Metric

1 square kilometer (km^2) = 1,000,000 square meters (m^2)
1 square meter (m^2) = 1,000,000 square millimeters (mm^2)
1 hectare (ha) = 10,000 square meters (m^2)
1 hectare (ha) = 0.01 square kilometer (km^2)

English

1 square foot (ft^2) = 144 square inches (in^2)
1 square yard (yd^2) = 9 square feet (ft^2)
1 square mile (mi^2) = 27,880,000 square feet (ft^2)
1 acre (ac) = 43,560 square feet (ft^2)

Metric–English

1 hectare (ha) = 2.471 acres (ac)
1 square kilometer (km^2) = 0.386 square mile (mi^2)
1 square meter (m^2) = 1.196 square yards (yd^2)
1 square meter (m^2) = 10.76 square feet (ft^2)
1 square centimeter (cm^2) = 0.155 square inch (in^2)

VOLUME

Metric

1 cubic kilometer (km^3) = 1,000,000,000 cubic meters (m^3)
1 cubic meter (m^3) = 1,000,000 cubic centimeters (cm^3)
1 cubic meter (m^3) = 1,000 liters
1 liter (L) = 1,000 milliliters (mL) = 1,000 cubic centimeters (cm^3)
1 milliliter (mL) = 0.001 liter (L)
1 milliliter (mL) = 1 cubic centimeter (cm^3)

English

1 gallon (gal) = 4 quarts (qt)
1 quart (qt) = 2 pints (pt)

Metric–English

1 liter (L) = 0.265 gallon (gal)
1 liter (L) = 1.06 quarts (qt)
1 liter (L) = 0.0353 cubic foot (ft^3)
1 cubic meter (m^3) = 35.3 cubic feet (ft^3)
1 cubic meter (m^3) = 1.30 cubic yards (yd^3)
1 cubic kilometer (km^3) = 0.24 cubic mile (mi^3)
1 barrel (bbl) = 159 liters (L)
1 barrel (bbl) = 42 U.S. gallons (gal)

MASS

Metric

1 kilogram (kg) = 1,000 grams (g)
1 gram (g) = 1,000 milligrams (mg)
1 gram (g) = 1,000,000 micrograms (μg)
1 milligram (mg) = 0.001 gram (g)
1 microgram (μg) = 0.000001 gram (g)
1 metric ton (mt) = 1,000 kilograms (kg)

English

1 ton (t) = 2,000 pounds (lb)
1 pound (lb) = 16 ounces (oz)

Metric–English

1 metric ton (mt) = 2,200 pounds (lb) = 1.1 tons (t)
1 kilogram (kg) = 2.20 pounds (lb)
1 pound (lb) = 454 grams (g)
1 gram (g) = 0.035 ounce (oz)

ENERGY AND POWER

Metric

1 kilojoule (kJ) = 1,000 joules (J)
1 kilocalorie (kcal) = 1,000 calories (cal)
1 calorie (cal) = 4.184 joules (J)

Metric–English

1 kilojoule (kJ) = 0.949 British thermal unit (Btu)
1 kilojoule (kJ) = 0.000278 kilowatt-hour (kW-h)
1 kilocalorie (kcal) = 3.97 British thermal units (Btu)
1 kilocalorie (kcal) = 0.00116 kilowatt-hour (kW-h)
1 kilowatt-hour (kW-h) = 860 kilocalories (kcal)
1 kilowatt-hour (kW-h) = 3,400 British thermal units (Btu)
1 quad (Q) = 1,050,000,000,000,000 kilojoules (kJ)
1 quad (Q) = 293,000,000,000 kilowatt-hours (kW-h)

Temperature Conversions

Fahrenheit (°F) to Celsius (°C):
$°C = (°F - 32.0) \div 1.80$
Celsius (°C) to Fahrenheit (°F):
$°F = (°C \times 1.80) + 32.0$

Reading Graphs and Maps (Chapters 1–17)

SUPPLEMENT 2

Graphs and Maps Are Important Visual Tools

A **graph** is a tool for conveying information that we can summarize numerically by illustrating that information in a visual format. This information, called *data*, is collected in experiments, surveys, and other information-gathering activities. Graphing can be a powerful tool for summarizing and conveying complex information.

In this textbook and the accompanying web-based Active Graphing exercises, we use three major types of graphs: *line graphs, bar graphs*, and *pie graphs*. Here, you will explore each of these types of graphs and learn how to read them. In the web-based Active Graphing exercises, you can try your hand at creating some graphs.

An important visual tool used to summarize data that vary over small or large areas is a **map**. We discuss some aspects of reading maps relating to environmental science at the end of this supplement.

Line Graphs

Line graphs usually represent data that fall in some sort of sequence such as a series of measurements over time or distance. In most such cases, units of time or distance lie on the horizontal *x-axis*. The possible measurements of some quantity or variable such as temperature that changes over time or distance usually lie on the vertical *y-axis*.

In Figure 1, the x-axis shows the years between 1950 and 2009, and the y-axis displays the possible values for the annual amounts of oil consumed worldwide during that time in millions of tons, ranging from 0 to 4,000 million (or 4 billion) tons. Usually, the y-axis appears on the left end of the x-axis, although y-axes can appear on the right end, in the middle, or on both ends of the x-axis.

The curving line on a line graph represents the measurements taken at certain time or distance intervals. In Figure 1, the curve represents changes in oil consumption between 1950 and 2009. To find the oil consumption for any year, find that year on the x-axis (a point called the *abscissa*) and run a vertical line from the axis to the curve. At the point where your line intersects the curve, run a horizontal line to the y-axis. The value at that point on the y-axis, called the *ordinate*, is the amount you are seeking. You can go through the same process in reverse to find a year in which oil consumption was at a certain point.

Questions

1. What was the total amount of oil consumed in the world in 1990?
2. In about what year between 1950 and 2000 did oil consumption first start declining?
3. About how much oil was consumed in 2009? Roughly how many times more oil was consumed in 2009 than in 1970? How many times more oil was consumed in 2009 than in 1950?

Line graphs have several important uses. One of the most common applications is to compare two or more variables. Figure 2 compares two variables: monthly temperature and precipitation (rain and snowfall) during a typical year in a temperate deciduous forest. However, in this case the variables are measured on two different scales, so there are two y-axes. The y-axis on the left end of the graph shows a Centigrade temperature scale, while the y-axis on the right shows the range of precipitation measurements in millimeters. The x-axis displays the first letters of each of the 12 months' names.

Questions

1. In which month does most precipitation fall? Which is the driest month of the year? Which is the hottest month?
2. If the temperature curve were almost flat, running throughout the year at roughly its highest point of about 30°C, how do you think this forest would change from what it is now (see Figure 7-14, p. 134)? If the annual precipitation suddenly dropped and remained under 25 centimeters all year, what do you think would eventually happen to this forest?

It is also important to consider what aspect of a set of data is being displayed on a graph. The creator of a graph can take two different aspects of one data set and create two very different-looking graphs that would give two different interpretations of the same phenomenon. For example, when talking about any type of

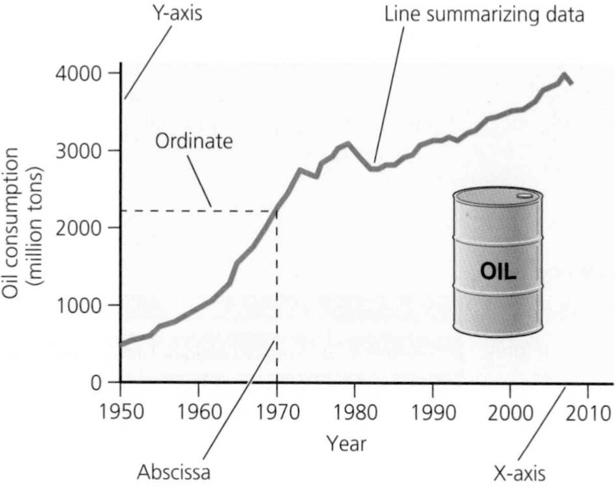

Figure 1 This graph tracks world oil consumption, 1950–2009. (Data from U.S. Energy Information Administration, British Petroleum, International Energy Agency, and United Nations)

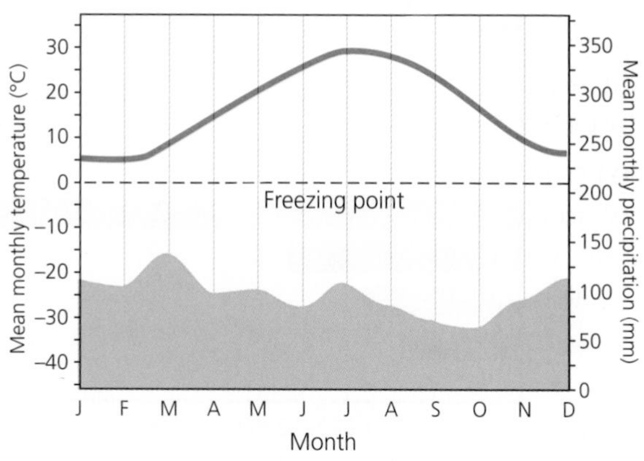

Figure 2 This climate graph shows typical variations in annual temperature (red) and precipitation (blue) in a temperate deciduous forest.

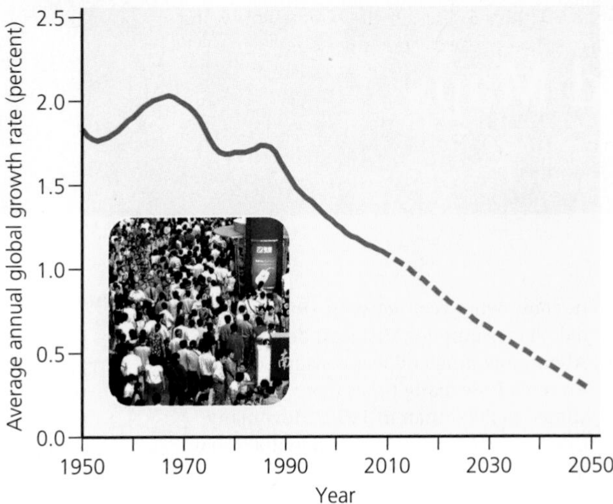

Figure 3 This graph tracks the annual growth rate in world population, 1950–2010, with projections to 2050. (Data from UN Population Division and U.S. Census Bureau)

growth we must be careful to distinguish the question of whether something is growing from the question of how fast it is growing. While a quantity can keep growing continuously, its rate of growth can go up and down.

One of many important examples of growth used in this book is human population growth. For example, the graph in Figure 1-11 (p. 16) gives you the impression that human population growth has, for the most part, been continuous and uninterrupted. However, consider Figure 3, which plots the rate of growth of the human population since 1950. Note that all of the numbers on the y-axis, even the smallest ones, represent growth. The lower end of the scale represents slower growth and the higher end faster growth. Thus, while one graph tracks population growth in terms of numbers of people, the other tracks the rate of growth.

Questions

1. If this graph were presented to you as a picture of human population growth, what would be your first impression?
2. Do you think that reaching a growth rate of 0.5% would relieve those who are concerned about overpopulation? Why or why not?

Bar Graphs

The *bar graph* is used to compare measurements for one or more variables across categories. Unlike the line graph, a bar graph typically does not involve a sequence of measurements over time or distance. The measurements compared on a bar graph usually represent data collected at some point in time or during a well-defined period. For instance, we can compare the *net primary productivity (NPP)*, a measure of chemical energy produced by plants in an ecosystem, for different ecosystems, as represented in Figure 4.

In most bar graphs, the categories to be compared are laid out on the x-axis, while the range of measurements for the variable under consideration lies along the y-axis. In our example in Figure 4, the categories (ecosystems) are on the y-axis, and the variable range (NPP) lies on the x-axis. In either case, reading the graph is straightforward. Simply run a line perpendicular to the bar you are reading from the top of that bar (or the right or left end, if it lies horizontally) to the variable value axis. In Figure 4, you can see that the NPP for continental shelf, for example, is close to 1,600 kcal/m2/yr.

Questions

1. What are the two terrestrial ecosystems that are closest in NPP value of all pairs of such ecosystems? About how many times greater is the NPP in a tropical rain forest than the NPP in a savannah?
2. What is the most productive of aquatic ecosystems shown here? What is the least productive?

An important application of the bar graph used in this book is the *age structure diagram* (see Figure 6-7, p. 102), which describes a population by showing the numbers of males and females in certain age groups (see Chapter 6, pp. 102–103).

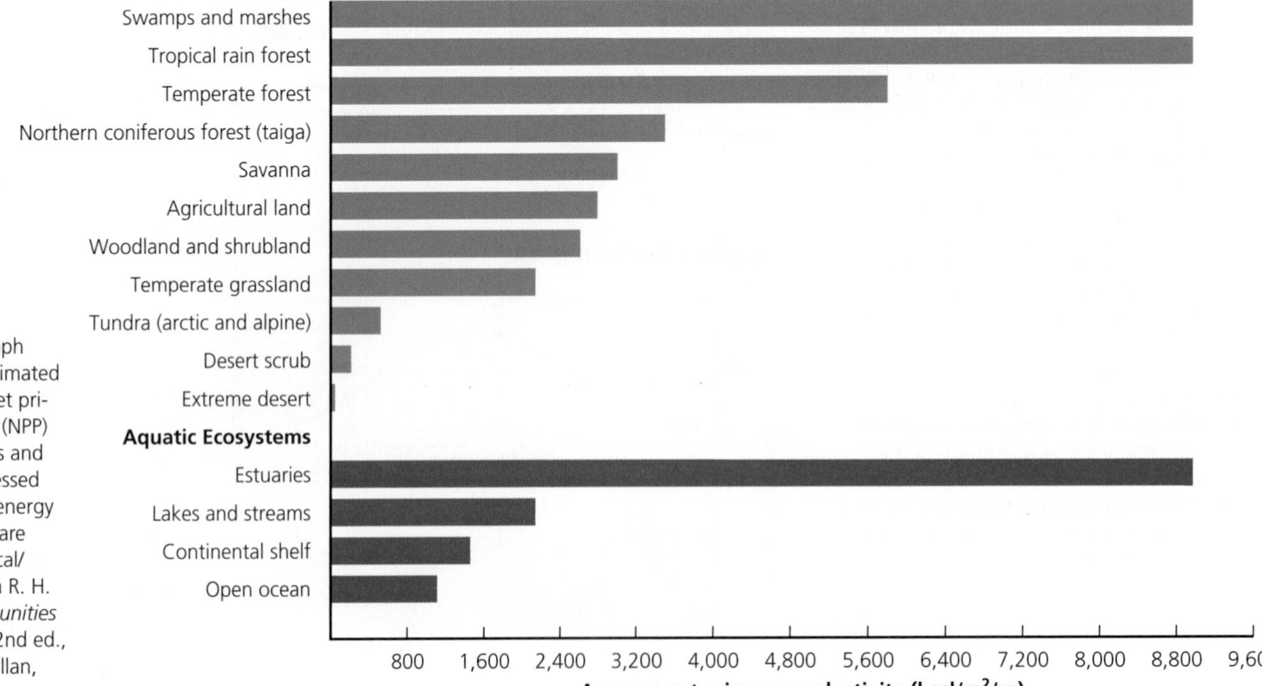

Figure 4 This graph represents the estimated annual average net primary productivity (NPP) in major life zones and ecosystems, expressed as kilocalories of energy produced per square meter per year (kcal/m^2/yr). (Data from R. H. Whittaker, *Communities and Ecosystems*, 2nd ed., New York: Macmillan, 1975)

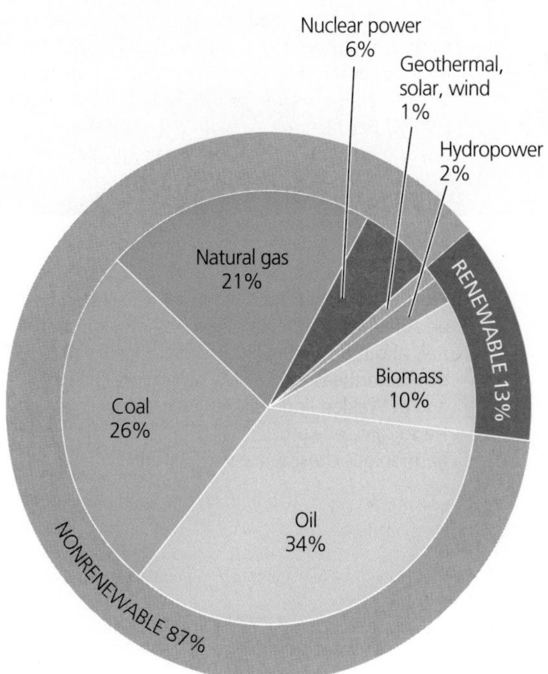

Figure 5 This pie graph represents world energy use by source in 2009.

Pie Graphs

Like bar graphs, *pie graphs* illustrate numerical values for two or more categories. But in addition to that, they can also show each category's proportion of the total of all measurements. The categories are usually ordered on the graph from largest to smallest, for ease of comparison, although this is not always the case. Also, as with bar graphs, pie graphs are generally snapshots of a data set at a point in time or during a defined time period. Unlike line graphs, one pie graph cannot show changes over time.

For example, Figure 5 shows how much each major energy source contributes to the world's total amount of energy used. This graph includes the numerical data used to construct it: the percentages of the total taken up by each part of the pie. But we can use pie graphs without including the numerical data and we can roughly estimate such percentages. The pie graph thereby provides a generalized picture of the composition of a data set.

Figure 13-3 (p. 303) shows this and other data in more detail, and illustrates how we can use pie graphs to compare different groups of categories and different data sets. Also, see the Active Graphing exercise for Chapter 13 on the website for this book.

Questions

1. Do you think that the data showing growth in the use of natural gas, coal, and oil over the years means that the use of other energy categories on this graph has shrunk in that time? Explain.
2. About how many times bigger is oil use than biomass use?

Reading Maps

We can use maps for considerably more than showing where places are relative to one another. For example, in environmental science, maps can be very helpful in comparing how people or different areas are affected by environmental problems such as air pollution and acid deposition (a form of air pollution). Figure 6 is a map of the United States showing the relative numbers of premature deaths due to air pollution in the various regions of the country.

Questions

1. Which part of the country generally has the lowest level of premature deaths due to air pollution?
2. Which part of the country has the highest level? What is the level in the area where you live or go to school?

The Active Graphing exercises available for various chapters on this textbook's website will help you to apply this information on graphs. Go to **www.cengagebrain.com** (see p. xvi in the Preface for details). Choose a chapter with an Active Graphing exercise, click on the exercise, and begin learning more about graphing. There is also a data analysis exercise at the end of each chapter in this book. Some of these exercises involve analysis of various types of graphs and maps.

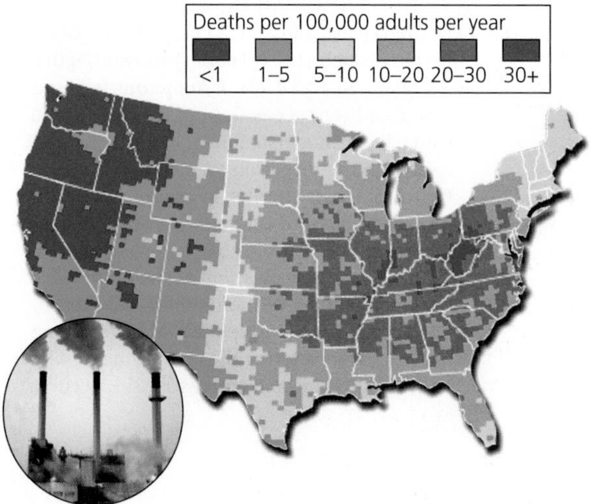

Figure 6 This map compares rates of death due to health effects of air pollution in the United States. (Data from U.S. Environmental Protection Agency)

SUPPLEMENT

3 Environmental History of the United States (Chapters 4, 11, 17)

The Four Major Eras of U.S. Environmental History

We can divide the environmental history of the United States into four eras. During the *tribal era*, people (Native Americans or members of the First Nations) representing several hundred tribes distinguished by language and culture occupied North America for at least 13,000 years before European settlers began arriving in the early 1600s. Some of the tribes were made up of nomadic hunter–gatherers, while others settled into communities centered around agriculture. Some of the more populous Native American societies were more sophisticated than their counterparts in Europe in terms of agricultural practices and other cultural aspects. Yet these North American tribes generally had more sustainable, low-impact ways of life because of their relatively limited numbers and modest rates of resource use per person.

Next was the *frontier era* (1607–1890) when European colonists settled in North America. Faced with a continent offering seemingly inexhaustible resources, the early colonists developed a **frontier environmental worldview**. They saw a wilderness to be conquered and managed for human use.

The *early conservation era* (1832–1870) overlapped the end of the frontier era. During this period, some people became alarmed at the scope of resource depletion and degradation in the United States. They argued that part of the unspoiled wilderness on public lands should be protected as a legacy to future generations. Most of these warnings and ideas were not taken seriously.

This period was followed by an era—lasting from 1870 to the present—featuring an increased role by the federal government and private citizens in resource conservation, public health, and environmental protection.

The Frontier Era (1607–1890)

During the frontier era, European settlers spread across the land by clearing forests for cropland and settlements. In the process, they displaced the Native Americans who, for the most part, had lived on the land sustainably for thousands of years.

The U.S. government accelerated this settling of the continent and the use of its resources by transferring vast areas of public land to private interests. To encourage settlement, between 1850 and 1890 more than half of the country's public land was given away or sold cheaply by the government to railroad, timber, and mining companies, land developers, states, schools, universities, and homesteaders. This era came to an end when the government declared the frontier officially closed in 1890.

Figure 1 Henry David Thoreau (1817–1862) was an American writer and naturalist who kept journals about his excursions into wild areas in parts of the northeastern United States and Canada, and at Walden Pond in Concord, Massachusetts. He sought self-sufficiency, a simple lifestyle, and a harmonious coexistence with nature.

Early Conservationists (1832–1870)

Between 1832 and 1870, some citizens became alarmed at the scope of resource depletion and degradation in the United States. They urged the government to preserve part of the unspoiled wilderness on public lands owned jointly by all people (but managed by the government), and to protect it as a legacy to future generations.

Two of these early conservationists were Henry David Thoreau (1817–1862) and George Perkins Marsh (1801–1882). Thoreau (Figure 1) was alarmed at the loss of numerous wild species from his native eastern Massachusetts. To gain a better understanding of nature, he built a cabin in the woods on Walden Pond near Concord, Massachusetts, lived there alone for 2 years, and wrote *Life in the Woods*, an environmental classic.*

In 1864, Marsh, a scientist and member of Congress from Vermont, published *Man and Nature*, an extensive study of how human activities were altering the environment, which helped legislators and citizens see the need for resource conservation. Marsh questioned the idea that the country's resources were inexhaustible. He also used scientific studies and case studies to show how the rise and fall of past civilizations were linked to the use and misuse of their soils, water supplies, and other resources. Some of his resource conservation principles are still used today.

*I (Miller) can identify with Thoreau. I spent 10 years living in the deep woods studying and thinking about how nature works and writing early editions of the book you are reading. See the whole story of this adventure on p. xxiv.

What Happened between 1870 and 1930?

Between 1870 and 1930, a number of events increased the role of the federal government and private citizens in resource conservation and public health. The *Forest Reserve Act of 1891* was a turning point in establishing the responsibility of the federal government for protecting public lands from resource exploitation.

In 1892, nature preservationist and activist John Muir (1838–1914) (Figure 2) founded the Sierra Club. He had been deeply troubled by the rapid deforestation that he had witnessed in large areas of North America. By 1870, the vast forests of the Northeast had been cut down, and between then and 1920, loggers essentially clear-cut the great pine forests of the upper Midwest—an area the size of Europe.

Muir became the leader of the *preservationist movement*, which called for protecting large areas of wilderness on public lands from human exploitation, except for low-impact recreational activities such as hiking and camping. However, this idea was not enacted into law until 1964. Muir also proposed and lobbied for creation of

Figure 2 John Muir (1838–1914) was a Scottish geologist, explorer, and naturalist. He spent 6 years studying, writing journals, and making sketches in the wilderness of California's Yosemite Valley and then went on to explore wilderness areas in Utah, Nevada, the Northwest, and Alaska. He was largely responsible for establishing Yosemite National Park in 1890. He also founded the Sierra Club and spent 22 years lobbying actively for conservation laws.

a national park system on public lands. In addition, he founded the Sierra Club, which is to this day a political force working on behalf of the environment.

Primarily because of political opposition, effective protection of forests and wildlife on federal lands did not begin until Theodore Roosevelt (1858–1919) (Figure 3), an ardent conservationist, became president. His term of office, 1901–1909, has been called the country's *Golden Age of Conservation*.

Figure 3 Theodore (Teddy) Roosevelt (1858–1919) was a writer, explorer, naturalist, avid birdwatcher, and twenty-sixth president of the United States. He was the first national political figure to bring conservation issues to the attention of the American public. According to many historians, Theodore Roosevelt contributed more than any other U.S. president to natural resource conservation in the United States.

While in office, Roosevelt persuaded Congress to give the president power to designate public land as federal wildlife refuges. He was motivated partly by the fates of prominent wildlife species such as the passenger pigeon (extinct by 1900) and the North American bison, or American buffalo, which was nearly wiped out. In 1903, Roosevelt established the first federal refuge at Pelican Island (see Figure 8-17, p. 170) off the east coast of Florida for preservation of the endangered brown pelican, and he added 35 more wildlife reserves by 1904. He also more than tripled the size of the national forest reserves.

In 1905, Congress created the U.S. Forest Service to manage and protect the forest reserves. Roosevelt appointed Gifford Pinchot (1865–1946) as its first chief. Pinchot pioneered scientific management of forest resources on public lands. In 1906, Congress passed the *Antiquities Act,* which allows the president to protect areas of scientific or historical interest on federal lands as national monuments. Roosevelt used this act to protect the Grand Canyon and other wilderness areas that would later become national parks.

Congress became upset with Roosevelt in 1907, because by then he had added vast tracts to the national forest reserves. Congress passed a law banning further executive withdrawals of public forests. However, on the day before the bill became law, Roosevelt defiantly reserved another large block of land. Most environmental historians view Roosevelt as the country's best environmental president.

Early in the 20th century, the U.S. conservation movement split into two factions over how public lands should be used. The *wise-use,* or *conservationist,* school, led by Roosevelt and Pinchot, believed all public lands should be managed wisely and scientifically to provide needed resources. The *preservationist* school, led by Muir, wanted wilderness areas on public lands to be left untouched. This controversy over use of public lands continues today.

In 1916, Congress passed the *National Park Service Act,* which declared that parks are to be maintained in a manner that leaves them unimpaired for future generations. The act also established the National Park Service (within the Department of the Interior) to manage the park system. Under its first head, Stephen T. Mather (1867–1930), the dominant park policy was to encourage tourist visits by allowing private concessionaires to operate facilities within the parks.

After World War I, the country entered a new era of economic growth and expansion. To stimulate economic growth during the administrations of Presidents Harding, Coolidge, and Hoover, the federal government promoted the increased sales, at low prices, of timber, energy, mineral, and other resources found on public lands.

President Herbert Hoover went even further and proposed that the federal government return all remaining federal lands to the states or sell them to private interests for economic development. But the Great Depression (1929–1941) made owning such lands unattractive to state governments and private investors. The Depression was bad news for the country. But some say that without it we might have little if any of the public lands that today make up about one-third of the total land area of the United States (see Figure 17-13, p. 448).

What Happened between 1930 and 1960?

Along with a second wave of national resource conservation, improvements in public health also began in the early 1930s as President Franklin D. Roosevelt (1882–1945) strove to bring the country out of the Great Depression. He persuaded Congress to enact federal programs to provide jobs and to help restore the country's degraded environment.

During this period, the government purchased large tracts of land from cash-poor landowners, and established the *Civilian Conservation Corps* (CCC) in 1933. According to the U.S. Forest Service, it eventually put more than 3 million unemployed people to work planting trees and developing and maintaining parks and recreation areas. During its 10 years of existence, the CCC planted an estimated 3 billion trees. It also restored silted waterways and built levees and dams for flood control.

The government also built and operated many large dams in the Tennessee River Valley, as well as in the arid western states, including Hoover Dam on the Colorado River (see Figure 11-1, p. 237). The goals were to provide jobs, flood control, cheap irrigation water, and cheap electricity for industry.

In 1935, Congress passed the Soil Conservation Act. During the Great Depression, erosion problems had ruined many farms in the Great Plains states and created a large area of degraded land known as the *Dust Bowl.* To correct these devastating erosion problems, the new law established the *Soil Erosion Service* as part of the Department of Agriculture. Its name was later changed to the *Soil Conservation Service,* and now it is called the *Natural Resources Conservation Service.* Many environmental historians praise Franklin D. Roosevelt for his efforts to get the

Figure 4 Aldo Leopold (1887–1948) worked in game management early in his career, but then grew interested in the emerging scientific field of ecology. He became a leading conservationist and his book, *A Sand County Almanac* (published after his death), is considered an environmental classic that helped to inspire the modern environmental and conservation movements.

Tom Coleman/Aldo Leopold Foundation

country out of a major economic depression, partly by helping to restore environmentally degraded areas.

Also, in 1935, Aldo Leopold (1887–1948) (Figure 4), wildlife manager, professor, writer, and conservationist, helped to found the U.S. Wilderness Society. Largely through his writings, especially his 1949 essay *The Land Ethic*, and his 1949 book *A Sand County Almanac*, he became one of the foremost leaders of the *conservation* and *environmental movements*. His energy and foresight helped to lay the critical groundwork for the field of environmental ethics. Leopold contended that the role of the human species should be to protect nature, not to conquer it.

Federal resource conservation policy changed little during the 1940s and 1950s, mostly because of preoccupation with World War II (1941–1945) and economic recovery after the war.

Between 1930 and 1960, improvements in public health included establishment of public health boards and agencies at the municipal, state, and federal levels; increased public education about health issues; introduction of vaccination programs; and a sharp reduction in the incidence of waterborne infectious diseases, mostly because of improved sanitation and garbage collection.

What Happened during the 1960s?

A number of milestones in American environmental history occurred during the 1960s. In 1962, biologist Rachel Carson (1907–1964) published *Silent Spring*, which documented the pollution of air, water, and wildlife from the use of pesticides such as DDT (see Chapter 10, Individuals Matter, p. 221). This influential book helped to broaden the concept of resource conservation to include preservation of the *quality* of the planet's air, water, soil, and wildlife.

Many environmental historians mark Carson's wake-up call as the beginning of the modern **environmental movement** in the United States. It consists of citizens organized to demand that political leaders enact laws and develop policies to curtail pollution, clean up polluted environments, and protect unspoiled areas from environmental degradation.

In 1964, Congress passed the *Wilderness Act*, inspired by the vision of John Muir more than 80 years earlier. It authorized the government to protect undeveloped tracts of public land as part of the National Wilderness System, unless Congress later decides they are needed for the national good. Land in this system is to be used only for nondestructive forms of recreation such as hiking and camping.

Between 1965 and 1970, the emerging science of *ecology* received widespread media attention. At the same time, the popular writings of biologists such as Paul Ehrlich, Barry Commoner, and Garrett Hardin awakened Americans to the interlocking relationships among population growth, resource use, and pollution. During that period, a number of events increased public awareness of pollution, habitat loss, and other forms of environmental degradation. For example, many people learned about these problems when well-known wildlife species such as the American bald eagle, the grizzly bear, the whooping crane, and the peregrine falcon became endangered. Also, the stretch of the Cuyahoga River running near Cleveland, Ohio, was so polluted with oil and other flammable pollutants that it caught fire several times. Another major event was a devastating oil spill off the California coast in 1969.

In 1968, when U.S. astronauts aboard Apollo 8 became the first humans to fly to the moon, they photographed the entire earth for the first time from lunar orbit. This allowed people to see the earth as a tiny blue and white planet in the black void of space, and it led to the development of the *spaceship-earth environmental worldview*. According to this view, we live on a planetary spaceship that we should not harm because it is the only home we have.

What Happened during the 1970s? The Environmental Decade

During the 1970s, media attention, public concern about environmental problems, scientific research, and action to address environmental concerns grew rapidly. This period is sometimes called the *environmental decade*, or the *first decade of the environment*.

During the 1970s, several major U.S. environmental laws were passed. The first and possibly most important of these laws, signed by President Richard M. Nixon on January 1, 1970, was the *National Environmental Policy Act* (NEPA). It established the *environmental impact statement* (EIS) process. An EIS is now required for any action taken by the federal government that could significantly affect on the environment. The EIS process forces the government and businesses that receive government contracts to carefully evaluate environmental impacts and avoid those with the potential for significant environmental damage. Many activities, including road building, dam construction, and logging on federal lands, are affected. NEPA set in motion a legislative process that produced several major environmental laws (see Figure 17-17, p. 452).

The first annual *Earth Day* was held on April 20, 1970. This event was proposed by Senator Gaylord Nelson (1916–2005) and organized by then Harvard graduate student Denis Hayes (see Individuals Matter, p. 451). Some 20 million people in more than 2,000 U.S. communities took to the streets to heighten the nation's environmental awareness and to demand improvements in environmental quality. There were also many environmental efforts that took place on a smaller scale that were greatly instructive for the environmental movement.

The *Environmental Protection Agency* (EPA) was established in 1970. In addition, the *Endangered Species Act of 1973* greatly strengthened the role of the federal government in protecting endangered species and their habitats. In addition, Congress created the Department of Energy in 1977, charging this new agency with developing a long-range energy strategy to help reduce the country's heavy dependence on imported oil.

During the 1970s, the area of land in the National Wilderness System tripled and the area in the National Park System doubled (primarily because vast tracts of land in the state of Alaska were added to these systems). In 1978, the *Federal Land Policy and Management Act* gave the *Bureau of Land Management* (BLM) its first real authority to manage the public land under its control, 85% of which is in 12 western states. This law angered a number of western interests whose use of these public lands was restricted for the first time.

In response, a coalition of ranchers, miners, loggers, developers, farmers, some elected officials, and other citizens in the affected states launched a political campaign known as the *sagebrush rebellion*. It had two major goals. *First*, sharply reduce government regulation of the use of public lands. *Second*, remove most public lands in the western United States from federal ownership and management and turn them over to the states. After that, the plan was to persuade state legislatures to sell or lease the resource-rich lands at low prices to ranching, mining, timber, land development, and other private interests. This represented a return to President Herbert Hoover's plan to get rid of all public land, which had been thwarted by the Great Depression. This political movement continues to exist.

What Happened during the 1980s? Environmental Backlash

In 1980, Congress created the *Superfund* as part of the *Comprehensive Environmental Response, Compensation, and Liability Act* (CERCLA, see Chapter 16, p. 426). Its goal was to clean up abandoned hazardous waste sites, including the Love Canal housing development in Niagara Falls, New York, which had to be abandoned when hazardous wastes from the site of a former chemical company began leaking into school grounds, yards, and basements.

During the decade of the 1980s, farmers, ranchers, and leaders of the oil, coal, automobile, mining, and timber industries strongly opposed many of the environmental laws and regulations developed in the 1960s and 1970s. They organized and funded multiple efforts to defeat environmental laws and regulations. In 1988, these industries backed a new coalition called the *wise-use movement*. Its major goals were to weaken or repeal most of the country's environmental laws and regulations, and destroy the effectiveness of the environmental movement in the United States. Backers of this movement argued that environmental laws had gone too far and were hindering economic growth.

Congress, influenced by this growing backlash, allowed for much more private energy and mineral development, and timber cutting on public lands. Congress also lowered automobile gas mileage standards and relaxed federal air and water quality pollution standards.

In 1980, the United States had led the world in research and development of wind and solar energy technologies. Between 1981 and 1983, however, Congress slashed by 90% government subsidies for renewable energy research and for energy efficiency research, and eliminated tax incentives for the residential solar energy and energy conservation programs enacted in the late 1970s. As a result, the United States lost its lead in developing and selling the wind turbines and solar cells to Denmark, Germany, Japan, and China. These businesses are rapidly becoming two of the biggest and most profitable industries in the world.

At the same time, the 1980s saw rising public interest in some environmental issues, largely due to the influence of some prominent environmental scientists. For example, oceanographer Jacques Cousteau continued to tell the story of his four decades of undersea explorations and his concerns for the ocean environment through dozens of TV programs, books, and films. In Africa, zoologist Dian Fossey studied mountain gorillas and raised public awareness about these endangered animals, their shrinking habitats, and the threat of poachers. Her murder in 1985, possibly at the hands of poachers, made her story even more compelling.

What Happened from 1990 to 2011?

Between 1990 and 2011, opposition to environmental laws and regulations gained strength. This occurred because of continuing political and economic support from corporate backers, who not only argued that environmental laws were hindering economic growth, but also helped elect many members of Congress who were generally unsympathetic to environmental concerns. Since 1990, leaders and supporters of the environmental movement have had to spend much of their time and funds fighting efforts to discredit the movement and efforts to weaken or eliminate most environmental laws passed during the 1960s and 1970s.

During the 1990s, many small and mostly local grassroots environmental organizations sprang up to help deal with environmental threats in their local communities. Interest in environmental issues increased on many college and university campuses, resulting in the expansion of environmental studies courses and programs at these institutions. In addition, there was growing awareness of critical and complex environmental issues, such as sustainability, population growth, biodiversity protection, and threats from atmospheric warming and projected climate change. Whereas 20 million Americans took part in the first Earth Day in 1970, an estimated 200 million people from more than 140 countries participated in Earth Day 1990.

During the first decade of the 21st century, some environmental progress was made as a number of business leaders began urging the United States to begin shifting from fossil fuels to a mix of renewable energy resources such as solar energy, wind power, and geothermal energy. Many of these leaders also put a high priority on saving their companies money by improving their energy efficiency and reducing their unnecessary energy waste.

In addition, during this decade many analysts as well as business and government leaders have called for developing an effective U.S. policy for slowing projected climate change caused mostly by the burning of fossil fuels. However, the politically and economically powerful fossil fuel industries have been able to block efforts to develop a comprehensive energy policy as well as any U.S. policy for dealing with the projected threat of climate disruption.

SUPPLEMENT

4 Some Basic Chemistry (Chapters 1–5 and 10–16)

Chemists Use the Periodic Table to Classify Elements on the Basis of Their Chemical Properties

The basic unit of each element is a unique *atom* that is different from the atoms of all other elements. Each atom consists of an extremely small and dense center called its *nucleus,* which contains one or more protons and, in most cases, one or more neutrons, as well as one or more electrons moving rapidly somewhere around the nucleus (Figure 1). Each atom has equal numbers of positively charged protons and negatively charged electrons. Because these electrical charges cancel one another, *atoms as a whole have no net electrical charge*.

We cannot determine the exact location of the electrons around any nucleus. Instead, we can estimate the *probability* that they will be found at various locations outside the nucleus—sometimes called an *electron probability cloud*. This is somewhat like saying that there are six airplanes flying around inside a cloud. We do not know their exact location, but the cloud represents an area in which we can probably find them.

Chemists have developed a way to classify the elements according to their chemical behavior in what is called the *periodic table of elements* (Figure 2). Each horizontal row in the table is called a *period*. Each vertical column lists elements with similar chemical properties and is called a *group*.

The periodic table in Figure 2 shows how the elements can be classified as *metals, nonmetals,* and *metalloids*. Examples of metals are sodium (Na), calcium (Ca), aluminum (Al), iron (Fe), lead (Pb), silver (Ag), and mercury (Hg).

Atoms of metals tend to lose one or more of their electrons to form positively charged ions such as Na^+, Ca^{2+}, and Al^{3+}. For example, an atom of the metallic element sodium (Na, atomic number 11) with 11 positively charged protons and 11 negatively charged electrons can lose one of its electrons. It then becomes a sodium ion with a positive charge of 1 (Na^+) because it now has 11 positive charges (protons) but only 10 negative charges (electrons). Examples of *nonmetals* are hydrogen (H), carbon (C), nitrogen (N), oxygen (O), phosphorus (P), sulfur (S), chlorine (Cl), and fluorine (F). Atoms of some nonmetals such as chlorine, oxygen, and sulfur tend to gain one or more electrons lost by metallic atoms to form negatively charged ions such as O^{2-}, S^{2-}, and Cl^-. For example, an atom of the nonmetallic element chlorine (Cl, atomic number 17) can gain an electron and become a chlorine ion. The ion has a negative charge of 1 (Cl^-) because it has 17 positively charged protons and 18 negatively charged electrons.

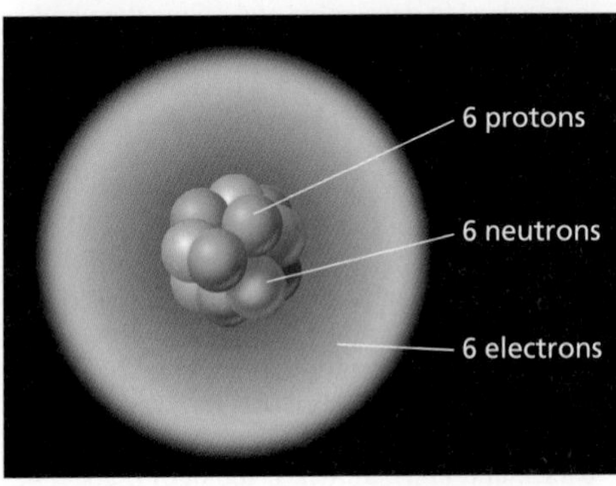

Figure 1 This is a greatly simplified model of a carbon-12 atom. It consists of a nucleus containing six protons with a positive electrical charge and six neutrons with no electrical charge. Six negatively charged electrons are found outside its nucleus.

Atoms of nonmetals can also combine with one another to form molecules in which they share one or more pairs of their electrons. Hydrogen, a nonmetal, is placed by itself above the center of the table because it does not fit very well into any of the groups.

The elements arranged in a diagonal staircase pattern between the metals and nonmetals have a mixture of metallic and nonmetallic properties and are called *metalloids*. Examples are germanium (Ge) and arsenic (As).

Figure 2 also identifies the elements required as *nutrients* (black squares) for all or some forms of life, and elements that are moderately or highly toxic (red squares) to all or most forms of life. Six nonmetallic elements—carbon (C), oxygen (O), hydrogen (H), nitrogen (N), sulfur (S), and phosphorus (P)—make up about 99% of the atoms of all living things.

THINKING ABOUT
The Periodic Table

Use the periodic table to identify by name and symbol two elements that should have chemical properties similar to those of **(a)** Ca, **(b)** potassium, **(c)** S, **(d)** lead.

Ionic and Covalent Bonds Hold Compounds Together

Sodium chloride (NaCl) consists of a three-dimensional network of oppositely charged *ions* (Na^+ and Cl^-) held together by the forces of attraction between opposite charges (Figure 3). The strong forces of attraction between such oppositely charged ions are called *ionic bonds.* Because ionic compounds consist of ions formed from atoms of metallic (positive ions) and nonmetallic (negative ions) elements (Figure 2), they can be described as *metal–nonmetal compounds.*

Sodium chloride and many other ionic compounds tend to dissolve in water and break apart into their individual ions (Figure 4, p. S12).

Water, a *covalent compound,* consists of molecules made up of uncharged atoms of hydrogen (H) and oxygen (O). Each water molecule consists of two hydrogen atoms chemically bonded to an oxygen atom, yielding H_2O molecules. The bonds between the atoms in such molecules are called *covalent bonds* and form when the atoms in the molecule share one or more pairs of their electrons. Because they are formed from atoms of nonmetallic elements (Figure 2), covalent compounds can be described as *nonmetal–nonmetal compounds*. Figure 5 (p. S12) shows the chemical formulas and shapes of the molecules that are the building blocks for several common *covalent compounds*.

What Makes Solutions Acidic? Hydrogen Ions and pH

The *concentration*, or number of hydrogen ions (H^+) in a specified volume of a solution (typically a liter), is a measure of its acidity. Pure water (not tap water or rainwater) has an equal number of hydrogen (H^+) and hydroxide (OH^-) ions. It is called a **neutral solution**. An **acidic solution** has more hydrogen ions than hydroxide ions per liter. A **basic solution** has more hydroxide ions than hydrogen ions per liter.

Scientists use **pH** as a measure of the acidity of a solution based on its concentration of hydrogen ions (H^+). By definition, a neutral solution has a pH of 7, an acidic solution has a pH of less than 7, and a basic solution has a pH greater than 7.

Group 1	2	3	4	5	6	7	8	9	10	11	12	13	14	15	16	17	18
1 H hydrogen																	2 He helium
3 Li lithium	4 Be beryllium											5 B boron	6 C carbon	7 N nitrogen	8 O oxygen	9 F fluorine	10 Ne neon
11 Na sodium	12 Mg magnesium											13 Al aluminum	14 Si silicon	15 P phosphorus	16 S sulfur	17 Cl chlorine	18 Ar argon
19 K potassium	20 Ca calcium	21 Sc scandium	22 Ti titanium	23 V vanadium	24 Cr chromium	25 Mn manganese	26 Fe iron	27 Co cobalt	28 Ni nickel	29 Cu copper	30 Zn zinc	31 Ga gallium	32 Ge germanium	33 As arsenic	34 Se selenium	35 Br bromine	36 Kr krypton
37 Rb rubidium	38 Sr strontium	39 Y yttrium	40 Zr zirconium	41 Nb niobium	42 Mo molybdenum	43 Tc technetium	44 Ru ruthenium	45 Rh rhodium	46 Pd palladium	47 Ag silver	48 Cd cadmium	49 In indium	50 Sn tin	51 Sb antimony	52 Te tellurium	53 I iodine	54 Xe xenon
55 Cs cesium	56 Ba barium	57–71	72 Hf hafnium	73 Ta tantalum	74 W tungsten	75 Re rhenium	76 Os osmium	77 Ir iridium	78 Pt platinum	79 Au gold	80 Hg mercury	81 Tl thallium	82 Pb lead	83 Bi bismuth	84 Po polonium	85 At astatine	86 Rn radon
87 Fr francium	88 Ra radium	89–103	104 Rf rutherfordium	105 Db dubnium	106 Sg seaborgium	107 Bh bohrium	108 Hs hassium	109 Mt meitnerium	110 Ds darmstadtium	111 Rg roentgenium	112 Cn copernicium	113 Uut ununtrium	114 Uuq ununquatium	115 Uup ununpentium	116 Uuh ununhexium	117 Uus ununseptium	118 Uuo ununoctium

Key: Atomic number — Symbol — Name

Metals

Nonmetals

Metalloids

Required for all or some life-forms

Moderately to highly toxic

Lanthanides (Rare Earth Elements)	57 La lanthanum	58 Ce cerium	59 Pr praeseidymium	60 Nd neodmium	61 Pm promethium	62 Sm samarium	63 Eu europium	64 Gd gadolinium	65 Tb terbium	66 Dy dysprosium	67 Ho holmium	68 Er erbium	69 Tm thulium	70 Yb ytterbium	71 Lu lutetium
Actinides	89 Ac actinium	90 Th thorium	91 Pa protactinium	92 U uranium	93 Np neptunium	94 Pu plutonium	95 Am americium	96 Cm curium	97 Bk berkelium	98 Cf californium	99 Es einsteinium	100 Fm fermium	101 Md mendelevium	102 No nobelium	103 Lr lawrencium

Figure 2 This is the periodic table of elements. Elements in the same vertical column, called a *group*, have similar chemical properties.

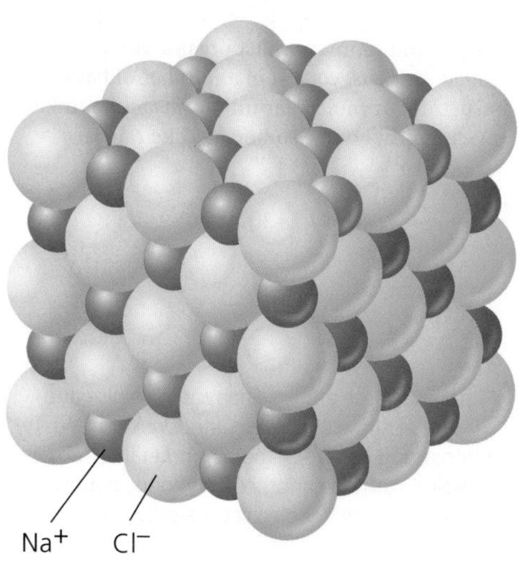

Figure 3 A solid crystal of an ionic compound such as sodium chloride consists of a three-dimensional array of oppositely charged ions held together by *ionic bonds* that result from the strong forces of attraction between opposite electrical charges. They are formed when an electron is transferred from a metallic atom such as sodium (Na) to a nonmetallic element such as chlorine (Cl).

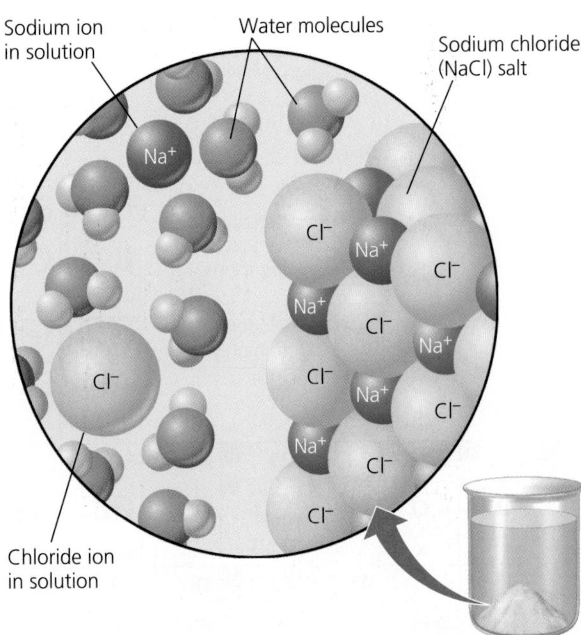

Figure 4 This model illustrates how a salt dissolves in water.

Each single unit change in pH represents a tenfold increase or decrease in the concentration of hydrogen ions per liter. For example, an acidic solution with a pH of 3 is ten times more acidic than a solution with a pH of 4. Figure 6 shows the approximate pH and hydrogen ion concentration per liter of solutions for various common substances.

THINKING ABOUT pH

A solution has a pH of 2. How many times more acidic is this solution than one with a pH of 6?

The measurement of acidity is important in the study of environmental science, as environmental changes involving acidity can have serious environmental impacts. For example, when coal and oil are burned they give off acidic compounds that can return to the earth as *acid deposition* (see Figures 15-5 and 15-6, p. 381), which has become a major environmental problem.

There Are Weak Forces of Attraction between Some Molecules

Ionic and covalent bonds form between the ions or atoms *within* a compound. There are also weaker forces of attraction *between* the molecules of covalent compounds (such as water) resulting from an unequal sharing of electrons by two atoms.

For example, an oxygen atom has a much greater attraction for electrons than does a hydrogen atom. Thus, the electrons shared between the oxygen atom and its two hydrogen atoms in a water molecule are pulled closer to the oxygen atom, but not actually transferred to the oxygen atom. As a result, the oxygen atom in a water molecule has a slightly negative partial charge and its two hydrogen atoms have a slightly positive partial charge (Figure 7).

The slightly positive hydrogen atoms in one water molecule are then attracted to the slightly negative oxygen atoms in another water molecule. These forces of attraction *between* water molecules are called *hydrogen bonds* (Figure 7). They account for many of water's unique properties (see Chapter 3, Science Focus, p. 52). Hydrogen bonds also form between other covalent molecules or between portions of such molecules containing hydrogen and nonmetallic atoms with a strong ability to attract electrons.

Four Types of Large Organic Compounds Are the Molecular Building Blocks of Life

Larger and more complex organic compounds, called *polymers,* consist of a number of basic structural or molecular units (*monomers*) linked by chemical bonds, somewhat like rail cars

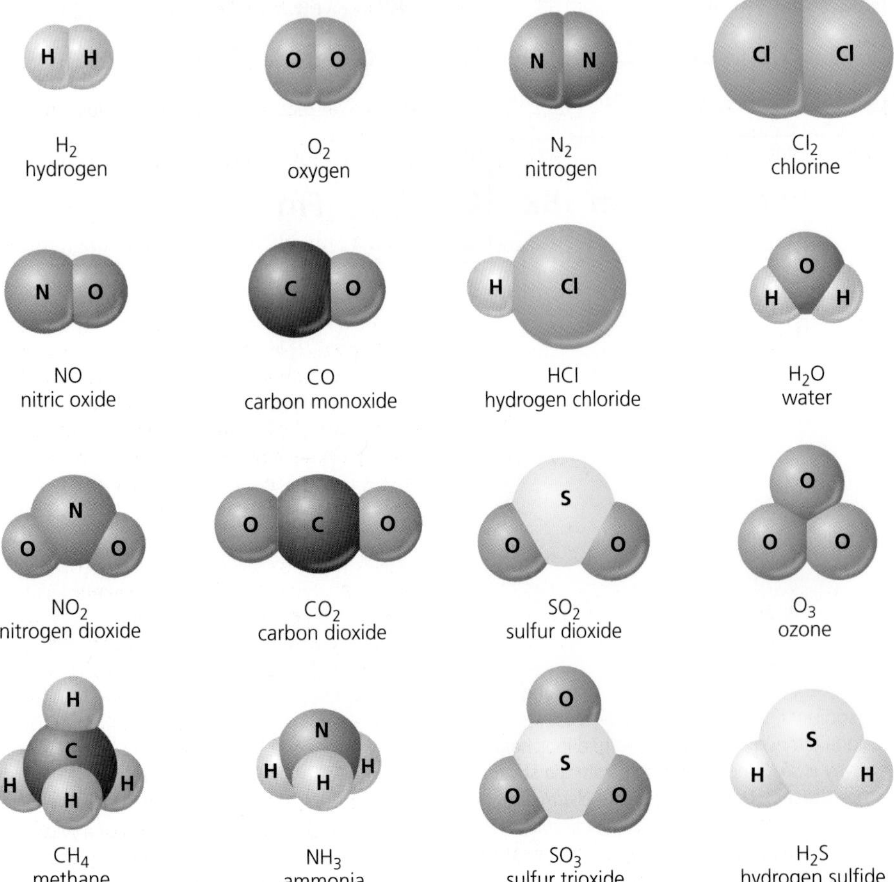

Figure 5 These are the chemical formulas and shapes for some *covalent compounds* formed when atoms of one or more nonmetallic elements combine with one another. The bonds between the atoms in such molecules are called *covalent bonds*.

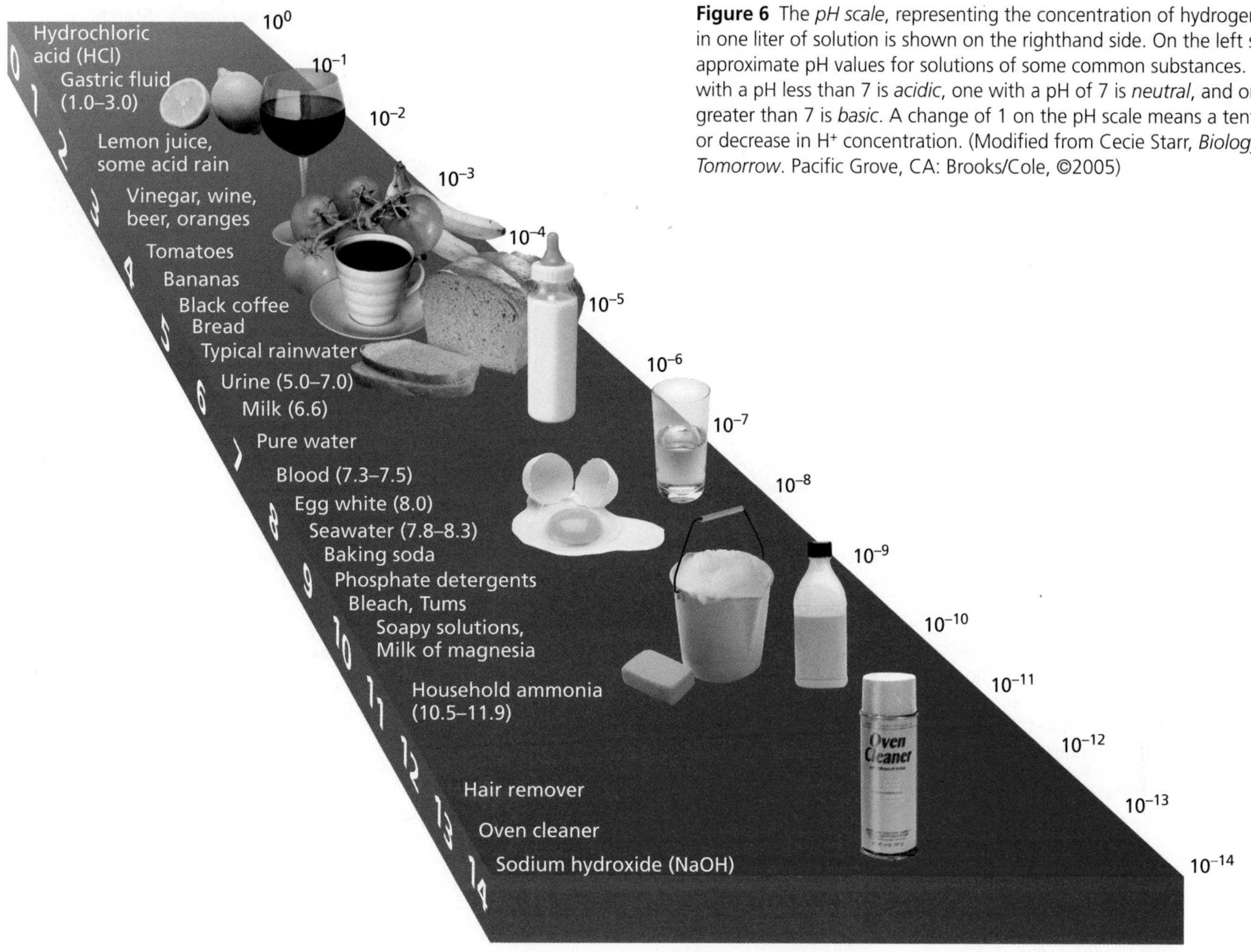

Figure 6 The *pH scale*, representing the concentration of hydrogen ions (H^+) in one liter of solution is shown on the righthand side. On the left side, are the approximate pH values for solutions of some common substances. A solution with a pH less than 7 is *acidic*, one with a pH of 7 is *neutral*, and one with a pH greater than 7 is *basic*. A change of 1 on the pH scale means a tenfold increase or decrease in H^+ concentration. (Modified from Cecie Starr, *Biology: Today and Tomorrow*. Pacific Grove, CA: Brooks/Cole, ©2005)

linked in a freight train. Four types of macromolecules—complex carbohydrates, proteins, nucleic acids, and lipids—are the molecular building blocks of life.

Slightly negative charge
δ−
O
H
δ+
Hydrogen bonds
Slightly positive charge

Figure 7 *Hydrogen bond:* The slightly unequal sharing of electrons in the water molecule creates a molecule with a slightly negatively charged end and a slightly positively charged end. Because of this electrical polarity, the hydrogen atoms of one water molecule are attracted to the oxygen atoms in other water molecules. These fairly weak forces of attraction *between* molecules (represented by the dashed lines) are called *hydrogen bonds*.

Complex carbohydrates consist of two or more monomers of *simple sugars* (such as glucose, Figure 8, p. S14) linked together. One example is the starches that plants use to store energy and also to provide energy for animals that feed on plants. Another is cellulose, the earth's most abundant organic compound, which is found in the cell walls of bark, leaves, stems, and roots.

Proteins are large polymer molecules formed by linking together long chains of monomers called *amino acids* (Figure 9, S14). Living organisms use about 20 different amino acid molecules to build a variety of proteins, which play different roles: Some help to store energy. Some are components of the *immune system* that protects the body against diseases and harmful substances by forming antibodies that make invading agents harmless. Others are *hormones* that are used as chemical messengers in the bloodstreams of animals to turn various bodily functions on or off. In animals, proteins are also components of hair, skin, muscle, and tendons. In addition, some proteins act as *enzymes* that catalyze or speed up certain chemical reactions.

Nucleic acids are large polymer molecules made by linking hundreds to thousands of four types of monomers called *nucleotides*. Two nucleic acids—DNA (**d**eoxyribo**n**ucleic **a**cid) and RNA (**r**ibo**n**ucleic **a**cid)—participate in the building of proteins and carry hereditary information used to pass traits from parent to offspring. Each nucleotide consists of a *phosphate group*, a *sugar molecule* containing five carbon atoms (deoxyribose in DNA molecules

Figure 8 Straight-chain and ring structural formulas of glucose, a simple sugar that can be used to build long chains of complex carbohydrates such as starch and cellulose.

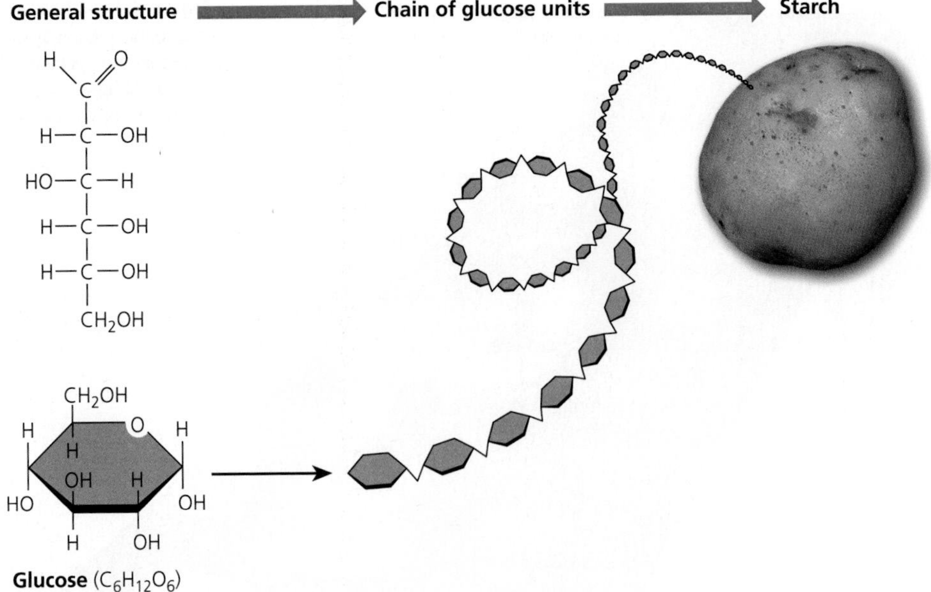

Figure 9 This model illustrates both the general structural formula of amino acids and a specific structural formula of one of the 20 different amino acid molecules that can be linked together in chains to form proteins that fold up into more complex shapes.

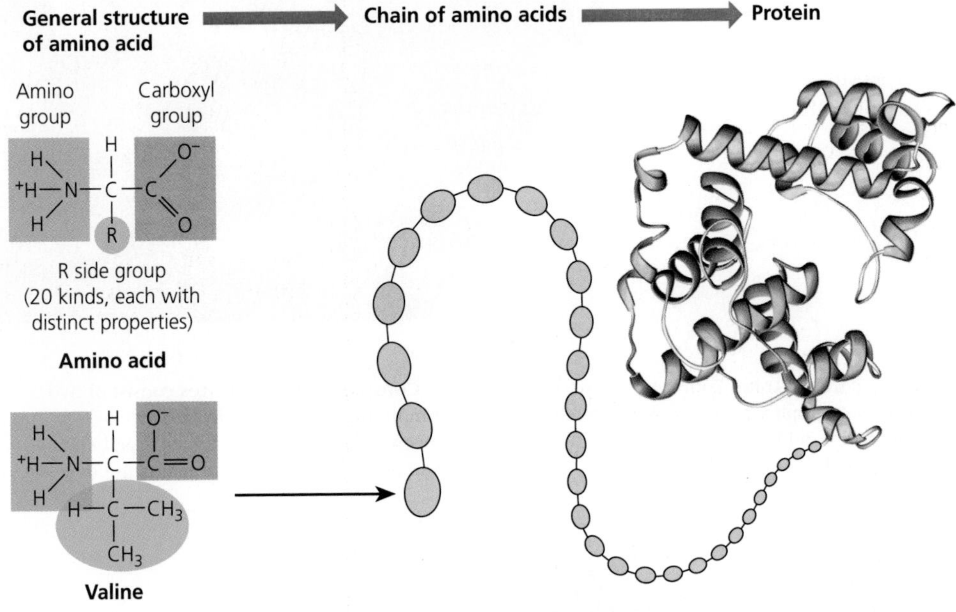

and ribose in RNA molecules), and one of four different *nucleotide bases* (represented by A, G, C, and T, the first letter in each of their names, or A, G, C, and U in RNA) (Figure 10). In the cells of living organisms, these nucleotide units combine in different numbers and sequences to form *nucleic acids* such as various types of DNA and RNA (Figure 11).

Hydrogen bonds formed between parts of the four nucleotides in DNA hold two DNA strands together like a spiral staircase, forming a double helix (Figure 11). DNA molecules can unwind and replicate themselves.

The total weight of the DNA needed to reproduce all of the world's people is only about 50 milligrams—the weight of a small match. If the DNA coiled in your body were unwound, it would stretch about 960 million kilometers (600 million miles)—more than six times the distance between the sun and the earth.

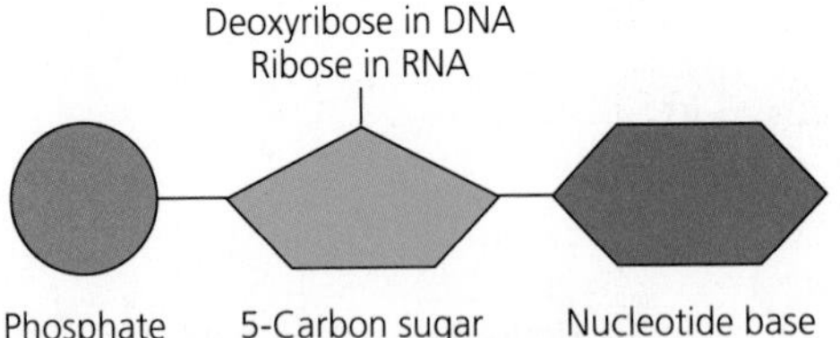

Figure 10 This diagram shows the generalized structures of the nucleotide molecules linked in various numbers and sequences to form large nucleic acid molecules such as various types of DNA (deoxyribonucleic acid) and RNA (ribonucleic acid). In DNA, the five-carbon sugar in each nucleotide is deoxyribose; in RNA it is ribose. The four basic nucleotides used to make various forms of DNA molecules differ in the types of nucleotide bases they contain—guanine (G), cytosine (C), adenine (A), and thymine (T). (Uracil, labeled U, occurs instead of thymine in RNA.)

The different molecules of DNA that make up the millions of species found on the earth are like a vast and diverse genetic library. Each species is a unique book in that library. The *genome* of a species is made up of the entire sequence of DNA "letters" or base pairs that combine to "spell out" the chromosomes in typical members of each species. In 2002, scientists were able to map out the genome for the human species by analyzing the 3.1 billion base sequences in human DNA.

Lipids, a fourth building block of life, are a chemically diverse group of large organic compounds that do not dissolve in water. Examples are *fats and oils* for storing energy (Figure 12), *waxes* for structure, and *steroids* for producing hormones.

Figure 13 shows the relative sizes of simple and complex molecules, cells, and multicelled organisms.

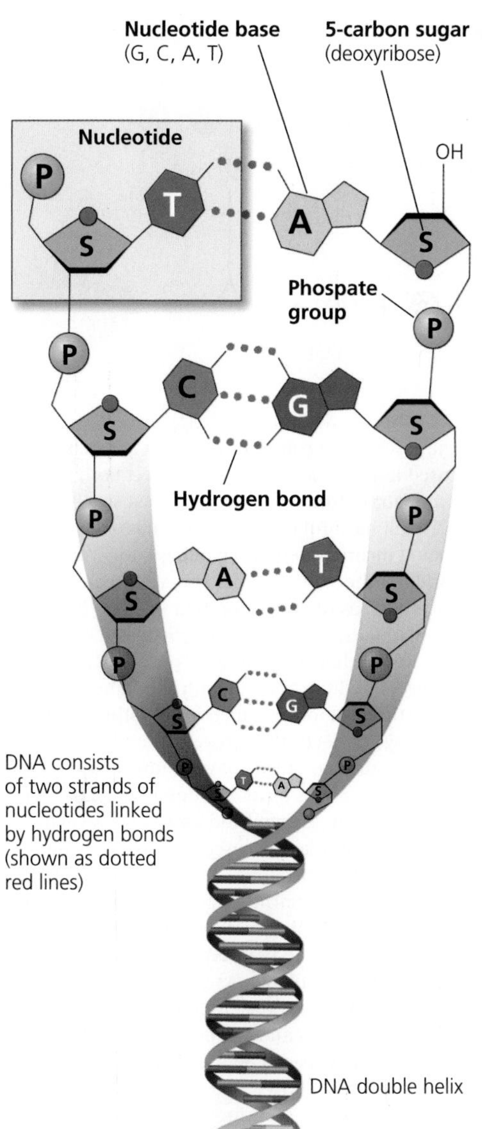

Figure 11 A portion of the double helix of a DNA molecule is shown here. The double helix is composed of two spiral (helical) strands of nucleotides. Each nucleotide contains a unit of phosphate (P), deoxyribose (S), and one of four nucleotide bases: guanine (G), cytosine (C), adenine (A), and thymine (T). The two strands are held together by hydrogen bonds formed between various pairs of the nucleotide bases. Guanine (G) bonds with cytosine (C), and adenine (A) with thymine (T).

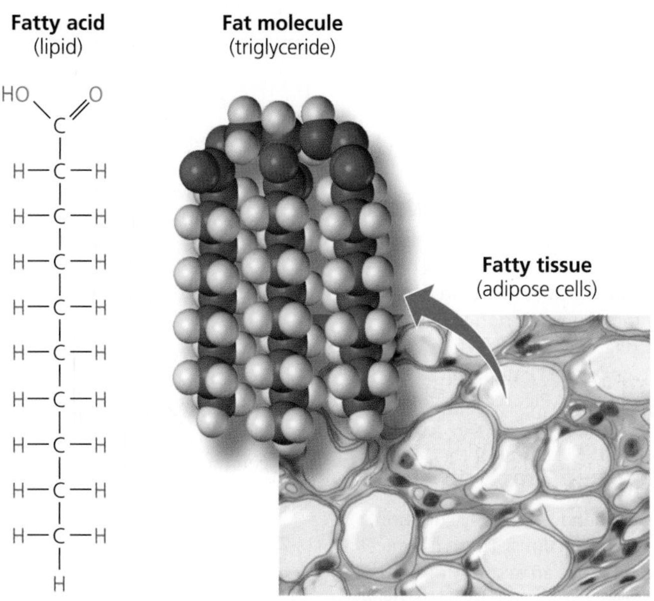

Figure 12 The structural formula of fatty acid that is one form of lipid (left) is shown here. Fatty acids are converted into more complex fat molecules (center) that are stored in adipose cells (right).

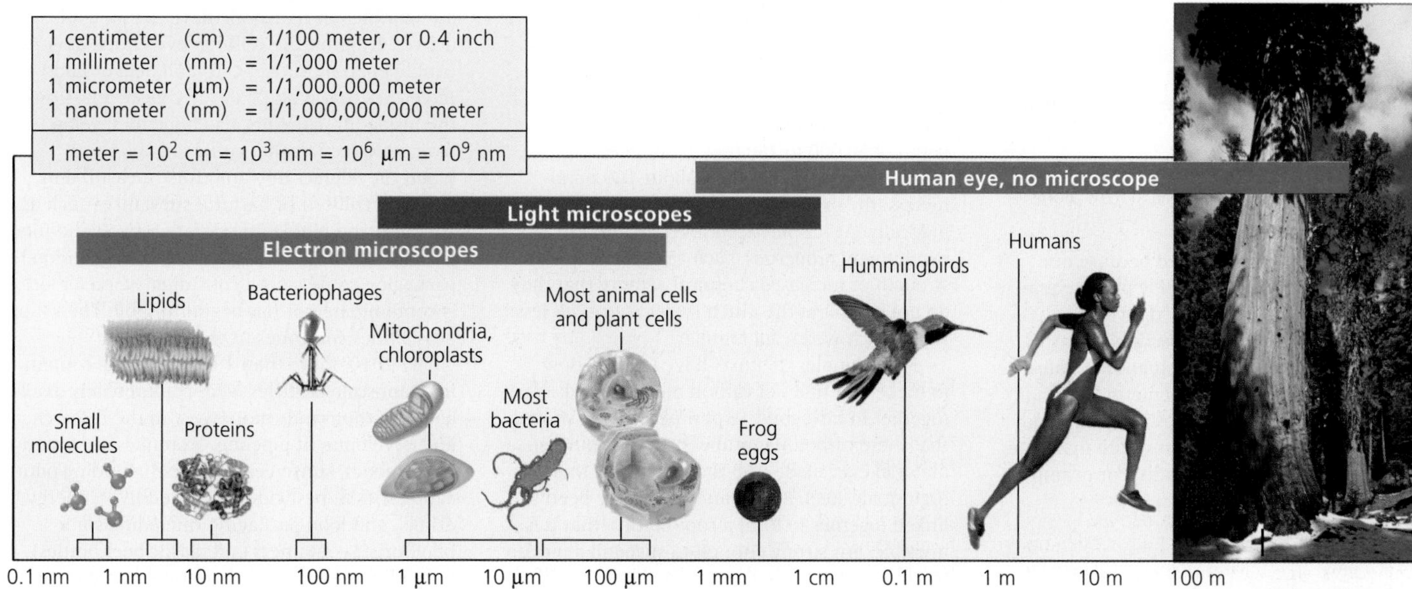

Figure 13 This chart compares the relative size of simple molecules, complex molecules, cells, and multicellular organisms. This scale is exponential, not linear. Each unit of measure is ten times larger than the unit preceding it. (Used by permission from Cecie Starr and Ralph Taggart, *Biology*, 11th ed. Belmont, CA: Thomson Brooks/Cole, © 2006)

Figure 14 These models represent energy storage and release in cells.

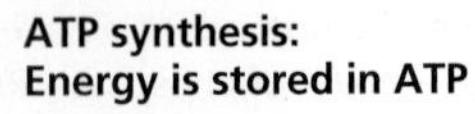

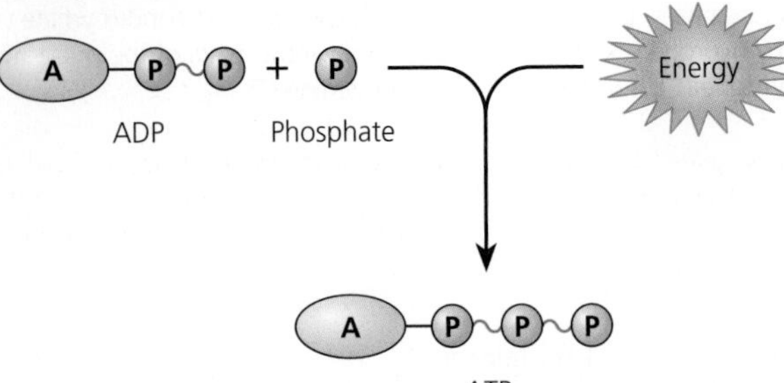

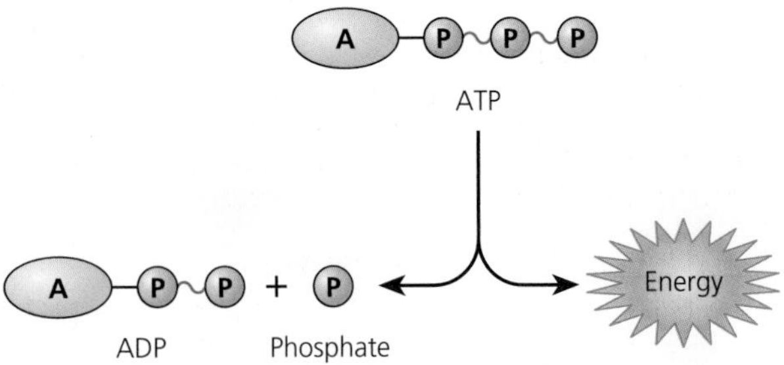

Certain Molecules Store and Release Energy in Cells

Chemical reactions occurring in plant cells during photosynthesis (see Chapter 3, p. 43) release energy that is absorbed by adenosine diphosphate (ADP) molecules and stored as chemical energy in adenosine triphosphate (ATP) molecules (Figure 14, left). When cellular processes require energy, ATP molecules release it to form ADP molecules (Figure 14, right).

Chemists Balance Chemical Equations to Keep Track of Atoms

Chemists use a shorthand system, or equation, to represent chemical reactions. These chemical equations are also used as an accounting system to verify that no atoms are created or destroyed in a chemical reaction as required by the law of conservation of matter (see Chapter 2, p. 34). As a consequence, each side of a chemical equation must have the same number of atoms or ions of each element involved. Ensuring that this condition is met leads to what chemists call a *balanced chemical equation*. The equation for the burning of carbon ($C + O_2 \longrightarrow CO_2$) is balanced because one atom of carbon and two atoms of oxygen are on both sides of the equation.

Consider the following chemical reaction: When electricity passes through water (H_2O), the latter can be broken down into hydrogen (H_2) and oxygen (O_2), as represented by the following equation:

$$H_2O \longrightarrow H_2 + O_2$$

2 H atoms → 2 H atoms, 2 O atoms
1 O atom

This equation is unbalanced because one atom of oxygen is on the left side of the equation but two atoms are on the right side.

We cannot change the subscripts of any of the formulas to balance this equation because that would change the arrangements of the atoms, leading to different substances. Instead, we must use different numbers of the molecules involved to balance the equation. For example, we could use two water molecules:

$$2\,H_2O \longrightarrow H_2 + O_2$$

4 H atoms → 2 H atoms, 2 O atoms
2 O atoms

This equation is still unbalanced. Although the numbers of oxygen atoms on both sides of the equation are now equal, the numbers of hydrogen atoms are not.

We can correct this problem by having the reaction produce two hydrogen molecules:

$$2\,H_2O \longrightarrow 2\,H_2 + O_2$$

4 H atoms → 4 H atoms, 2 O atoms
2 O atoms

Now the equation is balanced, and the law of conservation of matter has been observed. For every two molecules of water through which we pass electricity, two hydrogen molecules and one oxygen molecule are produced.

> **THINKING ABOUT**
> **Chemical Equations**
>
> Try to balance the chemical equation for the reaction of nitrogen gas (N_2) with hydrogen gas (H_2) to form ammonia gas (NH_3).

Scientists Are Learning How to Build Materials from the Bottom Up

Nanotechnology (see Chapter 12, Science Focus, p. 294) uses atoms and molecules to build materials from the bottom up using atoms of the elements in the periodic table as its raw materials. A *nanometer* (nm) is one-billionth of a meter—equal to the length of about 10 hydrogen atoms lined up side by side. A DNA molecule (Figure 10) is about 2.5 nanometers wide. A human hair has a width of 50,000 to 100,000 nanometers.

For objects smaller than about 100 nanometers, the properties of materials change dramatically. At this *nanoscale* level, materials can exhibit new properties, such as extraordinary strength or increased chemical activity that they do not exhibit at the much larger *macroscale* level with which we are all familiar.

For example, scientists have learned how to make tiny tubes of carbon atoms linked together in hexagons. Experiments have shown that these carbon nanotubes are the strongest material ever made—60 times stronger than high-grade steel. Such nanotubes have been linked together to form a rope so thin that it is invisible, but strong enough to suspend a pickup truck.

At the macroscale, zinc oxide (ZnO) can be rubbed on the skin as a white paste to protect against the sun's harmful UV rays; at the nanoscale it becomes transparent and is being used in invisible coatings to protect both skin and fabrics from UV damage. Because silver (Ag) can kill harmful bacteria, silver nanocrystals are being incorporated into bandages for wounds and in other antibacterial and antifungal products such as the imitation hair in some fuzzy animal toys, as well as in pet beds and clothing.

Researchers hope to incorporate nanoparticles of hydroxyapatite, with the same chemical structure as tooth enamel, into toothpaste to put coatings on teeth that prevent the penetration of bacteria. Nanotech coatings now being used on cotton fabrics form an impenetrable barrier that causes liquids to bead and roll off. Such stain-resistant fabrics, used to make clothing, rugs, and furniture upholstery, could eliminate the need to use harmful chemicals for removing stains. Self-cleaning window glass coated with a layer of nanoscale titanium dioxide (TiO_2) particles is now available. As the particles interact with UV rays from the sun, dirt on the surface of the glass loosens and washes off when it rains. Similar products can be used for self-cleaning sinks and toilet bowls.

Scientists are working on ways to replace the silicon in computer chips with carbon-based nanomaterials that greatly increase the processing power of computers. Biological engineers are working on nanoscale devices that could deliver drugs on the cellular level. Such devices could penetrate cancer cells and deliver nanomolecules that would kill the cancer cells from the inside. Researchers also hope to develop nanoscale crystals that could change color when they detect tiny amounts (measured in parts per trillion) of harmful substances such as chemical and biological warfare agents, and food pathogens. For example, a color change in food packaging could alert a consumer when a food is contaminated or has begun to spoil. The list of possibilities continues to grow.

By 2010, more than 1,000 products containing nanoscale particles were commercially available and thousands more were in the research and development pipeline. Examples are found in cosmetics, sunscreens, fabrics (including odor-eating socks), pesticides, food additives, energy drinks, and food packaging (including some hamburger containers and plastic beer bottles).

So far, these products are unregulated and unlabeled. In addition, consumers and government officials don't know what companies are using nanomaterials, what kinds and amounts they are using, and in what products they are using them. This concerns many health and environmental scientists because the tiny size of nanoparticles can allow them to penetrate the body's natural defenses against invasions by foreign and potentially harmful chemicals and pathogens. Nanoparticles of a chemical tend to be much more chemically reactive than macroparticles of the same chemical, largely because the tiny nanoparticles have relatively large surface areas for their small mass. This means that a chemical that is harmless at the macroscale may be hazardous at the nanoscale when the particles are inhaled, ingested, or absorbed through the skin.

We know little about such effects and risks at a time when the use of untested and unregulated nanoparticles is increasing exponentially. A few toxicological studies are sending up red flags:

- In 2004, Eva Olberdorster, an environmental toxicologist at Southern Methodist University in the U.S. state of Texas, found that fish swimming in water loaded with a type of carbon nanomolecules called buckyballs experienced brain damage within 48 hours.
- In 2005, NASA researchers found that injecting commercially available carbon nanotubes into rats caused significant lung damage.
- A 2005 study by researchers at the U.S. National Institute of Occupational Safety and Health found substantial damage to the hearts and aortas of mice exposed to carbon nanotubes.
- In 2008, a team of researchers, including Andrew Maynard and Ken Donaldson, injected short and long straight carbon nanotubes into mice tissues. They found that the short tubes did not cause lesions but the long tubes did produce lesions that could develop into mesothelioma, a deadly form of lung cancer caused by the inhalation of small needles of asbestos fibers. The study also suggested that curly carbon nanotubes did not have such harmful effects, suggesting that further research could lead to safe carbon nanotube products.

In 2004, the British Royal Society and Royal Academy of Engineering recommended that we avoid the environmental release of nanoparticles and nanotubes as much as possible until more is known about their potentially harmful impacts. As a precautionary measure, they recommended that factories and research laboratories treat manufactured nanoparticles and nanotubes as if they were hazardous to their workers and to the general public. Others have called for regulating the use of nanoparticle materials, labeling products with such materials, and greatly increasing research on the potential harmful health effects of nanoparticles. **GREEN CAREER:** nanotechnology

THINKING ABOUT
Nanotechnology

Do you think that the benefits of nanotechnology outweigh its potentially harmful effects? Explain. What are three things you would do to reduce its potentially harmful effects?

SUPPLEMENT

5 Weather Basics: El Niño, Tornadoes, and Tropical Cyclones (Chapters 3, 7, 10, 15)

Weather Is Affected by Moving Masses of Warm and Cold Air

Weather is the set of short-term atmospheric conditions—typically those occurring over hours or days—for a particular area. Examples of atmospheric conditions include temperature, pressure, moisture content, precipitation, sunshine, cloud cover, and wind direction and speed.

Meteorologists use equipment mounted on weather balloons, aircraft, ships, and satellites, as well as radar and stationary sensors, to obtain data on weather variables. They then feed these data into computer models to draw weather maps. Other computer models project the weather for a period of several days by calculating the probabilities that air masses, winds, and other factors will change in certain ways.

Much of the weather we experience results from interactions between the leading edges of moving masses of warm and cold air. Weather changes as one air mass replaces or meets another. The most dramatic changes in weather occur along a **front**, the boundary between two air masses with different temperatures and densities.

A **warm front** is the boundary between an advancing warm air mass and the cooler one it is replacing (Figure 1, left). Because warm air is less dense (weighs less per unit of volume) than cool air, an advancing warm front rises up over a mass of cool air. As the warm front rises, its moisture begins condensing into droplets, forming layers of clouds at different altitudes. Gradually, the clouds thicken, descend to a lower altitude, and often release their moisture as rainfall. A moist warm front can bring days of cloudy skies and drizzle.

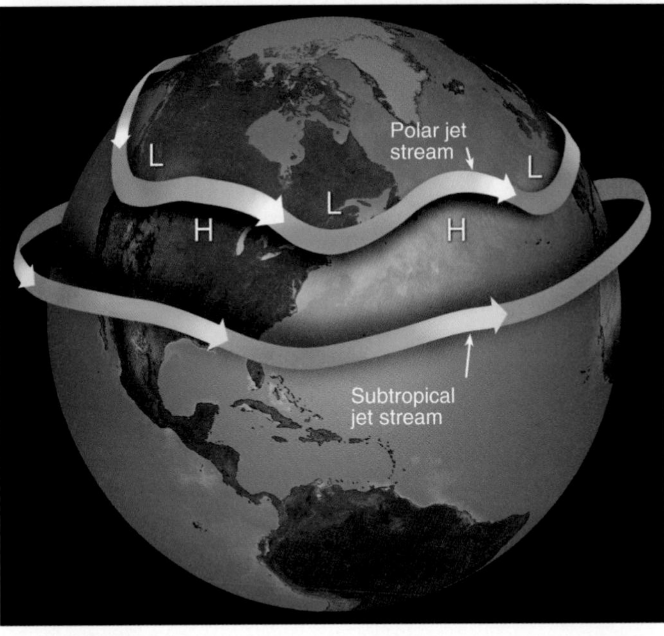

Figure 2 A *jet stream* is a rapidly flowing air current that moves west to east in a wavy pattern. This figure shows a polar jet stream and a subtropical jet stream in winter. In reality, jet streams are discontinuous and their positions vary from day to day. (Used by permission from C. Donald Ahrens, *Meteorology Today*, 8th ed. Belmont, CA: Brooks/Cole, 2006)

A **cold front** (Figure 1, right) is the leading edge of an advancing mass of cold air. Because cold air is denser than warm air, an advancing cold front stays close to the ground and wedges underneath less dense warmer air. An approaching cold front produces rapidly moving, towering clouds called *thunderheads*, with flat, anvil-like tops.

As a cold front passes through, it may cause high surface winds and thunderstorms. After it leaves the area, it usually results in cooler temperatures and a clear sky.

Near the top of the troposphere, hurricane-force winds circle the earth. These powerful winds, called *jet streams*, follow rising and falling paths that have a strong influence on weather patterns (Figure 2).

Weather Is Affected by Changes in Atmospheric Pressure

Changes in atmospheric pressure also affect weather. *Atmospheric pressure* results from molecules of gases (mostly nitrogen and oxygen) in the atmosphere zipping around at very high speeds and hitting and bouncing off everything they encounter.

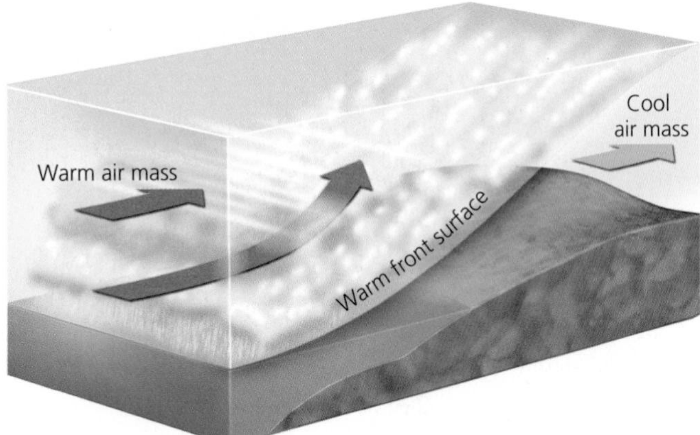

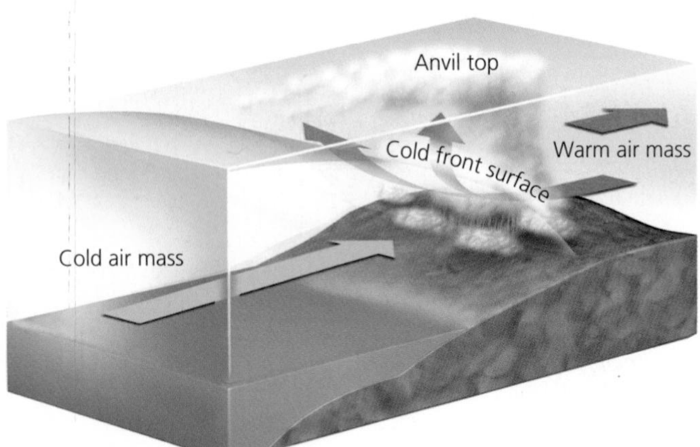

Figure 1 *Weather fronts:* A *warm front* (left) occurs when an advancing mass of warm air meets and rises up over a mass of denser cool air. A *cold front* (right) forms when a moving mass of cold air wedges beneath a mass of less dense warm air.

Atmospheric pressure is greater near the earth's surface because the molecules in the atmosphere are squeezed together under the weight of the air above them. An air mass with high pressure, called a **high**, contains cool, dense air that descends slowly toward the earth's surface and becomes warmer. Because of this warming, condensation of moisture usually does not take place and clouds usually do not form. Fair weather with clear skies follows as long as this high-pressure air mass remains over the area.

In contrast, a low-pressure air mass, called a **low**, produces cloudy and sometimes stormy weather. Because of its low pressure and low density, the center of a low rises, and its warm air expands and cools. When the temperature drops below a certain level where condensation takes place, called the *dew point*, moisture in the air condenses and forms clouds.

If the droplets in the clouds coalesce into larger drops or snowflakes heavy enough to fall from the sky, then precipitation occurs. The condensation of water vapor into water drops usually requires that the air contain suspended tiny particles of material such as dust, smoke, sea salts, or volcanic ash. These so-called *condensation nuclei* provide surfaces on which the droplets of water can form and coalesce.

Every Few Years Major Wind Shifts in the Pacific Ocean Affect Global Weather Patterns

An **upwelling**, or upward movement of ocean water, can mix upper levels of seawater with lower levels, bringing cool and nutrient-rich water from the bottom of the ocean to the warmer surface where it supports large populations of phytoplankton, zooplankton, fish, and fish-eating seabirds.

Figure 7-2, p. 123, shows the oceans' major upwelling zones. Upwellings far from shore occur when surface currents move apart and draw water up from deeper layers. Strong upwellings are also found along the steep western coasts of some continents when winds blowing along the coasts push surface water away from the land and draw water up from the ocean bottom (Figure 3).

Every few years, normal shore upwellings in the Pacific Ocean (Figure 4, left) are affected by changes in weather patterns called the *El Niño–Southern Oscillation*, or *ENSO* (Figure 4, right). In an ENSO, often called simply *El Niño*, prevailing winds called tropical trade winds blowing east to west weaken or reverse direction. This allows the warmer waters of the western Pacific

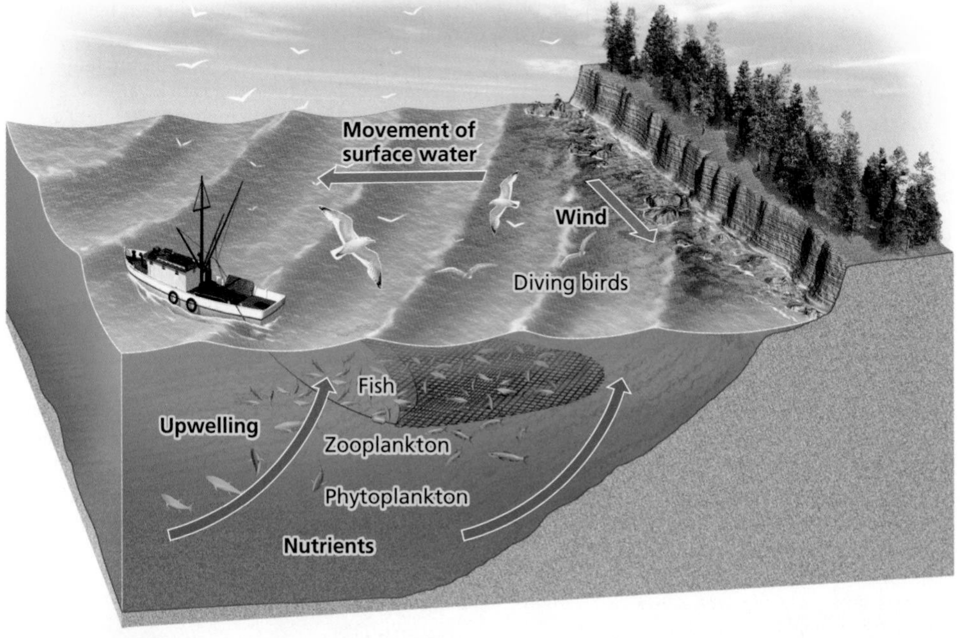

Figure 3 A shore upwelling occurs when deep, cool, nutrient-rich waters are drawn up to replace surface water moved away from a steep coast by wind flowing along the coast toward the equator.

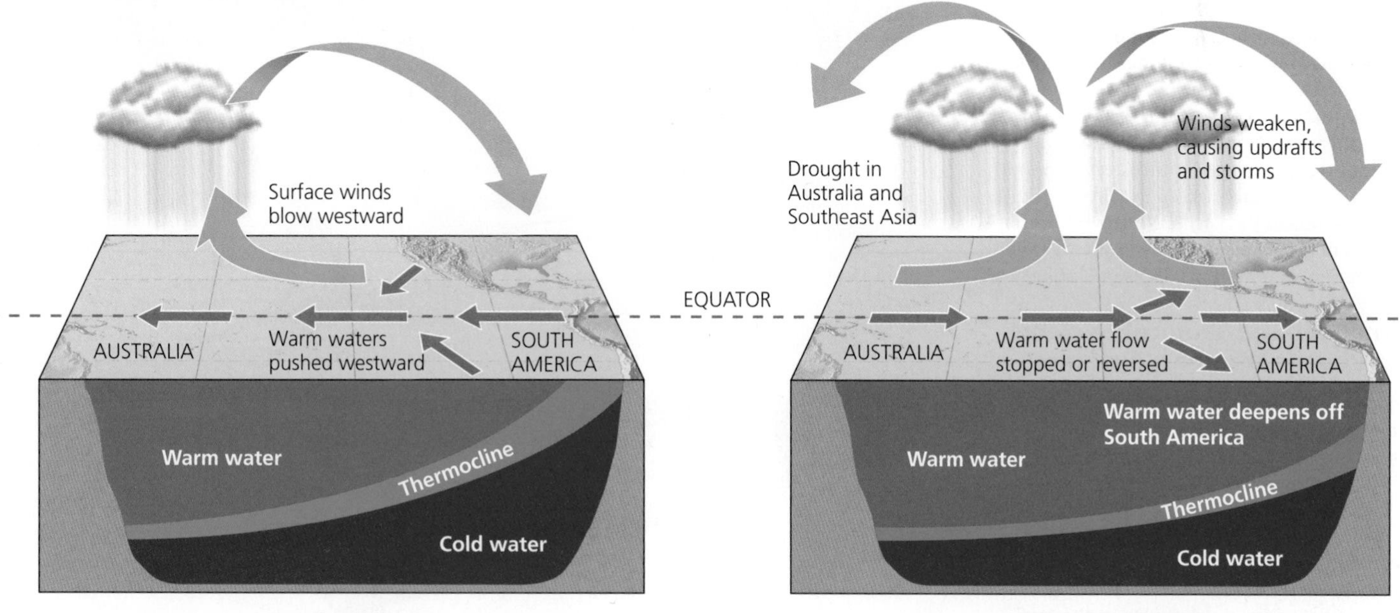

Figure 4 Normal trade winds blowing east to west cause shore upwellings of cold, nutrient-rich bottom water in the tropical Pacific Ocean near the coast of Peru (left). A zone of gradual temperature change called the *thermocline* separates the warm and cold water. Every few years, a shift in trade winds known as the *El Niño–Southern Oscillation (ENSO)* disrupts this pattern.

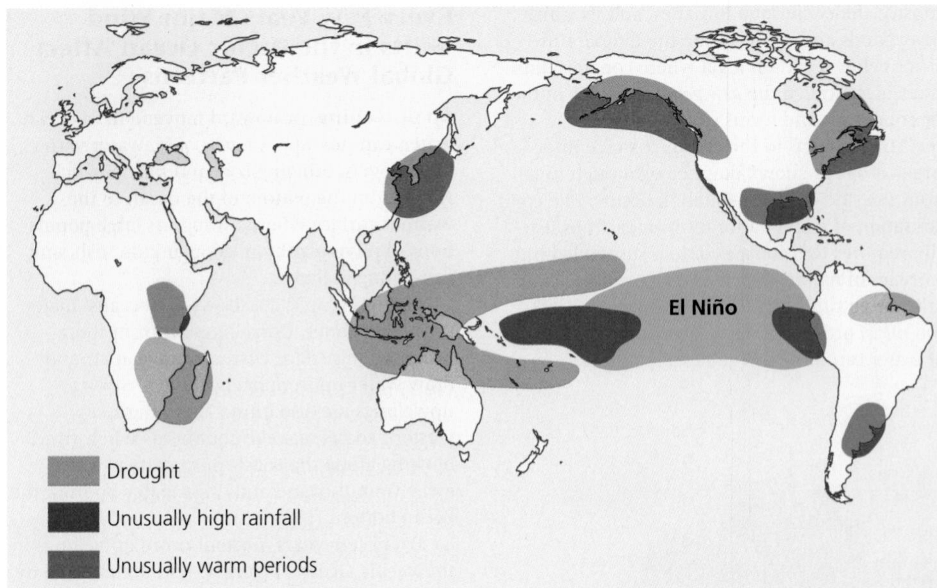

Figure 5 Typical global weather effects of an El Niño–Southern Oscillation. **Question:** How might an ENSO affect the weather where you live or go to school? (Data from United Nations Food and Agriculture Organization)

to move toward the coast of South America, which suppresses the normal upwellings of cold, nutrient-rich water (Figure 4, right). The decrease in nutrients reduces primary productivity and causes a sharp decline in the populations of some fish species.

When an ENSO lasts 12 months or longer, it severely disrupts populations of plankton, fish, and seabirds in upwelling areas. A strong ENSO can also alter weather conditions over at least two-thirds of the globe (Figure 5)—especially in lands along the Pacific and Indian Oceans. Scientists do not know exactly what causes an ENSO, but they do know how to detect its formation and track its progress.

La Niña, the reverse of El Niño, cools some coastal surface waters, and brings back upwellings. Typically, La Niña means more Atlantic Ocean hurricanes, colder winters in Canada and the northeastern United States, and warmer and drier winters in the southeastern and southwestern United States. It also usually leads to wetter winters in the Pacific Northwest, torrential rains in Southeast Asia, lower wheat yields in Argentina, and more wildfires in Florida.

Tornadoes and Tropical Cyclones Are Violent Weather Extremes

Sometimes we experience *weather extremes*. Two examples are violent storms called *tornadoes* (which form over land) and *tropical cyclones* (which form over warm ocean waters and sometimes pass over coastal areas).

Tornadoes, or *twisters*, are swirling, funnel-shaped clouds that form over land. They can destroy houses and cause other serious damage in areas where they touch down on the earth's surface. The United States is the world's most tornado-prone country, followed by Australia.

Tornadoes in the plains of the Midwestern United States usually occur when a large, dry, cold-air front moving southward from Canada runs into a large mass of warm humid air moving northward from the Gulf of Mexico. Most tornadoes occur in the spring and summer when fronts of cold air from the north penetrate deeply into the Great Plains and the Midwest.

As the large warm-air mass moves rapidly over the more dense cold-air mass, it rises swiftly and forms strong vertical convection currents that suck air upward, as shown in Figure 6. Scientists hypothesize that the interaction of the cooler air nearer the ground and the rapidly rising warmer air above causes a spinning, vertically rising air mass, or vortex.

Figure 7 shows the areas of greatest risk from tornadoes in the continental United States.

Tropical cyclones are spawned by the formation of low-pressure cells of air over warm tropical seas. Figure 8 shows the formation and structure of a tropical cyclone. *Hurricanes* are tropical cyclones that form in the Atlantic Ocean; those forming in the Pacific Ocean usually are called *typhoons*. Tropical cyclones take a long time to form and gain strength. As a result, meteorologists can track their paths and wind speeds, and warn people in areas likely to be hit by these violent storms.

For a tropical cyclone to form, the temperature of ocean water has to be at least 27°C (80°F) to a depth of 46 meters (150 feet). A tropical cyclone forms when areas of low pressure over the warm ocean draw in air from surrounding higher-pressure areas. The earth's rotation makes these winds spiral counterclockwise in the northern hemisphere and clockwise in the southern hemisphere (see Figure 7-3, p. 123). Moist air, warmed by the heat of the ocean, rises in a vortex through the center of the storm until it becomes a tropical cyclone (Figure 8).

The intensities of tropical cyclones are rated in different categories, based on their sustained wind speeds: *Category 1*, 119–153 kilometers

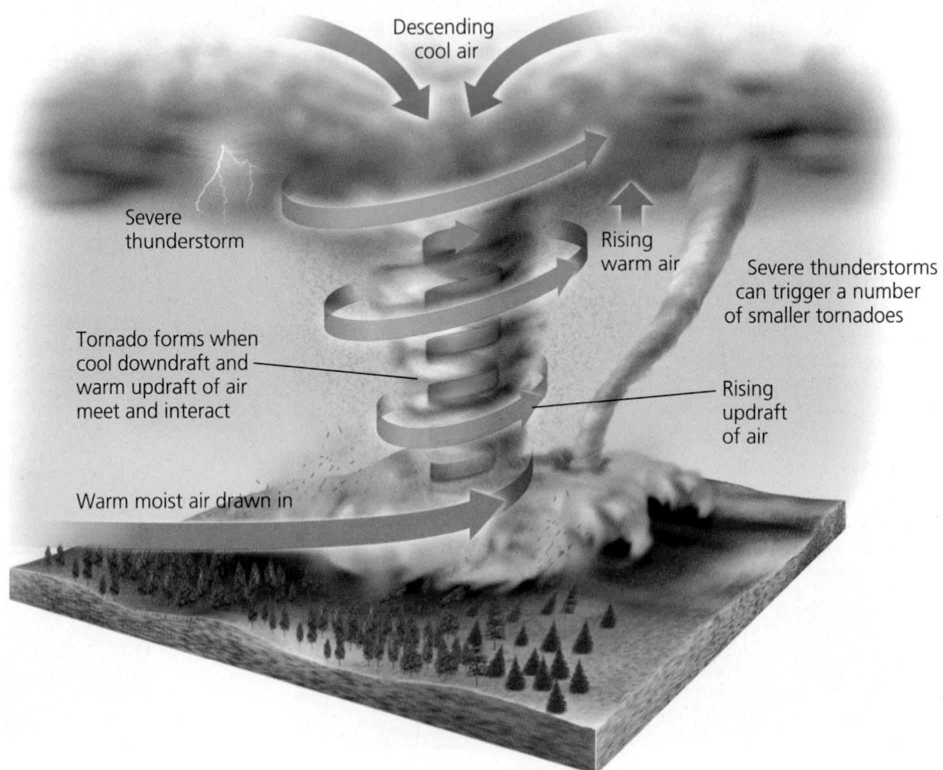

Figure 6 Formation of a *tornado*, or *twister*. Although twisters can form at any time of the year, the most active tornado season in the United States is usually March through August. Meteorologists cannot yet forecast exactly where tornadoes will form at any given time, but research on tornados and advanced computer modeling can help them to identify areas at risk each day for the formation of these deadly storms.

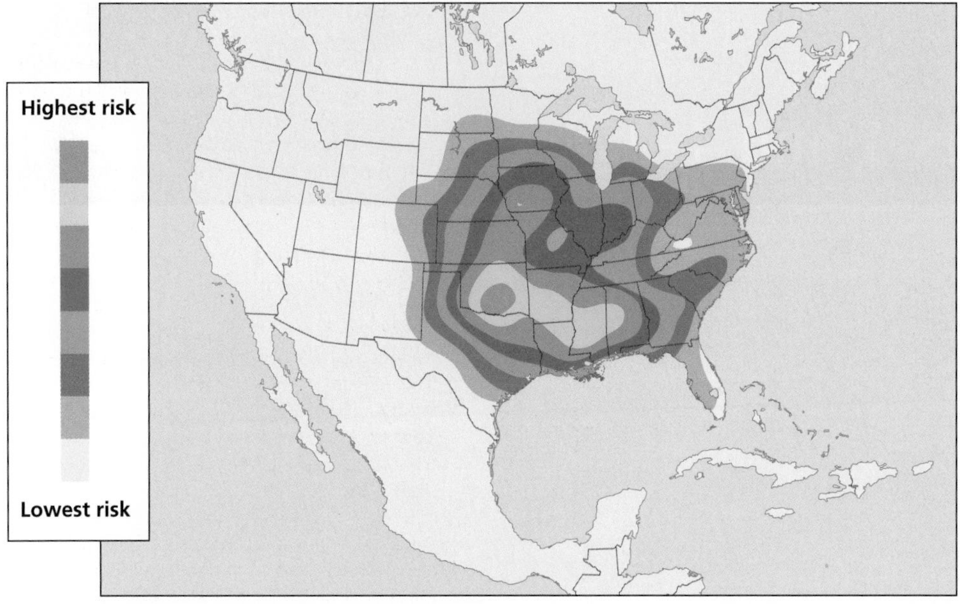

Figure 7 This map shows the relative risk of tornados across the continental United States. (Data from NOAA)

per hour (74–95 miles per hour); *Category 2,* 154–177 kilometers per hour (96–110 miles per hour); *Category 3,* 178–209 kilometers per hour (111–130 miles per hour); *Category 4,* 210–249 kilometers per hour (131–155 miles per hour); and *Category 5,* greater than 249 kilometers per hour (155 miles per hour). The longer a tropical cyclone stays over warm waters, the stronger it gets. Significant hurricane-force winds can extend 64–161 kilometers (40–100 miles) from the center, or eye, of a tropical cyclone.

Hurricanes and typhoons kill and injure people and damage property and agricultural production. Sometimes, however, the long-term ecological and economic benefits of a tropical cyclone exceed its short-term harmful effects.

For example, in parts of the U.S. state of Texas along the Gulf of Mexico, coastal bays and marshes, because of their unique natural formations and the barrier islands that protect them, normally receive very limited freshwater and saltwater inflows. In August 1999, Hurricane Brett struck this coastal area. According to marine biologists, the storm flushed out excess nutrients from land runoff and swept dead sea grasses and rotting vegetation from the coastal bays and marshes. It also carved out 12 channels through the barrier islands along the coast, allowing huge quantities of fresh seawater to flood the bays and marshes.

This flushing of the bays and marshes reduced brown tides consisting of explosive growths of algae feeding on excess nutrients. It also increased growth of sea grasses, which serve as nurseries for shrimp, crabs, and fish, and provide food for millions of ducks wintering in Texas bays. Production of commercially important species of shellfish and fish also increased.

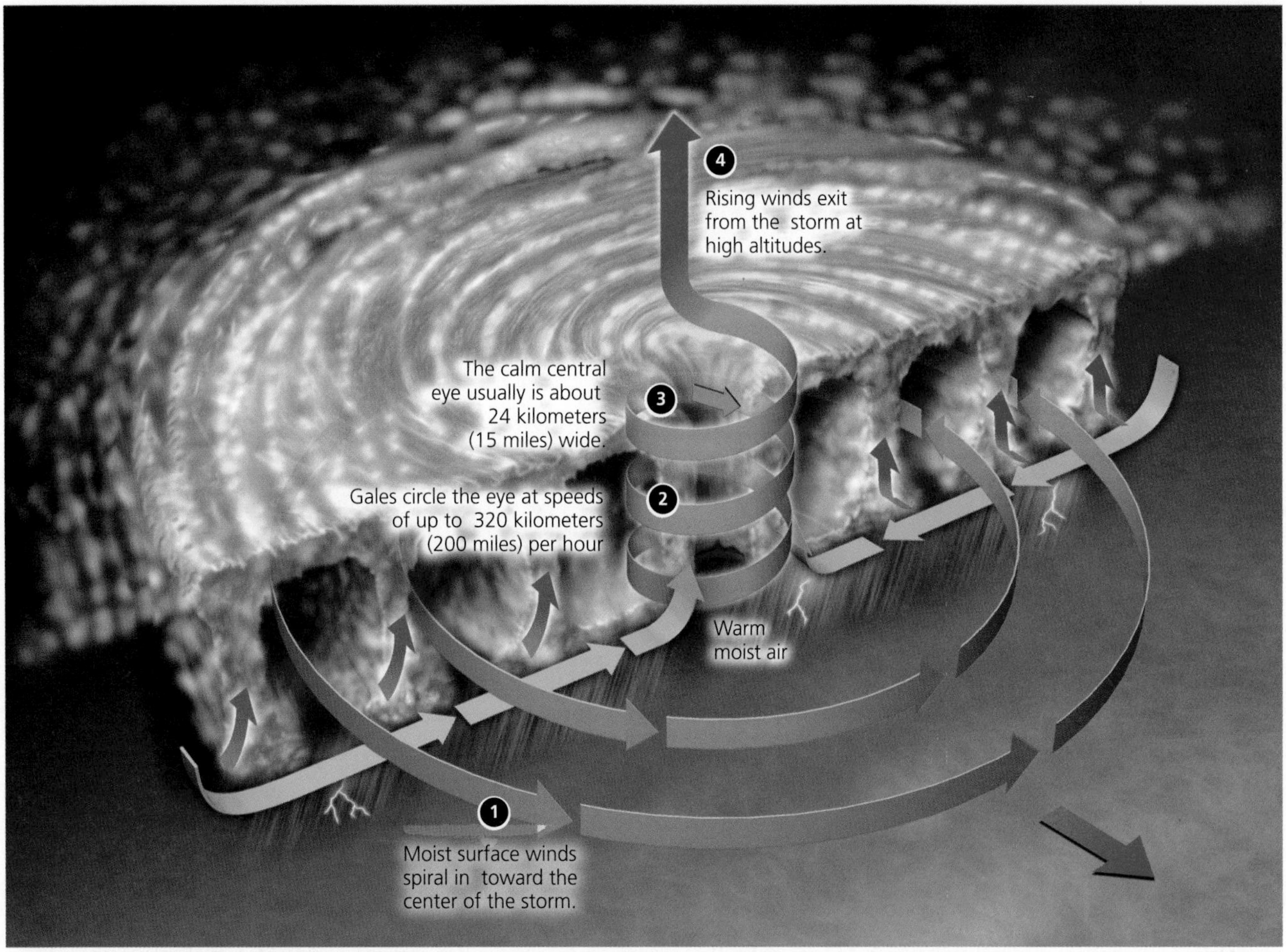

Figure 8 This diagram illustrates the formation of a *tropical cyclone*. Those forming in the Atlantic Ocean are called *hurricanes*; those forming in the Pacific Ocean are called *typhoons*.

SUPPLEMENT

6 Maps (Chapters 1 and 4–17)

Figure 1 The countries of the world.

Map Analysis

1. Which is the largest country in **(a)** North America, **(b)** Central America, **(c)** South America, **(d)** Europe, and **(e)** Asia?
2. Which countries surround **(a)** China, **(b)** Mexico, **(c)** Germany, and **(d)** Sudan?

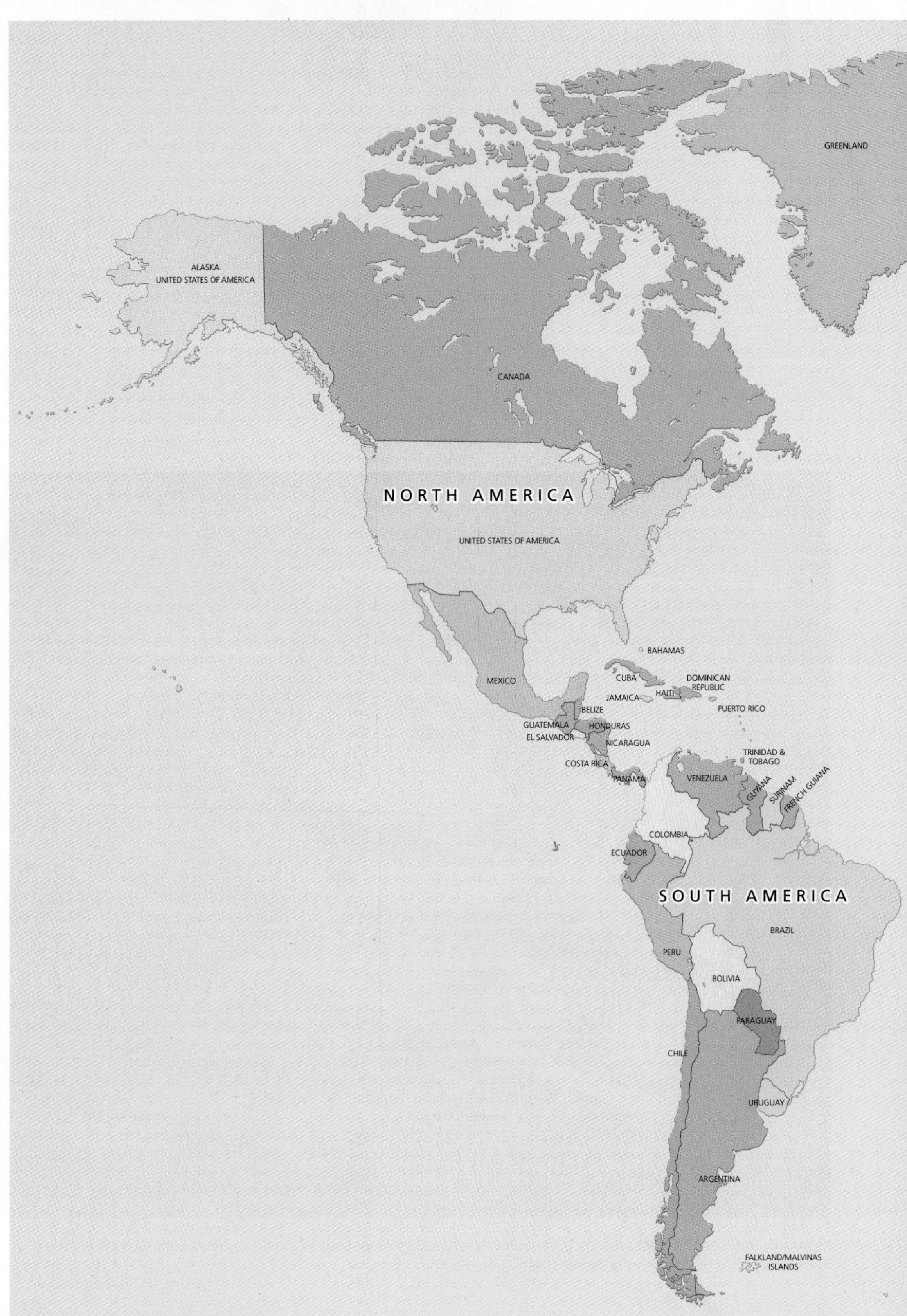

EUROPE
ASIA
AFRICA
AUSTRALIA
RUSSIA
NORWAY
SWEDEN
FINLAND
ESTONIA
LATVIA
LITHUANIA
DENMARK
UNITED KINGDOM
NETHERLANDS
BELARUS
GERMANY
POLAND
BELGIUM
LUXEMBOURG
CZECH REP.
SLOVAKIA
UKRAINE
AUSTRIA
SWITZERLAND
HUNGARY
MOLDOVA
SLOVENIA
CROATIA
ROMANIA
FRANCE
BOSNIA-HERZEGOVINA
YUGOSLAVIA
BULGARIA
ITALY
MACEDONIA
ALBANIA
GREECE
SPAIN
GEORGIA
ARMENIA
AZERBAIJAN
TURKEY
KAZAKHSTAN
UZBEKISTAN
KYRGYZSTAN
TURKMENISTAN
TAJIKISTAN
MONGOLIA
NORTH KOREA
SOUTH KOREA
JAPAN
CHINA
CYPRUS
SYRIA
LEBANON
ISRAEL
IRAQ
IRAN
AFGHANISTAN
JORDAN
KUWAIT
PAKISTAN
NEPAL
BHUTAN
SAUDI ARABIA
QATAR
UNITED ARAB EMIRATES
BANGLADESH
INDIA
TAIWAN
HONG KONG
MYANMAR
LAOS
OMAN
YEMEN
THAILAND
VIETNAM
PHILIPPINES
CAMBODIA
SRI LANKA
BRUNEI
MALAYSIA
INDONESIA
PAPUA NEW GUINEA
SOLOMON ISLANDS
VANUATU
FIJI
NEW CALEDONIA
NEW ZEALAND
TUNISIA
ALGERIA
LIBYA
EGYPT
MALI
NIGER
CHAD
SUDAN
ERITREA
BURKINA FASO
NIGERIA
GHANA
IVORY COAST
TOGO
BENIN
CENTRAL AFRICAN REPUBLIC
ETHIOPIA
CAMEROON
SOMALIA
EQUATORIAL GUINEA
GABON
CONGO
DEMOCRATIC REPUBLIC OF CONGO
UGANDA
KENYA
RAWANDA
BARUNDI
TANZANIA
ANGOLA
ZAMBIA
MALAWI
MOZAMBIQUE
MADAGASCAR
MAURITIUS
RÉUNION
NAMIBIA
ZIMBABWE
BOTSWANA
SWAZILAND
LESOTHO
SOUTH AFRICA

Figure 2 Natural capital degradation: This map shows the human footprint on the earth's land surface—in effect, the sum of all ecological footprints (see Figure 1-8, p. 14) of the human population. Colors represent the percentage of each area influenced by human activities. Excluding Antarctica and Greenland, human activities have, to some degree, directly affected about 83% of the earth's land surface and 98% of the area where it is possible to grow rice, wheat, or maize. (Data from Wildlife Conservation Society and the Center for International Earth Science Information Network at Columbia University)

Data and Map Analysis

1. What is the human footprint value for the area in which you live? List three other countries in the world that have about the same human footprint value as that of the area where you live. (See Figure 1 of this supplement, pp. S22–S23, for country names.)
2. Compare this map with that of Figure 5, which shows ecological debtors and creditors. Why do you think that some countries in Northern Africa are ecological debtors (Figure 5) while the human footprint value is low in large parts of these countries?

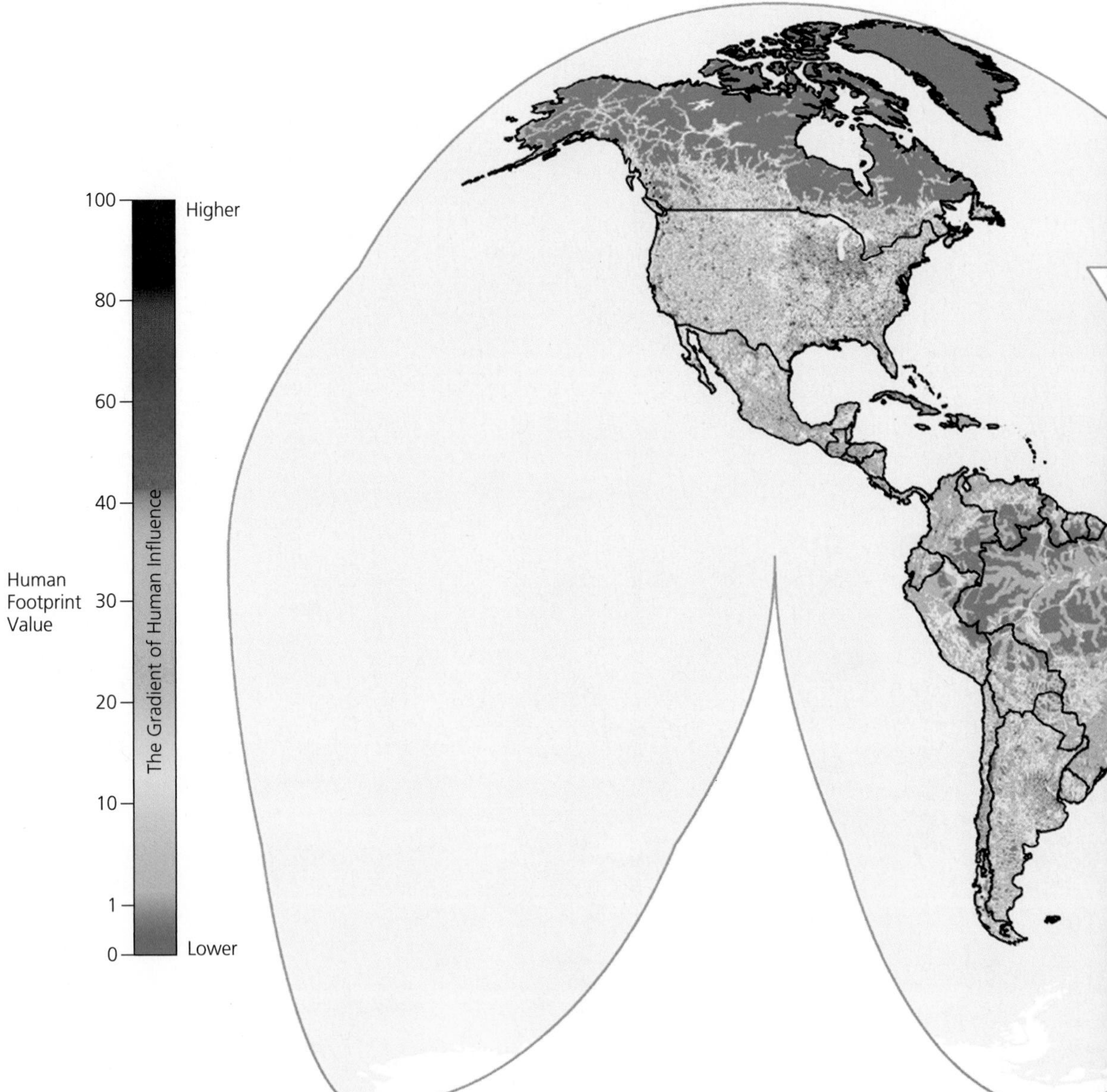

Figure 3 **Natural capital:** This map shows the major biomes found in North America.

Data and Map Analysis

1. Which type of biome occupies the most coastal area?
2. Which biome is the rarest in North America?

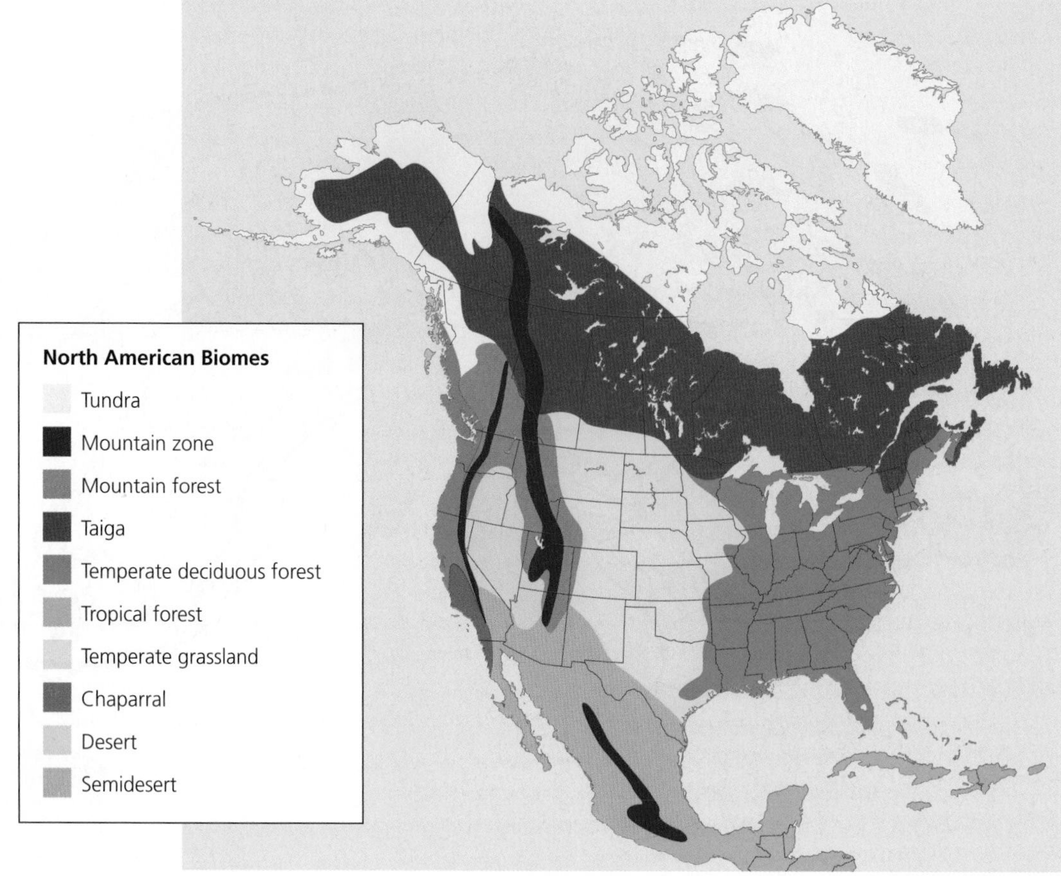

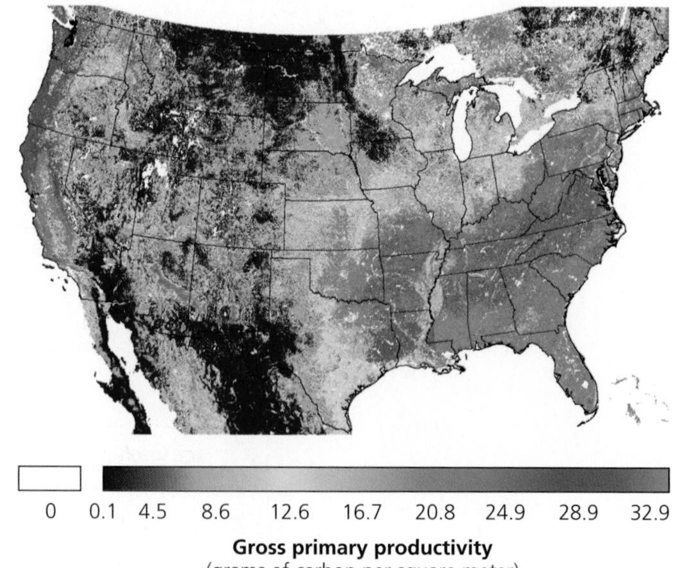

Figure 4 This map illustrates gross primary productivity across the continental United States, based on remote satellite data. The differences roughly correlate with variations in moisture and soil types. (NASA's Earth Observatory)

Data and Map Analysis

1. Comparing the five northwestern-most states with the five southeastern-most states, which of these regions has the greater variety in levels of gross primary productivity? Which of the regions has the highest levels overall?
2. Compare this map with that of Figure 3. Which biome in the United States is associated with the highest level of gross primary productivity?

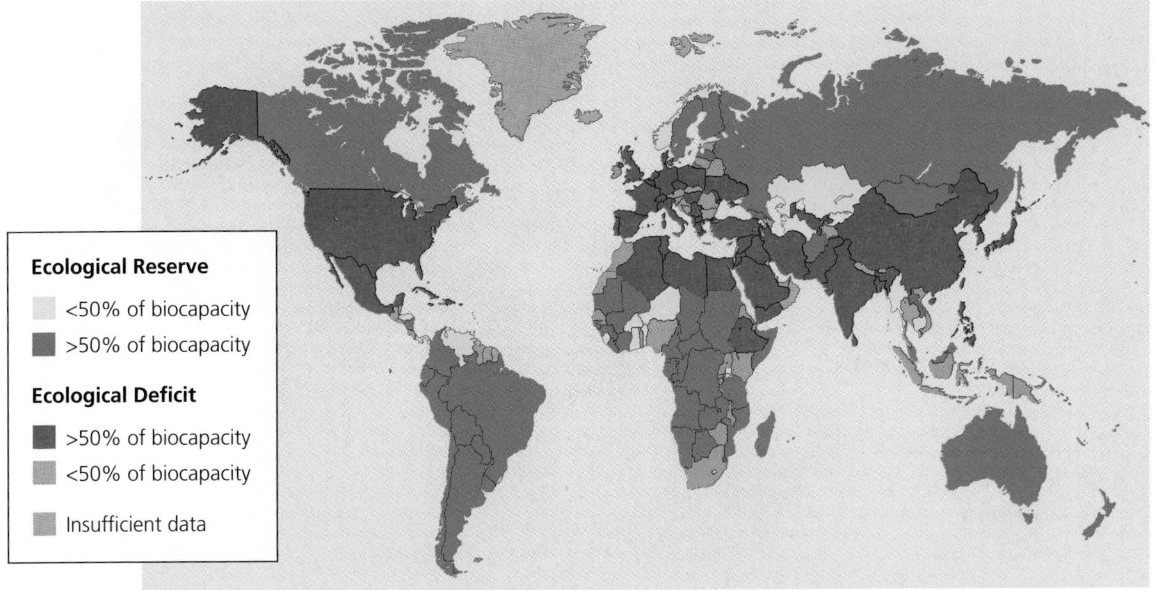

Figure 5 *Ecological Debtors and Creditors:* The ecological footprints of some countries exceed their biocapacity, while other countries still have ecological reserves. (Data from Global Footprint Network)

Data and Map Analysis

1. List five countries, including the three largest, in which the ecological deficit is greater than 50% of biocapacity. (See Figure 1 of this supplement, pp. S22–S23, for country names.)
2. On which two continents does land with ecological reserves of more than 50% of biocapacity occupy the largest percentage of total land area? (See Figure 1 of this supplement, pp. S22–S23, for continent names.) Look at Figure 2 and, for each of these two continents, list the highest human footprint value that you see on the map.

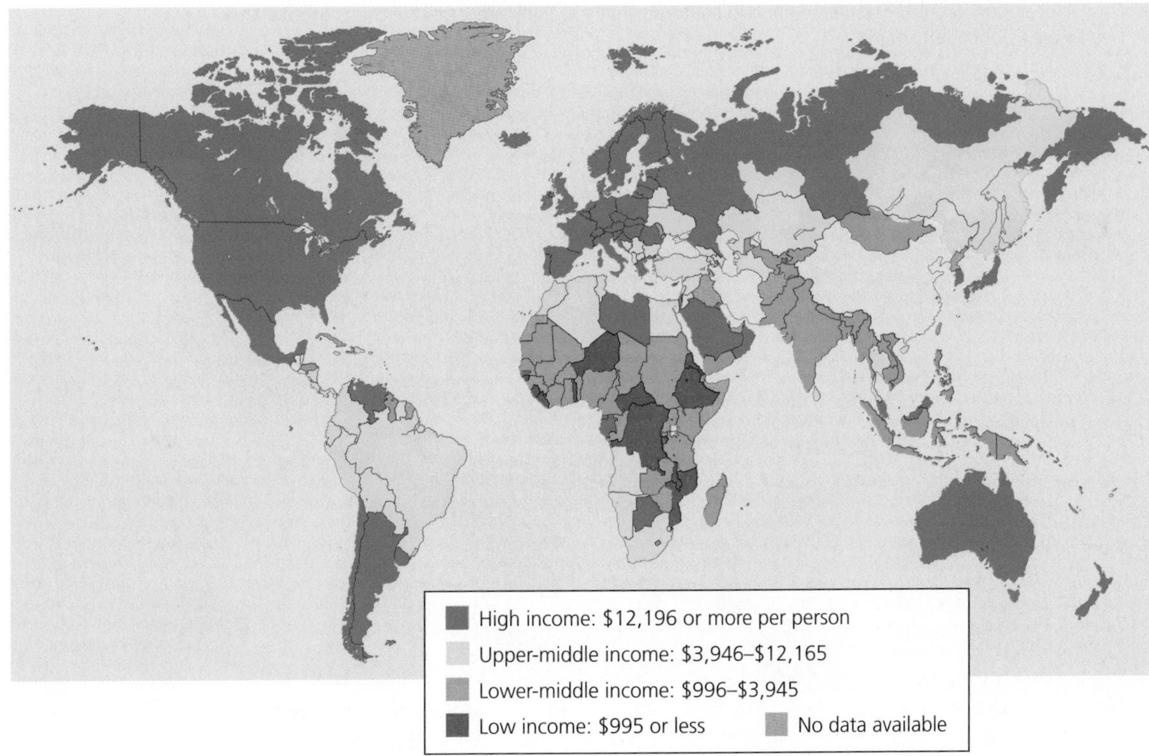

Figure 6 This map shows high-income, upper-middle-income, lower-middle-income, and low-income countries in terms of gross national income (GNI) PPP per capita (U.S. dollars) in 2009. (Data from World Bank and International Monetary Fund)

Data and Map Analysis

1. In how many countries is the per capita average income $995 or less? Look at Figure 1 and find the names of three of these countries.
2. In how many instances does a lower-middle- or low-income country share a border with a high-income country? Look at Figure 1 and find the names of the countries that reflect three examples of this situation.

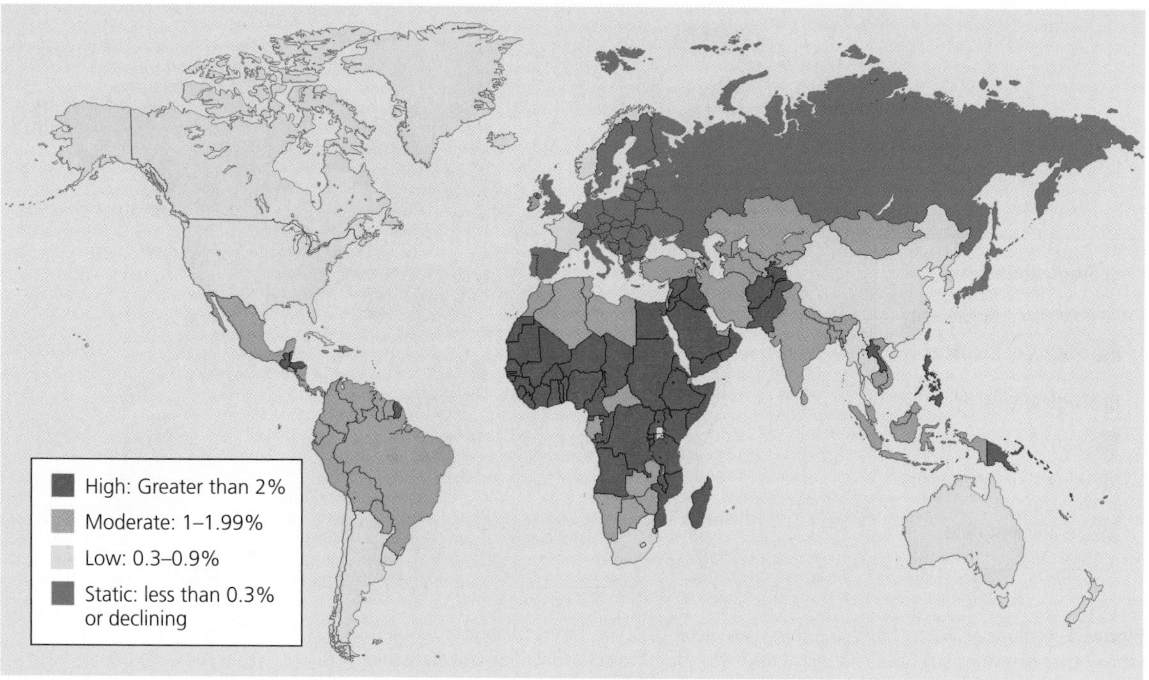

Figure 7 The rate of population increase (%) throughout the world in 2010 is shown here. (Data from Population Reference Bureau and United Nations Population Division)

Data and Map Analysis

1. Which continent has the greatest number of countries with high rates of population increase? Which continent has the greatest number of countries with static rates? (See Figure 1 of this supplement, pp. S22–S23, for continent names.)
2. For each category on this map, name the two countries that you think are largest in terms of total area (see Figure 1 for country names).

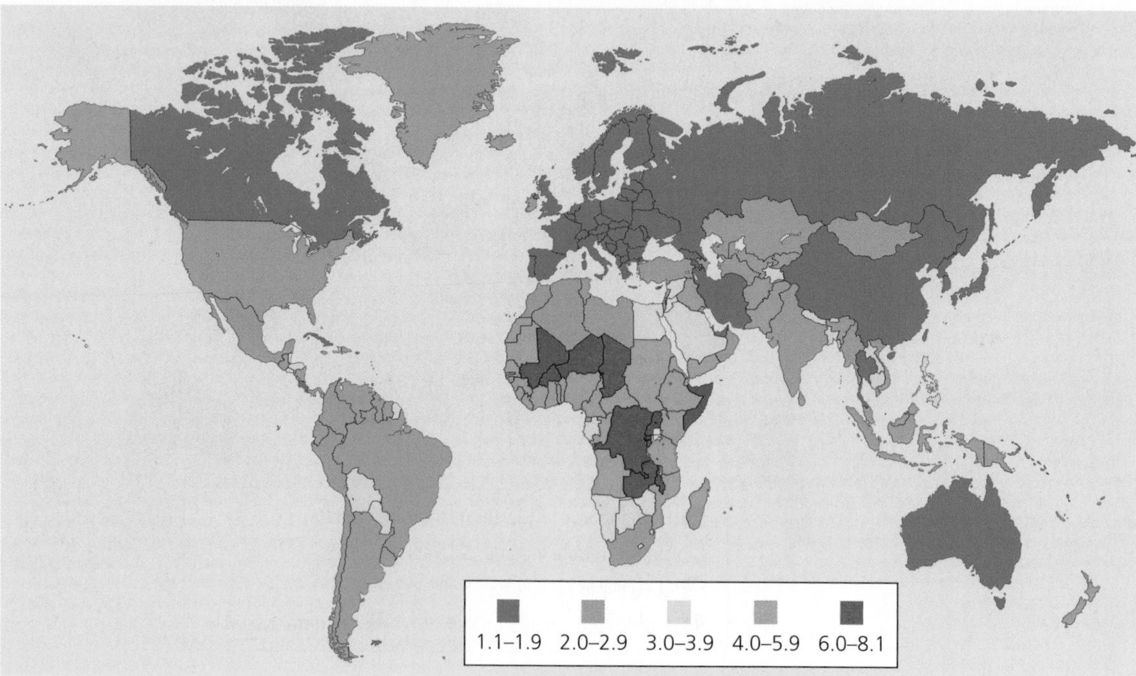

Figure 8 This map represents the total fertility rate (TFR), or average number of children born to the world's women throughout their lifetimes, as measured in 2010. (Data from Population Reference Bureau and United Nations Population Division)

Data and Map Analysis

1. Which country in the highest TFR category borders two countries in the lowest TFR category? What are those two countries? (See Figure 1 of this supplement, pp. S22–S23, for country names.)
2. Describe two geographic patterns that you see on this map.

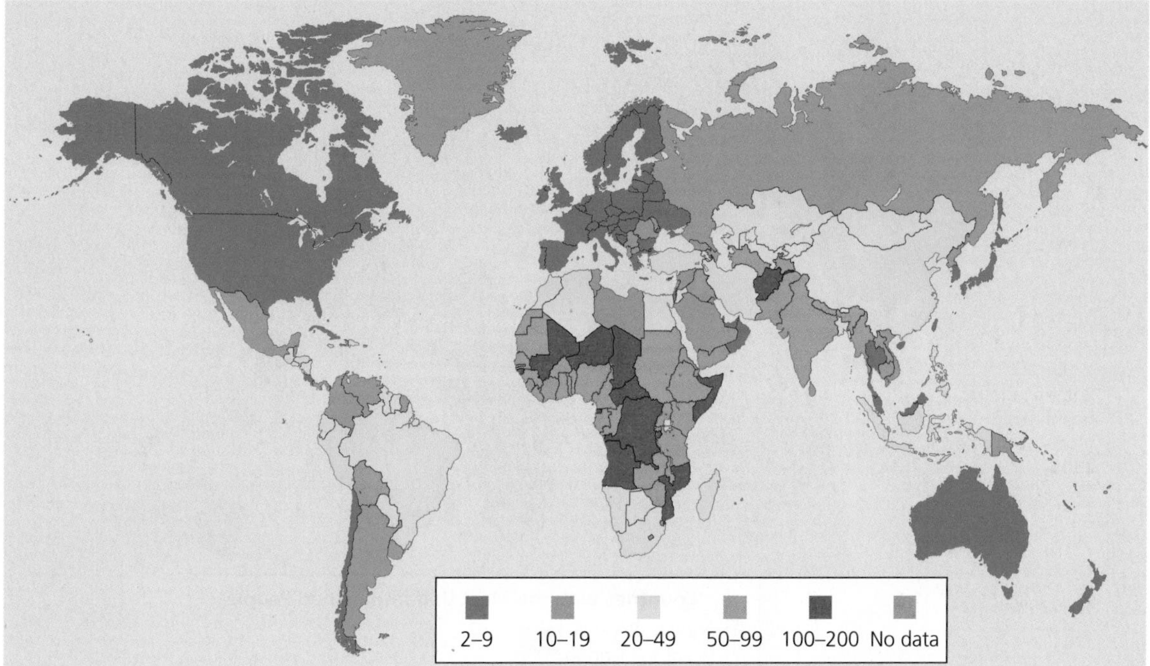

Figure 9 Infant mortality rates in 2010 are shown here. (Data from Population Reference Bureau and United Nations Population Division)

Data and Map Analysis

1. Describe a geographic pattern that you can see related to infant mortality rates as reflected on this map.
2. Describe any similarities that you see in geographic patterns between this map and the one in Figure 7.

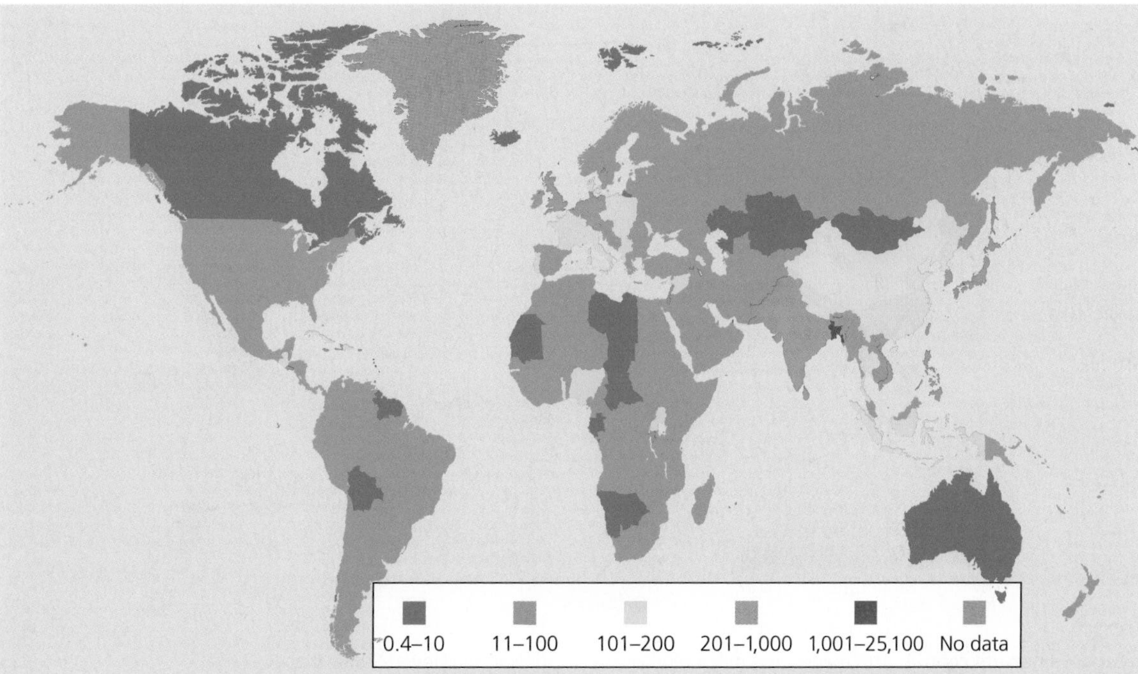

Figure 10 This map reflects global population density per square kilometer in 2008. (Data from Population Reference Bureau and United Nations Population Division)

Data and Map Analysis

1. Which country has the densest population? (See Figure 1 of this supplement, pp. S22–S23, for country names.)
2. List the continents in order from the most densely populated, overall, to the least densely populated, overall. (See Figure 1 of this supplement for continent names.)

Figure 11 World hunger is shown here as a population percentage that suffered from chronic hunger and malnutrition in 2005. (Data from Food and Agriculture Organization, United Nations)

Data and Map Analysis

1. List the continents in order, starting with the one that has the highest percentage of undernourished people and ending with the one that has the lowest such percentage. (See Figure 1 of this supplement, pp. S22–S23, for continent names.)
2. On which continent is the largest block of countries that suffer the highest levels of undernourishment? List five of these countries.

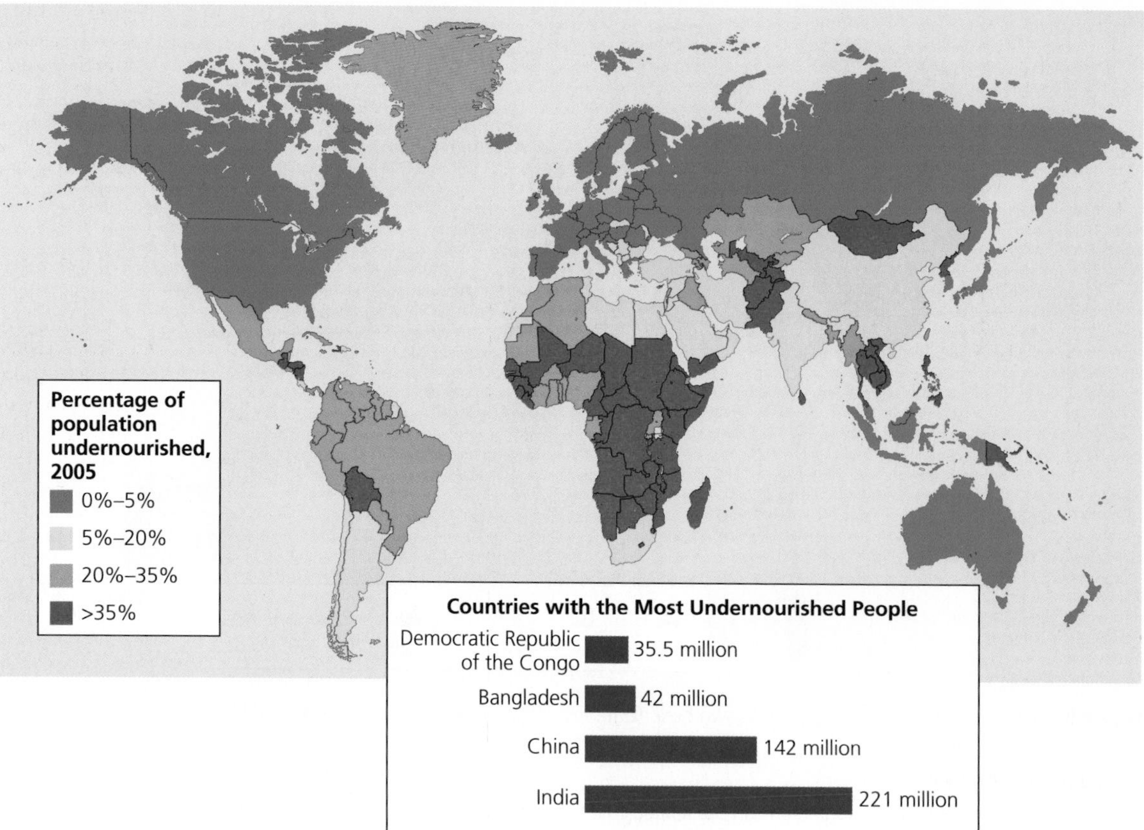

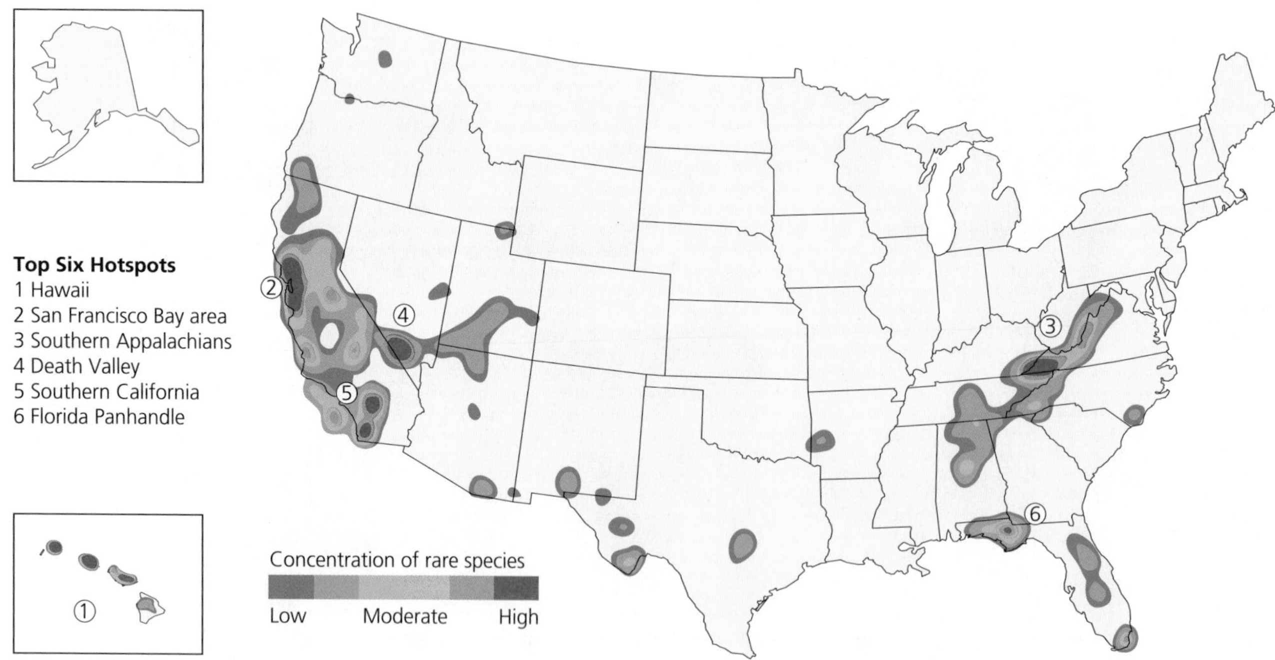

Figure 12 Endangered natural capital: This map shows major biodiversity hotspots in the United States that need emergency protection. The shaded areas contain the largest concentrations of rare and potentially endangered species. Compare these areas with those on the U.S. portion of the global map of the human ecological footprint shown in Figure 2 in this supplement. (Data from State Natural Heritage Programs, the Nature Conservancy, and Association for Biodiversity Information)

Data and Map Analysis

1. If you live in the United States, which of the top six hotspots is closest to where you live or go to school?
2. Do you think that hotspots near urban areas would be harder to protect than those in rural areas? Explain.
3. Which general part of the country has the highest overall concentration of rare species? Which part has the second-highest concentration?

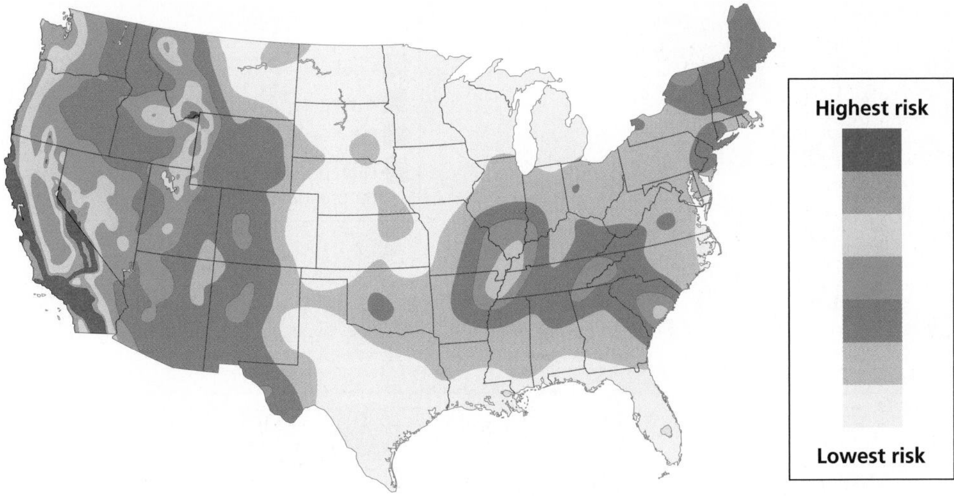

Figure 13 This map shows the earthquake (seismic) risk in various areas of the continental United States. In 2008, the U.S. Geological Survey estimated that the U.S. state of California has more than a 99% chance of experiencing a magnitude 6.7 earthquake within 30 years, and that Southern California has a 37% chance of experiencing a magnitude 7.5 earthquake during that period. (Data from U.S. Geological Survey)

Data and Map Analysis

1. Speaking in general terms (Northeast, Southeast, Central, West Coast, etc.), which area has the highest earthquake risk and which area has the lowest risk?
2. For each of the categories of risk, list the number of states that fall into the category.

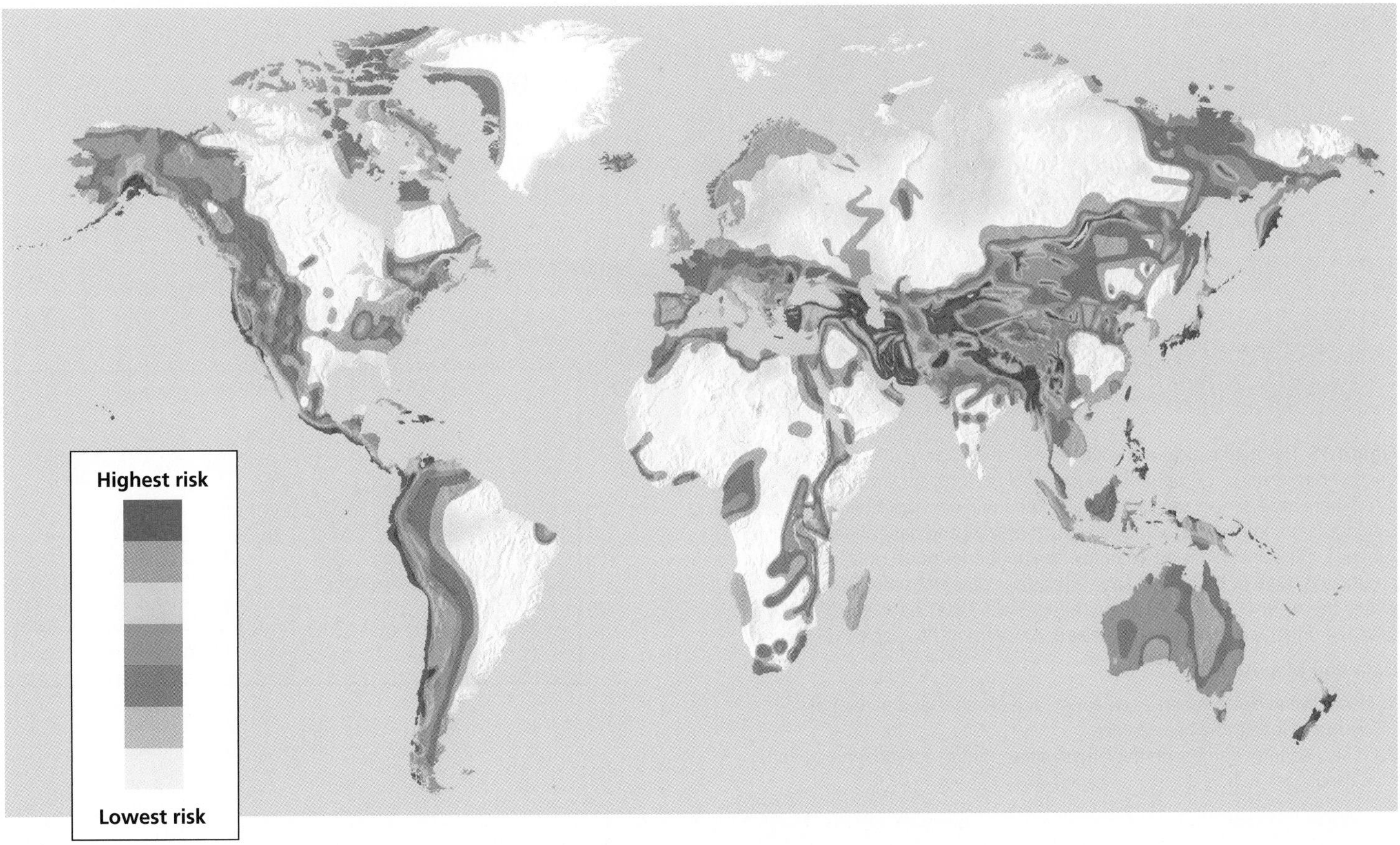

Figure 14 This maps shows earthquake (seismic) risk in the world. (Data from U.S. Geological Survey)

Data and Map Analysis

1. How are these areas related to the boundaries of the earth's major tectonic plates, as shown in Figure 12-4 (p. 278)?
2. Which continent has the longest coastal area subject to the highest possible risk? Which continent has the second longest such coastal area? (See Figure 1 of this supplement, pp. S22–S23, for continent names.)

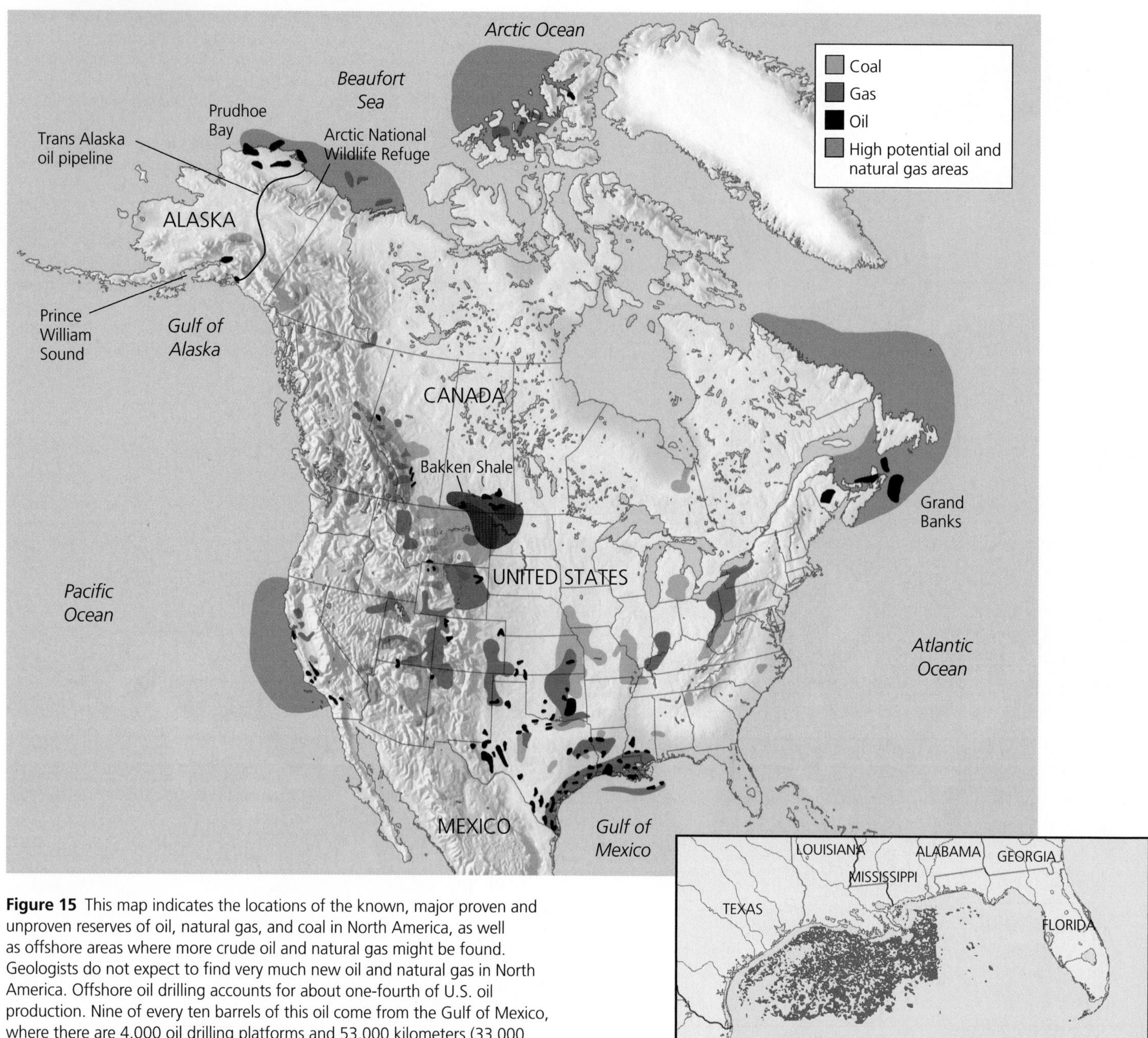

Figure 15 This map indicates the locations of the known, major proven and unproven reserves of oil, natural gas, and coal in North America, as well as offshore areas where more crude oil and natural gas might be found. Geologists do not expect to find very much new oil and natural gas in North America. Offshore oil drilling accounts for about one-fourth of U.S. oil production. Nine of every ten barrels of this oil come from the Gulf of Mexico, where there are 4,000 oil drilling platforms and 53,000 kilometers (33,000 miles) of underwater pipeline (see insert). (Data from U.S. Geological Survey)

Data and Map Analysis

1. If you live in North America, where are the oil, coal, and natural gas deposits closest to where you live?
2. Which country borders on the largest areas of high potential for oil and natural gas?

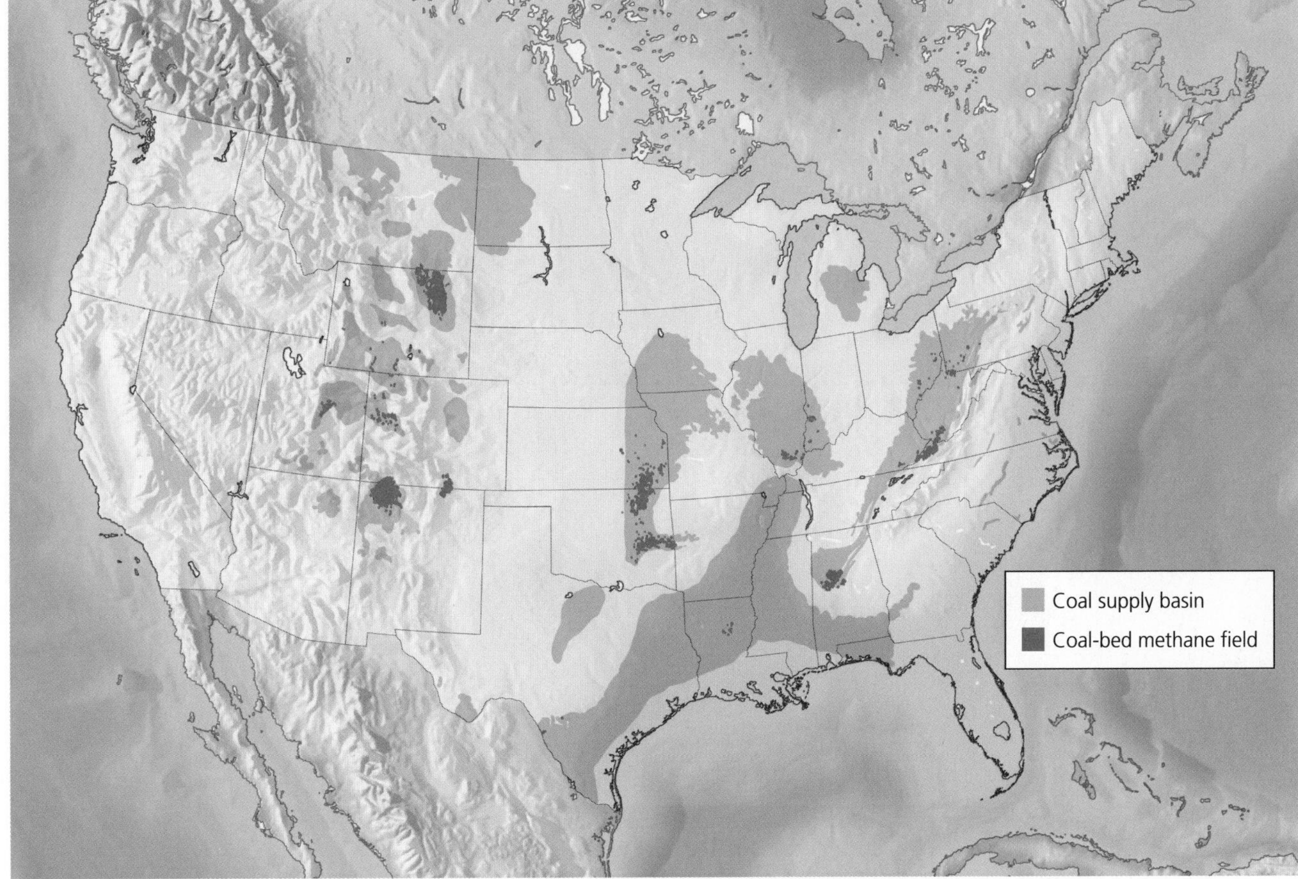

Figure 16 Major coal supply basins and coal-bed methane fields in the lower 48 states of the United States. (Data from U.S. Energy Information based U.S. Geological Survey and various other published studies)

Data and Map Analysis

1. If you live in the United States, where are the coal-bed methane deposits closest to where you live?
2. Removing these deposits requires lots of water. Compare the locations of the major deposits of coal-bed methane with water-deficit areas shown in Figure 11-4, p. 240, and Figure 11-5, p. 242.

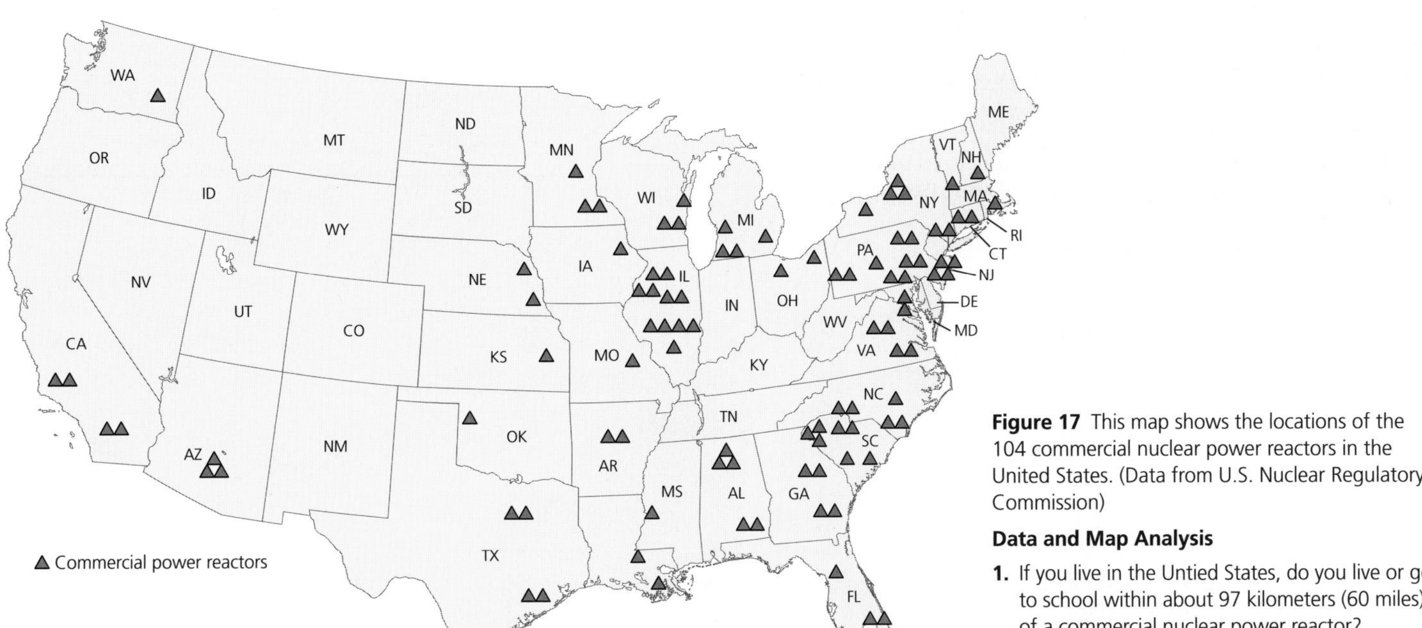

Figure 17 This map shows the locations of the 104 commercial nuclear power reactors in the United States. (Data from U.S. Nuclear Regulatory Commission)

Data and Map Analysis

1. If you live in the Untied States, do you live or go to school within about 97 kilometers (60 miles) of a commercial nuclear power reactor?
2. Which state has the largest number of commercial nuclear power reactors?

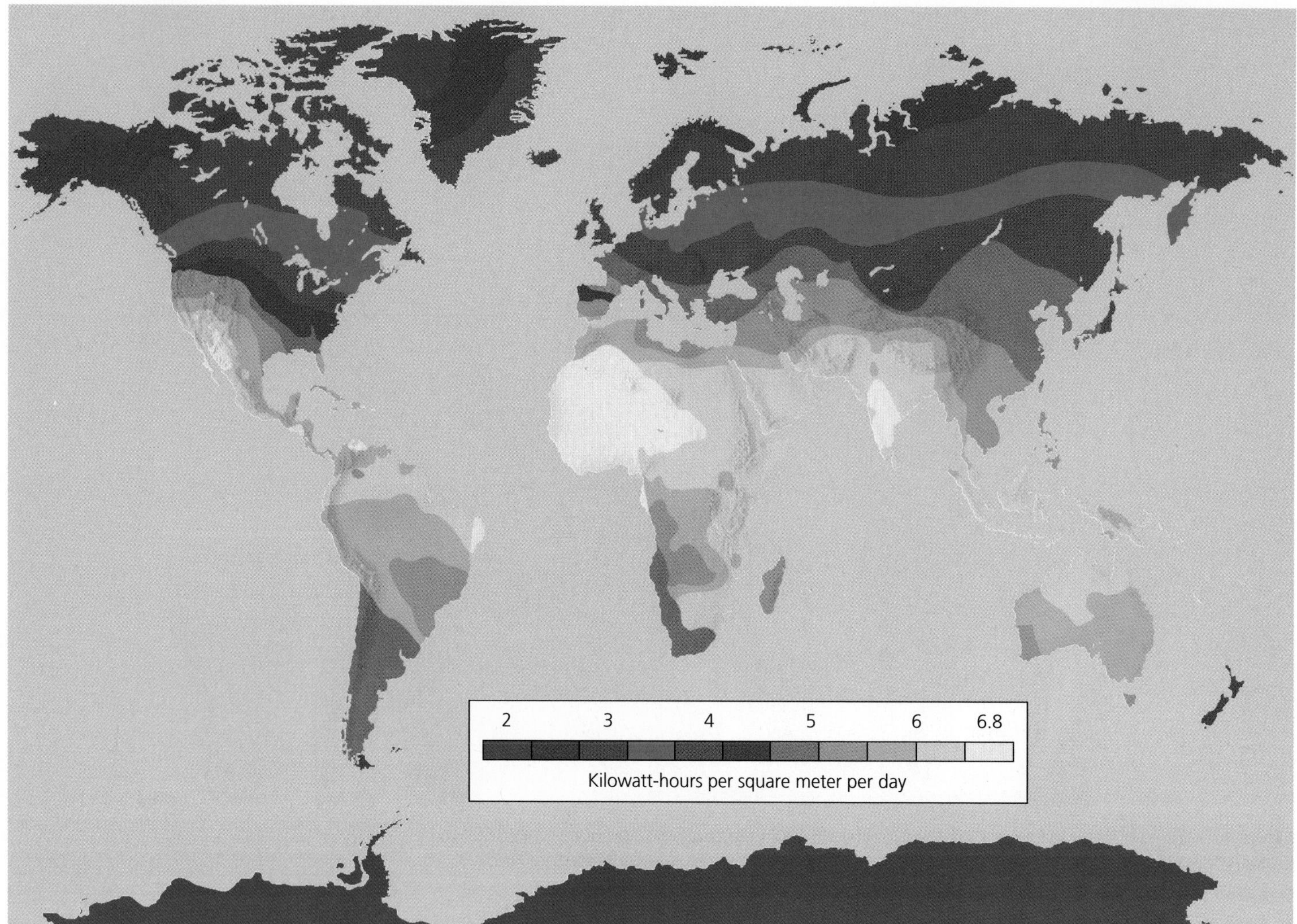

Figure 18 This map shows the global availability of direct solar energy. Areas with more than 3.5 kilowatt-hours per square meter per day (see scale) are good candidates for passive and active solar heating systems and use of solar cells to produce electricity. The United Nations is mapping the potential wind and solar energy resources of 13 developing countries in Africa, Asia, and South and Central America. (Data from U.S. Department of Energy)

Data and Map Analysis

1. What is the potential for making greater use of solar energy to provide heat and produce electricity (with solar cells) where you live or go to school?
2. List the continents in order of overall availability of direct solar energy, from those with the highest to those with the lowest. (See Figure 1 of this supplement, pp. S22–S23, for continent names.)

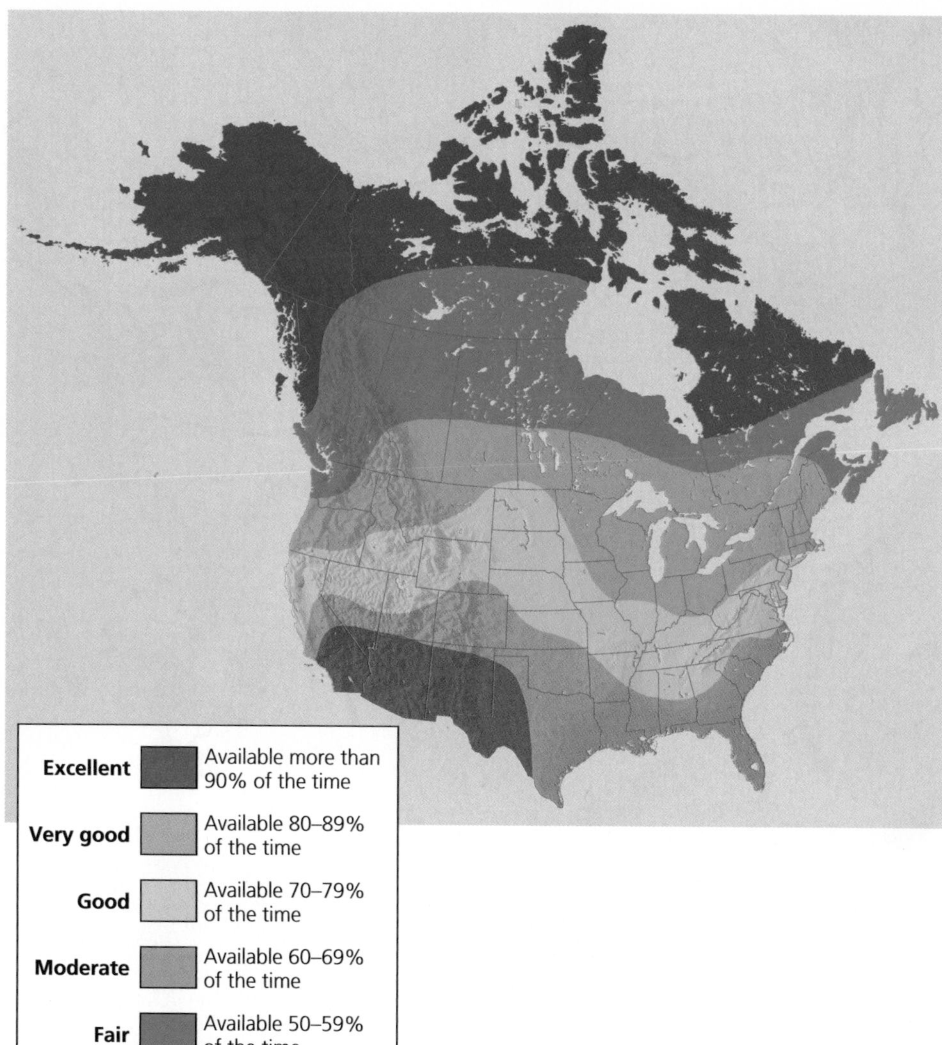

Figure 19 Availability of direct solar energy in the continental United States and Canada. If prices come down as expected, large banks of solar cells in desert areas in the southwestern United States could produce enough energy to meet all U.S. electricity needs. Electricity produced by such solar-cell power plants would be distributed through the country's electric power grid. (Data from the U.S. Department of Energy and the National Wildlife Federation).

Data and Map Analysis

1. If you live in the United States, what is the potential for making increased use of solar energy to provide heat and electricity (with solar cells) where you live or go to school?
2. How many states have areas with excellent, very good, or good availability of direct solar energy?

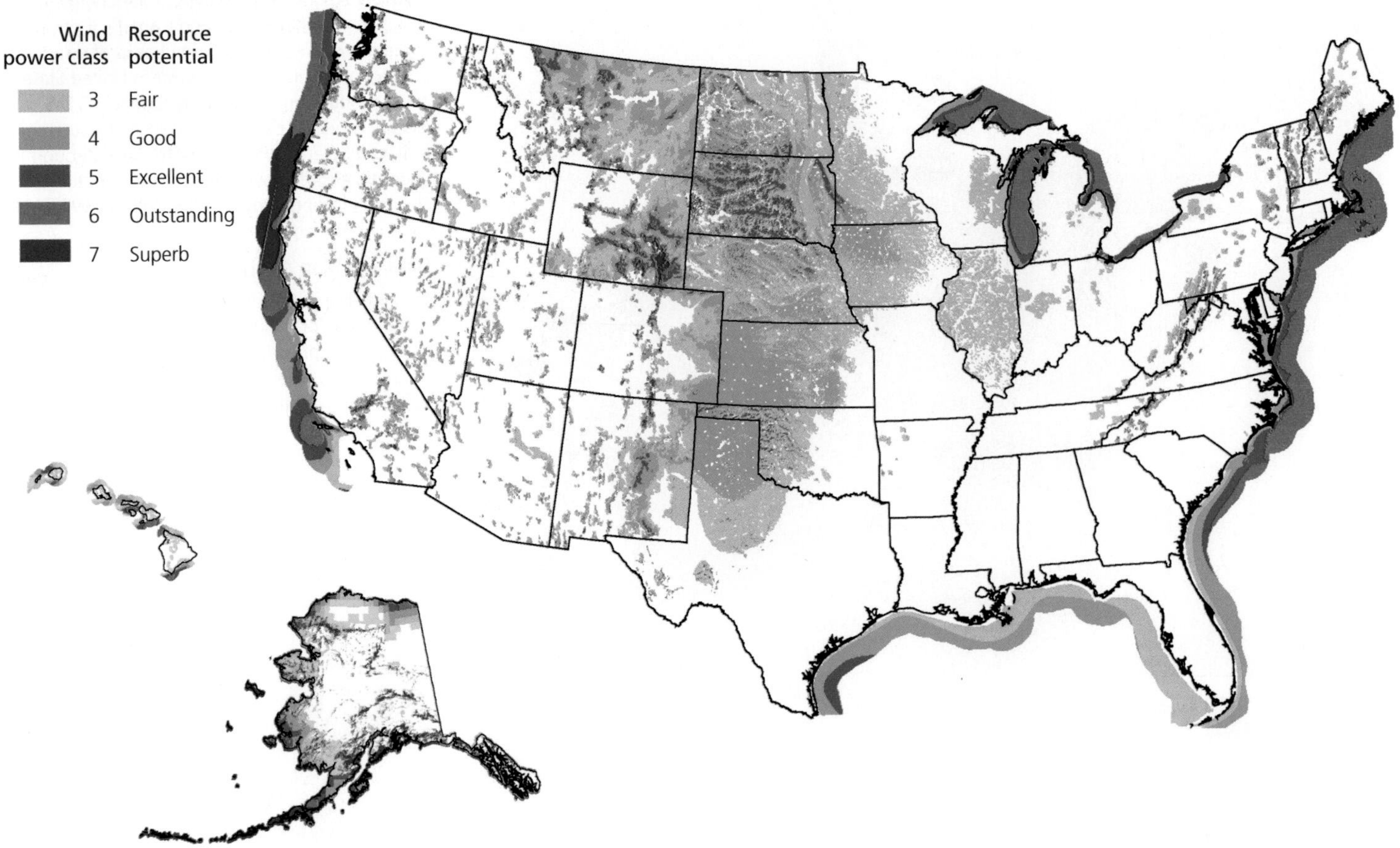

Wind power class	Resource potential
3	Fair
4	Good
5	Excellent
6	Outstanding
7	Superb

Figure 20 This map indicates the potential supply of land-based and offshore wind energy (an indirect form of solar energy) in the United States. Locate the land-based and offshore areas with the highest potential for wind power. For more detailed maps by state see the U.S. Department of Energy, National Renewable Energy Laboratory website at **http://www.nrel.gov/wind/resource_assessment.html**.

Data and Map Analysis

1. If you live in the United States, what is the general wind energy potential where you live or go to school?
2. How many states have areas with good or excellent potential for wind energy?

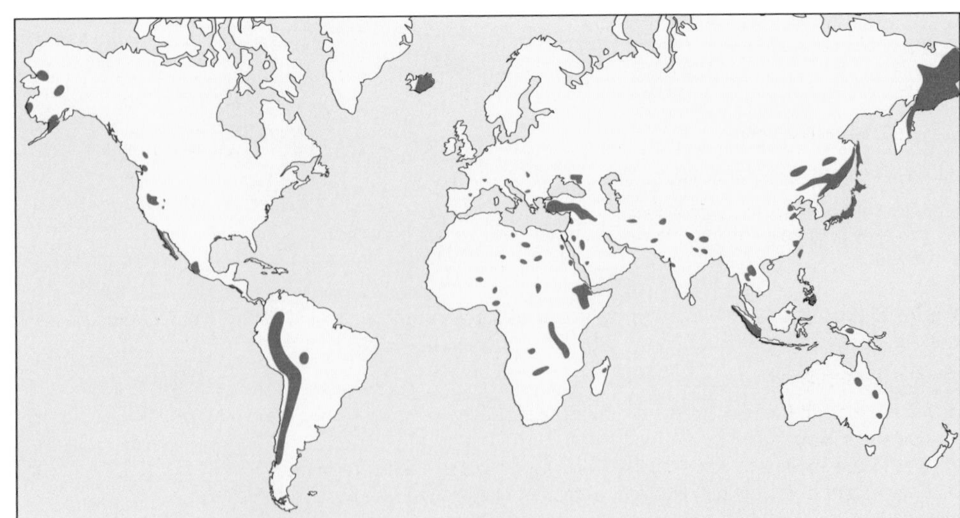

Figure 21 This map illustrates the known global reserves of moderate- to high-temperature geothermal energy. (Data from Canadian Geothermal Resources Council)

Data and Map Analysis

1. Between North and South America, which continent appears to have the greatest total potential for geothermal energy? (See Figure 1 of this supplement, pp. S22–S23, for continent names.)
2. Which country in Asia has the greatest known reserves of geothermal energy? (See Figure 1 of this supplement for country names.)

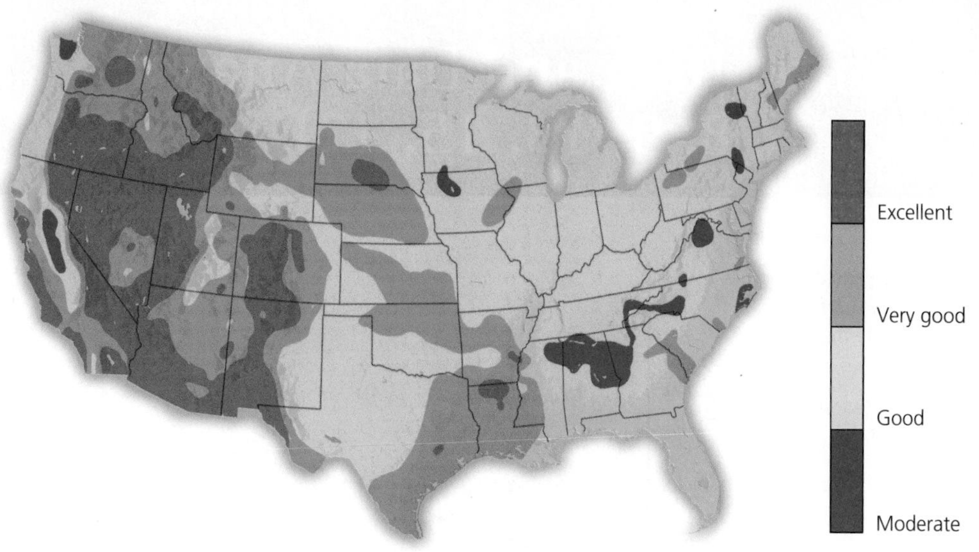

Figure 22 The potential geothermal energy resources in the continental United States are shown here. (Data from U.S. Department of Energy)

Data and Map Analysis

1. If you live in the United States, what is the potential for using geothermal energy to provide heat or to produce electricity where you live or go to school?
2. How many states have areas with very good or excellent potential for using geothermal energy?

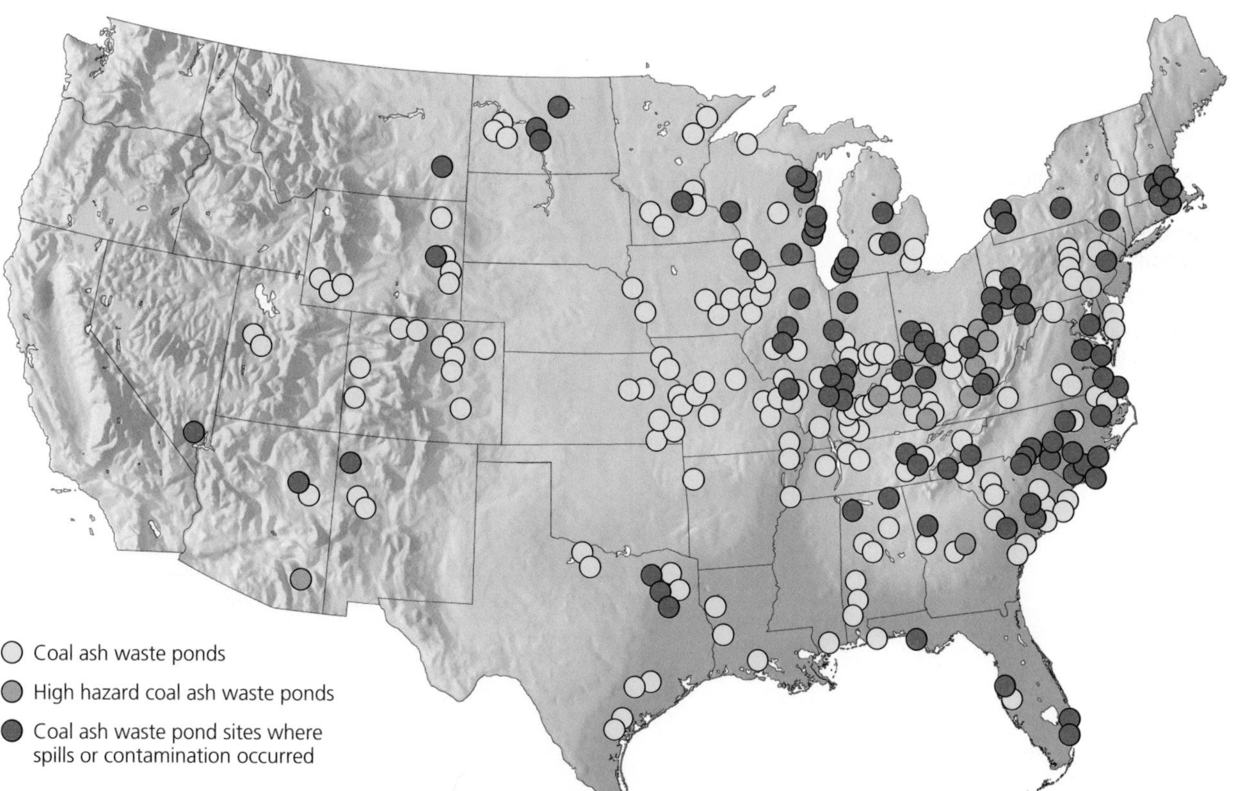

Figure 23 Coal ash waste pond sites in the United States are shown here. (Data from U.S. Environmental Protection Agency and the Sierra Club)

Data and Map Analysis

1. In what part of the country (eastern third, central third, or western third) have most of the cases of contamination and spills occurred?
2. What state appears to have had the most cases of spills and contamination?
3. In what part of the country are most of the high-hazard coal ash ponds located?

SUPPLEMENT

7 Environmental Data and Data Analysis (Chapters 2, 6, 8, 9, 13, 15, 17)

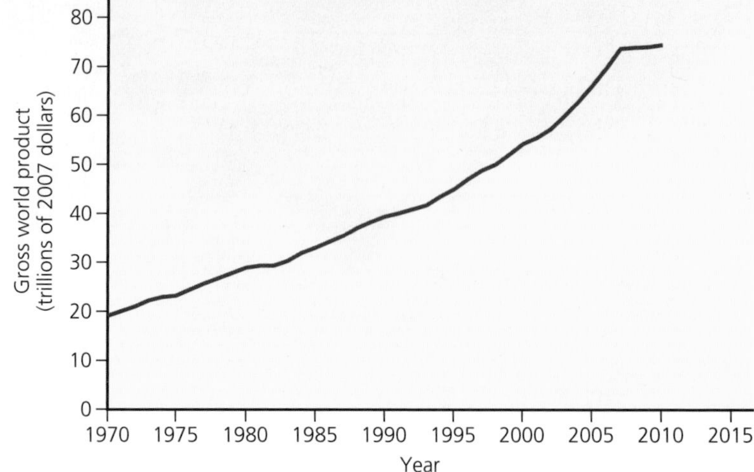

Figure 1 This graph tracks gross world product (GWP), 1970–2010. (Data from International Monetary Fund, the World Bank using purchasing power parity terms)

Data and Graph Analysis

1. Roughly how many times bigger than the gross world product of 1985 was the gross world product of 2010?
2. If the current trend continues, about what do you think the gross world product will be in 2013, in trillions of dollars?

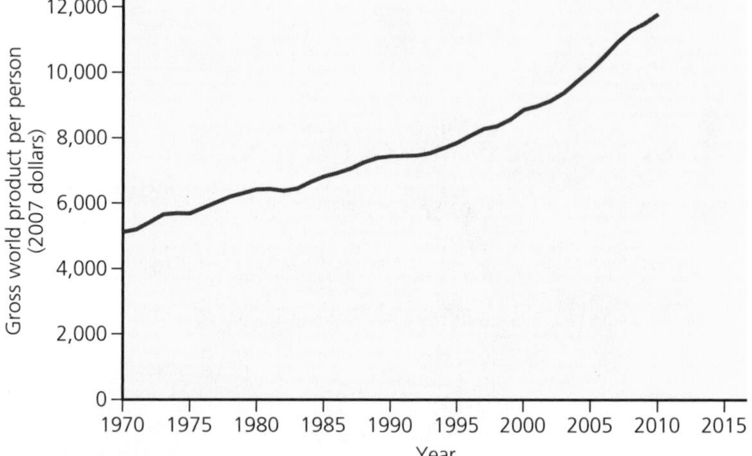

Figure 2 This graph tracks gross world product (GWP) per person, 1970–2010. (Data from International Monetary Fund and the World Bank using purchasing power parity terms)

Data and Graph Analysis

1. Roughly how many years did it take for the GWP per person value in 1970 to double?
2. If the current trend continues, about what do you think the GWP per person will be in 2013, in dollars?

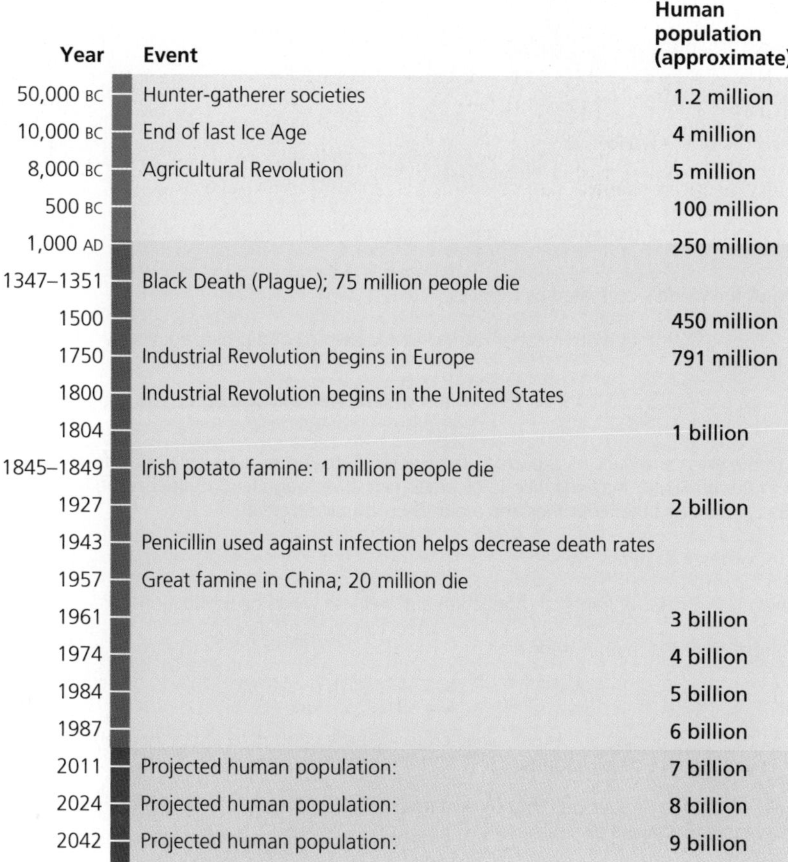

Figure 3 This is a population time line for the period 10,000 BC–2042 AD.

Data and Graph Analysis

1. About how many years did it take the human population to reach 1 billion? How long after that did it reach 2 billion?
2. In about what year was the population half of what it is projected to be in 2011?

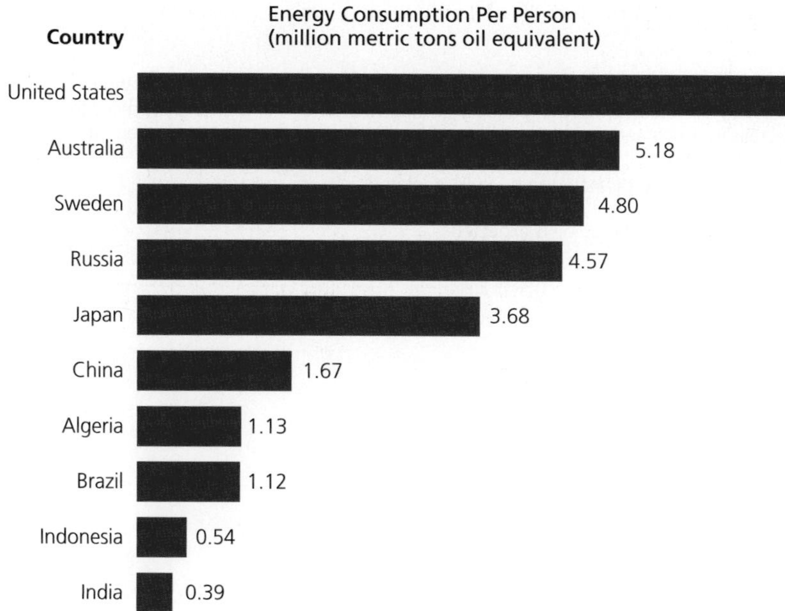

Figure 4 This bar graph represents energy consumption per person in 2009 for selected countries. (Data from World Resources Institute and International Energy Agency)

Data and Graph Analysis

1. On average, how many times more energy does an American consume per year than does a person in **(a)** China, **(b)** India, **(c)** Japan, and **(d)** Ethiopia?
2. On average, how many times more energy does a Chinese citizen consume per year than does a person in **(a)** India and **(b)** Algeria?

Year	Event
1857	First commercial oil well drilled near Titusville, Pennsylvania.
1905	Oil supplies 10% of U.S. energy.
1925	The United States produces 71% of the world's oil.
1930	Because of an oil glut, oil sells for 10¢ per barrel.
1953	U.S. oil companies account for about half the world's oil production, and the United States is the world's leading oil exporter.
1955	The United States has 20% of the world's estimated oil reserves.
1960	OPEC is formed so that developing countries, with most of the world's known oil and projected oil reserves, can get a higher price for their oil.
1973	The United States uses 30% of the world's oil, imports 36% of this oil, and has only 5% of the world's proven oil reserves.
1973–1974	OPEC reduces oil imports to the West and bans oil exports to the United States because of its support for Israel in the 18-day Yom Kippur War with Egypt and Syria. World oil prices rise sharply and lead to double-digit inflation in the United States and many other countries and a global economic recession.
1975	Production of estimated U.S. oil reserves peaks.
1979	Iran's Islamic Revolution shuts down most of Iran's oil production and reduces world oil production.
1981	The Iran-Iraq war pushes global oil prices to an historic high.
1983	Facing an oil glut, OPEC cuts its oil prices.
1985	U.S. domestic oil production begins to decline and is not expected to increase enough to affect the global price of oil or to reduce U.S. dependence on oil imports.
August 1990–June 1991	The United States and its allies fight the Persian Gulf War to oust Iraqi invaders of Kuwait and to protect Western access to Saudi Arabian and Kuwaiti oil supplies.
2003–2010	The United States and a small number of allies fight a second Persian Gulf War to oust Saddam Hussein from power in Iraq and to protect Western access to Saudi Arabian, Kuwaiti, and Iraqi oil.
2010	OPEC has 77% of the world oil reserves. The United States has only about 2% of world oil reserves, uses 22% of the world's oil production, and imports 57% of its oil.
2020	The United States could be importing at least 70% of the oil it uses, as consumption continues to exceed production.
2010–2030	Production of conventional oil from the world's estimated oil reserves is expected to peak as half of the world's oil reserves are used up. Oil prices are expected to increase gradually as the demand for oil increasingly exceeds the supply—unless the world decreases its demand by wasting less energy and shifting to other sources of energy.
2010–2048	Domestic U.S. conventional oil reserves are projected to be 80% depleted.
2042–2083	A gradual decline in dependence on conventional oil is expected.

Figure 5 This time line presents a brief history of the Age of Oil and some projections for the future.

Data and Graph Analysis

1. About what year did U.S. oil production begin to decline?
2. When might the U.S. share of only 1.5% of the world's proven conventional oil reserves be about 80% depleted and, therefore, too expensive to justify extracting what remains?

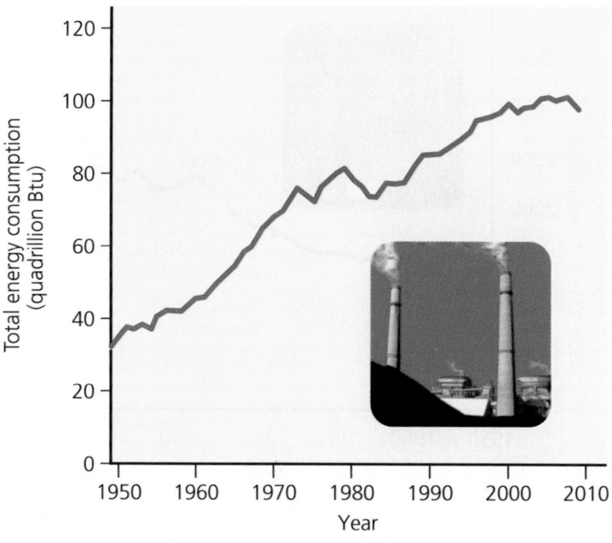

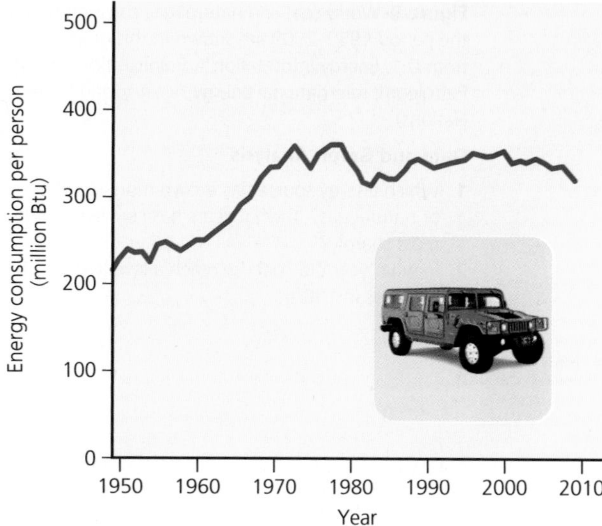

Figure 6 These graphs track total (left) and per capita (right) energy consumption in the United States, 1950–2009. (Data from U.S. Energy Information Administration/Annual Energy Review 2010)

Data and Graph Analysis

1. In what year or years did total U.S. energy consumption reach 80 quadrillion Btus?
2. In what year did energy consumption per person reach its highest level shown on this graph, and about what was that level of consumption?

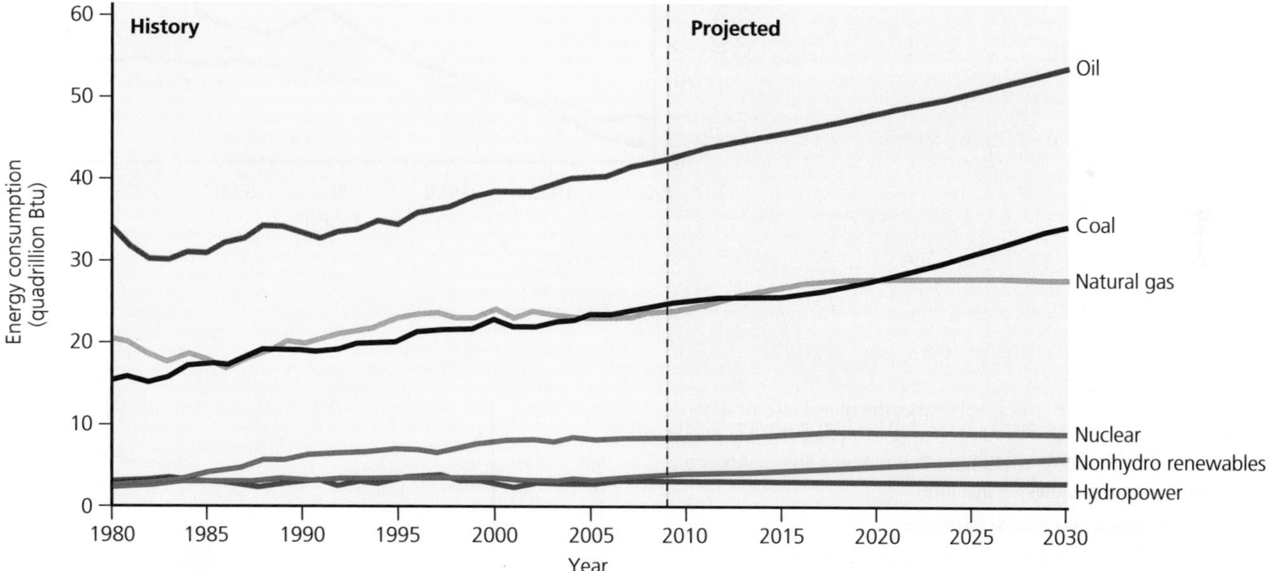

Figure 7 This graph represents energy consumption by fuel in the United States, 1980–2009, with projections to 2030. (Data from U.S. Energy Information Administration/Annual Energy Outlook 2010)

Data and Graph Analysis

1. Which energy source grew the most between 1990 and 2009? How much did it grow (in quadrillion Btus)?
2. In what years after 2007 is it projected that coal use would be higher than natural gas use?

Figure 8 World coal and natural gas consumption for the period 1950–2009 are shown in this graph. (Data from U.S. Energy Information Administration, British Petroleum, International Energy Agency, and United Nations)

Data and Graph Analysis

1. Which energy source has grown more steadily—coal or natural gas? In what years has coal use grown most sharply?
2. In what year did coal use reach a level twice as high as it was in 1960?

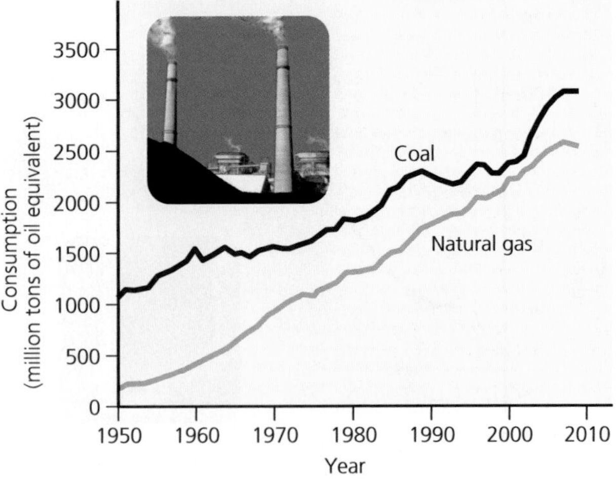

Figure 9 Coal consumption in China and the United States, 1980–2009, is represented here. (Data from British Petroleum, *Statistical Review of World Energy*, 2009)

Data and Graph Analysis

1. By what percentage did coal consumption increase in China between 1980 and 2009?
2. By what percentage did coal consumption increase in the United States between 1980 and 2009?

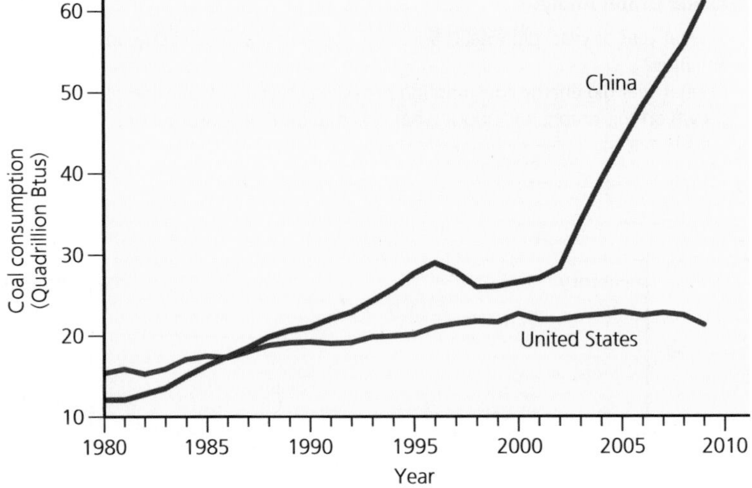

Figure 10 This graph tracks the global electrical generating capacity of nuclear power plants for the period 1960–2009. (Data from International Energy Agency and Worldwatch Institute)

Data and Graph Analysis

1. After 1980, how long did it take to double that year's generating capacity?
2. Considering the decades of the 1970s, 1980s, and 1990s, which decade saw the sharpest growth in generating capacity? During which decade did this growth level off?

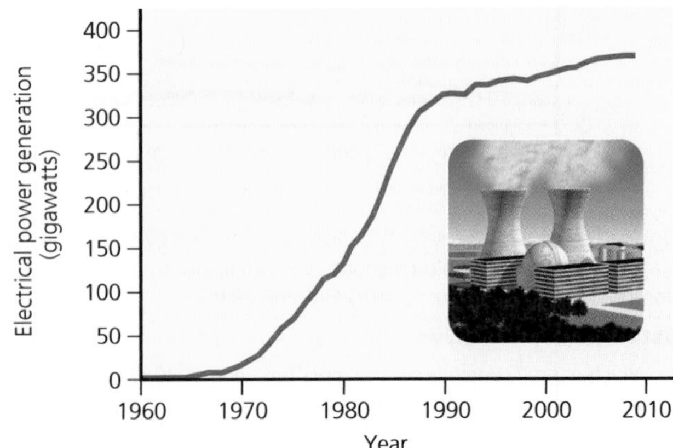

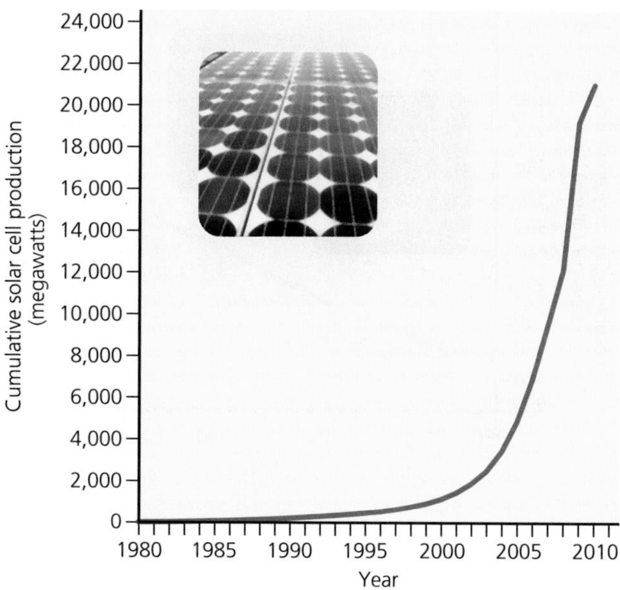

Figure 11 This graph tracks the global cumulative production of electricity by solar (photovoltaic) cells for the period 1980–2009. (Data from PV News, Maycock, Prometheous Institute, Worldwatch Institute, and Global Data 2009)

Data and Graph Analysis

1. About how many times more solar cell production had occurred by 2009 than had occurred by 2000?
2. How long did it take the world to go **(a)** from 0 to 6,000 megawatts in its cumulative production of electricity by solar cells and **(b)** from 6,000 to 19,200 megawatts?

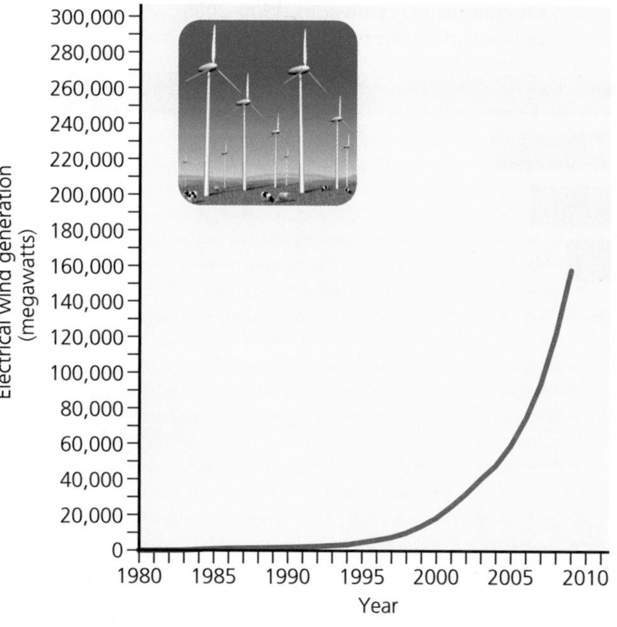

Figure 12 Global installed capacity for generation of electricity by wind energy for the period 1980–2009 is shown in this graph. (Data from Global Wind Energy Council, European Wind Energy Association, American Wind Energy Association, Worldwatch Institute, and World Wind Energy Association, 2010)

Data and Graph Analysis

1. How long did it take for the world to go from zero to 100,000 megawatts of installed capacity for wind-generated electricity?
2. In 2009, the world's installed capacity for generating electricity by wind power was about how many times more than it was in 1995?

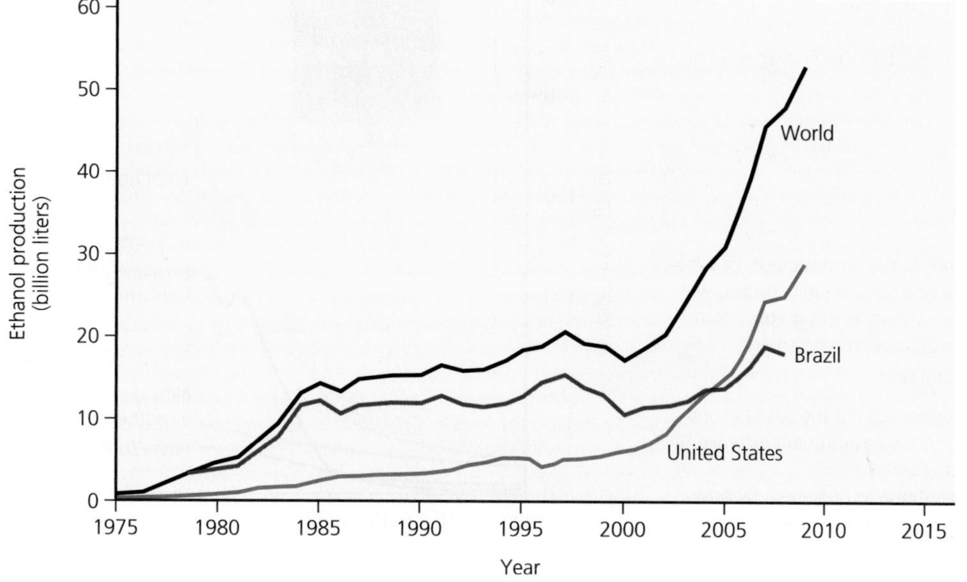

Figure 13 This graph tracks the production of ethanol motor fuel in the world, in Brazil, and in the United States for the period 1975–2009. (Data from F. O. Litch, USDA, and Earth Policy Institute)

Data and Graph Analysis

1. By roughly what percentage did global ethanol production increase between 2000 and 2009?
2. If U.S. production of ethanol continues at its current rate, when is it likely to reach about 50 billion liters?

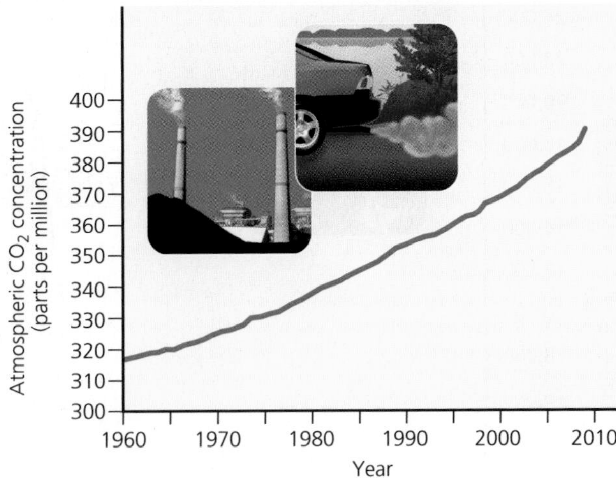

Figure 14 This graph shows the atmospheric concentration of carbon dioxide (CO_2) measured at a major atmospheric research center in Mauna Loa, Hawaii, 1960–2010. The annual fluctuation in CO_2 values occurs because land plants take up varying amounts of CO_2 in different seasons. (Data from Scripps Institute of Oceanography, 2010, and U.S. Energy Information Agency, 2010)

Data and Graph Analysis

1. By how much did atmospheric CO_2 concentrations grow between 1960 and 2010 (in parts per million)?
2. Assuming that atmospheric CO_2 concentrations continue growing as is reflected on this graph, estimate the year in which such concentrations will reach 450 parts per million.

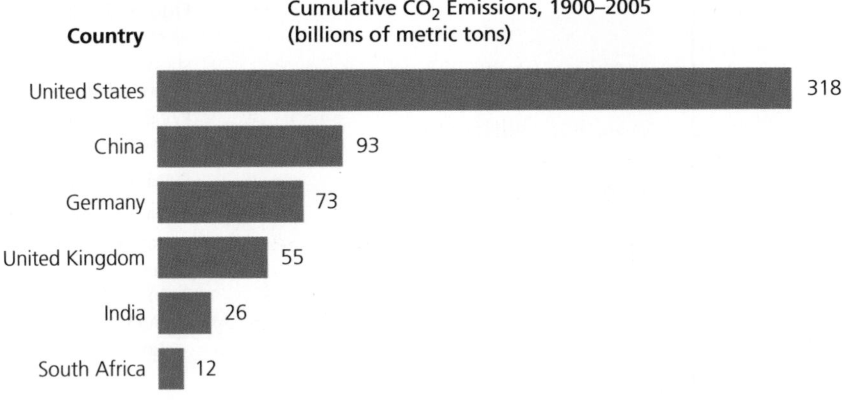

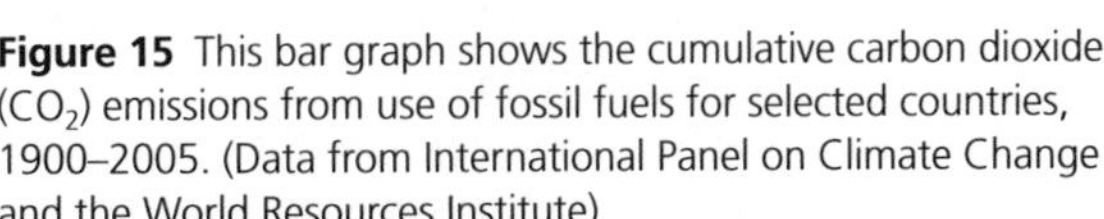

Figure 15 This bar graph shows the cumulative carbon dioxide (CO_2) emissions from use of fossil fuels for selected countries, 1900–2005. (Data from International Panel on Climate Change and the World Resources Institute)

Data and Graph Analysis

1. How many times higher are the cumulative CO_2 emissions of the United States than those of **(a)** China and **(b)** India?
2. How many times higher are the cumulative CO_2 emissions of China than those of **(a)** the United Kingdom and **(b)** India?

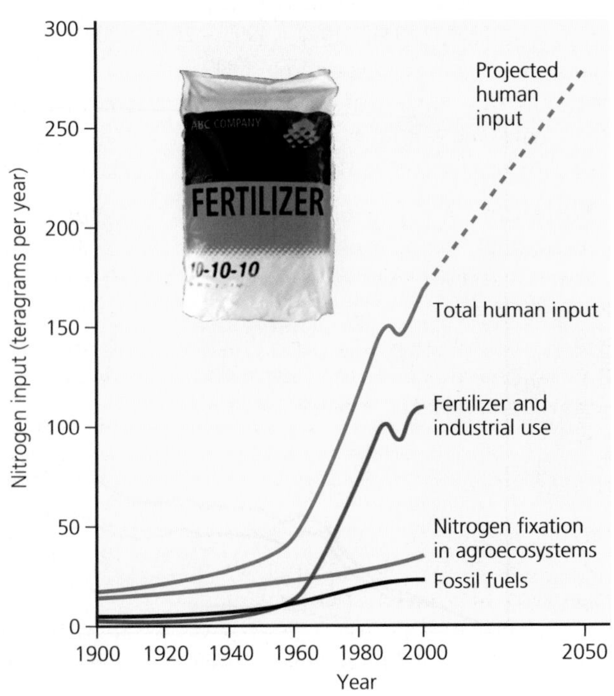

Figure 16 Global trends in the annual inputs of nitrogen into the environment from human activities, with projections to 2050, are shown in this graph. (Data from 2005 Millennium Ecosystem Assessment)

Data and Graph Analysis

1. By roughly what percentage did the world's total input of nitrogen into the environment increase between 1960 and 2000?
2. If nitrogen inputs continue as projected, by how many times will the inputs have increased between 1980 and 2050?

Glossary

acid deposition The falling of acids and acid-forming compounds from the atmosphere to the earth's surface. Acid deposition is commonly known as *acid rain*, a term that refers to the wet deposition of droplets of acids and acid-forming compounds.

acidic solution Any water solution that has more hydrogen ions (H^+) than hydroxide ions (OH^-); any water solution with a pH less than 7. Compare *basic solution, neutral solution*.

acidity Chemical characteristic that helps determine how a substance dissolved in water (a solution) will interact with and affect its environment; based on the comparative amounts of hydrogen ions (H^+) and hydroxide ions (OH^-) contained in a particular volume of the solution. See *pH*.

acid rain See *acid deposition*.

active solar heating system System that uses solar collectors to capture energy from the sun and store it as heat for space heating and water heating. Liquid or air pumped through the collectors transfers the captured heat to a storage system such as an insulated water tank or rock bed. Pumps or fans then distribute the stored heat or hot water throughout a dwelling as needed. Compare *passive solar heating system*.

adaptation Any genetically controlled structural, physiological, or behavioral characteristic that helps an organism survive and reproduce under a given set of environmental conditions. It usually results from a beneficial mutation. See *biological evolution, mutation, natural selection*.

adaptive trait See *adaptation*.

aerobic respiration Complex process that occurs in the cells of most living organisms, in which nutrient organic molecules such as glucose ($C_6H_{12}O_6$) combine with oxygen (O_2) to produce carbon dioxide (CO_2), water (H_2O), and energy. Compare *photosynthesis*.

affluence Wealth that results in high levels of consumption and unnecessary waste of resources, driven by the assumption that buying more and more material goods will bring fulfillment and happiness.

age structure Percentage of the population (or number of people of each gender) at each age level in a population.

air pollution One or more chemicals in high enough concentrations in the air to harm humans, other animals, vegetation, or materials. Such chemicals or physical conditions are called air pollutants. See *primary pollutant, secondary pollutant*.

alley cropping Planting of crops in strips with rows of trees or shrubs on each side.

anaerobic respiration Form of cellular respiration in which some decomposers get the energy they need through the breakdown of glucose (or other nutrients) in the absence of oxygen. Compare *aerobic respiration*.

animal manure Dung and urine of animals used as a form of organic fertilizer. Compare *green manure*.

aquaculture Growing and harvesting of fish and shellfish for human use in freshwater ponds, irrigation ditches, and lakes, or in cages or fenced-in areas of coastal lagoons and estuaries, or in the open ocean.

aquatic life zone Marine and freshwater portions of the biosphere. Examples include freshwater life zones (such as lakes and streams) and ocean or marine life zones (such as estuaries, coastlines, coral reefs, and the open ocean).

aquifer Porous, water-saturated layers of sand, gravel, or bedrock that can yield an economically significant amount of water.

area strip mining Type of surface mining used where the terrain is flat. An earthmover strips away the overburden and a power shovel digs a trench to remove the mineral deposit. The trench is then filled with overburden, and a new trench is made parallel to the previous one. The process is repeated over the entire site. Compare *mountaintop removal mining, open-pit mining, subsurface mining*.

artificial selection Process by which humans select one or more desirable genetic traits in the population of a plant or animal species and then use *selective breeding* to produce populations containing many individuals with the desired traits. Compare *genetic engineering, natural selection*.

asthenosphere Zone within the earth's mantle made up of hot, partly melted rock that flows and can be deformed like soft plastic.

atmosphere The whole mass of air surrounding the earth. See *stratosphere, troposphere*. Compare *biosphere, geosphere, hydrosphere*.

atom The smallest unit of an element that can exist and still have the unique characteristics of that element; made of subatomic particles, it is the basic building block of all chemical elements and thus all matter. Compare *ion, molecule*.

atomic number Number of protons in the nucleus of an atom. Compare *mass number*.

atomic theory Idea that all elements are made up of atoms; the most widely accepted scientific theory in chemistry.

autotroph See *producer*.

background extinction Normal extinction rate of various species as a result of changes in local environmental conditions. Compare *mass extinction*.

bacteria Prokaryotic, one-celled organisms, some of which transmit diseases. Most act as decomposers and get the nutrients they need by breaking down complex organic compounds in the tissues of living or dead organisms into simpler inorganic nutrient compounds.

basic solution Water solution with more hydroxide ions (OH^-) than hydrogen ions (H^+); water solution with a pH greater than 7. Compare *acid solution, neutral solution.*

bioaccumulation An increase in the concentration of a chemical in specific organs or tissues at a level higher than would normally be expected. Compare *biomagnification.*

biodegradable pollutant Material that can be broken down into simpler substances (elements and compounds) by bacteria or other decomposers. Paper and most organic wastes such as animal manure are biodegradable but can take decades to biodegrade in modern landfills. Compare *nondegradable pollutant.*

biodiversity Variety of different species (*species diversity*), genetic variability among individuals within each species (*genetic diversity*), variety of ecosystems (*ecological diversity*), and functions such as energy flow and matter cycling needed for the survival of species and biological communities (*functional diversity*).

biodiversity hotspot An area especially rich in plant species that are found nowhere else and are in great danger of extinction. Such areas suffer serious ecological disruption, mostly because of rapid human population growth and the resulting pressure on natural resources.

biofuel Gas (such as methane) or liquid fuel (such as ethyl alcohol or biodiesel) made from plant material (biomass).

biogeochemical cycle Natural processes that recycle nutrients in various chemical forms from the nonliving environment to living organisms and then back to the nonliving environment. Examples include the carbon, oxygen, nitrogen, phosphorus, sulfur, and hydrologic cycles.

biological community See *community.*

biological diversity See *biodiversity.*

biological evolution Change in the genetic makeup of a population of a species in successive generations. If continued long enough, it can lead to the formation of a new species. Note that populations, not individuals, evolve. See also *adaptation, natural selection, theory of evolution.*

biological extinction Complete disappearance of a species from the earth. It happens when a species cannot adapt and successfully reproduce under new environmental conditions or when a species evolves into one or more new species. Compare *speciation.* See also *endangered species, mass extinction, threatened species.*

biomagnification Increase in concentration of DDT, PCBs, and other slowly degradable, fat-soluble chemicals in organisms at successively higher trophic levels of a food chain or web. Compare *bioaccumulation.*

biomass Organic matter produced by plants and other photosynthetic producers; total dry weight of all living organisms that can be supported at each trophic level in a food chain or web; dry weight of all organic matter in plants and animals in an ecosystem; plant materials and animal wastes used as fuel.

biome A terrestrial region distinguished by the predominance of certain types of vegetation and other forms of life. Examples include various types of deserts, grasslands, and forests.

biomimicry Process of observing certain changes in nature, studying how natural systems have responded to such changing conditions over many millions of years, and applying what is learned to dealing with some environmental challenge.

biosphere Zone of the earth where life is found. It consists of parts of the atmosphere (the troposphere), hydrosphere (mostly surface water and groundwater), and lithosphere (mostly soil and surface rocks and sediments on the bottoms of oceans and other bodies of water). Compare *atmosphere, geosphere, hydrosphere.*

birth rate See *crude birth rate.*

broadleaf deciduous plants Plants such as oak and maple trees that survive drought and cold by shedding their leaves and becoming dormant. Compare *broadleaf evergreen plants, coniferous evergreen plants.*

broadleaf evergreen plants Plants that keep most of their broad leaves year-round. An example is the trees found in the canopies of tropical rain forests. Compare *broadleaf deciduous plants, coniferous evergreen plants.*

buffer Substance that can react with hydrogen ions in a solution and thus hold the acidity or pH of a solution fairly constant. See *pH.*

carbon capture and storage (CCS) Process of removing carbon dioxide gas from coal-burning power and industrial plants and storing it somewhere (usually underground or under the seabed). To be effective, it must be stored so that it cannot be released into the atmosphere, essentially forever.

carbon cycle Cyclic movement of carbon in different chemical forms from the environment to organisms and then back to the environment.

carbon oxides Collective term for the compounds *carbon monoxide* and *carbon dioxide,* which can act as pollutants.

carcinogen Chemicals, ionizing radiation, and viruses that cause or promote the development of cancer. Compare *mutagen, teratogen.*

carnivore Animal that feeds on other animals. Compare *herbivore, omnivore.*

carrying capacity (*K*) Maximum population of a particular species that a given habitat can support over a given period. Compare *cultural carrying capacity.*

CCS See *carbon capture and storage.*

cell Smallest living unit of an organism. Each cell is encased in an outer membrane or wall and contains genetic material (DNA) and other substances that enable it to perform its life function. Organisms such as bacteria consist of only one cell, but most organisms contain many cells.

cell theory The idea that all living things are composed of cells; the most widely accepted scientific theory in biology.

CFCs See *chlorofluorocarbons.*

chemical change Interaction between chemicals in which the chemical composition of the elements or compounds involved changes. Compare *nuclear change, physical change.*

chemical cycling The circulation of chemicals from the environment (mostly from soil and water) through organisms and back to the environment. See *nutrient cycling.*

chemical formula Shorthand way to show the number of atoms (or ions) in the basic structural unit of a compound. Examples include H_2O, NaCl, and $C_6H_{12}O_6$.

chemical reaction See *chemical change.*

chemosynthesis Process in which certain organisms (mostly specialized bacteria) extract inorganic compounds from their environment and convert them into organic nutrient compounds without the presence of sunlight. Compare *photosynthesis.*

chlorofluorocarbons (CFCs) Organic compounds, made up of atoms of carbon, chlorine, and fluorine, that can deplete the ozone layer when they slowly rise into the stratosphere and react with ozone molecules.

CHP (Combined heat and power) See *cogeneration.*

chromosome A grouping of genes and associated proteins in plant and animal cells that carry certain types of genetic information. See *genes.*

chronic malnutrition Faulty nutrition, caused by a diet that does not supply an individual with enough protein, essential fats, vitamins, minerals, and other nutrients needed for good health. Compare *overnutrition, chronic undernutrition.*

chronic undernutrition Condition suffered by people who cannot grow or buy enough food to meet their basic energy needs. Most chronically undernourished children live in developing countries and are likely to suffer from mental retardation and stunted growth, and to die from infectious diseases. Compare *chronic malnutrition, overnutrition.*

clear-cutting Method of timber harvesting in which all trees in a forested area are removed in a single cutting. Compare *selective cutting, strip cutting.*

climate General pattern of atmospheric conditions in a given area over periods ranging from at least 30 years to thousands of years. The two main factors determining an area's climate are its average temperature, with its seasonal variations, and the average amount and distribution of precipitation. Compare *weather.*

climate change tipping point Point at which an environmental problem reaches a threshold level where scientists fear it could cause irreversible climate disruption.

closed-loop recycling See *primary recycling.*

coal Solid, combustible mixture of organic compounds with 30–98% carbon by weight, mixed with various amounts of water and small amounts of sulfur and nitrogen compounds. It forms in several stages as the remains of plants are subjected to heat and pressure over millions of years.

coal gasification Conversion of solid coal to synthetic natural gas (SNG).

coal liquefaction Conversion of solid coal to a liquid hydrocarbon fuel such as synthetic gasoline or methanol.

coastal wetland Land along a coastline, extending inland from an estuary that is covered with salt water all or part of the year. Examples include marshes, tidal flats, and mangrove swamps. Compare *inland wetland.*

coastal zone Warm, nutrient-rich, shallow part of the ocean that extends from the high-tide mark on land to the edge of a shelf-like extension of continental land masses known as the continental shelf. Compare *open sea.*

coevolution Evolution in which two or more species interact and exert selective pressures on each other that can lead each species to undergo adaptations. See *evolution, natural selection.*

cogeneration Production of two useful forms of energy, such as high-temperature heat or steam and electricity, from the same fuel source.

cold front Leading edge of an advancing mass of cold air. Compare *warm front.*

combined heat and power (CHP) production See *cogeneration.*

commensalism An interaction between organisms of different species in which one type of organism benefits and the other type is neither helped nor harmed to any great degree. Compare *mutualism.*

commercial inorganic fertilizer Commercially prepared mixture of inorganic plant nutrients such as nitrates, phosphates, and potassium applied to the soil to restore fertility and increase crop yields. Compare *organic fertilizer.*

commercial forest See *tree plantation.*

common-property resource Resource that is owned jointly by a large group of individuals, such as land that belongs to a whole village and that can be used by anyone for grazing cows or sheep. Compare *open access renewable resource.* See *tragedy of the commons.*

community Populations of all species living and interacting in an area at a particular time.

complex carbohydrate Two or more monomers of simple sugars (such as glucose) linked together. One example is the starches that plants use to store energy and also to provide energy for animals that feed on plants. Another is cellulose, the earth's most abundant organic compound, which is found in the cell walls of bark, leaves, stems, and roots.

compost Partially decomposed organic plant and animal matter used as a soil conditioner or fertilizer.

compound Combination of atoms, or oppositely charged ions, of two or more elements held together by attractive forces called chemical bonds. Examples are NaCl, CO_2, and $C_6H_{12}O_6$. Compare *element.*

coniferous evergreen plants Cone-bearing plants (such as spruces, pines, and firs) that keep most of their narrow, pointed leaves (needles) all year. Compare *broadleaf deciduous plants, broad-leaf evergreen plants.*

coniferous trees Cone-bearing trees, mostly evergreens, that have needle-shaped or scale-like leaves. They produce wood known commercially as softwood. Compare *deciduous plants*.

conservation biology Multidisciplinary science created to deal with the crisis of accelerating losses of the genes, species, communities, and ecosystems that make up earth's biological diversity. Its goals are to investigate human impacts on biodiversity and to develop practical approaches to preserving it.

conservation-tillage farming Crop cultivation in which the soil is disturbed little (minimum-tillage farming) or not at all (no-till farming) in an effort to reduce soil erosion, lower labor costs, and save energy. Compare *conventional-tillage farming*.

consumer Organism that cannot synthesize the organic nutrients it needs and gets its organic nutrients by feeding on the tissues of producers or of other consumers; generally divided into *primary consumers* (herbivores), *secondary consumers* (carnivores), *tertiary* (*higher-level*) *consumers*, *omnivores*, and *detritivores* (decomposers and detritus feeders). In economics, one who uses economic goods. Compare *producer*.

contour farming Plowing and planting across the changing slope of land, rather than in straight lines, to help retain water and reduce soil erosion.

contour strip mining Form of surface mining used on hilly or mountainous terrain. A power shovel cuts a series of terraces into the side of a hill. An earth-mover removes the overburden, and a power shovel extracts the coal. The overburden from each new terrace is dumped onto the one below. Compare *area strip mining*, *mountaintop removal*, *open-pit mining*, *subsurface mining*.

conventional-tillage farming Crop cultivation method in which a planting surface is made by plowing land, breaking up the exposed soil, and then smoothing the surface. Compare *conservation-tillage farming*.

coral reef Formation produced by massive colonies containing billions of tiny coral animals, called polyps, that secrete a stony substance (calcium carbonate) around themselves for protection. When the corals die, their empty outer skeletons form layers; reefs grow with the accumulation of such layers. They are found in the coastal zones of warm tropical and subtropical oceans.

core Inner zone of the earth that consists of a solid inner core and a liquid outer core. Compare *crust*, *mantle*.

crop rotation Planting a field, or an area of a field, with different crops from year to year to reduce soil nutrient depletion. A plant such as corn, tobacco, or cotton, which removes large amounts of nitrogen from the soil, is planted one year. The next year a legume such as soybeans, which adds nitrogen to the soil, is planted.

crown fire Extremely hot forest fire that burns ground vegetation and tree-tops. Compare *ground fire*, *surface fire*.

crude birth rate Annual number of live births per 1,000 people in the population of a geographic area at the midpoint of a given year. Compare *crude death rate*.

crude death rate Annual number of deaths per 1,000 people in the population of a geographic area at the midpoint of a given year. Compare *crude birth rate*.

crude oil Gooey liquid consisting mostly of hydrocarbon compounds and small amounts of compounds containing oxygen, sulfur, and nitrogen. Extracted from underground accumulations, it is sent to oil refineries, where it is converted to heating oil, diesel fuel, gasoline, tar, and other materials.

crust Solid outer zone of the earth. It consists of oceanic crust and continental crust. Compare *core*, *mantle*.

cultural carrying capacity The limit on population growth that would allow most people in an area or in the world to live in reasonable comfort and freedom without impairing the ability of the planet to sustain future generations.

cultural eutrophication Overnourishment of aquatic ecosystems with plant nutrients (mostly nitrates and phosphates) resulting from human activities such as agriculture, urbanization, and discharges from sewage treatment plants. See *eutrophication*.

dam A structure built across a river to control the river's flow or to create a reservoir. See *reservoir*.

data Factual information collected by scientists.

death rate See *crude death rate*.

deciduous plants Trees, such as oaks and maples, and other plants that survive during dry or cold seasons by shedding their leaves. Compare *coniferous trees*, *succulent plants*.

decomposer Organism that digests parts of dead organisms, as well as cast-off fragments and wastes of living organisms, by breaking down the complex organic molecules in those materials into simpler inorganic compounds and then absorbing the soluble nutrients. Producers return most of these chemicals to the soil and water for reuse. Decomposers consist of various bacteria and fungi. Compare *consumer*, *detritivore*, *producer*.

deforestation Removal of trees from a forested area.

demographic transition Hypothesis that countries, as they become industrialized, have declines in death rates followed by declines in birth rates.

depletion time The time it takes to use a certain fraction (usually 80%) of the known or estimated supply of a nonrenewable resource at an assumed rate of use. Finding and extracting the remaining 20% usually costs more than it is worth.

desalination Purification of salt water or brackish (slightly salty) water by removal of dissolved salts.

desert Biome in which evaporation exceeds precipitation and the average amount of precipitation is less than 25 centimeters (10 inches) per year. Such areas have little vegetation or have widely spaced, mostly low vegetation. Compare *forest*, *grassland*.

desertification Conversion of rangeland, rain-fed cropland, or irrigated cropland to desert-like land, with a drop in agricultural productivity of 10% or more. It usually is caused by a combination of overgrazing, soil erosion, prolonged drought, and climate change.

detritivore Consumer organism that feeds on detritus, parts of dead organisms, and cast-off fragments and wastes of living organisms. Examples include earthworms, termites, and crabs. Compare *decomposer*.

detritus Parts of dead organisms and cast-off fragments and wastes of living organisms.

detritus feeder See *detritivore*.

dissolved oxygen (DO) content Amount of oxygen gas (O_2) dissolved in a given volume of water at a particular temperature and pressure, often expressed as a concentration in parts of oxygen per million parts of water.

dose Amount of a potentially harmful substance an individual ingests, inhales, or absorbs through the skin. Compare *response*. See *dose-response curve*.

dose-response curve Plot of data showing the effects of various doses of a toxic agent on a group of test organisms. See *dose, response*.

drainage basin See *watershed*.

drought Condition in which an area does not get enough water because of lower-than-normal precipitation or higher-than-normal temperatures that increase evaporation.

earthquake Shaking of the ground resulting from the fracturing and displacement of subsurface rock, which produces a fault, or from subsequent movement along the fault.

ecological diversity The variety of forests, deserts, grasslands, oceans, streams, lakes, and other biological communities. See *biodiversity*. Compare *functional diversity, genetic diversity, species diversity*.

ecological footprint Amount of biologically productive land and water needed to supply a population with the renewable resources it uses and to absorb or dispose of the wastes from such resource use. It is a measure of the average environmental impact of populations in different countries and areas. See *per capita ecological footprint*.

ecological niche Total way of life, or role of a species in an ecosystem. It includes all physical, chemical, and biological conditions that a species needs in order to live and reproduce in an ecosystem.

ecological restoration Deliberate alteration of a degraded habitat or ecosystem to restore as much of its ecological structure and function as possible.

ecological succession Process in which communities of plant and animal species in a particular area are replaced over time by a series of different and often more complex communities. See *primary ecological succession, secondary ecological succession*.

ecological tipping point Point at which an environmental problem reaches a threshold level, which causes an often irreversible shift in the behavior of a natural system.

ecologist Biological scientist who studies relationships between living organisms and their environment.

ecology Biological science that studies the relationships between living organisms and their environment; study of the structure and functions of nature.

economic depletion Exhaustion of 80% of the estimated supply of a nonrenewable resource. Finding, extracting, and processing the remaining 20% usually costs more than it is worth. May also apply to the depletion of a renewable resource, such as a fish or tree species.

economic development Improvement of human living standards by economic growth. Compare *economic growth*.

economic growth Increase in the capacity to provide people with goods and services; an increase in gross domestic product (GDP). Compare *economic development*. See *gross domestic product*.

economic resources Natural resources, capital goods, and labor used in an economy to produce material goods and services. See *natural resources*.

economics Social science that deals with the production, distribution, and consumption of goods and services to satisfy people's needs and wants.

ecosystem One or more communities of different species interacting with one another and with the chemical and physical factors making up their nonliving environment.

ecosystem services Natural services that support life on the earth and are essential to the quality of human life and the functioning of the world's economies. Examples are the chemical cycles, natural pest control, and natural purification of air and water. See *natural resources*.

electromagnetic radiation Forms of kinetic energy traveling as electromagnetic waves. Examples include radio waves, TV waves, microwaves, infrared radiation, visible light, ultraviolet radiation, X-rays, and gamma rays.

electron (e) Tiny particle moving around outside the nucleus of an atom. Each electron has one unit of negative charge and almost no mass. Compare *neutron, proton*.

element Chemical, such as hydrogen (H), iron (Fe), sodium (Na), carbon (C), nitrogen (N), or oxygen (O), whose distinctly different atoms serve as the basic building blocks of all matter. Two or more elements combine to form the compounds that make up most of the world's matter. Compare *compound*.

endangered species Wild species with so few individual survivors that the species could soon become extinct in all or most of its natural range. Compare *threatened species*.

endemic species Species that is found in only one area. Such species are especially vulnerable to extinction.

energy Capacity to do work or to transfer heat. Can involve mechanical, physical, chemical, or electrical tasks or heat transfers between objects at different temperatures.

energy efficiency Percentage of the total energy input that does useful work and is not converted into low-quality, generally useless heat in an energy conversion system or process. See *energy quality, net energy*.

energy quality Ability of a form of energy to do useful work. High-temperature heat and the chemical energy in fossil fuels are examples of concentrated high-quality energy. Low-quality energy such as low-temperature heat is dispersed or diluted and cannot do much useful work. See *high-quality energy, low-quality energy*.

environment All external conditions, factors, matter, and energy, living and nonliving, that affect any living organism or other specified system.

environmental degradation Depletion or destruction of a potentially renewable resource such as soil, grassland, forest, or wildlife that is used faster than it is naturally replenished. If such use continues, the resource becomes nonrenewable (on a human time scale) or nonexistent (extinct). See also *natural capital degradation, sustainable yield*.

environmental ethics Human beliefs about what is right or wrong with how we treat the environment.

environmentalism Social movement dedicated to protecting the earth's life support systems for us and other species.

environmental justice Fair treatment and meaningful involvement of all people, regardless of race, color, sex, national origin, or income, with respect to the development, implementation, and enforcement of environmental laws, regulations, and policies.

environmentally sustainable society Society that meets the current and future needs of its people for basic resources in a just and equitable manner without compromising the ability of future generations of humans and other species from meeting their basic needs.

environmental movement Citizens organized to demand that political leaders enact laws and develop policies to curtail pollution, clean up polluted environments, and protect unspoiled areas from environmental degradation.

environmental policy Laws, rules, and regulations related to an environmental problem that are developed, implemented, and enforced by a particular government body or agency.

environmental resistance All of the limiting factors that act together to limit the growth of a population. See *limiting factor*.

environmental science Interdisciplinary study that uses information and ideas from the physical sciences (such as biology, chemistry, and geology) as well as those from the social sciences and humanities (such as economics, politics, and ethics) to learn how nature works, how we interact with the environment, and how we can to help deal with environmental problems.

environmental scientist Scientist who uses information from the physical sciences and social sciences to understand how the earth works, learn how humans interact with the earth, and develop solutions to environmental problems. See *environmental science*.

environmental wisdom worldview Worldview holding that humans are part of and totally dependent on nature and that nature exists for all species, not just for us. Our success depends on learning how the earth sustains itself and integrating such environmental wisdom into the ways we think and act. Compare *frontier worldview, planetary management worldview, stewardship worldview*.

environmental worldview Set of assumptions and beliefs about how people think the world works, what they think their role in the world should be, and what they believe is right and wrong environmental behavior (environmental ethics). See *environmental wisdom worldview, frontier worldview, planetary management worldview, stewardship worldview*.

EPA U.S. Environmental Protection Agency; responsible for managing federal efforts to control air and water pollution, radiation and pesticide hazards, environmental research, hazardous waste, and solid waste disposal.

epidemiology Study of the patterns of disease or other harmful effects from exposure to toxins and diseases caused by pathogens within defined groups of people to find out why some people get sick and some do not.

erosion Process or group of processes by which loose or consolidated earth materials, especially topsoil, are dissolved, loosened, or worn away and removed from one place and deposited in another. See *weathering*.

estuary Partially enclosed coastal area at the mouth of a river from which freshwater, carrying fertile silt and runoff from the land, mixes with salty seawater.

eutrophication Physical, chemical, and biological changes that take place after a lake, estuary, or slow-flowing stream receives inputs of plant nutrients—mostly nitrates and phosphates—from natural erosion and runoff from the surrounding land basin. See *cultural eutrophication*.

eutrophic lake Lake with a large or excessive supply of plant nutrients, mostly nitrates and phosphates. Compare *oligotrophic lake*.

evaporation Conversion of a liquid into a gas.

evergreen plants Plants that keep some of their leaves or needles throughout the year. Examples include cone-bearing trees (conifers) such as firs, spruces, pines, redwoods, and sequoias. Compare *deciduous plants, succulent plants*.

evolution See *biological evolution*.

experiment Procedure a scientist uses to study some phenomenon under known conditions. Scientists conduct some experiments in the laboratory and others in nature. The resulting scientific data or facts must be verified or confirmed by repeated observations and measurements, ideally by several different investigators.

exponential growth Growth in which some quantity, such as population size or economic output, increases at a constant rate per unit of time. An example is the growth sequence 2, 4, 8, 16, 32, 64, and so on, which increases by 100% at each interval. When the increase in quantity over time is plotted, this type of growth yields a curve shaped like the letter J.

external cost Harmful environmental, economic, or social effect of producing and using an economic good that is not included in the market price of the good. Compare *internal cost*. See *full cost pricing*.

extinction See *biological extinction*.

extinction rate Percentage or number of species that go extinct within a certain period of time such as a year.

family planning Providing information, clinical services, and contraceptives to help people choose the number and spacing of children they want to have.

feedlot Confined outdoor or indoor space used to raise hundreds to thousands of domesticated livestock.

first law of thermodynamics Whenever energy is converted from one form to another in a physical or chemical change, no energy is created or destroyed, but energy can be changed from one form to another; you cannot get more energy out of something than you put in; in terms of energy quantity, you cannot get something for nothing. This law does not apply to nuclear changes, in which large amounts of energy can be produced from small amounts of matter. See *second law of thermodynamics*.

fishery Concentration of particular aquatic species suitable for commercial harvesting in a given ocean area or inland body of water.

fish farming See *aquaculture*.

fishprint Area of ocean needed to sustain the consumption of an average person, a nation, or the world. Compare *ecological footprint*.

floodplain Flat valley floor next to a stream channel. For legal purposes, the term often applies to any low area that has the potential for flooding, including certain coastal areas.

food chain Series of organisms in which each eats or decomposes the preceding one. Compare *food web*.

food insecurity Condition under which people live with chronic hunger and malnutrition that threatens their ability to lead healthy and productive lives. Compare *food security*.

food security Condition under which every person in a given area has daily access to enough nutritious food to have an active and healthy life. Compare *food insecurity*.

food web Complex network of many interconnected food chains and feeding relationships. Compare *food chain*.

forest Biome with enough average annual precipitation to support the growth of tree species and smaller forms of vegetation. Compare *desert, grassland*.

fossil fuel Product of partial or complete decomposition of plants and animals; occurs as crude oil, coal, natural gas, or heavy oil as a result of exposure to heat and pressure in he earth's crust over millions of years. See *coal, natural gas*.

fossils Skeletons, bones, shells, body parts, leaves, seeds, or impressions of such items that provide recognizable evidence of organisms that lived long ago.

fracking See *hydraulic fracturing*.

freshwater Relatively pure water containing few dissolved salts.

freshwater life zones Aquatic systems where water with a dissolved salt concentration of less than 1% by volume accumulates on or flows through the surfaces of terrestrial biomes. Examples include *standing* (lentic) bodies of freshwater such as lakes, ponds, and inland wetlands and *flowing* (lotic) systems such as streams and rivers. Compare *biome*.

front The boundary between two air masses with different temperatures and densities. See *cold front, warm front*.

frontier environmental worldview View held by European colonists settling North America in the 1600s that the continent had vast resources and was a wilderness to be conquered by settlers clearing and planting land.

frontier science See *tentative science*.

full cost pricing Setting market prices of goods and services to include hidden harmful environmental and health costs of producing and using them. See *external cost, internal cost*.

functional diversity Biological and chemical processes or functions such as energy flow and matter cycling needed for the survival of species and biological communities. See *biodiversity, ecological diversity, genetic diversity, species diversity*.

GDP See *gross domestic product*.

generalist species Species with a broad ecological niche. They can live in many different places, eat a variety of foods, and tolerate a wide range of environmental conditions. Examples include flies, cockroaches, mice, rats, and humans. Compare *specialist species*.

genes Coded units of information about specific traits that are passed from parents to offspring during reproduction. They consist of segments of DNA molecules found in chromosomes.

genetically modified organism (GMO) Organism whose genetic makeup has been altered by genetic engineering.

genetic diversity Variability in the genetic makeup among individuals within a single species. See biodiversity. Compare *ecological diversity, functional diversity, species diversity*.

genetic engineering Insertion of an alien gene into an organism to give it a beneficial genetic trait. Compare *artificial selection, natural selection*.

genuine progress indicator (GPI) GDP plus the estimated value of beneficial transactions that meet basic needs, but in which no money changes hands, minus the estimated harmful environmental, health, and social costs of all transactions. Compare *gross domestic product*.

geographic isolation Separation of populations of a species into different areas for long periods of time.

geology Study of the earth's dynamic history. Geologists study and analyze rocks and the features and processes of the earth's interior and surface.

geosphere Earth's intensely hot core, thick mantle composed mostly of rock, and thin outer crust that contains most of the earth's rock, soil, and sediment. Compare *atmosphere, biosphere, hydrosphere*.

geothermal energy Heat transferred from the earth's underground concentrations of dry steam (steam with no water droplets), wet steam (a mixture of steam and water droplets), or hot water trapped in fractured or porous rock.

global warming Warming of the earth's lower atmosphere (troposphere) because of increases in the concentrations of one or more greenhouse gases. It can result in climate change that can last for decades to thousands of years. See *greenhouse effect, greenhouse gases, natural greenhouse effect*.

GMO See *genetically modified organism*.

GPI See *genuine progress indicator*.

GPP See *gross primary productivity.*

graph A tool for conveying information that we can summarize numerically by illustrating that information in a visual format.

grassland Biome found in regions where enough annual average precipitation to support the growth of grass and small plants but not enough to support large stands of trees. Compare *desert, forest.*

greenhouse effect Natural effect that releases heat in the atmosphere near the earth's surface. Water vapor, carbon dioxide, ozone, and other gases in the lower atmosphere (troposphere) absorb some of the infrared radiation (heat) radiated by the earth's surface. Their molecules vibrate and transform the absorbed energy into longer-wavelength infrared radiation in the troposphere. If the atmospheric concentrations of these greenhouse gases increase and other natural processes do not remove them, the average temperature of the lower atmosphere will increase.

greenhouse gases Gases in the earth's lower atmosphere (troposphere) that cause the greenhouse effect. Examples include carbon dioxide, chlorofluorocarbons, ozone, methane, water vapor, and nitrous oxide.

green manure Freshly cut or still-growing green vegetation that is plowed into the soil to increase the organic matter and humus available to support crop growth. Compare *animal manure.*

green revolution Popular term for the introduction of scientifically bred or selected varieties of grain (rice, wheat, corn) that, with adequate inputs of fertilizer and water, can greatly increase crop yields.

gross domestic product (GDP) Annual market value of all goods and services produced by all firms and organizations, foreign and domestic, operating within a country. See *per capita GDP.* Compare *genuine progress indicator (GPI).*

gross primary productivity (GPP) Rate at which an ecosystem's producers capture and store a given amount of chemical energy as biomass in a given length of time. Compare *net primary productivity.*

ground fire Fire that burns decayed leaves or peat deep below the ground's surface. Compare *crown fire, surface fire.*

groundwater Water that sinks into the soil and is stored in slowly flowing and slowly renewed underground reservoirs called *aquifers;* underground water in the zone of saturation, below the water table. Compare *runoff, surface water.*

habitat Place or type of place where an organism or population of organisms lives. Compare *ecological niche.*

habitat fragmentation Breakup of a habitat into smaller pieces, usually as a result of human activities.

hazardous chemical Chemical that can cause harm because it is flammable or explosive, can irritate or damage the skin or lungs (such as strong acidic or alkaline substances), or can cause allergic reactions of the immune system (allergens). See also *toxic chemical.*

hazardous waste Any solid, liquid, or containerized gas that can catch fire easily, is corrosive to skin tissue or metals, is unstable and can explode or release toxic fumes, or has harmful concentrations of one or more toxic materials that can leach out. These substances are usually byproducts of manufacturing processes.

heat Total kinetic energy of all randomly moving atoms, ions, or molecules within a given substance, excluding the overall motion of the whole object. Heat always flows spontaneously from a warmer sample of matter to a colder sample of matter. This is one way to state the *second law of thermodynamics.*

herbivore Plant-eating organism. Examples include deer, sheep, grasshoppers, and zooplankton. Compare *carnivore, omnivore.*

heterotroph See *consumer.*

high Air mass with a high pressure. Compare *low.*

high-grade ore Ore containing a large amount of a desired mineral. Compare *low-grade ore.*

high-input agriculture See *industrialized agriculture.*

high-quality energy Energy that is concentrated and has great ability to perform useful work. Examples include high-temperature heat and the energy in electricity, coal, oil, gasoline, sunlight, and nuclei of uranium-235. Compare *low-quality energy.*

high-quality matter Matter that is concentrated and contains a high concentration of a useful resource. Compare *low-quality matter.*

high-throughput economy Economic system in most advanced industrialized countries, in which ever-increasing economic growth is sustained by maximizing the rate at which matter and energy resources are used, with little emphasis on pollution prevention, recycling, reuse, reduction of unnecessary waste, and other forms of resource conservation. Compare *low-throughput economy.*

high-waste economy See *high-throughput economy.*

HIPPCO Acronym used by conservation biologists for the six most important secondary causes of premature extinction: **H**abitat destruction, degradation, and fragmentation; **I**nvasive (nonnative) species; **P**opulation growth (too many people consuming too many resources); **P**ollution; **C**limate change; and **O**verexploitation.

human capital People's physical and mental talents that provide labor, innovation, culture, and organization. Compare *manufactured capital, natural capital.*

human resources See *human capital.*

hydraulic fracturing Process of injecting water mixed with sand and some toxic chemicals underground through horizontal natural gas wells and then using explosives and high pressure to fracture the deep rock and free up the natural gas stored there. The gas flows out of the well along with much of the water and a mix of compounds pulled from the rocks, including salts, toxic heavy metals, and naturally occurring radioactive materials. Commonly referred to as *fracking.*

hydroelectric power plant Structure in which the energy of falling or flowing water spins a turbine generator to produce electricity.

hydrologic cycle Biogeochemical cycle that collects, purifies, and distributes the earth's fixed supply of water from the environment to living organisms and then back to the environment.

hydropower Electrical energy produced by falling or flowing water. See *hydroelectric power plant*.

hydrosphere Earth's *liquid water* (oceans, lakes, other bodies of surface water, and underground water), *frozen water* (polar ice caps, floating ice caps, and ice in soil, known as permafrost), and *water vapor* in the atmosphere. See also *hydrologic cycle*. Compare *atmosphere, biosphere, geosphere*.

igneous rock Rock formed when molten rock material (magma) wells up from the earth's interior, cools, and solidifies into rock masses. Compare *metamorphic rock, sedimentary rock*. See *rock cycle*.

immigration Migration of people into a country or area to take up permanent residence.

indicator species Species whose decline serves as early warnings that a community or ecosystem is being degraded. Compare *keystone species, native species, nonnative species*.

industrialized agriculture (high-input agriculture) Production of large quantities of crops and livestock for domestic and foreign sale; involves use of large inputs of energy from fossil fuels (especially oil and natural gas), water, fertilizer, and pesticides.

industrial smog Type of air pollution consisting mostly of a mixture of sulfur dioxide, suspended droplets of sulfuric acid formed from some of the sulfur dioxide, and suspended solid particles. Compare *photochemical smog*.

industrial solid waste Solid waste produced by mines, factories, refineries, food growers, and businesses that supply people with goods and services. Compare *municipal solid waste*.

inertia (persistence) The ability of a living system such as a grassland or a forest to survive moderate disturbances.

infant mortality rate Number of babies out of every 1,000 born each year who die before their first birthday.

infectious disease Disease caused when a pathogen such as a bacterium, virus, or parasite invades the body and multiplies in its cells and tissues. Examples are flu, HIV, malaria, tuberculosis, and measles. See *transmissible disease*. Compare *nontransmissible disease*.

inland wetland Land away from the coast, such as a swamp, marsh, or bog, that is covered all or part of the time with freshwater. Compare *coastal wetland*.

inorganic compounds All compounds not classified as organic compounds. See *organic compounds*.

instrumental value Value of an organism, species, ecosystem, or the earth's biodiversity based on its usefulness to humans. Compare *intrinsic value*.

integrated pest management (IPM) Combined use of biological, chemical, and cultivation methods in proper sequence and timing to keep the size of a pest population below the level that causes economically unacceptable loss of a crop or livestock animal.

integrated waste management Variety of strategies for both waste reduction and waste management designed to deal with the solid wastes we produce.

internal cost Direct cost paid by the producer and the buyer of an economic good. Compare *external cost*. See *full cost pricing*.

interspecific competition Attempts by members of two or more species to use the same limited resources in an ecosystem.

intrinsic value Value of an organism, species, ecosystem, or the earth's biodiversity based on its existence, regardless of whether it has any usefulness to humans. Compare *instrumental value*.

invasive species See *nonnative species*.

inversion See *temperature inversion*.

ion Atom or group of atoms with one or more positive (+) or negative (–) electrical charges. Examples are Na^+ and Cl^-. Compare *atom, molecule*.

IPM See *integrated pest management*.

irrigation Supplying water to crops by artificial means rather than by relying on natural rainfall.

isotopes Two or more forms of a chemical element that have the same number of protons but different mass numbers because they have different numbers of neutrons in their nuclei.

keystone species Species that play roles affecting many other organisms in an ecosystem. Compare *indicator species, native species, nonnative species*.

kinetic energy Energy that matter has because of its mass and speed, or velocity. Compare *potential energy*.

lake Large natural body of standing freshwater formed when water from precipitation, land runoff, or groundwater flow fills a depression in the earth created by glaciation, earth movement, volcanic activity, or a giant meteorite. See *eutrophic lake, oligotrophic lake*.

land degradation Decrease in the ability of land to support crops, livestock, or wild species in the future as a result of natural or human-induced processes.

landfill See *sanitary landfill*.

land-use planning Planning to determine the best present and future uses of each parcel of land.

law of conservation of energy See *first law of thermodynamics*.

law of conservation of matter In any physical or chemical change, matter is neither created nor destroyed but merely changed from one form to another; in physical and chemical changes, existing atoms are rearranged into different spatial patterns (physical changes) or different combinations (chemical changes).

law of nature See *scientific law*.

LD50 See *median lethal dose*.

less-developed country Country that has low to moderate industrialization and low to moderate per capita GDP. Most are located in Africa, Asia, and Latin America. Compare *more-developed country*.

life expectancy Average number of years a newborn infant can be expected to live.

limiting factor Single factor that limits the growth, abundance, or distribution of the population of a species in an ecosystem. See *limiting factor principle*.

limiting factor principle Too much or too little of any abiotic factor can limit or prevent growth of a population of a species in an ecosystem, even if all other factors are at or near the optimal range of tolerance for the species.

lipids A chemically diverse group of large organic compounds that do not dissolve in water. Examples are fats and oils for storing energy, waxes for structure, and steroids for producing hormones.

liquefied natural gas (LNG) Natural gas converted to liquid form by cooling it to a very low temperature.

liquefied petroleum gas (LPG) Mixture of liquefied propane (C_3H_8) and butane (C_4H_{10}) gas removed from natural gas and used as a fuel.

lithosphere Outer shell of the earth, composed of the crust and the rigid, outermost part of the mantle outside the asthenosphere. Compare *crust, geosphere, mantle.*

LNG See *liquefied natural gas.*

logistic growth Pattern in which exponential population growth occurs when the population is small, and population growth decreases steadily with time as the population approaches the carrying capacity.

low Air mass with a low pressure. Compare *high.*

low-grade ore Ore containing a small amount of a desired mineral. Compare *high-grade ore.*

low-quality energy Energy that is dispersed and has little ability to do useful work. An example is low-temperature heat. Compare *high-quality energy.*

low-quality matter Matter that is dilute or dispersed or contains a low concentration of a useful resource. Compare *high-quality matter.*

low-throughput economy Economy based on working with nature by recycling and reusing discarded matter; preventing pollution; conserving matter and energy resources by reducing unnecessary waste and use; and building things that are easy to recycle, reuse, and repair. Compare *high-throughput economy.*

low-waste economy See *low-throughput economy.*

LPG See *liquefied petroleum gas.*

malnutrition See *chronic malnutrition.*

mangrove forest Ecosystem, found on some coastlines in warm tropical climates, that may contain any of 69 species that can live partly submerged in the salty environment of coastal swamps.

mantle Zone of the earth's interior between its core and its crust. Compare core, crust. See *geosphere, lithosphere.*

manufactured capital See *manufactured resources.*

manufactured inorganic fertilizer Commercially prepared mixture of inorganic plant nutrients such as nitrates, phosphates, and potassium applied to the soil to restore fertility and increase crop yields. Compare *organic fertilizer.*

manufactured resources Manufactured items made from natural resources and used to produce and distribute economic goods and services bought by consumers. They include tools, machinery, equipment, factory buildings, and transportation and distribution facilities. Compare *human capital, natural resources.*

map An important visual tool used to summarize data that vary over small or large areas.

marine life zone See *saltwater life zone.*

mass extinction Catastrophic, widespread, often global event in which major groups of species are wiped out over a short time compared with normal (background) extinctions. Compare *background extinction.*

mass number Sum of the number of neutrons and the number of protons in the nucleus of an atom. It gives the approximate mass of that atom. Compare *atomic number.*

matter Anything that has mass (the amount of material in an object) and takes up space. On the earth, where gravity is present, we weigh an object to determine its mass.

matter quality Measure of how useful a matter resource is, based on its availability and concentration. See *high-quality matter, low-quality matter.*

median lethal dose (LD50) Amount of a toxic material per unit of body weight of test animals that kills half the test population in a certain time.

megacity City with 10 million or more people.

metamorphic rock Rock produced when a preexisting rock is subjected to high temperatures (which may cause it to melt partially), high pressures, chemically active fluids, or a combination of these agents. Compare *igneous rock, sedimentary rock.* See *rock cycle.*

microorganisms Organisms such as bacteria that are so small that it takes a microscope to see them.

migration Movement of people into and out of specific geographic areas.

mineral Any naturally occurring inorganic substance found in the earth's crust as a crystalline solid. See *mineral resource.*

mineral resource Concentration of mineral material in or on the earth's crust in a form and amount such that extracting and converting it into useful materials or items is currently or potentially profitable. Mineral resources are classified as *metallic* (such as iron and tin ores) or *nonmetallic* (such as sand and salt).

minimum-tillage farming See *conservation-tillage farming.*

model Approximate representation or simulation of a system being studied.

molecule Combination of two or more atoms of the same chemical element (such as O_2) or different chemical elements (such as H_2O) held together by chemical bonds. Compare *atom, ion.*

monoculture Cultivation of a single crop, usually on a large area of land. Compare *polyculture.*

more-developed country Country that is highly industrialized and has a high per capita GDP. Compare *less-developed country.*

mountaintop removal mining Type of surface mining that uses explosives, massive power shovels, and large machines called draglines to remove the top of a mountain and expose seams of coal underneath a mountain. Compare *area strip mining, contour strip mining.*

MSW See *municipal solid waste.*

municipal solid waste (MSW) Solid materials discarded by homes and businesses in or near urban areas. See *solid waste*. Compare *industrial solid waste*.

mutagen Agent such as a chemical or form of radiation that increases the frequency of mutations in the DNA molecules found in cells. See *carcinogen, mutation, teratogen*.

mutation Random change in DNA molecules making up genes that can alter anatomy, physiology, or behavior in offspring. See *mutagen*.

mutualism Type of species interaction in which both participating species generally benefit. Compare *commensalism*.

nanotechnology Uses science and engineering to manipulate and create materials out of atoms and molecules at the ultra-small scale of less than 100 nanometers.

native species Species that normally live and thrive in a particular ecosystem. Compare *indicator species, keystone species, nonnative species*.

natural capital Natural resources and natural services that keep us and other species alive and support our economies. See *natural resources, natural services*.

natural capital degradation The waste, depletion, or destruction of any of the earth's natural capital. See *environmental degradation*.

natural gas Underground deposits of gases consisting of 50–90% by weight methane gas (CH_4) and small amounts of heavier gaseous hydrocarbon compounds such as propane (C_3H_8) and butane (C_4H_{10}).

natural greenhouse effect See *greenhouse effect*.

natural income Renewable resources such as plants, animals, and soil provided by natural capital.

natural resources Materials, such as air, water, and soil, and forms of energy in nature that are essential or useful to humans. See *natural capital*.

natural selection Process by which a particular beneficial gene (or set of genes) is reproduced in succeeding generations more than other genes. The result of natural selection is a population that contains a greater proportion of organisms better adapted to certain environmental conditions. See *adaptation, biological evolution, mutation*.

natural services Processes of nature, such as purification of air and water and pest control, which support life and human economies. See *natural capital*.

net energy yield Total amount of useful energy available from an energy resource or energy system over its lifetime, minus the amount of energy *used* (the first energy law), *automatically wasted* (the second energy law), and *unnecessarily wasted* in finding, processing, concentrating, and transporting it to users.

net primary productivity (NPP) Rate at which all the plants in an ecosystem produce net useful chemical energy; it is equal to the difference between the rate at which the plants in an ecosystem produce useful chemical energy (gross primary productivity) and the rate at which they use some of that energy through cellular respiration. Compare *gross primary productivity*.

neutral solution Water solution containing an equal number of hydrogen ions (H^+) and hydroxide ions (OH^-); water solution with a pH of 7. Compare *acid solution, basic solution*.

neutron (n) Elementary particle in the nuclei of all atoms (except hydrogen-1). It has a relative mass of 1 and no electric charge. Compare *electron, proton*.

niche See *ecological niche*.

nitric oxide (NO) Colorless gas that forms when nitrogen and oxygen gas in air react at the high-combustion temperatures in automobile engines and coal-burning plants. Lightning and certain bacteria in soil and water also produce NO as part of the *nitrogen cycle*.

nitrogen cycle Cyclic movement of nitrogen in different chemical forms from the environment to organisms and then back to the environment.

nitrogen dioxide (NO_2) Reddish-brown gas formed when nitrogen oxide reacts with oxygen in the air.

nitrogen oxides (NO_x) The collective term for nitric oxide and nitrogen dioxide. See *nitric oxide* and *nitrogen dioxide*.

noise pollution Any unwanted, disturbing, or harmful sound that impairs or interferes with hearing.

nondegradable pollutant Material that is not broken down by natural processes. Examples include the toxic elements lead and mercury. Compare *biodegradable pollutant*.

nonnative species Species that migrate into an ecosystem or are deliberately or accidentally introduced into an ecosystem by humans. Compare *native species*.

nonpoint source Broad and diffuse area, rather than a specific point, from which pollutants enter bodies of surface water or air. Examples include runoff of chemicals and sediments from cropland, livestock feedlots, logged forests, urban streets, parking lots, lawns, and golf courses. Compare *point source*.

nonrenewable resource Resource that exists in a fixed amount (stock) in the earth's crust and has the potential for renewal by geological, physical, and chemical processes taking place over hundreds of millions to billions of years. Examples include copper, aluminum, coal, and oil. We classify these resources as exhaustible because we are extracting and using them at a much faster rate than they are formed. Compare *renewable resource*.

nontransmissible disease Disease that is not caused by living organisms and does not spread from one person to another. Examples include most cancers, diabetes, cardiovascular disease, and malnutrition. Compare *transmissible disease*.

no-till farming See *conservation-tillage farming*.

NPP See *net primary productivity*.

nuclear change Process in which nuclei of certain isotopes spontaneously change, or are forced to change, into one or more different isotopes. The three principal types of nuclear change are radioactive decay, nuclear fission, and nuclear fusion. Compare *chemical change, physical change*.

nuclear fission Nuclear change in which the nuclei of certain isotopes with large mass numbers (such as uranium-235 and plutonium-239) are split apart into lighter nuclei when struck by a neutron. This process releases more neutrons and a large amount of energy. Compare *nuclear fusion*.

nuclear fuel cycle Includes the mining of uranium, processing and enriching the uranium to make fuel, using it in the reactor, safely storing the resulting highly radioactive wastes for thousands of years until their radioactivity falls to safe levels, and retiring the highly radioactive plant by taking it apart and storing its high- and moderate-level radioactive material safely for thousands of years.

nuclear fusion Nuclear change in which two nuclei of isotopes of elements with a low mass number (such as hydrogen-2 and hydrogen-3) are forced together at extremely high temperatures until they fuse to form a heavier nucleus (such as helium-4). This process releases a large amount of energy. Compare *nuclear fission*.

nucleic acid Large polymer molecule made by linking hundreds to thousands of four types of monomers called nucleotides. Two nucleic acids—DNA (deoxyribonucleic acid) and RNA (ribonucleic acid)—participate in the building of proteins and carry hereditary information used to pass traits from parent to offspring.

nucleus Extremely tiny center of an atom, making up most of the atom's mass. It contains one or more positively charged protons and one or more neutrons with no electrical charge (except for a hydrogen-1 atom, which has one proton and no neutrons in its nucleus).

nutrient Any chemical an organism must take in to live, grow, or reproduce.

nutrient cycle See *biogeochemical cycle*.

nutrient cycling The circulation of chemicals necessary for life, from the environment (mostly from soil and water) through organisms and back to the environment.

ocean currents Mass movements of surface water produced by prevailing winds blowing over the oceans.

oil sand See *tar sand*.

old-growth (primary) forest Virgin and old, second-growth forests containing trees that are often hundreds—sometimes thousands—of years old. Examples include forests of Douglas fir, western hemlock, giant sequoia, and coastal redwoods in the western United States.

oligotrophic lake Lake with a low supply of plant nutrients. Compare *eutrophic lake*.

omnivore Animal that can use both plants and other animals as food sources. Examples include pigs, rats, cockroaches, and humans. Compare *carnivore, herbivore*.

open access renewable resource Renewable resource owned by no one and available for use by anyone at little or no charge. Examples include clean air, underground water supplies, the open ocean and its fish, and the ozone layer. Compare *common property resource*.

open dump Fields or holes in the ground where garbage is deposited and sometimes covered with soil. They are rare in developed countries, but are widely used in many developing countries, especially to handle wastes from megacities. Compare *sanitary landfill*.

open-pit mining Removing minerals such as gravel, sand, and metal ores by digging them out of the earth's surface and leaving an open pit behind. Compare *area strip mining, contour strip mining, mountaintop removal, subsurface mining*.

open sea Part of any ocean that lies beyond the continental shelf. Compare *coastal zone*.

ore Part of a metal-yielding material that can be economically extracted from a mineral; typically contains two parts: the ore mineral, which contains the desired metal, and waste mineral material (gangue). See *high-grade ore, low-grade ore*.

organic agriculture Growing crops with limited or no use of synthetic pesticides and synthetic fertilizers and no use of genetically modified crops; raising livestock without use of synthetic growth regulators and feed additives; and using organic fertilizer (manure, legumes, compost) and natural pest controls (bugs that eat harmful bugs, plants that repel bugs and environmental controls such as crop rotation).

organic compounds Compounds containing carbon atoms combined with each other and with atoms of one or more other elements such as hydrogen, oxygen, nitrogen, sulfur, phosphorus, chlorine, and fluorine. All other compounds are called *inorganic compounds*.

organic farming See *organic agriculture* and *sustainable agriculture*.

organic fertilizer Organic material such as animal manure, green manure, and compost applied to cropland as a source of plant nutrients. Compare *manufactured inorganic fertilizer*.

organism Any form of life.

overburden Layer of soil and rock overlying a mineral deposit. Surface mining removes this layer.

overgrazing Destruction of vegetation when too many grazing animals feed too long on a specific area of pasture or rangeland and exceed the carrying capacity of a rangeland or pasture area.

overnutrition Diet so high in calories, saturated (animal) fats, salt, sugar, and processed foods, and so low in vegetables and fruits that the consumer runs a high risk of developing diabetes, hypertension, heart disease, and other health hazards. Compare *chronic malnutrition* and *chronic undernutrition*.

ozone (O_3) Colorless and highly reactive gas and a major component of photochemical smog. Also found in the ozone layer in the stratosphere. See *photochemical smog*.

ozone depletion Decrease in concentration of ozone (O_3) in the stratosphere. See *ozone layer*.

ozone layer Layer of gaseous ozone (O_3) in the stratosphere that protects life on earth by filtering out most harmful ultraviolet radiation from the sun.

parasite Consumer organism that lives on or in, and feeds on, a living plant or animal, known as the host, over an extended period. The parasite draws nourishment from and gradually weakens its host; it may or may not kill the host. See *parasitism*.

parasitism Interaction between species in which one organism, called the parasite, preys on another organism, called the host, by living on or in the host.

particulates Also known as suspended particulate matter (SPM); variety of solid particles and liquid droplets small and light enough to remain suspended in the air for long periods. About 62% of the SPM in outdoor air comes from natural sources such as dust, wild fires, and sea salt. The remaining 38% comes from human sources such as coal-burning electric power and industrial plants, motor vehicles, plowed fields, road construction, unpaved roads, and tobacco smoke.

passive solar heating system System that, without the use of mechanical devices, captures sunlight directly within a structure and converts it into low-temperature heat for space heating or for heating water for domestic use. Compare *active solar heating system*.

pasture Managed grassland or enclosed meadow that usually is planted with domesticated grasses or other forage to be grazed by livestock.

pathogen Living organism that can cause disease in another organism. Examples include bacteria, viruses, and parasites.

PCBs See *polychlorinated biphenyls*.

peak production Point in time when the pressure in an oil well drops and its rate of conventional crude oil production starts declining, usually a decade or so; for a group of wells or for a nation, the point at which all wells on average have passed peak production.

peer review Process of scientists reporting details of the methods and models they used, the results of their experiments, and the reasoning behind their hypotheses for other scientists working in the same field (their peers) to examine and criticize.

per capita ecological footprint Amount of biologically productive land and water needed to supply each person in a population with the renewable resources he or she uses and to absorb or dispose of the wastes from such resource use. It measures the average environmental impact of individuals in populations in different countries and areas. Compare *ecological footprint*.

per capita GDP Annual gross domestic product (GDP) of a country divided by its total population at midyear. See *gross domestic product*. Compare *genuine progress indicator (GPI)*.

perennial Plant that can live for more than 2 years. Compare *annual*.

permafrost Perennially frozen layer of the soil that forms when the water there freezes. It is found in arctic tundra.

perpetual resource Resource that is essentially inexhaustible on a human time scale because it is renewed continuously. Solar energy is an example. Compare *nonrenewable resource, renewable resource*.

persistence See *inertia*.

pest Unwanted organism that directly or indirectly interferes with human activities.

pesticide Any chemical designed to kill or inhibit the growth of an organism that people consider undesirable. See *fungicide, herbicide, insecticide*.

petrochemicals Chemicals obtained by refining (distilling) crude oil. They are used as raw materials in manufacturing most industrial chemicals, fertilizers, pesticides, plastics, synthetic fibers, paints, medicines, and many other products.

petroleum See *crude oil*.

pH Numeric value that indicates the relative acidity or alkalinity of a substance on a scale of 0 to 14, with the neutral point at 7. Acid solutions have pH values lower than 7; basic or alkaline solutions have pH values greater than 7.

phosphorus cycle Cyclic movement of phosphorus in different chemical forms from the environment to organisms and then back to the environment.

photochemical smog Complex mixture of air pollutants produced in the lower atmosphere by the reaction of hydrocarbons and nitrogen oxides under the influence of sunlight. Especially harmful components include ozone, peroxyacyl nitrates (PANs), and various aldehydes. Compare *industrial smog*.

photosynthesis Complex process that takes place in cells of green plants. Radiant energy from the sun is used to combine carbon dioxide (CO_2) and water (H_2O) to produce oxygen (O_2), carbohydrates (such as glucose, $C_6H_{12}O_6$), and other nutrient molecules. Compare *aerobic respiration*.

photovoltaic (PV) cell Device that converts radiant (solar) energy directly into electrical energy. Also called a solar cell.

physical change Process that alters one or more physical properties of an element or a compound without changing its chemical composition. Examples include changing the size and shape of a sample of matter (crushing ice and cutting aluminum foil) and changing a sample of matter from one physical state to another (boiling and freezing water). Compare *chemical change, nuclear change*.

phytoplankton Small, drifting plants, mostly algae and bacteria, found in aquatic ecosystems. Compare *plankton, zooplankton*.

pioneer species First hardy species—often microbes, mosses, and lichens—that begin colonizing a site as the first stage of ecological succession. See *ecological succession, pioneer community*.

planetary management worldview Worldview holding that humans are separate from nature, that nature exists mainly to meet our needs and increasing wants, and that we can use our ingenuity and technology to manage the earth's life-support systems, mostly for our benefit. It assumes that economic growth is unlimited. Compare *environmental wisdom worldview, stewardship worldview*.

plankton Small plant organisms (phytoplankton) and animal organisms (zooplankton) that float in aquatic ecosystems.

plantation agriculture Growing specialized crops such as bananas, coffee, and cacao in tropical developing countries, primarily for sale to developed countries.

plate tectonics Theory of geophysical processes that explains the movements of lithospheric plates and the processes that occur at their boundaries. See *lithosphere, tectonic plates*.

point source Single identifiable source that discharges pollutants into the environment. Examples include the smokestack of a power plant or an industrial plant, drainpipe of a meatpacking plant, chimney of a house, or exhaust pipe of an automobile. Compare *nonpoint source.*

policies programs, and the laws and regulations through which they are enacted, that a government enforces and funds.

politics Process through which individuals and groups try to influence or control government policies and actions that affect the local, state, national, and international communities.

pollutant Particular chemical or form of energy that can adversely affect the health, survival, or activities of humans or other living organisms. See *pollution.*

pollution Undesirable change in the physical, chemical, or biological characteristics of air, water, soil, or food that can adversely affect the health, survival, or activities of humans or other living organisms.

pollution cleanup Device or process that removes or reduces the level of a pollutant after it has been produced or has entered the environment. Examples include automobile emission control devices and sewage treatment plants. Compare *pollution prevention.*

pollution prevention Device, process, or strategy used to prevent a potential pollutant from forming or entering the environment or to sharply reduce the amount entering the environment. Compare *pollution cleanup.*

polychlorinated biphenyls (PCBs) Group of 209 toxic, oily, synthetic chlorinated hydrocarbon compounds that can be biologically amplified in food chains and webs.

polyculture Complex form of intercropping in which a large number of different plants maturing at different times are planted together. Compare *monoculture.*

population Group of individual organisms of the same species living in a particular area.

population change Increase or decrease in the size of a population. It is equal to (Births + Immigration) − (Deaths + Emigration).

population crash Dieback of a population that has used up its supply of resources, exceeding the carrying capacity of its environment. See *carrying capacity.*

population density Number of organisms in a particular population found in a specified area or volume.

potential energy Energy stored in an object because of its position or the position of its parts. Compare *kinetic energy.*

poverty Inability of people to meet their basic needs for food, clothing, and shelter.

prairie See *grassland.*

precautionary principle When there is significant scientific uncertainty about potentially serious harm from chemicals or technologies, decision makers should act to prevent harm to humans and to the environment. See *pollution prevention.*

precipitation Water in the form of rain, sleet, hail, and snow that falls from the atmosphere onto land and bodies of water.

predation Interaction in which an organism of one species (the predator) captures and feeds on some or all parts of an organism of another species (the prey).

predator Organism that captures and feeds on some or all parts of an organism of another species (the prey).

predator–prey relationship Relationship that has evolved between two organisms, in which one organism has become the prey for the other, the latter called the predator. See *predator, prey.*

prey Organism that is killed by an organism of another species (the predator) and serves as its source of food.

primary consumer Organism that feeds on some or all parts of plants (herbivore) or on other producers. Compare *detritivore, omnivore, secondary consumer.*

primary ecological succession Ecological succession in an area without soil or bottom sediments. See *ecological succession.* Compare *secondary ecological succession.*

primary pollutant Chemical that has been added directly to the air by natural events or human activities and occurs in a harmful concentration. Compare *secondary pollutant.*

primary productivity See *gross primary productivity, net primary productivity.*

primary (closed-loop) recycling Process in which materials are recycled into new products of the same type—turning used aluminum cans into new aluminum cans, for example.

primary sewage treatment Mechanical sewage treatment in which large solids are filtered out by screens and suspended solids settle out as sludge in a sedimentation tank. Compare *secondary sewage treatment.*

principles of sustainability Principles by which nature has sustained itself for billions of years by relying on solar energy, biodiversity, and nutrient recycling.

probability Mathematical statement about how likely it is that something will happen.

producer Organism that uses solar energy (green plants) or chemical energy (some bacteria) to manufacture the organic compounds it needs as nutrients from simple inorganic compounds obtained from its environment. Compare *consumer, decomposer.*

protein Large polymer molecules formed by linking together long chains of monomers called amino acids.

proton (p) Positively charged particle in the nuclei of all atoms. Each proton has a relative mass of 1 and a single positive charge. Compare *electron, neutron.*

proven oil reserves Identified deposits from which conventional crude oil can be extracted profitably at current prices with current technology.

PV cell See *photovoltaic cell.*

pyramid of energy flow Diagram representing the flow of energy through each trophic level in a food chain or food web. With each energy transfer, only a small part (typically 10%) of the usable energy entering one trophic level is transferred to the organisms at the next trophic level.

radioactive decay Change of a radioisotope to a different isotope by the emission of radioactivity.

radioactive waste Waste products of nuclear power plants, research, medicine, weapons production, and other processes involving nuclear reactions. See *radioactivity*.

radioactivity Nuclear change in which unstable nuclei of atoms spontaneously shoot out "chunks" of mass, energy, or both at a fixed rate. The three principal types of radioactivity are gamma rays and fast-moving alpha particles and beta particles.

rain shadow effect Low precipitation on the leeward side of a high mountain range when prevailing winds flow up and over these mountains, dropping their moisture on the windward side and creating semiarid and arid conditions on their leeward side.

rangeland Land that supplies forage or vegetation (grasses, grass-like plants, and shrubs) for grazing and browsing animals. Compare *pasture*.

reconciliation ecology Science of inventing, establishing, and maintaining habitats to conserve species diversity in places where people live, work, or play.

recycle To collect and reprocess discarded materials so that they can be made into new products; one of the three R's of resource use. An example is collecting aluminum cans, melting them down, and using the aluminum to make new cans or other aluminum products. See *primary recycling, secondary recycling*. Compare *reduce and reuse*.

reduce To consume less and live a simpler lifestyle; one of the three R's of resource use. Compare *recycle and reuse*.

reliable science Concepts and ideas that are widely accepted by experts in a particular field of the natural or social sciences. Compare *tentative science, unreliable science*.

reliable surface runoff Surface runoff of water that generally can be counted on as a stable source of water from year to year. See *runoff*.

renewable resource Resource that can be replenished rapidly (in hours to several decades) through natural processes as long as it is not used up faster than it is replaced. Examples include trees in forests, grasses in grasslands, wild animals, fresh surface water in lakes and streams, most groundwater, fresh air, and fertile soil. If such a resource is used faster than it is replenished, it can be depleted. Compare *nonrenewable resource* and *perpetual resource*. See also *environmental degradation*.

replacement-level fertility Average number of children a couple must bear to replace themselves. The average for a country or the world usually is slightly higher than two children per couple (2.1 in the United States and 2.5 in some less developed countries) mostly because some children die before reaching their reproductive years. See also *total fertility rate*.

representative democracy A government by the people through elected officials and representatives.

reproductive isolation Long-term geographic separation of members of a particular sexually reproducing species.

reserves Resources that have been identified and from which a usable mineral can be extracted profitably at present prices with current mining or extraction technology.

reservoir Artificial lake created when a stream is dammed. See *dam*.

resilience The ability of a living system to be restored through secondary succession after a more severe disturbance.

resource Anything obtained from the environment to meet human needs and wants. It can also be applied to other species.

resource partitioning Process of dividing up resources in an ecosystem so that species with similar needs (overlapping ecological niches) use the same scarce resources at different times, in different ways, or in different places. See *ecological niche*.

response Amount of health damage caused by exposure to a certain dose of a harmful substance or form of radiation. See *dose, dose-response curve*.

restoration ecology Research and scientific study devoted to restoring, repairing, and reconstructing damaged ecosystems.

reuse To use a product over and over again in the same form. An example is collecting, washing, and refilling glass beverage bottles. One of the 3 Rs. Compare *reduce and recycling*.

riparian zone A thin strip or patch of vegetation that surrounds a streams. These zones are very important habitats and resources for wildlife.

risk Probability that something undesirable will result from deliberate or accidental exposure to a hazard. See *risk analysis, risk assessment, risk management*.

risk analysis Identifying hazards, evaluating the nature and severity of risks associated with the hazards (*risk assessment*), ranking risks (*comparative risk analysis*), and using this and other information to determine options and make decisions about reducing or eliminating risks (*risk management*).

risk assessment Process of gathering data and making assumptions to estimate short- and long-term harmful effects on human health or the environment from exposure to hazards associated with the use of a particular product or technology.

risk communication Communicating information about risks to decision makers and the public. See *risk, risk analysis*.

risk management Use of risk assessment and other information to determine options and make decisions about reducing or eliminating risks. See *risk, risk analysis*.

rock Any solid material that makes up a large, natural, continuous part of the earth's crust. See *mineral*.

rock cycle Largest and slowest of the earth's cycles, consisting of geologic, physical, and chemical processes that form and modify rocks and soil in the earth's crust over millions of years.

runoff Freshwater from precipitation and melting ice that flows on the earth's surface into streams, lakes, wetlands, and reservoirs. See *reliable surface runoff, surface runoff, surface water*. Compare *groundwater*.

salinity Concentration of various salts dissolved in a given volume of water.

salinization Accumulation of salts in soil that can eventually make the soil unable to support plant growth.

saltwater life zones Aquatic life zones associated with oceans: oceans and their accompanying bays, estuaries, coastal wetlands, shorelines, coral reefs, and mangrove forests.

sanitary landfill Waste disposal site on land in which waste is spread in thin layers, compacted, and covered with a fresh layer of clay or plastic foam each day. Compare *open dump*.

science Attempts to discover order in nature and use that knowledge to make predictions about what is likely to happen in nature. See *data, reliable science, scientific hypothesis, scientific law, scientific model, scientific theory, tentative science, unreliable science*.

scientific hypothesis An educated guess that attempts to explain a scientific law or certain scientific observations. Compare *scientific law, scientific methods, scientific model, scientific theory*.

scientific law Description of what scientists find happening in nature repeatedly in the same way, without known exception. See *first law of thermodynamics, law of conservation of matter, second law of thermodynamics*. Compare *scientific hypothesis, scientific theory*.

scientific model A simulation of complex processes and systems. Many are mathematical models that are run and tested using computers.

scientific theory A well-tested and widely accepted scientific hypothesis. Compare *scientific hypothesis, scientific law*.

secondary consumer Organism that feeds only on primary consumers. Compare *detritivore, omnivore, primary consumer*.

secondary ecological succession Ecological succession in an area in which natural vegetation has been removed or destroyed but the soil or bottom sediment has not been destroyed. See *ecological succession*. Compare *primary ecological succession*.

secondary pollutant Harmful chemical formed in the atmosphere when a primary air pollutant reacts with normal air components or other air pollutants. Compare *primary pollutant*.

secondary recycling A process in which waste materials are converted into different products; for example, used tires can be shredded and turned into rubberized road surfacing. Compare *primary (closed-loop) recycling*.

secondary sewage treatment Second step in most waste treatment systems in which aerobic bacteria decompose as much as 90% of degradable, oxygen-demanding organic wastes in wastewater. Compare *primary sewage treatment*.

second-growth forest A stand of trees resulting from secondary ecological succession; these forests develop after the trees in an area have been removed by human activities, such as clear-cutting for timber or cropland, or by natural forces such as fire, hurricanes, or volcanic eruptions.

second law of thermodynamics Whenever energy is converted from one form to another in a physical or chemical change, we end up with lower-quality or less usable energy than we started with. In any conversion of heat energy to useful work, some of the initial energy input is always degraded to lower-quality, more dispersed, less useful energy—usually low-temperature heat that flows into the environment. See *first law of thermodynamics*.

sedimentary rock Rock that forms from the accumulated products of erosion and in some cases from the compacted shells, skeletons, and other remains of dead organisms. Compare *igneous rock, metamorphic rock*. See *rock cycle*.

selective cutting Cutting of intermediate-aged, mature, or diseased trees in an uneven-aged forest stand, either singly or in small groups. This encourages the growth of younger trees and maintains an uneven-aged stand. Compare *clear-cutting, strip cutting*.

septic tank Underground tank for treating wastewater from a home, used in rural and suburban areas. Bacteria in the tank decompose organic wastes, and the sludge settles to the bottom of the tank. The effluent flows out of the tank into the ground through a field of drainpipes.

shale oil Slow-flowing, dark brown, heavy oil obtained when kerogen in oil shale is vaporized at high temperatures and then condensed. Shale oil can be refined to yield gasoline, heating oil, and other petroleum products.

smart growth Form of urban planning that recognizes that urban growth will occur but uses zoning laws and other tools to prevent sprawl, direct growth to certain areas, protect ecologically sensitive and important lands and waterways, and develop urban areas that are more environmentally sustainable and more enjoyable places to live.

smelting Process in which a desired metal is separated from the other elements in an ore mineral.

SNG See *synthetic natural gas*.

soil Complex mixture of inorganic minerals (clay, silt, pebbles, and sand), decaying organic matter, water, air, and living organisms.

soil conservation Methods used to reduce soil erosion, prevent depletion of soil nutrients, and restore nutrients previously lost by erosion, leaching, and excessive crop harvesting.

soil erosion Movement of soil components, especially topsoil, from one place to another, usually by wind, flowing water, or both. This natural process can be greatly accelerated by human activities that remove vegetation from soil. Compare *soil conservation*.

soil horizons Horizontal zones, or layers, that make up a particular mature soil. Each horizon has a distinct texture and composition that vary with different types of soils. See *soil profile*.

soil profile Cross-sectional view of the horizons in a soil. See *soil horizon*.

solar cell See *photovoltaic cell.*

solar energy Direct radiant energy from the sun and a number of indirect forms of energy produced by the direct input of such radiant energy. Principal indirect forms of solar energy include wind, falling and flowing water (hydropower), and biomass (solar energy converted into chemical energy stored in the chemical bonds of organic compounds in trees and other plants)—none of which would exist without direct solar energy.

solid waste Any unwanted or discarded material that is not a liquid or a gas. See *industrial solid waste, municipal solid waste.*

specialist species Species with a narrow ecological niche. They may be able to live in only one type of habitat, tolerate only a narrow range of climatic and other environmental conditions, or use only one type or a few types of food. Compare *generalist species.*

speciation Formation of two species from one species because of divergent natural selection in response to changes in environmental conditions; usually takes thousands of years. Compare *extinction.*

species Group of similar organisms, and for sexually reproducing organisms, a set of individuals that can mate and produce fertile offspring. Every organism is a member of a certain species.

species diversity Number of different species (species richness) combined with the relative abundance of individuals within each of those species (species evenness) in a given area. See *biodiversity, species evenness, species richness.* Compare *ecological diversity, genetic diversity.*

species evenness Degree to which comparative numbers of individuals of each of the species present in a community are similar. See *species diversity.* Compare *species richness.*

species richness Variety of species, measured by the number of different species contained in a community. See *species diversity.* Compare *species evenness.*

spoils Unwanted rock and other waste materials produced when a material is removed from the earth's surface or subsurface by mining, dredging, quarrying, or excavation.

stewardship worldview Worldview holding that we can manage the earth for our benefit but that we have an ethical responsibility to be caring and responsible managers, or stewards, of the earth. It calls for encouraging environmentally beneficial forms of economic growth and discouraging environmentally harmful forms. Compare *environmental wisdom worldview, planetary management worldview.*

stratosphere Second layer of the atmosphere, extending about 17–48 kilometers (11–30 miles) above the earth's surface. It contains a layer of gaseous ozone (O_3), which filters out about 95% of the incoming harmful ultraviolet radiation emitted by the sun. Compare *troposphere.* See *ozone layer.*

strip-cropping Planting regular crops and close-growing plants, such as hay or nitrogen-fixing legumes, in alternating rows or bands to help reduce depletion of soil nutrients.

strip-cutting Variation of clear-cutting in which a strip of trees is clear-cut along the contour of the land, with the corridor being narrow enough to allow natural regeneration within a few years. After regeneration, another strip is cut above the first, and so on. Compare *clear-cutting, selective cutting.*

strip-mining Form of surface mining in which bulldozers, power shovels, or stripping wheels remove large chunks of the earth's surface in strips. See *area strip mining, contour strip mining, surface mining.* Compare *subsurface mining.*

subsidence Slow or rapid sinking of part of the earth's crust that is not slope-related.

subsurface mining Extraction of a metal ore or fuel resource such as coal from a deep underground deposit. Compare *surface mining.*

sulfur cycle Cyclic movement of sulfur in various chemical forms from the environment to organisms and then back to the environment.

sulfur dioxide (SO_2) Colorless gas with an irritating odor. About one-third of the SO_2 in the atmosphere comes from natural sources as part of the sulfur cycle. The other two-thirds come from human sources, mostly combustion of sulfur-containing coal in electric power and industrial plants and from oil refining and smelting of sulfide ores.

sulfuric acid Compound containing hydrogen, sulfur, and oxygen; a hazardous chemical that is often a component of acid precipitation. See *acid deposition.*

surface fire Forest fire that burns only undergrowth and leaf litter on the forest floor. Compare *crown fire, ground fire.* See *controlled burning.*

surface mining Removing soil, subsoil, and other strata and then extracting a mineral deposit found fairly close to the earth's surface. See *area strip-mining, contour strip-mining, mountaintop removal, open-pit mining, strip-mining.* Compare *subsurface mining.*

surface runoff Water flowing off the land into bodies of surface water. See *reliable surface runoff.*

surface water Precipitation that does not infiltrate the ground or return to the atmosphere by evaporation or transpiration. Found in streams, rivers, lakes, and wetlands. See *runoff.* Compare *groundwater.*

suspended particulate matter See *particulates.*

sustainability Ability of earth's various systems, including human cultural systems and economies, to survive and adapt to changing environmental conditions indefinitely.

sustainable agriculture Method of growing crops and raising livestock by relying on organic fertilizers, soil conservation, water conservation, biological pest control, and minimal use of nonrenewable fossil-fuel energy.

sustainable yield Highest rate at which a potentially renewable resource can be used indefinitely without reducing its available supply. See also *environmental degradation.*

synfuels Synthetic gaseous and liquid fuels produced from solid coal or sources other than natural gas or crude oil.

synthetic natural gas (SNG) Gaseous fuel containing mostly methane produced from solid coal.

tailings Rock and other waste materials removed as impurities when waste mineral material is separated from the metal in an ore.

tar sand Deposit of a mixture of clay, sand, water, and varying amounts of a tar-like heavy oil known as bitumen. Bitumen can be extracted from tar sand by heating. It is then purified and upgraded to synthetic crude oil.

tectonic plates Various-sized pieces of the earth's lithosphere that move slowly around atop the mantle's flowing asthenosphere. Most earthquakes and volcanoes occur around the boundaries of these plates. See *lithosphere*.

temperature inversion Layer of dense, cool air trapped under a layer of less dense, warm air. It prevents upward-flowing air currents from developing. In a prolonged inversion, air pollution in the trapped layer may build up to harmful levels.

tentative science Preliminary scientific data, hypotheses, and models that have not been widely tested and accepted. Compare *reliable science, unreliable science*.

teratogen Chemical, ionizing agent, or virus that causes birth defects. Compare *carcinogen, mutagen*.

terracing Planting crops on a long, steep slope that has been converted into a series of broad, nearly level terraces with short vertical drops from one to another that run along the contour of the land to retain water and reduce soil erosion.

tertiary (higher-level) consumers Animals that feed on animal-eating animals. They feed at high trophic levels in food chains and webs. Examples include hawks, lions, bass, and sharks. Compare *detritivore, primary consumer, secondary consumer*.

theory of evolution Widely accepted scientific idea that all life forms developed from earlier life forms. It is the way most biologists explain how life has changed over the past 3.5 billion years and why it is so diverse today.

theory of island biogeography Widely accepted scientific theory holding that the number of different species (species richness) found on an island is determined by the interactions of two factors: the rate at which new species immigrate to the island and the rate at which species become extinct, or cease to exist, on the island. See *species richness*.

thermal inversion See *temperature inversion*.

threatened species Wild species that is still abundant in its natural range but is likely to become endangered because of a decline in numbers. Compare *endangered species*.

tipping point Threshold level at which an environmental problem causes a fundamental and irreversible shift in the behavior of a system. See *climate change tipping point, ecological tipping point*.

total fertility rate (TFR) Estimate of the average number of children who will be born alive to a woman during her lifetime if she passes through all her childbearing years (ages 15–44). More simply, it is an estimate of the average number of children that women in a given population will have during their childbearing years.

toxic chemical Chemical that can cause harm to an organism. See *carcinogen, mutagen, teratogen*.

toxicity Measure of the harmfulness of a substance.

toxicology Study of the adverse effects of chemicals on health.

toxic waste See *hazardous waste*.

traditional intensive agriculture Production of enough food for a farm family's survival and a surplus that can be sold. This type of agriculture uses higher inputs of labor, fertilizer, and water than traditional subsistence agriculture. See *traditional subsistence agriculture*. Compare *industrialized agriculture*.

traditional subsistence agriculture Production of enough crops or livestock for a farm family's survival and, in good years, a surplus to sell or put aside for hard times. Compare *industrialized agriculture, traditional intensive agriculture*.

tragedy of the commons Depletion or degradation of a potentially renewable resource to which people have free and unmanaged access. An example is the depletion of commercially desirable fish species in the open ocean beyond areas controlled by coastal countries. See *common-property resource, open access renewable resource*.

trait Characteristic passed on from parents to offspring during reproduction in an animal or plant.

transform fault Area where the earth's lithospheric plates move in opposite but parallel directions along a fracture (fault) in the lithosphere. Compare *convergent plate boundary, divergent plate boundary*.

transmissible disease Disease that is caused by living organisms (such as bacteria, viruses, and parasitic worms) and can spread from one person to another by air, water, food, or body fluids, or in some cases by insects or other organisms. Compare *nontransmissible disease*.

transpiration Process in which water is absorbed by the root systems of plants, moves up through the plants, passes through pores (stomata) in their leaves or other parts, and evaporates into the atmosphere as water vapor.

tree farm See *tree plantation*.

tree plantation Site planted with one or only a few tree species in an even-aged stand. When the stand matures it is usually harvested by clear-cutting and then replanted. These farms normally raise rapidly growing tree species for fuelwood, timber, or pulpwood. Compare *old-growth forest, second-growth forest*.

trophic level All organisms that are the same number of energy transfers away from the original source of energy (for example, sunlight) that enters an ecosystem. For example, all producers belong to the first trophic level and all herbivores belong to the second trophic level in a food chain or a food web.

troposphere Innermost layer of the atmosphere. It contains about 75% of the mass of earth's air and extends about 17 kilometers (11 miles) above sea level. Compare *stratosphere*.

tsunami Series of large waves generated when part of the ocean floor suddenly rises or drops, typically due to an earthquake.

undernutrition See *chronic undernutrition*.

unreliable science Scientific results or hypotheses presented as reliable science without having undergone the rigors of the peer review process. Compare *reliable science, tentative science*.

upwelling Movement of nutrient-rich bottom water to the ocean's surface. It can occur far from shore but usually takes place along certain steep coastal areas where the warm surface layer of ocean water is pushed away from shore and replaced by cold, nutrient-rich bottom water.

urban area Geographic area containing a community with a population of 2,500 or more. The minimum number of people used in this definition varies among countries, from 2,500 to 50,000.

urban growth Rate of growth of an urban population.

urbanization Creation or growth of urban areas, or cities, and their surrounding developed land. See *degree of urbanization, urban area*.

urban sprawl Growth of low-density development on the edges of cities and towns. See *smart growth*.

virtual water Water that is not directly consumed but is used to produce food and other products.

virus Infectious agent that is smaller than a bacterium; it works by invading a cell and taking over its genetic machinery to copy itself. It then multiplies and spreads throughout one's body, causing a viral disease such as flu or AIDS.

volatile organic compounds (VOCs) Organic compounds that exist as gases in the atmosphere and act as pollutants, some of which are hazardous.

volcano Vent or fissure in the earth's surface through which magma, liquid lava, and gases are released into the environment.

warm front Boundary between an advancing warm air mass and the cooler one it is replacing. Because warm air is less dense than cool air, an advancing warm front rises over a mass of cool air. Compare *cold front*.

waste management Managing wastes to reduce their environmental harm without seriously trying to reduce the amount of waste produced. See *integrated waste management*. Compare *waste reduction*.

waste reduction Reducing the amount of waste produced; wastes that are produced are viewed as potential resources that can be reused, recycled, or composted. See *integrated waste management*. Compare *waste management*.

water cycle See *hydrologic cycle*.

water footprint A rough measure of the volume of water used directly and indirectly to keep a person or group alive and to support their lifestyles.

waterlogging Saturation of soil with irrigation water or excessive precipitation so that the water table rises close to the surface.

water pollution Any physical or chemical change in surface water or groundwater that can harm living organisms or make water unfit for certain uses.

watershed (drainage basin) Land area that delivers water, sediment, and dissolved substances via small streams to a major stream (river).

water table Upper surface of the zone of saturation, in which all available pores in the soil and rock in the earth's crust are filled with water. See *zone of aeration, zone of saturation*.

weather Short-term changes in the temperature, barometric pressure, humidity, precipitation, sunshine, cloud cover, wind direction and speed, and other conditions in the troposphere at a given place and time. Compare *climate*.

wetland Land that is covered all or part of the time with salt water or freshwater, excluding streams, lakes, and the open ocean. See *coastal wetland, inland wetland*.

wilderness Area where the earth and its ecosystems have not been seriously disturbed by humans and where humans are only temporary visitors.

windbreak Row of trees or hedges planted to partially block wind flow and reduce soil erosion on cultivated land.

wind farm Cluster of wind turbines in a windy area on land or at sea, built to capture wind energy and convert it into electrical energy.

worldview How people think the world works and what they think their role in the world should be. See *environmental wisdom worldview, planetary management worldview, stewardship worldview*.

zone of aeration Zone in soil that is not saturated with water and that lies above the water table. See *water table, zone of saturation*.

zone of saturation Zone below the water table where all available pores in soil and rock in the earth's crust are filled by water. See *water table, zone of aeration*.

zoning Designating parcels of land for particular types of use.

zooplankton Animal plankton; small floating herbivores that feed on plant plankton (phytoplankton). Compare *phytoplankton*.

Index

Note: Page numbers in **boldface** type indicate key terms. Page numbers followed by italicized *f* or *t* indicate figures and tables.